AF759510

ROBOT MANIPULATORS, NEW ACHIEVEMENTS

Edited by

ALEKSANDAR LAZINICA AND HIROYUKI KAWAI

Robot Manipulators, New Achievements

Edited by Aleksandar Lazinica and Hiroyuki Kawai

Republished by InTech

Janeza Trdine 9, 51000 Rijeka, Croatia

Copyright © 2010 InTech

Abstracting and non-profit use of the material is permitted with credit to the source. Statements and opinions expressed in the chapters are these of the individual contributors and not necessarily those of the editors or publisher. No responsibility is accepted for the accuracy of information contained in the published articles. Publisher assumes no responsibility liability for any damage or injury to persons or property arising out of the use of any materials, instructions, methods or ideas contained inside. After this work has been published by the Intech, authors have the right to republish it, in whole or part, in any publication of which they are an author or editor, and the make other personal use of the work.

Technical Editor Sonja Mujacic

Additional hard copies can be obtained from orders@intechopen.com

Robot Manipulators, New Achievements,
Edited by Aleksandar Lazinica and Hiroyuki Kawai
p. cm.
ISBN 978-953-307-090-2

We are IntechOpen, the world's leading publisher of Open Access books

Built by scientists, for scientists

4,200+
Open access books available

116,000+
International authors and editors

125M+
Downloads

151
Countries delivered to

Our authors are among the
Top 1%
most cited scientists

12.2%
Contributors from top 500 universities

WEB OF SCIENCE™

Selection of our books indexed in the Book Citation Index
in Web of Science™ Core Collection (BKCI)

Interested in publishing with us?
Contact book.department@intechopen.com

Numbers displayed above are based on latest data collected.
For more information visit www.intechopen.com

For EU product safety concerns:
IN TECH d.o.o., Prolaz Marije Krucifikse Kozulić 3, 51000 Rijeka, Croatia,
info@intechopen.com or visit our website at intechopen.com.

Meet the editors

Alex Lazinica is the founder and CEO of IntechOpen. After obtaining a Master's degree in Mechanical Engineering, he continued his PhD studies in Robotics at the Vienna University of Technology. Here he worked as a robotic researcher with the university's Intelligent Manufacturing Systems Group as well as a guest researcher at various European universities, including the Swiss Federal Institute of Technology Lausanne (EPFL). During this time he published more than 20 scientific papers, gave presentations, served as a reviewer for major robotic journals and conferences and most importantly he co-founded and built the International Journal of Advanced Robotic Systems- world's first Open Access journal in the field of robotics. Starting this journal was a pivotal point in his career, since it was a pathway to founding IntechOpen - Open Access publisher focused on addressing academic researchers needs. Alex is a personification of IntechOpen key values being trusted, open and entrepreneurial. Today his focus is on defining the growth and development strategy for the company.

Hiroyuki Kawai, Ph.D., Professor of Kanazawa Institute of Technology, Japan, in Spring 2018. He received the B.Eng. degree in electrical and computer engineering in 1999 from Kanazawa University, Japan, and received the M.Eng. and Ph.D. degrees in the Graduate School of Natural Science and Technology from Kanazawa University, in 2001 and 2004, respectively. From 2004 until 2005, he was a Postdoctoral Scholar at the Information Technology Research Center, Hosei University, Japan. He was an Assistant Professor and an Associate Professor at Kanazawa Institute of Technology from 2005 to 2010 and 2010 to 2018, respectively. From 2013 to 2014, he was a Visiting Scholar at the University of Florida. Dr. Kawai's research focuses on the visual feedback control of robot systems, the robot control with antagonistic bi-articular muscles, and Functional Electrical Stimulation(FES) based human motion control. In particular, he is interested in the passivity based control. He received the IEEE Transactions on Control Systems Technology Outstanding Paper Award from the IEEE Control Systems Society (CSS) in 2008.

Preface

Robot manipulators are developing as industrial robots instead of human workers. Recently, the application fields of robot manipulators are increasing such as Da Vinci as a medical robot, ASIMO as a humanoid robot and so on. There are many research topics with respect to robot manipulators, e.g. motion planning, cooperation with a human, and fusion with external sensors like vision, haptic and force, etc. Moreover, these include both technical problems in the industry and theoretical problems in the academic fields. In this book we have collected the latest research issues from around the world. Thus, we believe that this book is useful and joyful for readers. We would like to thank all authors for their interesting contributions and the reviewers for their devoted works.

Editors:

Aleksandar Lazinica and Hiroyuki Kawai

Contents

Preface VII

1. Modeling and Control of a New Robotic Deburring System 001
Jae H. Chung

2. Trajectory tracking control for robot manipulators with no velocity measurement using semi-globally and globally asymptotically stable velocity observers 017
Farah Bouakrif

3. Robotic Machining from Programming to Process Control 035
Zengxi Pan and Hui Zhang

4. Fuzzy Optimal Control for Robot Manipulators 059
Basil M. Al-Hadithi, Agustín Jiménez and Fernando Matía

5. Development of Adaptive Learning Control Algorithm for a two-degree-of-freedom Serial Ball And Socket Actuator 081
Hayder M. A. A. Al-Assadi and Ahmed Jaffar

6. Singularity-Based Calibration – A Novel Approach for Absolute-Accuracy-Enhancement of Parallel Robots 093
Philipp Last, Annika Raatz and Jürgen Hesselbach

7. Advanced Nonlinear Control of Robot Manipulators 107
Adel Merabet and Jason Gu

8. Modelling of HDD head positioning systems regarded as robot manipulators using block matrices 129
Tomasz Trawiński and Roman Wituła

9. Mobile Manipulation: A Case Study 145
A. HENTOUT, B. BOUZOUIA, I. AKLI and R. TOUMI

10. A Concept for Isles of Automation 169
Mikko Sallinen and Tapio Heikkilä

11. Stiffness Analysis for an Optimal Design of Multibody Robotic Systems 185
Carbone Giuseppe

12. Concurrent Engineering of Robot Manipulators 211
M. Reza Emami and Robin Chhabra

13. Desktop Cartesian-Type Robot with Abilities of Compliant Motion and Stick-Slip Motion 241
Fusaomi Nagata, Shintaro Tani,Takanori Mizobuchi, Tetsuo Hase, Zenku Haga and Keigo Watanabe

14. Kinematic calibration of articulated arm coordinate measuring machines and robot arms using passive and active self-centering probes and multipose optimization algorithm based in point and length constrains 255
Jorge Santolaria and Juan José Aguilar

15. Two Cooperating Manipulators with Fractional Controllers 279
N. M. Fonseca Ferreira, J. A. Tenreiro Machado and József K. Tar

16. MFR (Multi-purpose Field Robot) based on Human-robot Cooperative Manipulation for Handling Building Materials 289
Seungyeol Lee

17. A Sensor Classification Strategy for Robotic Manipulators 315
Miguel F. M. Lima, J. A. Tenreiro Machado and António Ferrolho

18. Passivity-based Visual Force Feedback Control for Eye-to-Hand Systems 329
Hiroyuki Kawai, Toshiyuki Murao and Masayuki Fujita

19. Kinematic Analysis of 3-UCR Parallel Robot Leg 343
Cheng Gang and Ge Shi-rong

20. Digital Control of Free Floating Space Robot Manipulators Using Transpose of Generalized Jacobian Matrix 361
Shinichi Sagara and Yuichiro Taira

21. Kinematics, Singularity and Dexterity Analysis of Planar Parallel Manipulators Based on DH Method 387
Serdar Kucuk

22. Robot Manipulator Probabilistic Workspace Applied to Robotic Assistance 401
Fernando A. Auat Cheein, Fernando di Sciascio, Juan Marcos Toibero and Ricardo Carelli

23. On the Design of Human-Safe Robot Manipulators 419
Vincent Duchaine, Nicolas Lauzier and Clément Gosselin

24. Vibration Based Control for Flexible Link Manipulator 435
Tamer Mansour, Atsushi Konno and Masaru Uchiyama

25. Control of Robotic Systems with Flexible Components using Hermite Polynomial-Based Neural Networks 459
Gerasimos G. Rigatos

26. Dimensional optimization of completely restrained positioning cable driven parallel manipulator with large span 487
XiaoQiang Tang and Rui Yao

27. Multi-Criteria Optimization Manipulator Trajectory Planning 503
E. J. Solteiro Pires, P. B. de Moura Oliveira and J. A. Tenreiro Machado

28. On Designing Compliant Actuators Based On Dielectric Elastomers for Robotic Applications 523
Giovanni Berselli, Gabriele Vassura, Vincenzo Parenti Castelli and Rocco Vertechy

29. Hybrid Control Techniques for Static and Dynamic Environments: a Step towards Robot-Environment Interaction 551
Fabrizio Romanelli

30. Maximal Operational Workspace of Parallel Manipulators 577
E. Macho, O. Altuzarra and A. Hernandez

31. Kinematical and Dynamical Models of KR 6 KUKA Robot, including the kinematic control in a parallel processing platform 601
John Faber Archila Díaz, Max Suell Dutra and Fernando Augusto de Noronha Castro Pinto

32. Manipulator Design Strategy for a Specified Task Based on Human-Robot Collaboration 621
Seungnam Yu, Seungwhan Suh, Woonghee Son, Youngsoo Kim and Changsoo Han

33. PSPRD and PSPRD+I Control of Robot Manipulators and Redundant Manipulators 645
Kiyotaka Shimizu

34. 3D Imaging System for Tele-Manipulation 663
Hideki Kakeya

35. Experimental evaluation of output–feedback tracking controllers for robot manipulators 679
Javier Moreno–Valenzuela, Víctor Santibáñez and Ricardo Campa

36. Higher Dimensional Spatial Expression of Upper Limb Manipulation Ability based on Human Joint Torque Characteristics 693
Makoto Sasaki, Takehiro Iwami, Kazuto Miyawaki, Ikuro Sato, Goro Obinata and Ashish Dutta

1

Modeling and Control of a New Robotic Deburring System

Jae H. Chung
US Army RDECOM-ARDEC
Building 95N
Picatinny Arsenal, NJ 07806
USA

1. Introduction

A machining manipulator is subject to mechanical interaction with the object being processed. The robot performs the task in constrained work space. In constrained tasks, one is concerned with not only the position of the robot end-point, but also the contact forces, which are desired to be accommodated rather than resisted. Therefore, interaction force needs to be considered in designing and controlling deburring tools.
Many researchers have proposed automated systems for grinding dies, deburring casting, removing weld beans, etc [Bopp, 1983; Gustaffson, 1983]. Usually, a deburring tool is mounted on a NC machining center or a robot manipulator. Several control laws have been developed for simultaneous control of both motion and force [Whitney, 1987; Hogan, 1984] of robotic manipulators. Despite the diversity of approaches, it is possible to classify most of the control methods into two major approaches: impedance control [Wang & Cheah, 1996; Carelli & Kelly, 1991] and hybrid position/force control [Raibert & Craig, 1981; Yoshikawa et al., 1988]. However, these methods require an accurate model of force interaction between the manipulator and the environment and are difficult to implement on typical industrial manipulators that are designed for position control.
An active feedback control scheme was developed in order to supply compliance for robotic deburring as a means to accommodate the interaction force due to contact motion. Kuntze [Kuntze, 1984] suggested an active control scheme, in which the actuators are commanded to increase torques in the opposite direction of the deflections. Paul [Paul et al., 1982] applied an active isolator to a chipping robot, where the isolator attached to the arm tip reduces the vibration seen by the robot. Sharon and Hardt [Sharon and Hardt, 1984] developed a multi-axis local actuator, which compensates for positioning errors at the end point, in a limited range.
Asada [Asada & Sawada, 1984] developed passive tool support mechanisms, which couple the arm tip to the workpiece surface and bear large vibratory loads. These mechanisms allow the robot to compensate for the excessive deflection when the robot contacts the workpiece. These methods reduce dynamic deflection in a certain frequency range. However, it is difficult for these control schemes, which are employed for a robot with a

passive tool, to perform well over a wide frequency band because they must drive the entire, massive robot arm. In addition, unknown compliance from a passive tool makes it difficult to control the deburring robot.
In this paper, a robotic deburring method is developed based on an integrated pneumatic actuation system (IPAS), which considers the interaction among the tool, the manipulator, and the workpiece and couples the tool dynamics and a control design that explicitly considers deburring process information. A new active tool is developed based on two pneumatic actuators, which utilizes double cutting action - initial cut followed by fine cut. Then, a coordination based control method is developed for the robotic deburring system based on the active pneumatic deburring tool. The developed control method employs a hierarchical control structure based on a coordination scheme. Robust feedback linearization is utilized to minimize the restrained elements of the precision deburring such as static and Coulomb friction and nonlinear compliance of the pneumatic cylinder stemming from the compressibility of air.

2. Modeling of the Deburring Robot

In this section, a dynamic model of a robotic arm with the new deburring tool or IPAS is developed as a robotic deburring system as shown in Fig. 2. Fig. 1 shows the integrated cylinder, which is comprised of three chambers and actuated by a single valve connected to Chamber 3. Note that the IPAS is a single input system with two pistons. The pistons are not directly connected to the inner pistons, M_{t3} and M_{t4}, which create a unique configuration of three chambers connected in series. This configuration allows the chambers adjacent to the active chamber to act as vibration isolators. This feature enables the IPAS to damp out the chatter caused by external loads and air compressibility. Therefore, a double cutting action and chattering reduction can be achieved simultaneously.

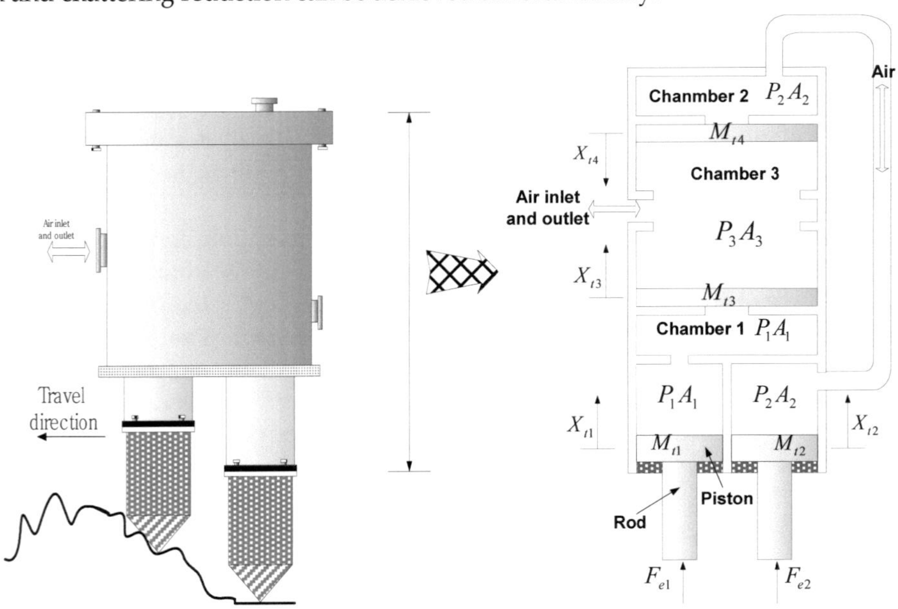

Fig. 1. Integrated double cylinder system

The dynamics of the chambers can be written as [Sorli et al., 1999]

$$G_3 = \rho_3 \frac{dV_3}{dt} + V_3 \frac{d\rho_3}{dt} \tag{1}$$

where G_3 is the entering air flow, ρ_3 the air density and V_3 the volume of Chamber 3. It is assumed that the condition of the air is ideal as following:

$$\rho_3 = \rho_{3j}\left(\frac{P_3}{P_{3j}}\right)^{1/n} = \frac{P_{3j}}{RT_{3j}}\left(\frac{P_3}{P_{3j}}\right)^{1/n} \tag{2}$$

where the subscript j indicates the initial conditions and n is the air transformation ratio. Now, V_3 is derived as

$$V_3 = A_3(L - X_{t4} - X_{t3}) \tag{3}$$

where A_3 denotes the area of Piston 3, and X_{ti} (i =4,3) is the position of Piston i . L denotes the length of Chamber 3 as shown in Fig. 3. By combining Eqs. (2) and (3) and their time derivatives in Eq. (1), the following expression is be obtained:

$$\begin{aligned} G_3 &= A_3(L - X_{t3})\frac{1}{nRT_{3j}}\left(\frac{P_3}{P_{3j}}\right)^{1/n-1}\frac{dP_3}{dt} + \frac{P_{3j}}{RT_{3j}}\left(\frac{P_3}{P_{3j}}\right)^{1/n} A_3\frac{dX_{t3}}{dt} \\ &+ A_3(L - X_{t4})\frac{1}{nRT_{3j}}\left(\frac{P_3}{P_{3j}}\right)^{1/n-1}\frac{dP_3}{dt} + \frac{P_{3j}}{RT_{3j}}\left(\frac{P_3}{P_{3j}}\right)^{1/n} A_3\frac{dX_{t4}}{dt} \end{aligned}. \tag{4}$$

Then, the pressure gradient is be written as

$$\begin{aligned} \frac{dP_3}{dt} &= \frac{nRT_{3j}}{A_3(L - X_{t3})(P_3/P_{3j})^{1/n-1}}G_3 - \frac{nP_3}{(L - X_{t3})}\frac{dX_{t3}}{dt} \\ &+ \frac{nRT_{3j}}{A_3(L - X_{t4})(P_3/P_{3j})^{1/n-1}}G_3 - \frac{nP_3}{(L - X_{t4})}\frac{dX_{t4}}{dt} \end{aligned}. \tag{5}$$

The dynamic equations are written as

$$\begin{bmatrix} M_{t3} & 0 \\ 0 & M_{t4} \end{bmatrix}\begin{bmatrix} \ddot{X}_{t3} \\ \ddot{X}_{t4} \end{bmatrix} + C\begin{bmatrix} (\dot{X}_{t1} - \dot{X}_{t3}) \\ (\dot{X}_{t1} - \dot{X}_{t4}) \end{bmatrix} + K\begin{bmatrix} (X_{t1} - X_{t3}) \\ (X_{t1} - X_{t4}) \end{bmatrix} = \begin{bmatrix} P_3A_3 - F_{f3} \\ P_3A_3 - F_{f4} \end{bmatrix} \tag{6}$$

where K and C are the stiffness and damping coefficients of the system, respectively, $\dot{X}_{ti}$ and $\ddot{X}_{ti}$ represent the velocity and the acceleration of each piston ($i = 1,2,3,4$). F_{fi} denotes the viscous friction force of the piston rod ($i = 1,2,3,4$), F_{ei} is the external force ($i = 1,2$), P_i and A_i ($i = 1,2,3$) denote the air pressure and the area of the piston, respectively, and M_{t1} and M_{t2} are the masses of each position rod.

2.3 Robotic Deburring System

Fig. 2 illustrates a multi-link rigid robot with the pneumatic deburring tool described earlier. Using the well-known Lagrangian equations, the following equations of motion of the deburring robot can be obtained:

$$\overline{m}(q)\ddot{q} + \overline{c}(q,\dot{q})\dot{q} + \overline{g}(q) = \tau \tag{7}$$

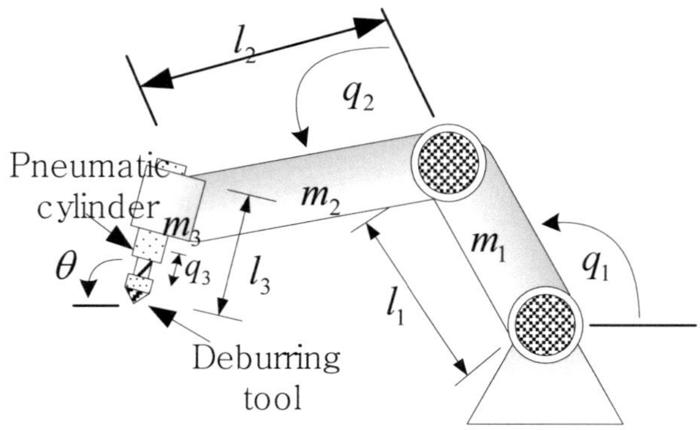

Fig. 2. Deburring robot with pneumatic tool

where $q, \dot{q}, \ddot{q}$ are the joint angle, the joint angular velocity, and the joint angular acceleration, respectively, $\overline{m}(q)$ is the 3×3 symmetric positive-definite inertia matrix, $\overline{c}(q,\dot{q})\dot{q}$ is the 3×1 vector of Coriolis and centrifugal torques, $\overline{g}(q)$ is the 3×1 gravitational torques, and τ is the 3×1 vector of the joint torques.
The mass of the links and pneumatic cylinder are considered as if they were rigidly attached. The relationship between the joint and the tip velocities can be written as

$$\dot{x} = J(q)\dot{q} \tag{8}$$

where $J(q)$ is the geometric Jacobian of the manipulator. By differentiating Eq. (8), the Cartesian acceleration term can be found as

$$\ddot{x} = J(q)\ddot{q} + \dot{J}\dot{q} \tag{9}$$

Then, the equations of motion of the robot are obtained as following:

$$m(x)\ddot{x} + c(x,\dot{x}) + g(x) = f \tag{10}$$

where $f = (J^T)^{-1}\tau$ is input expressed in task space and $m(x)$ is the inertia matrix, $c(x,\dot{x})$ is Coriolis and centrifugal forces, and $g(x)$ is gravitational forces.

Let the dynamic equation of the robot manipulator in the constraint coordinates be represented as

$$m(x)\ddot{x} + c(x,\dot{x})\dot{x} + g(x) = f + f_{rf} \tag{11}$$

where f denotes the input force and f_{rf} is the resultant force of the normal force f_n and the tangential force f_t exerted on the tool tip. The tangential force [18] can be represented as

$$f_t = \frac{bdv_t e_m}{V_t} \tag{12}$$

where V_t is the spindle speed of deburring tool; b is the tool width; d is the depth of cut; v_t is the feed rate (or the traveling speed of the end effector along the surface of the workpiece); e_m is the material-stiffness of the workpiece. The normal force f_n is assumed to be proportional to the tangential force f_t. Besides, the force angle of the deburring tool affects the tangential force. Although the value of the angle may vary substantially depending on the nature of the material flow at the tool-chip interface, as approximation 0.3 was used in these calculations [Raibert & Craig, 1981]. Therefore, the normal force f_n is considered to be smaller than the tangential force f_t in Eq. (12), where the ratio is $f_n / f_t \approx 0.3$ [Deccusse & Moog, 1985].

3. Control Design

The IPAS based deburring robot can be treated as a system that consists of two primary subsystems; the arm and the IPAS. The two subsystems differ substantially in their task assignments, dynamic characteristics and controller requirements. This physical interpretation provides an efficient approach to the control of the robotic deburring system. The control strategy for the deburring robot is illustrated in Fig. 3. The arm is commanded to follow the desired trajectory in task space, which is modified based on the position of the second piston due to varying length of the tool. In other words, the primary cutter at the front side cuts the burr first and the second cutter then attempts to eliminate the remaining burr. In case that the burr is not removed completely, the uncut depth is incorporated into the desired trajectory for compensation.

The developed control design is a decentralized control [Deccusse & Moog, 1985; Isidori, 1985], which consists of two independent controllers interacting based on the coordination scheme aforementioned for the manipulator and the IPAS, respectively. Constraint equations are derived in terms of position variables and are differentiated twice to lead to a relationship in terms accelerations, which integrate the separate controllers for stability proof. Feedback linearization is employed to design a coordination based controller. In what follows, it is shown that use of a nonlinear dynamic feedback achieves exact linearization and input-output decoupling for the robotic deburring system. However, pneumatic actuators typically have a limited bandwidth restricting the high gains which can be applied. Combined with their limited damping and low stiffness properties, which arise from the compressibility of air, the accuracy and repeatability of the performance can be limited under variable payload and supply pressure. Therefore, robust feedback linearization is employed to reduce the undesirable effect of nonlinear compliance of the pneumatic cylinder. The coordination control method is developed first and then its efficiency will be compared with the hybrid control method through simulation study.

3.1 Coordination Control

Shown in Fig. 3 is the control design for the deburring robot with the active pneumatic tool. Note that X_t denotes the position of the piston, respectively relative to their origins as described in Section 2.1. The desired trajectories of the robot wrist, denoted as $x_r^d, \dot{x}_r^d, \ddot{x}_r^d$, are modified to compensate the uncut depth based on the position of the second piston due to the varying length of the tool. Additionally, $X_t^d, \dot{X}_t^d$, and $\ddot{X}_t^d$ denote the desired trajectories for the IPAS. Feedback linearization [Isidori, 1985] is employed to design a coordination based controller. In what follows, it is shown that the use of a nonlinear dynamic feedback achieves exact linearization and input-output decoupling for the robotic deburring system.

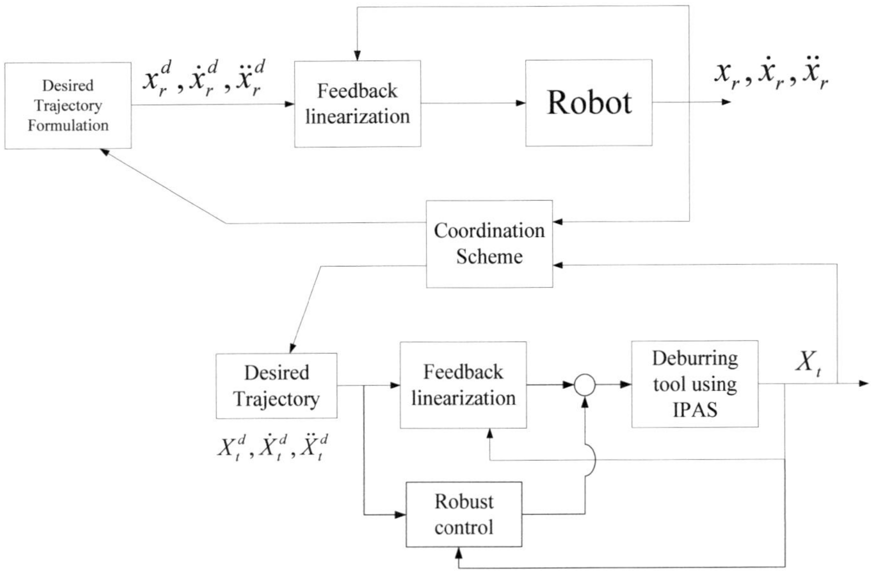

Fig. 3. Block diagram for coordinated control for robotic deburring

We assume that the robot has n links. The equations of motion of the arm are rewritten in a decentralized form as

$$m_r(x_r)\ddot{x}_r + c_r(x_r, \dot{x}_r) = f_r - R_r(x_r)\ddot{X}_t \tag{13}$$

where $x_r, \dot{x}_r$ and $\ddot{x}_r$ denote the displacement, velocity and acceleration matrix of the tip of the manipulator $n \times 1$, m_r is the inertia mass matrix $n \times n$, c_r is the matrix $n \times n$, which consists of Coriolis, centripetal, and gravity forces, f_r is the input force matrix acting on the tip of the manipulator $n \times 1$, R_r is the inertia matrix which reflects the dynamic effect of the deburring tool on the manipulator $n \times n$, and $\ddot{X}_t$ is the acceleration of IPAS $n \times 1$. Likewise, the equations of motion for deburring tool are written as

$$M_t\ddot{X}_t - C_t(\dot{X}_t, D(X_t), X_t, \text{sgn}(\dot{X}_t), \mu_c, \mu_u) + F_e = F_t(A, P) - R_t(x_r)\ddot{x}_r \tag{14}$$

where $\ddot{X}_t$ and $\dot{X}_t$ denote the acceleration and velocity matrix $n \times 1$ of the tool, M_t is the mass matrix $n \times n$ of the piston, C_t is a polynomial function of the nonlinear term $n \times 1$, μ_c is Coulomb term, μ_u is viscous coefficient [11], $D(X_t)$ is a polynomial function of the nonlinear spring caused by air compression in Eq. (14), F_t is the forces matrix $n \times 1$ acting on the piston, R_t is the inertia matrix $n \times n$ which represents the end effect of the manipulator on the tool, F_e is the external force matrix $n \times 1$ of the IPAS.

Let $p \in R^m$ denote the position vector of the tip of the robot in the fixed workspace coordinate system. The robotic deburring system is assumed to have the constraint surface defined in algebraic terms by

$$\phi(p) = \begin{bmatrix} \phi_1(p) \\ \vdots \\ \phi_{n+1}(p) \end{bmatrix} = 0 \tag{15}$$

where p is comprised of x_r and X_t. Now, the constraint Eq. (15) is differentiated once as following:

$$\dot{\phi}(p) = J_c(q)\dot{q} = 0 \tag{16}$$

where J_c denotes the geometric Jacobian matrix $n \times n$. The initial Lagrange coordinate q_0 satisfies the holonomic constraint $\phi(p_0) = 0$, where p_0 is the initial position of the robot. Then, Eq. (16) is differentiated once to produce $\ddot{\phi} = 0$, into which the subsystems, Eqs. (13) and (14) are incorporated. Then, feedback linearization can be applied to cancel the coupling terms and to design linear controllers as the outer feedback loop. Since the manipulator velocity is always in the null space of $\dot{\phi}(p)$, it is possible to define a vector of generalized velocities $\eta(t)$, which is the $n \times 1$ dimensional matrix as following:

$$\dot{x}_r = \zeta(x_r)\eta(t) = \begin{bmatrix} \dot{x}_{r1} \\ \vdots \\ \dot{x}_{rn} \end{bmatrix} = \begin{bmatrix} \zeta_1(x_{r1}) & \cdots & 0 \\ \vdots & \ddots & \vdots \\ 0 & \cdots & \zeta_n(x_{rn}) \end{bmatrix} \begin{bmatrix} \eta_1(t_1) \\ \vdots \\ \eta_n(t_n) \end{bmatrix} \tag{17}$$

where the columns of $\zeta(x_r)$ are in the $n \times n$ dimensional null space of $\phi(p)$. Differentiating Eq. (15), substituting the resulting expression for $\ddot{x}_r$ into Eq. (13), and premultiplying Eq. (13) by ζ^T, we obtain

$$\zeta^T(m_r\zeta\dot{\eta} + m_r\dot{\zeta}\eta + c_r) = \zeta^T f_r - \zeta^T R_r \ddot{X}_t \,. \tag{18}$$

Note that $\zeta^T\phi^T = 0$. Similarly substituting $\ddot{x}_r$ into Eq. (14), we have

$$M_t\ddot{X}_t - C_t + F_e = F_t - R_t\dot{\zeta}\eta - R_t\zeta\dot{\eta} \,. \tag{19}$$

Using the state vector $\chi = \begin{bmatrix} x_r^T & X_t^T & \eta^T & \dot{X}_t^T \end{bmatrix}^T$ and the block partition of the state vector

$$\chi = \begin{bmatrix} \chi_1 \\ \chi_2 \\ \chi_3 \end{bmatrix}, \text{ with } \chi_1 = x_r = \begin{bmatrix} x_{r1} \\ \vdots \\ x_{rn} \end{bmatrix}, \chi_2 = X_t = \begin{bmatrix} x_{t1} \\ \vdots \\ x_{tn} \end{bmatrix}, \chi_3 = \begin{bmatrix} \eta \\ \dot{X}_t \end{bmatrix} = \begin{bmatrix} \eta_1 \\ \vdots \\ \eta_n \\ \dot{X}_{t1} \\ \vdots \\ \dot{X}_{tn} \end{bmatrix} \tag{20}$$

the following expression is obtained:

$$\dot{\chi} = \begin{bmatrix} \dot{\chi}_1 \\ \dot{\chi}_2 \\ \dot{\chi}_3 \end{bmatrix} = \begin{bmatrix} \zeta\eta \\ \dot{X}_t \\ M^{-1}C \end{bmatrix} + \begin{bmatrix} 0 \\ 0 \\ M^{-1}E \end{bmatrix} \lambda \tag{21}$$

where

$$M = \begin{bmatrix} \zeta^T m_r \zeta & \zeta^T R_r \\ R_t \zeta & M_t \end{bmatrix}, \; E = \begin{bmatrix} \zeta^T & 0 \\ 0 & I \end{bmatrix}, \; C = \begin{bmatrix} -\zeta^T m_r \dot{\zeta}\eta & -\zeta^T c_r \\ C_t + F_e & -R_t \dot{\zeta}\eta \end{bmatrix}. \tag{22}$$

The system is input-output linearizable by using the following nonlinear feedback:

$$\begin{bmatrix} \lambda_1 \\ \vdots \\ \lambda_n \end{bmatrix} = \begin{bmatrix} E_1^{-1} \cdots 0 \\ \vdots \ddots \vdots \\ 0 \cdots E_1^{-1} \end{bmatrix} \left\{ \begin{bmatrix} M_1 \cdots 0 \\ \vdots \ddots \vdots \\ 0 \cdots M_n \end{bmatrix} \begin{bmatrix} u_1 \\ \vdots \\ u_n \end{bmatrix} - \begin{bmatrix} C_1 \\ \vdots \\ C_n \end{bmatrix} \right\} \tag{23}$$

which results in simpler state equations as following:

$$\dot{\chi} = \begin{bmatrix} \dot{\chi}_1 \\ \dot{\chi}_2 \\ \dot{\chi}_3 \end{bmatrix} = \begin{bmatrix} \begin{bmatrix} \zeta_1(x_{r1}) & \cdots & 0 \\ \vdots & \ddots & \vdots \\ 0 & \cdots & \zeta_n(x_{rn}) \end{bmatrix} \begin{bmatrix} \eta_1(t_1) \\ \vdots \\ \eta_n(t_n) \end{bmatrix} \\ \begin{bmatrix} \dot{X}_{t1} \\ \vdots \\ \dot{X}_{tn} \end{bmatrix} \\ \begin{bmatrix} 0_1 \\ \vdots \\ 0_n \end{bmatrix} \end{bmatrix} + \begin{bmatrix} \begin{bmatrix} 0_1 & \cdots & 0 \\ \vdots & \ddots & \vdots \\ 0 & \cdots & 0_n \end{bmatrix} \\ \begin{bmatrix} 0_1 & \cdots & 0 \\ \vdots & \ddots & \vdots \\ 0 & \cdots & 0_n \end{bmatrix} \\ \begin{bmatrix} I_1 & \cdots & 0 \\ \vdots & \ddots & \vdots \\ 0 & \cdots & I_n \end{bmatrix} \end{bmatrix} \begin{bmatrix} u_1 \\ \vdots \\ u_n \end{bmatrix}. \tag{24}$$

To derive the decoupling matrix, each component of the output equations is differentiated until the input appears explicitly in the derivative. In this case, the output equation is differentiated twice as following:

$$y = \begin{bmatrix} f_{11} \\ \vdots \\ f_{1n} \\ f_{21} \\ \vdots \\ f_{2n} \end{bmatrix} \tag{25}$$

$$\begin{bmatrix} \ddot{y}_1 \\ \vdots \\ \ddot{y}_n \end{bmatrix} = \dot{\Phi}(\chi) \begin{bmatrix} \begin{bmatrix} \eta_1(t_1) \\ \vdots \\ \eta_n(t_n) \end{bmatrix} \\ \begin{bmatrix} \dot{X}_{t1} \\ \vdots \\ \dot{X}_{tn} \end{bmatrix} \end{bmatrix} + \Phi(\chi) \begin{bmatrix} u_1 \\ \vdots \\ u_n \end{bmatrix} \tag{26}$$

where $\Phi(\chi)$ is the decoupling matrix of the system given by

$$\Phi(\chi) = \begin{bmatrix} \Phi_r(\chi) & 0 \\ 0 & \Phi_t(\chi) \end{bmatrix} \tag{27}$$

where

$$\Phi_t(\chi) = \begin{bmatrix} \Phi_{t1} & \cdots & 0 \\ \vdots & \ddots & \vdots \\ 0 & \cdots & \Phi_{tn} \end{bmatrix} = \begin{bmatrix} \frac{\partial f_{11}(X_{t1})}{\partial X_{t1}} & \cdots & 0 \\ \vdots & \ddots & \vdots \\ 0 & \cdots & \frac{\partial f_{1n}(X_{tn})}{\partial X_{tn}} \end{bmatrix}, \; f_{1n} = X_{tn} + f_{2n}$$

$$\Phi_r(\chi) = \begin{bmatrix} \Phi_{r1} & \cdots & 0 \\ \vdots & \ddots & \vdots \\ 0 & \cdots & \Phi_{rn} \end{bmatrix} = \begin{bmatrix} f_{21} & \cdots & 0 \\ \vdots & \ddots & \vdots \\ 0 & \cdots & f_{2n} \end{bmatrix}.$$

Applying the following nonlinear state feedback

$$\begin{bmatrix} u_1 \\ \vdots \\ u_n \end{bmatrix} = \begin{bmatrix} \Phi_r(\chi) & 0 \\ 0 & \Phi_t(\chi) \end{bmatrix}^{-1} \left(\begin{bmatrix} \upsilon_1 \\ \upsilon_2 \end{bmatrix} - \begin{bmatrix} \dot{\Phi}_r(\chi) & 0 \\ 0 & \dot{\Phi}_t(\chi) \end{bmatrix} \begin{bmatrix} \eta \\ \dot{X}_t \end{bmatrix} \right), \tag{28}$$

the input-output relationship is decoupled because each component of the auxiliary input, υ, controls one and only one component of the output, y. It is noted that the existence of the nonlinear feedback require the inverse of the decoupling matrix $\Phi(\chi)$. To complete the controller design, it is necessary to stabilize each of the above subsystem with constant state feedback. Then, the stability of the system is guaranteed by selecting appropriate constant feedback gains for the linearized system.

Now, robust feedback linearization is employed to minimize the undesirable effect of external disturbances such as static and Coulomb friction and nonlinear compliance of the pneumatic cylinder stemming from the compressibility of air as appeared in Eq. (14). Let the tracking error be defined $e_t = X_t - X_t^d$. From Eq. (14) the following expression can be obtained: one obtains

$$\ddot{X}_t = \frac{1}{M_t}(F_t(A,P) - R_t(x_r)\ddot{x}_r + C_t(\dot{X}_t, K_S(X_t), X_t, \mathrm{sgn}(\dot{X}_t), K_f) - F_e) \tag{29}$$

Then, the following error dynamics is employed:

$$(\ddot{X}_t - \ddot{X}_t^d) + \varsigma_1(\dot{X}_t^d - \dot{X}_t) + \varsigma_2(X_t^d - X_t) = 0 \tag{30}$$

Now, the feedback linearizing control P_{fl} is chosen to be

$$P_{fl} = \frac{1}{A}M_t(\ddot{X}_t^d - \varsigma_1(\dot{X}_t^d - \dot{X}_t) - \varsigma_2(X_t^d - X_t)) - \frac{1}{A}C_t + \frac{1}{A}F_e + \frac{1}{A}R_t(x_r)\ddot{x}_r \tag{31}$$

where $X_t^d, \dot{X}_t^d, \ddot{X}_t^d$ are the desired position, velocity, and acceleration and ς_1 and ς_2 are the control gains. In addition,Eq. (31) is uncertainty in the system, an auxiliary control input w can be injected as follows

$$\overline{P}_{fl} = P_{fl} + \frac{M_t}{A}w \tag{32}$$

Using $\overline{P}_{fl}$ Eq. (32) yields the error dynamics

$$(\ddot{X}_t - \ddot{X}_t^d) + \varsigma_1(\dot{X}_t - \dot{X}_t^d) + \varsigma_2(X_t - X_t^d) + \Theta(\cdot) - w = 0 \tag{33}$$

where $\Theta(\cdot)$ is lumped uncertainty originating from the bounded uncertainties in the plant. Here, a layer of sliding manifold and a switching law on the reduced order manifold are defined so as to compensate for the bounded lumped uncertainty stemming from the difference between the actual and the nominal plant parameters [Acarman et al., 2001]. Therefore the layer of sliding manifold can be defined as

$$S_w = \dot{e}_t + C_w e_t \tag{34}$$

where $\dot{e}_t$ and e_t denotes $\dot{X}_t - \dot{X}_t^d$ and $X_t - X_t^d$, respectively. It is noted that $C_w > 0$. Now, let

$$w = (\varsigma_1 - C_w)\dot{e}_t + \varsigma_2 e_t - \tilde{N}\,\mathrm{sgn}(S_w) \tag{35}$$

where $\tilde{N} > |\Theta(\cdot)|$. Then, $\dot{S}_w$ is expressed as

$$\dot{S}_w = \ddot{e}_t + C_w \dot{e}_t = -\Theta(\cdot) - \tilde{N}\,\mathrm{sgn}(S_w) \tag{36}$$

Therefore, $\dot{S}_w \cdot S_w < 0$ is achieved. In summary, the deburring system of interest is considered to have two subsystems as described. The interactive dynamics of the subsystems are decoupled in feedback sense by feedback linearization or Eq. (28) and suitable controllers are designed for the subsystems based on the motion coordination scheme as described. Then, a robust controller is designed for the tool subsystem to minimize the harmful effect of static and Coulomb frictions and nonlinear compliance of the pneumatic cylinder due to air compressibility. Therefore, the stability of the overall system can be achieved by properly selecting the feedback gains of each subsystem together with proper gains of the robust feedback for the tool as shown in Eqs. (33) – (36).

4. Simulation

Simulation study was performed to investigate the performance of the controllers developed for the robotic deburring systems with different tools: (1) the hybrid controller [12, 13, 20, 21, 22, 23, 24, 28, 39] for the rigid tool based system (2) the coordination controller for the single active pneumatic cylinder tool (3) the coordination controller for the double active pneumatic tool based system. Note that the mathematical models for the rigid and the single active pneumatic cylinder tools are not shown due to their nature of being a subset of the IPAS.

Fig. 4 shows the simulation results for the hybrid control system. The following parameters were used in simulation:

$$m_1 = 16\text{kg},\ m_2 = 12\text{kg},\ l_1 = 0.5\text{m, and } l_2 = 0.7\text{m}$$

where m_1 and m_2 are the masses of each link of the 2 DOF manipulator, l_1 and l_2 are the lengths of each link. The feedback gains of the controller were chosen as following: $f_d = 20N$, $k_{p1} =$ diag[150, 150, 150], $k_{d1} =$ diag[70, 70, 70], $k_{p2} =$ diag [750, 750, 750], and $k_{d2} =$ diag[230, 230, 230] where f_d is the desired force, and k_{pi} and k_{di} ($i = 1,2$) are the control PD gains.

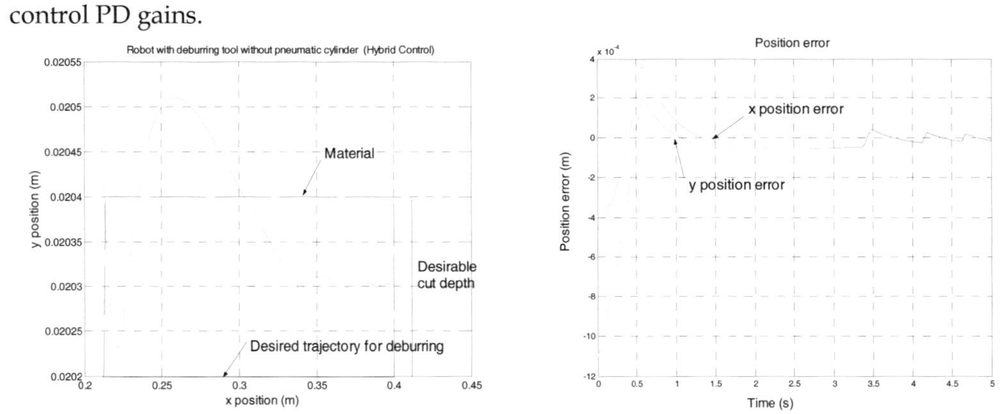

(a) (b)

Fig. 4. Rigid tool (a) tracking (b) position error

Fig. 4 (a) and (b) show the performance of the hybrid controller designed for the deburring robot with a rigid tool. In the simulation, the stiffness of the material was set to 500000 *N/m* and the desired cut depth was chosen to be 0.0002 *m*. The results show large deburring error, which remains oscillatory after large overshoot in the transient period due to chattering caused by the air compressibility and the contact motion between the robot and the workpiece. The following parameters were used for the tangential force as:

$$b = 16\ mm,\ v_t = 0.08\ m/s,\text{ and } V_t = 30{,}000\ RPM .$$

Fig. 5 depicts the deburring performance of the coordination controller designed for the robot with a single active pneumatic cylinder tool. The following parameters were used for simulation:

- Chamber pressures $P_{1s} = P_{2s} = 1\times10^5\,Pa$
- Piston areas $A_{s1} = A_{s1} = 0.000256m^2$,
- Piston mass M_{ts} = 0.01kg
- Chamber temperatures $T_{1in} = T_{2in} = 293^\circ K$.

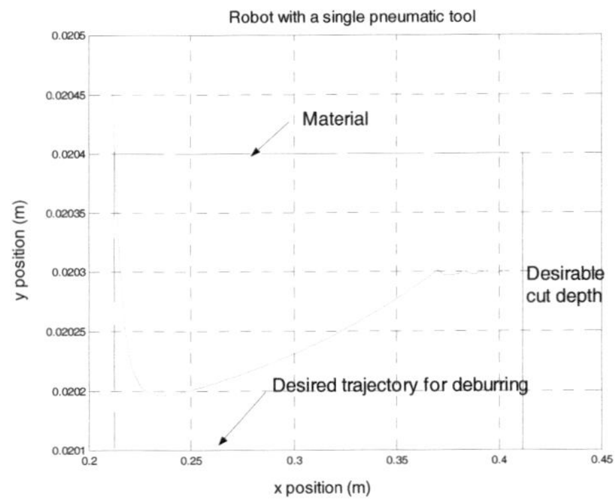

(a)

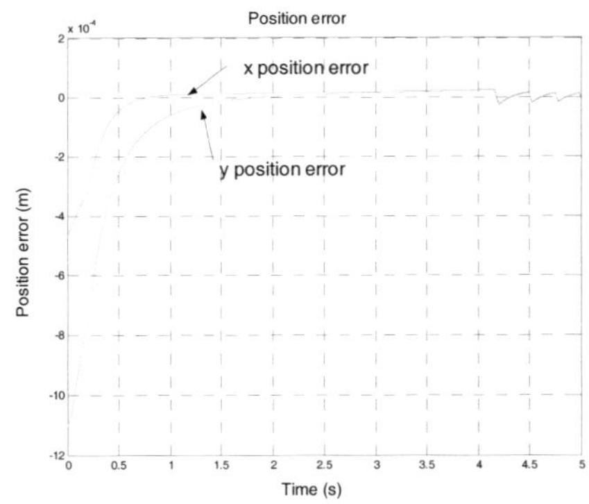

(b)

Fig. 5. Single pneumatic tool (a) tracking (b) position error

As shown in Fig. 5 (a) and (b), the transient performance is improved significantly with the single active pneumatic tool with the coordination controller in comparison to the previous case. However, the steady-state performance still remains unsatisfactory due to the chatter that appears in the response, which is caused by the compressibility of the air in the pneumatic cylinder and therefore requires repetitive deburring. Nevertheless, the simulation results demonstrate the potential of a pneumatic actuator as an efficient tool which can significantly enhance the performance of a deburring robot if the chattering effect can be eliminated or minimized by an improved design of the tool and/or an efficient control.

Fig. 6 demonstrates the deburring performance of the robot with the IPAS as shown in Fig. 1. The developed coordination control method by using feedback linearization was utilized for the IPAS based deburring system. It is noted that the initial position of X_{ti} (i =1, 2, 3, 4) is set to zero. The following is the additional parameters used for the integrated cylinder:

$P_{3j} = 1\times10^5\,Pa$, $A_1 = A_2 = 0.000256m^2$, $A_3 = A_4 = 0.00055m^2$, n=0.8, $F_{f1,2} = 10N$, $F_{f3,4} = 15N$, $M_{t1} = M_{t2}$ =0.01kg, $M_{t3} = M_{t4}$ =0.015kg, and $T_{3j} = 293^\circ K$.

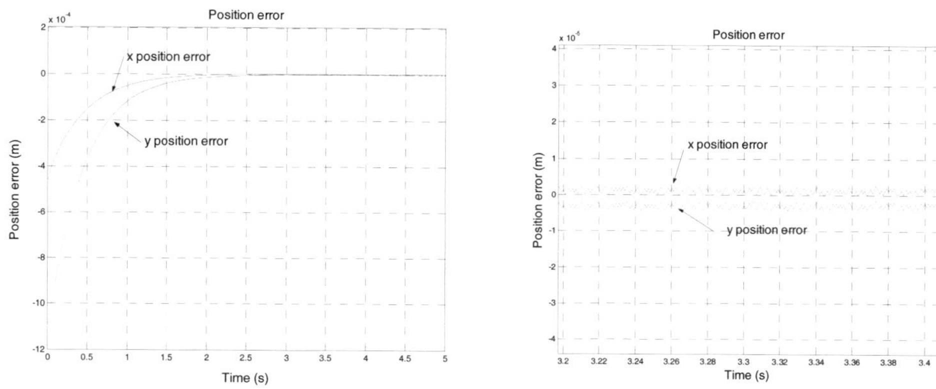

(a) (b)

Fig. 6. Integrated double pneumatic cylinder (Coordinated control without robust control) (a) Position error (b) Enlarged position error

It is evident as shown in Fig. 6 (a) that the deburring performance of the system is greatly improved with the IPAS and the coordination controller. The simulation results show quick and smooth transient response and nearly zero steady-state error. The integrated system particularly improves the transient behavior in comparison to the single cylinder system. However, Fig. 6 (b) shows the chatter of position error of the IPAS when Fig. 6 (a) is enlarged. The chatter is from the compressibility and flexibility of the air. Such fluctuating position error can occur in harm of the system. Also, Combined with their limited damping and low stiffness properties, which arise from the compressibility of air, the accuracy and repeatability of the performance can be limited under variable payload and supply pressure. To eliminated and/or reduce the undesirable effect of nonlinearity, in next simulation, robust feedback linearization is employed.

Fig. 7 depicts the deburring performance of the coordination controller based on robust feedback linearization. The following parameters were used for simulation:

$P_{3j} = 1\times10^5 Pa$, $A_1 = A_2 = 0.000256m^2$, $A_3 = A_4 = 0.00055m^2$, n=0.8, $F_{f1,2} = 10N$, $F_{f3,4} = 15N$, $M_{t1} = M_{t2}$ =0.01kg, $M_{t3} = M_{t4}$ =0.015kg, $T_{3j} = 293^\circ K$, $\varsigma_1 = 25$, $\varsigma_2 = 7$, $C_w = 7$, $\tilde{N} = 1$, and $\Theta(\cdot)$ =0.5,

Fig. 7 (b) shows the reduction of position error caused by the, which is caused by the compressibility of the air in the pneumatic cylinder. In this simulation, the oscillatory position errors are almost eliminated in difference with the previous results by using the robust feedback linearization. Through the robust feedback as shown in Fig. 3, the additional robust controller could soften the chatter by the air compressibility in pneumatic tool. The simulation results demonstrate the efficacy of the developed coordination control based on robust feedback linearization for the new deburring tool.

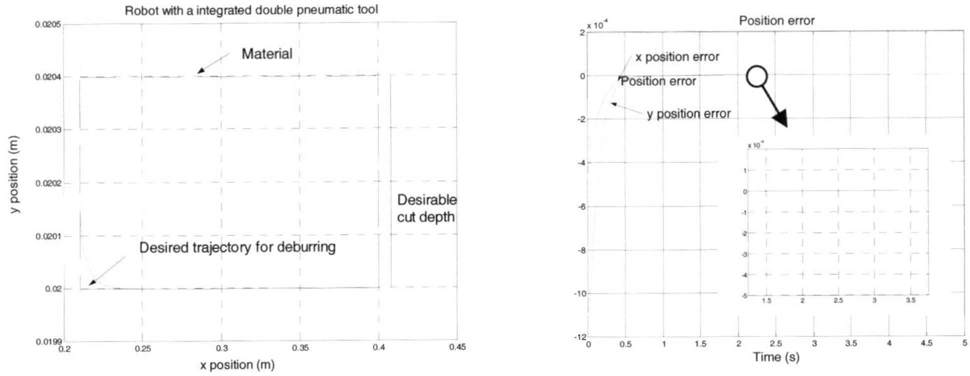

(a) (b)

Fig. 7. Integrated double pneumatic cylinder (a) tracking (b) position error (Robust Feedback linearization)

5. Conclusion

High-quality robotic deburring requires efficient control of the deburring path and contact forces, as well as optimal selection of a suitable feed-rate and tool design. In this paper, an efficient robotic deburring method was developed based on a new active pneumatic tool, which considers the interaction among the tool, the manipulator, and the workpiece and couples the tool dynamics and a control design that explicitly considers deburring process information. A new active pneumatic tool was developed by physically integrating two pneumatic actuators, which implements double cutting action - initial cut followed by fine cut. Then, a control method was developed for the robotic deburring system based on the active pneumatic tool, which utilizes coordinated control based on a feedback linearization for the manipulator and a robust feedback linearization for the deburring tool using a pneumatic cylinder. From the simulation results, robust feedback linearization achieved the smooth transient response and nearly zero steady-state error in spite of the undesirable effect of external disturbances. The developed control system employs the two-level hierarchical control structure based on a simple coordination scheme. Simulation results showed that the developed system significantly reduces the chattering of the deburring robot and improves the deburring accuracy. Implementation of the developed method is intended for experimental verification in the future.

6. References

[1] Kuntze, H. B. (1984). On the closed-loop control of an elastic industrial robot, *Proceedings: 1984 American Control Conference.*

[2] Sharon A. & Hardt D.E. (1984). Enhancement of robot accuracy using endpoint feedback and a macro-micro manipulator system, *Proceedings: 1984 American Control Conference*, pp.1836.

[3] Asada, H. & Sawada, Y. (1984). Design of an adaptable tool guide for grinding robot, *ASME Design Engineering Technical Conference paper*, No. 84-Det-41.

[4] Paul, F. W., Gettys, T.K. & Thomas, J.D. (1982). Defining of iron castings using a robot positioned chipper, *Proceedings: Robotics Research and Advanced Application, ASME*, pp. 269-278.

[5] Sorli, M., Gastaldi, L., Codina, E., & Heras, H. (1999). Dynamic analysis of pneumatic actuators, *Elsevier Science.*

[6]Armstmstrong-Helouvry, S., Dupont, P., & Canudas De Wit, C. (1994). A survey of models, analysis tools and compensation methods for the control of machines with friction, *Automatica*, pp.1083-1183.

[7] Deccusse, J. & Moog, C. H. (1985). Decoupling with dynamic compensation for strong invertible affine nonlinear systems, *International Journal of Control*, 42: 1387-1398.

[8] Isidori, A. (1985). Nonlinear control systems: An introduction, *Springer verlag*, Berlin, New york.

[9] Kazerooni, H. and Guo, J. (1987). Direct-drive, active compliant End-Effector, *IEEE Journal of Robotics and Automation.*

[10] Hollis, R.L. (1989). A planar XY robotic fine positioning device, *Proceedings of the IEEE International Conference on Robotics and Automation*, Raleigh, North Carolina.

[11]Raibert, M.H. & Craig J.J. (1981). Hybrid position/ force control of manipulator," *ASME Journal of Dynamics System, Measurements and Control*, Vol.102, pp.126-133.

[12] Whitney, D. E. (1987). Historical perspective and state of the art in robot force control, *International Journal of Robotics Research*, vol. 6, no. 1, pp. 3–14.

[13] Bopp, T. (1983). Robotic finishing applications: Polishing sanding, grinding, *Proceeding of the 13th International Symposium on Industrial Robots.*

[14] Gustaffson, L. (1983). Deburring with industrial, Robots, *Technical report*, Society of Manufacturing Engineers.

[15] Hogan, N. (1984). Impedance control of industrial robots, *Journal of Robotics and Computer Integrated Manufacturing*, Vol. 1, No. 1, pp.97-113.

[16] Wang, D. & Cheah, C. C. (1996). A robust learning control scheme for manipulators with target impedance at end-effectors, *Robotics and Manufacturing: Recent Trends in Research and Applications,* ASME Press, Vol. 6, pp. 851–856.

[17] R. Carelli, R. & Kelly, R. (1991). An adaptive impedance/force controller for robot manipulators, IEEE Transactions on Automatic Control, Vol. 36, pp. 967–972.

[18] Yoshikawa, T., Sugie, T. & Tanaka, M. (1988). Dynamic hybrid position/force control of robot manipulators: Controller design and experiment, *IEEE Journal of Robotics and Automation*, Vol. 4, pp. 699–705.

[19] Acarman, T., Hatipoglu, C. & Ozguner, U. (2001). A robust nonlinear controller design for a pneumatic actuator, *American Control Conference*, 2001. Proceedings of the 2001, 25-27 June 2001, Vol.6, pp.4490 - 4495 .

2

Trajectory tracking control for robot manipulators with no velocity measurement using semi-globally and globally asymptotically stable velocity observers

Farah Bouakrif
LAMEL Laboratory, University of Jijel
Algeria

1. Introduction

During the last decade the class of rigid robot systems has been the subject of intensive research in the field of systems and control theory, particularly owing to the inherent nonlinear nature of rigid robots. For the same reason, these systems have widely been used to exemplify general concepts in nonlinear control theory. As a result of this excessive research activity a large variety of control methods for rigid robot systems have been proposed, such as, proportional-integral-derivative (PID) control (Kelly, 1995), computed torque control (Luh et al., 1980), which achieve the trajectory tracking objective by feedback linearization of the nonlinear robot dynamics, adaptive control (Ortega & Spong, 1989), variable structure control (Slotine & Sastry, 1983), fuzzy control (Chang & Chen, 2000), passivity based control (Ryu et al, 2004; Bouakrif et al., 2010) and iterative learning control (Bouakrif et al., 2007; Tayebi, 2007).
Many of these previous controllers require the complete state measurements, that is position and velocity, is available for feedback. Unfortunately, in practice this assumption can only partially be fulfilled for two reasons. First, although robot systems generally are equipped with high precision sensors for position measurements, velocity measurements are often contaminated with a considerable amount of noise. This circumstance may reduce the dynamic performance of the manipulator, since in practice, the values of the controller gain matrices are limited by the noise present in the velocity measurements (Khosla & Kanade, 1988). Second, in robotic applications today velocity sensors are frequently omitted owing to the considerable savings in cost, volume and weight that can be obtained this way. A good solution of this problem is the use of the velocity observers to reconstruct the missing velocity signal starting from the available position measurements. Due to the nonlinear and coupled structure of the robot dynamical model, the problem of designing observers for robots is a very complex one. Recently, exploiting the structural properties of the robot dynamics, a number of conceptually different methods for both regulation and tracking control of robots equipped with only position sensors have been developed (Canudas dewit et al., 1992; Paden & Panja, 1988). (Berghuis & Nijmeijer, 1993) presented a controller-

observer scheme for the global regulation of robots using only position feedback. The PD control with high-gain observer was developed (Yu & Li, 2006), the authors propose to reconstruct the velocity signal via a high-gain observer, but a quite noisy movement of the manipulator, which may be undesirable for greater robots employed for industrial applications.

In this chapter, we want to solve the trajectory tracking problem of rigid robot manipulators which are not equipped with the tachometers (velocity sensors) to avoid the disadvantage mentioned in the previous paragraph. For this purpose, two velocity observers are presented to estimate the missing velocity. Using the first observer, an estimate region of attraction is given. It is important to observe that this region can be made arbitrarily large by increasing the observer gain. This kind of stability is called semi-global. The second is globally asymptotically stable. Thus, there is more freedom to choose the initial states. This presents an advantage of the second observer. Thereafter, these observers are integrated with a nonlinear controller by replacing the velocity in the control law with its estimation yielded by these observers, independently. Furthermore, the semi-global and global asymptotic stability conditions are established of the composite controller consisting of robot manipulator, nonlinear controller and the first and second velocity observer, respectively. This proof is based on Lyapunov theory and using saturation technique for the second observer. Finally, simulation results on two-link manipulator are provided to illustrate the effectiveness of the global velocity observer based trajectory tracking control.

2. Dynamic equation for robot manipulators

We consider a robot manipulator that is composed of serially connected rigid links. The motion of the manipulator with n-links is described by the following dynamic equation:

$$\tau = M(q)\,\ddot{q} + C(q,\dot{q})\,\dot{q} + G(q) \tag{1}$$

where $q(t)$, $\dot{q}(t)$, $\ddot{q}(t) \in R^n$ denote the link position, velocity, and acceleration vectors, respectively, $M(q(t)) \in R^{n\times n}$ represents the link inertia matrix, $C(q(t),\dot{q}(t)) \in R^{n\times n}$ represents centripetal-Coriolis matrix, $G(q(t)) \in R^{n\times 1}$ represents the gravity effects, and $\tau\ (t) \in R^{n\times 1}$ represents the torque input vector.

In the sequel, $q_d(t), \dot{q}_d(t), \ddot{q}_d(t) \in R^n$ denote the desired link position, velocity, and acceleration vectors, respectively.

The dynamic equation (1) has the following properties (Berghuis, 1993; Ortega & Spong, 1989) that will be used in the controller development and analysis.

P 1: The inertia matrix $M(q(t))$ is symmetric, positive definite and bounded as

$$0_n < M_m < \|M(q)\| < M_M \tag{2}$$

where $q \in R^n$, and $M_M > M_m > 0$.

P 2: $\forall i \in \{i,...,n\}$, the ith element of the vector $C(q,\dot{q})\dot{q}$ is equal to $\dot{q}^T N_i(q)\dot{q}$ with N_i symmetric, continuously differentiable, and such that $\exists \overline{N}_i > 0$ satisfies

$$\left\| N_i(q) \right\| \leq \overline{N}_i \qquad \forall q \in R^n \; . \tag{3}$$

P 3: Norm of the centripetal-Coriolis is bounded as follows

$$\left\| C(q,\dot{q} \right\| \leq C_m \left\| \dot{q} \right\| . \tag{4}$$

P 4: The matrix $\dot{M}(q,\dot{q}) - 2\,C(q,\dot{q})$ is skew-symmetric, i.e., for all $X \in \Re^n$,

$$X^T(\dot{M}(q,\dot{q}) - 2C(q,\dot{q}))\,X = 0\; . \tag{5}$$

P 5 : For all $x,y \in R^n$

$$C(q,x)y = C(q,y)x \tag{6}$$

$$C(q,z+\alpha x)y = C(q,z)y + \alpha C(q,x)y\; . \tag{7}$$

In this paper, the following lemmas are used.

Lemma 1 (Shim et al., 2001): Consider a C^1 function $f(x,y): R^p \times R^q \to R$ which is continuous and well defined on $X \times R^q$ where $X = \{x \in R^p \big| \; |x_i| \leq \rho_i, 1 \leq i \leq p\}$ with $\rho_i > 0$. Then $f(\sigma(x),y)$ is globally well defined and equal to $f(x,y)$ for $x \in X$, and where exists $L(y)$ such that

$$\left\| f(\sigma(\overline{x}),y) - f(\sigma(\widetilde{x}),y) \right\| \leq L(y) \left\| \overline{x} - \widetilde{x} \right\|, \forall \overline{x}, \widetilde{x} \in R^p, \quad \forall y \in R^q \tag{8}$$

where $\sigma(x)$ is an element-wise saturation function which is saturated outside X .

Proof

By the Mean Value Theorem, there exists $z \in R^p$ such that

$$f(\sigma(\overline{x}),y) - f(\sigma(\widetilde{x}),y) = \frac{\partial f}{\partial x}(z,y)(\sigma(\overline{x}) - \sigma(\widetilde{x})) \tag{9}$$

which implies

$$\left\| f(\sigma(\overline{x}),y) - f(\sigma(\widetilde{x}),y) \right\| \leq L(y) \left\| \sigma(\overline{x}) - \sigma(\widetilde{x}) \right\| \tag{10}$$

where $L(y)$ is the maximum of $\left\|\frac{\partial f}{\partial x}(z,y)\right\|$ with respect to z over the compact range of saturation function. Then the claim (8) follows from the fact that $\|\sigma(\bar{x})-\sigma(\tilde{x})\| \le \|\bar{x}-\tilde{x}\|$.
Lemma 2 "Barbalat's lemma" (Slotine & Li, 1991): If H is a continuous function, and it is bounded when $t \to \infty$, and if $\dot{H}$ is uniformly continuous in time, then $\dot{H}=0$.
In (2), (3), (4) and in the sequel the norm of a vector X is defined as

$$\|X\| = \sqrt{X^T X} \tag{11}$$

and the norm of a matrix A as

$$\|A\| = \sqrt{\lambda_{\max}(A^T A)} \tag{12}$$

with $\lambda_{\max}(.)$ denotes the maximum eigenvalue of A.
The following assumption is imposed.
Assumption: The robot velocity is bounded by a known constant V_m such that

$$\|\dot{q}(t)\| \le V_m \qquad \forall t \in R. \tag{13}$$

Remark 1. This assumption is definitively realistic. In fact, it is reasonable to expect that the joint velocities of a robot will not exceed certain a priori bounds that come from the mechanic limitations of the robot and/or from the way the robot operates. Moreover, this assumption is recurrent in the literature on control for robotic manipulators, for example (Berghuis & Nijmeijer, 1993; Nicosia & Tomei, 1990; Xian et al., 2004).

3. Controller-observers design

In this section we present the main results of this chapter, formulated in a lemma and two theorems and their proofs. Indeed, we want to solve the trajectory tracking problem of robot manipulators without using the velocity signal. This signal is reconstructed, firstly by a semi-globally stable velocity observer and secondly by a globally stable velocity observer.

3.1 Semi-globally asymptotically stable observer

Consider the following velocity observer

$$\dot{z} = M^{-1}\left[\tau - C(q,\dot{\hat{q}})\dot{\hat{q}} - G(q)\right] - L\dot{\hat{q}} \tag{14}$$

$$\dot{\hat{q}} = z + Lq \tag{15}$$

Where $\dot{\hat{q}}$ represents the estimated velocity, z is the observer state. $L = lI_n$, where $l > 0$ and $I_n \in R^{n\times n}$ is an identity matrix.

Lemma 3

If $L_m > \dfrac{2C_m V_m}{M_m}$ then $\lim_{t\to\infty} \dot{\tilde{q}} = 0$, and the initial error $\dot{\tilde{q}}(0)$ belongs to the ball B defined by: $B = \left\{ \dot{\tilde{q}} \in R^n \,\middle|\, \left\|\dot{\tilde{q}}(0)\right\| < \left(\dfrac{M_m L_m}{C_m} - V_m\right)\sqrt{\dfrac{M_m}{M_M}} \right\}$. Where L_m denotes the minimum eigenvalue of L and $\dot{\tilde{q}} = \dot{q} - \dot{\hat{q}}$.

Proof

The time-derivative of (15) gives us

$$\ddot{\hat{q}} = M^{-1}\left[\tau - C(q,\dot{\hat{q}})\dot{\hat{q}} - G(q)\right] + L(\dot{q} - \dot{\hat{q}}). \tag{16}$$

From (1), we can write

$$\ddot{q} = M^{-1}\left[\tau - C(q,\dot{q})\dot{q} - G(q)\right]. \tag{17}$$

Subtracting (16) from (17), we have

$$\ddot{\tilde{q}} = M^{-1}\left[-C(q,\dot{q})\dot{q} + C(q,\dot{\hat{q}})\dot{\hat{q}}\right] - L\dot{\tilde{q}}. \tag{18}$$

Using the property 5, we obtain

$$\ddot{\tilde{q}} = M^{-1}\left[-2C(q,\dot{q})\dot{\tilde{q}} + C(q,\dot{\tilde{q}})\dot{\tilde{q}}\right] - L\dot{\tilde{q}}. \tag{19}$$

Consider the following Lyapunov function

$$V(\dot{\tilde{q}}) = \frac{1}{2}\dot{\tilde{q}}^T M(q)\dot{\tilde{q}}. \tag{20}$$

Thus

$$\frac{1}{2}M_m\left\|\dot{\tilde{q}}(t)\right\|^2 \le V(\dot{\tilde{q}}(t)) \le \frac{1}{2}M_M\left\|\dot{\tilde{q}}(t)\right\|^2. \tag{21}$$

The time-derivative of (20) gives us

$$\dot{V}(\dot{\tilde{q}}) = \dot{\tilde{q}}^T M(q)\ddot{\tilde{q}} + \frac{1}{2}\dot{\tilde{q}}^T \dot{M}(q)\dot{\tilde{q}}. \tag{22}$$

From (19) and (22), we obtain

$$\dot{V}(\tilde{q}) = \dot{\tilde{q}}^T \left(\frac{1}{2}\dot{M}(q) - C(q,\dot{q}) \right)\dot{\tilde{q}} + \dot{\tilde{q}}^T \left(C(q,\dot{\tilde{q}}) - C(q,\dot{q}) \right)\dot{\tilde{q}} - \dot{\tilde{q}}^T ML\dot{\tilde{q}} \, . \tag{23}$$

Using the properties 1, 3, 4 and the assumption (13), we have

$$\dot{V} \le -\left[M_m L_m - C_m \left(V_m + \left\| \dot{\tilde{q}} \right\| \right) \right] \left\| \dot{\tilde{q}} \right\|^2 . \tag{24}$$

If

$$L_m > \frac{C_m \left(V_m + \left\| \dot{\tilde{q}} \right\| \right)}{M_m} \tag{25}$$

thus

$$L_m > \frac{2C_m V_m}{M_m} \tag{26}$$

then

$$\dot{V} \le 0 \, . \tag{27}$$

From (25), it comes

$$\left\| \dot{\tilde{q}} \right\| < \frac{M_m L_m}{C_m} - V_m \, . \tag{28}$$

From (21), (24) and (28), it follows that if

$$\left\| \dot{\tilde{q}}(0) \right\| < \left(\frac{M_m L_m}{C_m} - V_m \right) \sqrt{\frac{M_m}{M_M}} \, . \tag{29}$$

Then, we have a semi-global asymptotic stability.
Remark 2. It is important to observe that the region of attraction can be made arbitrarily large by increasing the observer gain L . As this region can be increased systematically by the gain L , we have semi-global asymptotic stability.
Now, we integrate this observer with a nonlinear controller and the semi-global asymptotic stability condition of the closed loop system is given in the following theorem.

Theorem 1
Given the robot dynamics (1), and let assumption (13) be satisfied. Under the following control law

$$\tau = M(q)\,\ddot{q}_d + C(q,\dot{\hat{q}})\dot{q}_d + G(q) + K_v \left(\dot{q}_d - \dot{\hat{q}} \right) + K_p\, E \tag{30}$$

with $\dot{\hat{q}}$ is given by (14) and (15). If

$$L_m > \frac{\left(C_m V_m + K_{vM}\right)^2}{2K_{vm}} M_m^{-1} + 2C_m V_m M_m^{-1} . \tag{31}$$

Then, the closed-loop system is semi-globally asymptotically stable. Hence

$$\lim_{t\to\infty} E(t) = \lim_{t\to\infty} \dot{E}(t) = \lim_{t\to\infty} \dot{\tilde{q}}(t) = 0 . \tag{32}$$

Moreover a region of attraction is given b

$$B = \left\{ y \in R^{3n} \middle| \left\| y(0) \right\| < \sqrt{\frac{q_m}{q_M}} \left(\frac{M_m L_m}{C_m} - \frac{\left(C_m V_m + K_{vM}\right)^2}{2K_{vm} C_m} - V_m \right) \right\} . \tag{33}$$

Where K_v and K_p are symmetric, positive definite matrices. $E(t) = q_d(t) - q(t)$, $\dot{E}(t) = \dot{q}_d(t) - \dot{q}(t)$. $y^T = \left[\dot{E}^T \; E^T \; \dot{\tilde{q}}^T \right]$, $Q = diag\{M(q), K_p, M(q)\}$, $q_m = \lambda_{\min}(Q) = \min\{M_m, K_{pm}\}$, $q_M = \lambda_{\max}(Q) = \max\{M_M, K_{pM}\}$, $K_{pM} = \left\| K_p \right\|$ and $K_{vM} = \left\| K_v \right\|$. L_m, K_{pm} and K_{vm} denote the minimum eigenvalue of L, K_p and K_v, respectively.

Proof

The analysis of asymptotic stability is in two parts. In the first part, we demonstrate that the closed-loop system is stable. In the second part, we demonstrate that it is semi-globally asymptotically stable.

Part 1

Let $\dot{\bar{q}} = \dot{q}_d - \dot{\hat{q}} = \dot{\tilde{q}} + \dot{E}$.

From (18), we can write

$$M\ddot{\tilde{q}} + C(q,\dot{q})\dot{q} - C(q,\dot{\hat{q}})\dot{\hat{q}} + ML\dot{\tilde{q}} = 0 . \tag{34}$$

Subtracting (1) from (30), we find

$$M\ddot{E} + C(q,\dot{\hat{q}})\dot{q}_d - C(q,\dot{q})\dot{q} + K_v \dot{\bar{q}} + K_p E = 0 . \tag{35}$$

The sum of (34) and (35) gives us

$$M\ddot{\bar{q}} + C(q,\dot{\hat{q}})\dot{\bar{q}} + K_v \dot{\bar{q}} + ML\dot{\tilde{q}} + K_p E = 0 . \tag{36}$$

Thus

$$M\ddot{E} + \left(C(q,\dot{\hat{q}}) + K_v\right)\dot{E} + K_p E = -M\ddot{\tilde{q}} - \left(C(q,\dot{\hat{q}}) + K_v\right)\dot{\tilde{q}} - ML\dot{\tilde{q}}\ . \tag{37}$$

From (34), (37) and using the property 5, we obtain

$$M\ddot{E} + \left(C(q,\dot{\hat{q}}) + K_v\right)\dot{E} + K_p E = C(q,\dot{q})\dot{\tilde{q}} - K_v\dot{\tilde{q}}\ . \tag{38}$$

Consider the following Lyapunov function

$$H(\dot{E},E,\dot{\tilde{q}}) = \frac{1}{2}\dot{E}^T M\dot{E} + \frac{1}{2}E^T K_p E + \frac{1}{2}\dot{\tilde{q}}^T M\dot{\tilde{q}}\ . \tag{39}$$

Hence

$$H(\dot{E},E,\dot{\tilde{q}}) = \frac{1}{2}y(t)^T Q\ y(t)\,. \tag{40}$$

It follows that

$$\frac{1}{2}q_m\|y(t)\|^2 \le H(y(t)) \le \frac{1}{2}q_M\|y(t)\|^2\,. \tag{41}$$

The time derivative of (39) evaluated along (34), (38) and using the properties 4 and 5,is

$$\dot{H} = -\dot{E}^T K_v\dot{E} - \dot{E}^T K_v\dot{\tilde{q}} + \dot{E}^T C(q,\dot{q})\dot{\tilde{q}} - \dot{\tilde{q}}^T ML\dot{\tilde{q}} - \dot{\tilde{q}}^T C(q,\dot{q})\dot{\tilde{q}} + \dot{\tilde{q}}^T C(q,\dot{\tilde{q}})\dot{\tilde{q}}\ . \tag{42}$$

Using the properties 1, 3 and the assumption (13), we find

$$\dot{H} \le -K_{vm}\|\dot{E}\|^2 - M_m L_m\|\dot{\tilde{q}}\|^2 + C_m\left(V_m + \|\dot{\tilde{q}}\|\right)\|\dot{\tilde{q}}\|^2 + \dot{E}^T C(q,\dot{q})\dot{\tilde{q}} - \dot{E}^T K_v\dot{\tilde{q}}\ . \tag{43}$$

Thus

$$\dot{H} \le -K_{vm}\|\dot{E}\|^2 - \left(M_m L_m - C_m\left(V_m + \|\dot{\tilde{q}}\|\right)\right)\|\dot{\tilde{q}}\|^2 + \dot{E}^T C(q,\dot{q})\dot{\tilde{q}} - \dot{E}^T K_v\dot{\tilde{q}}\ . \tag{44}$$

We note that

$$\dot{E}^T C(q,\dot{q})\dot{\tilde{q}} - \dot{E}^T K_v \dot{\tilde{q}} \leq 2 \left\|\dot{E}\right\| \left(\frac{C_m V_m + K_{vM}}{2}\right) \left\|\dot{\tilde{q}}\right\|$$

$$\leq 2 \left\|\dot{E}\right\| \left(\frac{C_m V_m + K_{vM}}{2}\right) \left[\frac{2}{M_m L_m - C_m\left(V_m + \left\|\dot{\tilde{q}}\right\|\right)}\right]^{\frac{1}{2}} \left[\frac{M_m L_m - C_m\left(V_m + \left\|\dot{\tilde{q}}\right\|\right)}{2}\right]^{\frac{1}{2}} \left\|\dot{\tilde{q}}\right\| \tag{45}$$

$$\leq \left\|\dot{E}\right\|^2 \left(\frac{C_m V_m + K_{vM}}{2}\right)^2 \left[\frac{2}{M_m L_m - C_m\left(V_m + \left\|\dot{\tilde{q}}\right\|\right)}\right] + \left\|\dot{\tilde{q}}\right\|^2 \left[\frac{M_m L_m - C_m\left(V_m + \left\|\dot{\tilde{q}}\right\|\right)}{2}\right].$$

From (44) and (45), we have

$$\dot{H} \leq -\left(K_{vm} - \frac{\left(C_m V_m + K_{vM}\right)^2}{2\left(M_m L_m - C_m\left(V_m + \left\|\dot{\tilde{q}}\right\|\right)\right)}\right) \left\|\dot{E}\right\|^2 - \frac{1}{2}\left(M_m L_m - C_m\left(V_m + \left\|\dot{\tilde{q}}\right\|\right)\right). \tag{46}$$

K_{vm} in (46) is chosen as follows

$$K_{vm} > \frac{\left(C_m V_m + K_{vM}\right)^2}{2\left(M_m L_m - C_m\left(V_m + \left\|\dot{\tilde{q}}\right\|\right)\right)}. \tag{47}$$

Since the assumption (13) is verified, we have

$$L_m > \frac{\left(C_m V_m + K_{vM}\right)^2}{2K_{vm}} M_m^{-1} + 2C_m V_m M_m^{-1}. \tag{48}$$

Thus

$$\dot{H} \leq -\frac{1}{2}\left(M_m L_m - 2C_m V_m\right) \left\|\dot{\tilde{q}}\right\|^2. \tag{49}$$

If we choose $L_m > 2C_m V_m M_m^{-1}$ (it is verified by (48)), then $\dot{H}$ is a negative semi-definite function, this result is not sufficient to demonstrate the asymptotic stability, and we can conclude only the stability of the system. Nevertheless, it is straightforward to verify that the equilibrium $(E, \dot{E}, \dot{\tilde{q}}) = (0,0,0)$ is the largest invariant set within the set $\dot{H} = 0$. Hence, using La Salle's invariance principle the asymptotic stability of the equilibrium can be proved. Therefore, one must insure that

$$\text{if} \quad \dot{\tilde{q}}=0 \quad \text{then} \quad E=0 \text{ and } \dot{E}=0 . \tag{50}$$

Part 2

When $\dot{H}=0$, it is necessary that $\dot{\tilde{q}}=0$, in addition $\ddot{\tilde{q}}=0$, therefore (38) will be

$$M\ddot{E}+\left(C(q,\dot{\hat{q}})+K_v\right)\dot{E}+K_pE=0 . \tag{51}$$

Choosing the following Lyapunov function candidate

$$W(\dot{E},E)=\frac{1}{2}\dot{E}^T M\dot{E}+\frac{1}{2}E^T K_p E . \tag{52}$$

Using the property 4, the time-derivative of (52) is

$$\dot{W}=-\dot{E}^T K_v \dot{E} . \tag{53}$$

Hence

$$\dot{W}\leq -K_{vm}\left\|\dot{E}\right\|^2 . \tag{54}$$

$\dot{W}$ is a negative semi-definite function, this result is not sufficient to demonstrate that $E\rightarrow 0$. Therefore, the Barbalat's lemma is required to complete the proof of asymptotic stability.

We note that, it is sufficient to show that $\ddot{W}$ is bounded to conclude that $\dot{W}$ is uniformly continuous. Indeed, the time-derivative of (53) is

$$\ddot{W}=-2\dot{E}^T K_v \ddot{E} . \tag{55}$$

From (48) and (49), we demonstrated the stability of the system (E and $\dot{E}$ are bounded). In addition, from (51), we can conclude that $\ddot{E}$ is bounded. Then W and $\ddot{W}$ are bounded. This result implies that $\dot{W}$ is uniformly continuous. Therefore, the Barbalat's lemma permits us to conclude that $\dot{W}=0$, then $\dot{E}=0$, $\ddot{E}=0$, and from (51) we find that $E=0$. Finally, we demonstrated that (50) is verified. Hence, the La Salle's invariance principle is applied, consequently, the equilibrium $(E,\dot{E},\dot{\tilde{q}})=(0,0,0)$ is the largest invariant set within the set $\dot{H}=0$. And the asymptotic stability of the equilibrium is proved.

Since $\|y\| > \|\dot{\tilde{q}}\|$, (47) holds if

$$\|y\| < \frac{M_m L_m}{C_m} - \frac{\left(C_m V_m + K_{vM}\right)^2}{2K_{vm} C_m} - V_m . \tag{56}$$

From (41), (46) and (56), it follows that if

$$\|y(0)\| < \sqrt{\frac{q_m}{q_M}}\left(\frac{M_m L_m}{C_m} - \frac{\left(C_m V_m + K_{vM}\right)^2}{2K_{vm} C_m} - V_m\right). \tag{57}$$

Then, the closed-loop system is semi-globally asymptotically stable. This completes the proof.

3.2 Globally asymptotically stable observer based controller

Now, a second observer is presented to reconstruct the velocity signal in the control law. Hence, the global asymptotic stability of the whole control system (robot plus controller plus observer) is guaranteed. This proof is based on Lyapunov theory and using saturation technique. This result is given in theorem 2.

Theorem 2

Given the robot dynamics (1), and let assumption (13) be satisfied. Under the following control law

$$\tau = M(q)\ddot{q}_d + C(q, sat(\dot{\hat{q}}))sat(\dot{\hat{q}}) + G(q) + K_v\left(\dot{q}_d - \dot{\hat{q}}\right) + K_p E \tag{58}$$

with

$$\dot{\hat{q}} = z + Lq \tag{59}$$

$$\dot{z} = \ddot{q}_d - L\dot{\hat{q}} + M^{-1} K_p E . \tag{60}$$

If

1. $L_m \geq \dfrac{4\psi^2}{M_m K_{vm}} + 2\left(K_{vM} M_m^{-1}\right)$
2 $L_m \geq 2\dfrac{\psi}{M_m}$.

Then, the closed-loop system is globally asymptotically stable. Hence

$$\lim_{t\to\infty} E(t) = \lim_{t\to\infty} \dot{E}(t) = \lim_{t\to\infty} \dot{\tilde{q}}(t) = 0 . \tag{61}$$

Where sat(V) represents the saturation for a vector V, this function is to be defined. $K_{vM} = \|K_v\|$. L_m and K_{vm} denote the minimum eigenvalue of L and K_v respectively. ψ is positive scalar constant such that $\left\|\frac{\partial}{\partial \dot{q}}\left(C(q,\dot{q})\dot{q}\right)\right\| \leq \psi$, $\forall(q,\dot{q}) \in R^n \times \overline{V}$, with $\overline{V} = \left\{\dot{q} \in R^n \middle| \ |\dot{q}_i| \leq V_i \quad i = 1,...,n\right\}$.

Proof
Since assumption (13) holds, in the rest of the proof regard

$$\tau = M(q)\ddot{q} + C(q, sat(\dot{q}))sat(\dot{q}) + G(q) \tag{62}$$

as the given dynamic equation instead of (1).
Where the saturation for a vector $V = \left[v_1, \ \dots \ , v_n\right]^T \in R^n$ is defined as

$$Sat(V) = \left[sat(v_1), \ \dots \ , sat(v_n)\right]^T \tag{63}$$

with

$$sat(v_i) = \begin{cases} v_i & if \left|v_i\right| \leq \overline{v}_i \\ \overline{v}_i & if \ v_i > \overline{v}_i \\ -\overline{v}_i & if \ v_i < -\overline{v}_i \end{cases} \qquad \text{for } i \in \{1 \ \dots \ n\} \tag{64}$$

and $V_m = \left\|\left[sat(v_1), \ \dots \ , sat(v_n)\right]^T\right\|$.
From (58), (59) and (60) we can eliminate the state z and write

$$\ddot{\hat{q}} = M^{-1}\left[\tau - C(q, sat(\dot{\hat{q}}))sat(\dot{\hat{q}}) - G(q) - K_v\dot{\tilde{q}} - K_v\dot{E}\right] + L\dot{\tilde{q}} . \tag{65}$$

From (62), we can write

$$\ddot{q} = M^{-1}\left[\tau - C(q, sat(\dot{q}))sat(\dot{q}) - G(q)\right]. \tag{66}$$

Subtracting (65) from (66), we obtain

$$M\ddot{\tilde{q}} = -C(q, sat(\dot{q}))sat(\dot{q}) + C(q, sat(\dot{\hat{q}}))sat(\dot{\hat{q}}) + K_v(\dot{E} + \dot{\tilde{q}}) - ML\dot{\tilde{q}}. \tag{67}$$

Subtracting (62) from (58), we have

$$M\ddot{E} - C(q, sat(\dot{q}))sat(\dot{q}) + C(q, sat(\dot{\hat{q}}))sat(\dot{\hat{q}}) + K_v(\dot{\tilde{q}} + \dot{E}) + K_p E = 0. \tag{68}$$

Consider the Lyapunov function

$$H(\dot{E}, E, \dot{\tilde{q}}) = \frac{1}{2}\dot{E}^T M\dot{E} + \frac{1}{2}E^T K_p E + \frac{1}{2}\dot{\tilde{q}}^T M\dot{\tilde{q}} \ . \tag{69}$$

The time-derivative of (69), evaluated along (67) and (68), is

$$\begin{aligned} \dot{H} &= \dot{E}^T\left(C(q, sat(\dot{q}))sat(\dot{q}) - C(q, sat(\dot{\hat{q}}))sat(\dot{\hat{q}})\right) \\ &- \dot{\tilde{q}}^T ML\dot{\tilde{q}} + \dot{\tilde{q}}^T\left(C(q, sat(\dot{\hat{q}}))sat(\dot{\hat{q}}) - C(q, sat(\dot{q}))sat(\dot{q})\right) \\ &- \dot{E}^T K_v \dot{E} + \dot{\tilde{q}}^T K_v \dot{\tilde{q}}. \end{aligned} \tag{70}$$

To simplify the notation, let

$$D(q, \dot{q}) = C\ \ (q, \dot{q})\dot{q}\ . \tag{71}$$

Using (P2), it follows that $\exists \psi > 0$ such that

$$\left\| \frac{\partial}{\partial \dot{q}}\left(D(q, \dot{q})\right) \right\| \le \psi \quad \forall (q, \dot{q}) \in R^n \times \overline{V} \tag{72}$$

with

$$\frac{\partial}{\partial \dot{q}}\left(D(q, \dot{q})\right) = 2\begin{pmatrix} \dot{q}^T N_1(q) \\ \vdots \\ \dot{q}^T N_n(q) \end{pmatrix} \tag{73}$$

and

$$\overline{V} = \left\{\dot{q} \in R^n \middle| \ |\dot{q}_i| \le V_i \qquad i = 1, \dots, n\right\}. \tag{74}$$

Then, using Lemma 1, we have

$$\dot{H} \le -K_{vm}\left\|\dot{E}\right\|^2 - \left(M_m L_m - K_{vM}\right)\left\|\dot{\tilde{q}}\right\|^2 + \psi\left\|\dot{\tilde{q}}\right\|^2 + \psi\left\|\dot{E}\right\|\left\|\dot{\tilde{q}}\right\|. \tag{75}$$

Hence

$$\dot{H} \le -K_{vm}\left\|\dot{E}\right\|^2 - \left(\frac{M_m L_m}{2} - K_{vM}\right)\left\|\dot{\tilde{q}}\right\|^2 - \left(\frac{M_m L_m}{2} - \psi\right)\left\|\dot{\tilde{q}}\right\|^2 + \psi\left\|\dot{E}\right\|\left\|\dot{\tilde{q}}\right\|. \tag{76}$$

We note that

$$\psi\|\dot{E}\|\;\|\dot{\tilde{q}}\| \leq 2\|\dot{E}\|\psi\|\dot{\tilde{q}}\|$$
$$\leq 2\;\|\dot{E}\|\psi\left[\frac{2}{\frac{M_mL_m}{2}-K_{vM}}\right]^{\frac{1}{2}}\left[\frac{\frac{M_mL_m}{2}-K_{vM}}{2}\right]^{\frac{1}{2}}\|\dot{\tilde{q}}\| \tag{77}$$
$$\leq \|\dot{E}\|^2\psi^2\left[\frac{2}{\frac{M_mL_m}{2}-K_{vM}}\right]+\|\dot{\tilde{q}}\|^2\left[\frac{\frac{M_mL_m}{2}-K_{vM}}{2}\right].$$

Therefore, we have

$$\dot{H} \leq -\left(K_{vm} - \frac{4\psi^2}{M_mL_m - 2K_{vM}}\right)\|\dot{E}\|^2 - \frac{1}{2}\left(\frac{M_mL_m}{2} - K_{vM}\right)\|\dot{\tilde{q}}\|^2 - \left(\frac{M_mL_m}{2} - \psi\right)\|\dot{\tilde{q}}\|^2. \tag{78}$$

Choosing K_{vm} and L_m as follows

$$L_m \geq 2\frac{\psi}{M_m} \tag{79}$$

and

$$K_{vm} \geq 4\frac{\psi^2}{M_mL_m - 2K_{vM}}. \tag{80}$$

Hence

$$L_m \geq \frac{4\psi^2}{M_mK_{vm}} + 2\left(K_{vM}M_m^{-1}\right). \tag{81}$$

Then, we have

$$\dot{H} \leq -\left(K_{vm} - \frac{4\psi^2}{M_mL_m - 2K_{vM}}\right)\|\dot{E}\|^2 - \frac{1}{2}\left(\frac{M_mL_m}{2} - K_{vM}\right)\|\dot{\tilde{q}}\|^2. \tag{82}$$

If we choose $L_m > 2\left(K_{vM}M_m^{-1}\right)$, (it is verified by (81)), then $\dot{H}$ is a negative semi-definite function, this result is not sufficient to demonstrate the asymptotic stability, and we can conclude only the stability of the system. Therefore, the lemma 2 is required to complete the proof of asymptotic stability.

In our case, H and $\dot{H}$ are given by (69) and (70) respectively. To conclude that $\dot{H}$ is uniformly continuous, it is sufficient to show that $\ddot{H}$ is bounded.

The time-derivative of (70) is

$$\begin{aligned}\ddot{H} &= \ddot{E}^T\left(C(q,sat(\dot{q}))sat(\dot{q}) - C(q,sat(\dot{\hat{q}}))sat(\dot{\hat{q}})\right)\\ &- 2\dot{\tilde{q}}^T ML\ddot{\tilde{q}} + \ddot{\tilde{q}}^T\left(C(q,sat(\dot{\hat{q}}))sat(\dot{\hat{q}}) - C(q,sat(\dot{q}))sat(\dot{q})\right)\\ &\left(\dot{E}^T + \dot{\tilde{q}}^T\right)\left(\frac{d}{dt}(C(q,sat(\dot{q}))sat(\dot{q})) - \frac{d}{dt}(C(q,sat(\dot{\hat{q}}))sat(\dot{\hat{q}}))\right)\\ &- 2\dot{E}^T K_v \ddot{E} + 2\dot{\tilde{q}}^T K_v \ddot{\tilde{q}}.\end{aligned} \tag{83}$$

From (79), (80), (81) and (82), we demonstrated the stability of the system (E, $\dot{E}$ and $\dot{\tilde{q}}$ are bounded). Therefore, from (69) H is bounded. In addition, from (67) and (68) we can conclude that $\ddot{E}$ and $\ddot{\tilde{q}}$ are bounded, then $\ddot{H}$ is bounded. This result implies that $\dot{H}$ is uniformly continuous. Hence, the Barbalat's lemma permits us to conclude that $\dot{H} = 0$. Thus, from (82), we have $\dot{E} = 0, \dot{\tilde{q}} = 0$, and necessary $\ddot{E} = 0, \ddot{\tilde{q}} = 0$. Finally from (67) and (68) we find that $E = 0$.
Then, the closed-loop system is globally asymptotically stable. This completes the proof.

4. Simulation results

In order to illustrate by simulation the efficiency of our design, we apply in this section the observer-controller laws (55-60) on two-link robot manipulator. The objective of our simulation work is to show that the tracking objective is achieved when an estimated velocity vector is used in the tracking control law.

Consider a two-link manipulator with masses m_1, m_2, lengths l_1, l_2, and angles q_1, q_2; then the model equations can be written as (1). $M(q)$, $C(q,\dot{q})$ and $G(q)$ are given by (Bouakrif et al., 2008):

$m_{11} = m_2 l_2^2 + 2m_2 l_1 l_2 \cos(q_2) + (m_1 + m_2) l_1^2$, $m_{12} = m_{21} = m_2 l_2^2 + m_2 l_1 l_2 \cos(q_2)$, $m_{22} = m_2 l_2^2$.

$C_{11} = -m_2 l_1 l_2 \sin(q_2)\dot{q}_2$, $C_{12} = -m_2 l_1 l_2 \sin(q_2)\dot{q}_2$, $C_{21} = m_2 l_1 l_2 \sin(q_2)\dot{q}_1$, $C_{22} = 0$.

$G_1 = m_2 l_2 g \cos(q_1 + q_2) + (m_1 + m_2) l_1 g \cos(q_1)$, $G_2 = m_2 l_2 g \cos(q_1 + q_2)$.

The desired trajectories are chosen as:

$q_{d1}(t) = 2\cos(4\pi t/3) + \sin(2\pi t/3)$ (rad), with $0 \le t \le 5$.

$q_{d2}(t) = 1 - 2\cos(4\pi t/3) - \sin(2\pi t/3)$ (rad), with $0 \le t \le 5$.

Simulation parameters:

$K_p = \{5000,5000\}, K_v = \{15,15\}, L = \{550,550\}$,

$m_1 = 0.5[kg]$, $m_2 = 0.7[kg]$, $l_1 = 1[m]$, $l_2 = 1.5[m]$.

Therefore, we find that $V_m = 10[rad/s]$, $M_m = 1[kgm^2]$.

The simulation results of the proposed scheme on two-link robot manipulator along a trajectory are shown below. Figure 1 show the observer result, where we can see the convergence of the observed velocity to real velocity, of each joint, in a minimum time.

Figure 2 show the simulation results for real and desired position trajectories, of each joint, when the velocity given by the observer (59) and (60) is used in the control law (58). We can see that the real trajectory follows the desired trajectory without error through time axis. Therefore, it is clear that the control algorithm works well.

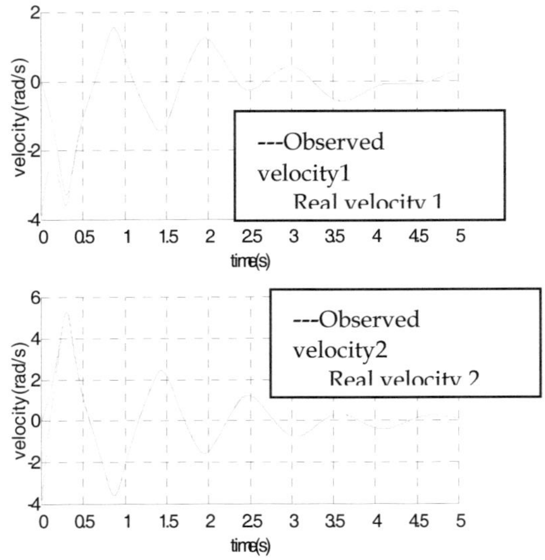

Fig.1. Real and observed velocities of two-link manipulator.

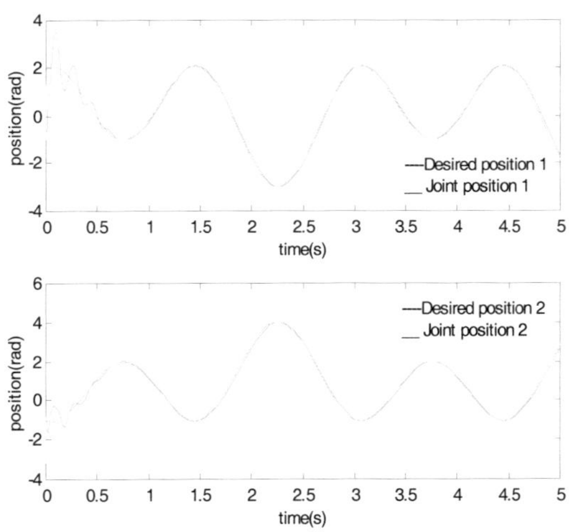

Fig.2. Real and desired position trajectories of two-link manipulator.

5. Conclusion

This chapter has presented two motion control schemes to solve the trajectory tracking problem of rigid-link robot manipulators, when the manipulator's joint velocities cannot be measured by the control system. The necessity of velocity measurements in the controllers can be removed by replacing the actual velocity signal by an estimate obtained from two observer systems. The whole control system consisting of robot manipulator, controller and the first observer is semi-globally asymptotically stable and a region of attraction is also given. Using the second observer, the global asymptotic stability of the closed loop system is guaranteed. Hence, there is more freedom to choose the initial states. These proofs are based on Lyapunov theory. Finally, simulation results on two-link manipulator are provided to illustrate the effectiveness of the global velocity observer based trajectory tracking control.

6. References

Berghuis, H. & Nijmeijer, H. (1993). Global regulation of robots using only position measurements, *Syst. and Contr. Lett.*, 21, 1993, pp. 289-293, ISSN 0167-6911.

Berghuis, H. (1993). Model based control: from theory to practice, Ph.D. thesis, University of Twent, ISBN 90-9006110-X, the Netherlands.

Bouakrif, F.; Boukhetala, D. & Boudjema, F. (2007). Iterative learning control for robot manipulators, *Archives of Control Sciences*, Vol. 17, No. 1, 2007, pp. 57-69.

Bouakrif, F.; Boukhetala, D. & Boudjema, F. (2008). Global asymptotic stability of controller-observer for robot manipulators using saturation technique, *The Mediterranean Journal of Measurement and Control*, Vol. 4, No. 1, 2008, ISSN 1743-9310.

Bouakrif, F.; Boukhetala, D. & Boudjema, F. (2010). Passivity based controller-observer for robot manipulators, *Int. J. Robotics and Automation*, Vol. 25, No. 1, 2010, ISSN 0826-8185.

Canudas de wit, C.; Fixot, N. & Aström, K.J. (1992). Trajectory tracking in robot manipulators via nonlinear estimated state feedback, *IEEE Trans. Robot. Automat.*, Vol. 8, No. 1, 1992, pp. 138-144, ISSN 1552-3098.

Chang, Y.C. & Chen, B.S. (2000). Robust tracking designs for both holonomic and nonholonomic constrained mechanical systems: adaptive fuzzy approach, *IEEE Trans. Fuzzy Syst.*, Vol. 8, 2000, pp. 46–66, ISSN 1063-6706.

Kelly, R. (1995). A tuning procedure of PID control for robot manipulators. *Robotica*, Vol. 13, 1995, pp. 141-148, ISSN 0263-5747.

Khosla, P.K. & Kanade, T. (1988). Experimental evaluation of nonlinear feedback and feedforward control schemes for manipulators, *Int. J. Robot. Res.*, Vol. 7, 1988, pp. 18-28, ISSN 0278-3649.

Luh, J.Y.S.; Walker, M.W. & Paul, R.C.P. (1980). Resolved-acceleration control of mechanical manipulators, *IEEE Trans. on Automatic Control, AC-25*(3), pp. 468-474, 1980, ISSN 0018-9286.

Nicosia, S. & Tomei, P. (1990). Robot control by using only joint position measurements, *IEEE Trans. Automat. Contr.*, Vol. 35, No. 9, 1990, pp. 1058–10 61, ISSN 0018-9286.

Ortega, R. & Spong, M.W. (1989). Adaptive motion control of rigid robots: a tutorial, *Automatica*, Vol. 25, no. 6, 1989, pp. 877-888, ISSN 0005-1098.

Paden, B & Panja, R. (1988). Globally asymptotically stable 'PD+' controller for robot manipulators, *Int. J. Control*, Vol. 47, No. 6, 1988, pp. 1697-1712, ISSN 0020-7179.

Ryu, J.H.; Kwon, D.S. & Hannaford, B. (2004). Stable teleoperation with time domain passivity control, *IEEE Trans. Robot. Automat.*, Vol. 20, 2004, pp. 365–373, ISSN 1552-3098.

Shim, H.; Son, Y. & Seo, J. (2001). Semi-global observer for multi-output nonlinear systems, *Systems and Control Letters*, Vol. 42, No. 3, 2001, pp. 233–44, ISSN 0167-6911.

Slotine, J.J.E. & Sastry, S.S. (1983). Tracking control of nonlinear systems using sliding surface with application to robot manipulators, *Int. J. Control*, Vol. 38, 1983, pp. 465-492, ISSN 0020-7179.

Slotine, J.J.E & Li, W. (1991). Applied Nonlinear Control", Prentice-Hall. Englewood Cliffs, 1991, ISBN: 0-13-040890-5.

Tayebi A. (2007). Analysis of two particular iterative learning control schemes in frequency and time domains, *Automatica*, Vol. 43, 2007, pp. 1565-1572, ISSN 0005-1098.

Xian, B.; Queiroz, M.; Dawson, D. & McIntyre, M. (2004). A discontinuous output feedback controller and velocity observer for nonlinear mechanical systems, *Automatica*, Vol. 40, No. 4, 2004, pp. 695–700, ISSN 0005-1098.

Yu, W. & Li, X. (2006). PD Control of robot with velocity estimation and uncertainties compensation, *Int. J. of Robot. and Aut.*, Vol. 21, no. 1, 2006, ISSN 0826-8185.

Robotic Machining from Programming to Process Control

Zengxi Pan[1] and Hui Zhang[2]
[1]*Faculty of Engineering, University of Wollongong, Australia*
[2]*ABB Corporate Research China*

1. Introduction

Cleaning and pre-machining operations are major activities and represent a high cost burden for casting producers. Robotics based flexible automation is considered as an ideal solution for its programmability, adaptivity, flexibility and relatively low cost, especially for the fact that industrial robot is already applied to tend foundry machines and transport parts in the process. Nevertheless, the foundry industry has not seen many success stories for such applications and installations due to the several major difficulties involved in robotic machining process with a conventional industrial robot. (Pan, 2006)
The first difficulty is the generation of robot motion for a complex workpiece. Secondly, the lower stiffness of articulated robot manipulator presents a unique disadvantage for machining of casting parts with complex geometry, which has non-uniform cutting depth and width. As a result, the machining force will vary dramatically, which induces uneven robot deformation. The third difficulty is the deformation caused by the interaction force between the tool and the workpiece, especially for milling process, which generates large cutting forces. The stiffness for a typical articulated robot is usually less than 1 N/μm, while a standard CNC machine very often has stiffness greater than 50 N/μm. As a result, force induced deformation is the major source of the inaccuracy of finished surface. The fourth difficulty is chatter/vibration occurred during the machining process.
Most of the existing literature on machining process, such as process force modelling (Kim, Landers & Ulsoy, 2003), accuracy improvement (Yang, 1996) and vibration suppression (Budak, & Altintas, 1998) are based on the CNC machine. Research in the field of robotic machining is still focused on accurate off-line programming and calibration. As the chatter analysis was discussed in a separate paper (Pan & Zhang et, al, 2006), our focus here is to address the first three major issues in robotic machining process.
This chapter is organized in six sections. Following this introduction section, section two presents an active force control platform, which is the foundation of various control strategies for solving difficulties in robotic machining processes. Section three addresses the programming issues for a part with complex contour. With two force control strategies, lead-through and path-learning, robot programming is made easy and efficient. Section four and five present two real-time process control techniques. The Controlled Material Removal Rate (CMRR) greatly reduces the process cycle time of the robotic machining operation,

while the real-time deformation compensation improves the quality and accuracy. The focus of these two sections will be the implementation of advanced control strategies and further analysis of robot stiffness modelling, as the preliminary research outcomes for CMRR and deformation compensation have been already introduced in (Wang, Zhang, & Pan, 2007). Experimental results are presented at the end of these sections. A summary and discussion is provided in section six.

2. Force Control Platform

The active force control platform is the foundation of the strategies adopted to address various difficulties in robotic machining processes. It is implemented on the most recent ABB IRC5 industrial robot controller which is a general controller for a series of ABB robots. The IRC5 controller includes a flexible teach pedant with a colourful graphic interface and touch screen which allows user to create customized Human Machine Interface (HMI) very easily. It only takes several minutes for a robot operator to learn the interface for a specific manufacturing task and it is programming free. An ATI 6 DOF force/torque sensor is equipped on the wrist of the robot to close the outer force loop and realize implicit hybrid position/force control scheme. The system setup for robotic machining with force control is shown in Fig. 1.

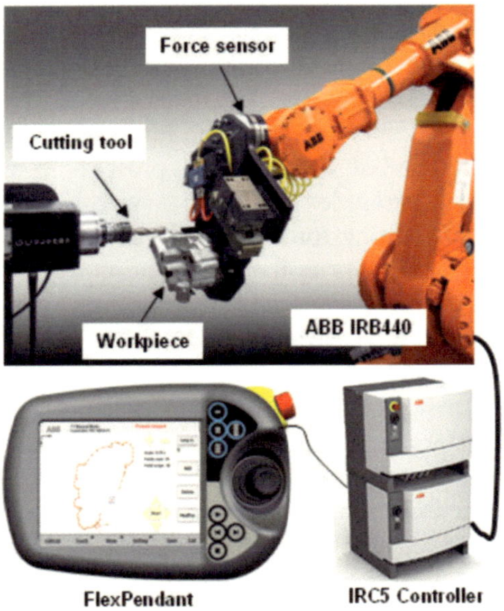

Fig. 1. System setup for robotic machining with force control

The force controller provides two major functions to make the entire programming process collision free and automatic. First function is lead-through, in which robot is compliant in selected force control directions and stiff in the rest of the position control directions. To change the position or orientation of the robot, the robot operator could simply push or drag the robot with one hand. The second function is called path-learning, in which robot is compliant in normal to the path direction to make the tool constantly contact with the workpiece. Thus, an accurate path could be generated automatically.

During the machining process, the force controller provides two more functions to achieve deformation compensation and CMRR. In both case, robot is still under position control, that is, stiff at all directions. Deformation compensation is achieved by update the target position of position loop based on the measured process force and robot stiffness model, while robot feed speed is adjusted to maintain constant spindle power consumption for CMRR. These two strategies are complementary to each other since CMRR adjusts robot speed at feed direction and deformation compensation adjusts the reference target at the rest of the directions. The detailed control strategies for process control of robotic machining will be explained in section four and five respectively.

3. Rapid Robot Programming

Although extensive research efforts have been carried out on the methodologies for programming industry robots, still only two methods are realistic in practical industrial application, which are, on-line programming (jog-and-teach method) and off-line programming (Basanez & Rosell, 2005)(Pires, et al., 2004). On-line programming relies on the experience of robot operators to teach robot motions by jogging the robot to the desired positions using teaching device (usually teach pendent) in real setup. Off-line programming generates the robot path from a CAD model of the workpiece in a computer simulated setup. The idea of programming by demonstration (PbD) has been proposed long time ago, while requirement of additional hardware devices and complicated calibration process make it unattractive in practical applications. The major advantage of the PbD method proposed here is that no additional devices and calibration procedures are required. The only sensor implemented for force feedback is an ATI 6 DOF force/torque sensor. This simple configuration will minimize the cost and simplify the complexity of the programming process greatly.

3.1 Lead-Through

Lead-though is the only step requires human intervention through the entire PbD process. The purpose of lead-through is to generate a few gross guiding points, which will be used to calculate the path frame in path-learning as shown in Fig. 2. The position accuracy of these guiding points is not critical because these guiding points are not the actual points/targets in the final program and they will be updated in automatic path-learning. However the orientation of these points should be carefully taught since it will determine the path frame and will be kept in the final program.

Theatrically all six DOFs could be released under force control and the user can adjust both position and orientation of the robot tool at the same time. In practice, we found it is almost impossible to adjust the tool orientation accurately by push/pull with a single hand. Thus, a force control jogging mode is created, under which the operator could push/pull the robot tool to any position easily and change the robot tool orientation using the joystick on the teach pendent. Since this jogging is under force control, collision is avoided even when the tool is in contact with the workpiece. As the instant position and orientation of the robot tool is displayed on the teach pendant, the operator could make very accurate adjustment on each independent rotation axis.

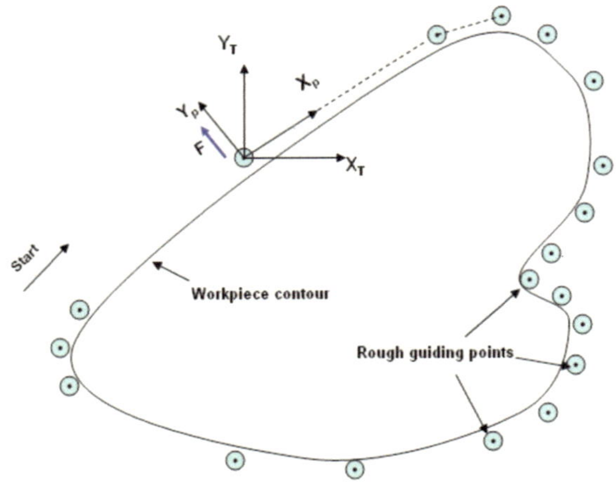

Fig. 2. Lead-through and path learning

3.2 Automatic Path-Learning

A robot program based on gross guiding points taught in lead-through is then generated. This program path, consisted of a group of linear movements from one guiding point to the next, is far different from the actual workpiece contour. The tool fixture would either move into the part or too far away from it.

During the automatic path-learning, the robot controller is engaged in a compliant motion mode, such that only in direction Yp, (Fig. 2.) which is perpendicular to path direction Xp, robot motion is under force control, while all other directions and orientations are still under position control. Further, it can be specified in the controller that a constant contact force in Yp direction (e.g., 20 N) is maintained. Because of this constrain, if the program path is into in the actual workpiece contour, the tool tip will yield along the Y axis until it reaches the equilibrium of 20N, resulting a new point which is physically on the workpiece contour. On the other hand, if the program path is away from the workpiece, the controller would bring the tool tip closer to the workpiece until the equilibrium is reached of 20N.

While the robot holding the tool fixture is moving along the workpiece contour, the actual robot position and orientation are recorded continuously. As described above, the tool tip would always be in continuous contact with the workpiece, resulting a recorded spatial relationship that is the exact replicate between the tool fixture and the workpiece. A robot program generated based on recorded path can be directly used to carry out the actual process.

3.3 Post Processing

After tracking the workpiece contour, the data from logging the robot position have to be filtered and reduced to generate a robot program. The measurements around sharp corners are often influenced by noise due to high dynamic forces, which has influence on the contact force. By using a threshold for the maximum and minimum acceptable contact force, the measurements influenced by this type of noise are removed. This is called force threshold filtering.

The amount of the targets from automatic path-learning are disproportionately large since the robot controller can recorded the points as fast as every 4 ms. An approach, namely deviation height method, is used to approximating the contour by straight-line segments.
As shown in Fig. 3, a straight line is made from a certain starting point on the contour to the current point. The deviation height is calculated between the line and each of the intermediate points. The deviation height is the length of the normal vector between the point and the line. The current point is displaced along the contour until the deviation height exceeds a certain limit. The previous point is then used as starting point for the next line segment. This continues until the whole contour is approximated with straight-line segments. From the reduced data, a robot program is generated in a standard format. The user could specify tool definitions, desired path velocity and orientation of the tool.

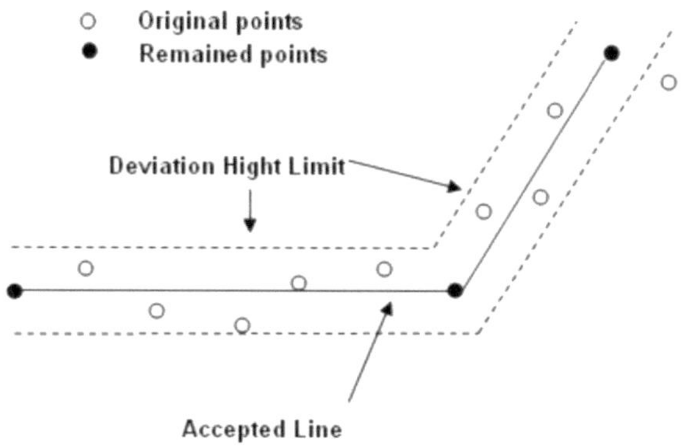

Fig. 3. Deviation height method

3.4 Experimental Results for PbD

With force control integrated in IRC5 controller, PbD method is available for a group of ABB industrial manipulators. An automatic deburring system using IRB 4400 manipulator is designed to clean the groove of a water pump to guarantee a seamless interface between two pump surfaces, as shown in Fig. 4.
A 2 mm cutting tool, driven by ultra high speed (~18,000rpm) air spindle is adopted to achieve this task. Since the groove is only about 5 mm wide and has contoured 2D shape, manually teaching a high quality program to clean the complete groove is almost impossible. Due to the process requirement, the cutting tool is always perpendicular to the surface of water pump. During path-learning, a contact force normal to the edge of 10 N is used, while the robot path learning velocity is set at 5mm/s. As shown in Fig. 5, the curvature of recorded targets changes dramatically along the path. The blue points represent the targets in the final cutting program, while the read points represent the offset targets in the test program. The average robot feed speed during the cutting process is about 10 mm/s, while the exact feed speed is determined by the local curvature, which is slower at sharp corner, to ensure a smooth motion throughout the path. The point reduction technique is performed on the filtered measurements. A deviation height of 0.2mm reduced the thousands of points recorded by the robot controller every 4 ms to about 300 points.

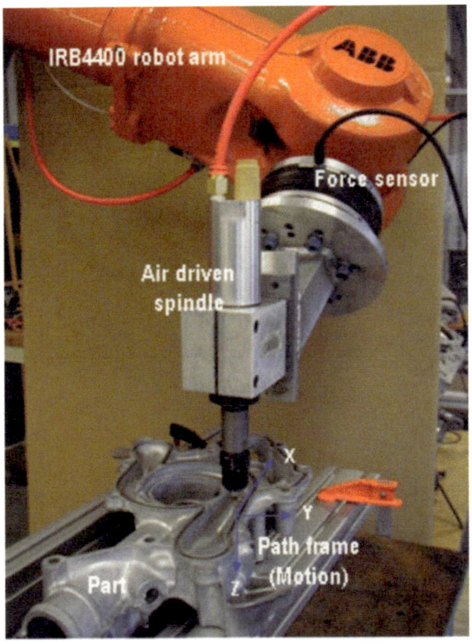

Fig. 4. Experimental setup for PbD

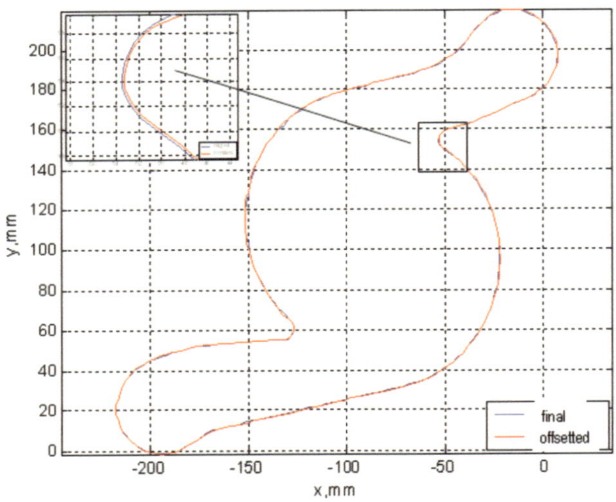

Fig. 5. Results from path-learning

With this programming strategy, generating a program for a water pump with complex contour, including more than three hundred robot target points, could be completed within one hour instead of several weeks by an experienced robot programmer. During this programming procedure, the operator is only involved with the first step of teaching the gross movement of the robot, while the bulk of the procedure is automated by the robot controller.

4. Controlled Material Removal Rate

The MRR in machining process is usually controlled by adjusting the tool feedrate. In robotic machining process, this means regulating robot feed speed to maintain a constant MRR. Machining force and spindle power are two variables proportional to MRR, which could be used to control robot feed speed. With 6-DOF force sensor fixed on robot wrist, the cutting force is available on real-time. Most spindles have an analog output whose value is proportional to the spindle current. With force feed back or spindle current feed back, MRR could be regulated to avoid tool damage and spindle stall.

In most cases, the relationship between process force and tool feedrate is nonlinear, and the process parameters, which describe the nonlinear relationship, are constantly changing due to the variations of the cutting conditions, such as, depth-of-cut , width-of-cut, spindle motor speed, and tool wearing condition, etc. Most of the time, conservative gains have to be chosen in order to maintain the stability of the close-loop system, while trading off the control performances.

Three different control strategies, PI control, adaptive control and fuzzy control, are designed to satisfy various process requirements. PI control is easy to tune and is very reliable. Adaptive control provides a more stable solution for machining process. Fuzzy control, which provides a much faster response by sacrificing control accuracy, is the best method for applications require fast robot feed speed

Fig. 6. Robotic end milling process setup

4.1 Robot Dynamic Model

A robotic milling process using industrial robot is shown in Fig. 6. The cutting force of this milling process is regulated by adjusting the tool feedrate. Since the tool is mounted on the robot end-effector, the tool feedrate is controlled by commanding robot end-effector speed. Thus, the robot dynamic model for this machining process is the dynamics from the command speed to the actual end-effector speed. The end-effector speed is controlled by the robot position controller. A model is identified via experiments for this position controlled close-loop system, which represents the dynamics from command speed to actual end-effector speed.

The dynamic model identified is given as

$$\frac{f(s)}{f_c(s)} = \frac{63s^2 - 45800s + 4330000}{s^3 + 575s^2 + 98670s + 4313000} \tag{1}$$

Where f(s) is the actual end-effector speed, $f_c(s)$ is the commanded end-effector speed.

The dynamic model Eq. (1) is a stable non-minimum phase system, and its root locus is shown in Fig. 7.

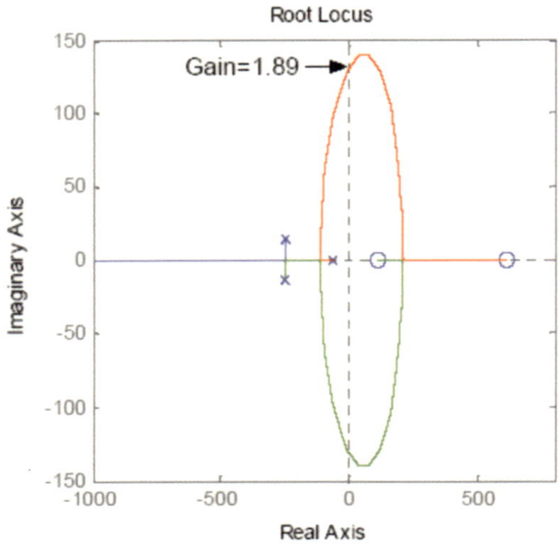

Fig. 7.Root locus of robot dynamic model

4.2 Process Force Model

MRR is a measurement of how fast material is removed from a workpiece; it can be calculated by multiplying the cross-sectional area (width of cut times depth of cut) by the linear feed speed of the tool:

$$MRR = w \cdot d \cdot f \tag{2}$$

Where w is width-of-cut (mm), d is depth-of-cut (mm), f is feed speed (mm/s).

Since it is difficult to measure the value of MRR directly, MRR is controlled by regulating the cutting force, which is readily available in real-time from a 6-DOF force sensor fixed on the robot wrist. The relationship between the machining process force and the tool feed speed is nonlinear and time-varying, as shown in the following dynamic model (Landers & Ulsoy, 2000)

$$F = K_C d^{\alpha} f^{\beta} w^{\gamma} \frac{1}{\tau_m s + 1} \tag{3}$$

Where K_C is the gain of the cutting process; α , β and γ are coefficients, and their values are usually between 0 and 1. τ_m is the machining process time constant. Since one spindle revolution is required to develop a full chip load, τ_m is 63% of the time required for a spindle revolution. (Daneshmend & Pak, 1986) Since τ_m is much smaller than the time constant of robot system, it is ignored here in the MRR controller design. Let,

$$K = K_C w^{\gamma} \tag{4}$$

K is considered as a varied process gain. Then, the force model is rewritten as a static model:

$$F = Kd^{\alpha} f^{\beta} \tag{5}$$

The depth-of-cut, d , depends on the geometry of the workpiece surface. It usually changes during the machining process, and is difficult to be measured on-line accurately. The cutting depth is the major contributor that causes the process parameter change during the machining process. K , α and β depend on those cutting conditions, such as, spindle speed, tool and workpiece material, and tool wearing condition, etc, which are pretty stable during the cutting process. If the tool and/or the workpiece are changed, these parameters could change dramatically. But they are not changing as quickly as the depth-of-cut d does during the machining process as explained above. A force model, which is only valid for the specific tool and workpiece setup in ABB robotics lab is identified from experiments as

$$F = 23d^{0.9} f^{0.5} \tag{6}$$

Eq. (6) models the process force very well from milling experimental data. The tool feedrate f is chosen as the control variable, i.e., to control the process force by adjusting the feed speed.

4.3 MRR Control Strategy

In roughing cycles, maximum material removal rates are even more critical than precision and surface finish. Conventionally, feed speed is kept constant in spite of variation of depth-of-cut during the pre-machining process of foundry part. This will introduce a dramatic change of MRR, which induces a very conservative selection of machining parameters to avoid tool breakage and spindle stall. The idea of MRR control is to adjust the feed speed to keep MRR constant during the whole machining process. As a result, a much faster feed speed, instead of conservative feed speed based on maximal depth-of-cut position, could be adopted. Fig. 8 illustrates the idea of MRR control while depth-of-cut changes during milling operation. (Pan, 2006)

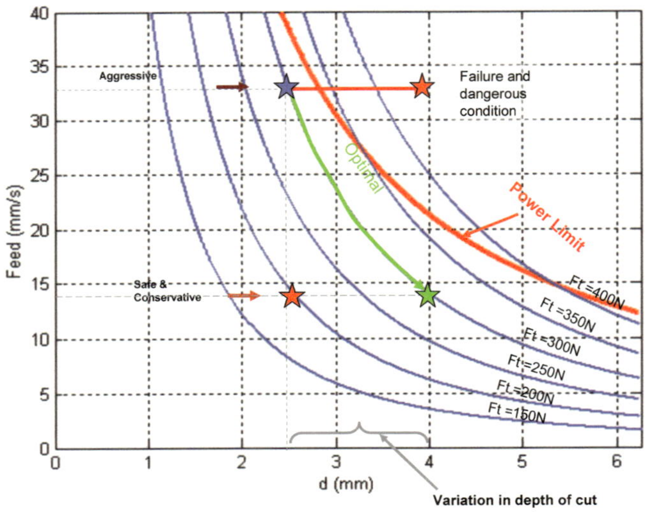

Fig. 8. Controlled material removal rate

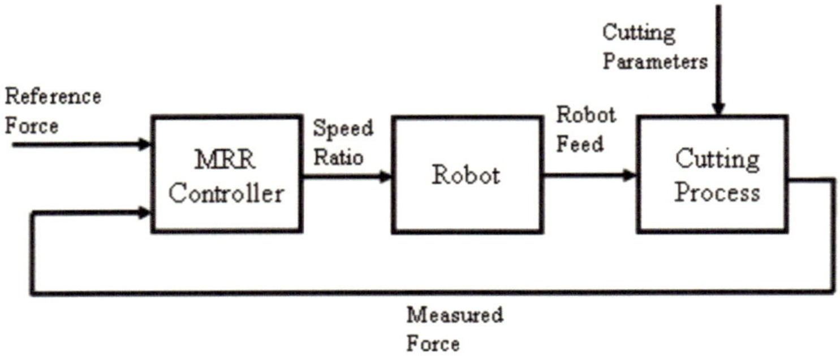

Fig. 9. The force control loop for CMRR

4.3.1 Force Control Sturcture

The block diagram of CMRR is shown in Fig. 9. The cutting force is controlled by varying the robot end-effecter speed in tool feed direction. The difference between the reference force and the measured cutting force is input to the MRR controller. In actual implementation, the robot motion is planned in advance based on a pre-selected command speed. The output of MRR controller is a term called speed_ratio, which is a ratio (e.g. from 0 to 1) of the actual robot feed speed to interpolate the reference trajectory in order to adjust the tool feedrate. Thus the command speed is the greatest speed robot can move. If the measured cutting force is larger than reference force, robot will slow down; otherwise robot will speed up until it reaches command speed. The CMRR function may implement several control approaches under the indirect force control framework. Three different control strategies, classical control (PI), adaptive control, and fuzzy logic control, will be introduced bellow.

4.3.2 PI Control

The cutting force model is nonlinear as described in Eq. (5), for controller design, it can be rewritten as

$$F = Kd^{\alpha} f^{\beta} = K_f f^{\beta} \tag{7}$$

Where $K_f = Kd^{\alpha}$. The effects of parameters K, d, and α to the process force are lumped into one parameter, force process gain K_f.

Define

$$F' = (F)^{1/\beta} \tag{8}$$

Together with Eq. (7), we get

$$F' = (F)^{1/\beta} = (K_f)^{1/\beta} f = kf \tag{9}$$

Where $k = (K_f)^{1/\beta}$ is time-varying. Instead of controlling cutting force F, we control F' to follow the new command force, i.e., $F_r' = (F_r)^{1/\beta}$, which is equivalent as controlling F to follow the original reference force F_r. By using Eq. (9), the nonlinear system is exactly linearized, and the linear system design technique can be applied to design a controller for the nonlinear system. PI type control is selected to achieve null steady-state error. The derivative term is not desirable due to the large noise associated with force readings.

The PI control in is given as

$$G_c = K_p + \frac{K_i}{s} \tag{10}$$

We put the zero of PI controller at -66.5 to cancel the slow stable pole of the robotic dynamic model. Since the zero of the PI controller is fixed, the proportional and integral gains will be given as

$$K_p = 0.015\alpha \text{ , } K_i = \alpha \tag{11}$$

Where α will be chosen to make the open loop gain of the whole system at the desired value. The magnitude of open loop gain, defined as kK_p determines the stability of the system. Conservative K_p and K_i are selected to ensure system still stable while the force process gain k takes the maximal value. The desired system response is that small overshot for command feed speed.

4.3.3 Adaptive Control

Since depth-of-cut and width-of-cut are likely to change dramatically due to the complex shape of workpiece and varied bur size, the force process gain k will vary dramatically during the machining process. The fixed-gain PI control will surely have problems to

maintain the stability and consistent system performance for wide range of cutting conditions. From Fig. 7, the close loop system becomes unstable when the open loop gain is greater than 1.89, which is consistent with our observations in machining experiments. So it is very important to adjust controller gains to compensate process parameter changes, in order to maintain close-loop system stability during the machining process.
A self-tuning mechanism is proposed here to adaptively adjust the gain of PI controller to maintain a stable machining process. The self-tuning PI controller is shown in Fig. 10. There is low positive speed_ratio output limit (because negative or larger than 1 speed_ratio is meaningless) assigned for tool feedrate command to avoid "stop and go" situation. So saturation nonlinearity is introduced into the control system. The anti-windup scheme is also necessary for the PI control to avoid the integration windup.
Let V_r be the maximum feed speed that the tool can be commanded. The saturation nonlinearity is defined as

$$sat(u) = \begin{cases} 1 & u \geq 1 \\ u & \delta < u < 1 \\ \delta & u \leq \delta \end{cases} \tag{12}$$

Where $\delta \geq 0$ and δV_r is the minimum feedrate command for the machining process.

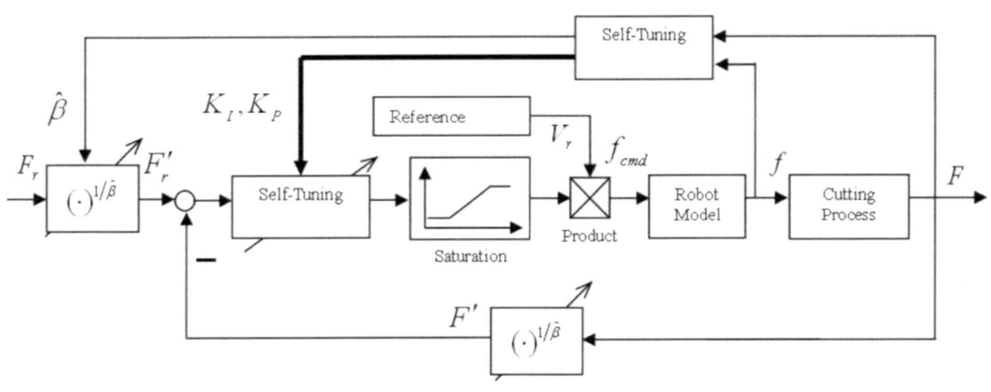

Fig. 10. Robotic machining system with self-tuning PI control

Without considering the saturation nonlinearity in the system block shown in Fig. 10, we set the open loop gain at 28.84, and the close loop system will have a dominant conjugate pair of poles with a damping factor around 0.7. The close loop system will have a quick response and very small overshoot, with the above damping factor. From Eq. (1), (9), (10), and (11), the open loop gain of the system is calculated as

$$\alpha \cdot V_r \cdot k = 28.84 \tag{13}$$

Combine Eq. (11) and Eq. (13), the proportional and integral gains can be given as

$$K_I = \frac{28.84}{V_r \hat{k}}, \; K_P = \frac{0.432}{V_r \hat{k}} \tag{14}$$

Where $\hat{k}$ is the on-line estimation of k in Eq. (9). Eq. (14) is used as the self-tuning rules for the PI controller, which aims to maintain the open loop gain at 28.84.

The following standard recursive linear least square (RLS) method is used to identify k and β of Eq. (9)

$$k(t) = \frac{P(t-1)x(t)}{\lambda + x^T(t)P(t-1)x(t)}$$

$$\hat{\theta}(t) = \hat{\theta}(t-1) + k(t)[y(t) - \hat{\theta}(t-1)x(t)]$$

$$P(t) = \frac{1}{\lambda}[I - K(t)x^T(t)]P(t-1) \tag{15}$$

Where $\hat{\theta}(t) = (\ln \hat{k}(t) \quad \hat{\beta})$; $y(t) = \ln F(t)$; $x(t) = (1 \quad \ln f(t))^T$; $t = 1,2,3,...$ is the sampling point; λ is the forgetting factor, which is usually chosen between 0.95 and 0.99. The on-line identified $\hat{k}$ and $\hat{\beta}$ are used in Eq. (9) and Eq. (14) respectively as the adaptive rules.

4.3.4 Fuzzy Logic Control

Although PI control and adaptive control provide stable and zero static error solutions for MRR control, they are only feasible for applications with slow feed speed, such as end milling and grinding. Their response is limited by the open loop gain to maintain a stable performance. For deburring applications, where the cycle time is critical, faster feed speed up to 200 mm/s is usually required. Also, the variation of material to be removed (bur size) is more dramatic in deburring process. Even with the largest stable gain, the PI and adaptive controller could not response fast enough to prevent spindle stall or robot vibration. Derivative term (change of force) must be included in the controller to predict the force trend and achieve faster response. Since the force/spindle current signal is very noisy, it is not practical to expand the PI control to a complete PID controller. A more intuitive control method must be adopted here to address this problem since the change of force information is only critical at the moment when the cutting tool start to engage a large bur.

Fuzzy control is a very popular approach for performing the task of controller design because it is able to transfer human skills to some linguistic rules. Therefore, fuzzy control is often applied to some ill-defined systems or systems without mathematical models. In this robotic machining situation we use a Mamdani type fuzzy PD control law to regulate the machining force. In Mamdani method, fuzzy logic controller (FLC) is viewed as directly translating external performance specifications and observations of plant behavior into a rule-based linguistic control strategy.

A FLC is a control law described by a knowledge base (defined with simple IF . . . THEN type rules over variables vaguely defined -- fuzzy variables) and an inference mechanism to obtain the current output control value. The designed FLC has three inputs, force difference, filtered change of force difference, and previous output speed_ratio, and one output change of speed_ratio. The inputs are divided in levels in accordance with the observed sensor characteristics and fuzzyfied using triangular membership functions.(Galichet & Foulloy, 1995) The output is fuzzyfied in the same way. The rule base is constructed using a methodology similar to that in the work of (Li, & Gatland, 1996). The rule base consist three groups of rules:

1) Force limit rule: Basic rules to speed up or slow down robot based on the difference of measured force and reference force. This group of rules perform similarly to classical control method.
2) Force trend rule: This group of rules are specially implemented to detect the large burs by evaluate the trend of force difference. Proper set of force trend rule could reduce overshoot of cutting force and achieves fast response.
3) System failure protection rule: Used for safety purpose. When speed_ratio is already on lowest stage and process force is still high, robot will stop to avoid motor overload and robot vibration.

FLC generates change of speed_ratio through evaluating various rules. Instead of changing speed_ratio continuously as in classical PID control, speed_ratio is set to several stages. The reason behind this is that continuously adjusting feed speed is not desirable for machining process because it increase tool wear and deteriorate surface quality. Since a too slow feed speed will change the chip generation mechanism, that is, tool becomes rubbing instead of cutting the workpiece; the minimal feed speed is also set. Although ideally more stages means more control accuracy, five stages (0.2, 0.4, 0.6 0.8, 1.0) would be enough for most applications. A special case is two-stage switching control which has only low or full speed. Two-stage switching control, which sacrifices control accuracy to achieve faster response, is a very attractive control method for many deburring process. One such example will be presented in the next session.

4.4 Experimental Results for CMRR

Experimental studies are conducted for an end milling process to verify the stability and performance of the proposed PI control and adaptive control algorithm. The robot used in the milling process is the ABB IRB 6400, the same robot on which we have done the parameter identification. The setup of robotic end milling process is shown as Fig. 6.

During the end milling experiment, a spindle was hold by the robot arm, and an aluminum block (AL2040) is fixed on a steel table. The cutting depth of the process was changed from 1 mm to 3 mm with a step of 1 mm. Both fixed gain PI control algorithm and self-tuning PI control algorithm, proposed, were tested with the same experimental setup. The control system performance and stability are compared for these two controllers. The experiment results for fixed-gain PI controller and for self-tuning PI controller are shown in Fig. 11 and Fig. 12, respectively.

The reference force was set at 250 N for the experiments. When the cutting depth is 1mm, both controllers are saturated with a full command speed at 30 mm/s. When the cutting depth changed to 2 mm, the fixed-gain PI controller started to vibrate, but still stable. When the cutting depth changed to 3 mm, the fixed-gain PI controller became unstable, just as predicted in the simulation results. On the other hand, the self-tuning adaptive controller maintained the stability and performance for all the cutting depths.

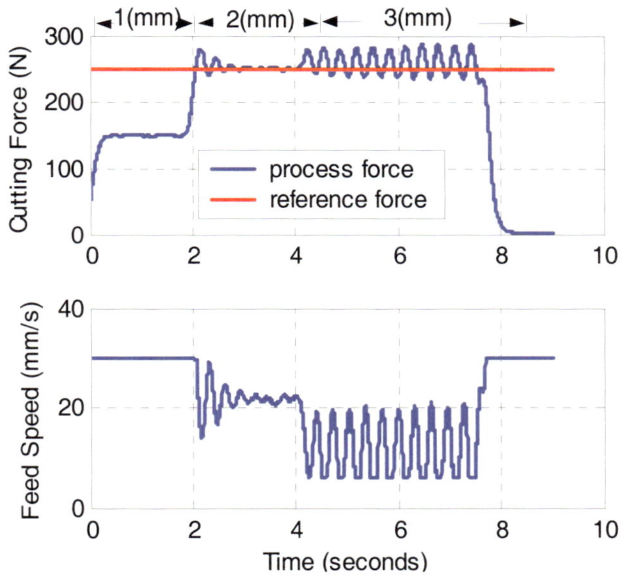

Fig. 11. Fixed-gain PI control experiment result

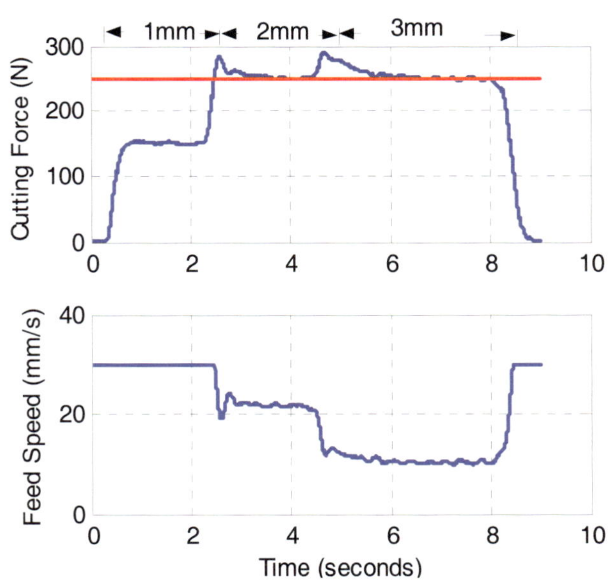

Fig. 12. Self-tuning PI control experiment result

The FLC is tested in another setup for robotic deburring with a grinder. There two burs located in the middle of the cast steel workpiece. A single cut with straight path is supposed to remove the burs. The limit of this system is the spindle power, which is equivalent to about 300 N. Without the CMRR function, the spindle will stall at the bur location and the entire system setup will be damaged. Since the bur location and size are not predicable, normally the command feed speed is set to be a very conservative value, such as 30~40 mm/s. With FLC MRR control, command feed speed is set to 100ms. Two-stage switch control (0.5, 1.0) is sufficient to keep the system under spindle limit. The motor current signal (blue) is also recorded for comparison purpose. It could be shown that after a linear conversion (a gain and an offset), spindle current is equivalent to machining force signal. Either signal could be used for feedback here. Note that the force measurements in the experiments were filtered with a low-pass filter before used. (Fig. 13)

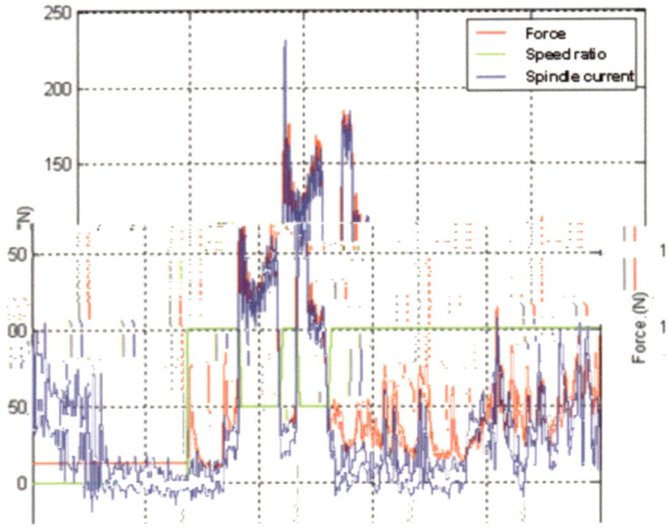

Fig. 13. FLC MRR control result

5. Robot Deformation Compensation

Among the many sources of errors of machine tools, thermal deformation and geometric errors are traditionally known as key contributors. For example, by studying a large amount of data, (Bryan, 1990) reported that thermal errors could contribute as much as 70% of workpiece errors in precision machining. RTEC techniques for geometric and thermal errors have successfully improved machine tool accuracy up to one order of magnitude (Donmez, 1986) (Chen, 1993).

After the geometric and thermal errors are compensated for, cutting force induced errors become the major source of machine tool errors. (Bajpai, 1972) and (Kops, et al., 1994) attempted to overcome the errors due to deflection using the relationship between workpiece deflection and the depth-of-cut applied at the final pass. However, most of the current error compensation research has not considered the cutting force induced errors. The following argument has been used to justify the neglect of the cutting force induced errors: in finish machining, the cutting force is small and the resulting deflection can be neglected.

However, in robotic machining process, due to the low stiffness of the industrial robot, the force induced deformation of the robot structure is the single most dominant source of workpiece surface error. Offline calibration strategies are often used to improve accuracy while sacrificing operation cycle time. The workpiece is calibrated with a distance sensor, usually LVDT or laser sensor before and after the machining process. The surface error is measured and calculated to update the tool/workpiece data of the next cut. Although offline calibration could improve robot path error as well as force induced error, the process cycle time is increased, mostly doubled. With force sensor attached on the robot wrist, force information is ready on real time. If an accurate stiffness model could be established, the force induced error could be compensated online by updating the robot targets.

5.1 Robot Stiffness Modeling

A robot stiffness model, which relates the force applied on the robot tool end point to the deformation of the tool end point in Cartesian space, is crucial for robot deformation compensation, since the force measurement and control is fulfilled in Cartesian space while the robot position control is implemented in joint space.

The proposed model must be accurate enough for a great improvement of the surface error, as well as simple enough for real-time implementation. Detailed modelling of all the mechanical components and connections will bring a too complicated model for real-time control; and difficulties for accurate parameter identification.

The sources of the stiffness of a typical robot manipulator are the compliance of its joints, actuators and other transmission elements, geometric and material properties of the links, base, and the active stiffness provided by its position control system (Alici & Shirinzadeh, 2005).. As commercial robotic systems are designed to achieve high positioning accuracy, elastic properties of the arms are insignificant. The dominant influence on a large deflection of the manipulator tip position is joint compliance, e.g., due to reducer elasticity (Pan et al., 2006).

The conventional formulation for the mapping of stiffness matrices between the joint and Cartesian spaces, was first derived by (Salisbury, 1980) and generally has been accepted and applied.

$$K_x = J(Q)^{-T} K_q J(Q)^{-1} \tag{16}$$

Where K_q is a 6×6 diagonal joint stiffness matrix, which relates the motor torque load τ on six joints to the 6×1 joint deformation vector ΔQ,

$$\tau = K_q \cdot \Delta Q \tag{17}$$

$J(Q)$ is the Jacobian matrix of the robot;

K_x is a 6×6 Cartesian stiffness matrix, which relates the 6 D.O.F. force vector in Cartesian space F to the 6 D.O.F. deformation of robot in Cartesian space ΔX

$$F = K_x \cdot \Delta X \tag{18}$$

Eq. (16) can be derived from the definition of Jacobian matrix in Eq. (19) and the principle of virtual work in Eq. (20).

$$\Delta X = J(Q) \cdot \Delta Q \tag{19}$$

$$F^T \cdot \Delta X = \tau^T \cdot \Delta Q \tag{20}$$

For articulated robot, K_x is not a diagonal matrix and it is configuration dependent. This means: first, the force and deformation in Cartesian space is coupled, the force applied in one direction will cause the deformation in all directions; second, at different positions, the stiffness matrix will take different values.
(Chen & Kao, 2000) introduced a more complex model using a new conservative congruence transformation as the generalized relationship between the joint and Cartesian stiffness matrices in order to preserve the fundamental properties of the stiffness matrices.

$$K_x = J(Q)^{-T}(K_q - K_g)J(Q)^{-1} \tag{21}$$

With

$$K_g = \left[\frac{\partial J^T(Q)}{\partial Q} \cdot F \right] \tag{22}$$

where K_g is a 6×6 matrix defining the changes in geometry via the differential Jacobian; F is external applied force.
The second model is more difficult to implement as the differential Jacobian is not available in the robot controller. The difference between these two models is the additional K_g in the second model. K_g accounts for the change in geometry under the presence of external load. IRB6400, a typical large sized industrial robot has a payload of 150kg, which will cause about 3 mm deformation considering its stiffness is around 0.5N/μm. From our calculation, K_g is negligible compared to K_q as this is a relative small deformation compared to the scale of robot structure.
Thus, the conventional formulation is selected in this research for stiffness modelling. In this model, robot stiffness is simplified to six rotational stiffness coefficients, that is, equivalent torsional spring with stiffness K as each joint is actuated directly with AC motor. Also from the control point of view, this model is the easiest to implement, since these are the 6 degree of freedom of the robot, which could be directly compensated by joint angles. Since the axis of force sensor is coincide with the axis of joint six, the stiffness of force sensor and its connection flange could be modelled into joint six.

5.2 Parameter Identification of the Stiffness Model

Experimental identification of the robot stiffness model parameters, joint stiffness of six joints, is critical in fulfilling real-time position compensation. In our model, the joint stiffness is an overall effect contributed by motor, joint link, and gear reduction units. It is not realistic and accurate to identify the stiffness parameter of each joint directly by dissembling the robot as the assembly process will affect the stiffness of the robot arm. The practical method is to measure it in Cartesian space.

The setup of robot stiffness measurement is shown in Fig. 14. The cutting tool at the end-effector is replaced by a sphere-tip. When robot is driven to a fixed position in the workspace, the joint angles of the robot are recorded. A weight is applied on the tool tip to generate a deformation. The position of the sphere-tip is measured by ROMOR CMM machine before and after the weight is applied to and the 3-DOF translational deformation is calculated. The applied force is measured by 6 DOF ATI force/torque sensor. A pulley is used to generate force on other directions than vertical down direction.

Fig. 14. Methodology of robot stiffness measurement

Given the kinematic parameters of the robot, the Jacobian matrix at any robot position could be calculated using robotics toolbox for MATLAB. Table 1 shows the IRB6400 kinematic model in Denavit-Hartenberg parameters.
The same procedure is repeated at multiple positions in the robot workspace and with different loads. From the relationship of

$$F = J(Q)^{-T} K_q J(Q)^{-1} \cdot \Delta X \tag{23}$$

K_q could be solved by least square method, given F, $J(Q)$ and ΔX. Only the first three equations from Eq. (23) are used in calculation as the orientation and torque are hard to measure accurately in the setup. The calibration results show that the standard deviation of the stiffness data is small, which means constant model parameter is adequate to model the deformation of robot. The deviation in the entire work space is less than 0.04mm.

Axis	θ	d	a	α	Home
1	q_1	900	188	-90°	0
2	q_2	0	950	0	-90°
3	q_3	0	225	-90°	0
4	q_4	1300	0	-90°	180°
5	q_5	0	0	90°	0
6	q_6	200	0	0	0

Table 1. DH model of IRB 6400

5.3 Real-time robot deformation compensation

The major sources of position error in robotic machining process can be classified into two classes, 1) machining force oriented error, and 2) motion error (kinematic, measurement and servo errors, etc.). The motion error is inherent from robot position controller and will appear even in non-contact movement. While the machining force in the milling process will typically over several hundreds of Newton, the force oriented error, which will easily go up to 0.5mm, is the dominant factor of surface error. Our objective here is to measure the deformation through a viable way and compensate it online to improve the overall machining accuracy.

To our best knowledge, none of the existing research has addressed the topic of online compensation of process force oriented robot deformation due to the lack of real-time force information and limited access to the controller of industrial robot.

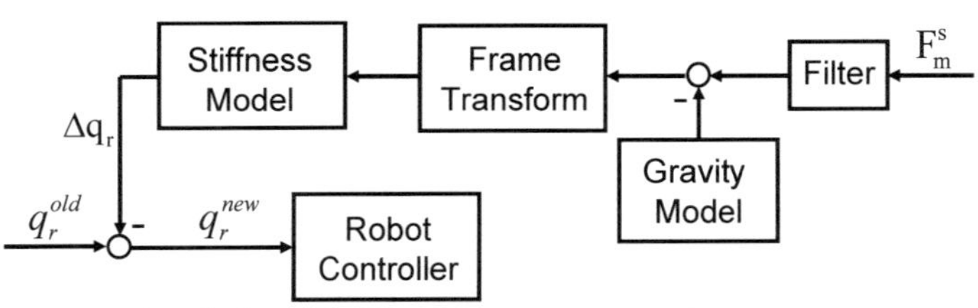

Fig. 15. Block diagram of real-time deformation compensation

The block diagram of real time deformation compensation algorithm is shown in Fig. 15. After the force sensor noise is filtrated, gravity compensation must be conducted to remove the force reading from the weight of spindle and tool. Since the robot may not always maintain a wrist down position, a general gravity compensation algorithm is developed to remove the gravity effects for any robot configuration. The algorithm takes measurement of gravity force at 15 distinctive robot configurations and uses least square method to calculate the mass and center of mass coordinates. This information is then updated to the robot tool data and the robot will always offset the gravity from the force reading at any robot configurations.

The force signal read from the sensor frame is then translated into the robot tool frame. Based on the stiffness model identified before, the deformation due to machining force is calculated online and the joint reference for robot controller is updated accordingly.

5.4 Experimental Results

The experimental tests on both standard aluminum block and real cylinder head workpiece have been conducted to verify the results of proposed real-time deformation compensation method.

5.4.1 Aluminum block end milling test

A 150mm×50mm 6063 aluminum alloy block is used for end milling test. Table 2 lists the detailed parameters for the experiment.

Test	End milling
Spindle	SETCO,5HP, 8000RPM
Tool type	SECO Φ75mm, Square insert×6
Cutting fluid	- (Dry cutting)
Feed rate	20 mm/s
Spindle speed	3600 RPM
DOC	3 mm

Table 2. Parameters for end milling

Fig. 16. Setup of aluminum end milling and surface scan

A laser distance sensor is used to measure the finished surface of aluminum block as shown in Fig. 16. The surface error without deformation compensation demonstrates anti-intuitive results, on average extra 0.4mm material was removed from the aluminum block, which is not possible for a CNC machine since the cutting force normal to the workpiece surface will always push the cutter away from the surface and cause negative surface error (cut less).

The coupling of robot stiffness model explains this phenomenon. When end milling using square inserts, the machining force in the robot feed direction and the cutting direction (around 300N each) are much larger than the force in the normal direction (around 50N). At this specific robot configuration, the force in feed and cutting direction will both push the cutter into the workpiece, which results in positive surface error (cut more). Since the feed force and cutting force are the major components in this setup, the overall effect is that the surface is removed 0.4 mm more than commanded depth. On the other hand, the result after deformation compensation shows a less than 0.1 mm surface error, which is in the range of robot path accuracy.

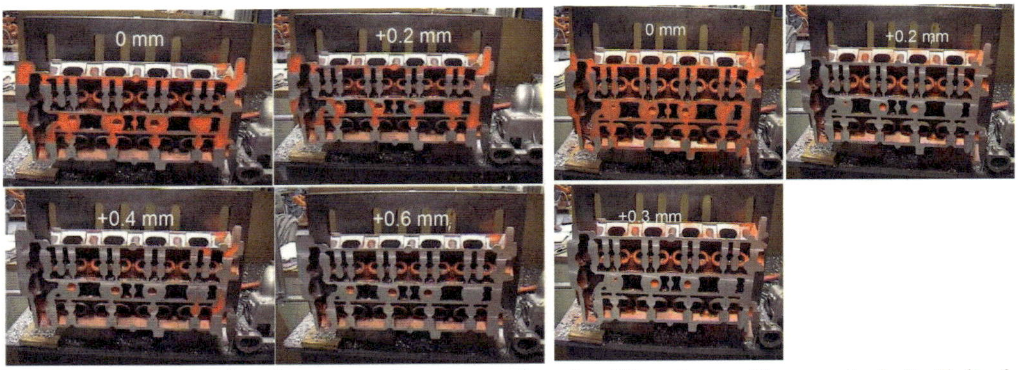

Fig. 17. A. Cylinder head part, surface error of end milling in position control; B. Cylinder head part, surface error of end milling in force control

5.4.2 Cylinder Head End Milling Test

A real cylinder head workpiece is also utilized here for deformation compensation test, using the same end milling parameters as listed in Table 2. To better visualize the surface error, the surface is covered by orange paint after the end milling. Then the tool is moved 0.1mm closer to the workpiece surface each time, until all the paint on the surface are cleaned. As shown in Fig.17A, under position control, the tool touches the surface at -0.3mm, and clean the surface at 0.6mm, the total surface error is 0.9mm. Under the force control, the tool touches the surface at -0.1mm, and clean the surface at 0.3mm, the total surface error reduced to 0.4mm, as shown in Fig. 17B.

6. Conclusion

This chapter has addressed the critical issues in robotic machining process from programming to process control. Three major contributions, including rapid robot programming, controlled material removal rate, and online deformation compensation have been introduced in detail. The complete solution is achieved with force control strategy based on ABB IRC5 robot controller.

Rapid robot programming is characterized by two main modules: lead-through and automatic path-learning. Lead-through gives robot operator the freedom to adjust the spatial relationship between the robot tool fixture and the workpiece easily, while robot automatically follow the workpiece contour, record the targets and generate the process program in path-learning. Since the robot programming is generated at actual process setup, no additional calibration is required.

Online deformation compensation is realized based on a robot structure model. Since force induced deformation is the major source of inaccuracy in robotic machining process, the surface quality is improved greatly adopting the proposed method. This function is especially important in milling applications, where cutting force could be as large as 1000 N.

Regulating machining forces provides significant economic benefits by increasing operation productivity and improving part quality. CMRR control the machining force by realtime adjusting the robot feed speed. Various control strategy, including PID, adaptive control and fuzzy logic controller were implemented on different cutting situations
Including the chatter and vibration analysis presented in (Pan & Zhang, et al, 2006), these complete set of solutions will greatly benefit the foundry industry with small to medium batch sizes. Dramatic increase of successful setups of industrial robots in foundry cleaning and pre-machining applications will be seen in the very near future.

7. References

Alici, G.; Shirinzadeh, B. (2005). "Enhanced Stiffness Modeling Identification and Characterization for Robot Manipulators", *IEEE Transactions on Robotics,* Vol. 21, No. 4, August 2005.

Donmez, M. A. et al. (1986). "A General Methodology for Machine Tool Accuracy Enhancement y Error Compensation", *Precision Engineering,* Vol. 8, No. 4, pp. 187-196.

Bajpai, S. (1972) "Optimization of Workpiece Size for Turning Accurate Cylindrical Parts", *International Journal of Machine Tool Design and Research,* Vol. 12, pp. 221-228.

Basanez, L. & Rosell, J. (2005) "Robotic Polishing Systems—From graphical task specification to automatic programming", *IEEE Robotics & Automation magazine,* Sep, 2005.

Bryan, J. B. (1990). "International Status of Thermal Error Research", *Annals of the CIRP,* Vol. 39, No. 2, pp. 645-656.

Budak, E.; Altintas, Y. (1998) "Analytical Prediction of Chatter Stability Conditions for Multi-Degree of Systems in Milling. Part I: Modeling, Part II: Applications," *Transactions of ASME, Journal of Dynamic Systems, Measurement and Control,* vol.120, pp.22-36

Chen, J. S. et al. (1993). "Real Time Compensation of Time-variant Volumetric Error on a Machining Center", *ASME Journal of Engineering for Industry,* Vol. 114, pp. 472-479.

Chen, S.F.; Kao, I. (2000) "Conservative congruence transformation for joint and Cartesian stiffness matrices of robotic hands and fingers", *The International Journal of Robotics Research,* Vol. 19, No. 9, September 2000, pp. 835-847.

Daneshmend, L., & Pak, H., (1986) "Model Reference Adaptive Control of Feed Force in Turning," *ASME Journal of Dynamic Systems, Measurement, and Control,* Vol. 108, No. 3, pp. 215-222

Galichet, S., & Foulloy, L. (1995) "Fuzzy controllers: synthesis and equivalences", *IEEE Transactions on System, Man and Cybernetics.*

Kim, S., Landers, R., Ulsoy, A., 2003, "Robust Machining Force Control with Process Compensation," *Journal of Manufacturing science and engineering,* Vol 125, pp. 423-430

Kops, L.; Gould, M.; & Mizrach, M. (1994). "A Search for Equilibrium between Workpiece Deflection and Depth of Cut: Key to Predictive Compensation for Deflection in Turing", *Proceedings 1994 ASME Winter Annual Meeting,* Vol. 68-2, pp. 819-825.

Landers, R. & Ulsoy, A., (2000) "Model-based machining force control", ASME Journal of Dynamic Systems, Measurement, and Control, vol. 122, no. 3, 2000, pp. 521-527.

Li, H.,& Gatland, H., (1996) "Conventional fuzzy control and its enhancement", IEEE Transactions on System, Man and Cybernetics.

Pan, Z. (2006). "Intelligent robotic machining with force control", Ph.D. dissertation, Stevens Institute of Technology, NJ, USA, Jan 2006.

Pan, Z.; & Zhang, H.; et al. (2006). "Chatter analysis of robotic machining process", Journal of Material Processing Technology, Volume 173, Issue 3, Pages 301-309, April 2006.

Pettersen, T. et al. (2003) "Augmented Reality for Programming Industrial Robots", Proceedings of the Second IEEE and ACM International Symposium on Mixed and Augmented Reality (ISMAR'03), 2003.

Pires,J. et al, (2004) "CAD interface for automatic robot welding programming", Industrial Robot: An International Journal, Vol 31, pp.71-76, 2004.

Salisbury, K. (1980) "Active stiffness control of a manipulator in Cartesian coordinates", Proceedings of the 19th IEEE Conference on Decision and Control, Albuquerque, NM, pp. 87–97.

Schmid, D. (1990) "Sensor Simulate Tools", The industrial robot, pp. 97-9, June, 1990

Sugita, S. et, al, (2003) "Development of robot teaching support devices to automate deburring and finishing works in casting", The International Journal of Advanced Manufacturing Technology, Springer-Verlag London, Dec 2003.

Wang, J. Zhang, H. and Pan, Z. (2007) "Machining with Flexible Manipulators: Critical Issues and Solutions", Industrial Robotics: Programming, Simulation and Applications, ISBN 3-86611-286-6 Edited by Low Kin Huat, Feb 2007.

Yang, S., (1996) "Real-time compensation for geometric, thermal, and cutting force induced errors in machine tools," Ph.D. dissertation, The University of Michigan

4

Fuzzy Optimal Control for Robot Manipulators

Basil M. Al-Hadithi, Agustín Jiménez and Fernando Matía
Intelligent Control Group, Universidad Polietécnica de Madrid
J.Gutierrez Abascal, 2.28006-Madrid
Spain

1. Introduction

This chapter deals with the design of a Fuzzy Logic Controller based Optimal Linear Quadratic Regulator (FC-LQR) for the control of a robotic system. The main idea is to design a supervisory fuzzy controller capable to adjust the controller parameters in order to obtain the desired axes positions under variations of the robot parameters and payload variations.
In the advanced control of robotic manipulators, it is important for manipulators to track trajectories in a wide range of work place. If speed and accuracy is required, the control using conventional methods is difficult to realize because of the high nonlinearity of the robot system.
In control design, it is often of interest to design a controller to fulfil, in an optimal form, certain performance criteria and constraints in addition to stability. The theme of optimal control addresses this aspect of control system design. For linear systems, the problem of designing optimal controllers reduces to solving algebraic Riccati equations , which are usually easy to solve and detailed literature of their solutions can be found in many references . Nevertheless, for nonlinear systems, the optimization problem reduces to the so-called Hamilton-Jacobi (HJ) equations, which are nonlinear partial differential equations. Different from their counterparts for linear systems, HJ equations are usually difficult to solve both numerically and analytically. Improvements have also been carried out on the numerical solution of the approximated solution of HJ equations. But few results so far can provide an effective way of designing optimal controllers for general nonlinear systems.
In the past, the design of controllers based on a linearized model of real control systems. In many cases a good response of complex and highly non-linear real process is difficult to obtain by applying conventional control techniques which often employ linear mathematical models of the process. One reason for this lack of a satisfactory performance is the fact that linearization of a non-linear system might be valid only as an approximation to the real system around a determined operating point.
However, fuzzy controllers are basically non-linear, and effective enough to provide the desired non-linear control actions by carefully adjusting their parameters.
In this chapter, we propose an effective method to nonlinear optimal control based on fuzzy control. The optimal fuzzy controller is designed by solving a minimization problem that minimizes a given quadratic performance function.
Both the controlled system and the fuzzy controller are represented by the affine Takagi-Sugeno (T-S) fuzzy model taking into consideration the effect of the constant term. Most of the research works analyzed the T-S model assuming that the non-linear system is linearized

with respect to the origin in each IF-THEN rule (Tanaka and Sano 1994), (Tanaka et al. 1996), which means that the consequent part of each rule is a linear function with zero constant term. This will in turn reduce the accuracy of approximating non-linear systems. Moreover, in linear control theory, the independent term does not affect the dynamics of the system rather the input to it. In the case of fuzzy control, the fuzzy system is resulted from blending all the subsystems. The blending of the independent term of each rule will no longer be a constant but a function of the variables of the system and thus affects the dynamics of the resultant system. A necessary condition has been added to deal with the independent term. The final fuzzy system can be obtained by blending of these affine models. The control is carried out based on the fuzzy model via the so-called parallel distributed compensation scheme. The idea is that for each local affine model, an affine linear feedback control is designed. The resulting overall controller, which is also a non-linear one, is again a blending of each individual affine linear controller.

LQR is used to determine best values for parameters in fuzzy control rules in which the robustness is inherent in the LQR thereby robustness in fuzzy control can be improved. With the aid of LQR, it provides an effective design method of fuzzy control to ensure robustness.

In this chapter, we will show how the LQR, the structure of which is based on mathematical analysis, can be made more appropriate for actual implementation by introduction of fuzzy rules.

The motivation behind this scheme is to combine the best features of fuzzy control and LQR to achieve rapid and accurate tracking control of a class of nonlinear systems.

The results obtained show a robust and stable behavior when the system is subjected to various initial conditions, moment of inertia and to disturbances.

The content of this chapter is organized as follows. In section 2, an Overview of various control techniques for robot manipulators are presented. Section 3 presents the modelling of the robot manipulator. Section 4 demonstrates Takagi-Sugeno model for the robot manipulator under study. In section 5 a detailed mathematical description of the proposed optimal controller is presented. Section 6 entails the application of the proposed FC-LQR on a robot manipulator to demonstrate the validity of the proposed approach. This example shows that the proposed approach gives a stable and well damped response infront of various initial conditions, moment of inertia and a robust behaviour in the presence of disturbances. The conclusion of the effectiveness and validity of the proposed apprach is explained in section 7.

2. Overview of Control Techniques for Robot Manipulators

It is well known that robotic manipulators are complicated, dynamically coupled, highly time-varying, highly nonlinear systems that are extensively used in tasks such as welding, paint spraying, accurate positioning systems and so on. In these tasks, end-effectors of robotic manipulators are commended to move from one place to another, or to follow some given trajectories as close as possible. Therefore, trajectory tracking problem is the most significant and fundamental task in control of robotic manipulators.

Motivated by requirements such as a high degree of automation and fast speed operation from industry, in the past decades, various control methods are introduced in the publications such as proportional, integration, derivative (PID) control (Luh 1983), feed-forward compensation control (Khosla and Kanade 1988), adaptive control (Slotine and Li 1988), variable structure control (Slotine et al. 1983), neural networks control (Purwar et al. 2005), fuzzy control (Chen et al. 1998) and so on.

As a predominant method in industrial robotic manipulators, traditional PID control has simple structure and convenient implementation (Luh 1983). However, some strong assumptions are required to be made, which involve that each joint of robotic manipulators is decoupled from others and the system has to be in the status of slow motion. Control performance degrades quickly as operating speed increases. Therefore, a robotic manipulator controlled in this way is only appropriate for relatively slow motion.

Robotic manipulator systems are inevitably subject to structured and unstructured uncertainty. Structured uncertainty is characterized by a correct dynamical model with parameters variations, which results from difference in weights, sizes and mass distributions of payloads manipulated by robotic manipulators, difference in links properties of robotic manipulators, difference in inaccuracies on torque constants of actuators and so on. Unstructured uncertainty is characterized by unmodeled dynamics, which is due to the presence of external disturbances, high-frequency modes ofrobotic manipulators, neglected time-delays and nonlinear frictions and so on.

Structured uncertainty can result in imprecision of dynamical models of robotic manipulators, and controllers designed for nominal parameters may not properly work for all changes in parameters. Adaptive control techniques (Slotine and Li 1988), can be used in this case. However, adaptive control law is unable to handle unstructured uncertainty. To overcome this difficulty, variable structure control (Slotine et al. 1983) that can simultaneously attenuate influences ofboth structured and unstructured uncertainty is employed. Unfortunately, undesirable chattering on sliding surface due to high frequent switching can deteriorate system performances, which cannot be eliminated completely.

For practical and complex control problem of robotic manipulators, traditional and effective schemes also cannot be ignored. Computed Torque Control (CTC) (Middleton and Goodwin 1988) is worth noting, because CTC is easily understood and of good performances. Briefly speaking, CTC is a linear control method to linearize and decouple robotic dynamics by using perfect dynamical models of robotic manipulator systems in order that motion ofeach joint can be individually controlled using other well-developed linear control strategies.

However, CTC method for robotic manipulators suffers from two difficulties. First, CTC requires exact dynamical knowledge of robotic manipulators, which is apparently impossible in practical situations. Second, CTC is not robust to structured uncertainty and/or unstructured uncertainty, which may result in performance devaluation.

One of successful fuzzy systems' (FS) applications is to model complex nonlinear systems by a set of fuzzy rules. One important property of fuzzy modeling approaches is that FS is a universal approximator (Wang and mendel 1992). In other words, FS can approximate virtually any nonlinear functions to arbitrary accuracy provided that enough rules are given. FS for control, i.e. Fuzzy Controller (FC) can integrate expertise of skilled personnel into control procedure and mathematical model is not required. Over the last few years, FC for complex nonlinear systems have been developed extensively (Hua et al. 2004), (Kim and Lewis 1999).

Recently, much attention has been devoted to FC for robotic manipulators. The latest survey on FC for robotic manipulators can be found in (Purwar et al. 2005) and references cited therein. Sun (Luh 1983) combined FC and variable structure control to construct a controller, where FS was greatly simplified by using system representative point and its derivative as inputs. Control laws designed by Hsu (Sun et al. 1999) consisted of a regular fuzzy controller and a supervisory control term, which ensured stability of closed-loop systems. In (Labiod et al. 2005), two FC schemes for a class of uncertain continuous-time multi-input

multi-output nonlinear dynamical systems were derived. Satisfactory performances were achieved by applying them to robotic manipulators (Song et al. 2006).
In (Song et al. 2006), it is supposed that robotic manipulator systems with structured uncertainty and/or unstructured uncertainty can be separated as two subsystems: nominal system with precise dynamical knowledge and uncertain system with unknown knowledge. An approach of CTC plus FC compensator is proposed.
The nominal system is controlled using CTC and for uncertain system, a fuzzy controller is designed. Here the fuzzy controller acts as compensator for CTC. Parameters updating laws of the fuzzy controller are derived using Lyapunov stability theorem.
FS have also been extensively adopted in adaptive control of robot manipulators (Berstecher et al. 2001), (Chuan-Kai Lin 2003), (Li et al. 2001), (Tzes et al. 1993), (Tong et al. 2000), (Tsai et al. 2000), (Yi and Chung 1997), (Yoo and Ham 2000), (Zhou et al. 1992), (Fukuda et al. 1992), (Meslin et al. 1992), (Sylvia et al. 2003).
In (Berstecher et al. 2001), Berstecher develops a linguistic heuristic-based adaptation algorithm for a fuzzy sliding mode controller. The algorithm relies on the linguistic knowledge in the form of fuzzy IF-THEN rules. Tsai et al. (Tsai et al. 2000) propose a robust multilayer fuzzy controller for the model following control of robot manipulators with torque disturbance and measurement noise.
Yi and Chung (Yi and Chung 1997) define a set of fuzzy rules based on the knowledge of error and derivative of error for designing the controller. Yoo and Ham (Yoo and Ham 2000) exploit the function approximation capabilities of FS to compensate for the parametric uncertainties of the robot manipulator. Chuan-Kai Lin (Chuan-Kai Lin 2003) proposes reinforcement learning systems combined with fuzzy control for robot arms. Here the reinforcement learning signal is used to update the weights of a fuzzy logic system which is used to approximate an unknown nonlinear function. This approximated function is then used for computing the control law. In (Li et al. 2001) Li presents a hybrid control scheme for tracking control of a manipulator which consists of a fuzzy logic proportional controller and a conventional integral and derivative controller.
Moreover, this controller was compared to a conventional PID controller and the performance of the fuzzy P+ID controller was found superior to conventional PID controller. In (Sylvia et al. 2003) Sylvia Kohn-Rich and Henryk Flashner present tracking control problem of mechanical systems based on Lyapunov stability theory and robust control of nonlinear systems. The control law has a two-component structure conventional PD control and a fuzzy component of robust control which is aimed at minimizing the chattering effect. Tong Shaocheng et al. (Tong et al. 2000) develops a robust fuzzy adaptive controller for a class of unknown nonlinear systems. In the control procedure, FS are implemented to estimate the unknown functions and robust compensators are designed in H_∞ sense for attenuating the unmatched uncertainties. In (Zhang et al. 2000), Rainer palm develops a mamdani fuzzy controller following the pattern of suboptimal control. The proposed controller in the paper is compared and found to have higher tracking quality than a conventional PD controller. In (Fuchun et al. 2003), Fuchun Sun et al. propose a nuero fuzzy adaptive control methodology for trajectory tracking of robotic manipulators. Here the fuzzy dynamic model of the manipulator is established using the Tagaki-Sugeno fuzzy framework. Based on the derived fuzzy dynamics of the manipulator, the neuro fuzzy adaptive controller is developed to improve the system performance by adaptively modifying the fuzzy model parameters. All these methods require both the position and velocity measurements, which can be problematic in practice (Purwar et al. 2005).

Applications in tracking control problems of robot manipulators are also available (Commuri et al. 1996), (Jagannathan et al. 1996), (Llama et al. 1998).
In (Commuri et al. 1996) an adaptive fuzzy logic controller is proposed. The structure of this controller is based on the so-called SlotineŰ Li controller (a PD term plus a model-based nonlinear compensation term using Hltered tracking errors). A framework that can approximate any nonlinear function with arbitrary accuracy is designed using a fuzzy logic system. By using this technique an estimate of the nonlinear compensation term of the control law is obtained. A learning algorithm that learns the membership function is developed, and the stability of the closed-loop system is demonstrated. In (Jagannathan et al. 1996) a tracking control system of a class of feedback linearizable unknown nonlinear dynamical systems, such as robotic systems, using a discrete time fuzzy logic controller, is presented.
Unlike (Commuri et al. 1996), instead of using fuzzy adaptation of the nonlinear compensation terms, in this paper the potential of a gain scheduling fuzzy self-tuning scheme is used in order to design a methodology for on-line parameter tuning of a robot motion controller. Particular attention is paid to provide a rigorous stability analysis including the robot nonlinear dynamics.
A basic problem in controlling robots is the so-called motion control formulation where a manipulator is requested to track a desired position trajectory. A number of such robot motion controllers having rigorous stability proofs have been reported in the literature and robotics textbooks (Lewis et al. 1994), (Sciavicco et al. 1996). Most of these stability results have been obtained provided that the controller parameters are constant and they belong to well-defined intervals (Llama et al. 2001).
In (Purwar et al. 2005), a stable fuzzy adaptive controller for trajectory tracking is developed for robot manipulators without velocity measurements, taking into account the actuator constraints. The controller is based on structural knowledge of the dynamics of the robot and measurements of link positions only. The gravity torque including system uncertainty like payload variation, etc., is estimated by FS. The proposed controller ensures the local asymptotic stability and the convergence of the position error to zero. The proposed controller is robust not only to structured uncertainty such as payload parameter variation, but also to unstructured one such as disturbances. The validity of the control scheme is shown by simulations on a two-link robot manipulator.
In (Llama et al. 2001) a motion control scheme based on a gain scheduling fuzzy self-tuning structure for robot manipulators is presented. They demonstrate, by taking into account the full non-linear and multivariable nature of the robot dynamics, that the overall closed-loop system is uniformly asymptotically stable. Besides the theoretical result, the proposed control scheme shows two practical characteristics. First, the actuators torque capabilities can be taken into account to avoid torque saturation, and second, undesirable e8ects due to Coulomb friction in the robot joints can be attenuated. Experimental results on a two degrees-of-freedom direct-drive arm show the usefulness of the proposed control approach.

3. Modelling of Robot Manipulators

The robot under study is characterized by having six rotational joints driven by hydraulic actuators(motors for the first joint and the robot wrist, and cylinders for other axes).
The main problem in controlling such processes is the nonlinearity. This makes it very difficult the use of conventional control techniques to implement the control job.
In this chapter, the robot which is a highly non-linear system is represented by affine T-S model, where the consequent part of each rule represents an affine model of the original sys-

tem in a certain operating point. The final fuzzy system can be obtained by blending of these affine models. The control is carried out based on the fuzzy model via the so-called parallel distributed compensation scheme. The idea is that for each local affine model, an affine variable structure controller is designed. the resulting overall controller, which is also a non-linear one, is again a blending of each individual affine linear controller.
The behaviour of the robot depends upon the robot working conditions, in particular the axes positions and the payload which are considered as the premise part of the fuzzy rule (Purwar et al. 2005), (Song et al. 2006).
The suggested fuzzy control considers every axis as a system whose control variables has to be tuned. It is necessary to establish differences between the first axis, which implies a rotation in the horizontal plane, and the axes 2,3 and 4, which imply rotations in the vertical plane. In the case of the latter two axes, which drive the robot wrist, it is not necessary to adjust the control parameters in real time, and they are automatically adjusted when the robot payload changes. For the latter two axes, due to the short length of the driven links and the robot kinematic configuration, their angular position doesnot have a significant amount of influence on their dynamic behaviour, which is mainly determined by the payload. All this means that these two axes are considered independientes with repsect to their control and influence on the adjustment of the other previous axes.
The variables that define the behaviour of each one of the axes are the angular values in each joint and the extreme payload. We should mention that not all the robot joints will influence the dynamic behaviour. The first axis position does not influence the others.
The angular values of the vertical joints that are placed behined the joint we are considering along the robot kinematic chain, and which influence the dynamic behaviour, can be combined in one fuzzy variable. Denoting the angular value for the joint j by θ_j, the effective angular value θ_{ia} to be considered as a fuzzy input variable for axes 2, 3 and 4 is:

$$\theta_{ia} = \sum_{j=2}^{i} \theta_j, \quad i = 2, 3, 4$$

Similarly, considering one particular axis, the angular axis, the angular values of the vertical joints that ar placed in front of it, as well as the robot payload, can be combined in the other fuzzy input variable, namely the effective moment of inertia from the considered axis J_i. This can be represented as:

$$J_i = f(\theta_{j>i}, M_{j>i}, M)$$

Where

- J_i represents the effective moment of inertia from axis i
- $\theta_{j>i}$ represents the angular values of the axes after i
- M_j represents the mass of the link j including its actuator
- M represents the mass of payload.

Figure 1 shows the scheme for the fuzzy input variable for axes 2,3 and 4.

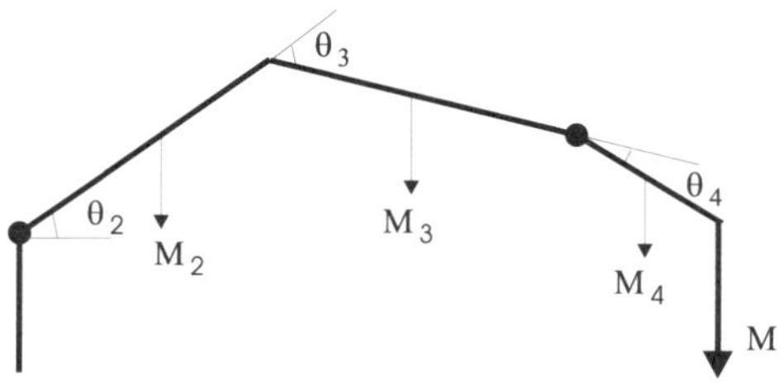

Fig. 1. scheme for the fuzzy input variable for axes 2,3 and 4

4. Takagi-Sugeno Model of Robot Manipulators

Consider the following system:

$$\acute{x} = f(x, u)$$

where

$$x = (x_1, x_2, \ldots, x_n)^t$$
$$u = (u_1, u_2, \ldots, u_m)^t$$

The local dynamics in various equilibrium states are represented by affine subsystems as follows:
Both the fuzzy system and the fuzzy controller are represented by the affine T-S fuzzy model. Let the $(i_1 \ldots i_n)^{th}$ rule of the T-S model be represented as:

$$S^{(i_1 \ldots i_n)} : \textit{If } x \textit{ is } M_1^{i_1} \textit{ and } \acute{x} \textit{ is } M_2^{i_2} \textit{ and} \ldots$$
$$\textit{and } x^{(n-1)} \textit{ is } M_n^{i_n} \textit{ then}$$
$$\acute{x} = a_0^{(i_1 \ldots i_n)} + A^{(i_1 \ldots i_n)}\mathbf{x} + b^{(i_1 \ldots i_n)}u \tag{1}$$

where $M_1^{i_1}$ $(i_1 = 1, 2, \ldots, r_1)$ are fuzzy sets for $\mathbf{x}$, $M_2^{i_2}$ $(i_2 = 1, 2, \ldots, r_2)$ are fuzzy sets for $\dot{x}$, $M_n^{i_n}$ $(i_n = 1, 2, \ldots, r_n)$ are fuzzy sets for $x^{(n-1)}$. Therefore the complete fuzzy system has $r_1 \times r_2 \times \ldots r_n$ rules.
We will adapt the affine T-S model to our robotic system. The premise part of each rule depends on the effective angular value and the effective moment of inertia. Both of them are linearized in three operating points. Table 1 shows the variables of each rule of the robotic system represented by T-S model. The input fuzzy variable which represent the angular axis position is linearized in three operating points. The moment of inertia is linearized in three operating points (Ishikawa 1988). The results were obtained from several tens of experiments of the real system (Gamboa 1996). The system has been approximated in each operating point by a linearized mathematical model looking for a suitable model that coincides with the nonlinear system.
Figure 2 shows the following triangular fuzzy sets of the angular position of the second axis:

$$\begin{aligned} \theta_{2a}^1 &= \{-\infty, 0, 55\} \\ \theta_{2a}^2 &= \{0, 55, 115\} \\ \theta_{2a}^3 &= \{55, 115, \infty\} \end{aligned} \tag{2}$$

Variable	Universe	Label
θ_{2a}	$[0°, 115°]$	$\{M^1_{\theta 2}, M^2_{\theta 2}, M^3_{\theta 2}\}$
θ_{3a}	$[-120°, 90°]$	$\{M^1_{\theta 3}, M^2_{\theta 3}, M^3_{\theta 3}\}$
θ_{3a}	$[-240°, 90°]$	$\{M^1_{\theta 4}, M^2_{\theta 4}, M^3_{\theta 4}\}$
J_2	$[5000, 51540]$	$\{M^1_{J2}, M^2_{J2}, M^3_{J2}\}$
J_3	$[1500, 18564]$	$\{M^1_{J3}, M^2_{J3}, M^3_{J3}\}$
J_4	$[140, 5093]$	$\{M^1_{J4}, M^2_{J4}, M^3_{J4}\}$

Table 1. Input fuzzy variables

$$\begin{aligned} \theta^1_{2a} &= \{-\infty, 0, 55\} \\ \theta^2_{2a} &= \{0, 55, 115\} \\ \theta^3_{2a} &= \{55, 115, \infty\} \end{aligned}$$

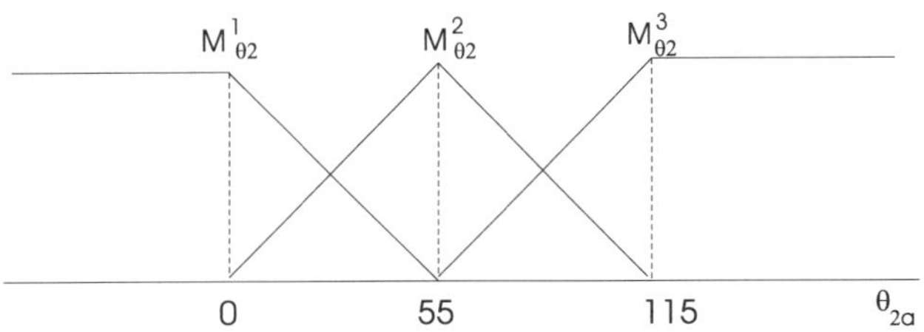

Fig. 2. Fuzzy sets of angular position of the second axis

Figure 3 shows the following triangular fuzzy sets of the moment of inertia of the second axis:

$$\begin{aligned} J^1_{2a} &= \{-\infty, 5000, 25000\} \\ J^2_{2a} &= \{5000, 25000, 51540\} \\ J^3_{2a} &= \{25000, 51540, \infty\} \end{aligned} \quad (3)$$

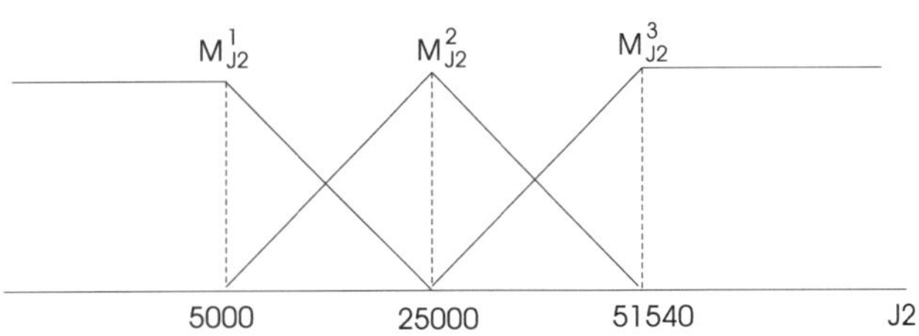

Fig. 3. Fuzzy sets of the moment of inertia of the second axis

Firstly, The model of the robotic model is linearized in three operation points for both the angular postion and its moment of interita. The universe of discourse of the anglular position is $[0, 115]$ rad. and the one of the moment of inertia is $[5000, 51540]$. The resultant identified fuzzy system is described as follows:

$$
\begin{aligned}
S_2^{11} &: If \left(\theta_{2a} \text{ is } M_{\theta 2}^1\right) \text{ and } (J_2 \text{ is } M_{J2}^1) \text{ then} \\
\ddot{\theta}_{2a}(t) &= -77.4\dot{\theta}_{2a}(t) - 3947.5\theta_{2a}(t) + 66150u(t) \\
S_2^{12} &: If \left(\theta_{2a} \text{ is } M_{\theta 2}^1\right) \text{ and } (J_2 \text{ is } M_{J2}^2) \text{ then} \\
\ddot{\theta}_{2a}(t) &= -43.8\dot{\theta}_{2a}(t) - 3276.4\theta_{2a}(t) + 48391u(t) \\
S_2^{13} &: If \left(\theta_{2a} \text{ is } M_{\theta 2}^1\right) \text{ and } (J_2 \text{ is } M_{J2}^3) \text{ then} \\
\ddot{\theta}_{2a}(t) &= -49.2\dot{\theta}_{2a}(t) - 1754.5\theta_{2a}(t) + 24964u(t) \\
S_2^{21} &: If \left(\theta_{2a} \text{ is } M_{\theta 2}^2\right) \text{ and } (J_2 \text{ is } M_{J2}^1) \text{ then} \\
\ddot{\theta}_{2a}(t) &= -74.4\dot{\theta}_{2a}(t) - 3452.4\theta_{2a}(t) + 59525u(t) \\
S_2^{22} &: If \left(\theta_{2a} \text{ is } M_{\theta 2}^2\right) \text{ and } (J_2 \text{ is } M_{J2}^2) \text{ then} \\
\ddot{\theta}_{2a}(t) &= -41.7\dot{\theta}_{2a}(t) - 3007.6\theta_{2a}(t) + 1.65 + 45907u(t) \\
S_2^{23} &: If \left(\theta_{2a} \text{ is } M_{\theta 2}^2\right) \text{ and } (J_2 \text{ is } M_{J2}^3) \text{ then} \\
\ddot{\theta}_{2a}(t) &= -51.1\dot{\theta}_{2a}(t) - 1832.8\theta_{2a}(t) + 3.3 + 26471u(t) \\
S_2^{31} &: If \left(\theta_{2a} \text{ is } M_{\theta 2}^3\right) \text{ and } (J_2 \text{ is } M_{J2}^1) \text{ then} \\
\ddot{\theta}_{2a}(t) &= -74.1\dot{\theta}_{2a}(t) - 3540.3\theta_{2a}(t) + 63995u(t) \\
S_2^{32} &: If \left(\theta_{2a} \text{ is } M_{\theta 2}^3\right) \text{ and } (J_2 \text{ is } M_{J2}^2) \text{ then} \\
\ddot{\theta}_{2a}(t) &= -33.4\dot{\theta}_{2a}(t) - 2379\theta_{2a}(t) + 11.74 + 39647u(t) \\
S_2^{33} &: If \left(\theta_{2a} \text{ is } M_{\theta 2}^3\right) \text{ and } (J_2 \text{ is } M_{J2}^3) \text{ then} \\
\ddot{\theta}_{2a}(t) &= -50.7\dot{\theta}_{2a}(t) - 1777.6\theta_{2a}(t) + 23.43 + 28130u(t)
\end{aligned} \tag{4}
$$

5. Design of an Optimal Controller

In this section, a design of a fuzzy optimal controller based on linear quadratic regulator is carried out for a robotic manipulator whose model can be described in the following form:

$$x^{(n)} = f(x, \acute{x}, \ldots, x^{(n-1)}, u)$$

The T-S model can be adjusted as follows:
The IF-THEN rules are as follows:

$$
\begin{aligned}
&S^{(i_1 \cdots\cdots i_n)} : \textit{If } x \textit{ is } M_1^{i_1} \textit{ and } \acute{x} \textit{ is } M_2^{i_2} \textit{ and} \ldots \textit{ and } x^{(n-1)} \textit{ is } M_n^{i_n} \\
&\textit{then } x^{(n)} = a_o^{(i_1 \ldots i_n)} + a_1^{(i_1 \ldots i_n)} x + a_2^{(i_1 \ldots i_n)} \acute{x} + \ldots + a_n^{(i_1 \ldots i_n)} x^{(n-1)} + b^{(i_1 \ldots i_n)} u
\end{aligned} \tag{5}
$$

where $M_1^{i_1}$ $(i_1 = 1, 2, \ldots, r_1)$ are fuzzy sets for x, $M_2^{i_2}$ $(i_2 = 1, 2, \ldots, r_2)$ are fuzzy sets for $\acute{x}$ and $M_n^{i_n}$ $(i_n = 1, 2, \ldots, r_n)$ are fuzzy sets for $x^{(n-1)}$.

The fuzzy system is described as:

$$x^{(n)} = \frac{\sum_{i_1=1}^{r_1}\cdots\sum_{i_n=1}^{r_n} w^{(i_1\ldots i_n)}(x)\left[a_0^{(i_1\ldots i_n)} + a_1^{(i_1\ldots i_n)}x\right]}{\sum_{i_1=1}^{r_1}\cdots\sum_{i_n=1}^{r_n} w^{(i_1\ldots i_n)}(x)} + \frac{\sum_{i_1=1}^{r_1}\cdots\sum_{i_n=1}^{r_n} w^{(i_1\ldots i_n)}(x)\left[a_2^{(i_1\ldots i_n)}\acute{x} + a_n^{(i_1\ldots i_n)}x^{(n-1)} + b^{(i_1\ldots i_n)}u\right]}{\sum_{i_1=1}^{r_1}\cdots\sum_{i_n=1}^{r_n} w^{(i_1\ldots i_n)}(x)} \tag{6}$$

The controller fuzzy rule is represented in a similar form:

$$\begin{gathered} C^{(i_1\cdots\cdots i_n)} : \textit{If } x \textit{ is } M_1^{i_1} \textit{ and } \acute{x} \textit{ is } M_2^{i_2} \textit{ and} \ldots \textit{ and } x^{(n-1)} \textit{ is } M_n^{i_n} \\ \textit{then } u = k_r^{(i_1\ldots i_n)}r - (k_0^{(i_1\ldots i_n)} + k_1^{(i_1\ldots i_n)}x + k_2^{(i_1\ldots i_n)}\acute{x} + \ldots + k_n^{(i_1\ldots i_n)}x^{(n-1)}) \end{gathered} \tag{7}$$

The closed-loop system is obtained substituting (7) in (5) as follows:

$$\begin{gathered} SC^{(i_1\cdots\cdots i_n)} : \textit{If } x \textit{ is } M_1^{i_1} \textit{ and } \acute{x} \textit{ is } M_2^{i_2} \textit{ and} \ldots \textit{ and } x^{(n-1)} \textit{ is } M_n^{i_n} \\ \textit{then } x^{(n)} = a_o^{(i_1\ldots i_n)} + a_1^{(i_1\ldots i_n)}x + \cdots + a_n^{(i_1\ldots i_n)}x^{(n-1)} + \\ b^{(i_1\ldots i_n)}[k_r^{(i_1\ldots i_n)}r - (k_0^{(i_1\ldots i_n)} + k_1^{(i_1\ldots i_n)}x + k_2^{(i_1\ldots i_n)}\acute{x} + k_n^{(i_1\ldots i_n)}x^{(n-1)})] \end{gathered} \tag{8}$$

5.1 Calculation of the Affine Term

The proposed methodology of design is based on the possibility of formulate the feedback system as shown previously in (8),
The affine term of the control action is used to eliminate the affine term of the system:

$$a_o^{(i_1\ldots i_n)} + b^{(i_1\ldots i_n)}k_0^{(i_1\ldots i_n)} = 0$$

$$k_0^{(i_1\ldots i_n)} = \frac{-a_o^{(i_1\ldots i_n)}}{b^{(i_1\ldots i_n)}}$$

and the feedback system is rewritten as follows:

$$\begin{gathered} SC^{(i_1\cdots\cdots i_n)} : \textit{If } x \textit{ is } M_1^{i_1} \textit{ and } \acute{x} \textit{ is } M_2^{i_2} \textit{ and} \ldots \textit{ and } x^{(n-1)} \textit{ is } M_n^{i_n} \\ \textit{then } x^{(n)} = a_1^{(i_1\ldots i_n)}x + \ldots + a_n^{(i_1\ldots i_n)}x^{(n-1)} + \\ b^{(i_1\ldots i_n)}[k_r^{(i_1\ldots i_n)}r - k_1^{(i_1\ldots i_n)}x + k_2^{(i_1\ldots i_n)}\acute{x} + \ldots + k_n^{(i_1\ldots i_n)}x^{(n-1)})] \end{gathered} \tag{9}$$

5.2 State Space Feedback Control based Linear Quadratic Regulator

Any control methodology by state feedback design can be applied to calculate the rest of control coefficients as pole assignments for example. The well known Linear Quadratic Regulator (LQR) method might be an appropriate choice. The system can be represented in state space form:

$$\acute{x} = Ax + Bu$$

$$x \in \Re^n, u \in \Re^m, A \in \Re^{n\times n}, B \in \Re^{n\times m}$$

The objective is to find the control action u(t) to transfer the system from any initial state $x(t_0)$ to some final state $x(\infty) = 0$ in an infinite time interval, minimizing a quadratic performance index of the form:

$$J = \int_{t_0}^{\infty} (x^t Qx + u^t Ru)dt$$

where $Q \in \Re^{n\times n}$ is a symmetric matrix, at least positive a semidefinite one and $R \in \Re^{m\times m}$ is also a symetric positive definite matrix and K is referred to as the state feedback gain matrix. The optimal control law is then computed as follows:

$$u(t) = -Kx(t) = -R^{-1}B^t Lx(t) \tag{10}$$

where the matrix $L \in \Re^{n\times n}$ is a solution of the Riccati equation:

$$0 = -Q + LBR^{-1}B^t L - LA - A^t L$$

The objective can be generalized to find the control action u(t) to transfer the system from any initial state $x(t_0)$ to any reference state $x(\infty) = x_r$ in an infinite time interval, minimizing a quadratic performance index of the form:

$$J = \int_{t_0}^{\infty} ((x - x_r)^t Q(x - x_r) + (u - u_r)^t R(u - u_r))dt$$

where u_r is the necessary input required to keep the system stable in the equilibrium state x_r, which can be calculated as follows:

$$0 = Ax_r + Bu_r \Longrightarrow u_r = -B^+ Ax_r$$

where B^+ is the pseudo inverse of B.
The solution in this case is:

$$u(t) - u_r = -K(x(t) - x_r) = -R^{-1}B^t L(x(t) - x_r) \tag{11}$$

$$u(t) = (K - B^+ A)x_r - Kx(t) \tag{12}$$

where L is the solution of the previously mentioned Riccati equation. Figure 4 shows a block diagram of the proposed optimal controller.
The design algorithm includes firstly the cancelation of the affine term in each subsystem of the form:

$$x^{(n)} = a_0^{(i_1 \ldots i_n)} + a_1^{(i_1 \ldots i_n)} x + a_2^{(i_1 \ldots i_n)} \acute{x} + \ldots + a_n^{(i_1 \ldots i_n)} x^{(n-1)} + b^{(i_1 \ldots i_n)} u \tag{13}$$

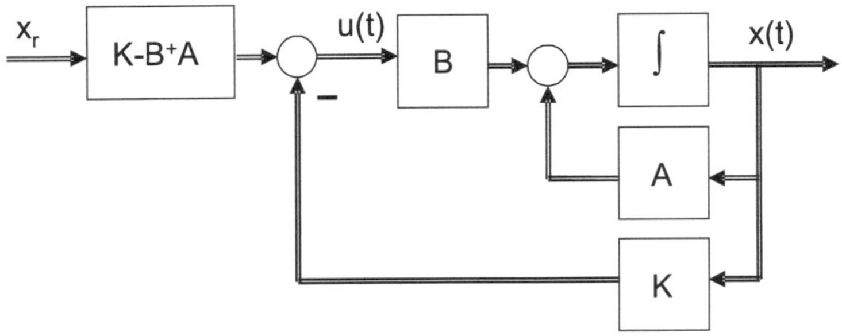

Fig. 4. A block diagram of the proposed optimal controller

The system is then represented in state space form as:

$$A^{(i_1 \ldots i_n)} == \begin{bmatrix} 0 & 1 & 0 & \ldots & 0 \\ 0 & 0 & 1 & \ddots & \vdots \\ \vdots & \vdots & \ddots & \ddots & 0 \\ 0 & 0 & \ldots & 0 & 1 \\ a_1^{(i_1 \ldots i_n)} & a_2^{(i_1 \ldots i_n)} & a_3^{(i_1 \ldots i_n)} & \ldots & a_n^{(i_1 \ldots i_n)} \end{bmatrix}, \quad B^{(i_1 \ldots i_n)} = \begin{bmatrix} 0 \\ \vdots \\ 0 \\ b^{(i_1 \ldots i_n)} \end{bmatrix},$$

$$x = \begin{bmatrix} x & x' & \ldots & x^{n-1} \end{bmatrix}^t$$

$$x_r = \begin{bmatrix} r & 0 & \ldots & 0 \end{bmatrix}^t$$

Secondly, the LQR methodology is applied for each subsystem using a common state weighting matrix Q and input matrix R for all the rules. Thus, Riccati equation is solved for each subsystem as follows:

$$0 = -Q + L^{(i_1 \ldots i_n)} B^{(i_1 \ldots i_n)} R^{-1} B^{(i_1 \ldots i_n)^t} L^{(i_1 \ldots i_n)} - L^{(i_1 \ldots i_n)} A^{(i_1 \ldots i_n)} - A^{(i_1 \ldots i_n)^t} L^{(i_1 \ldots i_n)}$$

Then the the state feedback gain vector can be obtained from (10):

$$K^{(i_1 \ldots i_n)} = \begin{bmatrix} k_1^{(i_1 \ldots i_n)} & k_2^{(i_1 \ldots i_n)} & \ldots & k_n^{(i_1 \ldots i_n)} \end{bmatrix} = R^{-1} B^{(i_1 \ldots i_n)^t} L^{(i_1 \ldots i_n)}$$

and finally,

$$u(t) = (K^{(i_1 \ldots i_n)} - B^{(i_1 \ldots i_n)^+} A^{(i_1 \ldots i_n)}) x_r - K^{(i_1 \ldots i_n)} x(t)$$

6. Application of the Proposed FC-LQR for Robotic Manipulator

A FC-LQR is designed which meets the requirements of small overshoot in the transient response and a well damped oscilations with no steady state error.
For example, in the first rule of the robot model described in (4), we have:

$$S_2^{11} : If \left(\theta_{2a} \text{ is } M_{\theta 2}^1 \right) \text{ and } (J_2 \text{ is } M_{J2}^1) \text{ then}$$
$$\ddot{\theta}_{2a}(t) = -77.4\dot{\theta}_{2a}(t) - 3947.5\theta_{2a}(t) + 66150u(t)$$

As the robot model in this rule has no affine term, there will be no affine term in the controller rule, this means that,

$$k_0^{11} = 0$$

and the state space model for this subsystem is:

$$A^{(11)} = \begin{bmatrix} 0 & 1 \\ -3947.5 & -77.4 \end{bmatrix}, \quad B^{(11)} = \begin{bmatrix} 0 \\ 66150 \end{bmatrix}$$

$$x = \begin{bmatrix} \theta_{2a} & \dot{\theta}_{2a} \end{bmatrix}^t$$

$$x_r = \begin{bmatrix} \theta_r & 0 \end{bmatrix}^t$$

If the weighting state and input matrices are:

$$Q = \begin{bmatrix} 1 & 0 \\ 0 & 50 \end{bmatrix}, \quad R = \begin{bmatrix} 3.10^4 \end{bmatrix}$$

the resultant state feedback gain vectors are:

$$K^{(11)} = \begin{bmatrix} 0.2786.10^{-3} & 0.3967.10^{-2} \end{bmatrix}$$

$$K^{(11)} - B^{(11)+}A = \begin{bmatrix} 0.0600 & 0.0408 \end{bmatrix}$$

Thus, the control action can be calculated as follows:

$$u(t) = 0.0600\theta_r - 0.2786.10^{-3}\theta_{2a} - 0.3967.10^{-2}\dot{\theta}_{2a}$$

Following the same procedure, we can calculate the control action for the rest of the subsystems.

The design parameters in this case are Q and R matrices whose values can be adjusted by trial and error. The objective should be the adjustment of the system with sufficiently fast response under admissible control action u(t). Taking into consideration that the range of possible values for θ_{2a} is $0 \div 115$, while the range for the control action is $\pm 3\ V$, it seems reasonable weight the input signal more than the output. In fact, we found that the admissible results can be obtained for the input action are:

$$q_{11} = 1 \ \ R = [10^3]$$

and better results can be obtained with:

$$q_{11} = 1 \ \ R = [10^4]$$

With respect to the weighting of the angular velocity, it has been found that with $q_{22} = 1$, the response peaks approach $160^\circ/s$ which is superior than the admissible range and with $q_{22} = 20$, the peaks are below $40^\circ/s$ which are within the admissible range. To get the optimal response, we have chosen:

$$Q = \begin{bmatrix} 1 & 0 \\ 0 & 20 \end{bmatrix}, \quad R = \begin{bmatrix} 10^4 \end{bmatrix}$$

and the control action for each subsystem is:

$$C_2^{11}: If \left(\theta_{2a} \; is \; M_{\theta 2}^1\right) \; and \; (\; J_2 \; is \; M_{J2}^1 \;) \; then$$

$$u(t) = 0.0605\theta_r - 0.8321.10^{-3}\theta_{2a} - 0.0436\dot{\theta}_{2a}$$

$$C_2^{12}: If \left(\theta_{2a} \; is \; M_{\theta 2}^1\right) \; and \; (\; J_2 \; is \; M_{J2}^2 \;) \; then$$

$$u(t) = 0.0684\theta_r - 0.7345.10^{-3}\theta_{2a} - 0.0438\dot{\theta}_{2a}$$

$$C_2^{13}: If \left(\theta_{2a} \; is \; M_{\theta 2}^1\right) \; and \; (\; J_2 \; is \; M_{J2}^3 \;) \; then$$

$$u(t) = 0.0710\theta_r - 0.7079.10^{-3}\theta_{2a} - 0.0428\dot{\theta}_{2a}$$

$$C_2^{21}: If \left(\theta_{2a} \; is \; M_{\theta 2}^2\right) \; and \; (\; J_2 \; is \; M_{J2}^1 \;) \; then$$

$$u(t) = 0.0589\theta_r - 0.8558.10^{-3}\theta_{2a} - 0.0434\dot{\theta}_{2a}$$

$$C_2^{22}: If \left(\theta_{2a} \; is \; M_{\theta 2}^2\right) \; and \; (\; J_2 \; is \; M_{J2}^2 \;) \; then$$

$$u(t) = 0.0663\theta_r - 0.0359.10^{-3} - 0.7588.10^{-3}\theta_{2a} - 0.0438\dot{\theta}_{2a}$$

$$C_2^{23}: If \left(\theta_{2a} \; is \; M_{\theta 2}^2\right) \; and \; (\; J_2 \; is \; M_{J2}^3 \;) \; then$$

$$u(t) = 0.0700\theta_r - 0.1247.10^{-3} - 0.7184.10^{-3}\theta_{2a} - 0.0428\dot{\theta}_{2a}$$

$$C_2^{31}: If \left(\theta_{2a} \; is \; M_{\theta 2}^3\right) \; and \; (\; J_2 \; is \; M_{J2}^1 \;) \; then$$

$$u(t) = 0.0562\theta_r - 0.8276.10^{-3}\theta_{2a} - 0.0436\dot{\theta}_{2a}$$

$$C_2^{32}: If \left(\theta_{2a} \; is \; M_{\theta 2}^3\right) \; and \; (\; J_2 \; is \; M_{J2}^2 \;) \; then$$

$$u(t) = 0.0608\theta_r - 0.2961.10^{-3} - 0.8276\theta_{2a} - 0.0439\dot{\theta}_{2a}$$

$$C_2^{33}: If \left(\theta_{2a} \; is \; M_{\theta 2}^3\right) \; and \; (\; J_2 \; is \; M_{J2}^3 \;) \; then$$

$$u(t) = 0.0640\theta_r - 0.8329 - 0.7863.10^{-3}\theta_{2a} - 0.0430\dot{\theta}_{2a}$$

Figure 5 shows the evolution of the angle θ_{2a} from an initial condition of 25° and zero reference signal It also shows the step response with reference input of 50° and a constant value of moment of inertia igual to $J_2 = 25000$. The step response has a settling time of 3 seconds.
Figure 6 shows the response with various initial conditions $10°, \ldots, 50°$ and zero reference input signal. After five seconds, the system is excited with various step reference inputs $10°, \ldots, 50°$ with a constant moment of inertia $J_2 = 25000$. It can be clearly observed that well damped and fast response is obtained in all the range of possible values of the output.

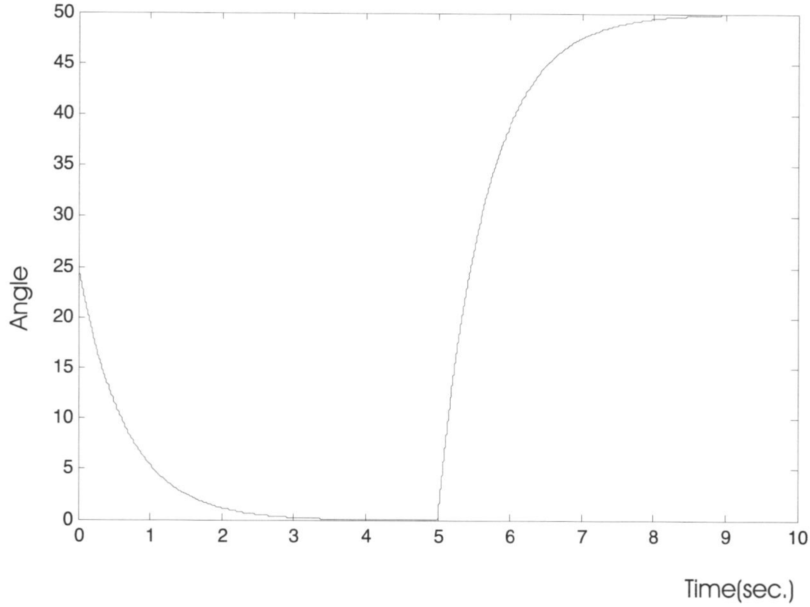

Fig. 5. Transient response of the robotic system with initial condition of 25° and moment of inertia $J_2 = 25000$

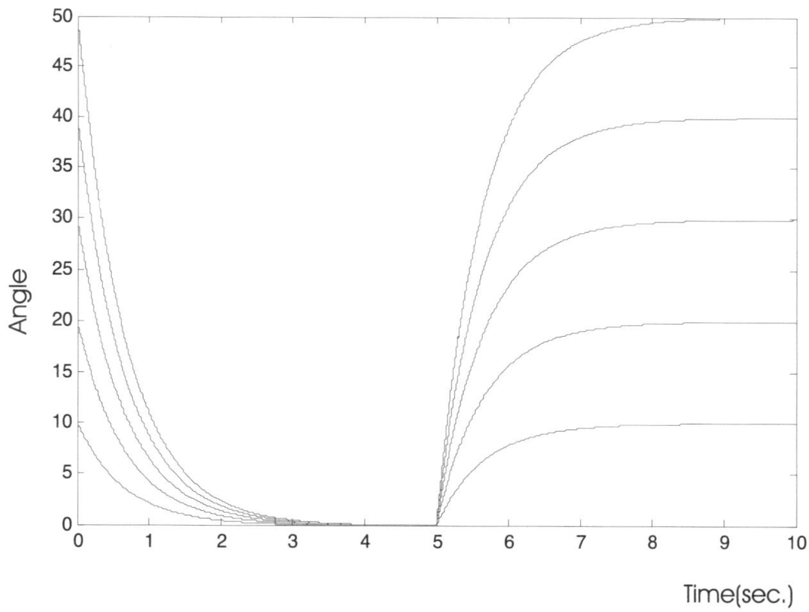

Fig. 6. Transient response of the robotic system with various initial conditions and reference input signals and constant moment of inertia of $J_2 = 25000$

Nevertheless, figure 7 shows the response with an intial condition and reference input signal of 25°. The response is initiated with moment of inertia $J_2 = 25000$ and after five seconds an abrupt change is applied in the moment of inertia to $J_2 = 50000$.

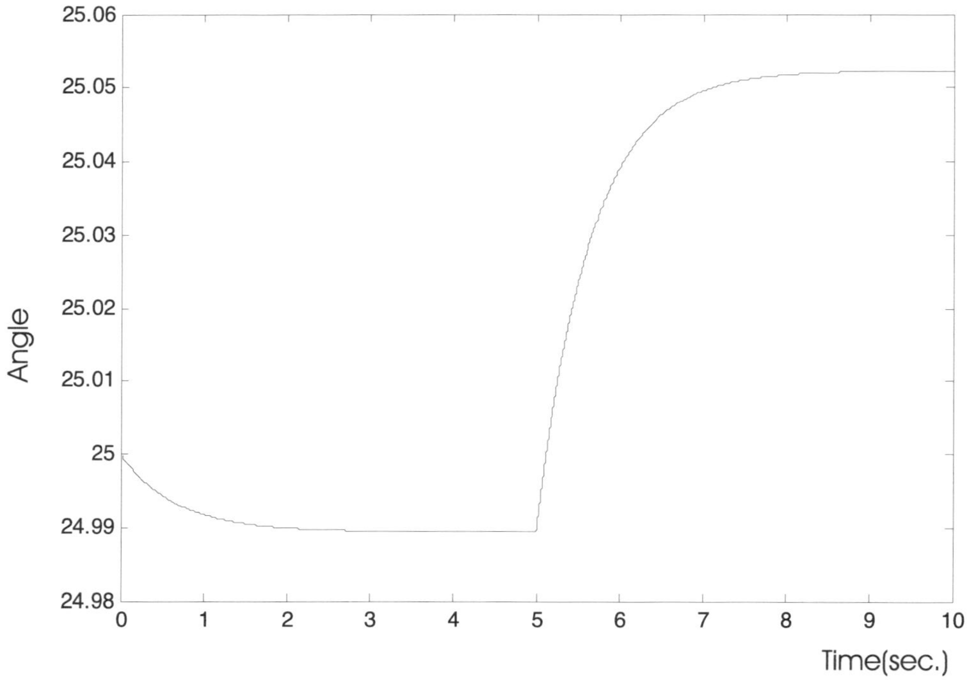

Fig. 7. Transient response of the robotic system with initial condition and reference input signal of 25°. An abrupt change is applied in moment of inertia from $J_2 = 25000$ to $J_2 = 50000$

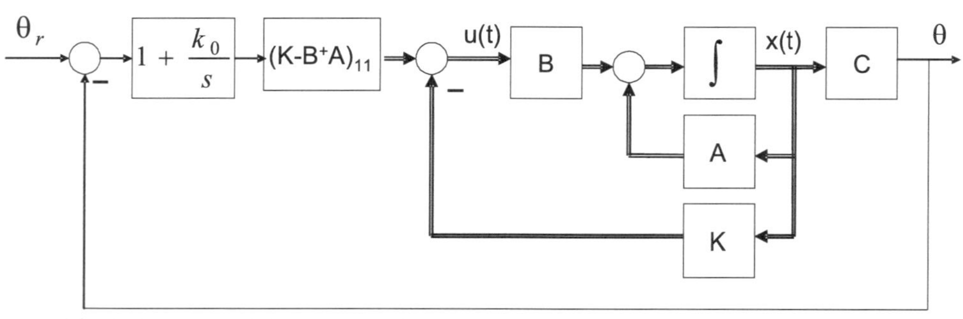

Fig. 8. A block diagram of the proposed controller with a PI controller to eliminate the steady state error

As can be seen in figure 7, the lack of precision in the model leads to a steady state error in the transient response. We propose a solution to eliminate this error. A simple but effective solution is realized by adding a feedback loop and including a PI controller as shown in figure 8.

$$e(t) = \theta_r(t) - \theta(t)$$

$$u = (K - B^{+}A)_{11}(e(t) + k_0 \int_{t_0}^{t} e(\tau)d\tau) + Kx(t)$$

Using the design shown in figure 8 and repeating the same experiment explained before with $k_0 = 1.5$ initial condition and reference input signal of 25°, keeping the moment of inertia constant with $J_2 = 25000$ and after five seconds an abrupt change is applied in the moment of inertia to $J_2 = 50000$. The result is shown in figure 9. It can be observed that a small disturbance effect is occurred in the output angle but it is immediately corrected resulting in a smooth response with zero steady state error. Figure 10 shows the response with an intial condition and reference input signal of 25°. The response is initiated with moment of inertia $J_2 = 25000$ and after five seconds an abrupt change is applied in the moment of inertia to $J_2 = 50000$. It can be easily noticed that the response has not been affected with the modification made to the propsed controller shown in figure 8 and the response is exactly similar to that shown in figure 6.

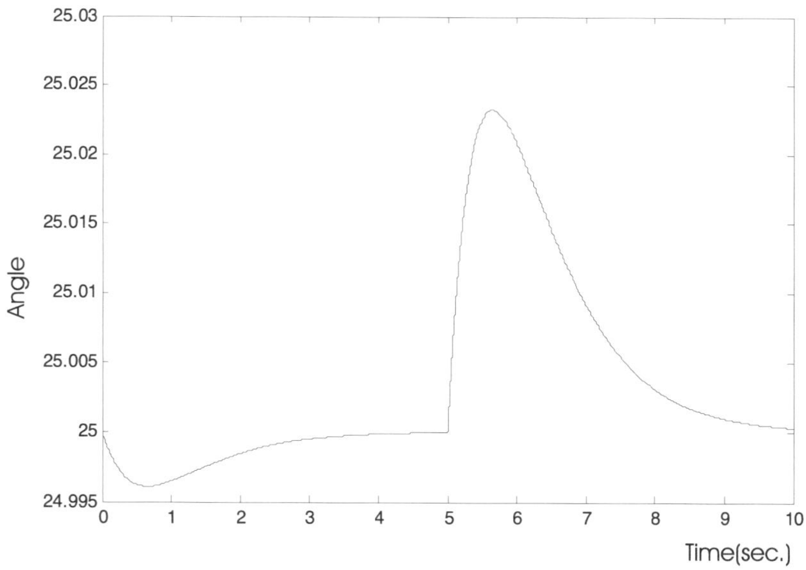

Fig. 9. Transient response of the robotic system by adding a PI controller to the proposed FC-LQR with initial condition and reference input signal of 25°. An abrupt change is applied in moment of inertia from $J_2 = 25000$ to $J_2 = 50000$

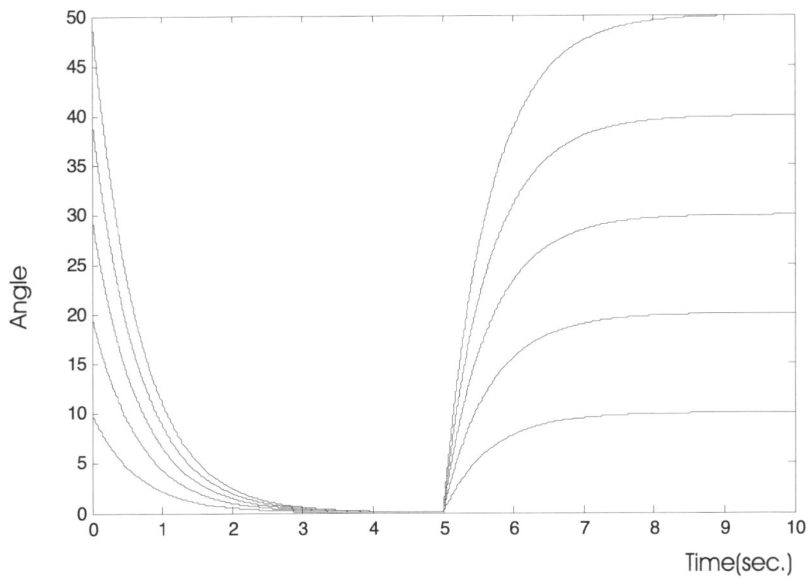

Fig. 10. Transient response of the robotic system by adding a PI controller to the proposed FC-LQR with various initial conditions and reference input signals and constant moment of inertia of $J_2 = 25000$

7. Conclusion

A robust FC-LQR for the control of a robotic system has been designed. The main idea is to design a supervisory fuzzy controller capable to adjust the controller parameters in order to obtain the desired axes positions under variations of the robot parameters and payload variations. The motivation behind this scheme is to combine the best features of fuzzy control and that of the optimal LQR.

Both the controlled system and the fuzzy controller are represented by the affine T-S fuzzy model taking into consideration the effect of the constant term. In the case of fuzzy control, the fuzzy system is resulted from blending all the sub-systems. The blending of the independent term of each rule will no longer be a constant but a function of the variables of the system and thus affects the dynamics of the resultant system. A necessary condition has been added to deal with the independent term.

In this chapter, we have demonstrated that the LQR, can be made more appropriate for actual implementation by introduction of fuzzy rules. The results obtained show a robust and stable behavior when the system is subjected to various initial conditions, moment of inertia and to disturbances.

8. References

[Berstecher et al. 2001] R.G. Berstecher, R. Palm, H.D. Unbehauen, An adaptive fuzzy sliding mode controller, IEEE Trans. Indust. Electron. 8 (1) (2001) 18Ű31.

[Burg et al. 2001] T. Burg, D. Dawson, M.D. Queiroz, An adaptive partial state feedback controller for RLED robot manipulators, IEEE Trans. Automat. Control 41 (7) (1996) 1024Ű1030.

[Canudas and Fixot 1991] C. Canudas de Wit, N. Fixot, Robot control via robust estimated state feedback, IEEE Trans. Automat. Control 36 (12) (1991) 1497Ű1501.

[Chen et al. 1998] B.S. Chen, H.J. Uang, C.S. Tseng, Robust tracking enhancement of robot systems including motor dynamics: a fuzzy-based dynamic game approach, IEEE Trans. Fuzzy Systems 6 (1998) 538Ű552.

[Chuan-Kai Lin 2003] Chuan-Kai Lin, A reinforcement learning adaptive fuzzy controller for robots, Fuzzy Sets and Systems 137 (3) (2003) 339Ű352.

[Commuri et al. 1996] S. Commuri, F.L. Lewis, Adaptive-fuzzy logic control of robots manipulators, Proc. IEEE Internat. Conf. on Robotics and Automation, Mineapolis, MI, 1996, pp. 2604Ű2609.

[Freund. 1982] Freund, E. Fast Nonlinear Control with Arbitrary Pole-placement for Industrial Robots and Manipulators, Robotics Research. pages 65-78 **1** (1982).

[Fuchun et al. 2003] Fuchun Sun, Zengqi Sun, Lei Li, Han-Xiong Li, Neuro-fuzzy adaptive control based on dynamic inversion for robotic manipulators, Fuzzy Sets and Systems 134 (1) (2003) 117Ű133.

[Fukuda et al. 1992] T. Fukuda, T. Shibata, Hierarchical intelligent control for robotic motion by using fuzzy, artificial intelligence, and neural networks, Proc. Internat. J. Conf. on Neural Networks, Vol. 1, Baltimore, 1992, pp. 269Ű274.

[Gamboa 1996] Gamboa, E., Control de robots con accionamiento hidráulicos. Aplicación a robots de grandes dimensiones para la construcción. Tesis doctoral. Universidad Politécnica de Madrid (1996).

[Hua et al. 2004] C. C. Hua, X. Guan, G. Duan, Variable structure adaptive fuzzy control for a class of nonlinear time delay systems, Fuzzy Sets and Systems 148 (2004) 453Ű468.

[Ishikawa 1988] Ishikawa, T. A study on fuzzy control of an arm robot. B. E. thesis, Dept. Contr. Eng., Tokyo Inst. Technol(1988).

[Jagannathan et al. 1996] S. Jagannathan, F.L. Lewis, Discrete-time adaptive fuzzy logic control of robotic systems, Proc. IEEE Internat. Conf. on Robotics and Automation, Mineapolis, MI, 1996, pp. 2586Ű2591.

[Khosla and Kanade 1988] P.K. Khosla, T. Kanade, Experimental evaluation of nonlinear feedback and feedforward control schemes for manipulator, Int. J. Robot. Res. 7 (1988) 18Ű28.

[Kim and Lewis 2000] Y.H. Kim, F.L. Lewis, Intelligent optimal control of robotic manipulators using neural networks, Automatica 36 (2000) 1355Ű1364.

[Kim and Lewis 1999] Y.H. Kim, F.L. Lewis, Neural network output feedback control of robot manipulators, IEEE Trans. Robot. Automat 15 (1999) 301Ű309.

[Koditschek. 1984] D. Koditschek, Natural motion for robot arms, Proc. 1984 IEEE Conf. on Decision and Control, Las Vegas, 1984, pp. 733Ű735.

[Labiod et al. 2005] S. Labiod, M.S. Boucherit, T.M. Guerra, Adaptive fuzzy control of a class of MIMO nonlinear systems, Fuzzy Sets and Systems 15 (1) (2005) 59Ű77.

[Lewis et al. 1994] F.L. Lewis, C.T. Abdallah, D.M. Dawson, Control of Robot Manipulators, MacMillan Publishing Company, New York, 1994.

[Li et al. 2001] W. Li, X.G. Chang, F.M. Wahl, J. Farrell, Tracking control of a manipulator under uncertainty by FUZZY P+ID controller, Fuzzy Sets and Systems 122 (1) (2001) 125Ű137.

[Llama et al. 2001] Miguel A. Llama, Rafael Kelly, Victor Santibañez. A stable motion control system for manipulators via fuzzy self-tuning. In *Fuzzy Sets ans Systems*, volume 124, pages 133 Ű154, 2001.

[Llama et al. 1998] M.A. Llama, V. Santibanez, R. Kelly, J. Flores, Stable fuzzy self-tuning computed-torque control of robot manipulators, Proc. IEEE Internat. Conf. on Robotics and Automation, Leuven, 1998, pp. 2369Ű2374. B 29 (1999) 371Ű388.

[Luh 1983] J.Y.S. Luh, Conventional controller design for Industrial robots-a tutorial, IEEE Trans. Systems Man Cybernet. 13 (1983) 298Ű316.

[Meslin et al. 1992] J.M. Meslin, J. Zhou, P. Coiffet, Fuzzy dynamic control of robot manipulators: a scheduling approach, Proc. IEEE Internat. Conf. on Systems Man and Cybernetics, Le Tourquet, 1993, pp. 69Ű73.

[Middleton and Goodwin 1988] R.H. Middleton, G.C. Goodwin, Adaptive computed torque control for rigid link manipulators, System Control Lett. 10 (1988) 9Ű16.

[Purwar et al. 2005] Purwar, S., Kar I. N., Jha A. N.: Adaptive control of robot manipulators using fuzzy logic systems under actuator constraints. In *Fuzzy Sets and Systems*, volume 152, pages 651 Ű664, 2005.

[Sciavicco et al. 1996] L. Sciavicco, B. Siciliano, Modeling and Control of Robot Manipulators, McGraw-Hill Co., New York, 1996.

[Slotine and Li 1988] J.J.E. Slotine, W. Li, Adaptive manipulator control: a case study, IEEE Trans. Automat. Control 33 (1988) 995Ű1003.

[Slotine et al. 1983] J.J.E. Slotine, S.S. Sastry, Tracking control of nonlinear systems using sliding surface with application to robot manipulator, Int. J. Control 38 (1983) 465Ű492.

[Song et al. 2006] Song, Z., Yi J., Zhao D., Li X. A computed torque controller for uncertain robotic manipulator systems: Fuzzy approach. In *Fuzzy Sets ans Systems*, volume 154, pages 208 Ű226, 2006.

[Spong and Vidyasagar 1989] M.W. Spong, M. Vidyasagar, Robot Dynamics and Control, Wiley, New York, 1989.

[Sugeno and Kang 1988] Structure identification of fuzzy model, *Fuzzy Sets Syst.*, vol. 28, pp. 15-33, 1988.

[Sugeno and Tanaka 1988] Successive identification of a fuzzy model and its applications to prediction of a complex system, *Fuzzy Sets and Systems* 42 (1991) 315-334, 1991.

[Sun et al. 1999] F. Sun, Z. Sun, G. Feng, An adaptive fuzzy controller based on sliding mode for robot manipulator, IEEE Trans. Systems Man Cybernet. 29 (1999) 661Ű667.

[Sylvia et al. 2003] Sylvia Kohn-Rich, Henryk Flashner, Robust fuzzy logic control of mechanical systems, Fuzzy Sets and Systems 133 (1) (2003) 77Ű108.

[Takagi and Sugeno 1985] Takagi, T., Sugeno M.: Fuzzy Identification of Systems and Its Applications to Modeling and Control. In *IEEE Transactions on Systems, Man and Cybernetics*, volume SMC-15, pages 116–132, 1985.

[Tanaka et al. 1996] K. Tanaka and T. Ikeda and H. O. Wang, Robust Stabilization of a Class of Uncertain Nonlinear Systems via Fuzzy Control: Quadratic Stabilizability, H_∞ Control Theory, and Linear Matrix Inequalities, IEEE Transactions on Fuzzy Systems, February, number 1, 1996, volume 4, pages 1-13.

[Tanaka and Sano 1994] K. Tanaka and M. Sano, A Robust Stabilization Problem of Fuzzy Control Systems and Its Application to Backing up Control of a Truck-Trailer, IEEE Transactions on Fuzzy Systems, May, number = "2", 1994, volume 2, pages 119-134.

[Tong et al. 2000] Tong Shaocheng, Tang Jiantao, Wang Tao, Fuzzy adaptive control of multivariable nonlinear systems, Fuzzy Sets and Systems 111 (2) (2000) 153Ű167.

[Tsai et al. 2000] C.H. Tsai, C.H.Wang, W.S. Lin, Robust fuzzy model following control of robot manipulators, IEEE Trans. Fuzzy Systems 8 (4) (2000) 462Ű469.

[Tzes et al. 1993] A. Tzes, K. Kyriakides, Adaptive fuzzy-control for Uexible-link manipulators: a hybrid frequency-time domain scheme, Proc. 2nd IEEE Internat. Conf. on Fuzzy Systems, San Francisco, 1993, pp. 122Ű127.

[Wang and mendel 1992] L.X. Wang, J.M. Mendel, Fuzzy basis function universal approximation, and orthogonal least square learning, IEEE Trans. Neural Networks 3 (1992) 807Ű814.

[Wijesoma and Richards 1990] S.W. Wijesoma, R.J. Richards, Robust trajectory following of robots using computed torque structure with VSS, Int. J. Control 52 (1990) 935Ű962.

[Yi and Chung 1997] S.Y. Yi, M.J. Chung, A robust fuzzy logic controller for robot manipulators with uncertainties, IEEE Trans. Systems Man Cybernet. Part B 27 (4) (1997) 706Ű713.

[Yoo and Ham 2000] B.K. Yoo, W.C. Ham, Adaptive control of robot manipulators using fuzzy compensator, IEEE Trans. Fuzzy Systems 8 (2) (2000) 186Ű199.

[Zhang et al. 2000] F. Zhang, D.M. Dawson, M.S. Queiroz de, W.E. Dixon, Global adaptive output feedback tracking control of robot manipulators, IEEE Trans. Automat. Control 45 (6) (2000) 1203Ű1208.

[Zhou et al. 1992] J. Zhou, P. Coi8et, Fuzzy control of robots, Proc. lst Internat. Conf. on Fuzzy Systems, San Diego, 1992, pp. 1357Ű1364.

5

Development of Adaptive Learning Control Algorithm for a two-degree-of-freedom Serial Ball And Socket Actuator

Hayder M. A. A. Al-Assadi and Ahmed Jaffar
Faculty of Mechanical Engineering, Universiti Teknologi MARA
40450,Shah Alam, Selangor
Malaysia

1. Introduction

For any newly developed mechanism the most challenging task is the controller. The controller is an algorithm that organizes the mechanism input energy to perform a specified task. Robot control usually requires the directing of signals or fluid paths of power elements to indicate robot end-effector dynamic behaviour. Furthermore, robot control can be divided into two main areas: kinematic control (the coordination of the links of the kinematic chain to produce desired motions of the robot) and dynamic control (driving the actuators of the mechanism to follow the commanded position velocities). These control strategies are widely used in most robots involving position coordination in Cartesian space by a direct kinematic method (Karlik and Aydin, 2000).

In this chapter, an artificial neural network (ANN) adaptive learning algorithm has been implemented for dynamic behaviour control of a new two-degree- of-freedom (2DOF) serial ball-and-socket actuator. The ANN provides computer simulation of human brain activity that gives computers the ability to learn and predict a decision for a specific task. The ANN requires a specific network design followed by a training process. A variety of modifications could be carried out for the network design during the training process.

For a robot control scheme there are many uncertainties in the parameters of both the actuators (hydraulic, pneumatic, and electric drivers) and the mechanical parts of the manipulators (Cheah et al., 2003, Tso & Law, 1993, Mills, 1994, Yang & Chu, 1993, Tsao & Tomizuka, 1994, Park & Cho, 1992, Ambrosino et al., 1998). Therefore, to cover the overall complexity of the robot control problem and the quest for a truly autonomous robot system, the application of an ANN to the robot control scheme has been considered (Ananthraman, 1991, Cruse & Bruwer, 1990, Kuperstein & Wang, 1990, Miller et al., 1990, Hasan et al., 2007, Abdelhameed, 1999, Sharkey & Noel, 1997, Brun-Picard et al., 1999). In addition, the proposed hydraulic power system is one with highly non-linear behaviour. Variations in parameters affect the hydraulic system operation and performance, e.g. the laminar and turbulent flows, channel geometry, friction results in the system equation, the relation between flow velocity and pressure, and oil viscosity (Knohl & Unbehauen, 2000). Hence, to cope with situations of this kind, the hydraulic system required a non-linear controller such situation such as an ANN, which has been the focus of work by various researchers (Mills et al., 1994, Chen & Billings, 1992).

In robotics, the revolute joint has one-degree-of-freedom and, because of its simplicity, is by far the most used joint. In order to imitate the shoulder or hip joint, two revolute actuators are required to provide the necessary 2DOF motion. In the biomedical literature, the representation of the human arm as three rigid segments connected by frictionless joints with a total of seven degrees of freedom is the generally accepted model (Desmurget & Prablanc, 1997, Lemay & Cragi, 1996, Raikova, 1992). In the 7DOF arm models the shoulder joint is usually considered as a ball-and-socket joint and the axes in the elbow and wrist joints are assumed to be orthogonal and intersecting (Perokopenko et al., 2001).

Consequently, a new 2DOF serial ball-and-socket actuator has been fabricated to replace the two revolute actuators in the serial robot manipulator. The fabricating process has been done by combining actuator elements such as the actuator mechanism, the electrohydraulic powering system, the communication interface board, and the adaptive learning algorithm.

The ball-and-socket joint, used in engineering as a mechanical connection between parts that must be allowed some relative angular motion in nearly all directions, represents articulation with two rotational degrees of freedom. Ball-and-socket joints are successfully used for parallel robots and simulators powered by pneumatic or hydraulic cylinders. The available basic methods to transmit the power are electrical, mechanical, and fluid drivers. Most applications are a combination of these three methods. Each of these methods has advantages and disadvantages, so the use of a particular method depends on the application and environment (McKerrow, 1991). Among the power transmission systems, the hydraulic system will be recommended for use in the developed actuator on account of its ability to store energy when no power supply is offered by keeping the pressurized fluid inside the cylinder. This is a necessary step to stabilize the ball-and-socket actuator. Therefore, two electrohydraulic cylinders have been developed; each will perform one degree of freedom with the other supporting, and vice versa.

An ANN model has been developed and trained to build control knowledge that covers all the control parameters for the ball-and-socket actuator. This control knowledge will function from digital signals, extracted by computer, to the target end-effector dynamic behaviour, without any involvement of actuator mechanism behaviour, with the flexibility to cover any modification without changing the control scheme. The ANN model has been simulated using C++ programming language. The completed system has been run and tested successfully in the laboratory. The remainder of this chapter will demonstrate the basic elements of the ball-and-socket actuator, and will examine the control approach and the process of development and training of the ANN model.

2. Actuator Design Specifications

The proposed ball-and-socket actuator comprised an actuator mechanism, a power system, and a communication interface board. The actuator mechanism represents the mechanical elements and comprises the base, ball-and-socket joints, two double-acting electrohydraulic cylinders, and the end-effector rod. A diagram of the ball-and-socket actuator is shown in Fig. 1, while Fig. 2 shows the fabricated actuator mechanism built to represent the developed ball-and-socket actuator.

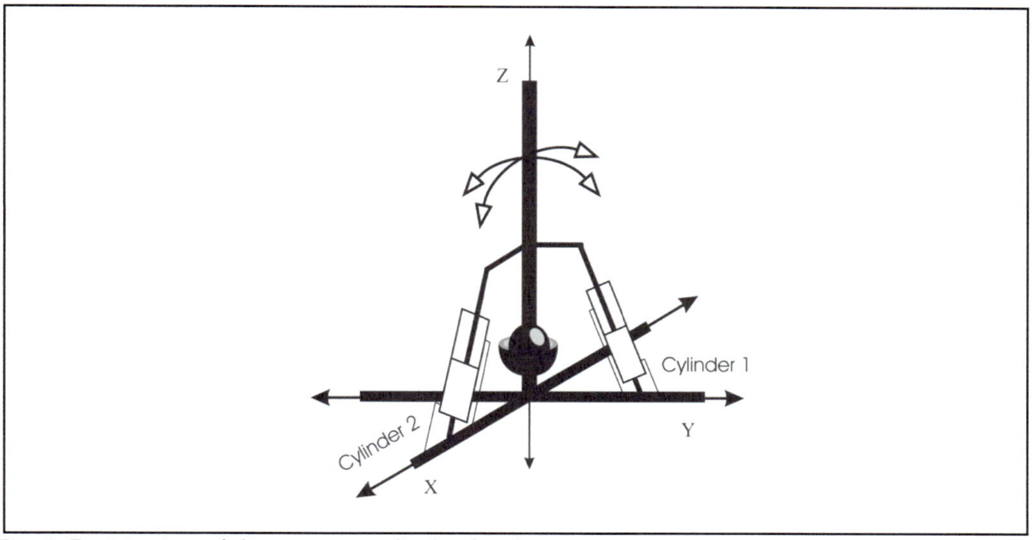

Fig. 1. Positioning of the support cylinder for the actuator

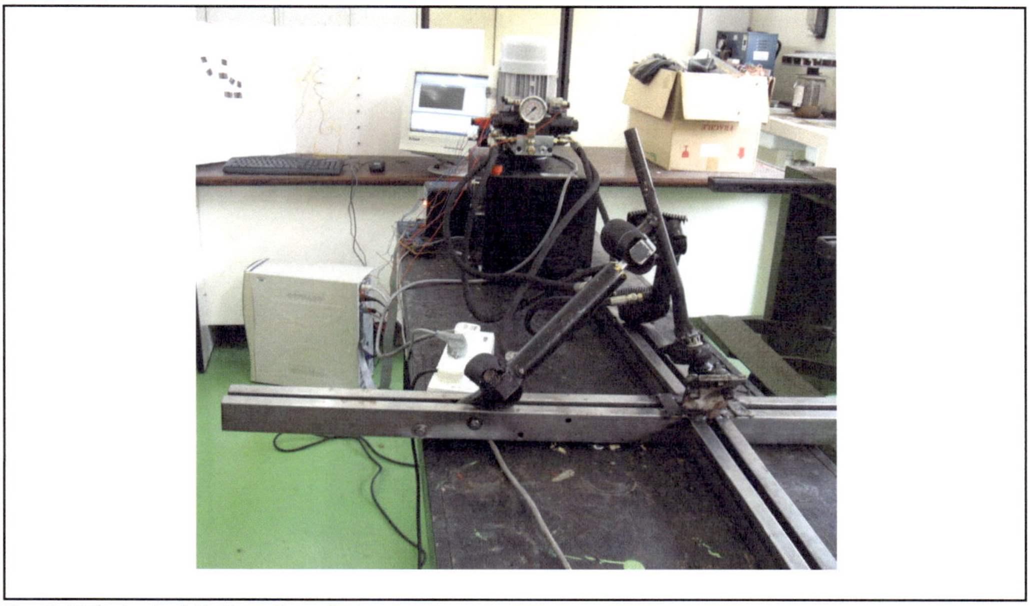
Fig. 2. Fabricated Ball and socket actuator

The power transmission system is complicated by the characteristics of the joints which must be free to rotate in all directions and need a dual-tasking power system. Therefore, an electrohydraulic cylinder powered by a 1 hp pump was used. The system consists of two double-acting electrohydraulic cylinders that are capable of maintaining their position when the pressurized fluid is kept inside them. This is a very necessary step to ensure sufficient actuator stability for the other cylinder when operating to the desired direction and is an advantage of the ball-and-socket actuator. The double-acting electrohydraulic cylinders have a two-direction movement scheme that provides an inward and outward motion for

the end-effector rod. Moreover, the deployment of double-acting electrohydraulic cylinders reduces the number of supporting points that are necessary to run and stabilize the actuator mechanism to 2 instead of 4 as in the case of single-acting cylinders.
A communication interface board has been designed and fabricated to establish the necessary signals to operate the actuator. Basically it is a transistor relay driver circuit converting a 5 V digital signal from the computer mother board operating the learning algorithm (ANN) to the necessary 24 d.c. signals required to operate the electrohydraulic cylinders.

3. Actuator Controlling Approach

To plan a controller it is necessary to understand the system behaviour and characteristics. The equations

$$\text{(x)} \qquad 2r_1 \sin\theta_1 \cos\phi_1 = r_2 \sin\theta_2 \cos\phi_2 + r_3 \sin\theta_3 \cos\phi_3 \tag{1}$$

$$\text{(y)} \qquad 2r_1 \sin\theta_1 \sin\phi_1 = r_2 \sin\theta_2 \sin\phi_2 + r_3 \sin\theta_3 \sin\phi_3 \tag{2}$$

$$\text{(z)} \qquad 2r_1 \cos\theta_1 = r_2 \cos\theta_2 + r_3 \cos\theta_3 \tag{3}$$

illustrate the relationship between angles θ_1 and ϕ_1, representing the angular displacement of the end effector, and $\theta_2, \theta_3, \phi_2, \phi_3, r_2,$ and r_3 for kinematic analysis on the *x*, *y*, and *z* axes, where $\theta_2, \theta_3, \phi_2, \phi_3$ are the angular displacements of cylinders 1 and 2, and r_2 and r_3 are the lengths of cylinders 1 and 2 respectively.

The equations

$$\theta_1 = \sin^{-1}\left(\frac{r_2 \sin\theta_2 \cos\phi_2 + r_3 \sin\theta_3 \cos\phi_3}{2r_1 \cos\phi_1}\right) \tag{4}$$

$$\phi_1 = \cos^{-1}\left(\frac{r_2 \sin\theta_2 \cos\phi_2 + r_3 \sin\theta_3 \cos\phi_3}{2r_1 \sin\theta_1}\right) \tag{5}$$

$$\theta_1 = \cos^{-1}\left(\frac{r_2 \sin\theta_2 \cos\varphi_2 + r_3 \sin\theta_3 \cos\varphi_3}{2r_1 \sin\theta_1}\right) \tag{6}$$

$$\phi_1 = \sin^{-1}\left(\frac{r_2 \sin\theta_2 \sin\phi_2 + r_3 \sin\theta_3 \sin\phi_3}{2r_1 \sin\theta_1}\right) \tag{7}$$

$$\theta_1 = \cos^{-1}\left(\frac{r_2 \cos\theta_2 + r_3 \cos\theta_3}{2r_1}\right) \tag{8}$$

represent the solutions for finding the angles θ_1 and ϕ_1 from equations (1) to (3).

Finding the solution for θ_1 and ϕ_1 as illustrated in the above equations will depend on $(\sin^{-1})$ and $(\cos^{-1})$ which are not single-value functions. Furthermore, equations (4), (6), and (8) can be used to find θ_1 values that are not unique.
In this chapter, an ANN adaptive learning algorithm has been proposed for controlling a 2DOF serial actuator. In this approach, the adaptive learning algorithm finds an alternative

solution of the kinematic relation for the ball-and-socket actuator. Therefore, all parameters operating the actuator will be considered as target learning input data for the ANN model, while the output target data will be the angular displacement, angular velocity, and angular acceleration of the actuator end-effector.
The shape of the actuator mechanism, as shown in Fig. 1, can be controlled by varying the length of the electrohydraulic cylinders. The hydraulic cylinders operate as a result of allowing pressurized fluid to run them. All the parameters affecting this operation, such as the valve order, time, flow-rate, pump pressure, and the fluid head losses, will have been incorporated as inputs for the ANN model. After running the cylinder length, the output for the ANN will be the dynamic behaviour of the actuator end-effector.
The workspace, the region that can be reached by the end-effector, is considered to be an important performance indicator. Therefore, the control approach is to drive the actuator to reach a point from any point within the desired workspace area. Experimental operation shows a square workspace for the fabricated actuator mechanism, as illustrated in Fig. 3. As can be seen from Fig. 3, the workspace is divided into nine points within the x–y plane. Therefore, experimental operation has been carried out to estimate and collect the control parameters that drive the actuator from one specific point to another individual point. These collected control data have been arranged as datasets. Each set represents input control data to drive the actuator mechanism and outputs as angular displacement, angular velocity, and angular acceleration of the end-effector. All the datasets were used as target learning data by the ANN to build the control knowledge required to operate the ball-and-socket actuator.

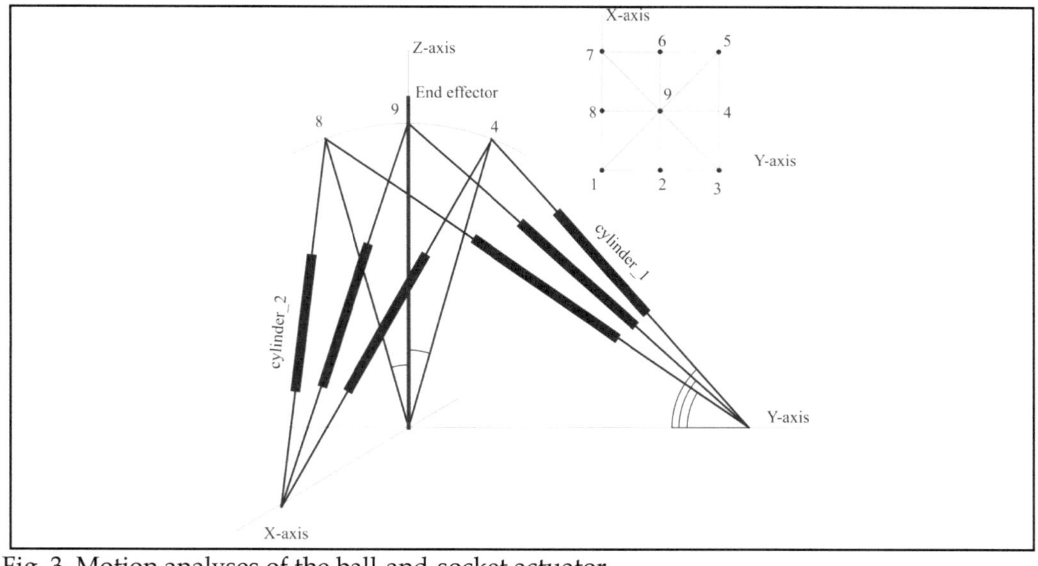

Fig. 3. Motion analyses of the ball-and-socket actuator

4. Adaptive Learning Algorithm

ANN adaptive learning algorithm computer software was proposed to learn and adopt the control parameters to provide the necessary digital signal from the computer main board to operate the actuator mechanism. These digital signals could be extracted through various

computer outputs such as serial, parallel, and USB ports. In this chapter, the parallel port (printer port) has been chosen to extract +5 V digital signals from the computer.
Although the ANN method is being implemented to learn a set of information, a specific network design is required to cover each individual dataset and application. Consequently, a special network has been designed to adopt the control parameters for the ball-and-socket actuator that consists of an input layer (valve order, time, pump power, flow-rate, output pressure, and head losses for the system), one hidden layer, and an output layer (angular displacement 1, angular displacement 2, angular velocity 1, angular velocity 2, angular acceleration 1, and angular acceleration 2), as shown in Fig. 4.

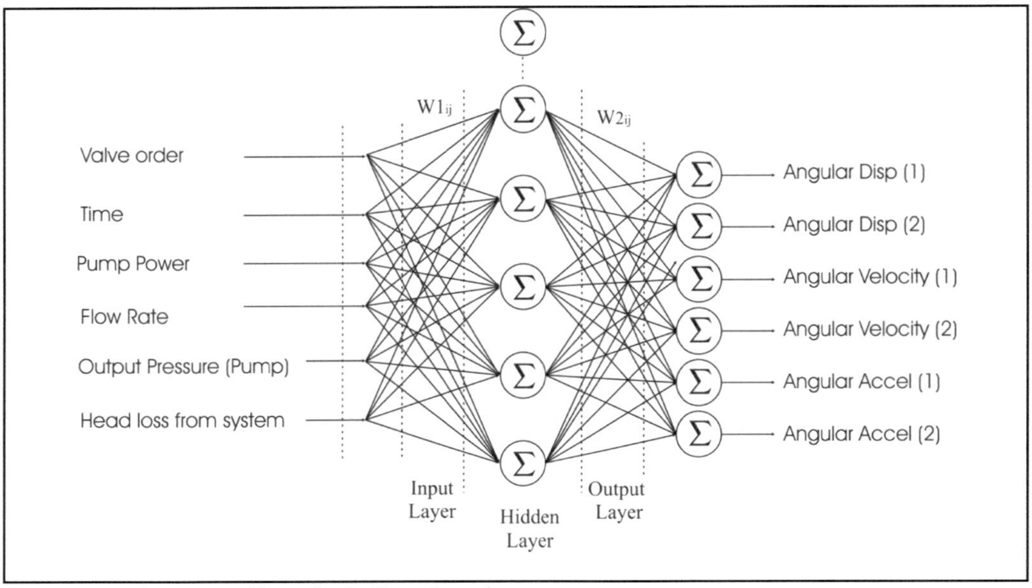

Fig. 4. ANN for controlling the ball-and-socket actuator

After designing the network, a training process had to be accomplished to build control knowledge, which is considered to be the most important step in designing ANN algorithms. A neural network was trained by presenting several target data that the network had to learn according to a learning rule. The training rule indicated transfer of a function such as the binary sigmoid transfer function (equation (9)), forward learning for the input layer (equation (10)), forward learning for the hidden layer (equation (11)), backward learning for the output layer (equation (12)), and backward learning for the hidden layer (equation (13))

$$\mathbf{y}_q = f(\mathbf{u}_q) = \frac{1}{1+\exp(-\mathbf{u}_q)} \tag{9}$$

$$\mathbf{h} = f^{(1)}\left[\mathbf{W}^{(1)}\mathbf{x}\right] = \frac{1}{1+\exp(-\mathbf{W}^{(1)}\mathbf{x})} \tag{10}$$

$$\mathbf{o} = f^{(2)}\left[\mathbf{W}^{(2)}\mathbf{h}\right] = \frac{1}{1+\exp(-\mathbf{W}^{(2)}\mathbf{h})} \tag{11}$$

$$\boldsymbol{\delta}_2 = \mathbf{o}\ (\mathbf{1}-\mathbf{o}\ )(\mathbf{y}\ -\mathbf{o}\ ) \tag{12}$$

$$\delta_1 = h\ (1 - h\ ) \sum \delta_2 W^{(2)} \tag{13}$$

The training process also indicates weight adjustments for each node of the network with adjustment of the hidden neuron numbers and learning factor. In this chapter, ten hidden neurons were assigned. This type of training process formally was known as the back-propagation learning algorithm or delta learning rule. The back-propagation for the output layer is represented by the equation

$$W^{(2)}(t + i) = W^{(2)}(t) + \mu\delta_2 h \tag{14}$$

and for the hidden layer by the equation

$$W^{(1)}(t + i) = w^{(1)} + \mu\delta_1 x \tag{15}$$

In addition, a learning factor μ of 0.7 was assigned to adjust the training process. The effectiveness and convergence of the error back-propagation learning algorithm depends significantly on the value of the learning factor. In general, the optimum value of μ depends on the problem being solved, and there is no signal learning factor value suitable for different training cases. This leads to the conclusion that μ should indeed be chosen experimentally for each problem (Zurda, 1992). The training process will be continued until the network is able to learn all the target data. The accuracy of the learning process depends on the type of data to be learned and the application of the network.

5. Results and Discussion

The ANN was trained with predefined target control datasets. C++ programming language was developed to simulate the ANN control algorithm with the necessary arrangement of output signals operating the electrohydraulic power system. All control datasets values had been scaled individually so that the overall difference in the dataset was maximized; this was due to the sigmoid transfer function employed with a learning range from 0 to 1. Training sets were taken by manually driving the actuator to follow a desired path.

The training control data were broken up into 64 input–output sets, which covered the entire motion range of the ball-and-socket actuator. Each set represented the valve order with the time needed to move the actuator from a desired point to another with the incorporated parameters. These control data were used to drive the actuator to follow a desired path and to move the actuator through all intermediate points. The neural network was trained repeatedly for 300 000 iterations with the predefined datasets. To validate the design of the network, predicted output sets for angular displacement 1, angular displacement 2, angular velocity 1, angular velocity 2, angular acceleration 1, and angular acceleration 2 were compared with values from experimental data collected.

The average absolute errors are summarized in Table 1. Figure 5 illustrates the deviation between predicted outputs and the data obtained from the ANN. The results show that the design network is capable of learning and predicting the control parameters as shwon in Figures 6, 7,and 8.

Parametrs	Percenatge of Error
Angular Disp_1	3.86
Angular Disp_2	5.23
Angular Velocity_1	6.35
Angular Velocity _2	4.36
Angular Accel_1	3.98
Angular Accel_2	2.77

Table 1. Mean absolute percentage error

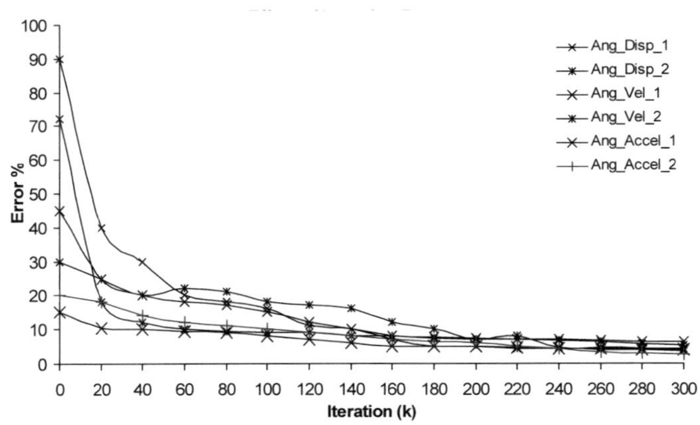

Fig. 5. Process of building knowledge for the learning Algorithm

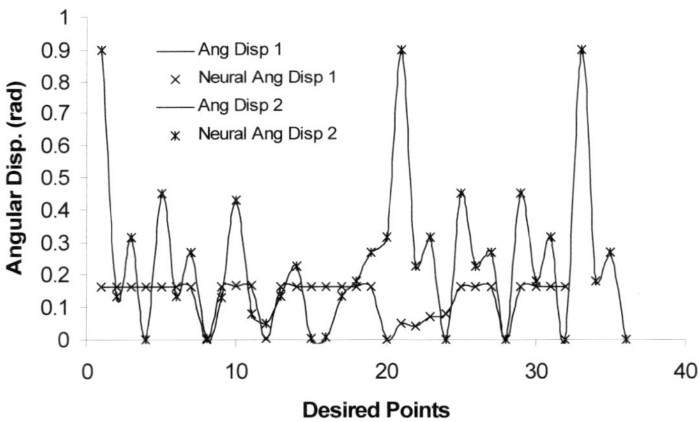

Fig. 6. Predicted angular displacements

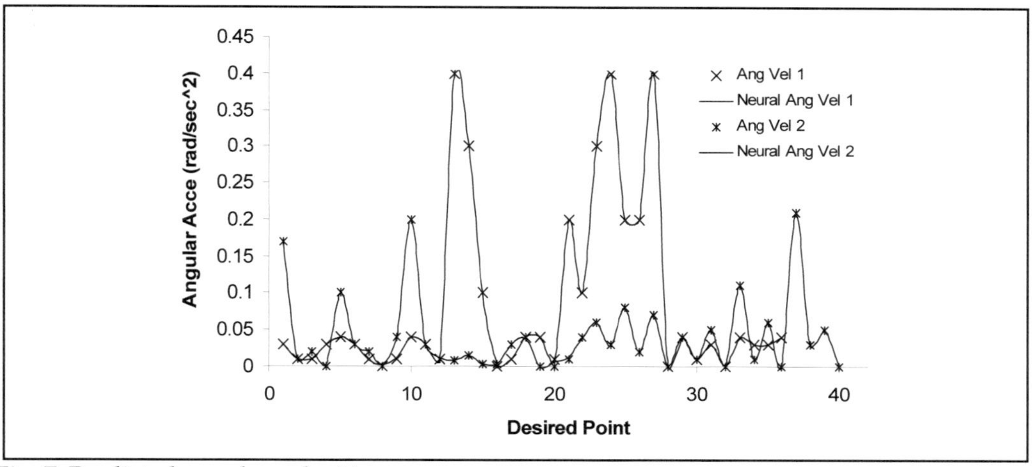

Fig. 7. Predicted angular velocities

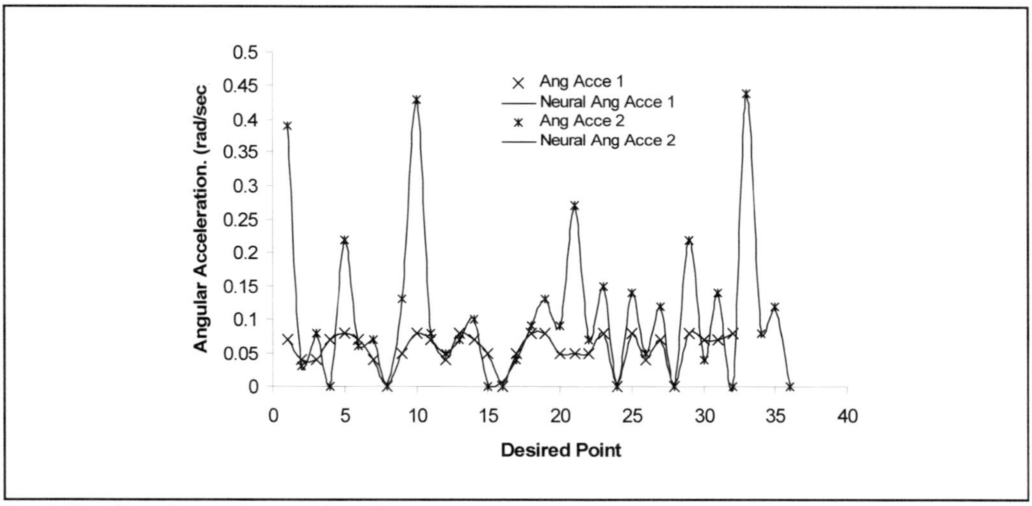

Fig. 8. Predicted angular acceleration

6. Conclusion

The ANN adaptive learning algorithm developed has been implemented successfully on a new 2DOF ball-and-socket actuator. The algorithm has the capability of getting round the drawback of some control schemes that depend on modelling the system being controlled. An actuator has been fabricated to replace the two revolute actuators in serial robot manipulators. The trained ANN showed the ability to operate the ball-and-socket actuator properly in real time by achieving angular displacement, angular network velocity, and angular acceleration.

7. References

Abdelhameed M. M. (1999). Adaptive neural network based controller for robots. *Journal of Mechatronics*, 9, (1999) 147-162, ISSN 0957 4158.

Ambrosino, G.; Celentano, G. & Garofalo, F. (1988). Adaptive tracking control of industrial robots. *ASME, Journal of Dynamic Systems, Measurment and Control,* 110, (1988) 215–220, ISSN 0022-0434.

Ananthraman S.; Nagchaudhuri A.; & Garg DP. (1991). Control of a robotic manipulator using a neural controller. *Proceedings of 22nd Modeling and Simulation Conference.* pp. 1846–1853, ISBN, Pittsburgh, PA, USA.

Brun-Picard, Daniel, & Sousa João S.S. (1999). Design of machines and robots endowed with a permanent learning ability. *J. of control engineering*, 7, (1999) 565-571, ISSN 0010-8049.

Cheah, C.C.; Kawamura, S. & Arimoto S. (2003). Stability of hybrid position and force control for robotics manipulators with kinematics and dynamics uncertainties. *Journal of Automatica.* 39(2003), 847–855, ISSN 0005-1098.

Chen S, & Billings SA. (1992). Neural networks for nonlinear dynamic system modeling and identification. *Int J. of Control*, 56, 2, (1992) 319– 46, ISSN 0020-7179.

Cruse H, & Bruwer M. (1990). A simple network controlling the movement of a three joint planar manipulator. *International Conference on Parallel Processing in Neural Systems Computers*. pp. 409–412, ISBN 0-444-88390-8, Elsevier Science Inc., North-Holland.

Desmurget, M. & Prablanc C. (1997). Postural control of three-dimensional perhension movement. *J. of Neurophysiology*, 77, 1, (1997) 452-464, ISSN 0022-3077.

Hasan, Ali T.; Hamouda A.M.S.; Ismail N.; & Al-Assadi H.M.A.A. (2007). A New Adaptive Learning Algorithm for Robot Manipulator Control. *Journal of Systems and Control Engineering*, 221, (2007) 663-672, ISSN 0959-6518.

Karlik, Bekir & Aydin, Serkan (2000). An improved approach to the solution of inverse kinematics problems for robot manipulator, *Journal Engineering Applications of Artificial Intelligence*, 13, (2000) 159-164, ISSN 0952-1976

Knohl, T. & Unbehauen H. (2000). Adaptive position control of electrohydraulic servo systems using ANN. *Mechatroincs*. 10, (2000) 127–143, ISSN 0957 4158.

Kuperstein M, & Wang J. (1990). Neural controller for adaptive movements with unforeseen payloads. *IEEE Trans Neural Network.* 1, (1990) 137–42, ISSN 1045-9227.

Lemay, M.A. & Cragi, P.E. (1996). A dynamic model for simulating movements of the elbow, forearm, and wrist. *J. of Biomechanics*, 29, (1996) 1319-1330, ISSN 0021-9290.

McKerrow, Phillip John. (1991). *Introduction to Robotics*, Addison-Wesley, ISBN 0-201-182240-8, Sydny, Austrlia.

Miller III WT; Hewes RP; Glanz FH; & Kraft III LG. (1990). Real-time dynamic control of an industrial manipulator using a neural-network-based learning controller. *IEEE Trans Robot Autom*, 6, (1990) 1–9, ISSN 1070-9932.

Mills PM; Zomaya AY; & Tade MO. (1994). Adaptive modle-based control using neural networks. *Int J. of Control,* 60, 6, (1994) 1163–92, ISSN 0020-7179.

Mills, K.J. (1994). Robotic manipulator control of generalized contact force and position. *IEEE Trans on SMAC*, 24, 3, (1994) 523–531, ISSN 1740-8865.

Park, H.J. & Cho, H.S. (1992). On the realization of an accurate hydraulic servo through an iterative learning control. *Journal of Mechatronics*, 2, (1992) 77–88, ISSN 0957-4158.

Perokopenko, R.A.; Frolov, A.A.; Briyukova E.V. & Roby-Brami, A. (2001). Assessment of the accuracy of a human arm model with seven degrees of freedom. *J. of Biomechanics* 32, (2001) 177-185, ISSN 0021-9290.

Raikova, R. (1992). A general approach of modeling and mathematical investigation of the human upper limb. *J. of Biomechanics*, 25, (1992) 857-867, ISSN 0021-9290.

Sharkey, & Noel E. (1997). Artificial neural networks for coordination and control: The portability of experiential representations. *Journal of Robotics and Autonomous Systems*, 22, (1997) 345-359, ISSN 0921-8890.

Tsao, T.C. & Tomizuka, M. (1994). Robust adaptive and repetitive digital tracking control and application to a hydraulic servo for noncircular machining. *ASME, Journal of Dynamic Systems, Measurment and Control*, 116, (1994) 24–32, ISSN 0022-0434.

Tso, S.K. & Law P.L. (1993). Implementing model-based variable-structure controllers for robot manipulators with actuator modeling. *Proceedings of the 1993 IEEE/RSJ International Conference on Intelligent Robot and Systems*, pp. 701–707, ISBN 078-030-8239, Yokohama, Japan. Publisher

Yang, Y.P. & Chu, J.S. (1993). Adaptive velocity control of DC motors with Coulomb friction identification. *ASME, Journal of Dynamic Systems, Measurment and Control*, 115, (1993) 95–102, ISSN 0022-0434.

Zurda, M. J. (1992). *Introduction to Artificial Neural System Network*. West Publishing Companies, ISBN 0-314-93397-3, St. Paul, MN, USA.

6

Singularity-Based Calibration – A Novel Approach for Absolute-Accuracy-Enhancement of Parallel Robots

Philipp Last
Siemens AG, Energy Sector, Power Distribution Division, Medium Voltage (E D MV C R&D 4)
Germany

Annika Raatz and Jürgen Hesselbach
Technical University Braunschweig
Institute of Machine Tools and Production Technology (IWF)
Germany

1. Introduction

Due to different reasons such as manufacturing and assembly errors, the real geometry of a robot manipulator usually deviates from that defined by the kinematic model which is used for robot control. As a consequence, the absolute accuracy of the robot is limited and offline generated robot programs cannot be executed with sufficient accuracy. With the help of robot calibration, these shortcomings can be overcome. The underlying concept is to estimate the parameters of the kinematic model based on redundant measurements. This data can be used to alter the kinematic model so that it matches the reality as closely as possible. This is ultimately used to correct the parameters in the controller model which thus improves the obtainable absolute accuracy.

This article covers a new and innovative approach for robot calibration which can be applied to parallel robots. In comparison to all known calibration techniques, this novel scheme has the advantage in that it does not rely on any additional calibration hardware. In addition to being cost effective, this method is simple to use as it can be completely automated.

The main aspect of the work at hand is an approach which allows the acquisition of redundant measurement-data required for calibration. Under consideration of special knowledge of the robot-structure's behaviour in certain configurations, so-called singularities of the second type, measurement information is gathered using the robot's actuator measurement systems only. The presented approach is thus denoted as singularity based calibration. In conjunction with qualified modelling and identification methods, the proposed measurement approach sets up a completely new robot calibration scheme. With

the exception of special cases, the proposed technique is principally applicable to all parallel kinematic structures.
The technique is first explained by means of a simple and comprehensible example. Subsequently successful implementation of the singularity-based calibration technique is exemplarily shown by practical experiments which are conducted on a parallel-robot with three degrees of freedom (dof), the so-called TRIGLIDE-robot system. The final results show that singularity based calibration is an adequate means to significantly improve a robot's absolute accuracy.

2. Kinematic Robot Calibration – Fundamental approach

The essential point in any robot calibration technique that follows the idea to replace the model parameters in the controller-model is to set up a residual of the form

$$r(\mathbf{k}) = \tilde{\mathbf{b}} - \mathbf{b}(\mathbf{k}) \tag{1}$$

with $\tilde{\mathbf{b}}$ some redundantly measurement information, $\mathbf{b}(\mathbf{k})$ a vector with corresponding information provided by the kinematic model and $\mathbf{k}$ the parameter vector which is supposed to be identified. If such a residual can be obtained at n different measurement configurations then it is possible to stack all the information in a residual vector $\mathbf{r}(\mathbf{k}) = [\mathrm{r}_{\mathbf{1}}^{\mathbf{T}}, \mathrm{r}_{\mathbf{2}}^{\mathbf{T}}, \ldots, \mathrm{r}_{\mathbf{n}}^{\mathbf{T}}]^{\mathbf{T}}$. Once this vector is available it is the goal to estimate the parameters in a way such that

$$\mathbf{r}(\mathbf{k}) = \mathbf{0} \tag{2}$$

Due to measurement noise and model simplifications this goal is, however, of theoretical nature and will never be exactly reached in reality. Instead one aims to minimize a cost function

$$F = \mathbf{r}^{\mathrm{T}}(\mathbf{k})\mathbf{r}(\mathbf{k}) \tag{3}$$

which if $\mathbf{r}(\mathbf{k}) = \mathbf{0}$ would be fulfilled equals zero as well and is otherwise bigger than zero. The minimization of F can be attained by any optimization method in principle. Usually, due to the special so-called least squares form of the function F, least-square algorithms such as the Levenberg-Marquardt approach (Scales, 1985) are applied. Minimization of F finally yields a parameter vector $\mathbf{k}^{kalib}$ which is then used to replace the original parameters that were used before calibration within the robot-controller.
Considering the aforementioned remarks, four essential steps can be identified which are existent in each model-based calibration approach. These are (Mooring et al., 1991):

1. **modelling**
   In the modelling phase a kinematic model is set up which includes a number of geometric parameters that are supposed to be identified by calibration
2. **measurement**
   The measurement step provides the redundant information required for calibration

3. **parameter identification**
   By means of feasible mathematical methods the model parameters are identified in a way so that model and measurements correspond to each other in a best possible manner
4. **parameter correction**
   Within the parameter correction step the identified parameters are transferred to the robot controller

3. Classification of Calibration techniques

A huge number of calibration methods already exist which follow the general scheme described in the preceding section. Differences between these techniques can be found in various aspects at different stages. The most obvious and most important differences, however, exist in the measurement phase. Based on this appraisal a general classification can be defined for the different calibration strategies which includes the two seperation criteria: First the degree of automation and second the data-aquisition method, both briefly explained in what follows.

- **degree of automation**
  In regard to the degree of automation autonomous and non-autonomous calibration techniques are distinguished. A calibration method is understood to work autonomously only if all steps of the overall procedure can be completely automated and absolutely no user interaction is required during calibration. If any effort is needed for preparation, for accomplishment or data-transfer during calibration then the corresponding technique is defined to be non-autonomous.
  It should be noted that the non-autonomous methods, although combined in one group, may drastically vary in the amount of required manual support.

- **data-aquisition method**
  Two fundamental data-aquisition methods may be used for robot calibration. The first one uses additional sensors (internal or external) which are not required for operating the robot but just in order to provide redundant information. The second method relies on kinematic constraints that are introduced in the system without raising the number of sensors. In this case due to constraints the actuator measurement systems which are already part of the robot system deliver enough information for robot calibration.

In combination of all possible classification attributes there are four principle types of calibration techniques, namely type A, type B, type C and type D (see Fig. 1). Whereas calibration methods of type A to type C are well established and intensively described in the literature (see Table 1), no methods of type D have been reported so far up to the authors knowledge. This gap is closed by the singularity based calibration strategy presented in this paper and in preliminary work (Last & Hesselbach, 2006; Last et al., 2006, Last et al., 2007a, Last et al., 2007b; O`Brien et al., 2007).

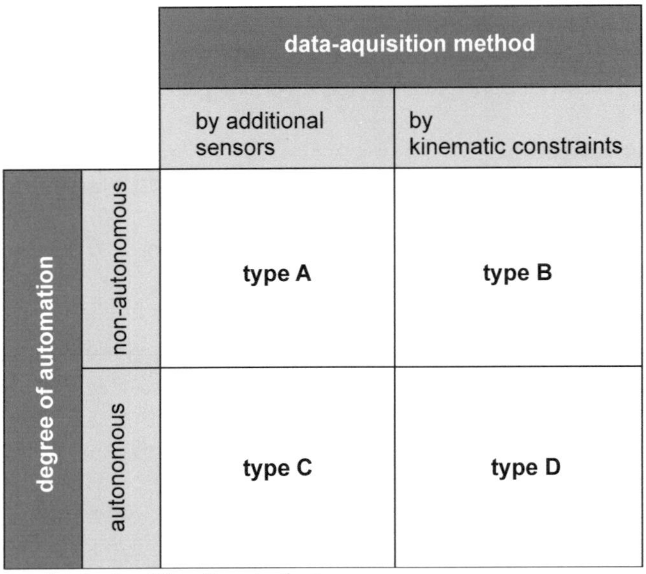

Fig. 1. Classification of robot calibration techniques

		exemplary approach	Source
Type A	-	Calibration by means of a lasertracker	(Corbel et al., 2006) (English et al., 2002)
	-	Calibration by means of camera-systems	(Beyer, 2004) (Nefzi et al., 2008)
	-	Calibration using a double-ball-bar	(Huang et al., 2006) (Ibaraki et al., 2004) (Ihara et al., 2000) (Takeda et al., 2004)
Type B	-	Calibration by contour tracking	(Ikits & Hollerbach, 1997) (Legnani et al., 2001) (Vischer, 1996) (Zhuang et al., 1999)
	-	Calibration by passive joint clamping	(Maurine et al., 1998) (Khalil & Besnard, 1999)
Type C	-	Calibration by passive joint sensors	(Hesselbach et al., 2005a) (Last et al., 2005) (Zhuang, 1997)
	-	Calibration with actuation redundancy	(Schönherr, 2002) (Zhang et al., 2007)

Table 1. Exemplarily chosen calibration strategies of different type

4. Singularity Based Calibration

The new calibration approach contributed here relies on passing singularities of type 2. Because these constitute structure configurations where several solutions of the direct kinematic problem (DKP) coincide, they are also called direct kinematic singularities. It is well known that a robot-structure is uncontrollable in this kind of configurations (Hesselbach et al., 2005b) and hence particular strategies need to be applied to savely guide a manipulator through singularities of type 2. Such a technique is described in section 4.1. Within the same section it is also shown how some specific measurement information is obtained during that process. Subsequently in section 4.2 it will be shown how to compute corresponding information from the kinematic model.

4.1 Passing singularities of type 2 as the basis of singularity based calibration

With the intention of workspace enlargement Helm has been the first who presented a technique to pass singularities of type 2 (Helm, 2003). It was experimentally proven at a planar robot-structure. Later the approach has been extended to spatial parallel structures in (Budde et al., 2005). Both methods rely on the basic idea which consists in temporarily underactuation of the robot system during passing the singular configuration and to use an additional driving force to guide the structure through the direct kinematic singularity. By means of the planar RRRRR-structure the approach is exemplarily summarized in Fig 2. In a pose near the singular configuration (a) the structure is underactuated by releasing one actuator (b). While the second actuator is kept at a constant motor-position the endeffector-point C passes the singularity (c) driven by gravity influence until it reaches a non-singular configuration (d) in which the released actuator can be activated again. Instead of exploiting gravity as the driving force which has been also done in (Budde et al., 2005), structure inertia may be used to pass the singularity as described in (Helm, 2003).
Performing the singularity passing while holding the motor that is not released, at a constant position it turns out that the released actuator changes its direction of movement (see dashed line) exactly in the point of the singularity that is reached if both rod elements of the robot manipulator build a common line. Consequently by observing the movement of the released actuator by its own motor-encoder it is possible to identify and save the actuator coordinate $\tilde{q}_{released}^{sing}$ of the released motor that corresponds to a singular configuration. Furthermore, since particular geometric conditions need to be fulfilled at a singular configuration of type 2, it is possible to compute the corresponding actuator coordinate $q_{released}^{sing}(\mathbf{k})$ from the kinematic model including the kinematic parameters $\mathbf{k}$. Comparing both information leads to a residual

$$r(\mathbf{k}) = \tilde{q}_{released}^{sing} - q_{released}^{sing}(\mathbf{k}) \tag{4}$$

corresponding to that in equation (1) which is the basis for singularity based calibration. Once such a residual can be conducted at a sufficiently high number of differing robot configurations the singularity based calibration procedure proceeds as described in section 2.
What is important to mention at this point is that the method is general for parallel robots and does not only apply to the RRRRR-structure. Independent on the robot structure a change of direction of the released actuator can be observed if a type 2 singularity is passed

while keeping all other actuators of the manipulator to be calibrated at a constant position during that approach. An application to serial robots is impossible because type 2 singularities only occur for parallel robots.

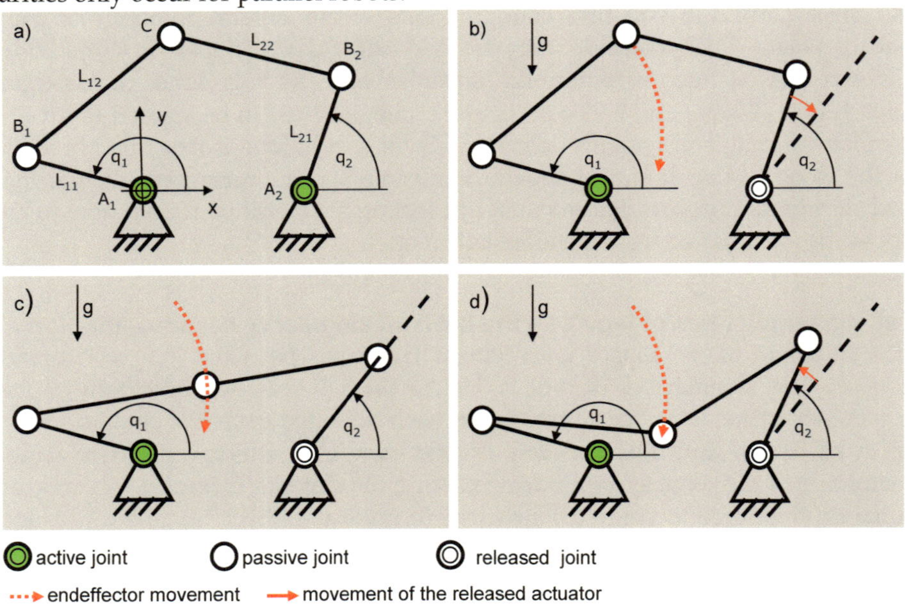

Fig. 2. Passing a singularity of type 2

4.2 Singular kinematic problem

In the preceding section it was assumed that the actuator coordinate $q_{released}^{sing}(\mathbf{k})$ which ideally corresponds to the measured value $\tilde{q}_{released}^{sing}$ can be computed from the kinematic model. Indeed this computation which is denoted as singular kinematic problem (SKP) is straightforward for the R̲RRRR̲-manipulator because due the simple kinematic structure an analytic solution exists. However, a closed form SKP-solution is not the general case. For more complex structures iterative numerical solutions need to be applied. Thus, in order to allow for a wide application of the singularity based calibration approach a general SKP-solution strategy is presented in this section.

A requirement for the application of the general SKP-solution-technique consists in a valid solution for the DKP that does not cause any numerical problems in or near singularities. Techniques which provide such a solution are presented in (Wang & Chen, 1991) and in (Last et al. 2007). Both methods follow an iterative numeric procedure and both methods return a loop closure error E_{total} with a clear geometrical meaning (see Fig. 3 (left)) that is zero if a valid DKP-solution exists and bigger than zero if no DKP-solution can be found.

The proposed SKP-solution exploits the fact that type 2 singularities define the boundary of the actuator space for parallel manipulators. This means that, when varying the released actuator coordinate $q_{released}$ away from the singular value $q_{released}^{sing}$, DKP solutions are found when moving $q_{released}$ in one direction while no DKP solutions are found when moving $q_{released}$ in the opposite direction. Without loss of generality this behaviour is illustrated in Fig. 3 for the the R̲RRRR̲ manipulator. Fig. 3 (left) shows the manipulator in three different situations. Situation B constitutes a "normal" configuration within the manipulator's actuator. In situation A, it is not possible to connect the loops of the mechanism, causing the DKP-solution to converge to an error $E_{total} > 0$. Hence the actuator position is not valid. Finally, a singular configuration is shown as situation S. A corresponding plot showing the loop-closure-error E_{total} vs. the released actuator value $q_{released}$ is illustrated in Fig. 3 (right). Obviously, for situation A, the DKP error E_{total} is greater than zero, while it holds that $E_{total} = 0$ for situation B. The singular situation corresponds to the value $q_{released}$ for which E_{total} starts deviating from zero. Based on the aforementioned observations, a simple bisection search can be applied to find $q_{released} = q_{released}^{sing}$. Its basic idea is to successively reduce an interval from which it is known that it includes $q_{released}^{sing}$. The method can be summarized as follows:

1) Provide an actuator value $q_{released}^{A}$ outside the workspace and a second actuator value $q_{released}^{B}$ inside the actuator space, both definig the initial search interval. Specify a termination threshold $\varepsilon_{q,diff}$ close to but bigger than zero, which defines the size of the search interval at which the algorithm terminates.
2) Compute an actuator coordinate $q_{released}^{C}$ located in the middle of $q_{released}^{A}$ and $q_{released}^{B}$
3) Solve the direct kinematics for $q_{released} = q_{released}^{C}$, thereby obtaining a loop closure error $E_{total}(q_{released}^{C})$.
4) If $E_{total}(q_{released}^{C}) > 0$ (to account for numerical deviations a value very close to but bigger than zero can be chosen instead of zero), then $q_{released}^{A} = q_{released}^{C}$, otherwise $q_{released}^{B} = q_{released}^{C}$.
5) If the difference $q_{released}^{B} - q_{released}^{A} > \varepsilon_{q,diff}$, repeat from step 3, otherwise terminate with $q_{released}^{sing} = q_{released}^{B}$.

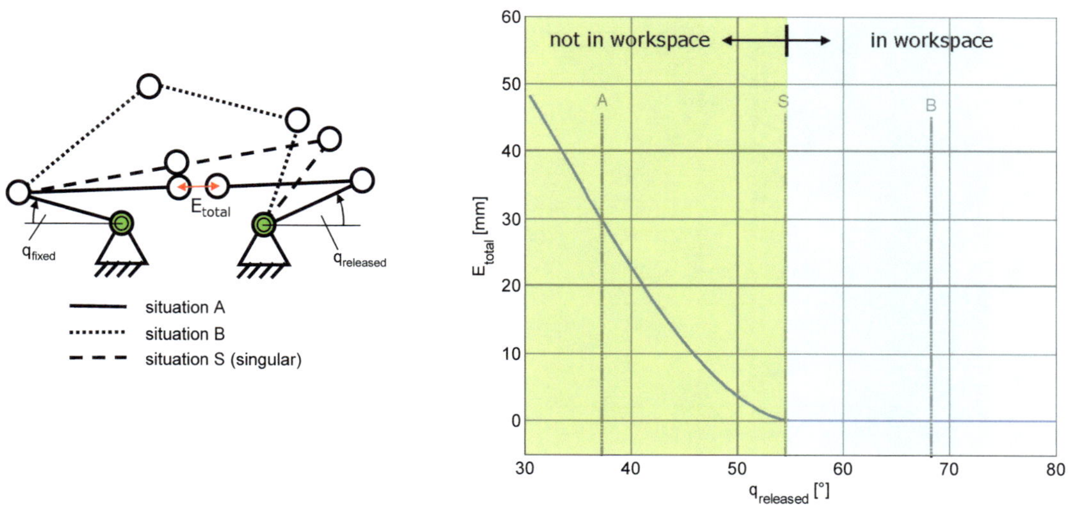

Fig. 3. Singularity as the boundary of the actuator space

4.3 Summary and review

The ideas presented in section 4.1 and in section 4.2 define the fundamental basis of the singularity based calibration. In conjunction with a suited modelling approach and an adequate parameter identification procedure (both described in the wide spread literature on robot calibration) the proposed methods build a general means to improve the absolute accuracy of parallel robot systems. In comparison to alternative techniques for robot calibration (type C - type singularity based calibration does neither rely on additional sensors (external or internal) nor requires the use of special hardware to constrain the robot motion. Due to the abandonment of particular calibration equipment singularity based calibration features the advantages of being cost effective and at the same time fully automatable. According to (Hidalgo & Brunn, 1998) these are aspects which are crucial for success of a calibration approach.

5. Application to the TRIGLIDE robot system

In order to validate the singularity based calibration method it is tested on a certain robot, the so-called TRIGLIDE structure (see Fig. 4) designed for high-speed handling and assembly tasks (Budde, 2007). Three equally designed kinematic chains connect the endeffector-platform of this robot with the base platform. Each chain is actuated by a linear drive. Due to the use of two parallel rods in the build-up of the three chains, the endeffector-platform is always kept at constant orientation. This fully parallel structure has three degrees of freedom allowing for free positioning of the endeffector in space. By attaching a serial rotational axis to the platform, an additional rotation around the z-axis can be accomplished, thus enabling the robot to perform Scara-type movements. Since the rotational axis is irrelevant to the calibration approach discussed here, it is neglected in the following.

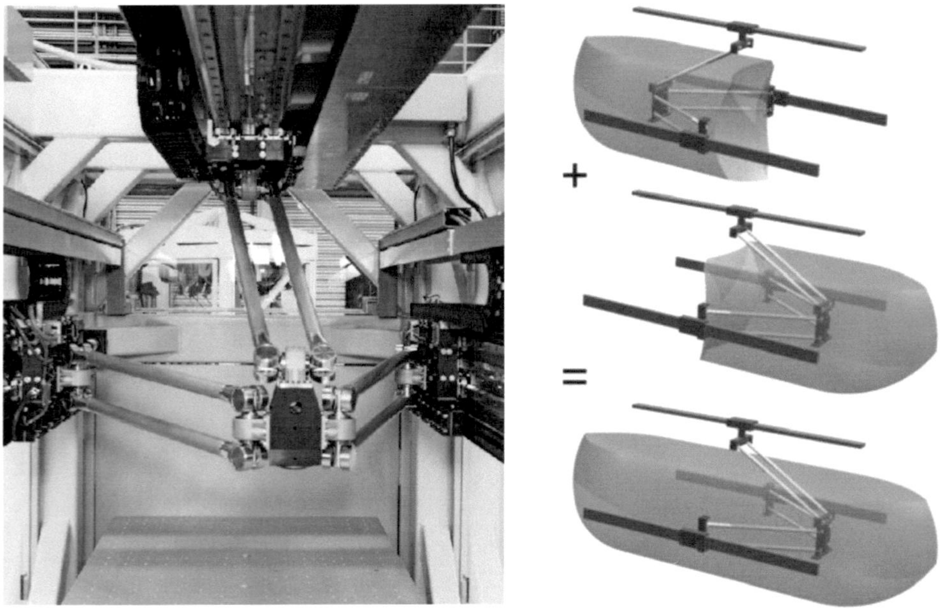

Fig. 4. TRIGLIDE robot system (left) and corresponding workspaces (right)

5.1. Passing Singularities

As stated above, singularity based calibration requires passing singularities of type 2. The proposed technique to realize such a passing (see section 4.1) has been successfully implemented on the TRIGLIDE structure - again with the original intention to enlarge its workspace. As can be seen in Fig. 4 (right) it allows to combine two symmetrical workspaces to an overall workspace which is almost twice as big as the single workspaces. Both of these workspaces are not diminished by direct kinematic singularities within them, allowing for their complete utilization. However, the transition between these two workspaces requires, that several singularities have to be passed and several other workspaces have to be crossed. Each of the workspaces corresponds to a specific working mode, also called IKP-configuration, where an IKP-configuration is characterized as follows: For a given position of the platform there are two possible positions of the carriage in each of the three kinematic chains $i = 1,2,3$ which will be described as $K_{IKP,i} = \{-1, +1\}$. They correspond to different solutions of the inverse kinematic problem. With this definition a complete IKP-configuration can be uniquely described using the vector $\mathbf{K} = [K_{IKP,1}, K_{IKP,2}, K_{IKP,3}]$. The two workspaces, the robot is going to be used in (Fig. 4) are based on the IKP-configurations $[-1,-1,-1]$ and $[+1,+1,+1]$ and are called the two working configurations. To switch between them several transition workspaces have to be passed. Due to the multitude of transition configurations there are several possibilities finding a way from one working configuration to the other one, of which one path is shown in Fig. 5 (a)-(d). In addition to the configurations to be passed, corresponding workspace sections parallel to the y-z-plane are shown in the figures.

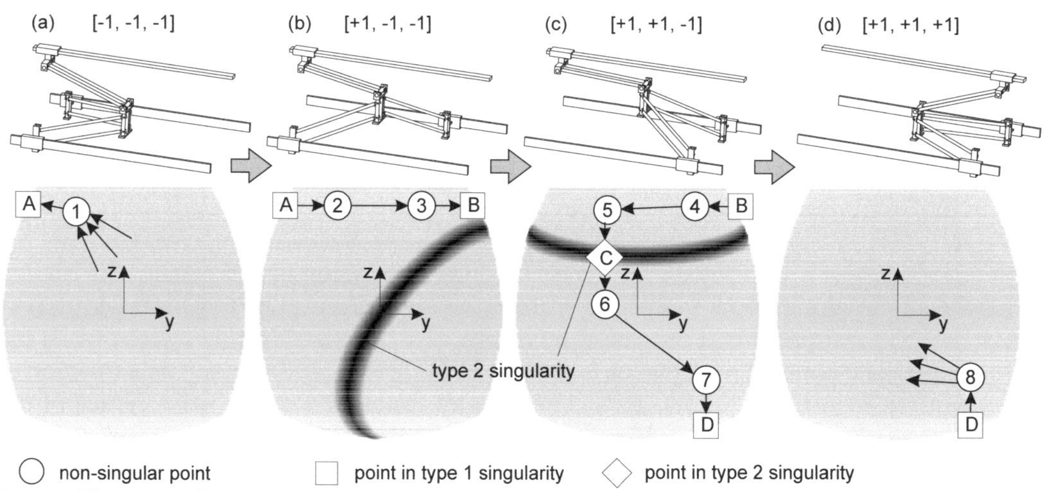

Fig. 5. Changing the working configuration

The total approach to switch between the two working configurations is explained in (Budde et al., 2008) and in (Budde, 2009) in more detail. For the calibration approach it is most essential to pass the singularity of type 2. Hence we focus on Fig. 5 (c) which shows the crossing of a type two singularity in position C. Similarly to the procedure described in section 4.1 the approach is as follows. First the endeffector is placed above the singularity (position 5) depicted by the black line within the workspace. At this point the robot system is underactuated by releasing the upper actuator. Forcing the other two actuators to remain at a constant position, the endeffector driven by gravity starts moving on a circular path. It passes the type 2 singularity in position C and finally reaches a non singular position 6 in which the released actuator can be activated again. Due to the temporary underactuation of the system, the risk of damage is avoided and the endeffector can reliably travel through the singularity. As for the simple manipulator shown above it becomes apparent that the released actuator changes its direction of movement exactly in the point of the singularity, thus allowing for experimental singularity detection. In the description of the proposed approach a particular IKP-configuration was chosen for passing direct kinematic singularities in the workspace of the TRIGLIDE robot and the upper actuator was released. In the same manner also one of the other actuators could have been released with the two remaining motors locked at the same time. Moreover, the singularity of type 2 not necessarily needs to be passed in the depicted position. Indeed, because the singularity builds a continuous surface in space (figure 3), it is possible to cross it at different positions even in other IKP-configurations and to collect enough information in order to allow for a complete calibration of the TRIGLIDE robot.

5.2 Experimental results

With the working configuration change procedure available as a robot command, singularity based calibration has been implemented on the TRIGLIDE robot system in a way so that calibration can be completely automatically performed. This means that once a special robot program is executed the whole calibration process is started and runs without a need for user interaction. First tests prove the principal functionality of the technique, however, it turns out that the absolute accuracy reached by the method is not sufficiently good. A critical review makes us believe that this is mainly due to elastic structure deformations occurring during singularity passing which finally result in disturbed measurement data. As a remedy the implementation is changed in a way that the singularity passing process is manually supported. By this means dynamics during singularity passing is significantly reduced thereby decreasing elastic deformation influence. Indeed, by this means the results can be drastically improved.

A typical calibration result is depicted in Fig. 6. It shows the position error, which is the difference between a computed and a measured target position at 125 equally distributed control configuration in each of the two working configurations [+1 +1 +1] and [-1 -1 -1] of the robot. The real position is measured by means of a Leica-lasertracker system while the computation of a corresponding value is accomplished by the DKP-solution as a function of the measured motor coordinates. As can be seen by the results, the initial accuracy of the robot is already quite good with maximum positioning deviations of approx. 0.6 mm. However the accuracy can be significantly improved by the proposed singularity based calibration method so that the remaining absolute positioning error after calibration is approx. 0.36 mm in maximum. Mean value as well as the standard deviation of the positioning error over the 250 control configurations also take better values after calibration compared to the uncalibrated case.

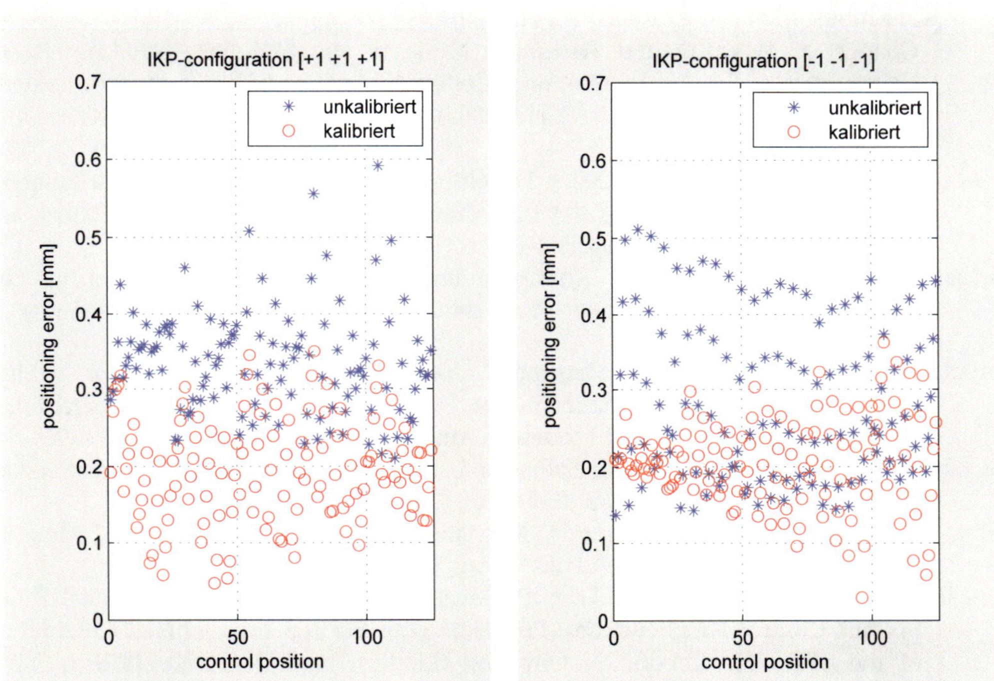

Fig. 6. positioning error at each control position of the validation routine.

6. Conclusion

For the first time a robot calibration approach has been presented that does exclusively rely on the information delivered by the robot-system itself. Hence, as neither additional sensors nor special constraint devices are required in order to apply parameter identification methods, the proposed technique is very economical and easy to use. It is thus especially convenient to be used in small and medium sized companies which do neither own special robot calibration equipment nor have professional skilled robot calibration experts. The basic idea of the new calibration scheme has been explained from a theoretical point of view by means of a simple example structure and subsequently validated through experiments by means of a more complex spatial parallel structure. The obtained results emphasize the promising potential of the approach.

7. Acknowledgements

The research work reported here was supported by the German Reserach Foundation (DFG) within the scope of the Collaborative Research Center SFB 562.

8. References

Beyer, L. (2004). Genauigkeitssteigerung von Industrierobotern - insbesondere mit Parallelkinematik, Helmut Schmidt Universität Hamburg, Dissertation

Budde, C.; Kolbus, M.; Last, P.; Raatz, A.; Hesselbach, J.; Schumacher, W. (2008). Optimized Change of Working and Assembly Mode of the SFB 562 TRIGLIDE-Robot, Proceedings of the 3rd International Colloquium of the SFB 562: Robotic Systems for Handling and Assembly (RSHA 2008), pp. 221-236, Braunschweig (Germany), Shaker Verlag, Aachen

Budde, C.; Last, P.; Hesselbach, J. (2007). Development of a Triglide-Robot with Enlarged Workspace. Proceedings of the IEEE International Conference on Robotics and Automation, Rome (Italy), pp. 543–548

Budde, C. (2009). Wechsel der Konfiguration zur Arbeitsraumvergrößerung bei Parallelrobotern, Technische Universität Braunschweig, Vulkan-Verlag , Dissertation

Corbel, D.; Company, O.; Nabat, V.; Maurine, P. (2006). Geometrical Calibration of the High Speed Robot Par4 using a Laser Tracker. Proceedings of the 12th IEEE International Conference on Methods and Models in Automation and Robotics, pp. 687–692

English, K.; Hayes, M.J.D.; Leitner, M.; Sallinger, C. (2002). Kinematic Calibration of Six-Axis Robots. Proceedings of the CSME Forum, Kingston (Canada)

Helm, M. B. (2003). Durchschlagende Mechanismen für Parallelroboter, Technische Universität Braunschweig, Vulkan Verlag, Dissertation

Hesselbach, J.; Bier, C.; Budde, C.; Last, P.; Maaß, J.; Bruhn, M. (20055). Parallel Robot Specific Control Functionalities. Proceedings of the 2nd International Colloquium of the SFB 562: Robotic Systems for Handling and Assembly (RSHA 2005). Braunschweig (Germany), pp. 93–108

Hesselbach, J.; Bier, C.; Campos, A.; Löwe, H. (2005). Direct Kinematic Singularity Detection of a Hexa Parallel Robot, Proceedings of the IEEE International Conference on Robotics and Automation, Barcelona (Spain), pp. 3507-3512

Hidalgo, F.; Brunn, P. (1998). Robot Metrology and Calibration Systems - A Market Review. Industrial Robot: An International Journal Vol. 25, No. 1, pp. 42–47

Huang, T.; Hong, Z.Y.; Mei, J.P.; Chetwynd, D.G. (2006). Kinematic Calibration of the 3-DOF Module of a 5-DOF Reconfigurable Hybrid Robot using a Double-Ball-Bar System. Proceedings of the IEEE/RSJ International Conference on Intelligent Robots and Systems, Peking (China), 2006, pp. 508–512

Ibaraki, S.; Yokawa, T.; Kakino, Y.; Nakagawa, M.; Matsushita, T. (2004). Kinematic calibration on a parallel kinematic machine tool of the Stewart platform by circular tests. Proceedings of the American Control Conference, Boston (USA), pp. 1394–1399

Ihara, Y.; Ishida, T.; Kakino, Y.; Li, Z.; Matsushita, T.; Nakagawa, M. (2000). Kinematic Calibration of a Hexapod Machine Tool by Using Circular Test. Proceedings of the 2000 JAPAN-USA Symposium on Flexible Automation, Ann Arbor (USA)

Ikits, M.; Hollerbach, J.M. (1997). Kinematic calibration using a plane constraint. Proceedings of the IEEE International Conference on Robotics and Automation, Albuquerque (USA), pp. 3191–3196

Khalil, W.; Besnard, S. (1999). Self Calibration of Stewart-Gough Parallel Robots Without Extra Sensors. IEEE Transactions on Robotics and Automation, Vol. 15, Nr. 6, pp. 1116-1121

Last, P.; Budde, C.; Hesselbach, J. (2005). Self-Calibration of the HEXA-Parallel-Structure. Proceedings of the IEEE International Conference on Automation Science and Engineering (CASE), Edmonton (Canada), pp. 393–398

Last, P.; Raatz, A.; Carretero, J.-A.; O`Brien, S. M. (2007). A General Approach to Solve the Singular Kinematic Problem, Proceedings of the IEEE/RSJ International Conference on Intelligent Robots and Systems (IROS), San Diego, CA, USA

Last, P.; Hesselbach, J. (2006). A new calibration strategy for a class of parallel mechanisms, Advances in Robot Kinematics (ARK) - Mechanisms and Motion, , Springer Verlag, Berlin , pp. 331-338

Last, P.; Budde, C.; Krefft, M.; Hesselbach, J. (2006). Parallel Robot Self-Calibration Without Additional Sensors or Constraint Devices, Proceedings of the ISR/Robotik, München

Last, P.; Gaggiotto, F.; Budde, C.; Raatz, A. (2007) Optimal Selection of Measurement Configurations for Singularity Based Calibration of Parallel Kinematic Robots. Proceedings of the IEEE/ASME Conference on Advanced and Intelligent Machines (AIM), Zurich (Switzerland)

Last, P.; Schütz, D.; Raatz, A.; Hesselbach, J. (2007). Singularity Based Calibration of 3-DOF Fully Parallel Planar Manipulators. Proceedings of the 12th IFToMM World Congress, Besancon (Frankreich)

Legnani, G.; Adamini, R.; Jatta, F. (2001). Calibration of a Scara Robot by Force-Controlled Contour Tracking. Proceedings of the 32nd International Symposium on Robotics, Seoul (South Corea)

Maurine, P.; Liu, D. M.; Uchiyama, M. (1998). Self Calibration of a New Hexa Parallel Robot. Proceedings of the 4th Japan-France Congress and 2nd Asia-Europe Congress on Mechatronics, Kitakyushu (Japan), pp. 290–295

Mooring, B. W.; Roth; Z. S.; Driels, M. R. (1991). Fundamentals of Manipulator Calibration, John Wiley and Sons INC.

O`Brien, S. M.; Carretero, J.-A.; Last, P. (2007) Self Calibration of 3-PRS Manipulator Without Redundant Sensors, Transactions of the Canadian Society for Mechanical Engineering, Vol. 31, No. 4, pp. 483-494

Scales, L. E. (1985). Introduction to Non-Linear Optimization. Macmillan Publishers Ltd.

Schönherr, J. (2000). Zur Selbstkalibrierung von Parallelmanipulatoren. Proceedings of the 2nd Chemnitz Parallel Kinematics Seminar, Chemnitz (Germany), pp. 113–123

Y.Takeda; Shen, G.; Funabashi, H. (2004). A DBB-Based Kinematic Calibration Method for In Parallel Actuated Mechanisms Using a Fourier Series. Journal of Mechanical Design, Vol. 126, Nr. 5, pp. 856–865

Vischer, P. (1996). Improving the Accuracy of Parallel Robots, Ecole Polytechnique Federale de Lausanne, Dissertation

Zhang, Y.; Cong, S.; Li, Z.; Jiang, S. (2007). Auto-calibration of a redundant parallel manipulator based on the projected tracking error. Archive of Applied Mechanics, Vol. 77, Nr. 10, pp. 697–706

Zhuang, H.; Motaghedi, S. H.; Roth, Z. S. (1999). Robot Calibration With Planar Constraints. Proceedings of the IEEE International Conference on Robotics and Automation, Detroit (USA), pp. 805–810

Zhuang, H. (1997). Self-Calibration of Parallel Mechanisms with a Case Study on Stewart Platforms. IEEE Transactions on Robotics and Automation, Vol. 13, Nr. 3, pp. 387–397

Advanced Nonlinear Control of Robot Manipulators

Adel Merabet[1] and Jason Gu[2]
[1]*Division of Engineering, Saint Mary's University, Halifax, NS*
[2]*Department of Electrical & Computer Engineering, Dalhousie University, Halifax, NS*
Canada

1. Introduction

In the development of modern robot manipulators, it is required that the robot controller has the capability to overcome unmodeled dynamics, variable payloads, friction torques, torque disturbances, parameter variations, measurement noises which can be often presented in the practical environment.
The objective of this chapter is to provide the reader with an overview on advanced nonlinear control techniques of a rigid robot manipulator. In nonlinear control field, a common strategy is called model based control, which can be derived from the mathematical model of the system. However, in case of robot manipulator, it is weakened by inaccuracies present in the robot model, where the performance of the control algorithm is not guaranteed. As mentioned above, these inaccuracies can be defined as parametric uncertainties, unmodeled dynamics, and unknown external disturbances. To overcome the uncertainties' drawback, robust nonlinear control can be a solution. The goal of robust control is to maintain performance in terms of stability, tracking error, or other specifications despite inaccuracies present in the system.
In this chapter we present two nonlinear model based control strategies: the feedback linearization control and a nonlinear model predictive control for rigid robot manipulator. We first consider the dynamic of the robot manipulator driven by the Euler-Lagrange equations. Based on this general representation, we are able to derive equations of the nonlinear controller for both strategies. Then, a robustness study is carried out through compensation of the system inaccuracies. Two methods are used; the first one is based on the theory of guaranteed stability of uncertain systems, while the second is figured out using the nonlinear control law.
The computation of the nonlinear model based control assumes that all state variables are available. In case of robot manipulators, it implies the presence of additional sensors in each joint such as velocity measurements. They are often obtained by means of tachometers, which are perturbed by noise, or moreover, velocity measuring equipment is frequently omitted due to the savings in cost, volume, and weight. Model-based observers are considered very well adapted for state estimation and allow, in most cases, a stability proof and a methodology to tune the observer gains, which guarantee a stable closed loop system.

In this chapter, nonlinear observer is discussed for state variables estimation. It is a powerful tool to handle nonlinear and uncertain systems, which is the case of the robot manipulator.
Finally, the coupling between the nonlinear model based control and the state observer is discussed and the global stability of the closed loop system is proven theoretically via Lyapunov stability theory.

2. Robot modeling

In this chapter, the nonlinear control laws will be developed for rigid robot manipulators. Therefore, the design and control of such robots require mathematical model of the process. The dynamic of n-link rigid robot manipulator is driven by the Euler-Lagrange equations as

$$\mathbf{D}(\mathbf{q})\ddot{\mathbf{q}} + \mathbf{C}(\mathbf{q},\dot{\mathbf{q}})\dot{\mathbf{q}} + \mathbf{G}(\mathbf{q}) = \mathbf{u} \tag{1}$$

where $\mathbf{q}(t)\in\Re^n$ is the vector of the angular joint positions, which are the generalized coordinates and assumed available by measurement. $\mathbf{u}(t)\in\Re^n$ is the vector of the driving torques, which are the control inputs. $\mathbf{D}(\mathbf{q})\in\Re^{n\times n}$, $\mathbf{D}(\mathbf{q}) = \mathbf{D}(\mathbf{q})^T > 0$ is the link inertia matrix. $\mathbf{C}(\mathbf{q},\dot{\mathbf{q}})\dot{\mathbf{q}}\in\Re^n$ is the vector of the Coriolis and centripetal torques. $\mathbf{G}(\mathbf{q})\in\Re^n$ is the vector of gravitational torques. The outputs to be controlled are the joint angles in the robot. For more detail about robot modeling, the reader can refer to (Spong et al., 2006; Kozłowski, 2004).
The practical implementation of the control law for robot manipulators requires consideration of various sources of uncertainties such as modeling errors, unknown loads, and computation errors. In order to get the real values of the system elements, the uncertainties of the system, error or mismatch represented by $\Delta(.)$, are added to the computed or nominal values represented by $(.)_0$. Therefore, the matrices are rewritten as

$$\begin{cases} \mathbf{D}(\mathbf{q}) = \mathbf{D}_0(\mathbf{q}) + \Delta\mathbf{D} \\ \mathbf{C}(\mathbf{q},\dot{\mathbf{q}}) = \mathbf{C}_0(\mathbf{q},\dot{\mathbf{q}}) + \Delta\mathbf{C} \\ \mathbf{G}(\mathbf{q}) = \mathbf{G}_0(\mathbf{q}) + \Delta\mathbf{G} \end{cases} \tag{2}$$

Moreover, the frictions $\mathbf{F_r}(t)\in\Re^n$, considered as unmodeled quantities, and the external disturbances $\mathbf{b}(t)\in\Re^n$ are added to the robot model (1), which becomes

$$\left(\mathbf{D}_0(\mathbf{q}) + \Delta\mathbf{D}\right)\ddot{\mathbf{q}} + \left(\mathbf{C}_0(\mathbf{q},\dot{\mathbf{q}}) + \Delta\mathbf{C}\right)\dot{\mathbf{q}} + \mathbf{G}_0(\mathbf{q}) + \Delta\mathbf{G} + \mathbf{F_r} = \mathbf{u} + \mathbf{b} \tag{3}$$

Then, after simplification, the model dynamic of the robot is given by

$$\mathbf{D}_0(\mathbf{q})\ddot{\mathbf{q}} + \mathbf{C}_0(\mathbf{q},\dot{\mathbf{q}})\dot{\mathbf{q}} + \mathbf{G}_0(\mathbf{q}) = \mathbf{u} + \eta(\ddot{\mathbf{q}},\dot{\mathbf{q}},\mathbf{q},\mathbf{b}) \tag{4}$$

η is called uncertainty, which is defined by

$$\eta = -\left\{\Delta\mathbf{D}\ddot{\mathbf{q}} + \Delta\mathbf{C}\dot{\mathbf{q}} + \Delta\mathbf{G} + \mathbf{F_r} - \mathbf{b}\right\} \tag{5}$$

It includes unmodeled quantities, parametric uncertainties, and external disturbances.
In a state space form, the nonlinear system of the robot model (4) can be written as

$$\dot{\mathbf{x}} = \mathbf{f}(\mathbf{x}) + \mathbf{g}(\mathbf{x})\mathbf{u} + \mathbf{g}(\mathbf{x})\eta \tag{6}$$

where, the state vector $\mathbf{x} = [\mathbf{x}_1 \quad \mathbf{x}_2]^T = [\mathbf{q} \quad \dot{\mathbf{q}}]^T$, and the vector functions $\mathbf{f}$: $\Re^n \rightarrow \Re^{2n}$ and $\mathbf{g}$:$\Re^n \rightarrow \Re^{2n}$ are vector fields and defined as follows

$$\mathbf{f}(\mathbf{x}) = \begin{bmatrix} \mathbf{x}_2 \\ -\mathbf{D}_0(\mathbf{x}_1)^{-1}\left(\mathbf{C}_0(\mathbf{x}_1, \mathbf{x}_2)\mathbf{x}_2 + \mathbf{G}_0(\mathbf{x}_1)\right) \end{bmatrix}, \quad \mathbf{g}(\mathbf{x}) = \begin{bmatrix} 0_{n \times n} \\ \mathbf{D}_0(\mathbf{x}_1)^{-1} \end{bmatrix} \tag{7}$$

The output vector of angular positions to be controlled is

$$\mathbf{y} = \mathbf{h}(\mathbf{x}) = C\,\mathbf{x} \tag{8}$$

where $C = [I_{n\times n} \quad 0_{n\times n}]$, and I: identity matrix. The functions $\mathbf{f}(\mathbf{x})$, $\mathbf{g}(\mathbf{x})$ and $\mathbf{h}(\mathbf{x})$: $\Re^n \rightarrow \Re^n$ are assumed to be continuously differentiable a sufficient number of time.
First, the development of the control laws will be carried out for the undisturbed system where the uncertainties are not included in the analysis. Then, a robust control is studied through a compensation of the uncertainties by estimation.

3. Nonlinear model based control of robot manipulators

3.1 Feedback linearization control

Feedback linearization is one of the most important strategies for nonlinear control design. There are two general types of linearization: input-state linearization and input-output linearization. Necessary and sufficient conditions have been established for each type of linearization. For a given nonlinear system, these conditions can be checked to determine if the system is linearizable (Corriou, 2004; Nijmeijer & Van der Schaft, 1990; Isidori, 1985; Isidori & Ruberti, 1984).
In this chapter, we will study the feedback linearization, based on input-output linearization, of a rigid robot manipulator. The idea is to differentiate the output $\mathbf{y}$, using Lie derivatives, to obtain an expression where the input $\mathbf{u}$ appears explicitly. The number of times of differentiation is called relative degree.
Definition 1: The Lie derivative of function $h_j(\mathbf{x})$ along a vector field $\mathbf{f}(\mathbf{x}) = (f_1(\mathbf{x}) \ldots f_n(\mathbf{x}))$ is given by

$$L_{\mathbf{f}} h_j(\mathbf{x}) = \sum_{i=1}^{n} \frac{\partial h_j}{\partial x_i} f_i(\mathbf{x}) \tag{9}$$

Iteratively, we have

$$\begin{cases} L_{\mathbf{f}}^0 h_j = h_j \\ L_{\mathbf{f}}^i h_j = L_f (L_{\mathbf{f}}^{i-1} h_j) \end{cases} \tag{10}$$

and, along another vector field $\mathbf{g}(\mathbf{x}) = (g_1(\mathbf{x}) \ldots g_n(\mathbf{x}))$,

$$L_{g_i} L_{\mathbf{f}} h_j = \frac{\partial L_{\mathbf{f}} h_j}{\partial x_i} g_i(\mathbf{x}) \tag{11}$$

Definition 2: The system is said to have a relative degree r if

$$L_{g_i} L_{\mathbf{f}}^{k-1} h_j(\mathbf{x}) = 0 \text{ and } L_{g_i} L_{\mathbf{f}}^{r-1} h_j(\mathbf{x}) \neq 0;\ k = 1,\ldots, r-1 \tag{12}$$

Then, r is the number of differentiation times to appear the input $\mathbf{u}$ in the expression of $\mathbf{y}$. Differentiating the output, using Lie derivatives and the nominal robot state model (6) (without uncertainties), we have

$$\begin{aligned} \dot{\mathbf{y}} &= L_{\mathbf{f}} h_1(\mathbf{x}) = \mathbf{x}_2 \\ \ddot{\mathbf{y}} &= L_{\mathbf{f}}^2 h_1(\mathbf{x}) + L_{\mathbf{g}_2} L_{\mathbf{f}} h_1(\mathbf{x})\mathbf{u} = -\mathbf{D}_0(\mathbf{x}_1)^{-1}\left(\mathbf{C}_0(\mathbf{x}_1,\mathbf{x}_2)\mathbf{x}_2 + \mathbf{G}_0(\mathbf{x}_1)\right) + \mathbf{D}_0(\mathbf{x}_1)^{-1}\mathbf{u} \end{aligned} \tag{13}$$

where, $\mathbf{y} = \mathbf{x}_1 = \mathbf{q}$, and the relative degree for input $\mathbf{u}$ is $r = 2$.
The principle of linearization control law is to get a linear system, where the output is influenced by an external input $\mathbf{v}$ only through a chain of two integrators as

$$\begin{aligned} \ddot{\mathbf{y}} &= \mathbf{v} \\ &= -\mathbf{D}_0(\mathbf{x}_1)^{-1}\left(\mathbf{C}_0(\mathbf{x}_1,\mathbf{x}_2)\mathbf{x}_2 + \mathbf{G}_0(\mathbf{x}_1)\right) + \mathbf{D}_0(\mathbf{x}_1)^{-1}\mathbf{u} \end{aligned} \tag{14}$$

Then, the control law is carried out as

$$\mathbf{u} = \mathbf{D}_0(\mathbf{x}_1)\left(\mathbf{v} + \mathbf{D}_0(\mathbf{x}_1)^{-1}\left(\mathbf{C}_0(\mathbf{x}_1,\mathbf{x}_2)\mathbf{x}_2 + \mathbf{G}_0(\mathbf{x}_1)\right)\right) \tag{15}$$

It is possible to realize a pole-placement by imposing $\mathbf{v}$ as

$$\mathbf{v} = \ddot{\mathbf{y}}_{\mathbf{r}} - K_2(\dot{\mathbf{y}} - \dot{\mathbf{y}}_{\mathbf{r}}) - K_1(\mathbf{y} - \mathbf{y}_{\mathbf{r}}) \tag{16}$$

where, $K_1 = diag(k_{1i})$, $K_2 = diag(k_{2i})$, $i = 1, \ldots, n$
Applying the control law (15) with the external input (16), the tracking error $\mathbf{e}_{\mathbf{y}}(t) = \mathbf{y} - \mathbf{y}_{\mathbf{r}}$ satisfies the second linear equation

$$\ddot{\mathbf{e}}_{\mathbf{y}}(t) + K_2\dot{\mathbf{e}}_{\mathbf{y}}(t) + K_1\mathbf{e}_{\mathbf{y}}(t) = 0 \tag{17}$$

and, hence, the error dynamics are determined by the choice of K_2 and K_1, so that the characteristic equation is Hurwitz.

3.2 Nonlinear model based predictive control

Model based predictive control (MPC) is considered an effective control method handling with constraints, nonlinear processes and disturbances. This control strategy requires an optimization method to solve for the control trajectory over a future time horizon based on a dynamic model of the process (Bordon & Camacho, 1998; Hedjar & Boucher, 2005; Hedjar, et al., 2002; Klančar & Škrjanc, 2007; Vivas & Mosquera, 2005).

The objective of the nonlinear model based predictive controller is to carry out a control law $\mathbf{u}(t)$ in order to track the desired output trajectory $\mathbf{y_r}$ at the next time $(t+\tau)$ through minimization of a general form of the cost function defined as

$$\Im = f\left(\mathbf{e}_{\mathbf{y}}(t+\tau), \mathbf{x}, \mathbf{u}\right) \tag{18}$$

where $\mathbf{e_y}(t+\tau)$ is a predicted error, $\mathbf{y}(t+\tau)$ is a τ-step ahead prediction of the output (angular positions) and $\tau > 0$ is a prediction horizon .

In order to minimize the cost function (18), it is needed to define a prediction model for the behavior of the output in the moving time frame. As the robot model (6) is known, a mathematical tool based on Taylor series expansion can be used to develop the prediction model.

By definition, the Taylor series expansion is carried out using Lie derivatives and given by

$$y_i(t+\tau) = h_i(\mathbf{x}) + \tau\, L_{\mathbf{f}} h_i(\mathbf{x}) + \frac{\tau^2}{2!} L_{\mathbf{f}}^2 h_i(\mathbf{x}) + \dots + \frac{\tau^{r_i}}{r_i!} L_{\mathbf{f}}^{r_i} h_i(\mathbf{x}) + \frac{\tau^{r_i}}{r_i!} L_{\mathbf{g}} L_{\mathbf{f}}^{(r_i-1)} h_i(\mathbf{x})\mathbf{u}(t) \tag{19}$$

where r_i is the relative degree.

Based on this expansion, the prediction model for robot model is expressed by

$$\mathbf{y}(t+\tau) = \mathbf{y}(t) + \tau\,\dot{\mathbf{y}}(t) + \frac{\tau^2}{2!}\ddot{\mathbf{y}}(t) \tag{20}$$

Using the output differentiations (13), we have

$$\mathbf{Y}(t) = \begin{bmatrix} \mathbf{y}(t) \\ \dot{\mathbf{y}}(t) \\ \ddot{\mathbf{y}}(t) \end{bmatrix} = \begin{bmatrix} \mathbf{x}_1 \\ \mathbf{x}_2 \\ -\mathbf{D}_0(\mathbf{x}_1)^{-1}\left(\mathbf{C}_0(\mathbf{x}_1,\mathbf{x}_2)\mathbf{x}_2 + \mathbf{G}_0(\mathbf{x}_1)\right) \end{bmatrix} + \begin{bmatrix} 0_{n\times 1} \\ 0_{n\times 1} \\ \mathbf{D}_0(\mathbf{x}_1)^{-1}\mathbf{u}(t) \end{bmatrix} \tag{21}$$

Then, the prediction model (20) is rewritten as

$$\mathbf{y}(t+\tau) = \mathbf{T}(\tau)\mathbf{Y}(t) \tag{22}$$

where,

$$\mathbf{T}(\tau) = [I_{n\times n} \quad \tau * I_{n\times n} \quad (\tau^2/2) * I_{n\times n}]$$

A similar analysis can be used to carry out the predicted reference trajectory $\mathbf{y_r}$

$$\mathbf{y_r}(t+\tau) = \mathbf{T}(\tau)\mathbf{Y_r}(t) \tag{23}$$

where,

$$\mathbf{Y_r}(t) = [\mathbf{y_r} \quad \dot{\mathbf{y}}_\mathbf{r} \quad \ddot{\mathbf{y}}_\mathbf{r}]^T$$

It is assumed that the information about the derivatives of the reference $\mathbf{y_r}$ is available.
The predicted error is given by

$$\mathbf{e_y}(t+\tau) = \mathbf{y}(t+\tau) - \mathbf{y_r}(t+\tau) = \mathbf{T}(\tau)\big(\mathbf{Y}(t) - \mathbf{Y_r}(t)\big) \tag{24}$$

The optimal control law can be carried out through minimization of a cost function with respect to the control input. In this work, two approaches to define the cost function will be studied:

1. A cost function based only on the tracking error. The goals are to show that the use of this cost function type allows realizing a pole-placement similar to feedback linearization, and designing an uncertainty estimator (section 5. 2) based on the control law derived from this specific cost function.
2. A general form of cost function where the tracking error and the control signal are defined over a future horizon.

3.2.1 Cost function based on the tracking error

The cost function is defined as a quadratic form of the tracking error over a future horizon

$$\Im = \frac{1}{2}\int_0^{\tau_r} \big(\mathbf{y}(t+\tau) - \mathbf{y_r}(t+\tau)\big)^T \big(\mathbf{y}(t+\tau) - \mathbf{y_r}(t+\tau)\big)\, d\tau \tag{25}$$

The control weighting term is not included in the cost function. However, the control effort can be achieved by adjusting τ_r (Chan et al., 1999; Merabet & Gu, 2008).
Using the prediction model of error (24), the cost function (25) can be simplified as

$$\begin{aligned}\Im &= \frac{1}{2}\int_0^{\tau_r} \mathbf{e_y}(t+\tau)^T \mathbf{e_y}(t+\tau)\, d\tau \\ &= \frac{1}{2}\int_0^{\tau_r} \big(\mathbf{T}(\tau)\big(\mathbf{Y}(t) - \mathbf{Y_r}(t)\big)\big)^T \big(\mathbf{T}(\tau)\big(\mathbf{Y}(t) - \mathbf{Y_r}(t)\big)\big)\, d\tau \\ &= \frac{1}{2}\big(\mathbf{Y}(t) - \mathbf{Y_r}(t)\big)^T \mathbf{\Pi}\big(\mathbf{Y}(t) - \mathbf{Y_r}(t)\big)\end{aligned} \tag{26}$$

where

$$\mathbf{\Pi} = \int_0^{\tau_r} \mathbf{T}(\tau)^T \mathbf{T}(\tau) d\tau = \left[\begin{array}{cc|c} \tau_r * I_{n\times n} & (\tau_r^2/2) * I_{n\times n} & (\tau_r^3/6) * I_{n\times n} \\ (\tau_r^2/2) * I_{n\times n} & (\tau_r^3/3) * I_{n\times n} & (\tau_r^4/8) * I_{n\times n} \\ \hline (\tau_r^3/6) * I_{n\times n} & (\tau_r^4/8) * I_{n\times n} & (\tau_r^5/20) * I_{n\times n} \end{array}\right] = \begin{bmatrix} \mathbf{\Pi}_1 & \mathbf{\Pi}_2 \\ \mathbf{\Pi}_2^T & \mathbf{\Pi}_3 \end{bmatrix}$$

The necessary and sufficient condition for cost function minimization is

$$\frac{\partial \Im}{\partial \mathbf{u}} = 0 \tag{27}$$

Using equations (21) and (26), the condition (27) can be rewritten as

$$\left(\frac{\partial(\mathbf{D}_0(\mathbf{x}_1)^{-1}\mathbf{u}(t))}{\partial \mathbf{u}(t)}\right)^T [\mathbf{\Pi}_2^T \quad \mathbf{\Pi}_3](\mathbf{M}(t) - \mathbf{Y}_\mathbf{r}(t)) + \left(\frac{\partial(\mathbf{D}_0(\mathbf{x}_1)^{-1}\mathbf{u}(t))}{\partial \mathbf{u}(t)}\right)^T \mathbf{\Pi}_3 \mathbf{D}_0(\mathbf{x}_1)^{-1}\mathbf{u}(t) = 0 \tag{28}$$

Therefore, the optimal control is

$$\mathbf{u}(t) = -\mathbf{D}_0(\mathbf{x}_1)\left\{[\mathbf{\Pi}_3^{-1}\mathbf{\Pi}_2^T \quad I_{n\times n}](\mathbf{M}(t) - \mathbf{Y}_\mathbf{r}(t))\right\} \tag{29}$$

where,

$$\mathbf{M}(t) = \begin{bmatrix} \mathbf{y}(t) \\ \dot{\mathbf{y}}(t) \\ -\mathbf{D}_0(\mathbf{x}_1)^{-1}(\mathbf{C}_0(\mathbf{x}_1,\mathbf{x}_2)\mathbf{x}_2 + \mathbf{G}_0(\mathbf{x}_1)) \end{bmatrix}$$

$$[\mathbf{\Pi}_3^{-1}\mathbf{\Pi}_2^T \quad I_{n\times n}] = [(10/(3\tau_r^2)) * I_{n\times n} \quad (5/(2\tau_r)) * I_{n\times n} \quad I_{n\times n}]$$

Finally, the control law (29) becomes

$$\mathbf{u}(t) = -\mathbf{D}_0(\mathbf{x}_1)\left\{K_1(\mathbf{y}-\mathbf{y}_\mathbf{r}) + K_2(\dot{\mathbf{y}}-\dot{\mathbf{y}}_\mathbf{r}) - \mathbf{D}_0(\mathbf{x}_1)^{-1}(\mathbf{C}_0(\mathbf{x}_1,\mathbf{x}_2)\mathbf{x}_2 + \mathbf{G}_0(\mathbf{x}_1)) - \ddot{\mathbf{y}}_\mathbf{r}\right\} \tag{30}$$

with, $K_1 = (10/(3\tau_r^2)) * I_{n\times n}$,
$K_2 = (5/(2\tau_r)) * I_{n\times n}$

From the form of the control law (30) and compared with the linearizing control law (15), it can be noticed that they are similar and allow realizing a pole-placement to have a linear dynamic of the tracking error of the closed loop system.

3.2.2 General form of the cost function

The cost function is defined as quadratic forms of the tracking error and weighting control over a future horizon

$$\Im = \frac{1}{2}\int_0^{\tau_r} \mathbf{e}_\mathbf{y}(t+\tau)^T \boldsymbol{Q}\, \mathbf{e}_\mathbf{y}(t+\tau) d\tau + \frac{1}{2}\int_0^{\tau_u} \mathbf{u}(t+\tau)^T \boldsymbol{R}\, \mathbf{u}(t+\tau) d\tau \tag{31}$$

where, $\boldsymbol{Q}\in\Re^{n\times n}$ is a positive semi-definite matrix and $\boldsymbol{R}\in\Re^{n\times n}$ is a positive definite matrix, τ_r and τ_u are respectively the observation horizon of the tracking error and the control horizon. We assume that the control signal is constant over the control horizon ($\mathbf{u}(t+\tau) = \mathbf{u}(t)$).
Using the same analysis, as in section 3.2.1, the cost function (31) can be rewritten as

$$\Im = \frac{1}{2}\left(\mathbf{Y}(t) - \mathbf{Y}_{\mathbf{r}}(t)\right)^T \mathbf{\Pi}\left(\mathbf{Y}(t) - \mathbf{Y}_{\mathbf{r}}(t)\right) + \frac{1}{2}\boldsymbol{R}\tau_u \mathbf{u}(t)^T \mathbf{u}(t) \tag{32}$$

where the new matrix $\mathbf{\Pi}$ is defined by

$$\mathbf{\Pi} = \int_0^{\tau_r} \mathbf{T}(\tau)^T \boldsymbol{Q}\mathbf{T}(\tau)\, d\tau = \begin{bmatrix} \mathbf{\Pi}_1 & \mathbf{\Pi}_2 \\ \mathbf{\Pi}_2^T & \mathbf{\Pi}_3 \end{bmatrix}$$

The necessary and sufficient condition (27) becomes

$$\left(\frac{\partial\left(\mathbf{D}_0(\mathbf{x}_1)^{-1}\mathbf{u}(t)\right)}{\partial \mathbf{u}(t)}\right)^T [\mathbf{\Pi}_2^T \quad \mathbf{\Pi}_3]\left(\mathbf{M}(t) - \mathbf{Y}_{\mathbf{r}}(t)\right) + \left(\frac{\partial\left(\mathbf{D}_0(\mathbf{x}_1)^{-1}\mathbf{u}(t)\right)}{\partial \mathbf{u}(t)}\right)^T \mathbf{\Pi}_3 \mathbf{D}_0(\mathbf{x}_1)^{-1}\mathbf{u}(t) + \boldsymbol{R}\tau_u \mathbf{u}(t) = 0 \tag{33}$$

$$\mathbf{D}_0(\mathbf{x}_1)^{-1^T} [\mathbf{\Pi}_2^T \quad \mathbf{\Pi}_3]\left(\mathbf{M}(t) - \mathbf{Y}_{\mathbf{r}}(t)\right) + \left(\mathbf{D}_0(\mathbf{x}_1)^{-1^T} \mathbf{\Pi}_3 \mathbf{D}_0(\mathbf{x}_1)^{-1} + \boldsymbol{R}\tau_u\right)\mathbf{u}(t) = 0$$

Then, the optimal control law is given by

$$\mathbf{u}(t) = -\left(\mathbf{D}_0(\mathbf{x}_1)^{-1^T} \mathbf{\Pi}_3 \mathbf{D}_0(\mathbf{x}_1)^{-1} + \boldsymbol{R}\tau_u\right)^{-1} \mathbf{D}_0(\mathbf{x}_1)^{-1^T} [\mathbf{\Pi}_2^T \quad \mathbf{\Pi}_3]\left(\mathbf{M}(t) - \mathbf{Y}_{\mathbf{r}}(t)\right) \tag{34}$$

4. Robust control based on uncertainties compensation

Robust control is considered among the high qualified methods in motion control. The goal of robust control is to maintain performance in terms of stability, tracking error, or other specifications despite inaccuracies present in the system. The robust motion control problem can be solved by designing an estimator to compensate the system uncertainties such as unknown external disturbances, unmodeled quantities and mismatched model (Spong et al., 2006; Kozłowski, 2004; Corriou, 2004; Feuer & Goodwin, 1989; Chen et al., 2000; Curk & Jezernik, 2001; Merabet & Gu, 2008; Curk & Jezernik, 2001).
The uncertainties compensation analysis will be developed for the linearizing control laws (15) and (30).
Using the Lie derivative analysis in (13) about the uncertainties, it can be verified that the relative degree for η is $r = 2$. Then, the control law becomes

$$\mathbf{u}(t) = -\mathbf{D}_0(\mathbf{x}_1)\left\{K_1(\mathbf{y} - \mathbf{y}_{\mathbf{r}}) + K_2(\dot{\mathbf{y}} - \dot{\mathbf{y}}_{\mathbf{r}}) - \mathbf{D}_0(\mathbf{x}_1)^{-1}\left(\mathbf{C}(\mathbf{x}_1, \mathbf{x}_2)\mathbf{x}_2 + \mathbf{G}(\mathbf{x}_1)\right) - \ddot{\mathbf{y}}_{\mathbf{r}}\right\} - \eta(t) \tag{35}$$

Usually the uncertainties η are unknown. Therefore, estimation is required to compute the control law and compensate their effects, and the robust control law is given by

$$\mathbf{u}(t) = -\mathbf{D}_0(\mathbf{x}_1)\left\{K_1(\mathbf{y}-\mathbf{y_r}) + K_2(\dot{\mathbf{y}}-\dot{\mathbf{y}}_\mathbf{r}) - \mathbf{D}_0(\mathbf{x}_1)^{-1}\left(\mathbf{C}(\mathbf{x}_1,\mathbf{x}_2)\mathbf{x}_2 + \mathbf{G}(\mathbf{x}_1)\right) - \ddot{\mathbf{y}}_\mathbf{r}\right\} - \eta_{est}(t) \quad (36)$$

There are several approaches to treat the robust control problem. In this chapter two methods will be discussed to design the uncertainties estimator; the first one is based on the theory of guaranteed stability of uncertain systems, while the second one is based on the model control law.

4.1 Estimator based on the theory of guaranteed stability of uncertain systems

In this section we will detail the so-called theory of guaranteed stability of uncertain systems, which is based on Lyapunov's second method (Spong et al., 2006).
Substituting the control input (36) in the robot model differentiation (13) plus uncertainties η, the tracking error dynamic of the closed loop system is given by

$$\ddot{\mathbf{e}}_\mathbf{y}(t) + K_2\dot{\mathbf{e}}_\mathbf{y}(t) + K_1\mathbf{e}_\mathbf{y}(t) = \mathbf{D}_0^{-1}(\mathbf{x}_1)\mathbf{e}_\eta(t) \quad (37)$$

where, $\mathbf{e}_\eta(t) = \eta(t) - \eta_{est}(t)$
In terms of tracking error, the state space model of the dynamic system (37) is given by

$$\dot{\bar{\mathbf{e}}} = A_1\bar{\mathbf{e}} + B\mathbf{D}_0^{-1}\mathbf{e}_\eta \quad (38)$$

where

$$\bar{\mathbf{e}} = \begin{bmatrix}\mathbf{e}_\mathbf{y} \\ \dot{\mathbf{e}}_\mathbf{y}\end{bmatrix}, \quad A_1 = \begin{bmatrix}0_{n\times n} & I_{n\times n} \\ -K_1 & -K_2\end{bmatrix}, \quad B = \begin{bmatrix}0_{n\times n} \\ I_{n\times n}\end{bmatrix}$$

Since $\{K_1, K_2\}>0$, the matrix A_1 is Hurwitz. Thus, for any symmetric positive define matrix Q, there exists a symmetric positive defined matrix P satisfying the Lyapunov equation

$$A_1^T P + PA_1 = -Q \quad (39)$$

Let define the Lyapunov function candidate

$$V = \bar{\mathbf{e}}^T P\bar{\mathbf{e}} + \mathbf{e}_\eta^T \Gamma \mathbf{e}_\eta \quad (40)$$

where Γ is a positive definite symmetric matrix.
Using equations (38) and (39), the time derivative of V is given by

$$\dot{V} = -\bar{\mathbf{e}}^T Q\bar{\mathbf{e}} + 2\mathbf{e}_\eta^T\{(\mathbf{D}_0^{-1})^T B^T P\bar{\mathbf{e}} + \Gamma\dot{\mathbf{e}}_\eta\} \quad (41)$$

If we define

$$\dot{\mathbf{e}}_{\eta} = -\Gamma^{-1}(\mathbf{D}_0^{-1})^T B^T P\overline{\mathbf{e}} \tag{42}$$

Since there is no information about uncertainties variations, it can be assumed that $\dot{\eta}$ (*t*) = 0 (Chan et al., 1999). This assumption does not necessarily mean a constant variable, but that the changing rate in every sampling interval should be slow.
From (42), the dynamics of the uncertainties estimation is given by

$$\dot{\eta}_{est} = \Gamma^{-1}(\mathbf{D}_0^{-1})^T B^T P\overline{\mathbf{e}} \tag{43}$$

Using the definition (42), it follows that the Lyapunov function V satisfies $\dot{V} < 0$ along solution trajectories of equation (6) because

$$\dot{V} = -\overline{\mathbf{e}}^T Q\overline{\mathbf{e}} \tag{44}$$

This guarantees that $\overline{\mathbf{e}}$ (*t*) and $\mathbf{e}_{\eta}(t)$, and therefore $\eta_{est}(t)$, are bounded.
The uncertainties estimation equation (42) can also be written as

$$\eta_{est}(t) = \int \Gamma^{-1}(\mathbf{D}_0^{-1})^T B^T P\overline{\mathbf{e}}(t)dt \tag{45}$$

4.2 Estimator based on the model control law

From the model dynamic of the robot (4), an estimator for uncertainties can be defined as

$$\begin{aligned}\dot{\eta}_{est} &= L\mathbf{D}_0^{-1}(\mathbf{x}_1)(\eta - \eta_{est}) \\ &= -L\mathbf{D}_0^{-1}(\mathbf{x}_1)\eta_{est} + L\left(\ddot{\mathbf{y}} + \mathbf{D}_0^{-1}(\mathbf{x}_1)\mathbf{C}_0(\mathbf{x}_1,\mathbf{x}_2)\dot{\mathbf{y}} + \mathbf{D}_0^{-1}(\mathbf{x}_1)\mathbf{G}_0(\mathbf{x}_1) - \mathbf{D}_0^{-1}(\mathbf{x}_1)\mathbf{u}\right)\end{aligned} \tag{46}$$

where, $L = \ell * I_{n\times n} \in \Re^{n\times n}$ is a matrix gain, and ℓ is a positive constant (Chen et al., 2000; Chen et al., 1999; Feng et al., 2002).
From the equation (40) and with the assumption $\dot{\eta}$ (*t*) = 0, the dynamic of the uncertainty estimator is given by

$$\dot{\mathbf{e}}_{\eta} + L\mathbf{D}_0^{-1}\mathbf{e}_{\eta} = 0 \tag{47}$$

Since $L > 0$ and $\mathbf{D}_0 > 0$, it can be easily verified that the tracking error of the estimation converge to zero.
Substituting the control law (30) in the observer equation (46), the dynamic of the uncertainties estimation is given by

$$\dot{\eta}_{est}(t) = L\left(\ddot{\mathbf{e}}_{\mathbf{y}}(t) + K_2\dot{\mathbf{e}}_{\mathbf{y}}(t) + K_1\mathbf{e}_{\mathbf{y}}(t)\right) \tag{48}$$

Integrating the equation (48), the uncertainties estimation is defined by

$$\eta_{est}(t) = L\left(\dot{\mathbf{e}}_{\mathbf{y}}(t) + K_2\mathbf{e}_{\mathbf{y}}(t) + K_1\int \mathbf{e}_{\mathbf{y}}(t)\,dt\right) \tag{49}$$

The advantage of the uncertainties estimator (49) compared to (45) is that it contains an integral action, which allows achieving zero steady state error for constant reference inputs and disturbances (Corriou, 2004; Cavallo et al., 1999; Feuer, & Goodwin 1989).

5. Nonlinear observer based state estimation

The computation of a model control law, such as linearization control and model predictive control, requires angular position and velocity measurements. In the practical robotic systems all the generalized coordinates can be precisely measured by the encoder for each joint, but the velocity measurements obtained through the tachometers are easily perturbed by noises. To overcome these physical constraints, a nonlinear observer can be used for state estimation (Kozłowski, 2004; Rodriguez-Angeles & Nijmeijer, 2004; Heredia & Yu, 2000).
The state space model of rigid robot (6), (7) can be reorganized as

$$\begin{cases} \dot{\mathbf{x}}_1 = \mathbf{x}_2 \\ \dot{\mathbf{x}}_2 = f(\mathbf{x}_1, \mathbf{x}_2) + g(\mathbf{x}_1)\mathbf{u} + g(\mathbf{x}_1)\eta \\ \mathbf{y} = \mathbf{x}_1 \end{cases} \tag{50}$$

where, $\mathbf{x}_1 = \mathbf{q}$; $\mathbf{x}_2 = \dot{\mathbf{q}}$, $\mathbf{y}$ is the measurable position vector.

$$\begin{cases} f(\mathbf{x}_1, \mathbf{x}_2) = -\mathbf{D}_0(\mathbf{x}_1)^{-1}\left(\mathbf{C}_0(\mathbf{x}_1, \mathbf{x}_2)\mathbf{x}_2 + \mathbf{G}_0(\mathbf{x}_1)\right) \\ g(\mathbf{x}_1) = \mathbf{D}_0(\mathbf{x}_1)^{-1} \end{cases} \tag{51}$$

The nonlinear state observer based on high gain for the system (50) can be designed, to estimate angular positions and velocities, as

$$\begin{cases} \dot{\hat{\mathbf{x}}}_1 = \hat{\mathbf{x}}_2 + H_1(\mathbf{y} - \hat{\mathbf{x}}_1) \\ \dot{\hat{\mathbf{x}}}_2 = f(\hat{\mathbf{x}}_1, \hat{\mathbf{x}}_2) + g(\hat{\mathbf{x}}_1)\mathbf{u} + g(\hat{\mathbf{x}}_1)\hat{\eta}_{est} + H_2(\mathbf{y} - \hat{\mathbf{x}}_1) \end{cases} \tag{52}$$

where, $\hat{\mathbf{x}}_i$ (i = 1, 2) are the estimated states; $\hat{\eta}_{est}$ is the estimated uncertainty carried out from (49) with estimated states.
The estimated nonlinear functions *f*(.) and *g*(.) are given by:

$$\begin{cases} f(\hat{\mathbf{x}}_1, \hat{\mathbf{x}}_2) = -\mathbf{D}_0(\hat{\mathbf{x}}_1)^{-1}\left(\mathbf{C}_0(\hat{\mathbf{x}}_1, \hat{\mathbf{x}}_2)\hat{\mathbf{x}}_2 + \mathbf{G}_0(\hat{\mathbf{x}}_1)\right) \\ g(\hat{\mathbf{x}}_1) = \mathbf{D}_0(\hat{\mathbf{x}}_1)^{-1} \end{cases} \tag{53}$$

From (50) and (52), the observer error dynamic is given, in matrix form, by

$$\dot{\tilde{\mathbf{e}}}(t) = H\tilde{\mathbf{e}}(t) + W\delta(t) \tag{54}$$

where,

$$\tilde{\mathbf{e}} = \begin{bmatrix} \tilde{\mathbf{e}}_y \\ \dot{\tilde{\mathbf{e}}}_y \end{bmatrix} = \begin{bmatrix} \mathbf{x}_1 - \hat{\mathbf{x}}_1 \\ \mathbf{x}_2 - \hat{\mathbf{x}}_2 \end{bmatrix}, H = \begin{bmatrix} -H_1 & I_{n\times n} \\ -H_2 & 0_{n\times n} \end{bmatrix}, W = \begin{bmatrix} 0_{n\times n} & I_{n\times n} \end{bmatrix},$$

$H_1 = h_1 * I_{n\times n}$, $H_2 = h_2 * I_{n\times n}$, and h_1, h_2 are positive constants.

$\delta(.)$ is the disturbance term in the state observer. It is given by

$$\delta(.) = f(.) - f(\hat{.}) + \left(g(\hat{.}) - g(\hat{.})\right)\mathbf{u} + g(.)\eta - g(\hat{.})\hat{\eta}_{est} \tag{55}$$

The observer gain H is chosen to be a Hurwitz matrix in order to guarantee the convergence. In the presence of δ, the observer gains are adjusted as

$$h_1 = \frac{\gamma_1}{\varepsilon},\ h_2 = \frac{\gamma_2}{\varepsilon^2} \tag{56}$$

where, $0 < \varepsilon << 1$, and γ_1, γ_2 are positive constants.
This adjustment allows making the transfer function from δ to the error small so that the estimation error is not sensitive to the modeling error (Wang & Gao, 2003; Khalil, 1999; Heredia & Yu, 2000).

6. Global stability of the closed loop system

This section aims to discuss the global convergence of the tracking error for the closed loop system. The theory of stability, based on Lyapunov method, is used to prove the global stability of the robot system controlled by the robust estimated nonlinear control law.
The propriety of boundedness of the model elements of the robot are given from (Spong et al., 2006).

- Since $\mathbf{D}_0(\mathbf{q}) > 0$, it can be assumed that $\underline{\mathbf{D}} \le \left\| \mathbf{D}_0(\mathbf{q})^{-1} \right\| \le \overline{\mathbf{D}}$, where $\underline{\mathbf{D}}$, $\overline{\mathbf{D}}$ are positive constants.
- The matrix $\mathbf{C}_0(\mathbf{q},\dot{\mathbf{q}})$ is linear on $\dot{\mathbf{q}}(t)$ and bounded on $\mathbf{q}(t)$. Therefore, $\left\| \mathbf{C}_0(\mathbf{q},\dot{\mathbf{q}}) \right\| \le \alpha_1 \left\| \dot{\mathbf{q}} \right\|;\ \alpha_1 \in \Re^+$.
- The vector $\mathbf{G}_0(\mathbf{q})$ satisfies $\| \mathbf{G}_0(\mathbf{q}) \| \le \alpha_2$; $\alpha_2 \in \Re^+$.
- All variations $\Delta(.)$ are bounded.
- The signals $\mathbf{q}_\mathbf{r}, \dot{\mathbf{q}}_\mathbf{r}, \ddot{\mathbf{q}}_\mathbf{r}$ are bounded, such as $\left\| \mathbf{q}_\mathbf{r}(t) \right\| \le r_0$, $\left\| \dot{\mathbf{q}}_\mathbf{r}(t) \right\| \le r_1$ and $\left\| \ddot{\mathbf{q}}_\mathbf{r}(t) \right\| \le r_2$.
- The disturbance term $\delta(.)$ is smaller than the state observer error. Thus, $\left\| \delta(t) \right\| \le \gamma \left\| \tilde{\mathbf{e}}(t) \right\|$.
- The vector function $f(\mathbf{x}_1, \mathbf{x}_2)$ is Lipschitz with respect to $\mathbf{x}_2$. Thus, there exists $\kappa > 0$ such that
  $\left\| f(\mathbf{x}_1, \mathbf{x}_2) - f(\mathbf{x}_1, \dot{\mathbf{q}}_{ref}) \right\| \le \kappa \left\| \mathbf{x}_2 - \dot{\mathbf{q}}_{ref} \right\| = \kappa \left\| \mathbf{e}_2 \right\|;$
  $\forall (\mathbf{x}_1, \mathbf{x}_2) \in \Re^n \times \Re^n$

Integrating the state observer in the control loop, the control law is carried out with the state estimation (Rodriguez-Angeles & Nijmeijer, 2004). Based on state observer (52), the model control law (36) becomes

$$\mathbf{u}(t) = -\mathbf{D}_0(\hat{\mathbf{x}}_1)\left\{K_1\hat{\mathbf{e}}_\mathbf{y} + K_2\dot{\hat{\mathbf{e}}}_\mathbf{y} - \mathbf{D}_0(\hat{\mathbf{x}}_1)^{-1}\left(\mathbf{C}(\hat{\mathbf{x}}_1,\hat{\mathbf{x}}_2)\hat{\mathbf{x}}_2 + \mathbf{G}(\hat{\mathbf{x}}_1)\right) - \ddot{\mathbf{y}}_{1\mathbf{r}}\right\} - \hat{\eta}_{est}(t) \tag{57}$$

where, $\hat{\mathbf{e}}_\mathbf{y} = \hat{\mathbf{y}} - \mathbf{y}_\mathbf{r}$

The disturbance estimator (49) is carried out with the state estimation, which is expressed by

$$\hat{\eta}_{est}(t) = L\left(\dot{\hat{\mathbf{e}}}_\mathbf{y}(t) + K_2\hat{\mathbf{e}}_\mathbf{y}(t) + K_1\int\hat{\mathbf{e}}_\mathbf{y}(t)\,dt\right) \tag{58}$$

Substituting the control law (57) in the equation (13) with estimated states from (52), we have the dynamic of the tracking error as

$$\ddot{\hat{\mathbf{e}}}_\mathbf{y}(t) + K_2\dot{\hat{\mathbf{e}}}_\mathbf{y}(t) + K_1\hat{\mathbf{e}}_\mathbf{y}(t) = H_2C\tilde{\mathbf{e}}(t) \tag{59}$$

Using the state space form, the tracking error system (59) can be written as

$$\dot{\hat{\mathbf{e}}}(t) = A\hat{\mathbf{e}}(t) + B\tilde{\mathbf{e}}(t) \tag{60}$$

where,

$$\hat{\mathbf{e}} = \begin{bmatrix}\hat{\mathbf{e}}_\mathbf{y}\\ \dot{\hat{\mathbf{e}}}_\mathbf{y}\end{bmatrix},\ A = \begin{bmatrix}0_{n\times n} & I_{n\times n}\\ -K_1 & -K_2\end{bmatrix},\ B = \begin{bmatrix}H_2 & 0_{n\times n}\\ 0_{n\times n} & 0_{n\times n}\end{bmatrix}$$

Using the estimated states of the robot model (52), the uncertainty estimator (46) can be re-designed as

$$\dot{\hat{\eta}}_{est} = -L\mathbf{D}_0^{-1}(\hat{\mathbf{x}}_1)\eta_{est} + L\left(\ddot{\hat{\mathbf{y}}} + \mathbf{D}_0^{-1}(\hat{\mathbf{x}}_1)\mathbf{C}_0(\hat{\mathbf{x}}_1,\hat{\mathbf{x}}_2)\dot{\hat{\mathbf{y}}} + \mathbf{D}_0^{-1}(\hat{\mathbf{x}}_1)\mathbf{G}_0(\hat{\mathbf{x}}_1) - \mathbf{D}_0^{-1}(\hat{\mathbf{x}}_1)\mathbf{u}\right) \tag{61}$$

where, $\hat{\eta}_{est}$ is the uncertainty estimation based on estimated states.

So, the new error dynamic of the uncertainty estimator, based on estimation state model (52), is given by

$$\dot{\hat{\mathbf{e}}}_\eta(t) = -LH_2\tilde{\mathbf{e}}(t) \tag{62}$$

where, $\hat{\mathbf{e}}_\eta = \eta - \hat{\eta}_{est}$ is uncertainty error

From (62), it can be noticed that the convergence of the uncertainty estimator is related to the convergence of the state observer.

Under the state space form, the tracking error of the global system (robot + state observer + controller) can be carried out using error models (54) and (60)

$$\dot{\mathbf{e}}(t) = \mathbf{A}\mathbf{e}(t) + \mathbf{B}\delta(t) \tag{63}$$

where,

$$\mathbf{e} = \begin{bmatrix} \hat{\mathbf{e}} \\ \tilde{\mathbf{e}} \end{bmatrix}, \mathbf{A} = \begin{bmatrix} A & B \\ 0_{2n\times 2n} & H \end{bmatrix}, \mathbf{B} = \begin{bmatrix} 0_{n\times 2n} & W \end{bmatrix}$$

By an appropriate choice of control parameters K_i (i=1,…,n) and state observer gain H, it can be ensured that the matrix **A** is Hurwitz. Therefore, for any symmetric positive define matrix **Q**, there exists a symmetric positive defined matrix **P** satisfying the Lyapunov equation

$$\mathbf{A}^T\mathbf{P} + \mathbf{P}\mathbf{A} = -\mathbf{Q} \tag{64}$$

Let define the Lyapunov function candidate

$$V = \mathbf{e}^T\mathbf{P}\mathbf{e} \tag{65}$$

Its derivative is given by

$$\dot{V} = -\mathbf{e}^T\mathbf{Q}\mathbf{e} + 2\mathbf{e}^T\mathbf{P}\mathbf{B}\,\delta \tag{66}$$

Using the relationship

$$\lambda_{\min}(\mathbf{Q})\|\mathbf{e}\|^2 \le \mathbf{e}^T\mathbf{Q}\mathbf{e} \le \lambda_{\max}(\mathbf{Q})\|\mathbf{e}\|^2 \tag{67}$$

where $\lambda_{\min}(\mathbf{Q})$, $\lambda_{\max}(\mathbf{Q})$ denote the minimum and the maximum eigenvalues, respectively, of the matrix **Q**.
We have

$$\dot{V} \le -\lambda_{\min}(\mathbf{Q})\|\mathbf{e}\|^2 + 2\|\delta\|\,\|\mathbf{B}\|\,\|\mathbf{P}\|\,\|\mathbf{e}\| \tag{68}$$

Using the last propriety of boundedness, we have

$$\begin{aligned} \dot{V} &\le -\lambda_{\min}(\mathbf{Q})\|\mathbf{e}\|^2 + 2\gamma\lambda_{\max}(\mathbf{P})\,\|\mathbf{e}\|^2 \\ &\le -\left(\lambda_{\min}(\mathbf{Q}) - 2\gamma\lambda_{\max}(\mathbf{P})\right)\|\mathbf{e}\|^2 \end{aligned} \tag{69}$$

The condition, $\dot{V}$ is definite negative, is held when $\gamma < \dfrac{\lambda_{\min}(\mathbf{Q})}{2\lambda_{\max}(\mathbf{P})}$. Therefore, by LaSalle's invariance theorem, the origin is asymptotically stable. The global asymptotic stability of the estimated closed loop system with uncertainties is guaranteed.

7. Simulation results and discussion

We consider the two-link rigid robot manipulator to illustrate the performances of the nonlinear model predictive controller (36) with uncertainties compensation expressed by the observers (45) and (49) respectively (Merabet & Gu, 2008). The structure of the robot system driven by nonlinear model based control law is shown in figure 1.

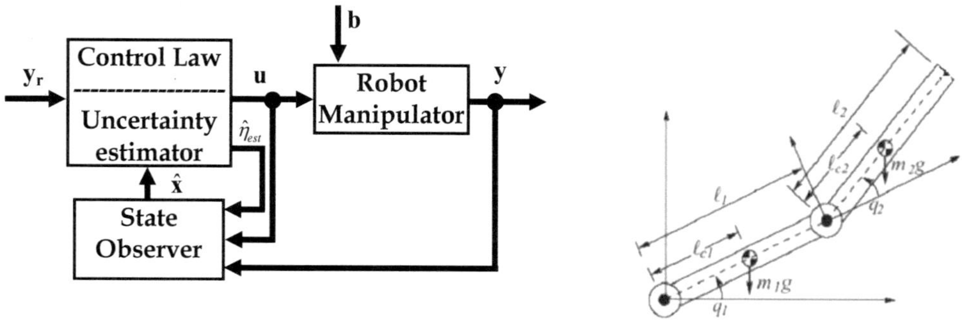

Fig. 1. Nonlinear model based control for two link rigid robot manipulator

The elements of the two-link robot model are given by

$$D_{11} = m_1 l_{c1}^2 + m_2(l_1^2 + l_{c2}^2 + 2l_1 l_{c2} \cos q_2) + I_1 + I_2;$$
$$D_{12} = D_{21} = m_2(l_{c2}^2 + l_1 l_{c2} \cos q_2) + I_2;$$
$$D_{22} = m_2 l_{c2}^2 + I_2$$
$$C_{11} = -(m_2 l_1 l_{c2} \sin q_2)\dot{q}_2;$$
$$C_{12} = -(\dot{q}_1 + \dot{q}_2) m_2 l_1 l_{c2} \sin q_2;$$
$$C_{21} = (m_2 l_1 l_{c2} \sin q_2)\dot{q}_1; C_{22} = 0$$
$$G_1 = (m_1 l_{c1} + m_2 l_1) g \cos q_1 + m_2 l_{c2} g \cos(q_1 + q_2);$$
$$G_2 = m_2 l_{c2} g \cos(q_1 + q_2)$$

For i = 1, 2, q_i denotes the joint angle; m_i denotes the mass of link i; l_i denotes the length of link i; l_{ci} denotes the distance from the previous joint to the center of mass of link i; and I_i denotes the moment of inertia of link i (Spong et al., 2006).
The nominal values of robot parameters are:

Link 1: m_1 = 10 kg, l_1 = 1 m, l_{c1} = 0.5 m, I_1 = 10/12 kg-m^2.
Link 2: m_2 = 5 kg, l_2 = 1 m, l_{c2} = 0.5 m, I_2 = 5/12 kg-m^2.

The model is simulated with a sample time of 10^{-4}s and the initial values of angular positions and velocities are $\hat{\mathbf{x}}$ = [0.1 rad 0 rad/s]^T for the state observer, and for the robot model $\mathbf{x}$ (0) = [0 rad 0 rad/s]^T. The parameters of the controller, uncertainties observers and state observer are chosen by trial and error in order to achieve accurate performances.
First, the tracking performance of robot system, driven by the nonlinear model predictive control law (36), is tested without the uncertainties observer. The robot system is affected by external disturbance **b**, which has the value 10 in the time interval [0.5s 4s]. The disturbance term is included in the robot model and the information about it is not taken into account when carrying out the control law. The value of prediction time is τ_r = 10^{-3}s. The state observer gain is taken as H_1=H_2= [10^4 0; 0 10^8]. Figure 2 shows the result for angular positions and tracking errors. It can be seen that small tracking errors, for both joints, are successfully achieved. However, steady errors occur in the system responses. The present

situation can be explained by the fact that the control law has no information about the external disturbances in order to compensate their effects. Figure 3 illustrates the induced control torque applied to robot manipulator. Note that the control torque lie inside the saturation limits. From figure 4, we can observe that the estimation errors are good although the presence of steady errors in the responses. As shown in the equation of state observer (52), the information about uncertainties is needed to have an accurate performance.

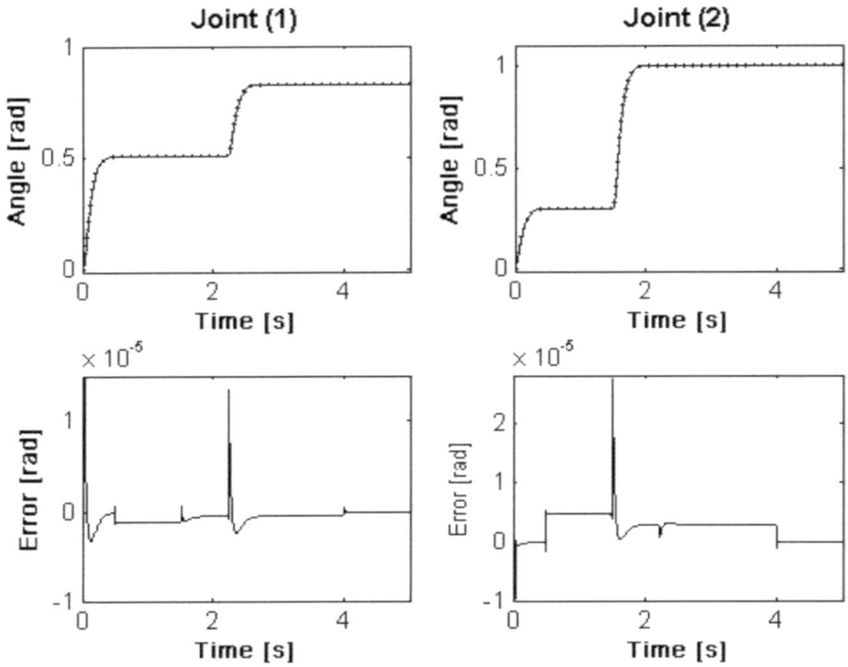

Fig. 2. Angular positions and tracking errors of distributed system without uncertainties compensator. .... reference, ——estimate

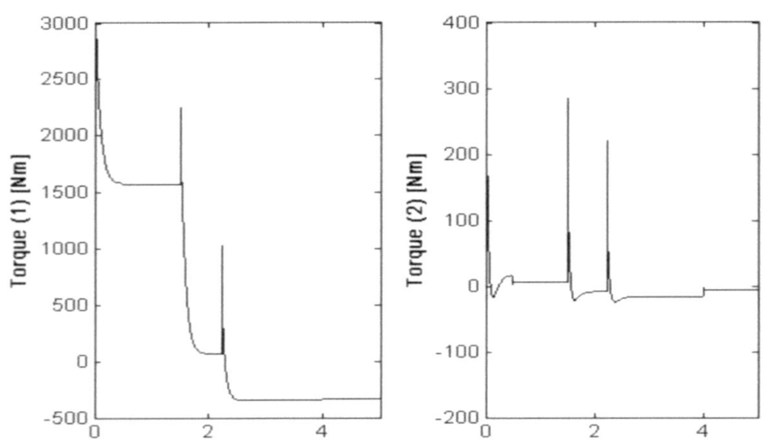

Fig. 3. Induced torque control produced from the nonlinear model predictive controller

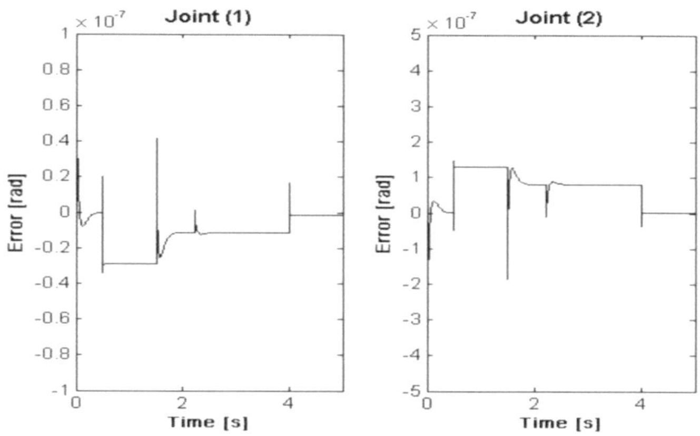

Fig. 4. Error estimation of the nonlinear state observer (controller without compensation)

Then, the uncertainties observers (45) and (49) are applied to the control law (36) respectively. The matrix P has the value [10^6 0; 0 10^6] and $\Gamma = I_{n\times n}$ for the observer (45), and L=[10^2 0; 0 10^2] for the observer (49). Figure 5 illustrates the angular positions and tracking errors of the system with uncertainties compensators. The steady error is vanished completely with the compensator (49), which means that the disturbance is well rejected. However, with the compensator (45), the steady error is only reduced compared with the results in figure (2).The elimination of steady errors by the compensator (49) can be explained by the presence of the integral action. It is known in control theory that an integral action achieves zero steady state error for constant reference inputs and disturbances. The same observation can be noticed in the result of state estimation errors shown in figure 6, where the uncertainties, carried out by the compensator (49), are included in state observer.

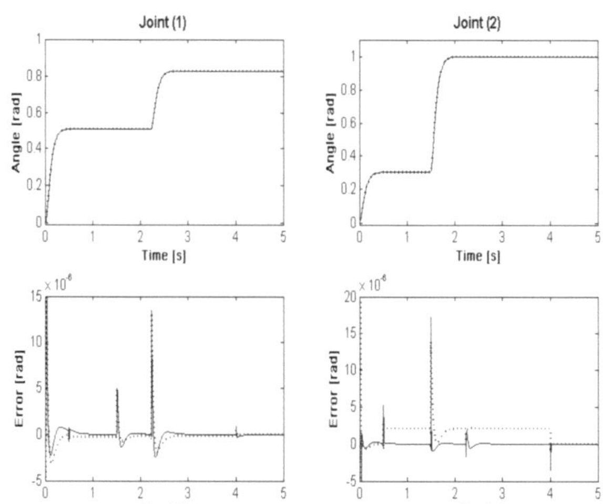

Fig. 5. Angular positions and tracking errors of distributed system with uncertainties compensator. ….. reference, —·—· estimate with compensator (45) , —— estimate with compensator (49) ….. tracking error with compensator (45), —— tracking error with compensator (49)

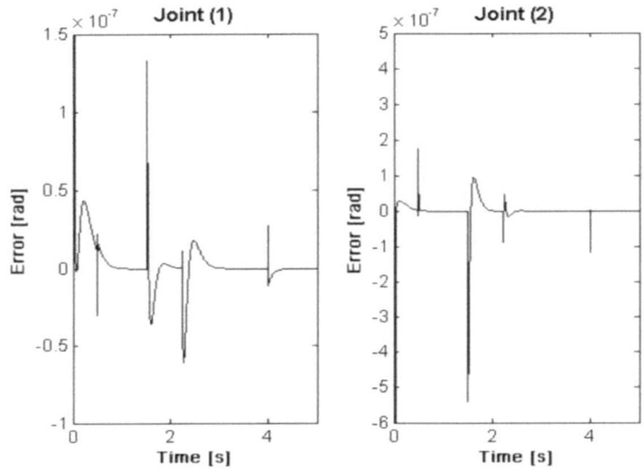

Fig. 6. Error estimation of the nonlinear state observer (controller with compensation)

In case of mismatched model, an unknown load carried by the robot is regarded as part of the second link, then the parameters m_2, l_{c2}, I_2 will change, $m_2+\Delta m_2$, $l_{c2}+\Delta l_{c2}$, $I_2+\Delta I_2$, respectively. The variations values are $\Delta m_2 = 1.5$, $\Delta l_{c2} = 0.125$, $\Delta I_2 = 1/12$. Also, the friction (Coulomb and viscous friction) given by $\mathbf{F}_r(\mathbf{x}_2)=\mathbf{F}_c\, sign(\mathbf{x}_2)+\mathbf{F}_v\mathbf{x}_2$, with values $\mathbf{F}_c=\mathbf{F}_v = diag(5, 5)$, are added to the robot model. The same parameters values of the controller, disturbance observer (45) and state observer are used as declared above. However, the gain of compensator (49) is decreased L = [70 0; 0 70]. As shown in figure 7, in case of the compensator (49), the errors occur in transient response, for this raison the gain is decreased, then reach zero. In case of the compensator (45), the errors in transient response are smaller than in the first case, but they do not reach zero like the other observer.

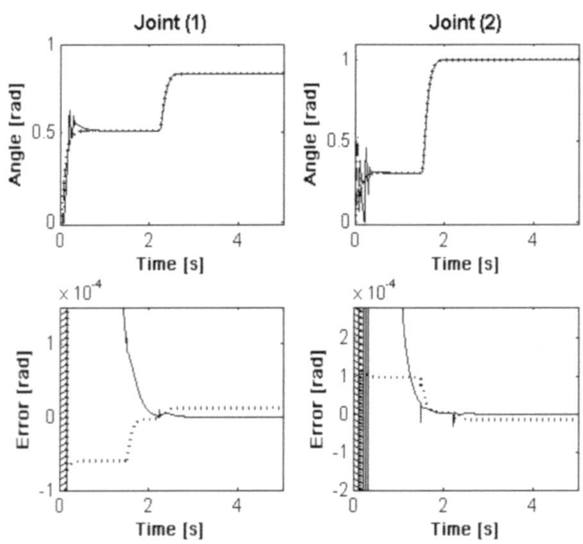

Fig. 7. Angular positions and tracking errors of mismatched model with uncertainties compensator. ….. reference, –·–·- estimate with compensator (45) , —— estimate with compensator (49). ….. tracking error with compensator (45), —— tracking error with compensator (49)

The results show that the tracking performance is successfully achieved and the effect of external disturbance is well rejected with the compensator (49). Concerning the unmodeled quantities and parametric uncertainties, the nonlinear model predictive controller, combined with uncertainties observer, deals well with their variations. It can be mentioned also that the state estimation, given by the nonlinear observer, is accurate for the tracking performance. The accuracy of the estimated nonlinear model predictive control combined with the compensator (49) is justified by the presence of the integral action, which eliminates steady state error.

8. Conclusions and future work

This chapter has presented some methods of advanced nonlinear control for robot systems. However, to cover all issues related to nonlinear control in detail will demand more than a chapter. The study has focused on model based control where a model dynamic of the process is needed to carry out the control law.

Two nonlinear control approaches have been detailed in this work. A feedback linearization control based on input-output linearization has been developed using differential-geometric methods for nonlinear systems. Then, a model based predictive control has been discussed for a nonlinear control design to robot manipulators. The predictive control law minimizes a cost function for the control trajectory over a future time horizon. The control solution has been analytically derived, with no need of an online optimization, which enables fast real-time implementation.

Because of the uncertainties present in the system, a robustness strategy has been studied to enhance the tracking response of the system. Two methods have been investigated to deal with system uncertainties. One method is based on the theory of guaranteed stability of uncertain systems, which results to an observer taking information from the system tracking errors. The other one is an observer derived from the nonlinear model control law. It contains an integral action on system tracking errors. This type of control strategy is robust with respect to modeling errors, very effective in disturbance rejection, and gives no steady error caused by either parameters uncertainties or external disturbances.

The development of these control strategies is related to the dynamic model of the process. In case of missing information about the system states, a version of control law based on state has been carried out with the quantities, angular positions and velocities, issued from a nonlinear state estimator. It has been shown that the tracking performance is achieved successfully when the uncertainties are well compensated.

The issue of global stability of the closed loop system has been proved analytically via Lyapunov stability theory.

The nonlinear control laws developed in this chapter are based on a dynamic model of the process. However, it is well known that mathematical representation of a dynamic model does not refer accurately to the reality. This is why it is very important to add to the control strategy a robustness analysis in order to compensate the uncertainties present in the dynamic model. As an alternative of this approach, intelligent control based on the process behavior can be considered as a solution for tracking motion of robot manipulators. Intelligent control achieves automation via the emulation of behavioral intelligence such as biological intelligence (e.g., the use of neural networks and genetics for control); the use of human's knowledge to design a smart control methodology (fuzzy control). This research area is very wide and the issues of modeling, mathematical stability, convergence and robustness analysis for learning systems must be investigated to design an accurate controller.

9. References

Bordon, C. & Camacho, EF. (1998). A generalized predictive controller for a wide class of industrial processes. *IEEE Transactions on Control Systems Technology 6*(3), pp. 372-387.

Cavallo, A.; De Maria, G. & Nistri, P. (1999). Robust control design with integral action and limited rate control. *IEEE Transactions on Automatic Control 44*(8), pp. 1569-1572.

Chen, W-H.; Balance, DJ.; Gawthrop, PJ. & O'Reilly, J. (2000). A nonlinear disturbance observer for robotic Manipulators. *IEEE Transactions on Industrial Electronics 47* (4), pp. 932-938.

Chen, W-H.; Balance, DJ.; Gawthrop, PJ.; Gribble JJ. & O'Reilly, J. (1999). Nonlinear PID predictive controller. *IEE Proceedings Control Theory application 146*(6), pp. 603-611.

Corriou, JP. (2004). *Process Control. Theory and Applications*. Springer, London, UK.

Curk, B. & Jezernik, K. (2001). Sliding mode control with perturbation estimation: Application on DD robot mechanism. *Robotica 19*(10), pp. 641-648.

Feng, W.; O'Reilly, J. & Balance, DJ. (2002). MIMO nonlinear PID predictive controller. *IEE Proceedings Control Theory application 149*(3), pp. 203-208.

Feuer, A. & Goodwin, GC. (1989). Integral Action in Robust Adaptive Control. *IEEE Transactions on Automatic Control 34*(10), pp. 1082-1085.

Hedjar, R. & Boucher, P. (2005). Nonlinear receding-horizon control of rigid link robot manipulators. *International Journal of Advanced Robotic Systems 2*(1), pp. 015-024.

Hedjar, R.; Toumi, R.; Boucher, P. & Dumur, D. (2002). Feedback nonlinear predictive control of rigid link robot manipulator. *Proceedings of the American Control Conference*, Anchorage AK, pp. 3594-3599.

Heredia, JA. & Yu, W. (2000). A high-gain observer-based PD control for robot manipulator. *Proceedings of the American Control Conference*, Chicago, Illinois, USA, pp. 2518-2522.

Isidori, A. (1985). *Nonlinear Control Systems: An Introduction*. Springer-Verlag, Berlin, New York.

Isidori, A., & Ruberti, A. (1984). On the synthesis of linear input output responses for nonlinear systems. Systems and Control Letters, 4(1), pp. 17-22.

Khalil, HK. (1999). High-gain observers in nonlinear feedback control. New Directions in nonlinear observer design. *Lecture Notes in Control and Information Sciences 24*(4), pp. 249-268.

Klančar, G. & Škrjanc, I. (2007). Tracking-error model-based predictive control for mobile robots in real time. *Robotics and Autonomous Systems 55*, pp. 460-469.

Kozłowski, K. (2004), *Robot motion and control. Recent developments*. Springer, London, UK.

Merabet, A. & Gu, J. (2008). Robust nonlinear predictive control based on state estimation for robot manipulator. *International Journal of Applied Mathematics and Mechanics*, Vol. 5, No. 1, 48-64.

Merabet, A. & Gu, J. (2008). Estimated feedback linearization controller with disturbance compensator for robotic applications. *The Mediterranean Journal of Measurement and Control*, Vol. 4, No. 3, 101-110

Nijmeijer, H., & A. J. van der Schaft. (1990). Nonlinear Dynamic Control Systems. Springer-Verlag, New York, 1990.

Richalet, J. (1993). Industrial Applications of Model Based Predictive Control. *Automatica 29*(5), pp. 1251-1274.

Rodriguez-Angeles, A. & Nijmeijer, H. (2004). Synchronizing Tracking Control for Flexible Joint Robots via Estimated State Feedback. *ASME Journal of Dynamic Systems, Measurement and Control 126*, pp. 162-172.

Spong, MW.; Hutchinson, S. & Vidyasagar, M. (2006). *Robot modeling and control*. John Wiley & Sons, USA.

Vivas, A. & Mosquera, V. (2005). Predictive functional control of a PUMA robot. *ICGST, ACSE 05 Conference*, Cairo, Egypt, pp. 372-387.

Wang, W. & Gao, Z. (2003). A comparison study of advanced state observer design techniques. *Proceedings of the American Control Conference*, Denver, Colorado, USA, pp. 4754-4759.

8

Modelling of HDD head positioning systems regarded as robot manipulators using block matrices

Tomasz Trawiński and Roman Wituła
Silesian University of Technology
Poland

1. Introduction

The modern hard disk drive (HDD) head positioning systems may be regarded as excellent example of mechatronics systems consisting of different components - subsystems: electrical (driving motors - actuators, flexible printed circuits, writing and reading heads etc.), mechanical (bearings, air bearings, swing arm, suspensions etc.) and electronics (power amplifiers, control system etc.). In this chapter we will focus only on the mechanical system of head positioning system, which usually consist of following components: main swing arm (so-called E-block) fixed with moving coil of the VCM (voice coil motor) motor, suspensions of the sliders, sliders with writing and reading heads. All of these elementary components (assumed to be stiff and rigid enough) are connected to each other and these connections may be treated as rotary or prismatic joints. Modern head positioning systems, beside fundamental VCM motor (which plays the role of fundamental source of driving torque), are equipped with additional micro-actuators for better track tracing or rejection of the internal and external disturbances. Usually the head positioning systems equipped with auxiliary micro-actuators are called as dual-stage (DS) positioning system. The dual-stage positioning systems may be classified according to kinds of auxiliary micro-actuators and place where the macro-actuators are attached to kinematic chain of head positioning system. For auxiliary micro-actuators very often the PZT (piezoelectric) micro-actuators or electrostatic MEMS (micro-electro-mechanical systems) micro-actuators are used. PZT micro-actuators are often placed between and tip of E-block and the beginning of slider and head suspension (Rotunno et al., 2006) and actuate the suspension or play the role of the sensor for vibration sensing (Huang et al., 2005), or they are placed between suspension and slider and drive slider directly (Hong et al., 2006). The MEMS micro-actuator in HDD head positioning systems, for the sake of relatively small dimensions and small generated forces (torque), are put between suspension and slider (drive slider directly) or they are placed between slider and heads (drive the heads directly). Some different and very interesting ideas for direct drives of HDD heads is presented in (Schultz, 2007), where thermal expansion of head pole tip is used for approaching the head to disk surface during write process. All presented mathematical models of head positioning systems are prepared for analysis of its cooperation only with one side of data disk. Some of proposed mathematical models take into account mutual interactions between auxiliary micro-actuator and main VCM motor, but they do not take into account this mutual interactions when positioning system is equipped with more then one micro-actuator. In this chapter mathematical model of head positioning system cooperating with more then one side of data disk will be derived. Firstly the real kinematic structure of HDD positioning system will be decomposed into elementary joints and links, that allows writing them in terms of open kinematics chain of small robot manipulators. Next the kinematic chains will be extended to multilayer kinematics chains. Secondly for multilayer kinematic chains of positioning system (using commonly known mathematical methods used in robot dynamics) mathematical model will be formulated and written in terms of Lagrange equations. During the mathematical model formulation the block matrix will be used for inverting the dynamics matrix of head positioning system. Finally the general method for dynamic matrix inversion for more complicated kinematic chains of positioning system will be given and carefully discussed.

2. Kinematic structure of HDD positioning system

2.1 Exemplary modern head positioning systems

The mechanical construction of head positioning system is strongly related with data areal density. Data areal density denotes the amounts of data which may be stored on unit area of data disk, and it is expressed in gigabits per square inch (Gb/in^2). Nowadays the data areal density in HDD reaches values up to several hundreds of Gb/in^2 (Trawiński & Kluszczyński 2008). For small areal densities (less then few tens of Gb/in^2) and resulting relatively width data track, the commonly used structures of HDD positioning systems were equipped with only one driving motor - VCM motor. Such a system forms one degree of freedom (1 DoF) mechanical system, usually equipped with massive E-block. Basic structure of positioning head system is presented in Fig.1; this positioning system operates with data areal densities reaching 15 Gb/in^2.

Fig. 1. Head positioning system for low data areal densities

In the Fig. 1 the numbers in the circles denote: (1) - E-block, (2) - sliders and heads suspensions, (3) - flexible printed circuit, (4) - VCM motor armature coil, (5) - pivot. This positioning system cooperated with spindle system consisting of set of three data discs. Another example of head positioning system which cooperates with data areal densities reaching 50 Gb/in^2 is presented in Fig. 2. Number in circles denotes this same part of positioning system like this presented in Fig. 1.

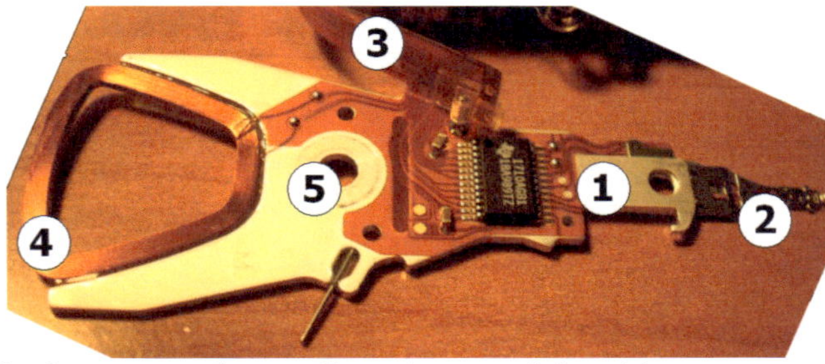

Fig. 2. Head positioning system for medium data areal densities

It is easy to spot that system presented in Fig. 2 is ready to cooperate only with one side of data disc. Basing on this two discussed positioning system it is very difficult to eliminate or suppress all internal disturbances such like: suspension air induced vibration, pivot nonlinearities, structural resonances of E-block, repeatable run-out (RRO) and non-repeatable run-out (NRRO) of data track due to rotation of spindle system (Wang & Krishnamurthy, 2006). This problem may be solved for example by utilising auxiliary macro-actuators or improvements in control system. (Chen & Horowitz, 2001) for this reason were proposed the silicon actuated suspension over PZT and achieved range of head motion (generated by PZT micro-actuator) about ±1.3 μm at ±30 V supply. In Fig. 3 exemplary and simplified view of PZT micro-actuator for suspension actuation (which is placed between end tip of E-block and beginning of suspension) is presented (Jiang et al., 2007), (Rotunno et al., 2006).

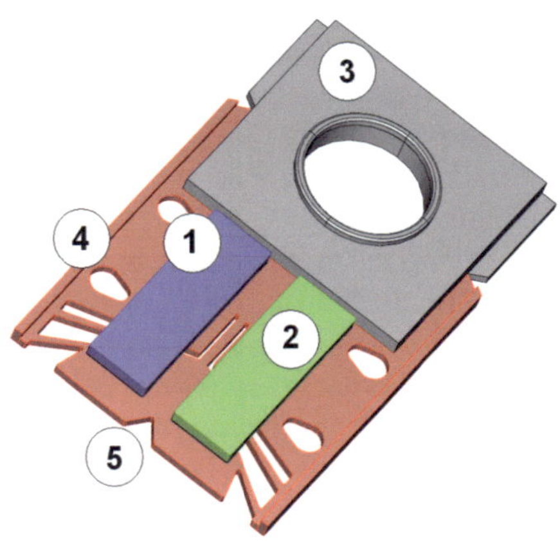

Fig. 3. Exemplary PZT micro-actuator for suspension actuation

In Fig. 3 the numbers in the circles denote: (1) and (2) - PZT stripes acting (extends) in opposite directions under voltage supply, (3) - end tip of E-block, (4) - flexible part - gimbals, (5) - place for suspension attaching.
Another example of PZT actuated suspension is presented in (Koganezawa & Hara, 2001) but this time the sheer-mode PZT where used to generate head motion. They achieved the motion of head in range of ± 0.5 μm at ± 30 V supply.
Placing the PZT micro-actuator between suspension and end tip of E-block may result (during PZT operation) in structural resonance excitation in suspension, thus certain

proposition in (Hong et al. 2006) was given for direct drive of the slider. Exemplary view of PZT actuated slider is presented in Fig. 4.

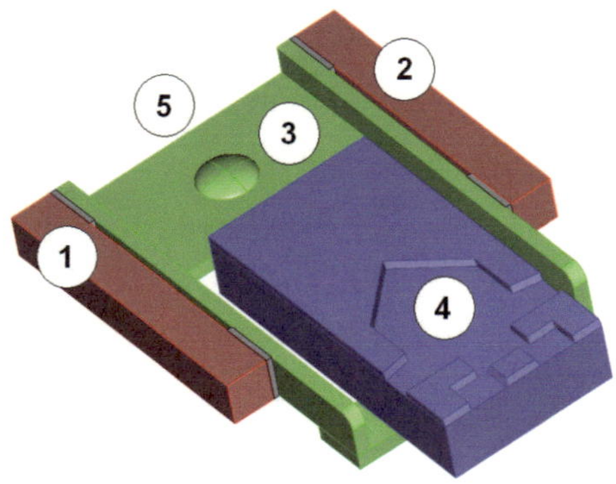

Fig. 4. Exemplary PZT actuated slider

In Fig. 4 the numbers in the circles denote: (1) and (2) - PZT stripes which are bending under voltage supply, (3) - flexible part - gimbals, (4) - slider, (5) - place for suspension attaching. Using higher rate of sampling frequencies in servo system, reducing NRRO and RRO, reducing air induced vibration due to spoiler (attached over spinning disk) is possible to push the border of areal density when the auxiliary actuation will be inevitable (Sugaya, 2006).

2.2 Decomposition of head positioning system into joints and links

The mechanical subsystem of head positioning system, as it was mentioned before, may be represented as a set of stiff links connected by rotary or prismatic joints with one degrees of freedom. In chosen joint may act torque (or forces) generated by main motor and auxiliary micro-actuators. Such a set of links and joints is very similar to kinematic chain of small robot manipulators. But the fundamental difference is in range of motions arising in every joints. In the robot manipulators joints the ranges of motion are usually high and almost equal to each other. In case of head positioning systems the angular rages of joint motions differ very much. Motion of the main joint usually covers the angle between 30 to 40 degrees for 3.5 inch disk drives, for smaller drives equipped with 2 inch disk (or smaller in diameter) the range of angular motion may by smaller then 30 degrees. For another joints the values for angular motion are small (usually few degrees or fraction of degree or micro-degree, except (Sarajlic et al., 2009)) and depending on kind of auxiliary micro-actuator and its place in kinematic chain (Sarajlic et al., 2009). For these reasons we may assume forgoing correlation between real parts of head positioning system and hypothetical robot manipulator kinematic chain:

- the fundamental kinematic pairs consist of HDD frame and housing, E-block and VCM armature coil which are connected by rotating joint (pivot). On this joint act torque generated by VCM motor and torque (force) generated by flexible printed circuit (this effects will be further omitted for simplicity). The first rotary joint will be treated as perfect rotary joint (with one degrees of freedom) without any nonlinearities (this is very serious simplify assumptions). Problem of pivot nonlinearities is discussed in

(Ohno & Horowitz, 2005). The fundamental link (HDD frame and housing) will be called as "base" and second link (E-block, VCM coil) will be called as "bough".

- The second kinematic pair consists of E-block and suspension connected with rotary joint. On this joint may acts torque (force) generated by PZT micro-actuator or alternatively spring torque (force), because connection between E-block and suspension is flexible in predominant cases.
- The third kinematic pair consists of suspension and slider which are connected by gimbals, but this kind of connections may be alternatively regarded as rotary or prismatic. Slider forms the fourth link.
- The fourth kinematic pair consists of slider and heads (reading head - magneto-resistive and writing heads - electromagnetic) connected to each other by means of prismatic joint. The set of heads forms the fifth link.

All links from third to fifth constitute the "branch" links. Number of links belonging to branch may vary and it depends on simplification made on kinematic chain of head positioning system. In illustrative way, the correlations between parts of real head positioning system and its robot manipulator kinematic chain equivalent representation is shown in Fig. 5.

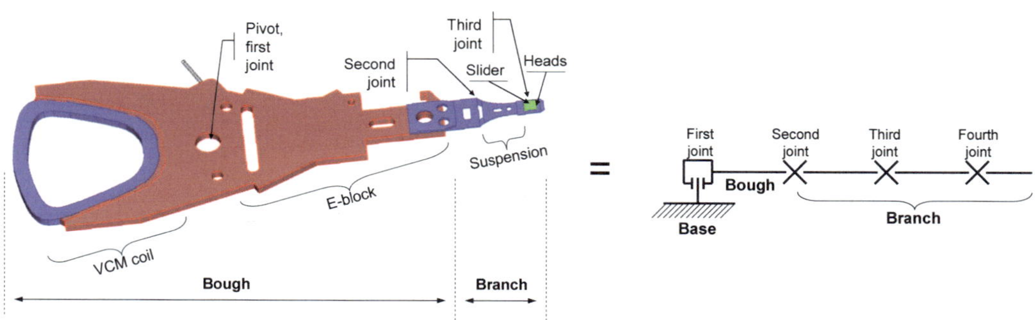

Fig. 5. Positioning system represented as manipulator

On the right side in Fig. 5 the simplified kinematic chain diagram is presented. The signs "x" denote joints which may be either rotating or prismatic. The first joint (in Fig.5) is rotating with rotating axis lie in the plain of drawings (and it is perpendicular to the bough). Basing on this schematic representation same kinematic chains of head positioning system presented in (Huang & Horowitz, 2005) may be represented in forthcoming figures. The head positioning system presented in (Huang & Horowitz, 2005) uses two sources of torque (force), one generated by VCM motor and the second (force) is generated by MEMS micro-generator (which drives directly the slider), so the simplified schematic representation of this manipulator is presented in Fig.6 and consists of two rotary joints (with rotating axis perpendicular to each other) and one prismatic joint (represented MEMS actuated slider). The second joint with rotating axis perpendicular to the plain of page is, in Fig.6, denoted by circle. The square with cross inside denotes, in Fig. 6, the prismatic joint.

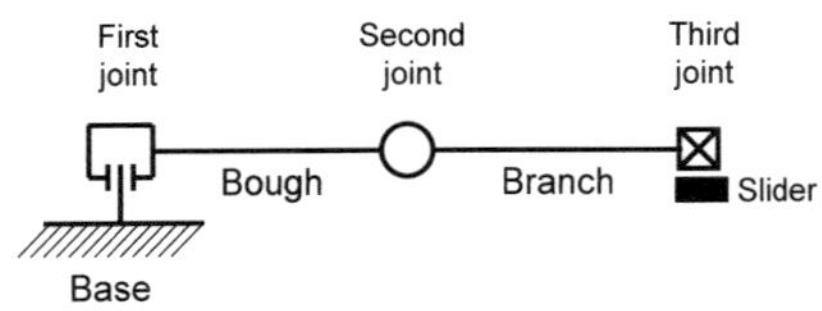

Fig. 6. Manipulator with 3 degrees of freedom

In Fig.6 in first joint acts VCM motor but second joint is not actuated - this is passive joint (Trawiński, 2007). The schematic representation of manipulator of positioning system which may be constructed basing on (Sarajlic et al., 2009) is presented in Fig. 7.

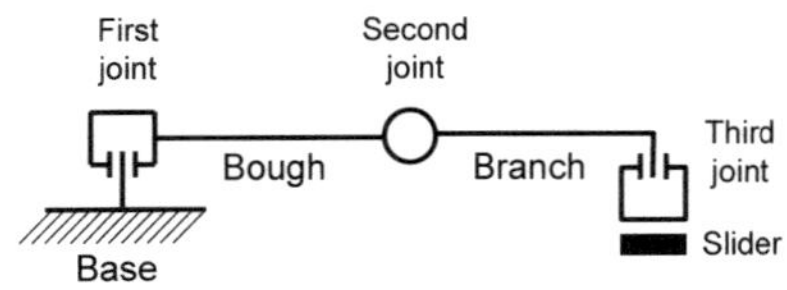

Fig. 7. Manipulator with 3 degrees of freedom

Kinematic chain of above mentioned manipulator consists of three rotating joints. The last rotating joint is driven by electrostatic MEMS 3-phase stepper motor (Sarajlic et al., 2009). This solution allows to compensate skew of reading and writing heads (Sarajlic et al., 2009). The second joint, as it was in previous case, is not actuated.

2.3 Multilayer head positioning system

Most of presented and known mathematical models of head positioning system assume its cooperation only with one side of data disk. It allows for analysis of internal dynamic interaction between parts of positioning systems, but does not take into consideration mutual interactions between multiple sets of suspensions and heads which cooperate with other sides of data disk. These mutual interactions may be shown only when the kinematics chain will be extended by another suspensions, sliders and heads which cooperate with the other sides of data disk. In our simplified schematic representation, presented in Figs. 6 & 7, for preparing them to cooperate with two sides of data disk, we have to add another branch. If it is done the schematic representation of kinematic chains look like these presented in Fig.8.

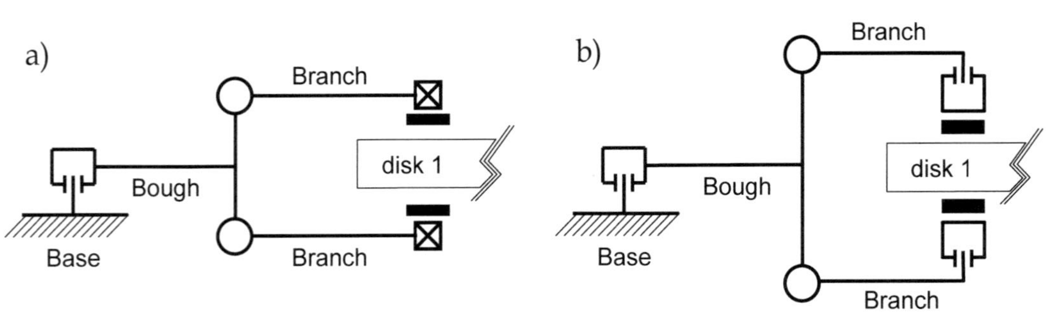

Fig. 8. Schematic view of positioning system manipulators capable of cooperation with two sides of data disk

When the head positioning system cooperates with set of two disk, and each side of disks is in use for data storing, then simplified kinematics chain will consist of four branches. Similarly for more additional disk the number of branches increases gradually for two branches for each disk. The positioning system now consists of multiple layer, one layer include single branch and one disk side. Such positioning system with multiple number of layers included branches, disk sides and bough will be further called as multilayer head positioning system. The individual branches, which belong to different layers, will be denoted by small letters starting from "a", every link of chosen branch will be assigned by number (starting form "2" upwards) and letter coincide with branch sign. The joints belonging to chosen branch will be denoted by letter coincide with the sign of branch and number (starting from "2" upwards). Bough link will be denoted by "1" and first joint by "(1)". The simplified schema of exemplary multilayer head positioning system, with symbols of branches etc., is presented in Fig. 9.

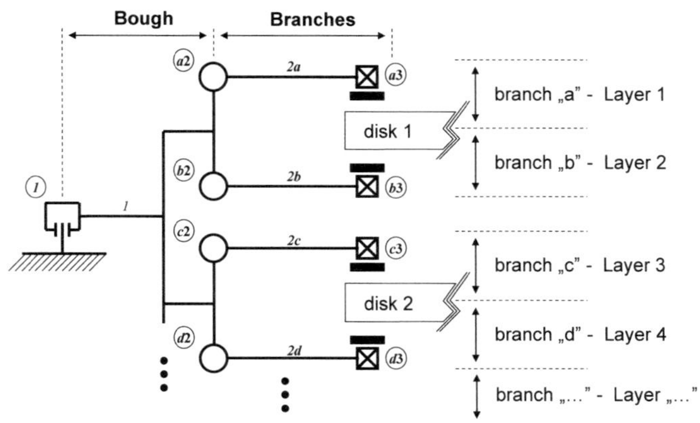

Fig. 9. Simplified schema of multilayer manipulator

For further consideration the multilayer kinematics chain presented in Figs. 8a) & 9 will be chosen, on its background the mathematical model will be formulated. The analysis of kinematics chains presented in Figs. 7 & 8.b) is discussed in (Trawiński & Kluszczyński, 2008).

3. Mathematical model of multilayer head positioning system

3.1 Dynamics matrix formulation

In matrix notation the Lagrange equations are given by:

$$\mathbf{D}\ddot{\mathbf{q}} + \mathbf{C}\dot{\mathbf{q}} + \mathbf{G} = \boldsymbol{\tau} \tag{1}$$

here and subsequently $\mathbf{D}$ - denotes dynamic matrix, $\mathbf{C}$ - centrifugal and Coriolis force matrix, $\mathbf{G}$ - gravitational forces and torque, τ - driving torque vector, $\mathbf{q}$ - vector of generalized displacements.

The Lagrange equation is a set of second order differential equation, and for convenient calculation should be rewritten into normal form:

$$\dot{\boldsymbol{\omega}} = \mathbf{D}^{-1}(\boldsymbol{\tau} - \mathbf{C}\boldsymbol{\omega} - \mathbf{G})$$
$$\dot{\mathbf{q}} = \boldsymbol{\omega} \tag{2}$$

which consist of the set of two first order differential equations (second set is related with generalized speeds). In equations (1) and (2) is present the dynamic matrix, which can be derived from kinetic energy of whole multilayer head positioning system. The kinetic energy of this system may be expressed in quadratic form which include the dynamic matrix, as follows:

$$T = \frac{1}{2}\dot{\mathbf{q}}^T\mathbf{D}\dot{\mathbf{q}} \tag{3}$$

When motion analysis of "j" – link will be carried out according to its centre of masses, then kinetic energy may be expressed in sun of two terms – translational and rotational terms of kinetic energy, therefore whole kinetic energy is equal to:

$$T = T_L + T_R = \frac{1}{2}\sum_j m_{cj}\mathbf{v}_{cj}^T\mathbf{v}_{cj} + \frac{1}{2}\sum_j \boldsymbol{\omega}_{cj}^T\mathbf{I}_{cj}\boldsymbol{\omega}_{cj} \tag{4}$$

where T_L, T_R – denote translational and rotational terms of kinetic energy respectively; m_{cj} – mass of the link concentrated in his mass centre; $\mathbf{v}_{cj}$ – vector of linear speed; $\boldsymbol{\omega}_{cj}$ – vector of angular speed of mass centre; $\mathbf{I}_{cj}$ – mass moment of inertia of link at mass centre.
The vector of linear speed occurring in equation (3) may be expressed in terms of the jacobian matrices, which describes the relation between joint generalised velocities and velocities of centre of masses expressed in coordinate system fixed with the base, thus:

$$\mathbf{v}_{cj} = \mathbf{J}_{vcj}\dot{\mathbf{q}} \tag{5}$$

where $\mathbf{J}_{vcj}$ – jacobian matrix of linear speed,
and for angular rotating speed, we have:

$$\boldsymbol{\omega}_{cj} = \mathbf{R}_{cj}^T\mathbf{J}_{\omega cj}\dot{\mathbf{q}} \tag{6}$$

where $\mathbf{J}_{wcj}$ – jacobian matrix of angular speed; $\mathbf{R}_{cj}$ – matrix of rotation (part of homogenous transformation matrices) of chosen link mass centre.
For jacobian matrices calculation and homogenous transformation matrices related with head positioning system refer to (Trawiński, 2007). Substituting (5) and (6) into equation (4), for "j" links of total "m" number of links (their centre of masses) of multilayer head positioning system, one may write:

$$T = \frac{1}{2}\dot{\mathbf{q}}^T\sum_{j=1}^{m}(m_{cj}\mathbf{J}_{vcj}^T\mathbf{J}_{vcj} + \mathbf{J}_{\omega cj}^T\mathbf{R}_{cj}\mathbf{I}_{cj}\mathbf{R}_{cj}^T\mathbf{J}_{\omega cj})\dot{\mathbf{q}} \tag{7}$$

The above derived equation may be expressed in terms of bough kinetic energy component and branches kinetic energy components:

$$T = \frac{1}{2}\dot{\mathbf{q}}_0^T\{m_{c1}\mathbf{J}_{vc1}^T\mathbf{J}_{vc1} + \mathbf{J}_{\omega c1}^T\mathbf{R}_{c1}\mathbf{I}_{c1}\mathbf{R}_{c1}^T\mathbf{J}_{\omega c1}\}\dot{\mathbf{q}}_0 + \frac{1}{2}\sum_g \dot{\mathbf{q}}_g^T\{\sum_{j=2}^{n}(m_{cj}\mathbf{J}_{vcgs}^T\mathbf{J}_{vcgs} + \mathbf{J}_{\omega cgs}^T\mathbf{R}_{cgs}\mathbf{I}_{cgs}\mathbf{R}_{cgs}^T\mathbf{J}_{\omega cgs})\}\dot{\mathbf{q}}_g \tag{8}$$

where m_{c1}, m_{cj} – masses of bough and "j" links of branches; $\mathbf{J}_{vc1}$, $\mathbf{J}_{vcj}$ – (3×n) dimensional jacobian matrices of linear speed; $\mathbf{J}_{wc1}$, $\mathbf{J}_{wcj}$ – (3×n) dimensional jacobian matrices of rotary speed; I_{c1}, I_{cgs} – (3×3) dimensional mass moment of inertia matrices of bough and branches "j" links mass centres respectively; g – subscript denotes branch sign; $\mathbf{q}_0$ – vector of generalised displacement of first joint – only one quotients q_1 is not equal zero; $\mathbf{q}_g$ – vector of generalised displacement of all branches joints (in this vector q_1 also is present); n – sum of number of degrees of freedom of bough and single branch respectively.
Now the expressions in curly bracket in equation (8) allow us to write the dynamic matrices in form:

$$\mathbf{D} = \left[\begin{array}{c|c|c|c} \mathbf{k} & \mathbf{a}_k & \mathbf{b}_k & \cdots \\ \hline \mathbf{a}_k^T & \mathbf{a} & \mathbf{0} & \cdots \\ \hline \mathbf{b}_k^T & \mathbf{0} & \mathbf{b} & \ddots \\ \hline \vdots & \vdots & \ddots & \ddots \end{array}\right] \tag{9}$$

Above presented matrix is a block – symmetric matrix, which consist of sub – matrices **k**, **a**, **b**, … and $\mathbf{a}_k$, $\mathbf{b}_k$, …. The physical interpretation of this sub – matrices is as follows:

- **k** – self inertial components of bough, it is (1×1) dimensional matrix, which k_{11} elements is expressed in form:

$$k_{(11)} = m_{c1}\sum_{i=1}^{3} J_{vc1\_i1}^2 + I_{zc1}(\sum_{i=1}^{3} J_{\omega c1\_i1} r_{c1\_i3})^2 + \sum_g\sum_{s=2}^{n} m_{cgs}\sum_{i=1}^{3} J_{vcgs\_i1}^2 + \sum_g\sum_{s=2}^{n} I_{zgs}(\sum_{i=1}^{3} J_{\omega cgs\_i1} r_{cgs\_i3})^2 \tag{10}$$

  where: J_{vc1_i1}, J_{wc1_i1}, J_{vcgs_i1}, J_{wgs_i1} – elements of jacobian matrices: linear and rotating speed of bough, linear and rotating speeds of branches respectively; $r_{c1\_i\,3}$, $r_{cgs\_i\,3}$ – $(i,3)$ elements of rotation matrix of homogenous transformation related to appropriate mass centre.
- **a**, **b**, … – square matrices in which internal diagonal quotients representing the self inertial components of branches. The quotients which lie above diagonal represent mutual inertial couplings between joints of chosen branch. This matrix components are given by:
  - diagonal components for c ≥ 2 (c – denotes columns of block matrix (9)):

$$g_{(c-1,c-1)} = \sum_{s=2}^{n}\left(m_{cgs}\sum_{i=1}^{3} J_{vcgs\_ic}^2 + I_{zgs}(\sum_{i=1}^{3} J_{\omega cgs\_ic} r_{cgs\_i3})^2\right) \tag{11}$$

- above diagonal components for r ≠ c and r ≥ 2 and c > 2 (r – denotes rows of block matrix (9)):

$$g_{(r,c)} = \sum_{s=2}^{n}\left(m_{cgs} \sum_{i=1}^{3} \prod_{j\in\{r,c\}} J_{vcgs\_ij} + I_{zgs} \prod_{j\in\{r,c\}} \sum_{i=1}^{3} J_{\omega cgs\_ij} r_{cgs\_i3} \right) \tag{12}$$

- **a**$_k$, **b**$_k$, … – row ((1×(n-1)) dimensional) matrices representing inertial mutual couplings between joints of branch and bough (the branch-bough inertial couplings). The components of this matrices are, for r = 1 and c ≥ 2, as follows:

$$g_{k(r,c-1)} = \sum_{s=2}^{n}\left(m_{cgs} \sum_{i=1}^{3} \prod_{j\in\{r,c\}} J_{vcgs\_ij} + I_{zgs} \prod_{j\in\{r,c\}} \sum_{i=1}^{3} J_{\omega cgs\_ij} r_{cgs\_i3} \right) \tag{13}$$

In the case of multilayer head positioning system, presented in Fig.9 (where it was assumed that it cooperates with set of two data disk), the dynamic block matrix is expressed by:

$$\mathbf{D} = \left[\begin{array}{c|cc|cc|cc|cc|c}
k_{11} & a_{k11} & a_{k12} & b_{k11} & b_{k12} & c_{k11} & c_{k12} & d_{k11} & d_{k11} & \cdots \\
\hline
 & a_{11} & 0 & 0 & 0 & 0 & 0 & 0 & 0 & \cdots \\
 & & a_{22} & 0 & 0 & 0 & 0 & 0 & 0 & \cdots \\
\hline
 & & & b_{11} & 0 & 0 & 0 & 0 & 0 & \cdots \\
 & & & & b_{22} & 0 & 0 & 0 & 0 & \cdots \\
\hline
 & & & & & c_{11} & 0 & 0 & 0 & \cdots \\
 & & & & & & c_{22} & 0 & 0 & \cdots \\
\hline
 & & & & & & & d_{11} & 0 & \\
 & sym. & & & & & & & d_{22} & \ddots \\
\hline
 & & & & & & & & & \ddots
\end{array}\right] \tag{14}$$

What is worth to underlining, there exists mutual inertial couplings between each joint of branches and bough – because the elements of row matrices are different from zero. However there is not inertial coupling between branches joints – this is a consequence that rotating axis of second joints (of every branch) is parallel to translation axes of third joints (of every branch). The self inertial components of bough, for multilayer kinematics chain at the point, is given by:

$$\begin{aligned} k_{11} = {} & m_{c1}a_{c1}^2 + I_{zc1} + \sum_{g\in\{a,b,c,d\}} m_{cg2}(a_1 + a_{cg2}\,c_{g2})^2 + \sum_{g\in\{a,c\}} m_{cg3}((a_1 + a_{g2}c_{g2} - a_{cg3}s_{g2})^2 + d_{g3}^2) + \\ & + \sum_{g\in\{b,d\}} m_{cg3}((a_1 + a_{g2}c_{g2} + a_{cg3}s_{g2})^2 + d_{g3}^2) \end{aligned} \tag{15}$$

where a_1, a_{c1}, a_{g2}, a_{cg2}, d_{g3} – denotes adequately: length of bough link, position of mass centre of bough, length of seconds and position of mass centre of thirds links and elongation of prismatic joints of adequate "g" branches; m_{cg2}, m_{cg3} – denotes masses of adequate branches links; s_{g2}, c_{g2} - denotes the abbreviated notation of cosine and sine functions of second joints angles of appropriate branches. The self inertial components of g matrices (for "a", "b", "c" and "d" branches) is expressed by:

$$\begin{cases} g_{11} = m_{cg2}a_{cg2}^2 + I_{zg2} + m_{cg3}(a_{g2}^2 + a_{cg3}^2) + I_{zg3} \\ g_{22} = m_{cg3} \end{cases} \tag{16}$$

where I_{zg2}, I_{zg3} – denotes adequately mass moment of inertia of appropriate branches links. The branch-bough inertial couplings matrices components are as follows:

- for "a" and "c" branches:

$$\begin{cases} g_{k11} = -d_{g3}m_{cg3}(a_{cg3}c_{g2} + a_{g2}s_{g2}) \\ g_{k12} = -m_{cg3}(a_1 + a_{g2}c_{g2} - a_{cg3}s_{g2}) \end{cases} \tag{17}$$

- for "b" and "d" branches:

$$\begin{cases} g_{k11} = -m_{cg3}d_{g3}(-a_{cg3}c_{g2} + a_{g2}s_{g2}) \\ g_{k12} = -m_{cg3}(a_1 + a_{g2}c_{g2} + a_{cg3}s_{g2}) \end{cases} \tag{18}$$

Components of matrices for "a", "c" and "b", "d" branches differs, because the first two cooperate with top part of data disk, but the other two with bottom part of data disk.

3.2 Dynamics matrix inversion using block matrices

For dynamics block matrix inversion, one advantages may be taken of hers block structure which allows for her inversion with the help of block matrices. According to the definition of inverse matrix, we have:

$$\mathbf{D}^{-1}\mathbf{D} = \mathbf{1} \tag{19}$$

where **1** – denotes unity matrix, or:

$$\frac{adj\mathbf{D}_r}{\det \mathbf{D}_r}\mathbf{D}_r = \mathbf{1} \tag{20}$$

Multiplying then both sides of equation (20) by determinant of dynamic matrix we get following matrix equation expressed in terms of elementary sub-matrices (corresponding with bough and branches):

$$\begin{bmatrix} \mathbf{A}_{11} & \mathbf{A}_{12} & \mathbf{A}_{13} & \mathbf{A}_{14} & \mathbf{A}_{15} \\ \mathbf{A}_{21} & \mathbf{A}_{22} & \mathbf{A}_{23} & \mathbf{A}_{24} & \mathbf{A}_{25} \\ \mathbf{A}_{31} & \mathbf{A}_{32} & \mathbf{A}_{33} & \mathbf{A}_{34} & \mathbf{A}_{35} \\ \mathbf{A}_{41} & \mathbf{A}_{42} & \mathbf{A}_{43} & \mathbf{A}_{44} & \mathbf{A}_{45} \\ \mathbf{A}_{51} & \mathbf{A}_{52} & \mathbf{A}_{53} & \mathbf{A}_{54} & \mathbf{A}_{55} \end{bmatrix} \begin{bmatrix} \mathbf{k} & \mathbf{a}_k & \mathbf{b}_k & \mathbf{c}_k & \mathbf{d}_k \\ \mathbf{a}_k^T & \mathbf{a} & \mathbf{0} & \mathbf{0} & \mathbf{0} \\ \mathbf{b}_k^T & \mathbf{0} & \mathbf{b} & \mathbf{0} & \mathbf{0} \\ \mathbf{c}_k^T & \mathbf{0} & \mathbf{0} & \mathbf{c} & \mathbf{0} \\ \mathbf{d}_k^T & \mathbf{0} & \mathbf{0} & \mathbf{0} & \mathbf{d} \end{bmatrix} = \mathbf{1}\det\mathbf{D} \tag{21}$$

This equation, after multiplication give us a five sets of five matrix equations, with unknown sub-matrices $\mathbf{A}_{ij}$ of adjunction matrix. This sets of equation should be solved according to unknown $\mathbf{A}_{ij}$ sub-matrices. The exemplary set of five matrix equation – related with the first row of adjunction matrix and first column of dynamics matrix is presented below:

$$\begin{cases} \mathbf{A}_{11}\mathbf{k} + \mathbf{A}_{12}\mathbf{a}_k^T + \mathbf{A}_{13}\mathbf{b}_k^T + \mathbf{A}_{14}\mathbf{c}_k^T + \mathbf{A}_{15}\mathbf{d}_k^T = \mathbf{1}\det\mathbf{D} \\ \mathbf{A}_{11}\mathbf{a}_k + \mathbf{A}_{12}\mathbf{a} = \mathbf{0} \\ \mathbf{A}_{11}\mathbf{b}_k + \mathbf{A}_{13}\mathbf{b} = \mathbf{0} \\ \mathbf{A}_{11}\mathbf{c}_k + \mathbf{A}_{14}\mathbf{c} = \mathbf{0} \\ \mathbf{A}_{11}\mathbf{d}_k + \mathbf{A}_{15}\mathbf{d} = \mathbf{0} \end{cases} \tag{22}$$

After solving of this matrix equation (22) and the rest similar, we get (here is only the one of five sets of solution presented):

$$\begin{cases} \mathbf{A}_{11} = \det\mathbf{D}\ (\mathbf{k} - \mathbf{a}_k\mathbf{a}^{-1}\mathbf{a}_k^T - \mathbf{b}_k\mathbf{b}^{-1}\mathbf{b}_k^T - \mathbf{c}_k\mathbf{c}^{-1}\mathbf{c}_k^T - \mathbf{d}_k\mathbf{d}^{-1}\mathbf{d}_k^T)^{-1} \\ \mathbf{A}_{12} = -\det\mathbf{D}\ (\mathbf{k} - \mathbf{a}_k\mathbf{a}^{-1}\mathbf{a}_k^T - \mathbf{b}_k\mathbf{b}^{-1}\mathbf{b}_k^T - \mathbf{c}_k\mathbf{c}^{-1}\mathbf{c}_k^T - \mathbf{d}_k\mathbf{d}^{-1}\mathbf{d}_k^T)^{-1}\mathbf{a}_k\mathbf{a}^{-1} \\ \mathbf{A}_{13} = -\det\mathbf{D}\ (\mathbf{k} - \mathbf{a}_k\mathbf{a}^{-1}\mathbf{a}_k^T - \mathbf{b}_k\mathbf{b}^{-1}\mathbf{b}_k^T - \mathbf{c}_k\mathbf{c}^{-1}\mathbf{c}_k^T - \mathbf{d}_k\mathbf{d}^{-1}\mathbf{d}_k^T)^{-1}\mathbf{b}_k\mathbf{b}^{-1} \\ \mathbf{A}_{14} = -\det\mathbf{D}\ (\mathbf{k} - \mathbf{a}_k\mathbf{a}^{-1}\mathbf{a}_k^T - \mathbf{b}_k\mathbf{b}^{-1}\mathbf{b}_k^T - \mathbf{c}_k\mathbf{c}^{-1}\mathbf{c}_k^T - \mathbf{d}_k\mathbf{d}^{-1}\mathbf{d}_k^T)^{-1}\mathbf{c}_k\mathbf{c}^{-1} \\ \mathbf{A}_{15} = -\det\mathbf{D}\ (\mathbf{k} - \mathbf{a}_k\mathbf{a}^{-1}\mathbf{a}_k^T - \mathbf{b}_k\mathbf{b}^{-1}\mathbf{b}_k^T - \mathbf{c}_k\mathbf{c}^{-1}\mathbf{c}_k^T - \mathbf{d}_k\mathbf{d}^{-1}\mathbf{d}_k^T)^{-1}\mathbf{d}_k\mathbf{d}^{-1} \end{cases} \tag{23}$$

The obtained results should be divided by determinant of dynamic matrix. It is easy to spot that in the set of solutions appears common quotients, which since then will be called as heading matrix (element) $\mathbf{k}_1$:

$$\mathbf{k}_1 = (\mathbf{k} - \mathbf{a}_k\mathbf{a}^{-1}\mathbf{a}_k^T - \mathbf{b}_k\mathbf{b}^{-1}\mathbf{b}_k^T - \mathbf{c}_k\mathbf{c}^{-1}\mathbf{c}_k^T - \mathbf{d}_k\mathbf{d}^{-1}\mathbf{d}_k^T)^{-1} \tag{24}$$

The heading matrix is present also in diagonal elements of calculated adjunction matrix:

$$\begin{cases} \mathbf{A}_{22} = \det\mathbf{D}\ (\mathbf{a} - \mathbf{a}_k^T(\mathbf{k}_1^{-1} + \mathbf{a}_k\mathbf{a}^{-1}\mathbf{a}_k^T)^{-1}\mathbf{a}_k)^{-1} \\ \mathbf{A}_{33} = \det\mathbf{D}\ (\mathbf{b} - \mathbf{b}_k^T(\mathbf{k}_1^{-1} + \mathbf{b}_k\mathbf{b}^{-1}\mathbf{b}_k^T)^{-1}\mathbf{b}_k)^{-1} \\ \mathbf{A}_{44} = \det\mathbf{D}\ (\mathbf{c} - \mathbf{c}_k^T(\mathbf{k}_1^{-1} + \mathbf{c}_k\mathbf{c}^{-1}\mathbf{c}_k^T)^{-1}\mathbf{c}_k)^{-1} \\ \mathbf{A}_{55} = \det\mathbf{D}\ (\mathbf{d} - \mathbf{d}_k^T(\mathbf{k}_1^{-1} + \mathbf{d}_k\mathbf{d}^{-1}\mathbf{d}_k^T)^{-1}\mathbf{d}_k)^{-1} \end{cases} \tag{25}$$

After division the equation (25) by determinant of dynamic matrix, the diagonal elements of inverted block matrix are as follows:

$$\begin{cases} \mathbf{a}_1 = (\mathbf{a} - \mathbf{a}_k^T(\mathbf{k}_1^{-1} + \mathbf{a}_k\mathbf{a}^{-1}\mathbf{a}_k^T)^{-1}\mathbf{a}_k)^{-1} \\ \mathbf{b}_1 = (\mathbf{b} - \mathbf{b}_k^T(\mathbf{k}_1^{-1} + \mathbf{b}_k\mathbf{b}^{-1}\mathbf{b}_k^T)^{-1}\mathbf{b}_k)^{-1} \\ \mathbf{c}_1 = (\mathbf{c} - \mathbf{c}_k^T(\mathbf{k}_1^{-1} + \mathbf{c}_k\mathbf{c}^{-1}\mathbf{c}_k^T)^{-1}\mathbf{c}_k)^{-1} \\ \mathbf{d}_1 = (\mathbf{d} - \mathbf{d}_k^T(\mathbf{k}_1^{-1} + \mathbf{d}_k\mathbf{d}^{-1}\mathbf{d}_k^T)^{-1}\mathbf{d}_k)^{-1} \end{cases} \tag{26}$$

When all results will be collected together and written in matrix form, it forms the inverted block matrix of multilayer head positioning system consisted of four branches (which is presented in Fig. 9):

$$\mathbf{D}^{-1}=\left[\begin{array}{c|c|c|c|c} \mathbf{k}_1 & -\mathbf{k}_1\mathbf{a}_k\mathbf{a}^{-1} & -\mathbf{k}_1\mathbf{b}_k\mathbf{b}^{-1} & -\mathbf{k}_1\mathbf{c}_k\mathbf{c}^{-1} & -\mathbf{k}_1\mathbf{d}_k\mathbf{d}^{-1} \\ \hline & \mathbf{a}_1 & \mathbf{a}^{-1}\mathbf{a}_k^T\mathbf{k}_1\mathbf{b}_k\mathbf{b}^{-1} & \mathbf{a}^{-1}\mathbf{a}_k^T\mathbf{k}_1\mathbf{c}_k\mathbf{c}^{-1} & \mathbf{a}^{-1}\mathbf{a}_k^T\mathbf{k}_1\mathbf{d}_k\mathbf{d}^{-1} \\ \hline & & \mathbf{b}_1 & \mathbf{b}^{-1}\mathbf{b}_k^T\mathbf{k}_1\mathbf{c}_k\mathbf{c}^{-1} & \mathbf{b}^{-1}\mathbf{b}_k^T\mathbf{k}_1\mathbf{d}_k\mathbf{d}^{-1} \\ \hline & & & \mathbf{c}_1 & \mathbf{c}^{-1}\mathbf{c}_k^T\mathbf{k}_1\mathbf{d}_k\mathbf{d}^{-1} \\ \hline sym. & & & & \mathbf{d}_1 \end{array}\right] \tag{27}$$

If head positioning system, in point, will be equipped with three branches the block matrix presented by equation (27) shrinks to the first four block rows and block columns. Similarly when head positioning system will consist only with two branches, then inverted block matrix will decreased to three block rows and block columns. The relation between numbers of branches of multilayer head positioning system and construction of inversed dynamic block matrix in illustrative way is presented in Fig. 10.

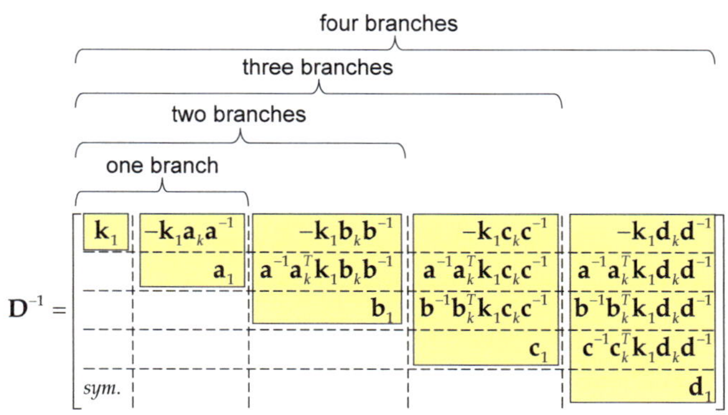

Fig. 10. Increase of inversed dynamic block matrix dimension v. branch number increase

And also the heading matrix change when number of branches will change. When multilayer head positioning system is equipped with three branches from equation (24) disappears the last term, but when he is equipped with two branches - disappears two last terms, etc. The relation between numbers of branches of multilayer head positioning system and construction of heading matrix (elements) is presented in Fig.11.

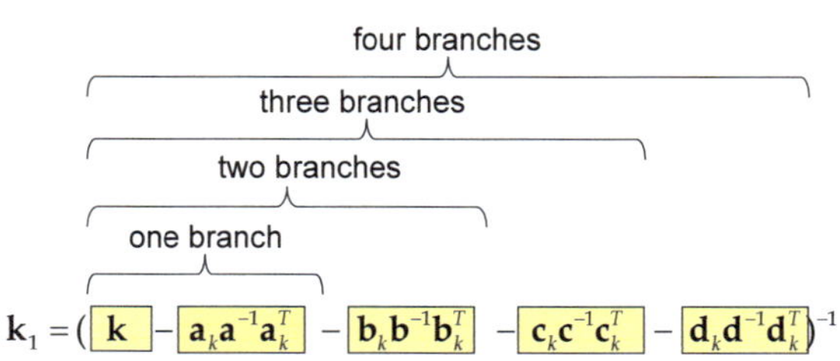

$$\mathbf{k}_1=(\mathbf{k}-\mathbf{a}_k\mathbf{a}^{-1}\mathbf{a}_k^T-\mathbf{b}_k\mathbf{b}^{-1}\mathbf{b}_k^T-\mathbf{c}_k\mathbf{c}^{-1}\mathbf{c}_k^T-\mathbf{d}_k\mathbf{d}^{-1}\mathbf{d}_k^T)^{-1}$$

Fig. 11. Increase of heading matrix components v. branch number increase

The dimension of heading matrix not changing versus increase of number of branches, her size is defined when dynamics block matrix is formulated and always is (1×1). Only numbers of components in equation (24) changes upon branch numbers change.
As my be observed in equation (27) and Fig.10 the rest elements of inversed dynamic matrix my by easily derived. The heading matrix in every block columns (except the first one) of inversed dynamic block matrix is right hand multiplied by product of two matrices - inversed self inertial components matrix of the branch (given by general equation (11) and (12)) and transposed branch-bough inertial couplings matrix (given by general equation (13)), which actually lie in desired column before dynamics matrix inversion.
Every block rows (except the first one) of inverted dynamic block matrix should be left hand multiplied by product of two matrices - inversed self inertial matrix of branch and transposed branch-bough inertial couplings matrix, which actually lie in desired column before dynamics matrix inversion. In illustrative way the deriving the rest components of inversed dynamics block matrix of multilayer head positioning system is presented in Fig.12.

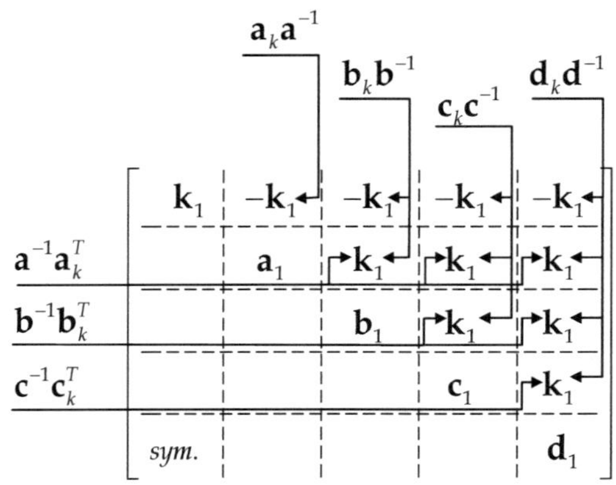

Fig. 12. Graphical method for deriving the components of inversed dynamics block matrix

4. Conclusion

The formulated dynamics block matrix of multilayer head positioning system consist of sub-matrixes which are related directly with structure of his kinematic chain. The dynamic block matrix consists of: bough self inertial matrix, self inertial matrix of branches, branch-bough inertial coupling matrix. The bough self inertial matrix is always one by one dimensional. But this matrix is very sensitive for increase of numbers of branches, adding one new branch into kinematic chains it result in two new components in equation (10). The self inertial matrices of branches are square, symmetrical matrices which may be very often diagonal matrices (Trawiński, 2007), (Trawiński, 2008). The dimension of these matrices always equals the branches number degrees of freedom. The branch-bough inertial couplings matrices are row matrices with numbers of elements equalling the numbers of degrees of freedom of chosen branches. The presented block matrices of multilayer head positioning system may be easily inverted by methods presented in chapter 3.2. In inverted form of dynamic block matrix the common heading matrix is present and the rest of inverted matrix element may be expressed in terms of them. Assumed and presented division of dynamics matrix into block matrix is natural and strictly related with structure of kinematic chain. In some special cases of multilayer head positioning system it is possible to divide dynamics matrix into very small block matrices - one by one dimensional. It usually happens when the number of degrees of freedom equals two. For highest numbers of branch degrees of freedom the division of dynamics matrix, which assure smallest possible dimensions of sub-matrices, is that presented in this chapter. One should be stressed that sizes of sub-matrices of dynamics block matrices influence on numbers of algebraic operations which have to be made during inversion process. This problem is discussed in (Trawiński, 2009).

5. References

Chen, T.-L. & Horowitz, R. (2001). Design, fabrication and dynamic analysis of a PZT-actuated silicon suspension, *Proceedings of American Control Conference*, pp. 1235 - 1240, Arlington, June 2001,

Hong, E.-J.; Kim, W.-S. & Lee, H. S. (2006). Design modification of micro-actuator to improve shock resistance of HDD, *Proceedings of Asia-Pacific Magnetic Recording Conference*, p.1-2, ISBN 1-4244-0863-6, Singapore, December 2006,

Huang, X.; Horowitz, R. & Li, Y. (2005). Design and analysis of robust track-following controllers for dual-stage servo system with an instrumented suspension, *Proceedings of American Control Conference*, pp.1126-1131, June 2005, Portland, USA,

Huang, X. & Horowitz, R. (2005). Robust controller design of a dual-stage disk drive servo system with an instrumented suspension, *IEEE Transacions on Magnetics*, Vol.41, No. 8, August 2005, pp. 2406-2413,

Jiang, M.; Bordson, T.; Gunderson, N. & Lawrence, B. (2007). HDD micro-actuator reliability study, *Proceedings of Reliability and Maintainability Symposium RAMS '07*, ISBN 0-7803-9766-5, pp. 254-258, January 2007, Orlando, FL,

Koganezawa, S. & Hara, T. (2001). Development of shear-mode piezoelectric microactuator for precise head positioning, *Fujitsu Scientific & Technical Journal*, Vol. 37, No.2, December 2001, pp. 212-219,

Ohno, K. & Horowitz, R. (2005). A pivot nonlinearity compensation by use of variable structure estimator for hard disk drives, *Microsystem Technologies*, Vol. 11, No. 8, August 2005, pp. 702–710, ISSN 0946-7076

Rotunno, M.; Oboe, R. & de Callafon, R. A. (2006). Modeling product variations in hard disk drive micro-actuator suspensions, *Microsystem Technologies*, Vol.12, No. 9, August 2006, pp.803-813, ISSN: 0946-7076,

Sarajlic, E.; Yamahata, C.; Cordero, M. & Fujita, H. (2009). Electrostatic rotary stepper micromotor for skew angle compensation in hard disk drive, *IEEE 22nd International Conference on Micro Electro Mechanical Systems*, pp.1079 - 1082, January 2009,

Sugaya, S. (2006). Trends in Enterprise hard disk drives, *Fujitsu Scientific & Technical Journal*, Vol. 42, No.1, January 2006, pp. 61-71,

Suh, S.-M.; Chung, C. C. & Lee, S.-H. (2001). Discrete-time LQG/LTR dual-stage controller design in magnetic disk drives, *IEEE Transactions on Magnetics*, Vol. 37, No.4, July 2001, pp.1891-1895,

Schultz, B. E. (2007). Thermal Fly-height Control (TFC) Technology in Hitachi Hard Disk Drives, White Paper, Hitachi Global Storage Technologies 2007,

Trawiński, T. (2007). Mathematical model of head actuator of hard-disk drive with passive joint, *Electromotion*, Vol.14, No.1, January-March 2007, p.32-37, ISSN 1223-057X,

Trawiński, T. & Kluszczyński, K. (2008). Mathematical modelling of double-shell hard disk drive positioning system regarded as manipulator, *Electrical Review*, Vol.84, No. 6, June 2008, pp.153-156, ISSN 0033-2097, (in polish),

Trawiński, T. (2008). Double layer head positioning system with five degrees of freedom, *XVIII Symposium PTZE*, pp.101-102, ISBN 978-83-7373-038-0, Zamość, Poland, Jun 2008, Polish Society of Applied Electromagnetism, Warszawa,

Trawiński, T. (2009). Inversion method of matrices with chosen structure with help of block matrices, *Electrical Review*, Vol. 85, No. 6, June 2009, pp. 98-101, ISSN 0033-2097, (in polish),

Wang, Z. & Krishnamurthy, P. (2006). A novel recursive filtering approach to estimate repeatable run-out (RRO) disturbance in HDD, *Proceedings of the 2006 American Control Conference*, pp. 2011-2015, ISBN 1-4244-0209-3, Minneapolis, Minnesota, USA, June 2006,

9

Mobile Manipulation: A Case Study

A. HENTOUT[1], B. BOUZOUIA[2], I. AKLI[3] and R. TOUMI[4]
[1, 2, 3] *Division of Computer-Integrated Manufacturing and Robotics (DPR)*
Advanced Technologies Development Centre (CDTA)
BP 17, Baba Hassen, Algiers 16303
Algeria
[4] *Laboratory of Robotics, Parallelism and Electro-energy (LRPE)*
University of Sciences and Technology Houari Boumediene (USTHB)
BP 32, El Alia, Bab Ezzouar, Algiers 16111
Algeria

1. Introduction

Classically, manipulators consist of several links connected together by joints. The main purpose in using these robots is to manumit the human from tedious, arduous and repetitive tasks. Nevertheless, the limited dimensions of the links and the morphology of the fixed-base manipulators, create, therefore, limited accessible workspaces.

To support the development and the new application fields of manipulators, the locomotion had to be combined to the manipulation creating, thus, mobile manipulators. This kind of robots consists of coupling manipulation (represented by a manipulator) and locomotion (represented by a mobile base). The conventional structure of this type of robots is a manipulator mounted upon a mobile base. The mobility extends the workspace of the manipulator and increments its operational capability and flexibility(Sugar & Kumar, 1998).

Mobile manipulators allow the most usual missions of robotics that require both abilities of locomotion and manipulation. They have applications in many areas such as grasping and transporting objects, mining, manufacturing, forestry, construction, etc. Recently, target environment for for activity of such robots has been shifting from factory environment to human environment (Nagatani et al., 2002) (offices, hospitals, homes, assistant for disabled and elderly persons, etc.) because they are particularly well suited for human-like tasks (Alfaro et al., 2004).

However, the motion study of these robots is different and more difficult than that of manipulators. Firstly, combining a mobile base and a manipulator creates redundancy. Secondly, the mobile base has a slower dynamic response than the manipulator. Thirdly, the mobile base is often subject to non-holonomic constraints while the manipulator is usually unconstrained. Finally, the task to be carried out by the robot must be decomposed into tiny movements to be executed by the manipulator and large movements to be carried out by the mobile base (Chen et al., 2006).

In recent years, there are a number of researchers studying mobile manipulators control. These studies led to different approaches.
One of the general approaches is to consider the locomotion as extra joints of the manipulator (Nagatani et al., 2002). In this case, the mobile manipulator is regarded as a redundant robot where the redundancy is introduced by the motion of the mobile base (Sasaki et al., 2001). Erden and colleagues (Erden et al., 2004) describe a multi-agent control system to a service mobile manipulator that interacts with human during an object delivery and hand-over task in two dimensions. The identified agents of the system are controlled using fuzzy control. The membership functions of the fuzzy controller are tuned by using genetic algorithms. The authors in (Chen et al., 2006) propose a three-level neural network-based hierarchical controller. The bottom-level controls each joint motor independently. The middle-level consists of a neural network and two sub-controllers. The high-level is a task-planning unit that defines the desired motion trajectories of each degree of freedom (dof). Colle et *al.* (Colle et al., 2006) propose a multi-agent system for controlling their mobile manipulator *ARPH*. For each articulation is affected a reactive agent that realize in parallel a local task without a priori knowledge on the actions of the other agents. Each agent computes the current position of the end-effector and attempts by tiny local movements to match that position with the desired one.
The other type of approaches controls separately the mobile base and the manipulator neglecting the dynamic interaction between the two sub-systems. Such strategies are appropriate when the coupled dynamics is not significant (ex. when the robot moves at low speed) (Chen et al., 2006). The authors in (Waarsing et al., 2003) implement a behaviour-based controller over a mobile manipulator to make it able to open a door. The locomotion control system, the manipulator control system and the sensor systems cooperate in order to realize such a behaviour. Petersson et *al.* (Petersson et al., 1999) propose an architecture that enables the integration of the manipulator into a behaviour-based control structure of the mobile base. This architecture combines existing techniques for navigation and mobility with a flexible control system for the manipulator.
The robot, as human, must have the ability to obtain information about its environment in order to achieve each step of the manipulation task. The most important sensor which provides rich and varied information on the environment is the vision sensor (the camera) (Trabelsi et al., 2005). Based on hand-eye relation, visual servo system has two types of camera configuration (*i*) Eye-in-hand configuration and (*ii*) Eye-to-hand configuration (Flandin et al., 2000). The manipulator behaves as a hand and the camera as its eye. The camera is said as Eye-in-hand when rigidly mounted on the end-effecter. Here, there exists a known, often constant relationship between the position of the camera and that of the end-effecter. The camera is said as Eye-to-hand when it observes both of the robot and the (Muis & Ohnishi, 2005). Visionbased servoing schemes are flexible and effective methods to control robot motion from camera observations (Hutchinson et al., 1996). Many applications in vision-based robotics, such as mobile robot localization (Blaer & Allen, 2002), object grasping (Muis & Ohnishi, 2005) (Janabi-Sharifi & Wilson, 1998) and manipulation (Trabelsi et al., 2005), handling and transporting objects from one place to another (Trabelsi et al., 2005), navigation (Winter et al., 2000), etc.
This chapter highlights several issues around mobile manipulation in indoor environments. The first aspect consists of planning a coordinated trajectory for the non-holonomic mobile base and the manipulator so that the end-effector of the robot can be as near as possible,

from a predefined operational trajectory. The second aspect deals with a position-based servoing control of mobile manipulators by using an eye-in-hand camera and a *LMS* sensor. These applications are developed within the framework of control architecture of such robots while taking into account the constraints and difficulties mentioned above. The architecture consists of a multi-agent system where each agent models a principal function and manages a different sub-system of the robot. The unified models of the mobile manipulator are derived from the sub-models describing the manipulator and the mobile base. These applications are considered in the case of the *RobuTER/ULM* mobile manipulator of the *Division of Computer-Integrated Manufacturing and Robotics* of the *Advanced Technologies Development Centre.*

The second section of the chapter describes the hardware and the software architecture of the experimental robotic system. Section three explains the multi-agent architecture proposed to control mobile manipulators. Section four describes the implementation of the control architecture. The agents are implemented as a set of concurrent threads communicating by *TCP/IP* sockets. In addition, the threads of each agent communicate by shared variables protected by semaphores. The autonomy of decision-making and the cooperation between the agents are presented in section five through two problems. The first one focuses on trajectory planning and control for mobile manipulators. The end-effector of the robot has to follow a predefined operational trajectory (given by a set of Cartesian coordinates (x, y, z)) while the mobile base avoids obstacles present in the environment. The second part of the experiments, in order to give the robot the ability to manipulate in an indoor environment, deals with position-based servoing control of mobile manipulators using a single camera mounted at its end-effector (eye-in-hand camera) and a *LMS* sensor. Conclusions and future works are presented at the end of the chapter.

2. Architecture of the experimental

The experimental robotic system, given by Fig. 1, consists of a *Local* (*Operator*) *site* and a *Remote site*, connected by wireless communication systems:

- *Local site*: it includes an off-board PC running under *Windows XP*, a wireless TCP/IP communication media, a wireless video reception system and input devices.
- *Remote site*: it includes the *RobuTER/ULM* mobile manipulator, a wireless TCP/IP communication media and a wireless video transmission system.

2.1 Architecture of the *RobuTER/ULM* mobile manipulator

RobuTER/ULM is composed of a rectangular differentially-driven mobile base on which is mounted a manipulator. The robot is controlled by an on-board *MMX* industrial *PC* and by four *MPC555* microcontroller cards communicating via a *CAN* bus. The on-board *PC* is running under *Linux* 6.2 with *RTAI* layer 1.3. This layer interfaces *C/C++* application with that developed under *SynDEx* (http://www.syndex.org). The first *MPC555* card controls the mobile base. The second and the third control the first three and the last three joints of the manipulator. The last *MPC555* controls the effort sensor.

The mobile base has two driven wheels ensuring its mobility and two free wheels to maintain its stability. The mobile base is equipped with a belt of 24 ultrasonic sensors, a laser measurement system at the front and an odometer sensor on each driven wheel.

The manipulator is a six-dof ultra-light manipulator (*ULM*) with two-finger electrical gripper. All of the joints are rotatable. The manipulator is equipped with incremental position sensor for each articulation and with a six-dof effort sensor integrated on the gripper.
The robot is also equipped with a monochrome *CCD* camera placed on the gripper (eye-in-hand camera) with an acquisition card. The resolution of the camera is 352*240 pixels. Images are directly transmitted to the off-board *PC* via the wireless video transmission system. The camera is maneuverable enough to explore the environment of the robot due to the six dof of the manipulator.

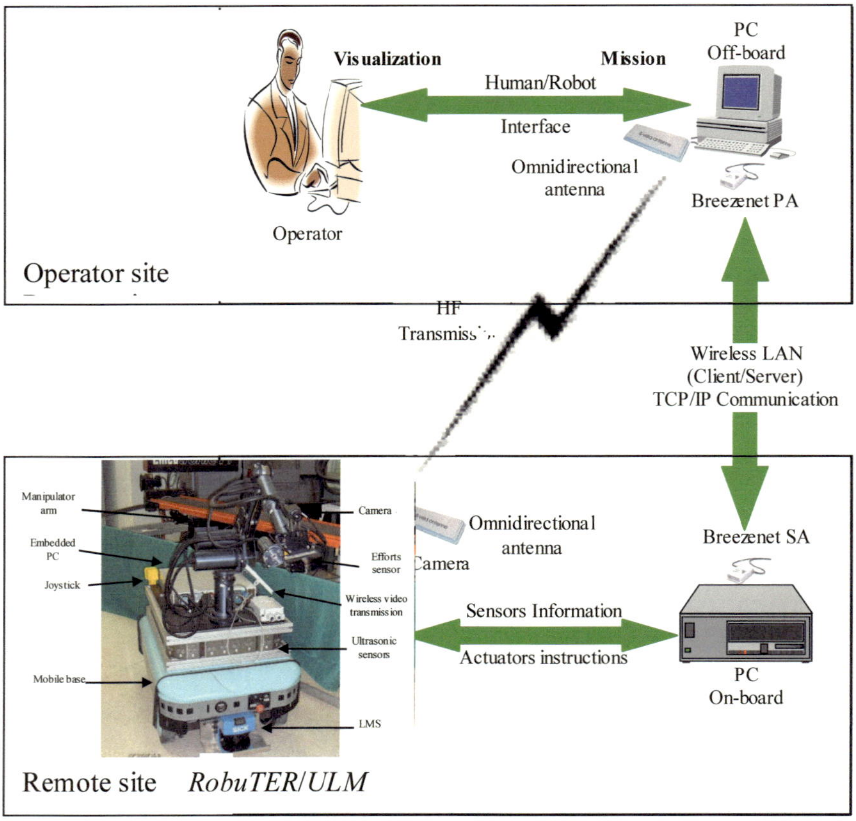

Fig. 1. Architecture of the experimental robotic system

2.2 Kinematic analysis of *RobuTER/ULM*

2.2.1 Main reference frames

The kinematic analysis of the robot needs to focus on the following main reference frames and transformation matrices (Fig. 2) (Hentout et al., 2009a):

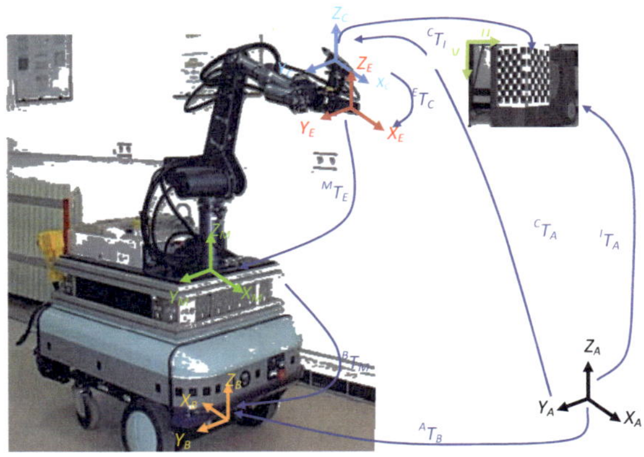

Fig. 2. Reference frames of the *RobuTER/ULM* and the transformation matrices

- $R_A = (O_A, \overrightarrow{x_A}, \overrightarrow{y_A}, \overrightarrow{z_A})$: The absolute reference frame.
- $R_B = (O_B, \overrightarrow{x_B}, \overrightarrow{y_B}, \overrightarrow{z_B})$: The mobile base reference frame.
- $R_M = (O_M, \overrightarrow{x_M}, \overrightarrow{y_M}, \overrightarrow{z_M})$: The manipulator reference frame.
- $R_E = (O_E, \overrightarrow{x_E}, \overrightarrow{y_E}, \overrightarrow{z_E})$: The end-effector reference frame.
- $R_C = (O_C, \overrightarrow{x_C}, \overrightarrow{y_C}, \overrightarrow{z_C})$: The camera reference frame.
- $R_I = (O_I, \overrightarrow{u_I}, \overrightarrow{v_I})$: The image reference frame.
- ${}^{M}T_E$: The transformation matrix defining R_E in R_M. It corresponds to the *Kinematic Model* of the manipulator(see 2.3.2).
- ${}^{B}T_M$: This matrix defines R_M in R_B(see 2.3.4).
- ${}^{A}T_B$: This matrix defines R_B in R_A(see 2.3.4).
- ${}^{A}T_E$: is the matrix defining R_E in R_A(see 2.3.4).
- ${}^{I}T_C$: The camera intrinsic parameters matrix (see 4.3.1).
- ${}^{C}T_A$: The camera extrinsic parameters matrix (see 4.3.1).
- ${}^{I}T_A$: The camera projection matrix (see 4.3.1).
- ${}^{E}T_C$: The Camera/Gripper transformation matrix (see 4.3.2).

2.2.2 Kinematic analysis of the *ULM* manipulator

The position coordinates and orientation angles of the end-effector are calculated in R_M by (1) following the *Modified Denavit-Hartenberg* (*MDH*) representation (Khalil & Kleinfinger, 1986) where ${}^{M}T_2$ defines R_2 in R_M, ${}^{k-1}T_k$ (k=3… 6) defines R_k in R_{k-1} and ${}^{6}T_E$definesR_E in R_6.

$$ {}^{M}T_E = {}^{M}T_2 * {}^{2}T_3 * {}^{3}T_4 * {}^{4}T_5 * {}^{5}T_6 * {}^{6}T_E \quad (1) $$

${}^{M}T_2$, ${}^{6}T_E$ and ${}^{k-1}T_k$ are given by (2) (Dombre & Khalil, 2007):

$$ {}^{k-1}T_k = \begin{bmatrix} \cos\theta_k & -\sin\theta_k & 0 & a_k \\ \cos\alpha_k * \sin\theta_k & \cos\alpha_k * \cos\theta_k & -\sin\theta_k & -d_k * \sin\alpha_k \\ \sin\alpha_k * \sin\theta_k & \sin\alpha_k * \cos\theta_k & \cos\alpha_k & d_k * \cos\alpha_k \\ 0 & 0 & 0 & 1 \end{bmatrix} \quad (2) $$

The different *MDH* parameters α_k, d_k, θ_k, a_k and the joints limits of the *ULM* manipulator are given in Table 1 (Hentout et al., 2009a).

k	α_k (rad)	d_k (mm)	θ_k	a_k(mm)	Q_{Min}(°)	Q_{Max}(°)
1	0	d_1=290	θ_1	0	-95	96
2	π/2	d_2=108.49	θ_2	0	-24	88
3	-π/2	d_3=113	0	a_3=402	-	-
4	π/2	0	θ_3	0	-2	160
5	π/2	d_4=389	θ_4	0	-50	107
6	-π/2	0	θ_5	0	-73	40
7	π/2	d_{eff}=220	θ_6	0	-91	91

Table 1. The *MDH* parameters and the joints limits of the *ULM* manipulator

2.2.3 Kinematic analysis of the mobile base

Assuming that the robot moves on the plane, the kinematic model of the non-holonomic mobile base can be decided, in R_A, by three parameters: X_B, Y_B and θ_B the Cartesian coordinates and the orientation angle. During its motion, the mobile base calculates, by odometry, its position coordinates and orientation angle in real time as shown in (Hentout et al., 2009a).

2.2.4 Kinematic analysis of the mobile manipulator

It involves the interaction between the mobile base and the manipulator. The location of the end-effector is given in R_A by:

$$ {}^{A}T_E = {}^{A}T_B * {}^{B}T_M * {}^{M}T_E \tag{3}$$

${}^{A}T_B$ and ${}^{B}T_M$ are given by (4) and (5) respectively. (X_B, Y_B, Z_B) are the Cartesian coordinates of O_B in R_A and (X_M, Y_M, Z_M) are the Cartesian coordinates of O_M in R_B.

$$ {}^{A}T_B = \begin{bmatrix} \cos\theta_B & -\sin\theta_B & 0 & X_B \\ \sin\theta_B & \cos\theta_B & 0 & Y_B \\ 0 & 0 & 1 & Z_B \\ 0 & 0 & 0 & 1 \end{bmatrix} \tag{4}$$

$$ {}^{B}T_M = \begin{pmatrix} 1 & 0 & 0 & X_M \\ 0 & 1 & 0 & Y_M \\ 0 & 0 & 1 & Z_M \\ 0 & 0 & 0 & 1 \end{pmatrix} \tag{5}$$

For *RobuTER/ULM*, as shown in Fig. 3, Z_B=120mm, X_M=30mm, Y_M=0mm and Z_M=520mm (Hentout et al., 2009a).

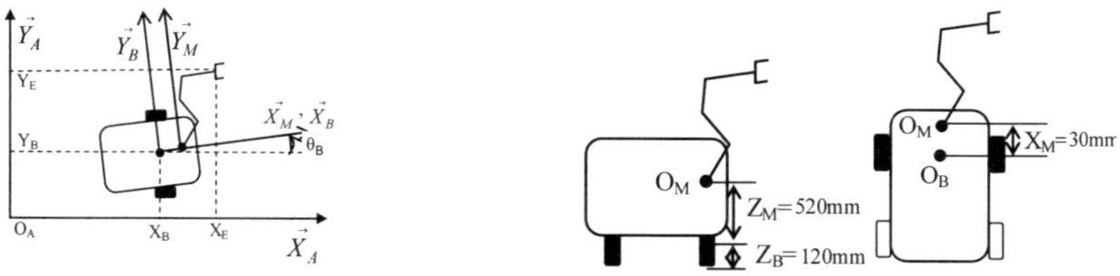

Fig. 3. Kinematics parameters of *RobuTER/ULM*

3. Multi-agent control architecture of mobile manipulators

Fig. 4 shows the proposed multiagent control architecture of mobile manipulators as proposed in (Hentout et al., 2008).The architecture consists of six agents: *Supervisory Agent* (*SA*), *Local Mobile Robot Agent* (*LMRA*), *Local Manipulator Robot Agent* (*LARA*), *Remote Mobile Robot Agent* (*RMRA*) and *Remote Manipulator Robot Agent* (*RARA*). Each agent models a principal function of the mobile manipulator and manages a different sub-system. In addition, for each agent corresponds a mechanism connecting the four capacities *Supervision, Perception, Decision* and *Action* explained in more details in (Hentout et al., 2008). The *Supervision* capacity is a virtual entity that select modules which result in the necessary behaviour facing a given situation.

The following are the basic functions of the architecture agents (Hentout et al., 2009b):

- *SA, Supervisory Agent*: *SA* receives the mission to be executed, decides on its feasibility according to the status and the availability (*Perception* + *Decision*) of the required equipments and resources of the robot (sensors, mobile base, manipulator, camera, etc.). If the mission is accepted, *SA* distributes it on on the corresponding agents for execution (*Action*).
- *LMRA, Local Mobile Robot Agent/LARA, Local Manipulator Robot Agent*: It receives the remote environment information of the mobile base/manipulator in order to build an up-to-date image on the environment where the robot evolves and, obtains feedback (reports) from *RMRA/RARA* on the execution of operations (*Perception*). In addition, the agent cooperates with the other agents (*LARA/LMRA, VSA*) in order to make a decision (*Decision*) according to the received information (sensors information, reports, etc.) and the status of the other agents of the architecture. At the end, it sends requests to *RMRA/RARA* for execution (*Action*).
- *VSA, Vision System Agent*: This agent observes the environment of the robot (*Perception*) by the vision system (the camera installed on the robot) and extracts useful and required information for the execution of the mission (*Decision* + *Action*) from captured images (images processing, localization and recognition of objects, etc.).
- *RMRA, Remote Mobile Robot Agent/RARA, Remote Manipulator Robot Agent:* This agent scans the various proprioceptif and exteroceptif sensors equipping the mobile base/manipulator (*Perception*) and sends the useful information to *LMRA/LARA* in order to maintain a correct representation of the environment. In addition, this agent ensures the local control of the mobile base/manipulator by sending instructions to its actuators and executing the multiple control strategies (navigation of the mobile base/motion of the manipulator) offered by *LMRA/LARA* (*Decision* + *Action*).

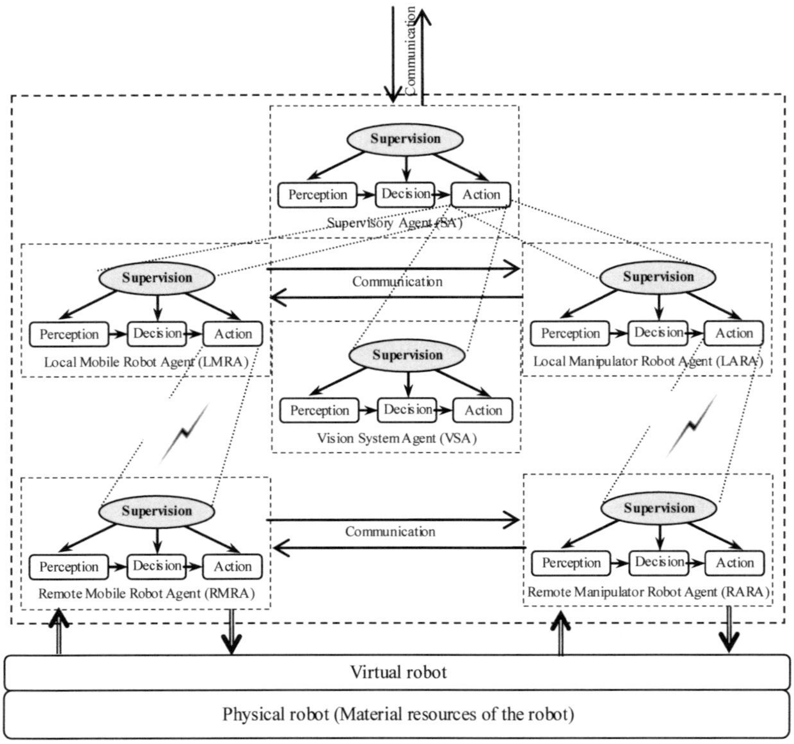

Fig. 4. Multi-agent control architecture

4. Implementation of the control architecture

The agents must be able to respond to asynchronous and external events, and to deal with requests, as soon as possible, according to the dynamics of the robot. Consequently, each agent is implemented as a set of concurrent communicating entities (a set of threads) executing autonomously and in parallel.

The agents communicate by sockets using *TCP/IP* protocol. Furthermore, *semaphores* are used to protect the access to the shared variables between the threads of the agent. *P*(*Variable*) to lock and *V*(*Variable*) to unlock the access to these variables. In addition, each agent has a *Knowledge Base* that describes its configuration. More details on the implementation of the multi-agent control architecture can be found in (Hentout et al., 2009c) (see Fig. 5 for the legend of the next figures of this section).

LMRA, *LARA*, *VSA* and *SA* agents are developed in *Visual studio C# 2008* and installed on the off-board *PC*. *RMRA* and *RARA* agents are installed on the off-board *PC*. They are developed in *C/C++* and *SynDEx*.

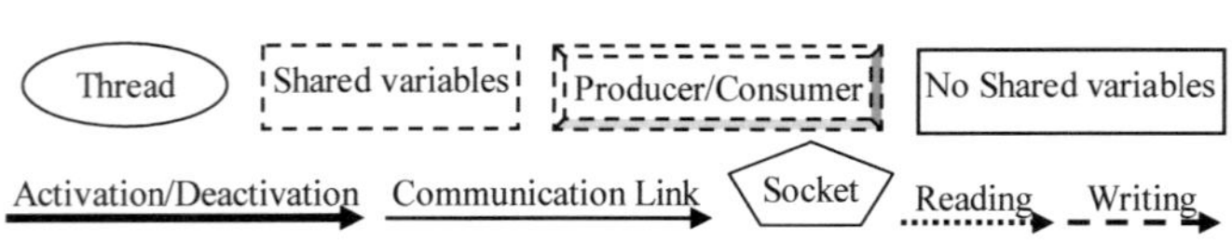

Fig. 5. Legend of the different components of the agents

4.1 Supervisory Agent

The *Supervisory Agent* (Fig. 6) consists of the following threads:

- *Configuration:* This thread allows the operator to configure the agent (ID, port, competences, etc.) and to introduce all the information on the knowledge of the agent, and the partial knowledge about its acquaintances and its environment.
- *Communication*: it interfaces the agent with its acquaintances. It contains traditional communication functions as *Connect*, *Disconnect*, *Send* and *Receive.*
- *Supervision*: Its role is to activate threads that result in the necessary behavior (facing a given situation) and deactivate the others.
- *Human/Robot Interface*: it displays data of odometer, LMS, US, effort and joints positions sensors, the state of the gripper, CCD camera images, etc. In addition, it allows the operator to introduce the mission to be executed by the robot and to dialog with the control architecture.
- *Mission Decision*: This thread decides either the mission to be carried out by the robot is accepted or not. For this aim, it checks the availability and status of all the required resources. If the mission is accepted, it is sent to the other agents for execution. Otherwise, this thread informs the operator of its incapacity to accomplish this mission.

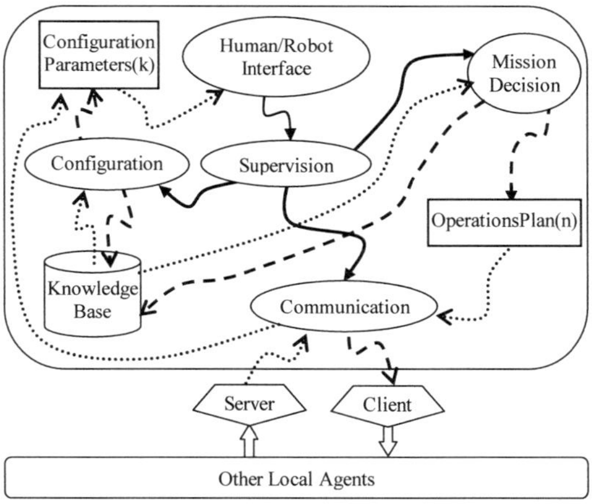

Fig. 6. Multithreading architecture of *SA*

4.2 Local Mobile/Manipulator Robot Agent

The *Local Mobile/Manipulator Robot Agent* (Fig. 7) consists of six threads:

- *Communication High-level*: communication with the other local agents.
- *Communication Low-Level*: it allows communicating with *RMRA/RARA* agent.
- *Configuration.*
- *Supervision.*
- *Sensors Processing*: it receives information on the environment of the remote robot. LMS, US and Odometer sensors for the mobile base; Positions sensors, Effort sensors and the State of the gripper for the manipulator.
- *Position Calculation*: for the *Mobile Robot* agent, this thread calculates the position of the mobile base on a plan relatively to any frame (R_B or R_A).

For the *Manipulator Robot* agent, this thread tests either a given Cartesian position belongs of the current workspace of the manipulator or not. In addition, this thread calculates the *Direct Kinematic Model* (*DKM*) and the *Inverse Kinematic Model (IKM)* of the manipulator.

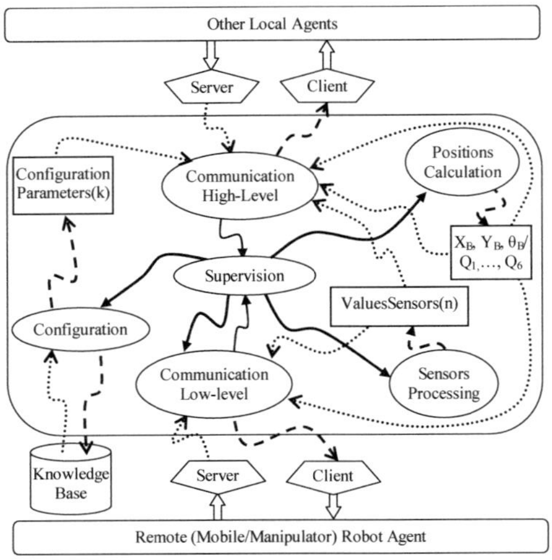

Fig. 7. Multithreading architecture of *LMRA/LARA*

4.3 Vision System Agent

The images captured by the *CCD* camera of the robot must undergo several operations to extract useful information and to calculate the coordinates of the objects of the scene. The *Vision System Agent* (Fig. 8) is composed of the following threads:

- *Communication.*
- *Supervision.*
- *Image Capturing*: The *CCD* camera of the robot delivers continuous video images of the scene. Images are stocked in *Image.*
- *Configuration*: the configuration of this agent consists of the *Camera calibration* (see 4.3.1) and *Camera/Gripper calibration* (see 4.3.2). These parameters are stored in the *knowledge base* of the *VSA* agent.
- *Image Processing*: Firstly, the *Median filter* is applied to remove the noise. It consists of replacing the value of a pixel by the median value of its neighbor pixels. Secondly, the resulting image is *binarised*. The *binarisation* consists of transforming the image into another format with two colors only: *black* for the *objects* and *white* for the *background*. Thirdly, *objects contours* are detected. The *contours* consist of finding pixels in the image that correspond to changes of the luminance intensity. The *algorithm of Canny* (Canny, 1986) has been used. Finally, the *forms recognition* (*characterization*) consists of identifying the forms and classifying them in the corresponding category (triangles, rectangles, circles, etc.). To this aim, the *Hough transformation* (Duda & Hart, 1972) has been used.The result is saved in *ProcessedImage.*
- *2D Extraction*: this thread calculates the *2D* coordinates of the gravity center (u_i, v_i) of all the objects. The results are stocked in *UV* array of *n* elements.

- *3D Extraction*: The aim of this thread is to compute the *3D* real coordinates (x, y, z) of the objects of the scene using the calibration parameters of the *CCD* camera.With a single camera, it is possible to estimate only two coordinates (y, z). Thus, to get the other one (x), another measurement system is needed. The used approach is that developed in (Bouzouia & Rahiche, 2009). It is as follows:
  - From the captured and processed image, the (y, z) coordinates of the selected object are calculated by using the camera model obtained by the calibration process.
  - The measure representing the other component (x) is obtained from the *LMS* sensor.

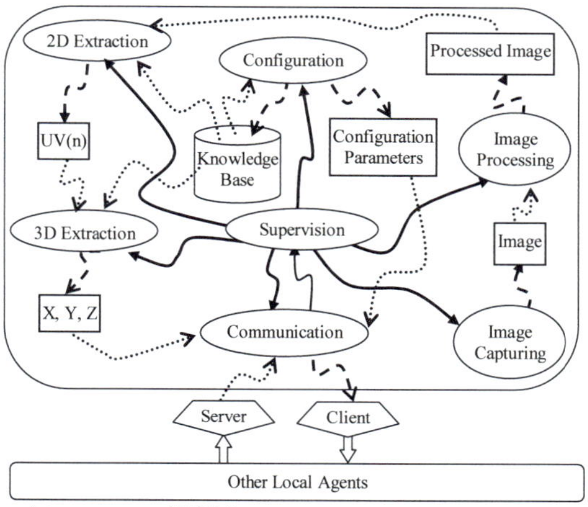

Fig. 8. Multithreading architecture of *VSA*

The *acquisition*, *Filtering*, *Segmentation* and *Characterization* steps are illustrated in Fig. 9 (Bouzouia &Rahiche, 2009).

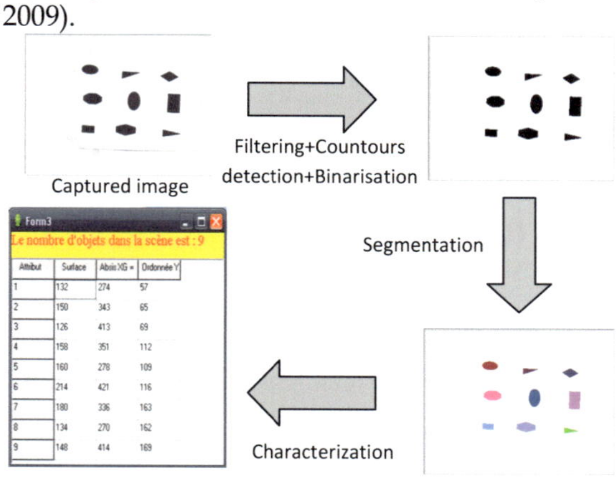

Fig. 9. Result of the image processing process

4.3.1 Camera calibration

Camera calibration consists in determining the 3×4 transformation matrix ${}^{I}T_{A}$ that maps a 3D coordinates of a point in the space $P\begin{bmatrix}x\\y\\z\end{bmatrix}$ expressed in R_A using a calibration grid, onto its 2D image projection whose coordinates $m\begin{bmatrix}u\\v\end{bmatrix}$ are expressed in pixel in R_I (Telle et al., 2003). The relation between P and m is given by (6) where s is an arbitrary scale factor (Muis & Ohnishi, 2005):

$$\begin{bmatrix}su\\sv\\s\end{bmatrix} = {}^{I}T_{A} * \begin{bmatrix}x\\y\\z\\1\end{bmatrix} \tag{6}$$

For the camera calibration, the method proposed in (Bénallal, 2002) is adopted. It consists of solving (7) with $n \geq 6$ (Hartley & Zisserman, 2001) and m_{11}, m_{12} ... m_{33} are the elements of the matrix ${}^{I}T_{A}$.

$$\begin{bmatrix}
x_1 & y_1 & z_1 & 1 & 0 & 0 & 0 & 0 & -u_1.x_1 & -u_1.y_1 & -u_1.z_1\\
0 & 0 & 0 & 0 & x_1 & y_1 & z_1 & 1 & -v_1.x_1 & -v_1.y_1 & -v_1.z_1\\
x_2 & y_2 & z_2 & 1 & 0 & 0 & 0 & 0 & -u_2.x_2 & -u_2.y_2 & -u_2.z_2\\
0 & 0 & 0 & 0 & x_2 & y_2 & z_2 & 1 & -v_2.x_2 & -v_2.y_2 & -v_2.z_2\\
. & . & . & . & . & . & . & . & . & . & .\\
. & . & . & . & . & . & . & . & . & . & .\\
x_n & y_n & z_n & 1 & 0 & 0 & 0 & 0 & -u_n.x_n & -u_n.y_n & -u_n.z_n\\
0 & 0 & 0 & 0 & x_n & y_n & z_n & 1 & -v_n.x_n & -v_n.y_n & -v_n.z_n
\end{bmatrix} * \begin{bmatrix}m_{11}\\m_{12}\\m_{13}\\m_{14}\\m_{21}\\m_{22}\\m_{23}\\m_{24}\\m_{31}\\m_{32}\\m_{33}\end{bmatrix} = \begin{bmatrix}u_1\\v_1\\u_2\\v_2\\.\\.\\u_n\\u_n\end{bmatrix} \tag{7}$$

The obtained matrix M is given by (8) (Hentout et al., 2009d):

$$M \approx \begin{pmatrix}-0.2057 & 0.8110 & -0.079 & 69.4449\\ 0.6498 & 0.1155 & -0.434 & 139.8394\\ 0.0005 & 0.0002 & 0.0006 & 1\end{pmatrix} \tag{8}$$

4.3.2 Camera/Gripper Calibration

Camera/Gripper calibration consists of finding the matrix ${}^{E}T_{C}$ defining R_C in R_E.

Let ${}^{C}T_{A1}$ and ${}^{C}T_{A2}$ be the transformation matrices defining a *first* and a *second position* of the camera in R_A. Let ${}^{M}T_{E1}$ and ${}^{M}T_{E2}$ be the transformation matrices defining the two positions of the end-effector in R_M corresponding to the first and the second position of the camera.

To find the Camera/Gripper calibration matrix ${}^{E}T_{C}$, the method developed in (Tsai & Lenz, 1989) is chosen. It is based on the *Least squares method* and consists of solving (9) where:

- $A=({}^{C}T_{A2})*({}^{C}T_{A1})^{-1}$: The measurable transformation matrix of the camera from its first to its second location (relative camera motion).
- $B=({}^{M}T_{E2})^{-1}*({}^{C}T_{E1})$: The measurable transformation matrix of the gripper from its first to its second location (relative robot gripper motion).

$$A * {}^{E}T_{C} = {}^{E}T_{C} * B \tag{9}$$

The obtained matrix ${}^{E}T_{C}$ is given by (10) (Hentout et al., 2009e):

$$
{}^{E}T_C \approx \begin{pmatrix} 0.7322 & -0.6573 & 0.1787 & -183.1608 \\ 0.6595 & 0.7497 & 0.0555 & -69.4161 \\ -0.1704 & 0.0772 & 0.9823 & 0 \\ 0 & 0 & 0 & 1 \end{pmatrix} \tag{10}
$$

4.4 Remote Mobile Robot Agent

The *Remote Mobile Robot Agent*, given by Fig. 10, consists of seven threads:

- *Configuration.*
- *Communication.*
- *Supervision.*
- *Reading LMS Sensors*: it scans continuously the serial port of the *LMS sensors* and stores data in *ValuesLMS* of 181 items.
- *Reading US Sensors*: This thread, in its turn, reads constantly the *US sensors* and stores *US* data in *ValuesUS* of 24 items.
- *Odometry*: This thread reads the values of the incremental encoders (*E_R*, *E_L*), installed on the driven wheels of the mobile base, and calculates its current position and orientation angle (*New_X*, *New_Y*, *New_θ*) as shown in (Hentout et al., 2009a).
- *Navigation*: it consists of the main role of this agent. It uses data of all the other threads. *Navigation* calculates velocities (*Spd_R*, *Spd_L*) to be sent to the actuators of the mobile base in order to move to a *Target position* given by (X_{Target}, Y_{Target}, θ_{Target}) while avoiding possible obstacles.

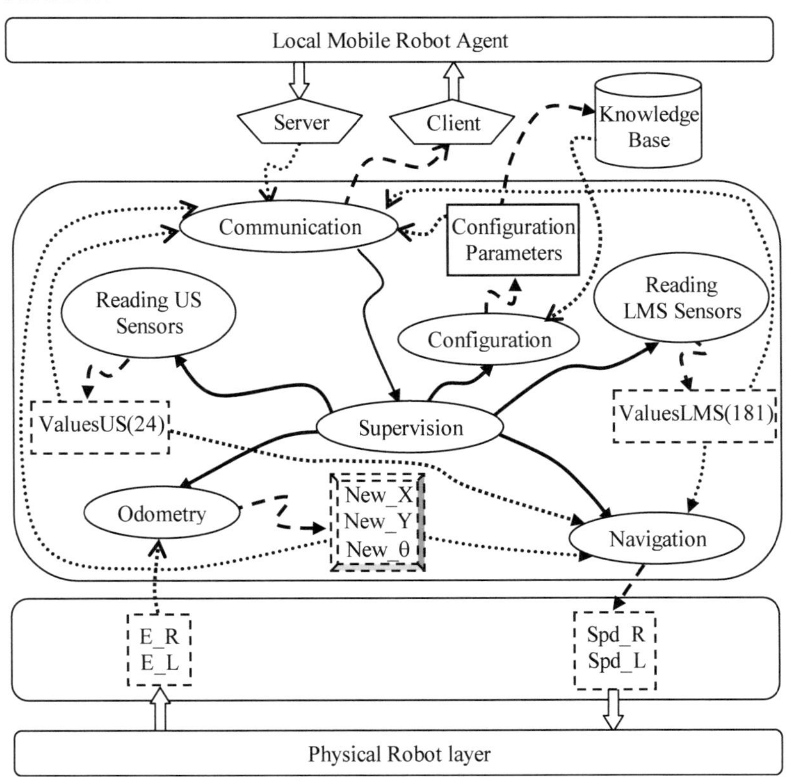

Fig. 10. Multithreading architecture of *RMRA*

4.5 Remote Manipulator Robot Agent

The *Remote Manipulator Robot Agent* (Fig. 11) is composed, in its turn, of seven threads:

- *Communication.*
- *Configuration.*
- *Supervision.*
- *Reading Positions Sensors*: it reads the incremental positions sensors (*PcIn*) installed on each articulation of the manipulator and saves data in *ValuesPos* of 6 elements.
- *Reading Effort Sensors/State Gripper*: this thread, in its turn, reads the effort sensor data (*Gripper*) and the state of the gripper (Opened or Closed). This thread stores read data in *ValuesGripper* of 7 items (*Forces (Fx, Fy, Fz), Torques (Tx, Ty, Tz), State of the Gripper*).
- *Open/Close gripper*: it opens or closes the gripper.
- *Movement*: This thread consists of the main role of this agent. It calculates orders (*PcOut*) to be sent to the actuators of the manipulator in order to move to a *Target position* given by ($Q_1 \ldots Q_6$) with a given velocity V_i (i=1…6) for each joint.

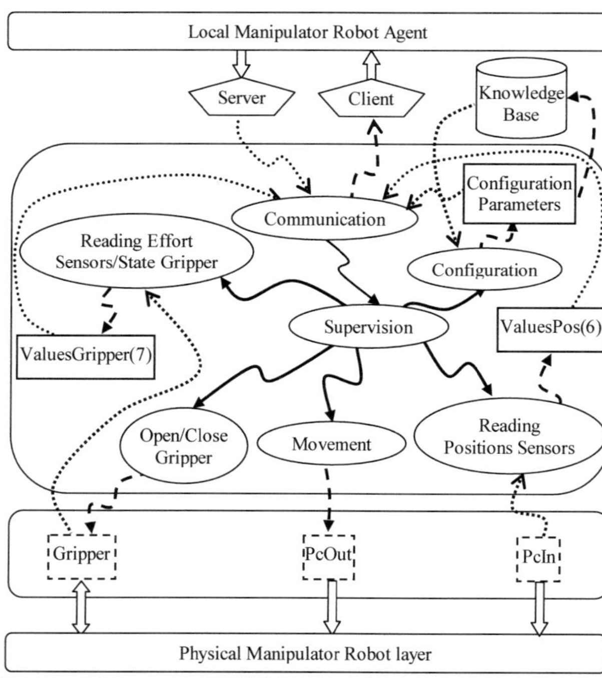

Fig. 11. Multithreading architecture of *RARA*

5. Experimental part

The core thinking of modeling and controlling mobile manipulators using a multi-agent system is that of realizing cooperation between the manipulator, the mobile base and the sensors system. In order to show the validity of the implementation of the *SA*, *LMRA*, *LARA*, *VSA*, *RMRA* and *RARA* agents, two different missions are considered in this section. For the envisaged experiments, all the positions and orientations are given in R_A. In addition, two cases are distinguished (Fig. 12):

- *The Target belongs of the current workspace of the robot*: this task requires only the motion of the manipulator.
- *The Target is outside the current workspace of the robot*: In this case, the mobile base moves until the *Target* is within the new workspace of the manipulator. Then, the robot manipulates the *Target* with its end-effector.

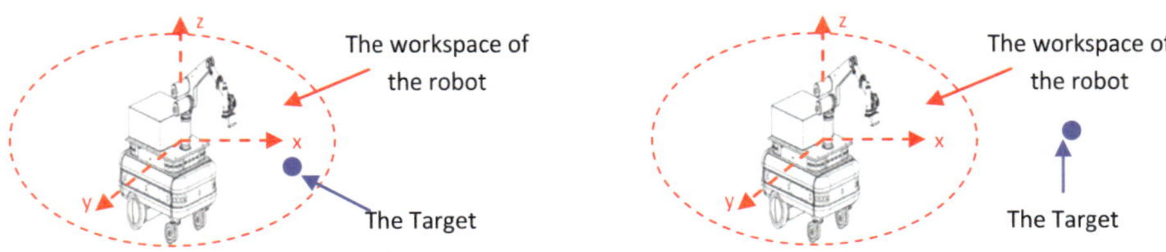

Fig. 12. The two possible cases for a given *Target*

5.1. Following a predefined operational trajectory

This experiment presents trajectory planning and control for mobile manipulators. The end-effector of the robot has to be, as near as possible, from a predefined operational trajectory (given by a set of Cartesian coordinates (*X*, *Y*, *Z*)) while the non-holonomic mobile base has to avoid the obstacles present in the environment.

5.1.1 Straight-line following

In this experiment, as shown in Fig. 13, the operational trajectory to be followed by the end-effector of the robot consists of a straight-line connecting an initial position $P_i(X_i, Y_i, Z_i)$ to a final position $P_f(X_f, Y_f, Z_f)$ (Hentout et al., 2009e).

To compute the imposed positions (*Targets*) to be reached by the end-effector, let's consider:

- $[P_i, P_f]$: the segment connecting P_i to P_f.
- $p(X_p, Y_p, Z_p)$: the current position coordinates of the end-effector.
- $h(X_h, Y_h, Z_h)$: the projection of p on the segment $[P_i, P_f]$.
- *New_X*, *New_Y*, *New_θ*: the current position coordinates and orientation angle of the mobile base.
- $Position_{Init}(X_{BInit}, Y_{BInit}, \theta_{BInit})$: the initial position coordinates and orientation angle of the mobile base.
- $Position_{Fin}(X_{BFin}, Y_{BFin}, \theta_{BFin})$: the final position coordinates and orientation angle of the mobile base.
- $m(X_m, Y_m, Z_m)$: a point in the space.

$$m \in [P_i, P_f] \Leftrightarrow \begin{cases} X_m = t * (X_f - X_i) + X_i \\ X_m = t * (Y_f - Y_i) + Y_i \\ Z_m = t * (Z_f - Z_i) + Z_i \\ t \epsilon [0, 1] \end{cases} \tag{11}$$

$\vec{hp}$ is orthogonal to $\vec{P_i P_f}$ so:

$$(X_h - X_p) * (X_f - X_i) + (Y_h - Y_p) * (Y_f - Y_i) + (Z_h - Z_p) * (Z_f - Z_i) = 0 \tag{12}$$

From (11) and (12), the position of h (the next *Target* to be reached by the end-effector of the robot) on the segment $[P_i, P_f]$ is given by (13):

$$\begin{aligned} X_h &= t*(X_f - X_i) + X_i \\ Y_h &= t*(Y_f - Y_i) + Y_i \\ Z_h &= t*(Z_f - Z_i) + Z_i \\ t &= \frac{(X_f - X_i)*(X_p - X_i) + (Y_f - Y_i)*(Y_p - Y_i) + (Z_f - Z_i)*(Z_p - Z_i)}{(X_f - X_i)^2 + (Y_f - Y_i)^2 + (Z_f - Z_f)^2} \end{aligned} \quad (13)$$

The positioning error of the end-effector is calculated by (14):

$$Err = \sqrt{(X_h - X_p)^2 + (Y_h - Y_p)^2 + (Z_h - Z_p)^2} \quad (14)$$

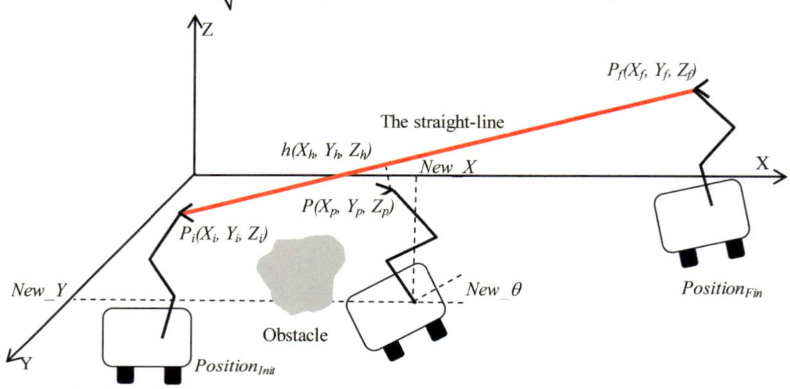

Fig. 13. The straight-line following mission and its parameters

The straight-line following algorithm for *LARA* is given as follows:

```
Straight_Line_Following_LARA(Pi, Pf){
        if(Pi ∉ Current workspace of the manipulator)
                wait for message (Mobile base in Position_Init) from LMRA;
        Generate the possible orientations for Pi using the IKM;
        Qi(i=1…6) = Choose the best configuration;
        Send Move(Qi(i=1…6)) to RARA;
        while (P != Pf){
                Receive (New_X, New_Y, New_θ) from LMRA;
                Calculate P in RA according to (New_X, New_Y, New_θ);
                if (P∉ [Pi, Pf]){
                        Calculate h, the projection of P on [Pi, Pf];
                        Generate the possible orientations for h using the IKM;
                        Qi(i=1…6) = Choose the best configuration;
                }
                Send Move(Qi(i=1…6)) to RARA;
        }
}
```

The straight-line following algorithm for *LMRA* is given here below:

```
Straight_Line_Following_LMRA(Pi, Pf){
```

```
        Calculate PositionInit corresponding to Pi;
        Send Move(PositionInit) to RMRA;
        wait for (Mobile base in PositionInit) from RMRA;
        Send (Mobile base in PositionInit) to LARA;
        Calculate PositionFin corresponding to Pf;
        Send Move(PositionFin) to RMRA;
        while ((New_X, New_Y, New_θ)!=PositionFin){
                Receive (New_X, New_Y, New_θ) from RMRA;
                Send (New_X, New_Y, New_θ) to LARA;
        }
}
```

These two previous algorithms are executed in parallel on the off-board *PC* by the corresponding agents. In addition, *LMRA* and *LARA* send requests and receive sensors data and reports from the corresponding agent (*RMRA* and *RARA*). At the same time, *RMRA* and *RARA* move towards the received positions: $Position_{Fin}$ for the mobile base and ($Q_1 \ldots Q_6$) for the manipulator.

5.1.2 Experimental result

The straight-line following algorithms proposed previously for *LMRA* and *LARA* are implemented to the *RobuTER/ULM*. (13) is used to generate the *Target* positions so that the end-effector of the mobile manipulator follows the desired line (Hentout et al., 2009e).
For this experiment $P_i(X_i, Y_i, Z_i)$ =(-691.72mm, -108.49mm, 1128.62mm) and $P_f(X_f, Y_f, Z_f)$ = (-2408.17mm, -108.49mm, 1472.30mm). Therefore, the operational trajectory consists of a straight-line with a slope of about 350mm (343.68mm).
The initial posture of the mobile base and that of the end-effector corresponding to P_i is $Target_{Init}(X_{BInit}, Y_{BInit}, \theta_{BInit}, X_{EInit}, Y_{EInit}, Z_{EInit}, \psi_{EInit}, \theta_{EInit}, \varphi_{EInit})$ = (0mm, 0mm, 0°, -691.72mm, -108.49mm, 1128.62mm, -90°, -90°, -90°). For this initial position, the initial joint angles (Q_{1Init}, Q_{2Init}, Q_{3Init}, Q_{4Init}, Q_{5Init}, Q_{6Init}) = (0°, 60°, 0°, 0°, 32°, 0°).The final position of the mobile base and that of the end-effector corresponding to $P_f$is $Target_{Fin}(X_{BFin}, Y_{BFin}, \theta_{BFin}, X_{EFin}, Y_{EFin}, Z_{EFin}, \psi_{EFin}, \theta_{EFin}, \varphi_{EFin})$ = (-1920mm, 2mm, 15°, -2408.17mm, -108.49mm, 1472.30mm, 0°, -90°, 0°). For this final position, the final joint angles (Q_{1Fin}, Q_{2Fin}, Q_{3Fin}, Q_{4Fin}, Q_{5Fin}, Q_{6Fin}) = (37°, 52°, 61°, 73°, -57°, 28°).
Two cases are tested for this example(Hentout et al., 2009e):

- The environment of the robot is free (no obstacles are considered). The motion of the mobile base consists also of a straight-line connecting $Position_{Init}$ to $Position_{Fin}$. For this case, the robot follows perfectly the imposed straight-line.
- The second case is more difficult. The non-holonomic mobile base has to avoid an obstacle present in the environment while the end-effector has to be always at the desired configuration (on the straight-line).For the second case of this experiment, the operational trajectory followed by the end-effector and the imposed trajectory for the end-effector are shown on Fig. 14.

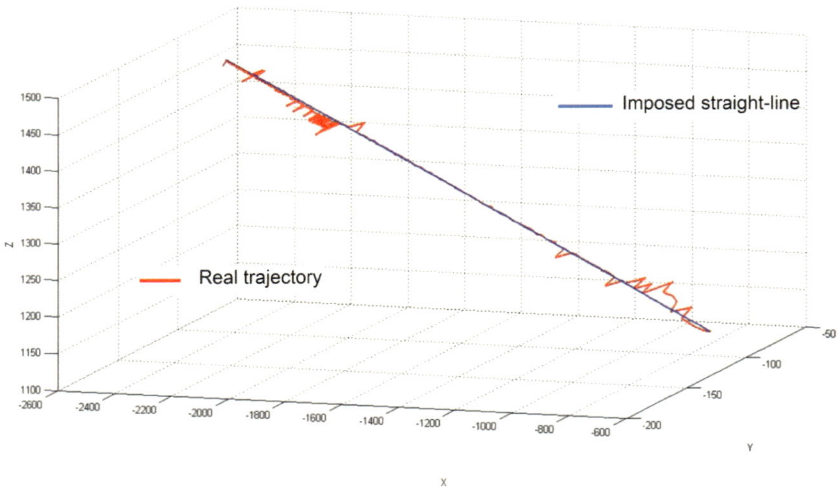

Fig. 14. Imposed operational trajectory and the real trajectory of the end-effector

The real joints variations rather than the desired trajectory for some joints (1, 2, 3 and 5 respectively) are shown on Fig. 15.

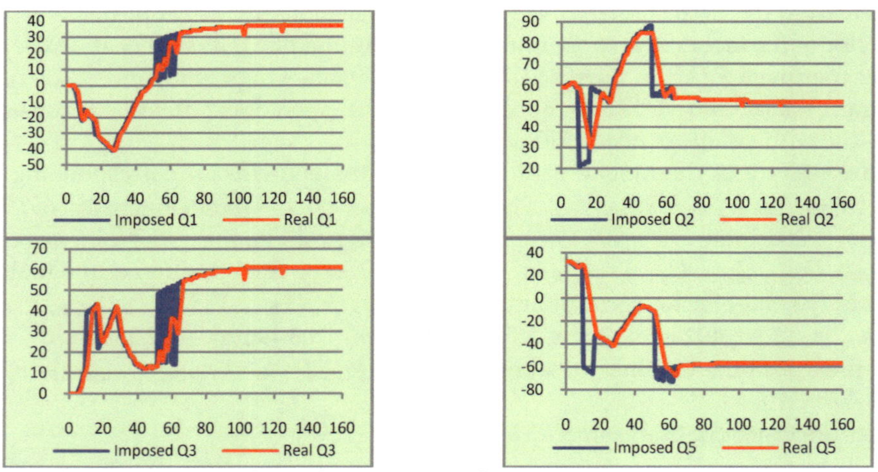

Fig. 15. Joints variations and desired trajectories of some joints

Fig.16 shows the trajectory followed by the mobile base and the avoidance of the obstacle.

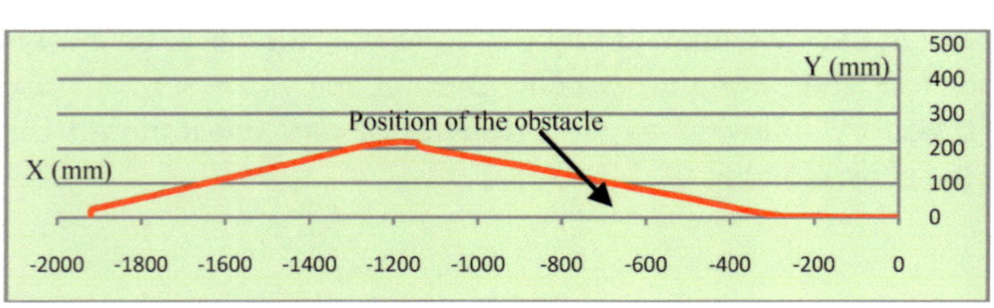

Fig. 16. Real trajectory followed by the mobile base

5.1.3 Discussion of results

Figs. 14, 15 and 16 showed the operational trajectory of the end-effector, the variations of some joints of the manipulator and the motion of the mobile base respectively. The mission took about 160 seconds.

The maximum positioning error calculated by (14) is 24.43mm while the average error is 3.41mm. The errors show that it is difficult to follow the desired straight-line.

The first reason for this error is the initial positioning error of the mobile base ($Position_{Init}$). It causes straying from the initial position for the end-effector in the trajectory. To solve this problem, the mobile manipulator must absorb this error by the motion of its manipulator. Secondly, an estimated positioning error of the mobile base, calculated by odometry (*New_X, New_Y, New_θ*), during its motion effects the tip position of the end-effector directly. To absorb this error, the manipulator should move quickly to adjust itself when the error is detected. Finally, the low velocity of the manipulator's motion during the motion of the mobile base causes a delay in the positioning of the end-effector. This problem can be solved by incrementing the velocity of the manipulator according to that of the mobile base.

5.2 Aligning the end-effector of the robot to different objects by using the eye-in-hand camera and the LMS sensor

A position-based servoing control of mobile manipulators by using the eye-inhand camera and the *LMS* sensor is considered. The working mission is to reach different positions (corresponding to various objects) by the end-effector of the robot.

To reach an object, it is necessary to capture an image of this object. *VSA* sends, for this aim, a request *Move Gripper* (*Position*) to *LARA* in order to position the manipulator. *Position* is read from the *Knowledge Base* of *VSA*. After the positioning of the manipulator, *VSA* captures an image and carries out the necessary processing to extract the *2D* coordinates (*u, v*) of the gravity center of the object in the image. At the end, *VSA* extract (*y, z*) coordinates and sends them to *LARA*. *LARA* always needsthe (*x*) coordinate. To this aim, a request is sent, in parallel, to *LMRA* (*Read LMS*) which transmits it to *RMRA*. Receiving this request, *RMRA* send *LMSValues* data to *LMRA*. This latter selects the minimum value from the 60th element to 120th element (corresponding to 60° to 120°). This value corresponds to the (*x*) coordinate. It is sent to *LMRA* which has now (*x, y, z*) coordinates of the position to be reached.

5.2.1 Experimental result

For this experiment, as shown in Fig. 18, the initial posture of the mobile base and that of the end-effector is $Target_{Init}(X_{BInit}, Y_{BInit}, \theta_{BInit}, X_{EInit}, Y_{EInit}, Z_{EInit}, \psi_{EInit}, \theta_{EInit}, \varphi_{EInit})$ = (0mm, 0mm, 0°, -546.62mm, -110.36mm, 1200.73mm, -90°, -90°, -90°). The position to be reached by the end-effector of the robot are at a distance x=-2470. For this initial position, the initial joint angles ($Q_{1Init}, Q_{2Init}, Q_{3Init}, Q_{4Init}, Q_{5Init}, Q_{6Init}$) = (0°, 87°, 0°, 0°, 5°, 0°). Table 2 shows the different parameters of this experiment.

Image Coordinates (u, v) (pixel)	Real Coordinates (x, y, z) (mm)	LMS value (x) (mm)	Calculated Coordinates (x, y, z) (mm)	Mobile Base coordinates (X_B, Y_B, θ_B)
166, 176	-2470, -63, 1325	2470	-2470, -63.01, 1328.56	-1670mm, 0mm, 0°
182, 228	-2470, 52, 1295	2470	-2470, 46.40, 1294.85	-1670mm, 0mm, 0°
228, 178	-2470, -58, 1200	2470	-2470, -58.93, 1197.94	-1670mm, 0mm, 0°
234, 227	-2470, 47, 1185	2470	-2470, 44.25, 1185.30	-1670mm, 0mm, 0°

Table 2. Different parameters of the experiment

The joints angles and the corresponding end-effector coordinates are given in Table 3.

Target	Joints angles Q_i (i=1..6) (°)	End-effector coordinates (X_E, Y_E, Z_E, ψ_E, θ_E, φ_E)
$Target_1$	(5, 49, 63, -13, -22, -78)	-2466.52mm, -60.45mm, 1335.20mm, -90°, -90°, 180°
$Target_2$	(16, 51, 54, -48, -22, -44)	-2469.86mm, 44.17mm, 1293.73mm, -90°, -90°, 180°
$Target_3$	(5, 53, 37, 89, 5, 1)	-2468.08mm, -60.95mm, 1199.39mm, -90°, -90°, 0°
$Target_4$	(16, 53, 35, 83, 16, 7)	-2469.41mm, 45.15mm, 1185.48mm, -90°, -90°, 0°

Table 3. Joints angles and end-effector postures for the different *Targets*

The following snapshots (Fig. 18) show the obtained result (Bouzouia & Rahiche, 2009):

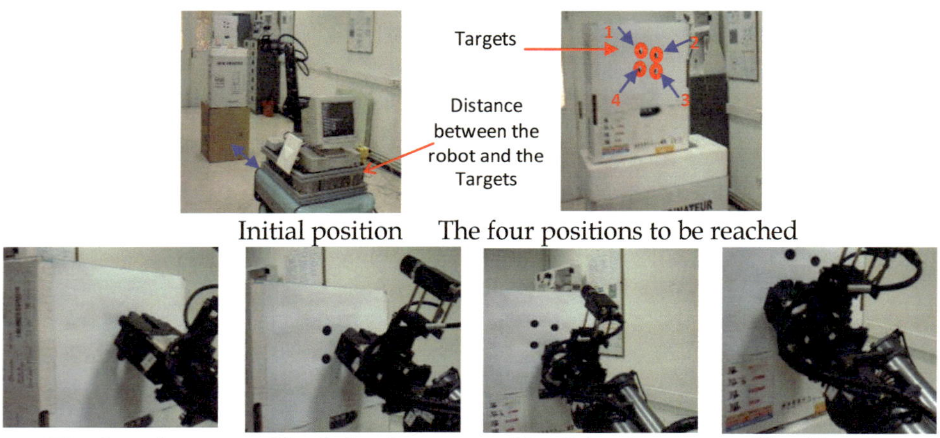

The 1st point The 2nd point The 3rd point The 4th point

Fig. 18. Position-based servoing control by using the camera and the LMS sensor of *RobuTER/ULM*

5.2.2 Discussion of results

The *VSA* agent uses the eye-in-hand camera to extract the two last coordinates (y and z) of the object to be manipulated by the robot. The *LMRA* and the *RMRA* agent use the *LMS* sensor to obtain the first coordinate (x).

The maximum 3D reconstruction error calculated by (14) is 5.60mm while the minimum error is 2.26mm. These errors are acceptable. They are due to the weak precision of the measured real values, to the low rigidity of the manipulator, to the accumulation errors of the calibration processand to the feeble precision of the *LMS* sensor (±15mm).

The maximum positioning error is 11.07mm while the minimum error is 2.00mm. The errors are principally due to the error in the estimated positioning of the mobile base, calculated by odometry (*New_X, New_Y, New_θ*), during its motion. These errors are also due the accumulated errors in the *IKM* of the manipulator.

6. Conclusions and future works

This chapter presented a multi-agent control architecture of mobile manipulators. The architecture consists of six agents: *Supervisory Agent* (*SA*), *Local Mobile Robot Agent* (*LMRA*), *Local Manipulator Robot Agent* (*LARA*), *Vision System Agent* (*VSA*), *Remote Mobile Robot Agent* (*RMRA*) and *Remote Manipulator Robot Agent* (*RARA*). The first four agents are installed on an off-board *PC* while the two other agents are installed on the on-board *PC* of the robot.
The controller was applied successfully to follow a predefined straight-line operational trajectory by the end-effector of a differentially-driven *RobuTER/ULM* mobile manipulator while considering obstacles in its environment. The controller was shown to be relatively effective when the robot moves with small velocities.
To realize the operational trajectory following, one of the biggest problems is that an accumulated error of the estimated position of the mobile base affects the position accuracy of the end-effector. Therefore, the manipulator should have a capability to adjust its position when the mobile base detects positioning errors.
The results obtained the position-based servoing control of the robot by using the eye-in-hand camera and the *LMS* sensor are satisfactory since the positioning error of the end-effector is less than 15mm. The calculation of the 3D coordinates is based on the eye-in-hand camera (for (y, z) coordinates) and on the *LMS* sensor (for (x) coordinate).
In future works, the performances and the robustness of the implemented agents of the control architecture should be shown and discussed through examples of other types of trajectories (circular, etc.). Furthermore, and especially for the *VSA* agent, a moving target tracking problem should be performed. In addition, the real time constraint for the *VSA* agent will be verified and discussed. Another extension of this work is to introduce a virtual reality system (a graphic simulator) to give more effective action for the developed architecture.

7. References

Alfaro, C., Ribeiro, M. I.& Lima, P. (2004). "Smooth Local Path Planning for a Mobile Manipulator". *Robotica 2004, The 4th Portuguese Robotics Festival*, Protugal, pp. 127-134.

Bénallal, M. (2002). "Système de Calibration de Caméra: Localisation de Forme Polyédrique par Vision Monoculaire", *Ph. D. Thesis*, Ecole des mines de Paris, France.

Blaer, P.& Allen, P.K. (2002). "Topological Mobile Robot Localization Using Fast Vision Techniques". *IEEE International Conference on Robotics and Automation*, pp. 1031-1036, Washington, USA.

Bouzouia, B. & Rahiche, A. (2009). "Teleoperation System of the Mobile Manipulator Robot RobuTER/ULM: Implementation Issues". *The XXII International Symposium on Information, Communication and Automation Technologies (ICAT 2009)*, Sarajevo, Bosnia & Herzegovina, October 29-31.

Canny, J. (1986). "A Computational Approach to Edge Detection", *IEEE Transactions on Pattern analysis and machine intelligence, 8(6)*, pp. 679-698.

Chen, Y., Liu, L., Zhang, M.& Rong, H. (2006). "Study on Coordinated Control and Hardware System of a Mobile Manipulator". *Proceedings of the 6th World Congress on Intelligent Control and Automation*, Dalian, China. pp. 9037-9041, June 21-23.

Colle, E., Nait Chabane, K., Delarue, S. & Hoppenot, P. (2006). "Comparaison d'une méthode classique et d'une méthode utilisant la coopération homme-machine pour exploiter la redondance de l'assistant robotisé". *HANDICAP 2006*, France, June 7-9.

Dombre, E.& Khalil, W. (2007). "Modeling, Performance Analysis and Control of Robot Manipulators". *Control Systems, Robotics and Manufacturing Series*. Hermes Science Publishing.

Duda, R.O.& Hart, P.E. (1972). "Use of the Hough transformation to detect lines and curves in pictures", *Communications of ACM, 15(1)*, pp. 11-15.

Erden, M. S., Leblebicioglu, K. & Halici, U. (2004). "Multi-Agent System-Based Fuzzy Controller Design with Genetic Tuning for a Mobile Manipulator Robot in the Hand Over Task". *Journal of Intelligent and Robotic Systems, 39(3)*, pp. 287-306.

Flandin, G., Chaumette, F. & Marchand, E. (2000). "Eye-in-hand/eye-to-hand cooperation for visual servoing," *Proceedings of the IEEE International Conference on Robotics and Automation*, pp. 2741–2746.

Hartley, R. & Zisserman, A. (2001). "Multiple View Geometry in Computer Vision". *Cambridge*.

Hentout, A., Bouzouia, B. & Toukal, Z. (2008). "Multi-agent Architecture Model for Driving Mobile Manipulator Robots". *International Journal of Advanced Robotic Systems, 5(3)*, pp. 257-269.

Hentout, A., Bouzouia, B., Akli, A., Toumi, R.,& Toukal, Z. (2009a). "Agent-Based Coordinated Control of Mobile Manipulators: Study on Hardware System of a Mobile Manipulator". *International Conference on Systems and Processing Information (ICSIP'09)*, Guelma, Algeria, May 2-4.

Hentout, A., Bouzouia, B., Toumi, R. & Toukal, Z. (2009b). "Agent-Based Control of Remote Mobile Manipulator Robotic System: Study of Cooperation". *Proceeding of theInternational Congress ofIndustrial Engineering 2009 (CIGI'09)*, Toulouse, France, June 10-12.

Hentout, A., Akli, I., Bouzouia, B., Ouzzane, M. E., Benbouali, R. & Bouskia, M. A. (2009c). "Implementation of a Multi-agent Architecture for Remote Control of Mobile Manipulators". *The IEEE International Conference on Applied Informatics (ICAI'09)*, Bordj Bou Arréridj, Algeria, November 15-17.

Hentout, A., Bouzouia, B., Akli, I., Bouskia, M. A., Benbouali, R. & Ouzzane, M. E. (2009d). "Multi-Agent Control Architecture of Mobile Manipulators: Extraction of 3D Coordinates of Object Using an Eye-in-Hand Camera", *The 3rd 2009 IEEE International Conference on Signals, Circuits and Systems (SCS'09)*, Jerba, Tunisia, November 6-8.

Hentout, A., Bouzouia, B., Akli, I., Ouzzane, E., Benbouali, R. & Bouskia, M. A. (2009e). "Multi-Agent Control Architecture of Mobile Manipulators: Following an Operational Trajectory". *The 3rd 2009 IEEE International Conference on Signals, Circuits and Systems (SCS'09)*, Jerba, Tunisia, November 6-8.

Hutchinson, S., Hager, G. D.& Corke, P.I. (1996). "A tutorial on visual servo control". *IEEE Transactions on Robotics and Automation, 12(5)*, pp. 651-670.

Janabi-Sharifi, F.& Wilson, W. J. (1998). "Automatic Grasp Planning for Visual-Servo Controlled Robotic Manipulators". *IEEE Transactions on Systems, Man, and Cybernetics-Part B: Cybernetics, 28(5).*

Katz, D., Horrell, E., Yang, Burns, Y., Buckley, B., Th., Grishkan, A., Zhylkovskyy, V., Brock, O.& Learned-Miller,E. (2006). "The UMass Mobile Manipulator UMan: An Experimental Platform for Autonomous Mobile Manipulation". *IEEE Workshop on Manipulation for Human Environments*, Philadelphia, USA.

Khalil, W. & Kleinfinger, J. F. (1986). "A new geometric notation for open and closed-loop robots". *The International Conference on Robotics and Automation*, USA, pp. 1175-1180.

Muis, A.& Ohnishi, K. (2005). "Eye-to-Hand Approach on Eye-in-Hand Configuration within Real-Time Visual Servoing". *IEEE/ASME Transactions on Mechatronics, 10(4).*

Nagatani, K., Hirayama, T., Gofuku, A. & Tanaka, Y. (2002). "Motion planning for mobile manipulator with keeping manipulability". *Proceeding of the 2002 IEEE/RSJ International conference on Intelligent Robots and Systems (IROS 2002)*, Switzerland.

Petersson, L., Egerstedt, M. & Christensen, H.I. (1999). "A hybridcontrol architecture for mobile manipulation". *Proceedings of IROS'99*, Kyongju, Korea, pp. 1285-1291.

Sasaki, H., Takahashi, T. & Nakano E. (2001). "A door opening method by a mobile manipulator with passive joints". *Journal of Robotic Society of Japan, 19(2)*, pp. 129-136.

Sugar, T. & Kumar, V., (1998). "Decentralized Control of Cooperating Mobile Manipulators," *Proceedings of the IEEE International Conference on Robotics and Automation*, Leuven, Belgium, pp. 2916-2921.

Telle, B., Aldon, M.-J. & Ramdani, N. (2003). "Guaranteed 3D Visual Sensing Based on Interval Analysis". *Proceedings 2003 IEEE/RSJ International Conference on Intelligent Robots and Systems (IROS 2003).* pp. 1566-1571, October 27-31.

Trabelsi, M., Aitoufroukh, N.& Lelandais, S. (2005). "Saisie d'objets à l'aide d'une caméra et des ultrasons pour la robotique de service". *The 3rd International Conference: Sciences of Electronic, Technologies of Information and Telecommunications (SETIT 2005)*, Tunisia, March 27-31.

Tsai, R.Y. & Lenz R.K. (1989). "A New Technique for Fully Autonomous and Effcient 3D Robotics Hand/Eye Calibration", *IEEE Transactions on robotics and automation, 5(3)*, pp. 345-358.

Waarsing, B. J. W., Nuttin, M.& Van Brussel,H. (2003) "Behaviour-based mobile manipulation: the opening of a door". *ASER'03, Bardolino (Italy)*, pp. 168-175.

Winter, N., Gaspar, J., Lacey, G.& Santos-Victor, J. (2000). "Omnidirectional Vision for Robot Navigation". *Proceedings of the IEEE Workshop on Omnidirectional Vision (OMNIVIS)*, pp. 21-28, South Carolina, USA.

10

A Concept for Isles of Automation

Mikko Sallinen and Tapio Heikkilä
VTT Technical Research Centre of Finland
Finland

1. Introduction

1.1 Approach to the topic

There has been a major change in the manufacturing after year 2000. At the same time when production is struggling with tougher price competition, the requirements for flexibility are increasing. Life cycles of the products are getting shorter, the variety of products is increasing, and production costs should be decreased at the same time when there is no good technical solution on the market to answer to this. Customer assumes to get a tailored solution with the same price and delivery as a mass product previously.

Using the current manufacturing technologies response to the needs of the market is getting increasingly difficult. Even if the modern manufacturing systems include a remarkable amount of ICT technologies, the flexibility of a human worker is very challenging to reach. Although robotic systems are classified as a flexible production technology, in practice the current robotic implementations are concentrated to high volume production (Naumann et. al 2006). Only a few industrial solutions have been installed for short and single series production. The main obstacle in installing robots for short series production is the amount of product-specific costs caused by the manual work still required for using the robot system (Sallinen et. al 2004). In most of the cases each work phase and process for each product has to be programmed, and the function of auxiliary equipment is usually based on part-specific geometry e.g. no exceptions beyond the designed parts are allowed. If the volumes of the product are low, the effective utilization of a robot assumes that it is applied to a large variety of parts, or to a large amount of different work phases, to bring the robot utilization rate to a decent level typical of the SME industry. In addition, parts to be processed in the production environment very often have complicated forms which make manipulation more challenging. This is the case especially in the short series production when most of the manufacturing processes are solved relatively well at the moment.

1.2 State-of-the-art

The development of manufacturing systems has had two main approaches (Maraghy 2006): Flexible Manufacturing Systems (FMS) and Reconfigurable Manufacturing Systems (RMS). The concept presented here has adapted features from both. Our approach is that the basic element of the automation island is an industrial robot equipped with different kinds of external sensors and auxiliary devices combining mechanics, sensor technology, tools and

software. This gives high level flexibility in terms of programmability, reconfigurability, reusability and price.
A lot of studies from different areas related to flexible robot automation can be found such as off-line programming (Brock 2000)(Burns & Brock 2005)(Sanchez & Latombe 2002). There are solutions for off-line programming where the first area has been welding (Dai & Kampker 2000). Another approach is on-line programming which modern version is called programming by demonstration (PbD) (Dillman et. al 1999)(Chen & McGarragher 2000)(Rogalla et. al 2002). Typical example of PbD is where human with ordinary tools shows the task by guiding the robot by hand (physical interaction) or human shows some task and the robot recognizes that by a vision system. System even try to learn tasks from the demonstation (Kang & Ikeuchi 1995). Learning can be divided to task- skill and body recognition which have been studied in (Nakaoka et. al 2006).
Studies about flexible robot system architectures can be found from mobile robotics. In the studies the task level control is common topic (Camargo et. al 1992)(Parker 1998). Compared with the architectures presented in the literature we present a solution called "Isle of Automation" which enables multiplying the workcell while the whole structure defines the overall architecture. Architecture of a holonic manufacturing system (Takamura et. al 2003) is close the approach presented in this chapter. The principle of operation of an automation island is that it is working autonomously but is not cooperative by default as holons are.
In this chapter, we present a concept for short series production based on backround system called *Engineering resources* and executive robot cell called *Production cell* consisting of industrial robots equipped with assistive manipulation systems, sensors and a control system. We call this complete system as an *Isle of automation*. The new concept is beyond the current robot workcells by the properties of flexibility, reconfigurability, context awareness and programmability. In the concept, there are modules for programming, sensing, material handling and flow as well as communication. The overall architecture defines how these modules are working together. In this chapter, we present these modules and illustrate the operation in pilot case.
This chapter has been organized as follows. In chapter 2 we introduce and define the overall concept of the isle of automation, in chapter 3 we introduce the architecture of production cells, in chapter 4 components of the isles of automation. Communication between the structures are explained in paragraph 5. Feasibility of our approach is shown in chapter 6 by a pilot case, discussion is given in chapter 7 and finally conclusions are drawn in chapter 8.

2. The Overall Concept for Isles of Automation

The concept of isles of automation for short series production consists of two main parts: Engineering Resources and Production cells. The overall system follows a modular structure and realizes highly flexible and controllable robotized system. It exploits the features of ubiquitous technology including flexibility, adaptivity, context awareness and reactivity, which are beyond the current automation solutions. The production system easily adapts to new products or product variants and to deviations in work pieces. In addition, data acquisition presents new possibilities, when open interfaces are offered down to the sensor level (e.g. measurement signals can be monitored). This means sensors offer services and are visible to the whole control system including user. Sensors can be used for on-line purposes such as control but also off-line monitoring such as quality control and prognosis of the

maintenance of the machines. This kind of features can not be found in the current systems such as presented in (Bloomenthal et. al 2002).
The operation of the automation island is managed by control software locating in Engineering Resources, called *Isle Manager*, which is controlling the execution of tasks in a high level. It also manages the sensing and reactions to non-expected situations in the robot cell. The work is carried out by communicating with distributed modules and providing the ways to carry out the tasks.
Figure 1 illustrates the concept of Isles of Automation. In the top there is the Engineering Resources and below that Production cells. More detailed description is given in next paragraphs.

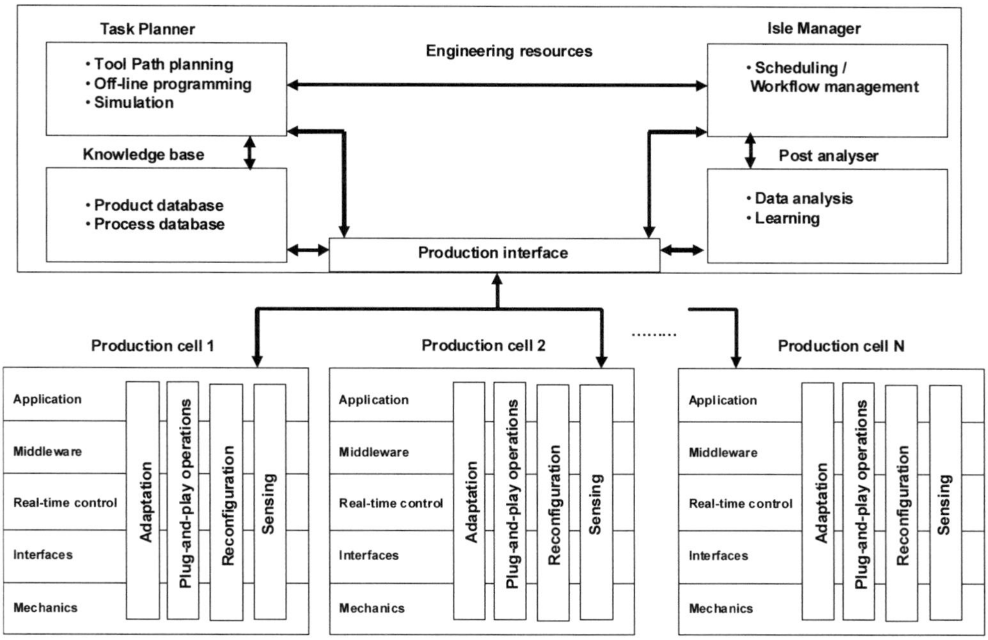

Fig. 1 The conceptual structure of isles of automation

The concept of the automation island is described in the figure 1. The main parts of the system are *Engineering Resources* and *Production Cells*. Interaction between these parts is described in paragraph 5.

2.1 Description of the concept: Engineering resources

The Engineering Resource is working as an autonomous backround system and operating mainly off-line. It consists of pre-processing and post-prosessing functions depending on the time of operation. Operation of the robot and auxiliary device is supported by pre-processing functions for which there are planning, off-line programming and simulation, see box in top left in the figure 1. To support the process, there is a knowledge base for product and process information. Knowledge base may locate in local engineering computer or the information can be obtained from enterprise factory database. For overall feedback and analysis, there are data analysis and learning modules, see boxes in right down in figure 1.

Isle manager takes care of the overall management and workflow in the whole system, see box right up in the figure 1.
If the factory contains a large number of production cells in many departments, there can be a pool of Engineering Resources which gives services for productions cells. Another option is that each cell has its own Engineering Resource which means they are operating autonomously like separate islands. This is a typical case when there are one or two production cells in the factory and they are operating in very different applications. In that case engineering resources are embedded in the Production Cell. In principle, an engineering resource of one production cell can offer services to other production cells as well.
The decision making is distributed in the Isle of Automation. There is a high level controller, which takes care of the high level production management. Flexibility in production also sets requirements to the managing and controlling of the island. To use hardware efficiently, flexible, modular and reconfigurable software must be used at every level to manage the whole system. Modular structure and re-programmable software means that operations and functions of the production cells can easily be configured and used on-line. This approach has several features of Service Oriented Architecture approach in production environment, see e.g. (Veiga et. al 2007).

2.2 Production cell

The concept of the Production cell has a layered structure for different response levels. These layers include hardware, interfaces, real-time control, middleware and application layers. The key-functions in the island are adaptation, reconfiguration, sensing and plug-and-play operations. These functions are operating vertically in the cell, see figure 1. Depending on the requirements of the applications, the properties, operation and status level of these key-functions are defined. They are explained more detailed in chapter 3.2.

2.3 Interaction between Engineering Resources and Production Cell

Communication between the cells and engineering resources is carried out through a production interface between the application layer and modules of the engineering resources system, see figure 1. Data exchange is not time critical and common formats are defined.
There can be several production cells in the system as illustrated in the figure 1. Each cell may have it's own function such as the first cell is making the cutting, the second cell is doing the welding and the third cell is doing the deburring. They can exchange and share information (e.g. updated product status data and geometrical information) and resources (e.g. sensors, devices, tools). Flexibility of production means that a product can be manufactured in any of the cells if the cells change the required tools and sensors guided by the Isle Manager.

3. Architecture of the production cells

The architecture of production cells. It is built based on layered structure consisting horizontal layers for required operations. In addition to horizontal layers, there are vertical functions called Key functions which use properties of different horizontal layers.

Architecture described in this chapter gives rules and methods for cross-operation of these layers and functions.

3.1 5-Layered structure

Layers of the production cell are described in the figure 2. All the units of the cell (e.g. robot manipulator, controller and device controller) will contain the same layered structure: application layer, middleware, real-time control, interface (API) and physical layer (e.g. mechanics). Each layer consists of operations to operate with other layers. Also if a new unit or device is connected and the operation should be transparent to the user, the layer structure should be the same. Depending on the functional requirements of each unit, different layers will have respective operations.

Communication between the vertical layers is carried out using interfaces suitable for each device (e.g. sockets, buffers or ethernet). Communication is recommended to be carried out between same layers to enable reliable and secure synchronization of the communication, especially in the real-time control layer. In the upper layers (application, middleware and real-time control) communication is carried out using textual structures, e.g. XML. In a time critical layer such as real-time control, interfaces and communication can be carried out using special real-time standards such as industrial Ethernet or digital or analog I/O or industrial field buses if very fast communication is required.

An exemplary content of each layer is described in table 1. In the application layer, there can be application or robot application program running in the cell computer or in a robot controller. The common property is that programs are not time-critical compared with programs in real-time control layer. In the case when programs run in cell computer, they may operate on Windows or Linux operating systems. In the middleware layer, there are services for application layer. Most of the services are built such that they are invisible to user.

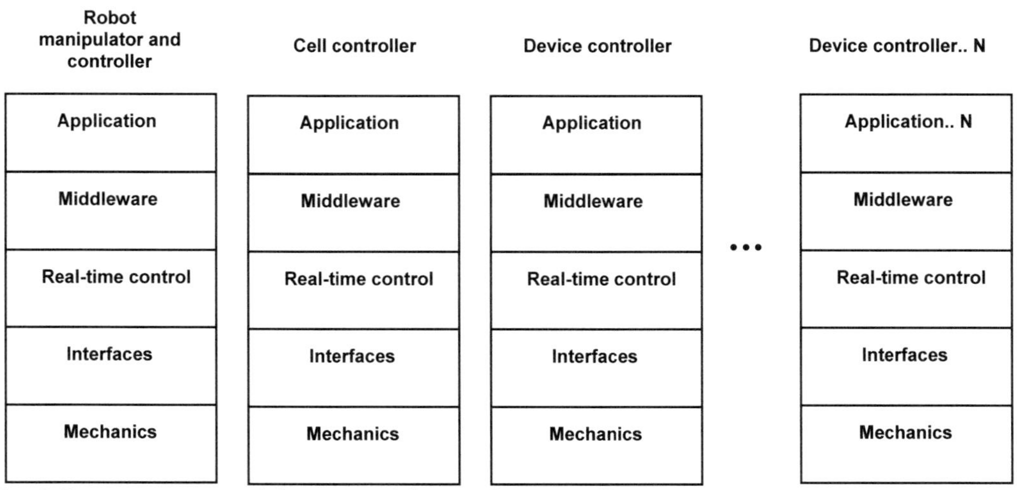

Fig. 2. Layer structure of the units of the production cell

The basis for key functions are in the middleware layer. Real-time control layer is established with user functions, upon the services of real-time operating system. In the robot controller, all the kinematic calculation and motion control is carried out in this layer. In this layer, there are often real-time operating systems such as real-time linux or embedded windows or KUKA's RT kernel. Interface layer has interfaces to external devices and communication networks using digital or analog lines or standard ethernet or industrial Ethernet. At the bottom, there are Mechanics layer which has physical devices, interface cards and tools, see table 1.

3.2 Key functions

Key functions are services available in the production island going through the layers as described in figure 3. Multi-layer operation means that they utilize each layer depending on the requirements. The purpose of the key functions is to carry out ubiquitous operations of automation island. It consist of intelligent, interactive and reactive operations of a cell can consist of one or several key functions.

There are four key functions which are adaptation, plug-and-play operations, reconfiguration and sensing. As layers described above, there do not have be fully operating key functions in every unit. Also, the architecture supports the operating principle where different units or devices can or do utilize key functions from each other. Example of this can be e.g. that operation of force sensor is utilized by both programming-by-demonstration and reactive execution. Operation for requirement of application of force sensor is provided by the co-operation of both Key-functions adaptation and sensing where adaptation includes operations for changing the robot motion paths and sensing includes properties for signal processing of low-level force sensor.

The operation principle of key functions are as follows: Adaptation function is on-line or off-line reaction to changes of product or production. It utilizes sensing -key-function to achieve the measurement data for the basis of the operation. Plug-and-play function enables easy connectivity of new sensors which can be used in the adaptation of the production system to new, different size of workobjects. In general, plug-and-play functions enable an easy way to connect and disconnect components such as sensors, actuators, tools and devices between production islands. Reconfiguration function enables making of structural changes in the production cell automatically or by physical assistance of operator.

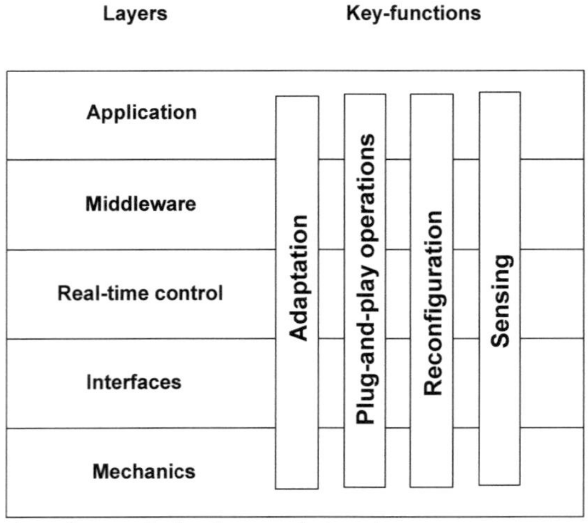

Fig. 3. Key functions going through the layered structure.

The changes are carried out such that all the required properties of the island will be achieved. Reconfiguration is also supported by plug-and-play operations. Sensing includes low-level signal processing properties and it also provides different kind of upper level sensing / measurement services for other functions and layers. It will utilize plug-and-play operations to easily change sensors between production cells.

Layer	Example of operation
Application	Application program, robot program
Middleware	Services for upper and lower layers including key functions
Real-time control / OS	RTOS: RTLinux, linux, embedded windows
Interfaces	Analog, digital, ethernet, device drivers
Mechanics	Manipulators, grippers, feeders, tools, sensors

Table 1. The content of the layers

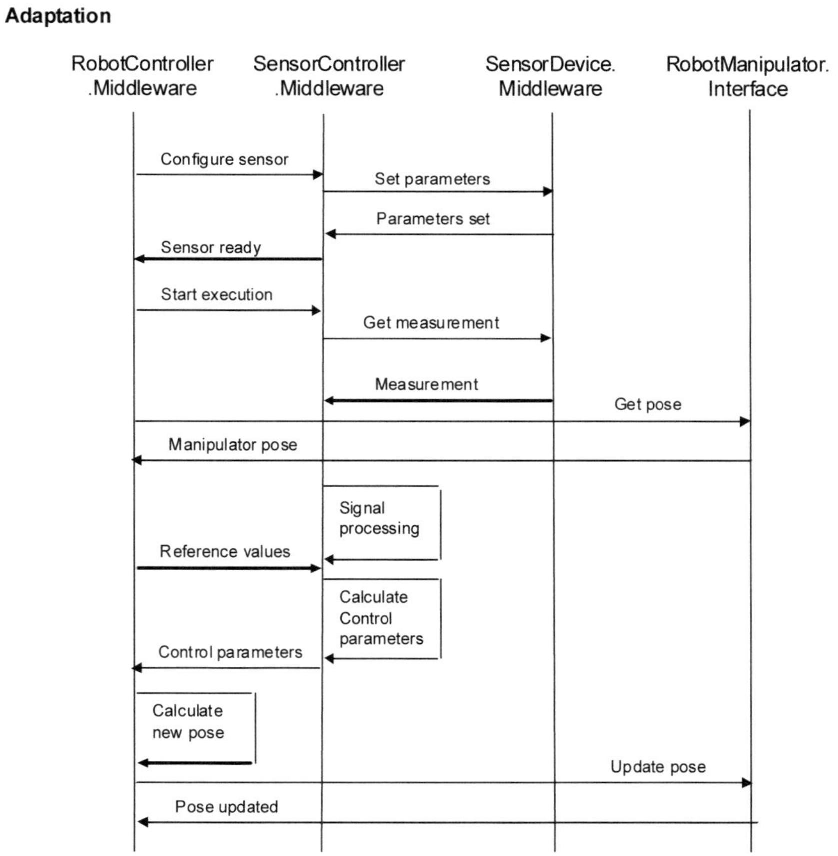

Fig. 4. Message sequence for adaptation –key function

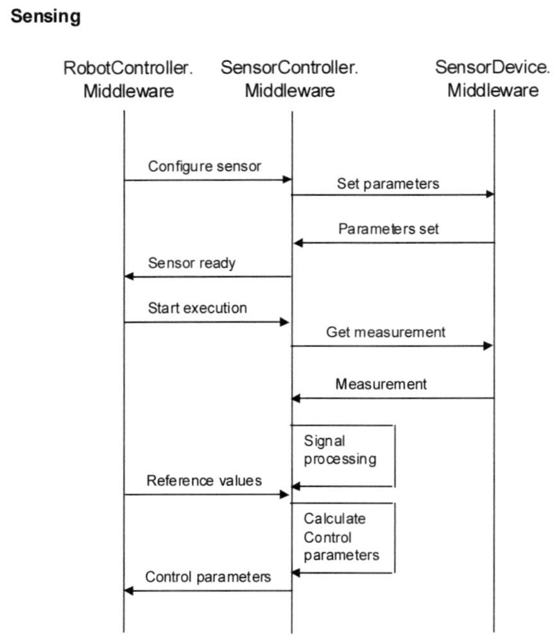

Fig. 5. Message sequence for sensing –key function

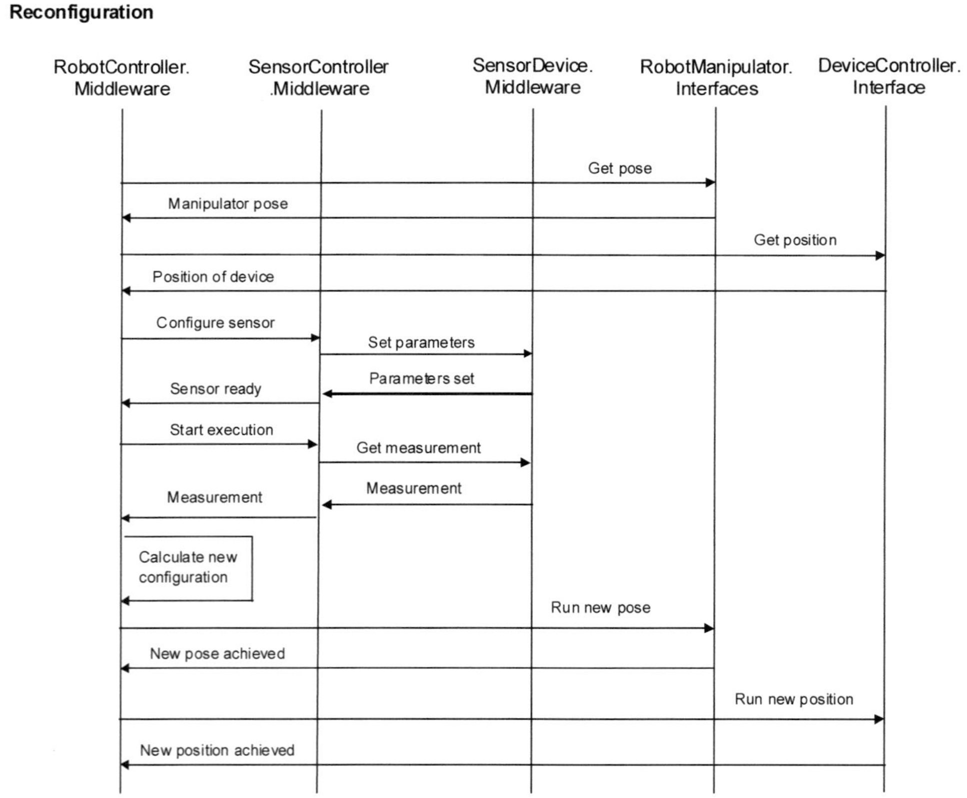

Fig. 6. Message sequence for reconfiguration –key function

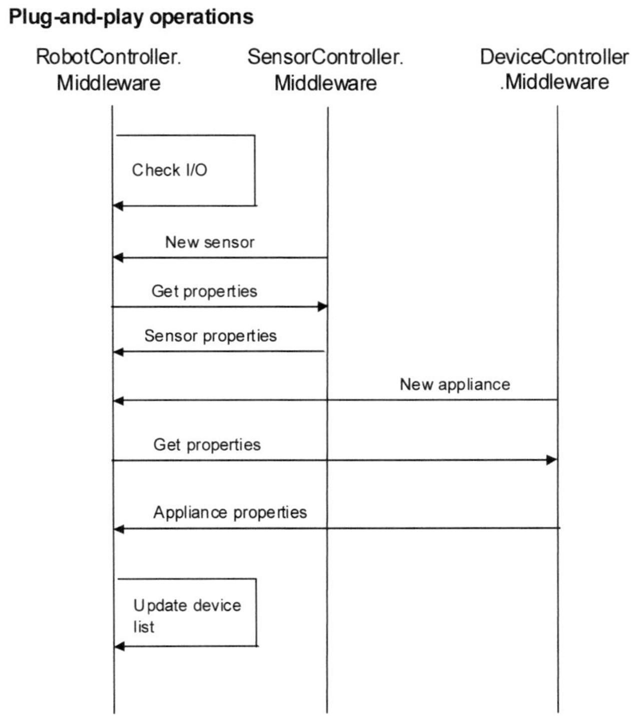

Fig. 7. Message sequence for plug-and-play -key function

4. Components Isles of Automation

Here we introduce components used in the Isles of Automation. The work operations of the Isles of Automation can be grouped and named as components and they are working in the layers and key functions described above. For this component-based approach for the Isles of Automation is given. Components are also in line with the architectural description given in chapters 2 and 3. Based on analyses of the current stage of the technology, technologies and methods are selected for the concept (Sallinen et. al 2006)(Salmi et. al 2007).

4.1 Description of the components

The main components of the automation island are 1) programming subsystem, 2) robot and external sensors, 3) material handling devices (e.g., grippers, feeders), 4) control system and 5) communication system. Simplified information flow of these is also described in figure 4. Programming tools include both off-line programming tools and on-line programming which is required in on-line reactivity. Robot and external sensors include robot manipulator and sensors like force, vision and laser rangefinders to observe the environment. The selection of these sensors depends on the requirements of the application. Material handling devices will make sure that the robot has pieces in the right position to be manipulated. Grippers and manipulators are specially designed or selected from the existing ones to manage flexible operations. Requirement of those is at least a low level programming to behave actively in the Automation Island. In that way they can support also reconfigurable operations such as modification to very different size of workobjects.

Workflow management software in Engineering Resources is above all and controls operations in the task level, e.g. how different phases of the workobject are carried out in the work flow. New tools and devices can be connected in a plug-and-play manner without parameter configuration. They utilize plug-and-play key functions. Communication and control system defines the information flow in the Isle of the Automation, where communication defines the protocols of the communication. All these components are designed to be built up using both commercial components available from the market as well as components built by ourselves. If the component available in the market fills the system requirement, it is the best selection for the use.

Component-based approach is a key element in achieving the desired flexibility and reconfigurability features. The components are spread out from the factory level down to the smallest functional units of devices such as sensors. It affects the physical structure, control devices, data transfer solutions and sensor utilization. The concept includes necessary modules for various purposes. The modularization also serves the aims of standardization and quality.

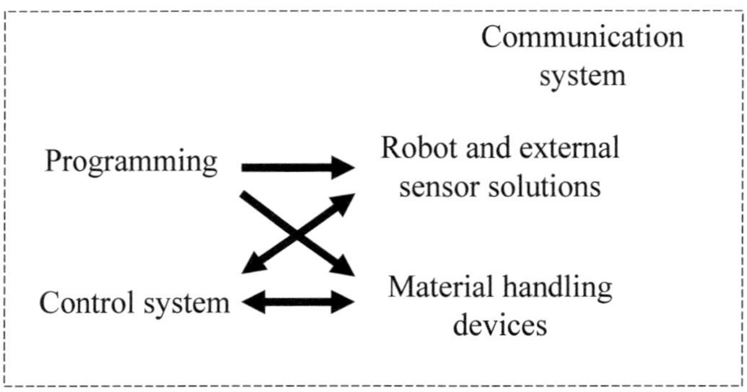

Fig. 8. The connectivity flow between the main components of the isles of automation.

5. Communication in the Isles of Automation

Here we explain the communication between the units in Isles of Automation. In the figures 5 and 6 there is a description of signal flows of in the case of task planning and task execution.

Task planning is operating in Engineering resources and is starting by order request from scheduler, see figure 5. It is requested from the task planner. Task planner is requesting a program from CAD tool. CAD tool will collect data from product database and process database. It has also information about the workcell environment including robots and all additional peripherals such as tools and sensors. Whet it receives this information it plans, simulates and makes a program ready-to-run in the robot. When program is ready, it's timing in the work line will be requested from the workflow manager and returned to schedule that task is in organized.

Task execution is operating in production cells, see figure 6. Task planner is sending the program to robot controller using ethernet or serial line. This can be done off-line. Scheduler will be responsible to start the execution of the program in the robot controller.

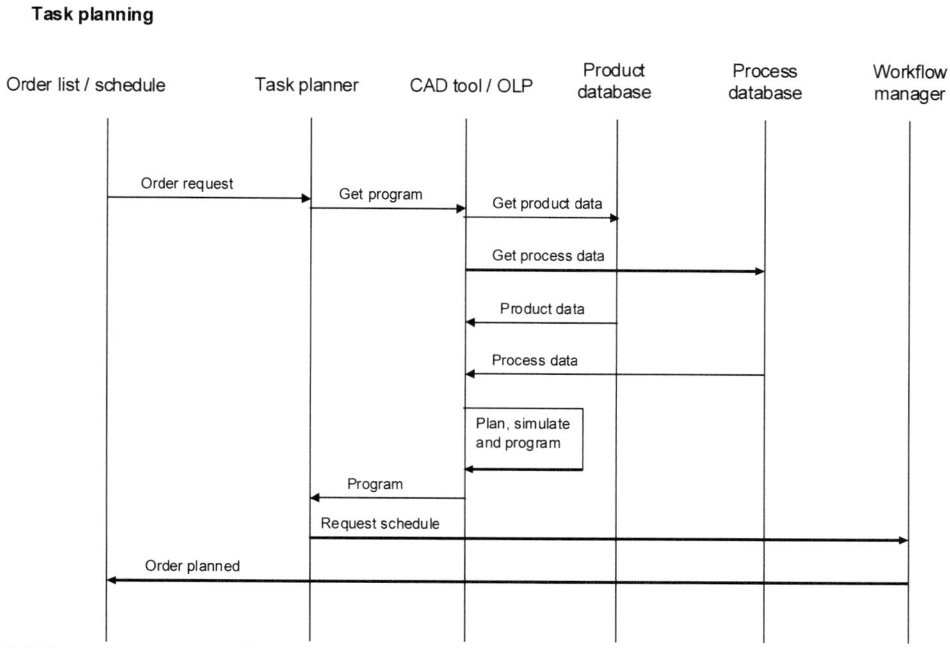

Fig. 9. Message sequence for the task planning

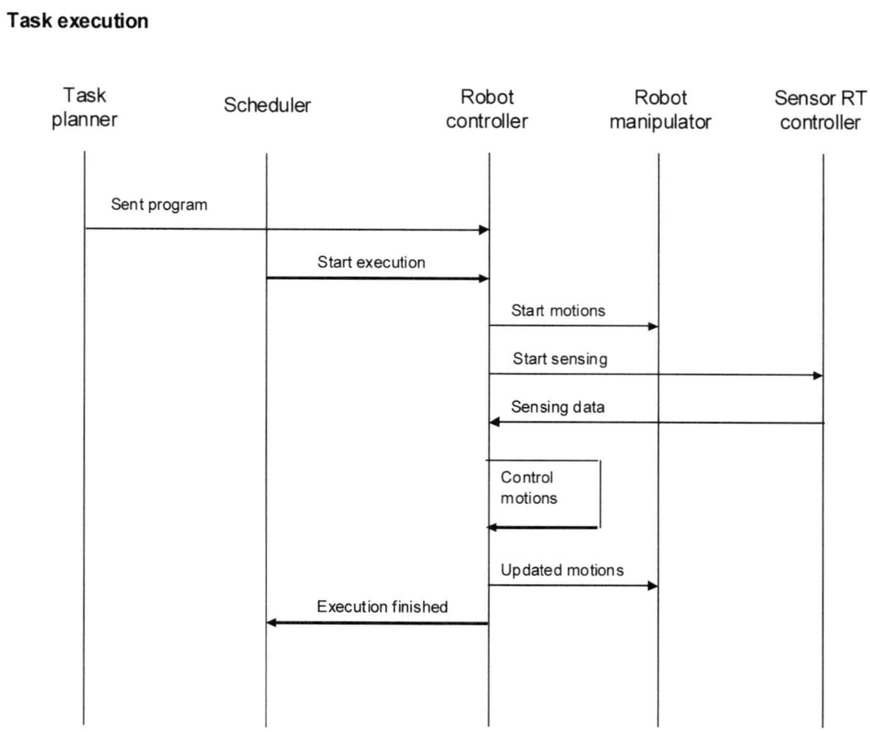

Fig. 10. Message sequence for the task execution.

Execution is carried out by first starting the motions in the robot manipulator and starting also the sensing of the external sensors by communicating with the sensor real-time controller. This sensor is typically force-torque sensor. During the execution, sensor returns the sensing data back to the robot controller. Based on the motions and pose of the robot and sensor measurements, motions for the robot manipulator will be calculated. Afterwards these updated motions will be sent to robot manipulator. When the execution is finished, information to the scheduler will be sent.

6. Demonstration

In this chapter, we give an example of applying the concept for Isles of Automation in a pilot case. The task of the demonstration was to deburr bevels of a sheet metal plate which was bent into 3D form. Input data for the system was a 2D-CAD drawing of the workobject and manufacturing data. The properties of the robot workcell (such as dimensions between the objects and reachability of the robot) was known.
In the engineering resources, off-line programming of the robot motion paths is based on 2D-CAD drawings made in Nestix2 (Nestix 2009) software. The software itself is designed for nesting 2D workobjects such as sheet metal plates and bewelling or deburring paths in 2D space. The drawings included both geometrical information and 2,5D milling paths for the deburring of the bevels. The 2,5D information of the paths included location in the 2D plane and angle of the bevel.

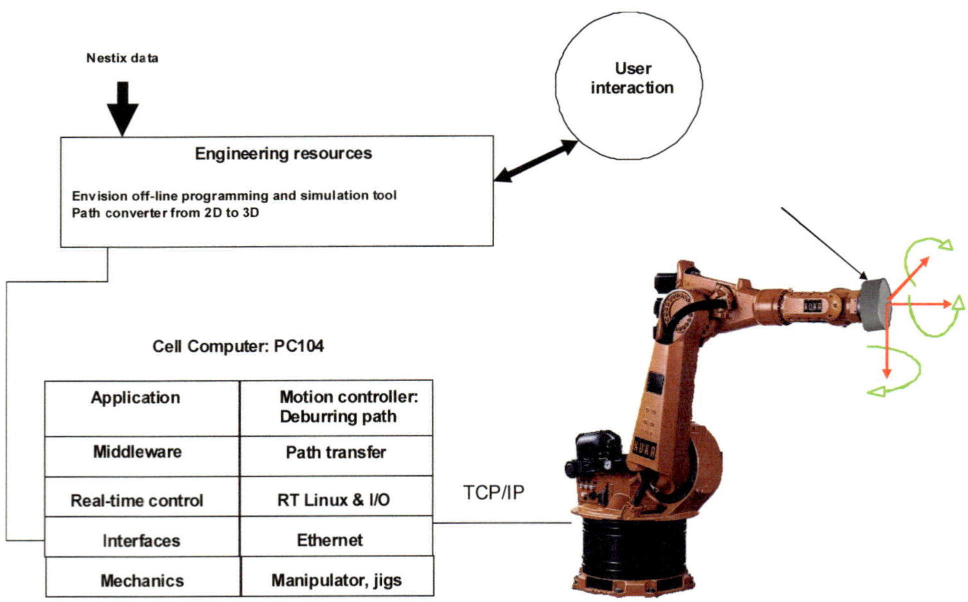

Fig. 11. Case implemented into Automation Island framework.

To fasten the programming of the robot, a converter to transform paths from 2D plane into 3D space based on the part 3D bending information was developed. After the transformation, there was a 3D model of the workobject and a 3D deburring paths (tags in the surface of the workobject). The robot motion paths were generated based on the 3D tags

in the surface of the workobject. This phase was supported by a robot motion path planner which calculated the paths for robot motion such that all points are reachable in a same joint configuration (for more information, see (Sallinen et. al 2006)).
The workflow of the demonstration task is illustrated in figure 8. In the workflow, first three operations are carried out by the engineering resources and the last one by the production cell. Scheduling / Workflow management is carried out manually by the shop floor operators. Engineering resources will generate programs to application layer in the production cell.
The robot programming was carried out using the ENVISION off-line programming tool by Delmia (Delmia 2009) for visualizing the virtual robot cell and transformation of workobject from 2D to 3D data. In the actual demonstration we used KUKA KR150-L110 industrial robot with KRC2 controller and deburring of the bevelling were done by a simple tool protype. Localization of the workobject was carried out using robot's own touching method where user shows axis in the workobject. In the demonstration, the purpose was to show the interfaces between the different parts of the system could be done easily. Generation of 3D model and paths from workobject 2D data succeed. In the demonstration, we did not consider any further process related issues such as tools and quality of the bevelling.
The implementation of the architecture into proposed framework is illustrated in figure 7. It also described the communication between cell computer and robot controller. Lines where data is transferred. Cell computer is PC104 –based solution with real-time linux which enables easy-to-integrate interfaces for sensors and actuators. There is not so much attention paid to workflow management because demonstration is not an industrial case or the productivity in the sense of workflow is not that important.
In the demonstration case there was no external sensors, especially which would need real-time communication and control. Therefore Ethernet communication was a proper solution for the communication.

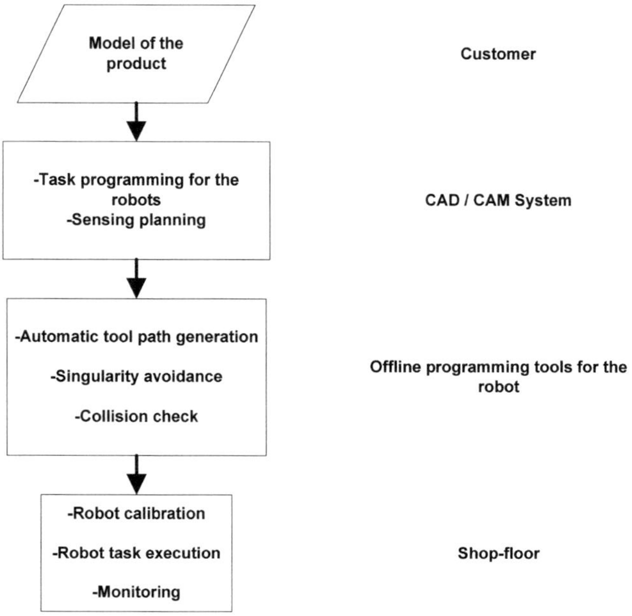

Fig. 12. Workflow in demonstration case.

7. Discussion

The proposed concept gives a framework for design of the robot workcells and different types of production units. The purpose has also been to give a design tool or guideline for making an efficient production unit. The proposed system do not necessary have to be completely implemented, there is possibility to also take part of the concept for the system. The demonstration gave very promising results about the usability of the concept. From 2D to 3D converter operated well and it fasten the programming. Off-line programming in short series production is cost effective when it is done half- or fully automatically. If the user have to make a lot of manual work, it may even take more time and is more expensive than on-line programming. Unfortunately that is the case in many real production cell. One solution for this is fully automatic off-line programming tool which optimises robot motion paths using tag point information (Simtech 2010).

8. Conclusions

In this chapter, we presented a novel concept for short series manufacturing. The concept is called Isles of Automation and it defines a system structure composed of engineering resources and production cells. System consists of key functions whose content we defined. Also communication between the functions and different tasks was described. System is scalable and can be implemented into several applications. We described the content of these parts in detail and how the whole system operates. In the chapter, we illustrated a demonstration case in laboratory where selected parts of the concept were implemented into a robot cell in the deburring application. The proposed concept showed to be efficient and easy-to-integrate into the different applications.

9. References

Bloomenthal M., Riesenfeld R., Cohen E., Fish R., "An Approach to Rapid Manufacturing with Custom Fixturing", IEEE Int. Conference on Robotics and Automation, San Francisco, USA, pp. 212-219. 2002.

Brock, O. Generating Robot Motion: The Integration of Planning and Execution. Ph.D. thesis, Stanford University, Stanford University, USA; 2000.

Burns B., and Brock, O., "Single-Query Entropy-Guided Path Planning" Proc. 2005 IEEE Int. Conf. on Robotics and Automation, Barcelona, Spain, April 2005.

Camargo R. F., Chatila R., Alami R.,. "Hardware and Software Architecture for Execution Control of an Autonomous Mobile Robot". Proc. Of IEEE Int. Conference on Industrial Electronics, Control, Instrumentation and Automation. 818-825. 1992.

Chen, J.R., McGarragher, B.J. Programming by Demonstration - constructing task level plans in hybrid dynamic framework. Robotics and Automation, 2000. Proceedings. ICRA '00. IEEE International Conference on, Volume 2, pp. 1402-1407. 24-28 April, 2000.

Dai W, Kampker M. (2000) User Oriented Integration of Sensor Operations in a Offline Programming System for Welding Robots. Proceedings of the 2000 IEEE International Conference on Robotics & Automation, San Francisco. CA April 2000.

Delmia. www.delmia.com/, 2009. Delmia, Digital 3D Manufacturing Solutions.

Dillman R., Rogalla M., Ehrenmann M., Zöllner R., Bordegoni M., Learning robot behavior and skills based on human demonstration and advice: the machine learning paradigm, 9th Int Symp. of Robotics Research, October 1999.

Kang S.B., Ikeuchi K., A Robot System that Observes and Replicates Grasping Tasks, Fifth Int. Conf. Computer Vision, 1995.

Maraghy, H.A., "Flexible and reconfigurable manufacturing systems paradigms". *International Journal of Flexible Manufacturing Systems* 17, Springer Science Business Media. pp. 261 - 276. 2006.

Nakaoka S., Nakazawa A., Kanehiro F., Kaneko K., Morisawa M., Hirukawa H., Ikeuchi K., "Leg Task Models for Reproducing Human Dance Motions on Biped Humanoid Robots", *Journal of the Robotics Society of Japan*, 24:3, pp. 388-399, April 2006

Naumann M., Wegener K., Schraft R., Lachello L., "Robot Cell Integration by Means of Application P'N'P.". 8p. ISR2006.

Nestix. http://www.nestix.com/, 2009.

Parker. "ALLIANCE: An Architecture for Fault Tolerant Multirobot Cooperation". *IEEE Transactions on Robotics and Automation*, Vol. 14, No 2. Pp. 220-240. 1998.

Rogalla O., Ehrenmann M., Zollner R., Becher R., Dillmann R: Using gesture and speech control for commanding a robot assistant; 11th IEEE Int. Workshop on Robot and Human Interactive Communicative, pp. 454-459, 2002

Sallinen M., Heikkilä T., Sirviö M. "Planning of sensory feedback in industrial robot workcells". Proc. of the IEEE Int. Conference on Robotics and Automation. pp. 675-680. 2006.

Sallinen M, Salmi T, Haataja K, Göös J, Voho P., "A Concept for Short Series Production: Isles of Automation". Proceeding of the Smart Systems 2006. 2006.

Sallinen M., Järviluoma M., Sirviö M., Väinölä J., Ruusu R.,. "A robotized system for prototype manufacturing of castings with various sizes of pieces". Proceedings of the 5th Int. Conf. on Machine Automation. ICMA2004., 517 - 522. 2004

Salmi T., Haataja K., Sallinen M., Göös J., Voho P. "Automation Islands - Requirements and Solutions for a Highly Flexible Concept of Robotic Systems". The 2nd International Conference on Changeable, Agile, Reconfigurable and Virtual Production (CARV 2007). 10p.

Sanchez G., and Latombe J.C., "A single-query bi-directional probabilistic roadmap planner with lazy collision checking", 2002

Simtech, http://www.easysimulation.com/,2010

Tamura S., Seki T., Hasegawa T., "HMS Development and Implementation Environments". Agent-Based Manufacturing, Advances in the Holonic Approach. Springler. 2003.

Veiga G.1, Pires J., Nilsson N.,. "On the Use of Service Oriented Platforms for Industrial Robotic Cells". IMS2007. 7p.

11

Stiffness Analysis for an Optimal Design of Multibody Robotic Systems

Carbone Giuseppe
LARM: Laboratory of Robotics and Mechatronics, University of Cassino
Via G. Di Biasio, 43 – 03043 Cassino (Fr)
Italy

1. Introduction

Robots are widely used to help human beings and/or to execute various manipulative tasks in industrial applications and even in non-industrial environments. Researchers are still widely investigating robotics with the aim to further improve a robot performance and/or to enlarge their fields of application. These tasks can be achieved only when the peculiarities in Kinematics and Dynamics behaviors are properly considered since the early design stage. Significant works on the topics can be considered the pioneer papers (Shimano & Roth, 1978), (Vijaykumar et al., 1986), (Paden & Sastry 1988), (Manoochehri & Seireg 1990), and more recently the papers (Angeles 2002), (Hao & Merlet, 2005), (Carbone et al. 2007), just to cite a few references in a very rich literature. Algorithms have been proposed, for example, as based on workspace characteristics (Schonherr, 2000), and global isotropy property (Takeda, & Funabashi,1999), separately. Several (often conflicting) criteria can be taken into account in the design process. Only recently, it has been possible to consider simultaneously several design aspects in design procedures for manipulators. Multi-criteria optimal designs have been proposed for example in (Ottaviano & Carbone 2003), (Hao & Merlet, 2005).
The significance of each design criterion is often strongly related with specific application task(s) and constraints. Therefore, in this chapter several design criteria are overviewed with specific numerical evaluation procedures for analytical definition of design optimization problems. But, among the design criteria special attention is addressed to stiffness, since it can be considered of primary importance in order to guarantee the successful use of any robotic system for a given task (Ceccarelli, 2004). Indeed, there are still open problems related with stiffness. Still an open issue can be considered, for example, the formulation of computationally efficient algorithms that can give direct engineering insight of the design parameter influence on stiffness response. There is also lack of a standard procedure for the comparison of stiffness performance for different multibody robotic architectures. Therefore, this chapter is also an attempt to propose a formulation for a reliable determination and comparison of the stiffness performance of multibody robotic systems by means of proper local and global stiffness performance indices. Then, the proposed numerical procedure is included into a multi-objective optimal design procedure, whose solution(s) can be achieved

even by taking advantage of solving techniques in commercial software packages. Illustrative examples are reported, also with the aim to clarify the computational efforts.

2. The optimal design problem and its formulation

The design problem for manipulators consists in several phases. The first phase is the type synthesis. In this phase a designer should select the type of kinematic architecture that can provide the desired stiffness, mobility, force, efficiency, size. For example, the architecture can be chosen as open chain or parallel structure, Fig.1. In addition, different solutions can be selected within each structure as depending on manipulative tasks.
After the type synthesis one should perform a dimensional synthesis aiming to compute values of design parameters that characterize and size the kinematic structure of a manipulator. Several aspects can be considered in a design procedure at this stage in order to achieve suitable performance for the desired application tasks.
Often performance improvements can be obtained from the point of view of a design criterion at the cost of worst performance in terms of other design criteria. Thus, it is very useful to develop computer aided procedures that can attempt to provide a design solution by considering more than one design criterion at the same time.
An optimization problem can be formulated in a very general form as

$$\min \mathbf{F}(\mathbf{X}) \tag{1}$$

subject to

$$\begin{aligned} \mathbf{G}(\mathbf{X}) &\le 0 \\ \mathbf{H}(\mathbf{X}) &= 0 \end{aligned} \tag{2}$$

where **X** is the vector whose components are the design parameters; **F** is the objective function vector, whose components are the expressions of mobility criteria. **G(X)** is the vector of inequality constraint functions that describes limiting conditions. **H(X)** is the vector of equality constraint functions that describes design prescriptions.

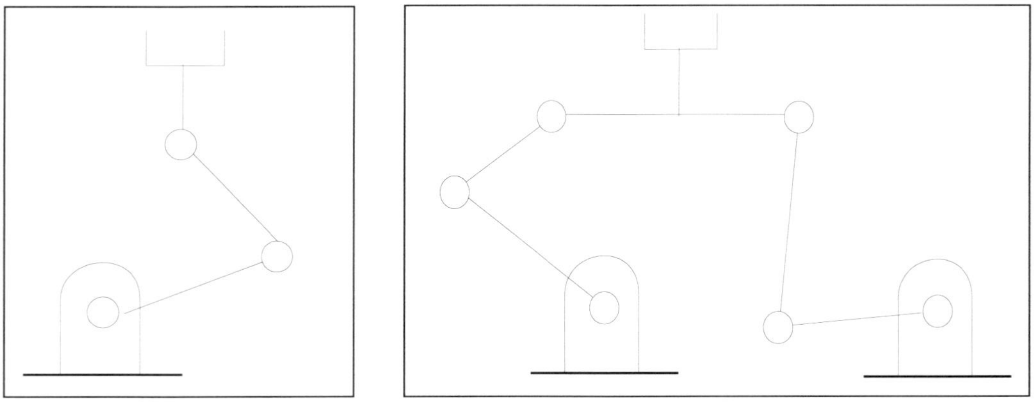

a) b)

Fig. 1. Planar examples of kinematic chains of manipulators, (Ceccarelli, 2004): a) serial chain as open type; b) parallel chain as closed type.

In general, the design parameters **X** in Eq.(1) are the sizes and mobility angles of manipulators architectures. Referring to Eq.(1), the main design issue is to properly define the objective function F(**X**) so that it can express the design criteria that have to be optimized in a computationally efficient form. Equation (1) can be modified to consider several design criteria, for example, by using a weighted sum such as

$$\min_{X} \left[\sum_{i} w_i F_i(X) \right] \tag{3}$$

where F_i is the mathematical expression of the i-th objective function; w_i is the i-th weight coefficient. The weighted sum in Eq.(3) has two main limits. The first limit of the weighted sum approach is related with the choice of numerical value for the weight coefficients w_i. In fact, even small changes in the weight coefficients w_i will lead to different results. Then, the choice of weight coefficient should be done according to the experience of a designer to a specific application. The second limit of the weighted sum approach is that a minimization of the weighed sum objective function does not guarantee that any of the objective function is minimized. Thus, one has no guarantee that the solution of the optimization process will lead to an optimal design solution from the point of view of any design criterion.
Another possible formulation for Eq.(1) can be

$$\min_{\mathbf{X}} [\mathbf{F}(\mathbf{X})] = \min_{\mathbf{X}} \left\{ \max_{i=1,\dots,N} [f_i(\mathbf{X})] \right\} \tag{4}$$

where min is the operator for calculating the minimum of a vector function **F**(**X**); similarly max determines the maximum value among the N functions [w_i f_i(**X**)] at each iteration; **G**(**X**) is the vector of constraint functions that describes limiting conditions, and **H**(**X**) is the vector of constraint functions that describes design prescriptions; **X** is the vector of design variables. The proposed optimization formulation uses the objective function **F**(**X**) at each iteration by choosing the worst-case value among all the scalar objective functions for minimizing it in the next iteration, as outlined in (Grace, 2002), (Mathworks, 2009). In particular, the worst-case value is selected in Eq.(4) at each iteration as the objective function with maximum value among the N available objective functions. This approach for solving multi-objective problems with several objective functions and complex tradeoffs among them is known as "minimax method", (Mathworks, 2009). The "minimax method" is widely indicated in the literature for many problems, like for example for estimating model parameters by minimizing the maximum difference between model output and design specification, (Pankov et al., 2000), (Eldar, 2006).
Optimal design of manipulators can be also formulated the form

$$\min_{\mathbf{X}} [\mathbf{F}(\mathbf{X})] = \min_{\mathbf{X}} \left\{ \max_{i=1,\dots,N} [w_i \, f_i(\mathbf{X})] \right\} \tag{5}$$

In this case, weighting factors w_i (with i=1, …,N) have been used in order to scale all the objective functions. In particular, weighting factors w_i are chosen so that each product w_i

$f_i(X)$ is equal to one divided by N for an initial guess of a design case. The above-mentioned conditions on the objective functions can be written in the form

$$\sum_{i=1}^{N}\left(w_i f_i\right)_0 = 1 \tag{6}$$

$$N\left(w_i f_i\right)_0 = 1 \tag{7}$$

where the subscript 0 indicates that the values are computed at an initial guess of the design case. Bigger/lower weighting factors can be chosen in order to increase/reduce the significance of an optimal criterion with respect to others.

Main aspects of the numerical procedure to solve the proposed multi-objective optimization are described in the flowchart of Fig. 2. The first step in the optimization process consists of selecting the design variables, which in this manuscript correspond to geometrical properties such as robot link lengths and equivalent areas. Then, robot constraints, and upper and lower limits of design variables must be identified. In this process, preliminary data on the kinematics and physical properties of the robot are needed in order to obtain computationally efficient expressions for the objective functions. In addition, the weighting factors have to be assumed as based also on the initial guess design variables that are used for the normalization process. On the other hand, the numerical minimax technique minimizes the worst-case value of a set of multivariable functions, starting at an initial estimate (vector X0). The minimax technique uses SQP (Sequential Quadratic Programming) to choose a merit function for the line search. The MATLAB SQP implementation consists of three main stages: Updating of the Hessian matrix of the Lagrangian function, Quadratic Programming problem Solution (QPS) and Line search and merit function calculation. First and second stages are explained in (Mathworks, 2009), the result of the QPS produces a vector Ψ_k which is used to obtain a new iteration ($X_{k+1}=X_k+ \Psi_k\ \delta_k$). The step length parameter δ_k is determined in order to produce a sufficient decrease in a merit function. The new design parameter value is used to compute again the normalized objective functions that are used to check if the objective functions reach an optimal solution and fulfil the constraints. In this case the algorithm stops with an optimal solution Otherwise, the loop starts again with a new iteration, as shown in Fig. 2.

Other search methods such as interval analysis (Merlet, 2004) can be also effectively used for an optimal design algorithm. Nevertheless, they have often too high computational costs. Therefore, numerical procedures are still widely used in optimisation processes even if they can suffer of known drawbacks. Some algorithms such as flooding techniques, simulated annealing, genetic algorithms can be faster in finding an optimal solution with a single objective function. But, they still cannot guarantee the convergence (Vanderplaats, 1984), (Branke 2008). Moreover, they cannot still guarantee that an optimal solution is a global optimum. In fact, one can be sure to reach a global optimum only for convex optimization problems (Boyd & Vandenberghe, 2004).

The formulation of the design problem as an optimization problem gives the possibility to consider contemporaneously several design aspects that can be contradictory for an optimal solution. Thus, optimality criteria are of fundamental interest even for efficient computations in solving optimization problems for manipulator design.

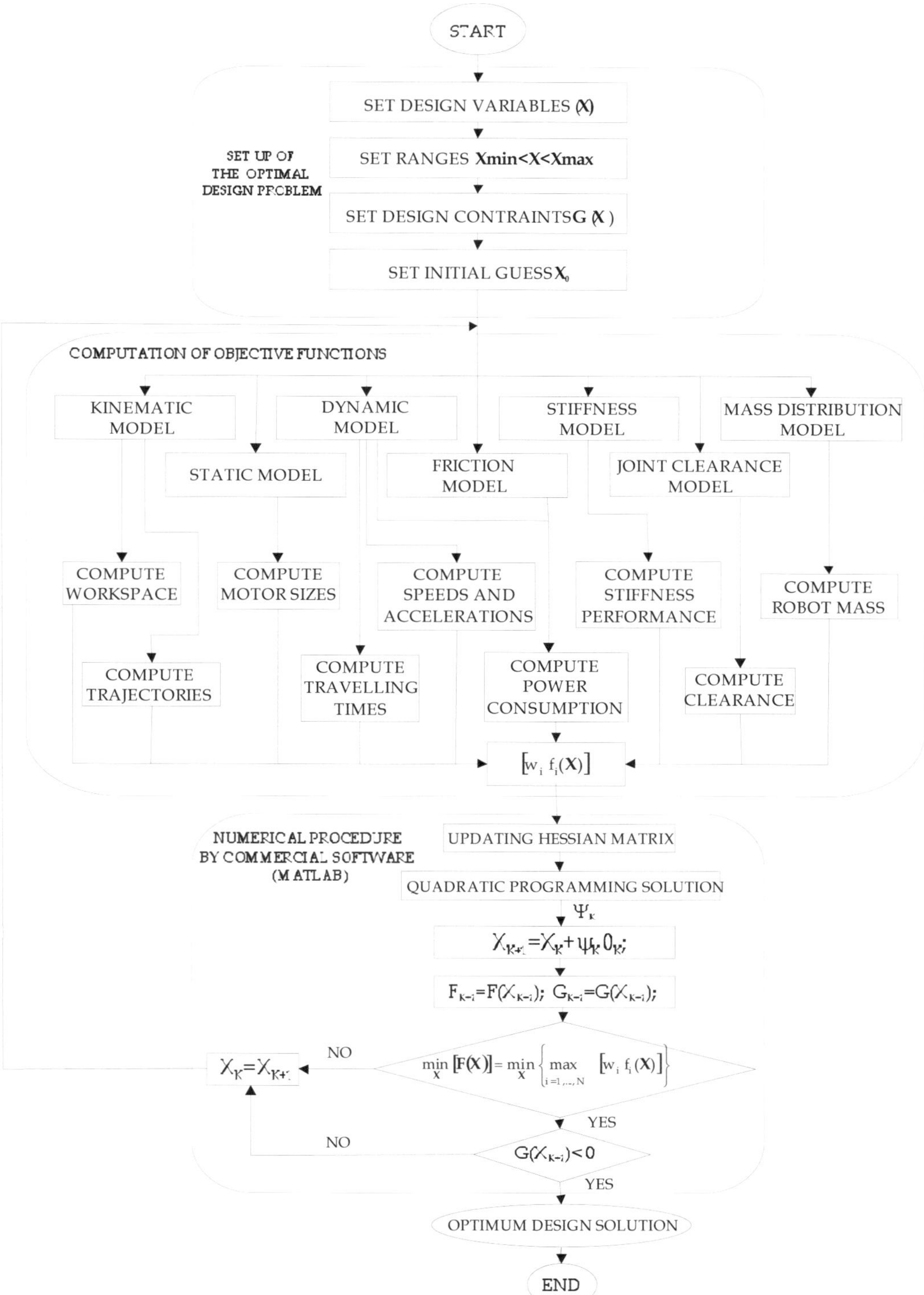

Fig. 2. A flow-chart for the proposed optimal design procedure by using MATLAB.

The analysis of manipulator performance should be aimed to computational algorithms that can be efficiently linked to the solving technique of highly non-linear optimal design of manipulators. Among the design criteria special attention should be addressed to stiffness, since it can directly affect the successful and efficient use of any robotic system for a given task as mentioned, for example, in (ANSI, 1990), (UNI, 1995), (Duffy, 1996), (Rivin, 1999).

3. Stiffness analysis for multibody robotic systems

A load applied on a body produces changes in the geometry of a body that are known as deformations or compliant displacements. Stiffness can be defined as the capacity of a mechanical system to sustain loads without excessive changes of its geometry (Rivin, 1999). Moreover, the stiffness of a body can be defined as the amount of force that can be applied per unit of compliant displacement of the body (Nof, 1985), or the ratio of a steady force acting on a deformable elastic medium to the resulting displacement. Compliant displacements in a multibody robotic system allow for mechanical float of the end-effector relative to the fixed base. This produces negative effects on static and fatigue strength, efficiency (friction losses), accuracy, and dynamic stability (vibrations) (Rivin, 1999). However, in some limited cases, compliant displacements can have even a positive effect if they are properly controlled. In fact, they can enable the correction of misalignment errors encountered for example when parts are mated during assembly operations (Nof, 1985), or in peg into hole tasks, (Tsumugiwa et al., 2002), or in deburring tasks (Schimmels, 2001), or in the operation of prosthetic limbs (English and Russell, 1999).

The analysis and evaluation of stiffness performances can be achieved by using finite element methods or lumped parameter models. The finite elements methods can provide accurate results but they require the simulation of a different model for each configuration assumed by a multibody robotic system. Therefore, models with lumped parameters are usually preferred in the literature since only one model is needed and since they require less computational efforts with respect to finite elements methods (Carbone, 2006).

The compliance of each component of a multibody robotic system can be modelled with lumped parameters by using linear and torsion springs as proposed for example in (Gosselin, 1990), (Duffy, 1996), (Tsai, 1999), (Ceccarelli, 2004). These lumped parameters are used for taking into account both stiffness properties of actuators and flexibility of links. Figures 3a) and b) show two models with lumped parameters for multibody robotic systems. In particular, Fig3a) shows a model of a 2R serial manipulator. Its links are elastically compliant and have been modelled as springs. Figure 3b) illustrates a planar parallel manipulator having three RPR legs connecting the movable plate to the fixed plate. Even in this scheme springs have been used to model the elastic compliance of the links. Schemes similar to Fig.3 can be defined for any multibody robotic system.

One can consider a compliant multibody robotic system in equilibrium with an externally applied wrench **W** that acts upon it in a point A. This point can be located on the robot end-effector and a reference frame $X_AY_AZ_A$ can be attached to point A as shown in Figs.3a) and b). In this condition, a change in the applied wrench **W** will cause a compliant displacement of the multibody robotic system. In particular, the reference frame attached to point A will change in $X'_AY'_AZ'_A$. In the most general case, a translation and rotation of the reference frame occurs.

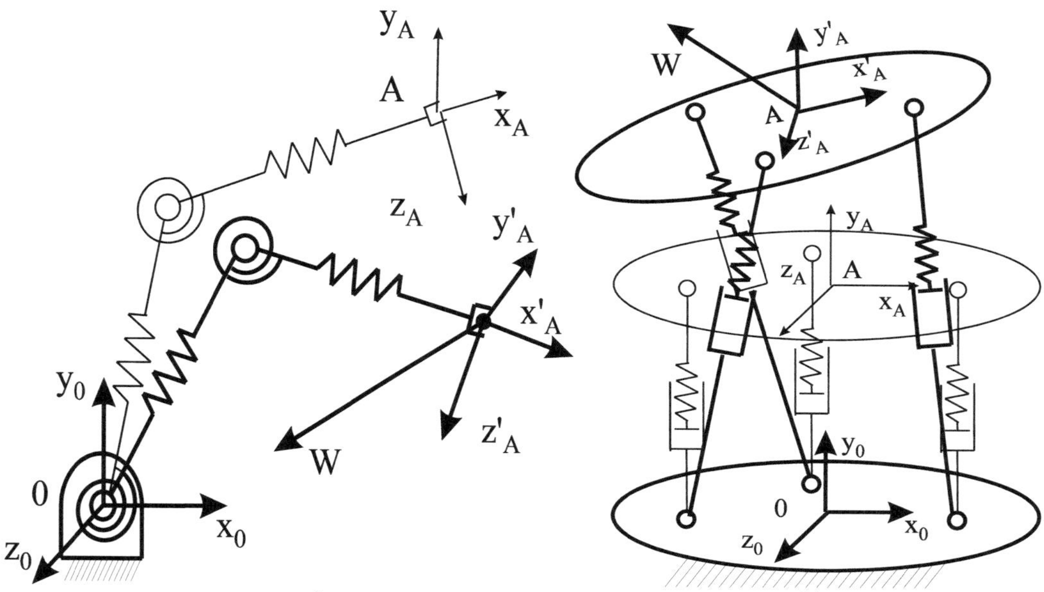

a) b)

Fig. 3. Schemes of elastically compliant multibody robotic systems: a) a 2R serial manipulator; b) a planar parallel manipulator with three RPR legs.

Usually the purpose of the stiffness analysis is the definition of the stiffness of the overall system through the derivation of a Cartesian stiffness matrix K. This stiffness matrix K express the relationship between the compliant displacements **ΔS** occurring to a frame fixed at the end of the kinematic chain when a static wrench **W** acts upon it and **W** itself. Considering Cartesian reference frames, 6x1 vectors can be defined for compliant displacements **ΔS** and external wrench **W** as

$$\mathbf{\Delta S} = (Ux, Uy, Uz, U\alpha, U\gamma, U\delta)^{\tau}$$
$$\mathbf{W} = (F_X, F_Y, F_Z, T_X, T_Y, T_Z)^{t} \qquad (8)$$

where Ux, Uy, and Uz are the differences between the coordinates and Uα, Uγ and Uδ are the differences between the Euler angles of the reference frames $X'_AY'_AZ'_A$ and $X_AY_AZ_A$ that are expressed with respect to the fixed reference frame $X_0Y_0Z_0$; F_X, F_Y and F_Z are the force components acting upon point A in X, Y and Z directions, respectively; T_X, T_Y and T_Z are the torque components acting upon point A along X, Y and Z directions, respectively.

The relationship between the vector s **ΔS** and **W** can be written in the form

$$K(\mathbf{q}) : \Re^r \rightarrow \Re^r \,, \;\; \mathbf{W} = K\,\mathbf{\Delta S} \qquad (9)$$

where K is the so-called 6x6 Cartesian stiffness matrix or spatial stiffness matrix.

Therefore, Eq.(9) defines K as a 6x6 matrix whose components are the amount of forces or torques that can be applied per unit of compliant displacements of the end-effector for the multibody robotic system. However, the linear expression in Eq.(9) is valid only for small

magnitude of the compliant displacements $\Delta\mathbf{S}$. Moreover, Eq.(9) is valid only in static conditions.
The entries in the 6x6 Cartesian stiffness matrix K depends on the configuration assumed by the robotic system, on the reference frame in which it is computed, and on the stiffness properties of each components of the multibody robotic system. A 6×6 stiffness matrix can be derived through the composition of suitable matrices.
A first matrix C_F gives all the wrenches $\mathbf{W}_L$, acting on manipulator links when a wrench $\mathbf{W}$ acts on the manipulator extremity according to the expression

$$\mathbf{W} = C_F \; \mathbf{W}_L \tag{10}$$

with the matrix C_F representing the force transmission capability of the manipulator mechanism.
A second matrix K_p gives the possibility to compute the vector $\Delta\mathbf{v}$ of all the deformations of the links when each wrench $\mathbf{W}_{Li}$ on a i-th link given by $\mathbf{W}_L$, acts on the legs according to

$$\mathbf{W}_L = K_p \; \Delta\mathbf{v} \tag{11}$$

with the matrix K_p grouping the spring coefficients of the deformable components of a manipulator structure.
A third matrix C_K gives the vector $\Delta\mathbf{S}$ of compliant displacements of the manipulator extremity due to the displacements of the manipulator links, as expressed as

$$\Delta\mathbf{v} = C_K \; \Delta\mathbf{S} \tag{12}$$

Therefore, the stiffness matrix K can be computed as

$$K = C_F \, K_p \, C_K \tag{13}$$

with matrix C_F giving the force transmission capability of the mechanism; K_p grouping the spring coefficients of the deformable components; C_K considering the variations of kinematic variables due to the deformations and compliant displacements of each compliant component.
Matrices C_K and C_F can be computed, for example, as a Jacobian matrix and its transpose, respectively, as proposed in (Tsai, 1999), (Tahmasebi, & Tsai, 1992), (Carbone et al., 2003). Nevertheless, this is only an approximate approach as pointed out, for example, in (Alici & Shirinzadeh, 2003). A more accurate computation of matrices C_K and C_F can be obtained as reported, for example in (Carbone, 2003). The K_P matrix can be computed as a diagonal matrix whose components are the lumped stiffness parameters of links, joints and motors that compose a multibody robotic system. The lumped stiffness parameters can be estimated by means of analytical and empirical expressions or by means of experimental tests. For example, the stiffness matrix of a generic beam element can be written as reported for example in (Kardestuncer, 1974),

$$K = \begin{bmatrix} \frac{EA}{L} & 0 & 0 & 0 & 0 & 0 \\ 0 & \frac{12EI_Z}{L^3} & 0 & 0 & 0 & -\frac{6EI_Z}{L^2} \\ 0 & 0 & \frac{12EI_Y}{L^3} & 0 & \frac{6EI_Y}{L^2} & 0 \\ 0 & 0 & 0 & \frac{GJ}{L} & 0 & 0 \\ 0 & 0 & \frac{6EI_Y}{L^2} & 0 & \frac{4EI_Y}{L} & 0 \\ 0 & -\frac{6EI_Z}{L^2} & 0 & 0 & 0 & \frac{4EI_Y}{L} \end{bmatrix} \tag{14}$$

where E is the Young modulus; A is the cross section area; L is the link length; I_Y and I_Z are the two principal moment of inertia of the cross sections; G is the shear modulus; J is the equivalent torsional moment of inertia. The stiffness of direct drive actuators can be computed by using an empirical expression as proposed in (Rivin, 1999) in the form

$$(k_m)^{-1} = \omega_0 \nu \tau_e \tag{15}$$

with

$$\begin{aligned} \tau_e &= L_r / R_r \\ \nu &= (((e/\Omega) - I_0) K_M) / \omega_0 \end{aligned} \tag{16}$$

where ω_0 is the no load angular velocity, R_r, L_r, e, Ω and K_M are the terminal resistance, inductance, voltage, resistance and torque constant of the motors, respectively. In presence of mechanical transmissions the values obtained by Eqs.(15) and (16) should be corrected by considering the transmission ratio and the stiffness properties of the transmission itself.

4. Stiffness as optimal design criterion

The stiffness matrix K can be computed numerically according with the flow chart that is proposed in Fig.4. A numerical algorithm can be composed of a first part in which the numerical values for the geometrical dimensions, masses and lumped stiffness parameters are defined. A second part defines the kinematic model, the force transmission model and the lumped parameter model through the matrices C_F, K_p, and C_K, respectively. Then, a third part can compute a close-form expression of the stiffness matrix K by means of Eq.(13). It is worth noting that the matrices C_F, and C_K are configuration dependant. Therefore, also the stiffness matrix K is configuration dependent. Thus, one should define configuration(s) of a multibody robotic system where the stiffness matrix will be computed. The configuration(s) should be carefully chosen in order to have significant information on the stiffness performance of the system in its whole workspace.

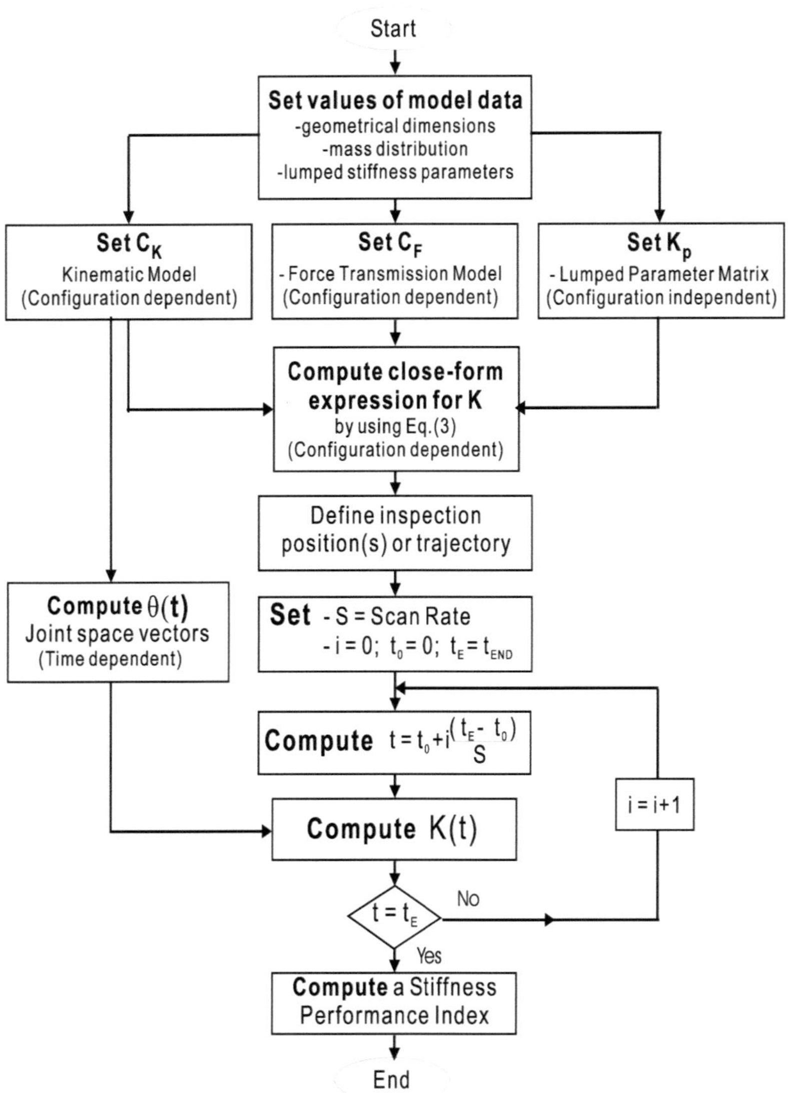

Fig. 4. A flow-chart for the proposed numerical computation of stiffness performance.

Then, the kinematic model can be used for computing the vector **θ** that express input angles and strokes in the joint space for any pose assumed by a multibody robotic system.
In some cases, a multibody robotic system can have few trajectories that are mostly used during its operation. In these cases, a kinematic model can be used together with a proper path planning strategy for computing a vector **θ(t)** that express input angles and strokes in the joint space as function of time for a given trajectory. Thus, the vector **θ(t)** can be used for computing the stiffness matrix as function of time for a given end-effector trajectory. However, it is necessary to define a scan rate S, and the time t_{END} in which the motion of the robotic system will be completed. Of course, the higher is the scan rate the higher is the number of configurations in which stiffness matrix K is computed. It is worth noting that the

accuracy in the estimation of model data such as geometrical dimensions, and values of lumped stiffness parameters can significantly affect the accuracy of the stiffness matrix that is computed through Eq.(13). Thus, experimental tests should be carried out in order to validate stiffness model and model data.

Once the stiffness matrix has been derived, it is necessary to be able to compare different stiffness matrices (for comparing local stiffness properties) and estimate the stiffness performance of the overall system (for comparing global stiffness properties). A local stiffness index can be directly related with the Cartesian stiffness matrix by means of different mathematical operators that can be applied to a matrix. Feasible choices can be the determinant, trace, norm, eigenvalues and eigenvectors at a given posture. In particular, the determinant of a stiffness matrix K is invariant in similarity transformations. Thus, it does not rely on the choice of reference frame. Moreover, it can be computed as

$$\det(K) = (-1)^6 + P_1(-1)^5 + P_2(-1)^4 + P_3(-1)^3 + P_4(-1)^2 + P_5(-1) + P_6 \quad (17)$$

where P_i (with i=1,2,…,6) is the sum of the principal minors of order i of the matrix K.

But the determinant can be expressed also as the product of matrix eigenvalues as given in Matrix Algebra. Each entry K_{ij}^{-1} of the inverse matrix of K can be computed as

$$K_{ij}^{-1} = \frac{(K)_{ji}}{\det(K)} \quad (18)$$

where $(K)_{ji}$ is the algebraic complement of the entry K_{ij} of the matrix K with i, j=1,2,…,6. Thus, if the determinant det(K) is zero, the Eq.(13) gives singular values and Eq.(12) cannot be computed. Therefore, the determinant of K can be used as a performance index to investigate synthetically the effect of the design parameters on the stiffness behaviour, since it is easy to compute and it is particularly significant for determining stiffness singularity properties. Merits and drawbacks of other local indices are summarized in (Carbone & Ceccarelli, 2007).

A local index of stiffness performance is neither suitable for an accurate design analysis nor useful for a comparison of different designs. In fact, even if a multibody robotic system has suitable stiffness for a given system posture it can have inadequate stiffness at other postures. Therefore, one should look at stiffness performance at all points of workspace or define a single global stiffness index over the whole workspace yet.

A global index of stiffness performance for a multibody robotic system can be defined by means of graphical methods that are based on plotting curves connecting postures having the same value of the local stiffness index (iso-stiffness curves or surfaces), as proposed for example in (Merlet, 2006). Nevertheless, the number of iso-stiffness curves or surfaces that one can plot is graphically limited. Moreover, few curves or surfaces usually do not provide sufficient insight of the overall stiffness behaviour of a multibody robotic system. These aspects significantly reduce the effectiveness of iso-stiffness curves or surfaces.

Global stiffness indices can be defined also in a mathematical form by using minimum, maximum, average or statistic evaluations of a local stiffness index. For example, one can compute a global index in the form

$$GI_d = \min \left|\det(K)\right| \tag{19}$$

It is worth noting that a GI_d index equal to zero means that at least one singular configuration is within the workspace of a multibody robotic system. This is a critical situation that should be avoided at the design stage.
Among the possible method the determinant of K and maximum values of compliant displacements can be most easily related with a physical meaning. However, one should note that the choice of a comparison method is strongly related with the application field. For example, eigenvalues and eigenvectors and the identification of a center of compliance are widely used for machine tools and grasping systems, respectively, as reported for example in (Gosselin & Angeles, 1991).

5. Other optimal design criteria

Alternatives in formulating and choosing optimality criteria are always possible depending on the designer experience, design goals, and manipulator applications. Many different indices and/or their computations have been proposed in a rich literature on manipulators in order to provide a numerical value of the performance of a manipulator. Those indices can be used and they have been used with proper formulation as optimality criteria in specific algorithms for optimal design of specific manipulators. Of course, any optimality criterion as well as its formulation can suffer drawbacks in terms of conceptual aim and numerical efficiency. Considering the above-mentioned aspects one can propose optimality criteria for taking into account, just to cite few examples, well known design aspects such as

- workspace,
- dynamic performance,
- lightweight design.

One can even define optimality criteria for other specific design aspects such as safety in robots for service tasks as proposed, for example, in (Castejón et al., 2007).

5.1 Workspace

The workspace is one of the most important kinematic properties of manipulators, because of its impact on manipulator design and its location in a work cell. A manipulator workspace can be identified as a set of reachable positions by a reference point at the manipulator's extremity. This is referred as position workspace. Similarly, orientation workspace can be identified as a set of reachable orientations by a reference point at the manipulator's extremity. Interpreting the orientation angles as workspace coordinates permits to treat the determination of the orientation workspace likewise the determination of the position workspace when a Cartesian space is considered in the computations. A general numerical evaluation of the workspace can be deduced by formulating a suitable binary representation of a cross-section in the task-space, as described, for example, in (Ottaviano & Ceccarelli, 2002).
The workspace volume V can be computed considering the cross-sections areas A_z and the number of slices n_z that have been considered for the workspace volume evaluation, according to scheme of Fig. 5, as

$$V = \sum_{z=1}^{n_z} \left(\sum_{i=1}^{i_{max}} \sum_{j=1}^{j_{max}} \left(P_{ij}\, \Delta x\, \Delta y \right) \right) \Delta z \tag{20}$$

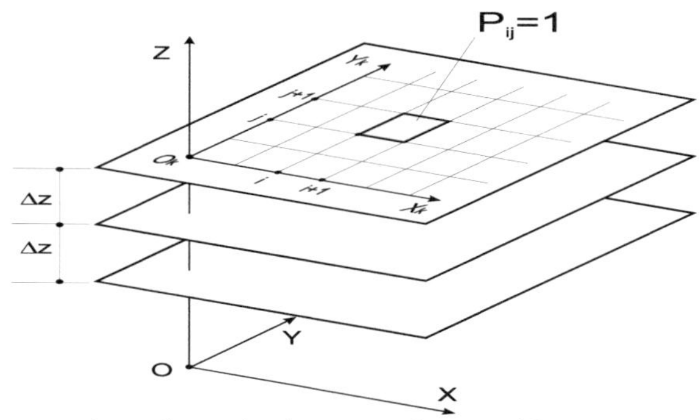

Fig. 5. A binary representation of manipulator workspace (Ottaviano & Ceccarelli, 2002).

Similarly, the orientation workspace can be analyzed by using a suitable binary representation with another binary matrix for a workspace region that can be described in term of orientation angles. Therefore, an optimum design problem with objective functions regarding workspace characteristics can be formulated as finding the optimal design parameters values to obtain the position and orientation workspace volumes that are as close as possible to prescribed ones in the form

$$f_{PW}(\mathbf{X}) = \left| 1 - \frac{V_{pos}}{V_{pos}'} \right| \tag{21}$$

$$f_{OW}(\mathbf{X}) = \left| 1 - \frac{V_{or}}{V_{or}'} \right|$$

where | . | is the absolute value; the subscripts pos and or indicate position and orientation, respectively; and prime refers to prescribed values.

An optimality criterion for addressing workspace performance could be defined also by taking into account several other aspects such as the shape of the workspace, the absence of singularities or voids within the desired workspace, isotropy of the workspace, manipulability index for specific manipulative tasks.

5.2 Dynamic performance

An optimality criterion concerning with dynamic performance, power consumption and energy aspects of the path motion can be conveniently expressed in terms of the work that is needed by the actuators. In particular, the work by the actuators is needed for increasing the kinetic energy of the system in a first phase from a rest condition to actuators states at which each actuator is running at maximum velocity. In a second phase bringing the system back to a rest condition, the kinetic energy will be decreased to zero through the actions of actuators and brakes. Thus, one can write the work W_{act} done by the actuators in the first phase of the path motion as an optimality criterion for optimal path generation as given by the expression

$$W_{act} = \sum_{k=1}^{3}\left[\int_{0}^{t_k} \tau_k \ \dot{\alpha}_k \ dt \right] \tag{22}$$

in which τ_k is the k-th actuator torque; α_k dot is the k-th shaft angular velocity of the actuator; and t_k is the time coordinate value delimiting the first phase of path motion with increasing speed of the k-th actuator. Therefore, trying to minimize the ratio W_{act} / W_{act0} with W_{act0} as a prescribed value, has the aim to size at the minimum level the design dimensions and operation actions of the actuators in generating a path between two given extreme positions. The prescribed value W_{act0} has to be chosen as referring to the power of a commercial actuator.

5.3 Lightweight design

Lightweight design is desirable in order to have a light mechanical structure for safety reasons and at the most for a general suitable maneuverability, installation, and location of the robot. A reasonable and computationally efficient expression of the lightweight design criterion can be given by

$$f_L(X) = \left| 1 - \frac{M_T}{M_d} \right| \tag{23}$$

as referred to M_T which is the overall mass of a robot and to M_d which is the desired overall mass of the same robot. The robot mass, M_T can be computed as the sum of the mass of links and joints M_i, the mass of actuators M_j, and the mass of cables and sensors M_k, in the form

$$M_T = \sum_{i=1}^{n_{link}} M_i + \sum_{j=1}^{m_{actuator}} M_j + \sum_{k=1}^{l_{component}} M_k \tag{24}$$

It is worth noting that the most critical aspect for obtaining a lightweight mechanical design is to reduce the weight of links and joints. In fact, cables and sensors are usually market components with given size and mass. Although actuators are usually market components their size and mass is mainly selected according to the desired output power and dynamics.

6. Cases of study

6.1 A Parallel Manipulator

The CaPaMan (Cassino Parallel Manipulator) manipulator has been considered to test the engineering feasibility of the above-mentioned formulation for optimal design of manipulators as specifically applied to parallel architectures. CaPaMan architecture has been conceived at LARM in Cassino since 1996, where a prototype has been built for experimental activity. Indeed, by using the existing prototype, simulations have been carried out also to validate the proposed optimum design by considering several guess solutions and imposing workspace and stiffness characteristics of the built prototype. According to those satisfactory results a numerical example has been proposed to obtain the same workspace characteristics but with enhanced stiffness and conditions for avoiding singularities. A schematic representation of the CaPaMan manipulator is shown in Fig.6a), and a photo of a prototype is shown in Fig.6b).

Position and orientation workspace volumes can be conveniently evaluated by using Eqs.(20-21) and the algebraic formulation for the Kinematics of CaPaMan manipulator that has has been reported, for example, in (Ottaviano & Ceccarelli, 2002). Similarly, singularity analysis for CaPaMan manipulator has been reported in (Ottaviano & Ceccarelli, 2002).

Stiffness analysis of CaPaMan has been reported in (Ceccarelli & Carbone, 2002). By modeling each leg of CaPaMan as shown in Fig.7, the stiffness matrix of CaPaMan can be derived as defined in Eq.(13) with

$$C_F = M_{FN}\,; \qquad\qquad C_K = C_P^{-1} A_d^{-1} \tag{25}$$

where M_{FN} is a 6x6 transmission matrix for the static wrench applied on H and transmitted to points H_1 H_2 and H_3 of each leg; K_p is a 6x6 matrix with the lumped stiffness parameters of the 3three legs; C_p is a 6x6 matrix giving the displacements of the links of each leg as a function of the displacements of points H_1, H_2 and H_2; A_d is a 6x6 matrix that has been obtained by using the Direct Kinematics of the CaPaMan to give the position of point H on the movable plate as function of the position of points H_1, H_2 and H_2 in the form

$$\mathbf{X_H} = A_d\ \mathbf{v} \tag{26}$$

with $\mathbf{v}=[y_1, z_1, y_2, z_2, y_3, \Delta z_3]^T$ and $\mathbf{X_H} = [x_H, y_H, z_H, \varphi, \theta, \psi]^T$. The derivation of matrices M_{FN}, K_p, A_d, and C_p for CaPaMan can be found in (Ceccarelli & Carbone, 2002).

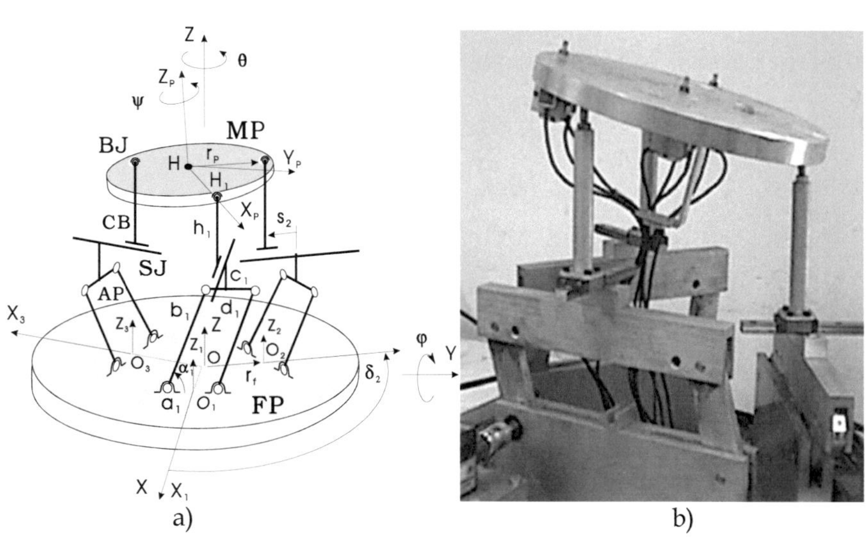

Fig. 6. CaPaMan (Cassino Parallel Manipulator) design: a) a kinematic diagram; b) a built prototype at LARM.

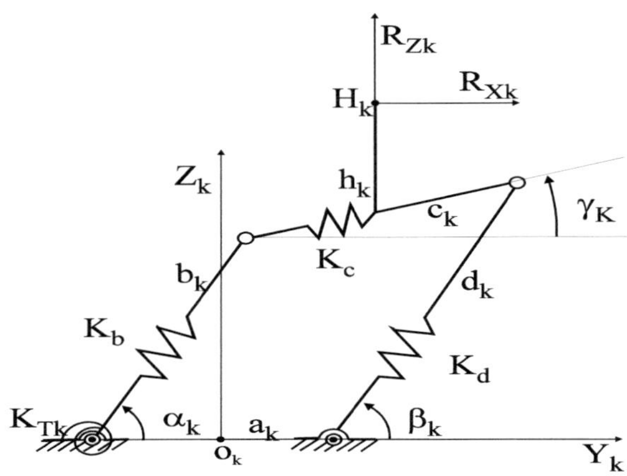

Fig. 7. A scheme for stiffness evaluation of a CaPaMan leg.

The lumped stiffness parameters has been assumed as $k_{bk}=k_{dk}=2.625\text{x}10^6$ N/m and $k_{Tk}=58.4\text{x}10^3$ Nm/rad; the couplers c_k have been assumed rigid bodies because of the massive design that has been imposed to have a fix position of the sliding joints. Further details on the derivation of the matrices in Eqs.(28) and (29) can be found in (Ceccarelli & Carbone, 2002). In the numerical example, for evaluation and design purposes we have assumed $r_p = r_f$, $a_k = c_k$, $b_k = d_k$.
Results of the proposed design procedure as applied to the CAPAMAN architecture are reported in Figs. 10, 11, and 12 and Table 1 and 2. In particular, the evolution of the objective functions is reported in Fig. 8, form which one can note that the numerical procedure takes 65 iterations to converge to the optimum values that are reported in Table 1. Evolution of design parameters and constraints are shown in Figs.9 and 10. Design characteristics for the optimum solution are reported in Table 2. For the proposed numerical example, the Inverse Kinematic singularities related to matrix A in Eq. (24) gives the condition that input crank angle α_i should be different from 90 deg, (for 1=1,2,3). This condition and Direct Kinematic singularities have been taken into account in the numerical procedure through a constraint equation.

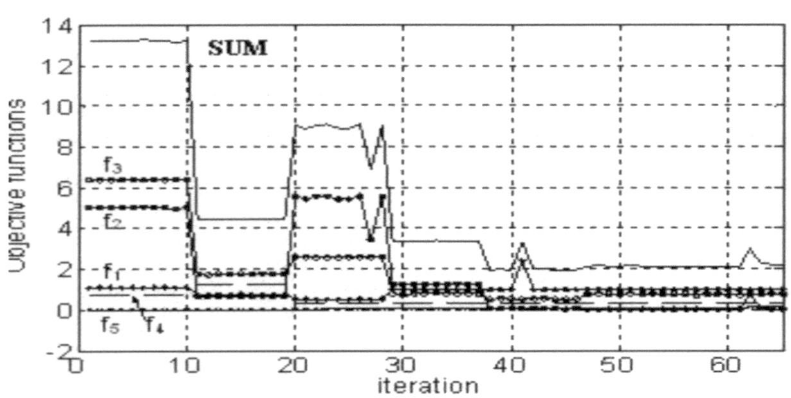

Fig. 8. Evolution of the objective functions versus number of iterations for the example of CaPaMan optimal design of Fig.6.

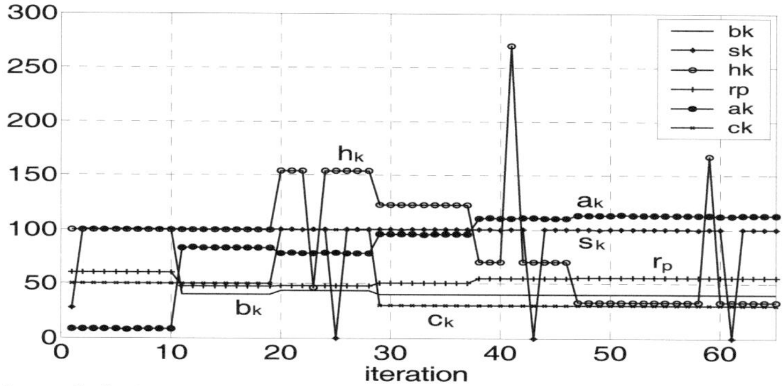

Fig. 9. Evolution of design parameters versus number of iterations for the example of CaPaMan optimal design of Fig.6.

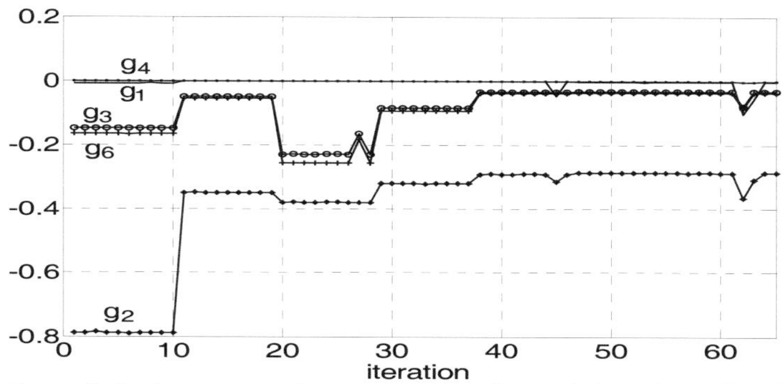

Fig. 10. Evolution of design constraint versus number of iterations for the example of CaPaMan optimal design of Fig.6 and Table 1.

Values	a_k (mm)	b_k (mm)	h_k (mm)	r_p (mm)	α_k (deg)	s_k (mm)
Initial Guess	27.85	100.0	100.0	60.0	45;135	50.0
Optimal	113.1	40.0	32.9	55.8	45;112	30.0

Table 1. Design parameters for optimal CaPaMan design of Figs.8 to 10.

Values of workspace ranges	Δx (mm)	Δy (mm)	Δz (mm)	$\Delta\varphi$ (deg)	$\Delta\psi$ (deg)	$\Delta\theta$ (deg)
Initial Guess	105.8	112.4	29.3	38.0	179.9	321.8
Optimal	48.6	55.9	11.7	16.1	179.9	212.4
Values of compliant displacements	Ux (mm)	Uy (mm)	Uz (mm)	$U\alpha$ (deg)	$U\gamma$ (deg)	$U\delta$ (deg)
Initial Guess	$5.5\ 10^{-4}$	$6.7\ 10^{-6}$	$3.2\ 10^{-4}$	$2.4\ 10^{-5}$	$2.4\ 10^{-5}$	$2.3\ 10^{-9}$
Optimal	0.002	$1.6\ 10^{-6}$	0.001	$6.0\ 10^{-4}$	$6.5\ 10^{-4}$	$2.3\ 10^{-8}$

Table 2. Design characteristics of optimum solution for optimal CaPaMan design of Figs.8 to 10 and Table 1.

The numerical example for the CaPaMan manipulator has been elaborated in an Intel

Pentium M 2.00 GHz. The algorithm takes 65 iterations to converge to an optimal solution with a computation time of 4 min and 8 sec. The accuracy for the objective function evaluations has been set equal to 1e-5 and the accuracy for the design parameters has been set equal to 1e-3. Numerical examples show satisfactory results with a quite rapid convergence to a feasible optimal solution. The robustness of the design algorithm is proved in some extent even by relatively large distance of the computed optimal design solutions from the guess values.

6.2 A robotic hand

LARM Hand as been considered to test the engineering feasibility of the above-mentioned formulation for optimal design of robotic hands. LARM Hand architecture has been conceived at LARM in Cassino in the second half of 90's. Four different design solutions have been developed and built at LARM as shown in Fig.11. Recently, special care has been addressed in designing a novel underactuated linkage mechanism with passive elements that can adjust the position of links and envelope object with only one motor as input actuator. A feasible design schemes has been defined as shown in Fig. 12.

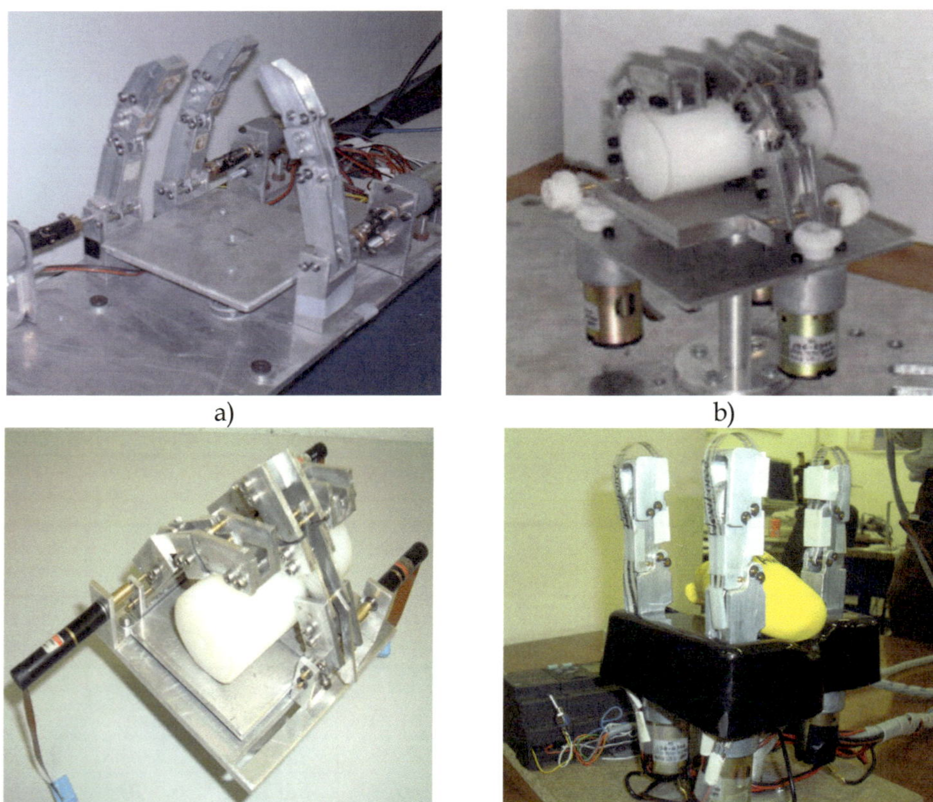

a) b) c) d)

Fig. 11. LARM Hand prototypes in Cassino: a) version I; b) version II; c) version III; d) version IV.

Link sizes of a first solution for the proposed finger mechanism are listed in Table 3 by referring to previous LARM Hand prototypes.

Initial values for coefficients of the springs have been determined as $k_1 = k_2 = 7.7\times10^{-2}$ Nm/rad. Referring to Fig.12 the design parameters can be considered as angles of the links, and the coefficients of the springs, namely L_i, α_i, δ_i, θ_i, for i=1,2,…9; θ_{s1}, s_2, k_1, k_2, c_1, c_2. A design for an anthropomorphic finger must fulfil basic features such as human-like contact forces, actuation efficiency, grasping capability, underactuated design, compact size, transmission efficiency.

The multi-objective design optimization problem has been solved by using a numerical procedure through Matlab Optimization Toolbox. For the numerical example the data have been given as reported in Table 3. The sizes and forces needed for the grasping have been defined by referring to experimental tests on a cylindrical object with a diameter 60 mm as reported in (Yao et al., 2009). Two main objective functions F1 and F2 have been defined. F1 combines together the optimal criteria for compact size, underactuated design, and stiffness performance. F2 combines together human-like contact forces, actuation efficiency, grasping capability, and transmission efficiency,(Yao et al., 2009).

Optimal solution is obtained after 81 iterations for F1 and 363 iterations for F2, with total 168 seconds of CPU computation with standard PC Genuine Intel(T2050). The accuracy for the objective function evaluations has been set equal to 1e-5 and the accuracy for the design parameters has been set equal to 1e-3. Results of optimal program are shown in Figs. 13 and 14 and numerical values are listed in the tables 4, and 5. The actuator torque f2 is obtained as about 0.15 Nm. This is due to an optimization of driving transmission efficiency and reduction of coefficients for springs and dampers. Additionally, it has been checked that the transmission angles are obtained in a reasonable range while the final grasping configuration occurs.

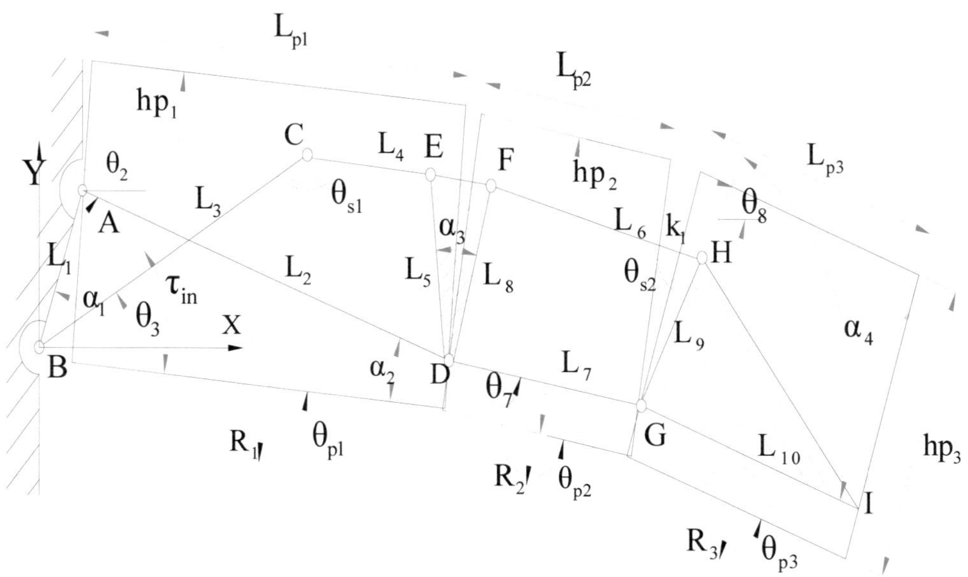

Fig. 12 The design parameters and phalanx bodies for a novel underactuated driving mechanism for LARM Hand.

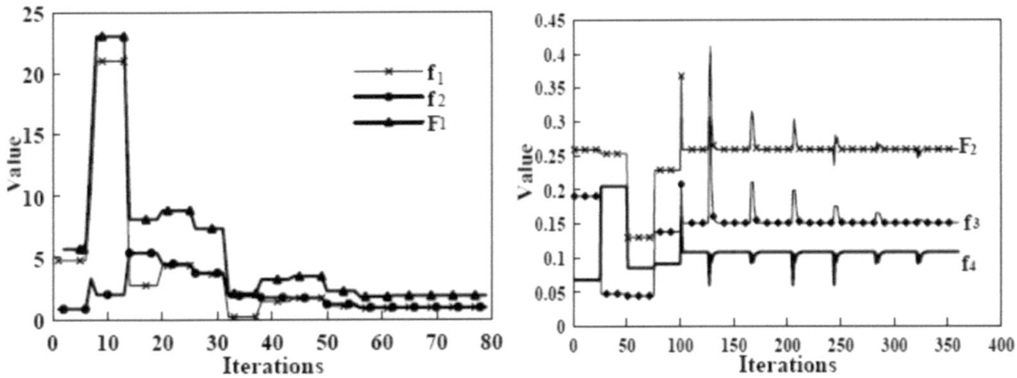

a) b)

Fig. 13. Evolution of the objective functions: a) evolution of the objective functions within F1; b) evolution of the objective functions within F2.

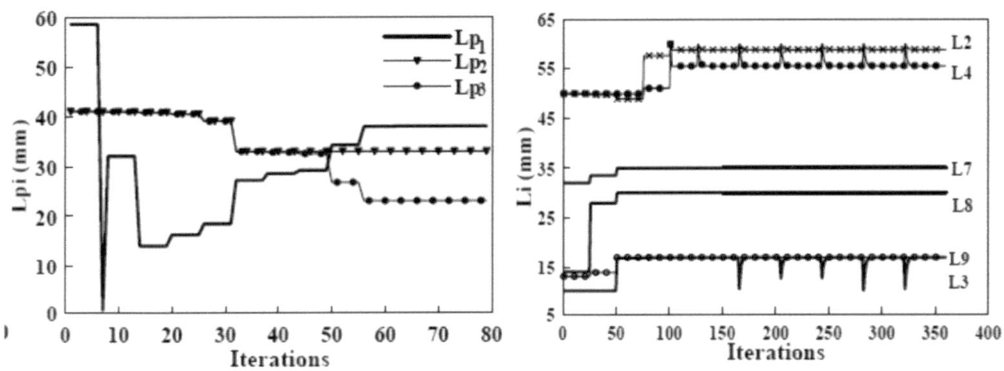

a) b)

Fig. 14. Results of the optimal design procedure: a) evolution of phalanx sizes; b) evolution of design parameters.

i	1	2	3	4	5	6	7	8	9	10
L_i(mm)	11.2	66.5	12.0	51.4	17.8	37.1	33.1	14.3	13.0	41.0
L_{pi}(mm)	60.0	41.5	41.0	-	-	-	-	-	-	-
h_{pi}(mm)	20.0	20.0	20.0	-	-	-	-	-	-	-
α_i (deg)	64.4	22.7	23.1	123.1	-	-	-	-	-	-
θ_{pi0}(deg)	0.0	0.0	0.0	-	-	-	-	-	-	-

Table 3. Initial guess design parameters for the proposed driving mechanism in Fig. 12.

Parameters	λtr1/λtr2/λtr3 (deg)	k_1/k_2(10□2 Nm/rad)	c_1/c_2 (Nms/deg)	h_{pi}/D (mm)	E_{spring}/τ_{in} (N/m)
Guess solution	140/153/85	0.210/0.008	0.25/0.25	20.0/89.8	0.19/0.07
Optimal solution	97/113/79	0.150/0.011	0.05/0.05	20.0/113.2	0.15/0.10

Table 4. Design parameters before and after optimality.

i	1	2	3	4	5	6	7	8	9	10
L_i(mm)	7.8	61.1	16.2	54.4	16.8.	33.7	35.0	30.0	17.1	23.2
L_{pi}(mm)	37.9	33.1	23.2	-	-	-	-	-	-	-
α_i (deg)	78.7	62.8	25.5	76.4	-	-	-	-	-	-

Table 5. Structure parameters of the optimal results for the underactuated finger mechanism.

6.3 A humanoid leg

The leg module of the humanoid robot WABIAN R-IV has been considered to test the engineering feasibility of the proposed formulation for optimal design of humanoid legs. WABIAN R-IV has been conceived at Waseda University within the series of WABIAN humanoid robots that started to walk on 1972. A collaboration has been established with LARM since 2001 aiming to investigate kinematics, stiffness, and dynamics aspects both from theoretical and experimental point of view.

Figures 15a), b) and c) show the humanoid robot WABIAN R-IV and a detailed kinematic model for its leg module, respectively. It is worthy to note that in this model the Denavit-Hartenberg convention has been used in order to define position and orientation of the link coordinate frames $X_iY_iZ_i$.

By considering the references frames of Fig.15c), the D-H link parameters for the kinematic chain of the leg can be computed as shown in Table 6. Then, the rotation matrices expressing the relation between the frames can be straightforward derived by using the D-H link parameters in in Table 6. Further details can be found in (Carbone et al. 2003).

A multi-objective optimization problem in the form of Eq.(23) can be defined also in order to find an optimum compromise between stiffness and lightweight design. It is worthy to note that the objective functions are affected by the choice of shape of links and material that is used. In this paper, hollow square sections are assumed, since they give high stiffness performances as pointed out in (Rivin, 1999). Moreover, it has been decided to use as material Extra Super Duralluminium having Young module E=70 GPa and specific weight ρ=3000 kg/m^3 as based on previous experiences at Waseda University.

Figures 16 shows the plot of the objective functions versus the number of iterations for a successful application of the proposed optimum design procedure. Figure 17 shows the plots of the compliant displacements versus the number of iterations. Tables 7 shows the optimum set of design sizes and evolution of the objective function that have been obtained as result of the proposed formulation to give an optimal compromise between stiffness and lightweight design. The objective function has evolved from an initial value of 37.028 to a final value of 0.3597. The numerical example for the leg module of the humanoid robot WABIAN R-IV has been elaborated in an Intel Pentium M 2.00 GHz. The algorithm takes 2600 iterations to converge to an optimal solution with a computation time of about 30 min. The accuracy for the objective function evaluations has been set equal to 1e-5 and the accuracy for the design parameters has been set equal to 1e-3.

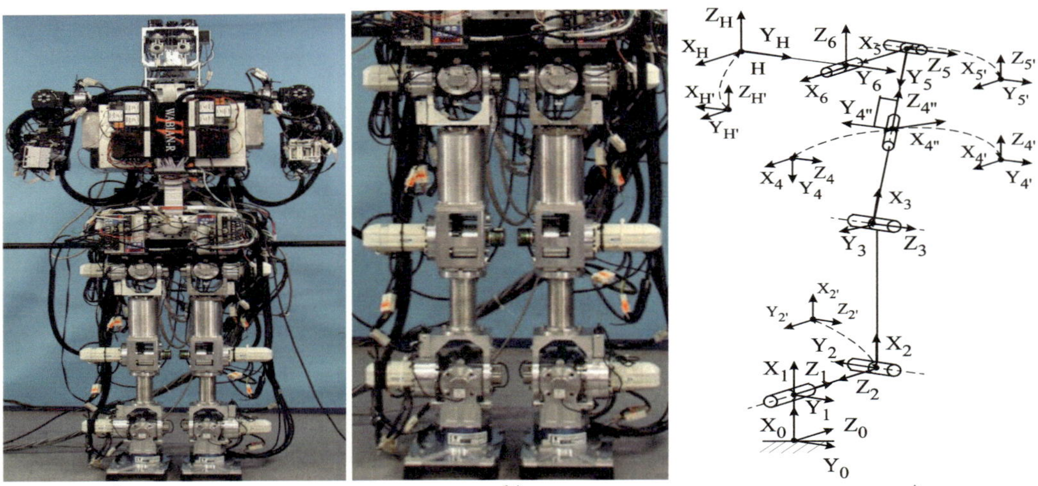

a) b) c)

Fig. 15. WABIAN-RIV: a) a photo of the built prototype ; b) a zoom view of the leg module; c) a kinematic scheme for the leg module.

D-H par. / Link No.	$\alpha_{(i-1)}$ deg]	$a_{(i-1)}$ [mm]	d_i [mm]	θ_i [deg]
1	0	a_0=185	0	θ_1
2	180	0	$-d_2$=0	0
2′	90	0	0	θ_2
3	0	a_2=300	0	θ_3
4	0	a_3=223.5	0	90
4′	90	0	0	0
4″	0	0	0	θ_4+180
5	90	0	$-d_5$=130	180
5′	90	0	0	θ_5
6	θ_6	a_5=0	0	0
H′	0	0	d_6=0	-90
H	0	0	0	90

Table 6. D-H parameters for the leg module of WABIAN-RIV in Fig.15.

Link N.	Length [m]		Cross-section Edge [m]	
	Initial	Final	Initial	Final
1	0.05	0.323	0.034	0.008
2	0.185	0.119	0.028	0.009
3	0.159	0.382	0.021	0.004
4	0.141	0.135	0.022	0.027
5	0.224	0.055	0.021	0.018
6	0.354	0.099	0.016	0.019

Table 7. Optimum set of design sizes.

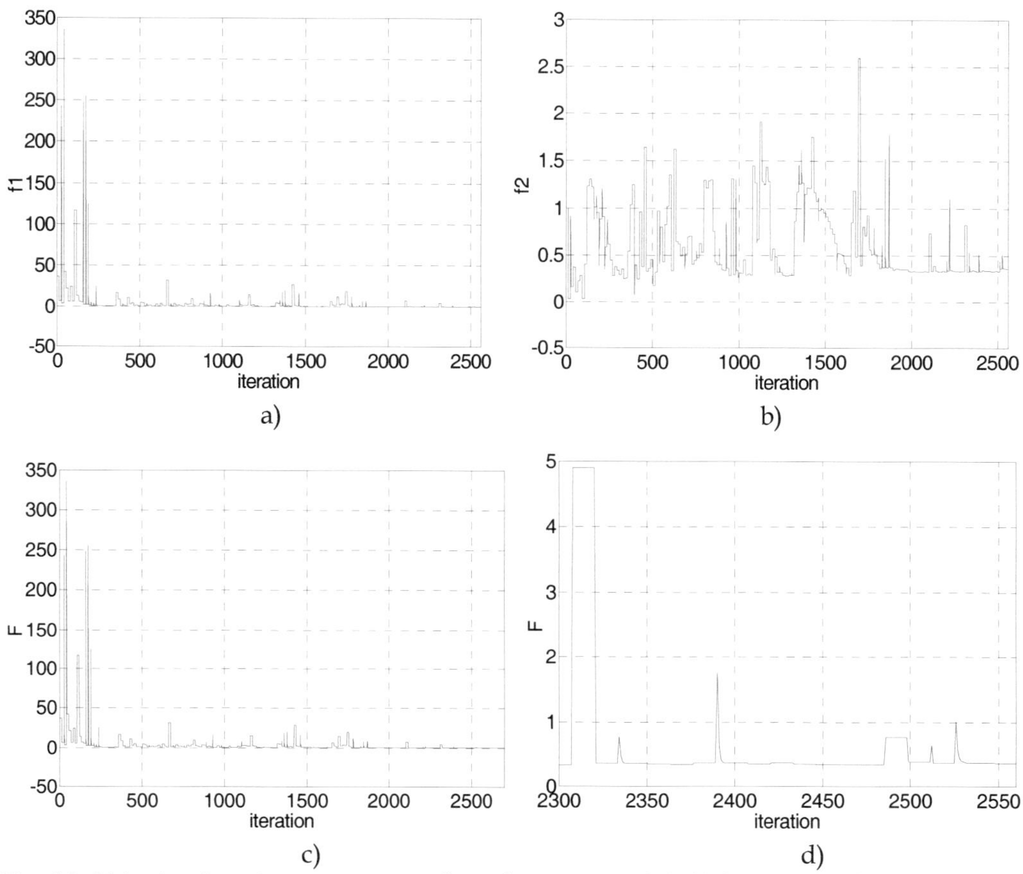

Fig. 16. Objective functions versus number of iterations: a) f_1; b) f_2; c) $F= f_1+ f_2$; d) zoom of F.

7. Conclusion

In this paper, a multi-objective optimal design procedure is outlined by discussing optimality criteria and numerical aspects. In particular, optimality criteria have been chosen according to the most common design requirements for robotic systems by paying special attention to stiffness, since it is of primary importance in order to guarantee the successful use of any robotic system for a given task. Additional alternative objective functions can be used to extend the proposed design procedure to more general design problems. The feasibility of such a complex design formulation for robotic manipulators has been illustrated by referring to experiences that have been developed at LARM in Cassino.

8. Aknowledgements

The author wishes to thank Prof. Marco Ceccarelli at LARM in Cassino for his valuable comments and support during the elaboration of this work.

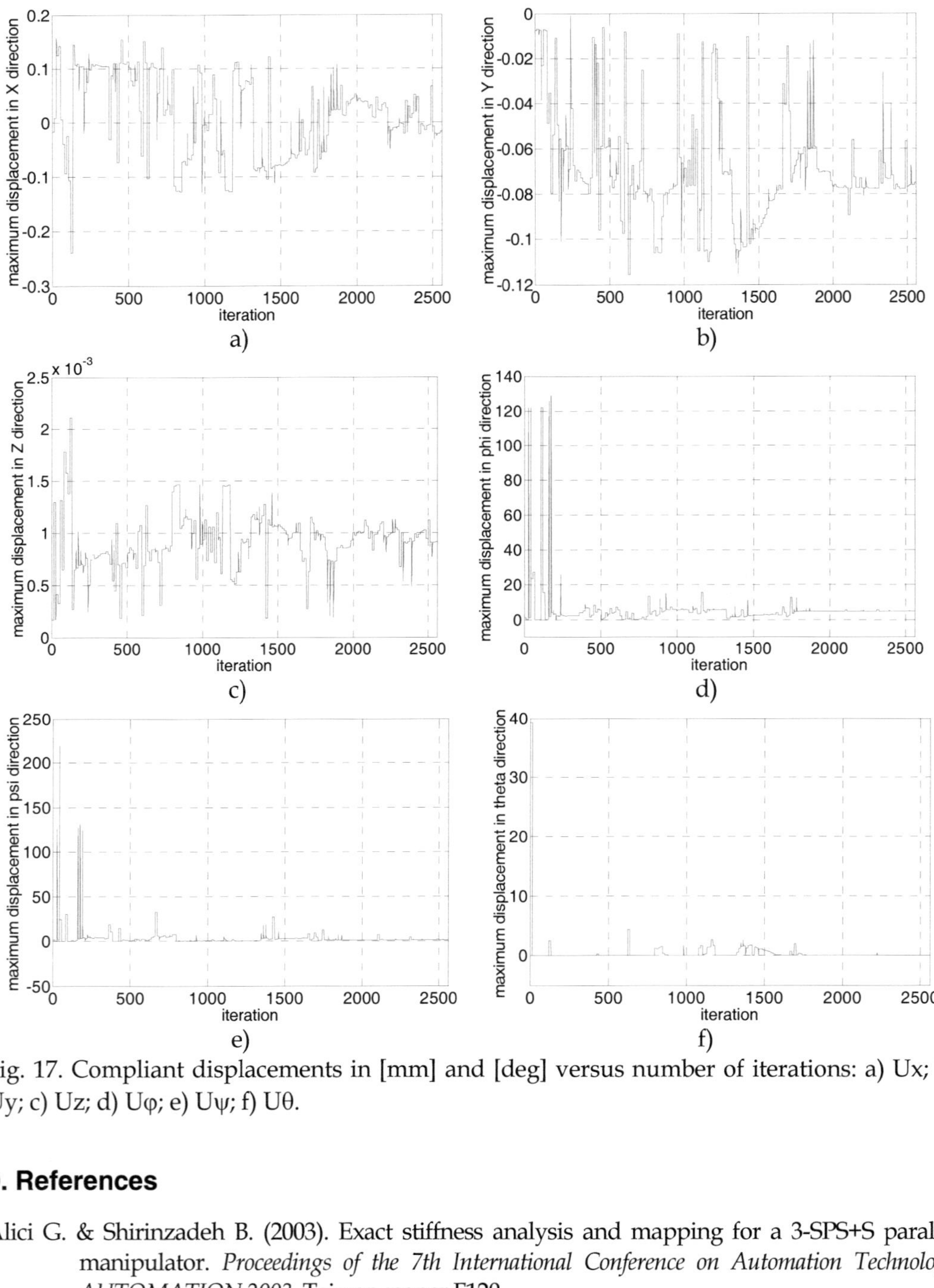

Fig. 17. Compliant displacements in [mm] and [deg] versus number of iterations: a) Ux; b) Uy; c) Uz; d) Uφ; e) Uψ; f) Uθ.

9. References

Alici G. & Shirinzadeh B. (2003). Exact stiffness analysis and mapping for a 3-SPS+S parallel manipulator. *Proceedings of the 7th International Conference on Automation Technology AUTOMATION 2003*, Taiwan, paper F120.

Angeles, J. (2002). The Robust Design of Parallel Manipulators. *Proceedings of 1st International Colloquium Collaborative Research Center 562*, Braunschweig, pp. 9-30.

ANSI, American National Standards Institute (1990). American National Standard for Industrial Robots and Robot Systems: Point-to-Point and Static Performance Characteristics - Evaluation. *ANSI/RIA 15.05-1-1990*, New York.

Boyd, S., & Vandenberghe, L. (2004). Convex Optimization. *Cambridge University Press*. Available on line at http://www.stanford.edu/~boyd/cvxbook/.

Branke J. (2008). Multiobjective optimization: interactive and evolutionary approaches, *Springer Verlag*.

Castejón C., Carbone G., García-Prada J.C., Ceccarelli M. (2007). A Multi-Objective Optimization Design for a 4R Service Robot, *International Journal of Mechanics and Control*, vol.08, n.02, pp.3-8.

Carbone G. (2003). Stiffness Evaluation of Multibody Robotic Systems. *PhD Dissertation, LARM, University of Cassino*, Cassino.

Carbone G. (2006). Stiffness Performance of Multibody Robotic Systems. *Proceedings of IEEE-TTTC International Conference on Automation, Quality&Testing, Robotics AQTR 2006*, vol.2, pp.219-224, Cluj-Napoca.

Carbone G., Ceccarelli M., Ogura Y., Lim H.O., Takanishi A. (2003). Numerical Simulation for an Optimum Design of a Humanoid Leg Through Stiffness Analysis. *Proceedings of 3rd IEEE International Conference on Humanoid Robots HUMANOIDS2003*, München and Karlsruhe, paper 3bWalking03.

Carbone G., Ottaviano E., Ceccarelli M. (2007). An Optimum Design Procedure for Both Serial and Parallel Manipulators", *IMechE Part C: Journal of Mechanical Engineering Science*, vol.221, n.7, pp.829-843.

Ceccarelli, M. (2004). *Fundamentals of Mechanics of Robotic Manipulation*, Kluwer, Dordrecht, 2004.

Ceccarelli, M., Carbone, G. (2002). A Stiffness Analysis for CaPaMan (Cassino Parallel Manipulator), *Mechanism and Machine Theory*, vol.37, n.5, pp.427-439.

Duffy, J. (1996). *Statics and Kinematics with Applications to Robotics*, Cambridge University Press, 153-169.

Eldar, Y.C. (2006). Minimax estimation of deterministic parameters in linear models with a random model matrix, *IEEE Transactions on Signal Processing*, Vol. 54, No. 2, pp. 601–612.

English C.E., Russell D. (1999). Mechanics and Stiffness Limitations of a Variable Stiffness Actuator for Use in Prosthetic Limbs. *Mechanism and Machine Theory*, vol.34, n.1, pp.7-25.

Gosselin, C. (1990). Stiffness Mapping for Parallel Manipulators", *IEEE Transactions on Robotics and Automation*, 6(3):.377-382.

Gosselin C., Angeles J. A Global Performance Index for the Kinematic Optimization of Robotic Manipulators, *Transaction of the ASME Journal of Mechanical Design*, 113: 220-226, 1991.

Grace, A. (1992). Optimization Toolbox User's Guide, *The Matlab Works Inc.*.

Hao, F., & Merlet, J.-P. (2005). Multi-Criteria Optimal Design of Parallel Manipulators Based on Interval Analysis, *Mechanism and Machine Theory*, Vol. 40, 2005, pp. 157-171.

Kardestuncer, H. (1974). *Elementary Matrix Analysis of Structures*, McGraw-Hill Kogakusha, Tokyo.

Manoochehri, S., & Seireg, A.A. (1990). A Computer-Based Methodology for the Form Synthesis and Optimal Design of Robot Manipulators, *ASME Journal of Mechanical Design*, Vol. 112, pp. 501-508.

Mathworks webpage (2009). Optimization toolbox™ 4. User guide. available on-line at http://www.mathworks.com/access/helpdesk/help/pdf_doc/optim/optim_tb.pdf.

Merlet, J-P. (2004). A C++ algorithms library of interval analysis for equation systems, version 2.3, *COPRIN project,* available on-line at http://www-sop.inria.fr/coprin/equipe/merlet/merlet.html
Merlet, J.-P. (2006). *Parallel Robots*. Springer Verlag.
Nof S.Y. (1985). Handbook of Industrial Robotics. *John Wiley & Sons*, New York.
Ottaviano, E., & Carbone G. (2003). A Procedure for the Multiobjective Design of Parallel Manipulators, *International Journal of Mechanics and Control*, Vol.04, No. 02, pp.57-62.
Ottaviano, E., & Ceccarelli, M. (2002). Optimum Design of Parallel Manipulators for Workspace and Singularity Performances, *Workshop on Fundamental Issues and Future Research Directions for Parallel Mechanisms and Manipulators*, Quebec City, pp.98-105.
Paden, B., & Sastry, S. (1988). Optimal Kinematic Design of 6R Manipulators, *International Journal of Robotics Research*, Vol.7, No. 2, pp. 43-61.
Pankov, A.R., Platonov, E.N., & Siemenikhin, K.V. (2000). On Minimax Identification: Method of Dual Optimization, *Proceedings of 39th IEEE Conference on Decision and Control*, Sydney, Vol. 5, pp. 4759-4764.
Rivin, E.I. (1999). *Stiffness and Damping in Mechanical Design,* Marcel Dekker.
Shimano, B.E., & Roth, B. (1978). Dimensional Synthesis of Manipulators, *Proceedings of 3rd RoManSy*, pp.166-187, Udine.
Schimmels J.M. (2001). Multidirectional Compliance and Constraint for Improved Robotic Deburring. Part 1: Improved Positioning. *Robotics and Computer-Integrated Manufacturing*, vol.17, n.4, pp.277-286.
Schonherr, J. (2000). Evaluation and Optimum Design of Parallel Manipulators Having Defined Workspace, *Proceedings of ASME 26th Biennial Mechanisms and Robotics Conference,* Baltimore, Paper DETC2000/MECH-14092.
Tahmasebi F., Tsai L.W. (1992). Jacobian and Stiffness Analysis of a Novel Class of Six-Dof Parallel Minimanipulators. *Proceedings of the ASME 22nd Biennial Mechanism Conference,* New York, vol.47, pp. 95-102.
Takeda, Y., & Funabashi, H. (1999). Kinematic Synthesis of In-Parallel Actuated Mechanisms Based on the Global Isotropy Index", Journal of Robotics and Mechatronics, Vol. 11, No. 5, pp. 404-410.
Tsai, L.W. (1999). *Robot Analysis: the Mechanics of Serial and Parallel Manipulators*, John Wiley & Sons, New York, pp.260-297.
Tsumugiwa T., Yokogawa R., Hara K. (2002). Variable Impedance Control with Virtual Stiffness for Human-Robot Cooperative Peg-in-Hole Task. *Proceedings of the IEEE/RSJ International Conference on Intelligent Robots ans Systems IROS'02*, Lausanne, pp.1075-1081.
Yao S., Zhan Q., Ceccarelli M., Carbone G., Lu Z. (2009). Analysis and Optimal Design of a Modular Underactuated Mechanism for Robot Fingers", IEEE/RSJ International Conference on Intelligent Robots and Systems IROS'09, St. Luis.
UNI, Italian National Institute for Standards (1995). Manipulating Industrial Robots: Performance Criteria and Related Test Methods. *UNI EN 29283 (= ISO 9283)*, Milan.
Vanderplaats, G. (1984). *Numerical Optimization Techniques for Engineers Design*, McGraw-Hill, New York.
Vijaykumar, R., Waldron, K.J., & Tsai, M.J. (1986). Geometric Optimization of Serial Chain Manipulator Structures for Working Volume and Dexterity, *International Journal of Robotics Research*, Vol.5, No. 2, pp. 91-103.

12

Concurrent Engineering of Robot Manipulators

M. Reza Emami and Robin Chhabra
University of Toronto Institute for Aerospace Studies
Canada

1. Introduction

Robot manipulators are good examples of complex engineering systems, where designers occasionally employ a subsystem-partitioning approach for their analysis and synthesis. The design methodology is traditionally based on the sequential decomposition of mechanical, electromechanical, and control/instrumentation subsystems, so that at each step a subset of design variables is considered separately (Castano et al., 2002). Although conventional *decoupled* or *loosely-coupled* approaches of design seem intuitively practical, they undermine the interconnection between various subsystems that may indeed play a crucial role in multidisciplinary systems. The necessity of communication and collaboration between the subsystems implies that such systems ought to be synthesized concurrently. In the concurrent design process, design knowledge is accumulated from all the participating disciplines, and they are offered equal opportunities to contribute to each state of design in parallel. The *synergy* resulting from integrating different disciplines in concurrent design has been documented in several case studies, to the effect that the outcome is a new and previously unattainable set of performance characteristics (Hewit, 1996). However, the challenge in a concurrent design process is that the multidisciplinary system model can become prohibitively complicated; hence computationally demanding. Plus, a large number of multidisciplinary objective and constraint functions must be taken into account, simultaneously, with a great number of design variables. As the complexity of the system model increases, in terms of the interactions between various subsystems, the coordination of all the constraints distributed in different disciplines becomes more difficult, in order to maintain the consistency between performance specifications and design variables.
Within the context of robotics, several *ad hoc* techniques of concurrent engineering have been reported in the literature. They are innovative design schemes for specific systems, such as Metamorphic Robotic System (Chirikjian, 1994), Molecule (Rus & McGray, 1998), Miniaturised Self-Reconfigurable System (Yoshida et al., 1999), Crystalline (Rus & Vona, 2000), and Semi-Cylindrical Reconfigurable Robot (Murata et al., 2000). But, more systematic approaches have been suggested by other researchers beyond the robotics community to tackle the challenge of high dimensionality in concurrent design. These approaches can be divided into two major groups. The first group translates the model complexity into a large volume of computations, and then attempts to find efficient algorithms or parallel

processing techniques to make these computations feasible. For example, parallel genetic algorithms were used for multi-objective optimizations (Coello, 1999), and later augmented with a penalty method to handle constraints (Kurapati et al., 2000). This approach was later adopted for the concurrent engineering of modular robotic systems (Bi & Zhang, 2001). Also, an integration of agent-based methods and simulated annealing was used for the modular configuration design (Ramachandran & Chen, 2000). The second group tries to alleviate the complexity by reducing the optimization space; either through breaking the optimization process into several stages (Paredis, 1996), or by approximating the space with the one with lower dimensions (Dhingra & Rao, 1995). Each group brings certain contributions to concurrent engineering, yet cannot avoid some drawbacks. While efficient algorithms, mostly taking advantage of parallel processing, can handle high computational demands in concurrent engineering, they tend to lose transparency, so that designers can no longer relate to the process. On the other hand, a better understanding of design may be achieved, should one be able to simplify the optimization model, but at a great cost of obtaining outcomes for an approximated version of the system that can be far from reality.

This chapter introduces a solution for the complexity of concurrent engineering, which in essence consists of two unique constituents, each relating to one of the above-mentioned groups. For the first part, it utilizes an efficient system modeling technique that not only does not compromise the transparency, but also accounts for complex phenomena such as sensor noise, actuator limitation, transmission flexibility, etc., which can hardly be captured by computational modeling. The model efficiency, in terms of both computation and accuracy, is due to the use of real hardware modules in the simulation loop and, hence, the real-time execution. In other words, the solution uses a Robotics Hardware-in-the-loop Simulation (RHILS) platform for "computing" the system model in the design process. And for the second part, the solution applies an alternative design methodology, namely Linguistic Mechatronics (LM), which not only formalizes subjective notions and brings the linguistic aspects of communication into the design process, but also transforms the multi-objective constrained optimization model into a single-objective unconstrained formulation. A combination of the above two techniques will ensure an efficient solution for concurrent engineering of robot manipulators, without simplifying the system model. Further, it facilitates communication between designers (of different background) and customers by including linguistic notions in the design process.

The chapter is organized as follows: Section 2 introduces Linguistic Mechatronics (LM). Section 3 details the Robotics Hardware-in-the-loop Simulation (RHILS) platform. Section 4 describes the LM-RHILS based concurrent engineering methodology and its application to an industrial robot manipulator. Some concluding remarks are made in Section 5.

2. Linguistic Mechatronics: An Alternative Approach to Concurrent Engineering

The premise of concurrent engineering is to provide a common language to fill in the communication gap between different engineering disciplines, and to devise a means for helping them collaborate towards a common goal. The need for communication and collaboration in concurrent engineering implies that, in addition to physical features, many subjective notions must be involved, which can hardly be captured by pure mathematical formulations. Both customers and designers need to communicate beyond the equations to convey design requirements and specifications. Hence, there is a need for a communication means in concurrent engineering that can convey qualitative and subjective notions that are used frequently in human interactions, in addition to holistic criteria that finalize the design process based on objective performances in the real physical world. A few methodologies of concurrent design have attempted to include subjective notions in the design process (e.x., Dhingra et al., 1990). Amongst them, Method of Imprecision (MoI) is a notable attempt to take into account imprecision in design (Otto & Antonsson, 1995). This approach defines a set of designer's preferences for design variables and performance parameters to model the imprecision in design. It determines and maximizes the global performance under one of the two conservative or aggressive design tradeoff strategies, and uses fuzzy-logic operators for tradeoff in the design space. This method offers a number of advantages that are crucial in concurrent engineering. However, it does not provide a systemic means to distinguish the constraints from the goals in the aggregation process; instead it simply offers two extreme designer's attitudes. Further, in the MoI methodology designer's attitudes are not justified with any objective performance criterion. While subjective notions can play a crucial role in concurrent engineering of multidisciplinary systems, their relevance must eventually be checked against the objective criteria of system performance.

This section introduces *Linguistic Mechatronics* (LM) as an alternative concurrent design framework, which emphasizes on the designer's satisfaction, instead of pure performance optimization, and brings the linguistic aspects of communication into the design process. It not only formalizes subjective notions of design and simplifies the complicated multi-objective constrained optimization, but also resolves the above-mentioned deficiencies of the MoI methodology through **a)** dividing the design attributes into two inherently-different classes, namely *wish* and *must* attributes; and **b)** aggregating satisfactions using *parametric* fuzzy-logic operators so that the designer's attitude can be adjusted based on an objective performance criterion. Linguistic Mechatronics involves three stages of system modeling. First, a fuzzy-logic model is developed in the *primary* phase of design; secondly, a software and/or hardware simulation of the system is used for the *secondary* phase. And lastly, a bond graph model of the system assigns appropriate *supercriteria* that finalize the design. In the following sub-sections, the foundations of linguistic mechatronics, namely fuzzy modeling and fuzzy operators, will be reviewed first, and then a step-by-step formulation of the LM methodology will be presented.

2.1 Fuzzy-Logic Modeling

Fuzzy-logic modeling is an approach to forming a system model by using a descriptive language based on fuzzy-logic with fuzzy propositions. In (Emami, 1997), a systematic approach of fuzzy-logic modeling is developed, which is adopted in this work. In general, the clustered knowledge of a system can be interpreted by fuzzy models consisting of IF-THEN rules with multi-antecedent and multi-consequent variables (n antecedents, s consequents, and r rules):

IF U_1 is B_{11} **AND**…**AND** U_n is B_{1n} **THEN** V_1 is D_{11} **AND**…**AND** V_s is D_{1s}
ALSO
… (1)
ALSO
IF U_1 is B_{r1} **AND**…**AND** U_n is B_{rn} **THEN** V_1 is D_{r1} **AND**…**AND** V_s is D_{rs}

where U_j (j=1,…,n) is the jth input variable and V_k (k=1,…,s) is the kth output variable, B_{ij} (i=1,...,r, j=1,…,n) and D_{ik} (i=1,…,r, k=1,…,s) are fuzzy sets over the input and output universes of discourse, respectively. Constructing a fuzzy model can be divided into two major steps: **a)** fuzzy rule-base generation, and **b)** fuzzy inference mechanism selection.

A. Fuzzy Rule-base Generation

Assuming the existence of sufficient knowledge of the system, the process of rule-base generation can be performed in the following sequence: **a)** clustering output data and assigning output membership functions, **b)** finding the non-significant input variables and assigning the membership functions to the rest of them, and **c)** tuning the input and output membership functions. Clustering methods are occasionally based on the optimization of an objective function to find the optimum membership matrix, $\boldsymbol{U}=[u_{ik}]$, that contains the membership value of the kth data point, $z_k \in \boldsymbol{Z}$, to the ith partition. In Fuzzy C-Means (*FCM*) clustering method, this function, J_m, is defined as the weighted sum of the squared errors of data points, and the minimization problem is formulated as:

$$\min_{(\mathbf{U},\mathbf{V})}\left[J_m(\boldsymbol{U},\boldsymbol{V};\boldsymbol{Z}) = \sum_{i=1}^{c}\sum_{k=1}^{N}(u_{ik})^m (\boldsymbol{z}_k-\boldsymbol{v}_i)^T(\boldsymbol{z}_k-\boldsymbol{v}_i)\right]; \tag{2}$$

where $\boldsymbol{V}=\{\boldsymbol{v}_1,\boldsymbol{v}_2,\ldots,\boldsymbol{v}_c\}$ is the set of unknown cluster centers, N and c are the number of data points and clusters, respectively, and m is the weighting exponent.

A prerequisite for *FCM* is assigning c and m. The optimal values of these numbers are calculated based on two requirements: **a)** maximum separation between the clusters; and **b)** maximum compactness of the clusters. Therefore, the *fuzzy within-cluster scatter matrix,*

$$\boldsymbol{S}_W = \sum_{i=1}^{c}\sum_{k=1}^{N}(u_{ik})^m(\boldsymbol{z}_k-\boldsymbol{v}_i)(\boldsymbol{z}_k-\boldsymbol{v}_i)^T \tag{3}$$

and *between-cluster scatter matrix,*

$$\boldsymbol{S}_B = \sum_{i=1}^{c}\left(\sum_{k=1}^{N}(u_{ik})^m\right)(\boldsymbol{v}_i-\bar{\boldsymbol{v}})(\boldsymbol{v}_i-\bar{\boldsymbol{v}})^T \tag{4}$$

are defined to reflect the two criteria (Emami et al., 1998). Note that the fuzzy total mean array, $\bar{\boldsymbol{v}}$, is defined as:

$$\bar{\boldsymbol{v}} = \frac{1}{\sum_{i=1}^{c}\sum_{k=1}^{N}(u_{ik})^m}\sum_{i=1}^{c}\sum_{k=1}^{N}(u_{ik})^m \boldsymbol{z}_k \,. \tag{5}$$

The matrix S_B represents the separation between the fuzzy clusters, and S_W is an index for the compactness of fuzzy clusters. For obtaining the best clusters the trace of matrix S_W, $tr(S_W)$, should be minimized to increase the compactness of clusters and $tr(S_B)$ should be maximized to increase the separation between clusters. Alternatively, $s_{cs} = tr(\boldsymbol{S_W}) - tr(\boldsymbol{S_B})$ can be minimized to identify the optimum number of clusters, *c*. The weighting exponent, *m*, varies in (1,+∞) and indicates the degree of fuzziness of the assigned membership functions. In order to have a reliable index for s_{cs}, *m* should be far enough from both extremes. Hence, the reliable value of *m* is what holds the trace of *fuzzy total scatter matrix* (S_T),

$$s_T = tr(\boldsymbol{S_T}) = tr(\boldsymbol{S_W} + \boldsymbol{S_B}), \tag{6}$$

somewhere in the middle of its domain. Since s_T and s_{cs} are both functions of *m* and *c*, the process of choosing the parameters should be performed by a few iterations.
In systems with a large number of variables, there occasionally exist input variables that have less effect on the output, in the range of interest. In order to have an efficient fuzzy-logic model, an index, π_j, is defined as an overall measure of the non-significance of input variable x_j as:

$$\pi_j = \prod_{i=1}^{n} \frac{\Gamma_{ij}}{\Gamma_j}; \qquad (j=1,\ldots,r) \tag{7}$$

where Γ_{ij} is the range in which membership function $B_{ij}(x_j)$ is one, and Γ_j is the entire range of the variable x_j. The smaller the value of π_j is, the more effect the jth variable has in the model, and vice versa.
Finally, to map the output membership functions onto the input spaces, a clustering method, called *line fuzzy clustering*, is employed. This method works based on the distance of each data point located on the axis x_j, to the interval of the jth input variable corresponding to the output membership function equal or close to one (Emami, 1997).

B. Fuzzy Reasoning Mechanism
To interpret connectives in fuzzy set theory, there exist a number of different classes of triangular norm (*t-norm*) and triangular conorm (*t-conorm*), such as *Max-Min Operators* (T_{min}, S_{max}), *Algebraic Product and Sum* (T_{prod} S_{sum}), and *Drastic Product and Sum* (T_W, S_W). Using the basic properties of these operators, it is shown in (Emami, 1997) that for any arbitrary *t-norm* (*T*) and *t-conorm* (*S*) and for all $a_i \in [0,1]$:

$$\begin{aligned} T_W(a_1,\ldots,a_n) \le T(a_1,\ldots,a_n) \le T_{\min}(a_1,\ldots,a_n), \\ S_W(a_1,\ldots,a_n) \le S(a_1,\ldots,a_n) \le S_{\max}(a_1,\ldots,a_n). \end{aligned} \tag{8}$$

Various types of parameterized operators have been suggested in the literature to cover this range. In particular, a class of operators for fuzzy reasoning is introduced in (Emami et al., 1999), which is adopted here for aggregating the satisfactions, as explained in the next sub-section:

$$S^{(p)}(b_1, b_2, \ldots, b_n) = [b_1^{\,p} + (1-b_1^{\,p})[\ldots[b_{n-2}^{\,p} + \ldots + (1-b_{n-2}^{\,p})[b_{n-1}^{\,p} + (1-b_{n-1}^{\,p})b_n^{\,p}]]\ldots]]]^{\,1/p}; \tag{9}$$

where $b_i \in [0,1]$ and $p \in (0,+\infty)$. Consequently, the corresponding *t-norm* operator is defined based on De Morgan laws using standard complementation operator, as:

$$T^{(p)}(a_1,a_2,...,a_n) = 1 - S^{(p)}((1-a_1),(1-a_2),...,(1-a_n)) \,. \tag{10}$$

In the extreme cases, this class of parameterized operators approaches (T_{min},S_{max}) as $p \to +\infty$, (T_{prod} S_{sum}) as $p \to 1$, and (T_W, S_W) as $p \to 0$.

The meaning of an aggregation operator is sometimes neither pure AND (*t-norm*) with its complete lack of compensation, nor pure OR (*t-conorm*). This type of operator is called *mean aggregation* operator. For example, a suitable parametric operator of this class, namely *generalized mean operator*, is defined in (Yager & Filev, 1994) as:

$$G^{(\alpha)}(a_1,a_2,...,a_n) = \left(\frac{1}{n}\sum_{i=1}^{n} a_i^{\alpha}\right)^{1/\alpha} ; \tag{11}$$

where $\alpha \in (-\infty,+\infty)$. It appears that this type of aggregation monotonically varies between *Min* operator while $\alpha \to -\infty$ and *Max* operator as $\alpha \to +\infty$. Subsequently, an appropriate inference mechanism should be employed to combine the rules and calculate the output for any set of input variables. Takagi-Sugeno-Kang (*TSK*) reasoning method is associated to a rule-base with functional type consequents instead of the fuzzy sets and the crisp output, y^*, is defined by the weighted average of the outputs of individual rules, y_i's, as:

$$y^* = \sum_{i=1}^{r} \frac{\tau_i}{\sum_{j=1}^{r}\tau_j} y_i = \sum_{i=1}^{r} \frac{\tau_i}{\sum_{j=1}^{r}\tau_j}(b_{i0} + b_{i1}x_1 + ... + b_{in}x_n) ; \tag{12}$$

where τ_i is the degree of fire of the ith rule:

$$\tau_{\mathrm{i}} = T(\mathrm{B}_{i1}(x_1),...,\mathrm{B}_{in}(x_n)) \,. \tag{13}$$

Since the *TSK* method of reasoning is compact and works with crisp values, it is computationally efficient; and therefore, it is widely used in fuzzy-logic modeling of engineering systems, especially when tuning techniques are utilized. Ultimately, the parameters of input membership functions and output coefficients are tuned by minimizing the mean square error of the output of the fuzzy-logic model with respect to the existing data points.

2.2 The LM Formulation

A design problem consists of two sets: *design variables* $\boldsymbol{X} \equiv \{X_j : \forall j = 1,..., n\}$ and *design attributes* $\boldsymbol{A} \equiv \{A_i : \forall i = 1,..., N\}$. Design variables are to be configured to satisfy the *design requirements* assigned for design attributes, subject to the design availability $\boldsymbol{D} \equiv \{D_j : \forall j = 1,...,n\}$. Each design attribute stands for a design function providing a functional mapping $F_i : \aleph \rightarrow \Im_i$ that relates a state of design configuration $\boldsymbol{X} \in \aleph$ to the attribute $A_i \in \Im_i$, i.e., $A_i = F_i(\boldsymbol{X})$ (*i*=1,...,*N*). These functional mappings can be of any form, such as closed-form equations, heuristic rules, or set of experimental or simulated data.

Given a set of design variables and a set of design attributes along with an available knowledge that conveys the relationship between them, the process of Linguistic Mechatronics is performed in two phases: **a)** *primary* phase in which proper intervals for the design variables are identified subject to design availability, and **b)** *secondary* phase in which design variables are specified in their intervals in order to maximize an overall design satisfaction based on the design requirements and designer's preferences. Thus, the secondary phase involves a single-objective optimization, yet it is critically dependant on the initial values of a large number of design variables. The primary phase makes the optimization more efficient by providing proper intervals for the design variables from where the initial values are selected. The *overall satisfaction* is an aggregation of satisfactions for all design attributes. The satisfaction level depends on the designer's attitude that is modeled by fuzzy aggregation parameters. However, different designers may not have a consensus of opinion on *satisfaction*. Therefore, the system performance must be checked over a holistic *supercriterion* to capture the objective aspects of design considerations in terms of physical performance. Designer's attitude is adjusted through iterations over both primary and secondary phases to achieve the enhanced system performance. Therefore, this methodology incorporates features of both human subjectivity (i.e., designer's intent) and physical objectivity (i.e., performance characteristics) in multidisciplinary system engineering.

Definition 1 - Satisfaction: A mapping μ such that $\mu : \boldsymbol{Y} \rightarrow [0,1]$ for each member of $\boldsymbol{Y}$ is called satisfaction, where $\boldsymbol{Y}$ is a set of available design variables or design attributes based on the design requirements. The grade one corresponds to the ideal case or the most satisfactory situation. On the other hand, the grade zero means the worst case or the least satisfactory design variable or attribute.

Satisfaction on a design attribute, $a_i \equiv \mu_{A_i}(\boldsymbol{X})$, indicates the achievement level of the corresponding design requirement based on the designer's preferences. The satisfaction for a design variable, $x_j \equiv \mu_{X_j}(\boldsymbol{X})$, reflects the availability of the design variable. In the conceptual phase, design requirements are usually subjective concepts that imply the costumer's needs. These requirements are naturally divided into *demands* and *desires*. A designer would use engineering specifications to relate design requirements to a proper set of design attributes. Therefore, in *LM* the design attributes are divided into two subsets, labeled *must* and *wish* design attributes.

Definition 2 - *Must* design attribute: A design attribute is called *must* if it refers to costumer's demand, i.e., the achievement of its associated design requirement is mandatory with no room for compromise. These attributes form a set coined $\boldsymbol{M}$.

Definition 3 - *Wish* design attribute: A design attribute is called *wish* if it refers to costumer's desire, i.e., its associated design requirement permits room for compromise and it should be achieved as much as possible. These attributes form a set coined **W**.
Therefore,

$$\boldsymbol{M} \cap \boldsymbol{W} = \phi, \qquad \boldsymbol{M} \cup \boldsymbol{W} = \boldsymbol{A}. \tag{14}$$

The satisfaction specified for *wish* attribute W_i is $w_i(\boldsymbol{X}) \equiv \mu_{W_i}(\boldsymbol{X})$ (i=1,...,N_W), and the satisfaction specified for *must* attribute M_i is $m_i(\boldsymbol{X}) \equiv \mu_{M_i}(\boldsymbol{X})$ (i=1,...,N_M). Therefore, for each design attribute A_i (corresponding to either M_i or W_i), there is a predefined mapping to the satisfaction a_i (m_i or w_i), i.e., $\{(A_i, a_i) : \forall i = 1,..., N\}$. Fuzzy set theory can be applied for defining satisfactions through fuzzy membership functions and also for aggregating the satisfactions using fuzzy-logic operators.
Remark: $[F_i(\boldsymbol{X}_1) \succeq F_i(\boldsymbol{X}_2)] \Leftrightarrow [a_i(\boldsymbol{X}_1) \geq a_i(\boldsymbol{X}_2)]$ for monotonically non-decreasing satisfaction. More specifically, if $0 < a_i(\bullet) < 1$ then $[F_i(\boldsymbol{X}_1) \succ F_i(\boldsymbol{X}_2)] \Leftrightarrow [a_i(\boldsymbol{X}_1) > a_i(\boldsymbol{X}_2)]$ and if $a_i(\bullet) = 0 \text{ or } 1$ then $[F_i(\boldsymbol{X}_1) \succ F_i(\boldsymbol{X}_2)] \Leftrightarrow [a_i(\boldsymbol{X}_1) = a_i(\boldsymbol{X}_2)]$, where $\succeq$ denotes loosely superior and $\succ$ represents strictly superior. In other words, the better the performance characteristic is the higher the satisfaction will be, up to a certain threshold.
Definition 4 - Overall satisfaction: For a specific set of design variables **X**, overall satisfaction is the aggregation of all *wish* and *must* satisfactions, as a global measure of design achievement.

A. Calculation of Overall Satisfaction
Must and *wish* design attributes have inherently-different characteristics. Hence, appropriate aggregation strategies must be applied for aggregating the satisfactions of each subset.

1) Aggregation of Must Design Attributes
Axiom 1: Given *must* design attributes, $\{(M_i, m_i) : \forall i = 1,..., N_M\}$, and considering component availability, $\{(D_j, x_j) : \forall j = 1,..., n\}$, the overall *must* satisfaction is the aggregation of all *must* satisfactions using a class of *t-norm* operators.

Must attributes correspond to those design requirements that are to be satisfied with no room of negotiation, and, linguistically, it means that all design requirements associated with *must* attributes have to be fulfilled simultaneously. Therefore, for aggregating the satisfactions of *must* attributes an AND logical connective is suitable. Considering satisfactions as fuzzy membership degrees, the AND connective can be interpreted through a family of *t-norm* operators. Thus, the overall *must* satisfaction is quantified using the *p*-parameterized class of *t-norm* operators, i.e.,

$$\mu_{\boldsymbol{M}}^{(p)}(\boldsymbol{X}) = T^{(p)}(m_1, m_2, ..., m_{N_M}, x_1, x_2, ..., x_n). \quad (p > 0) \tag{15}$$

The parametric *t-norm* operator $T^{(p)}$ is defined based on (9) and (10).

Parameter p can be adjusted to control the fashion of aggregation. Changing the value of p makes it possible to obtain different tradeoff strategies. The larger the p, the more pessimistic (conservative) designer's attitude to a design will be, and vice versa.

2) Aggregation of Wish Design Attributes

Definition 5 - Cooperative *wish* attributes: A subset of *wish* design attributes is called cooperative if the satisfactions corresponding to the attributes all vary in the same direction when the design variables are changed.

Therefore, *wish* attributes can be divided into two cooperative subsets:

a) Positive-differential *wish* attributes ($\boldsymbol{W}^+$): In this subset the total differential of the satisfactions for the *wish* attributes (with respect to design variables) are non-negative.

$$\boldsymbol{W}^+ = \{(W_i, w_i) : W_i \in \boldsymbol{W}, \quad dw_i(\boldsymbol{X}) \geq 0\}. \tag{16}$$

This subset includes all attributes that tend to reach a higher satisfaction when all design variables have an infinitesimal increment.

b) Negative-differential *wish* attributes ($\boldsymbol{W}^-$): In this subset the total differential of the satisfactions for the *wish* attributes (with respect to design variables) are negative.

$$\boldsymbol{W}^- = \{(W_i, w_i) : W_i \in \boldsymbol{W}, \quad dw_i(\boldsymbol{X}) < 0\}. \tag{17}$$

This subset includes all attributes that tend to reach a lower satisfaction when all design variables have an infinitesimal increment.

$$\boldsymbol{W}^+ \cap \boldsymbol{W}^- = \phi, \quad \boldsymbol{W}^+ \cup \boldsymbol{W}^- = \boldsymbol{W}. \tag{18}$$

Since in each subset all *wish* attributes are cooperative, their corresponding design requirements can all be fulfilled simultaneously in a linguistic sense. Hence, according to **Axiom 1**, similar to *must* satisfactions, a q-parameterized class of *t-norm* operators is suitable for aggregating satisfactions in either subsets of *wish* attributes.

$$\mu_{W^\pm}{}^{(q)}(\boldsymbol{X}) = T^{(q)}(w_1, w_2, \ldots, w_{N_{W^\pm}}) \quad (q > 0); \tag{19}$$

where $N_{W^\pm}$ are the number of positive-/negative-differential *wish* attributes.

Axiom 2: Given the satisfactions corresponding to positive- and negative-differential *wish* attributes, $\mu_{W^+}{}^{(q)}(\boldsymbol{X})$ and $\mu_{W^-}{}^{(q)}(\boldsymbol{X})$, the overall *wish* satisfaction can be calculated using an a-parameterized *generalized mean* operator.

The two subsets of *wish* attributes cannot be satisfied simultaneously as their design requirements compete with each other. Therefore, some compromise is necessary for aggregating their satisfactions, and the class of *generalized mean* operators in (11) reflects the averaging and compensatory nature of their aggregation.

$$\mu_W^{(\alpha,q)}(\boldsymbol{X})=\left[\frac{1}{2}\left(\left(\mu_{W^+}^{(q)}(\boldsymbol{X})\right)^{\alpha}+\left(\mu_{W^-}^{(q)}(\boldsymbol{X})\right)^{\alpha}\right)\right]^{1/\alpha}. \quad (20)$$

This class of *generalized mean* operators is monotonically increasing with respect to *a* between *Min* and *Max* operators; therefore, offers a variety of aggregation strategies from conservative to aggressive, respectively. The overall *wish* satisfaction is governed by two parameters *q* and *a,* representing subjective tradeoff strategies. They can be adjusted appropriately to control the fashion of aggregation. The larger the *a* or the smaller the *q*, the more optimistic (aggressive) one's attitude to a design will be, and vice versa.

3) Aggregation of Overall Wish and Must Satisfactions

Axiom 3: The overall satisfaction is quantified by aggregating the overall *must* and *wish* satisfactions, $\mu_M^{(p)}(\boldsymbol{X})$, and $\mu_W^{(q,\alpha)}(\boldsymbol{X})$, with the *p*-parameterized class of *t-norm* operators, i.e.,

$$\mu^{(p,q,\alpha)}(\boldsymbol{X})=T^{(p)}(\mu_M^{(p)}(\boldsymbol{X}),\mu_W^{(q,\alpha)}(\boldsymbol{X})).\quad (p>0). \quad (21)$$

The aggregation of all *wish* satisfactions can be considered as one *must* attribute, i.e., it has to be fulfilled to some extent with other *must* attributes with no compromise. Otherwise, the overall *wish* satisfaction can become zero and it means none of the *wish* attributes is satisfied, which is unacceptable in design. Therefore, the same aggregation parameter, *p,* that was used for *must* attributes should be used for aggregating the overall *wish* and *must* satisfactions. In (21), three parameters, i.e., *p, q* and *a,* called *attitude parameters*, govern the overall satisfaction.

B. Primary Phase of LM

Once the overall satisfaction is calculated, in order to obtain the most satisfactory design, this index should be maximized. The optimization schemes are critically dependent on the initial values and their search spaces. Therefore, to enhance the optimization performance, suitable ranges of design variables are first found in the primary phase of *LM*. In linguistic term, primary phase of *LM* methodology provides an *imprecise* sketch of the final product and illustrates the decision-making environment by defining some ranges of possible solutions. For this purpose, the mechatronic system is represented by a fuzzy-logic model based on (1). This model consists of a set of fuzzy IF-THEN rules that relates the ranges of design variables as fuzzy sets to the overall satisfaction; i.e.,

IF X_1 is B_{11} **AND…AND** X_n is B_{1n} **THEN** μ is D_1
ALSO
… (22)
ALSO
IF X_1 is B_{r1} **AND…AND** X_n is B_{rn} **THEN** μ is D_r

where μ is the overall satisfaction and B_{lj} and D_l (j=1,…,n and l=1,…,r) are fuzzy sets on X_j

and μ, respectively, which can be associated with linguistic labels.

The fuzzy rule-base is generated from the available data obtained from simulations, experimental prototypes, existing designs or etc., using fuzzy-logic modeling algorithm as detailed in the previous section. The achieved consequent fuzzy sets, D_l's, can be further defuzzified by (23) to crisply express the level of overall satisfaction corresponding to each rule.

$$\mu^*_l = \frac{1}{N}\sum_{i=1}^{N}\mu^i_l = \frac{1}{N}\sum_{i=1}^{N}(b_{l0} + b_{l1}X_1^i + \ldots + b_{ln}X_n^i); \tag{23}$$

where μ^i_l ($l=1,2,\ldots,r$, $i=1,2,\ldots,N$) is the overall satisfaction corresponding to the ith data point in l^{th} rule, N is the number of data points in the existing database, b_{lj} ($j=1,2,\ldots,n$) is the *TSK* consequent coefficient corresponding to the jth design variable in the lth rule, X_j^i is the jth design variable in the ith data point and μ^*_l corresponds to the overall satisfaction of rule l. The rule with the maximum μ_l^* is selected, and the set of its antecedents represents the appropriate intervals for the design variables. The set of these *suitable* intervals is denoted as $\boldsymbol{C} = \{C_j : \forall j = 1,\ldots,n\}$ and the corresponding fuzzy membership functions are labeled as $c_j(X_j)\,(j=1,\ldots,n)$. Finally, these fuzzy sets are defuzzified using *Centre of Area* (*CoA*) defuzzification method (Yager & Filev, 1994) to introduce the set of initial values $\boldsymbol{X_0} = \{X_{j0} : \forall j = 1,\ldots,n\}$ for design variables in the secondary phase of optimization process.

$$X_{j0} = \frac{\int_{C_j} X_j c_j(X_j)dX_j}{\int_{C_j} c_j(X_j)dX_j}. \quad (j = 1,\ldots,n) \tag{24}$$

C. Secondary Phase of LM

In the secondary phase, *LM* employs regular optimization methods to perform a single-objective unconstrained maximization of the overall satisfaction. The point-by-point search is done within the suitable intervals of design variables obtained from the primary phase. Therefore, the locally unique solution $\boldsymbol{X_s}$ is obtained through:

$$\mu^{(p,q,\alpha)}(\boldsymbol{X_s}) = \max_{\boldsymbol{X}\in\boldsymbol{C}} T^{(p)}(\mu_{\boldsymbol{M}}{}^{(p)}(\boldsymbol{X}), \mu_{\boldsymbol{W}}{}^{(q,\alpha)}(\boldsymbol{X})). \tag{25}$$

It can be shown that the pareto-optimality of the solution is a result of how the satisfactions are defined: Assume that $\boldsymbol{X_s}$ is not locally pareto-optimal. Then $\exists \boldsymbol{X_1} \in \boldsymbol{C}$ such that

$$F_i(\boldsymbol{X_1}) \succeq F_i(\boldsymbol{X_s}), \quad \forall i = 1,\ldots,N \tag{26}$$

particularly, there exists an i_0 that:

$$F_{i_0}(\boldsymbol{X_1}) \succ F_{i_0}(\boldsymbol{X_s}). \tag{27}$$

Thus, according to the **Remark**,

$$a_{i_0}(\boldsymbol{X}_l) > a_{i_0}(\boldsymbol{X}_s), \tag{28a}$$

or

$$a_{i_0}(\boldsymbol{X}_l) = a_{i_0}(\boldsymbol{X}_s) = 1. \tag{28b}$$

Hence, if F_{i_0} corresponds to a *must* attribute, due to the monotonicity of *t-norm* operator in (15),

$$\mu_M{}^{(p)}(\boldsymbol{X}_l) \geq \mu_M{}^{(p)}(\boldsymbol{X}_s). \tag{29}$$

And if F_{i_0} corresponds to a *wish* attribute, due to the monotonicity of both *t-norm* and *generalized mean* operators in (20),

$$\mu_W{}^{(q,\alpha)}(\boldsymbol{X}_l) \geq \mu_W{}^{(q,\alpha)}(\boldsymbol{X}_s). \tag{30}$$

Finally, the monotonicity of *t-norm* in (21) lead to:

$$\mu^{(p,q,\alpha)}(\boldsymbol{X}_l) \geq \mu^{(p,q,\alpha)}(\boldsymbol{X}_s). \tag{31}$$

Obviously, (31) contradicts the fact that $\boldsymbol{X}_s$ is a locally optimal solution. Note that in (29), (30) and (31) the equality holds when both satisfactions are 1. Thus, in order to avoid the equality, the satisfactions can be defined monotonically increasing or decreasing on the set of suitable intervals, **C**.

As indicated in (25), various attitude parameters, *p*, *q* and *a*, result in different optimum design values for maximizing the overall satisfaction. Consequently, a set of satisfactory design alternatives ($\boldsymbol{C}_s$) is generated based on subjective considerations, including designer's attitude and preferences for design attributes.

D. Performance Supercriterion

From the set of optimally satisfactory solutions, $\boldsymbol{C}_s$, the best design needs to be selected based on a proper criterion. In the previous design stages, decision making was critically biased by the designer's preferences (satisfaction membership functions) and attitude (aggregation parameters). Therefore, the outcomes must be checked against a supercriterion that is defined based on physical system performance. Indeed, such a supercriterion is used to adjust the designer's attitude based on the reality of system performance. A suitable supercriterion for multidisciplinary systems should take into account interconnections between all subsystems and consider the system holistically, as the synergistic approach of mechatronics necessitates.

Although mechatronic systems are multidisciplinary, the universal concept of energy and energy exchange is common to all of their subsystems. Therefore, an energy-based model can deem all subsystems together with their interconnections, and introduce generic notions that are proper for mechatronics. A successful attempt in this direction is the conception of *bond graphs* in the early 60's (Paynter, 1961). Bond graphs are domain-independent graphical descriptions of dynamic behaviour of physical systems. In this modeling strategy all components are recognized by the energy they supply or absorb, store or dissipate, and reversibly or irreversibly transform. In (Breedveld, 2004; Borutzky, 2006) bond graphs are utilized to model mechatronic systems. This generic modeling approach provides an efficient means to define holistic supercriteria for mechatronics based on the first and second laws of thermodynamics (Chhabra & Emami, 2009).

1) Energy Criterion

Any mechatronic system is designed to perform a certain amount of work on its environment while the input energy is supplied to it. Based on the first law of thermodynamics, this *supplied energy* (S) does not completely convert into the *effective work* (E) since portions of this energy are either stored or dissipated in the system by the system elements or alter the global state of the system in the environment. This *cost energy* (f) should be paid in any mechatronic system in order to transfer and/or convert the energy from the suppliers to the effective work. Therefore, a supercriterion, coined *energy criterion*, can be defined as minimizing $f(\boldsymbol{X})$ for a known total requested effective work from the system. Based on the principle of conservation of energy:

$$S(\boldsymbol{X}) = E + f(\boldsymbol{X}), \tag{32}$$

which shows that minimizing the supplied energy is equivalent to the energy criterion. Therefore, by minimizing the supplied energy or cost function, depending on the application, with respect to the attitude parameters the best design can be achieved in the set of optimally-satisfied solutions ($\boldsymbol{C}_s$).

$$S(\boldsymbol{X}^*) = \min_{\boldsymbol{X}_s \in \boldsymbol{C}_s} S(\boldsymbol{X}_s; p, q, \alpha). \tag{33}$$

In bond graphs the supplied energy is the energy that is added to the system at the source elements, which are distinguishable by S_e and S_f with the bonds coming out of them. Hence, by integrating the supplied power at all of the source elements during the simulation $S(\boldsymbol{X})$ can be calculated.

2) Entropy Criterion

Based on the second law of thermodynamics, after a change in supplied energy, a mechatronic system reaches its equilibrium state once entropy generation approaches its maximum. During this period the system loses its potential of performing effective work, constantly. Therefore, if the loss work of the system is less, available work from the system or, in other words, the aptitude of the system to perform effective work on the environment is more. This is equivalent to minimizing the entropy generation or the irreversible heat exchange at the dissipative elements of the bond graphs, i.e., $Q_{irr}(t; \boldsymbol{X})$, with respect to $\boldsymbol{X}$ and accordingly it is called *entropy criterion*. Given a unit step change of supplied energy, the equilibrium time, denoted by $t_{eq}(\boldsymbol{X})$, is the time instant after which the rate of change of dissipative heat remains below a small threshold, ε,

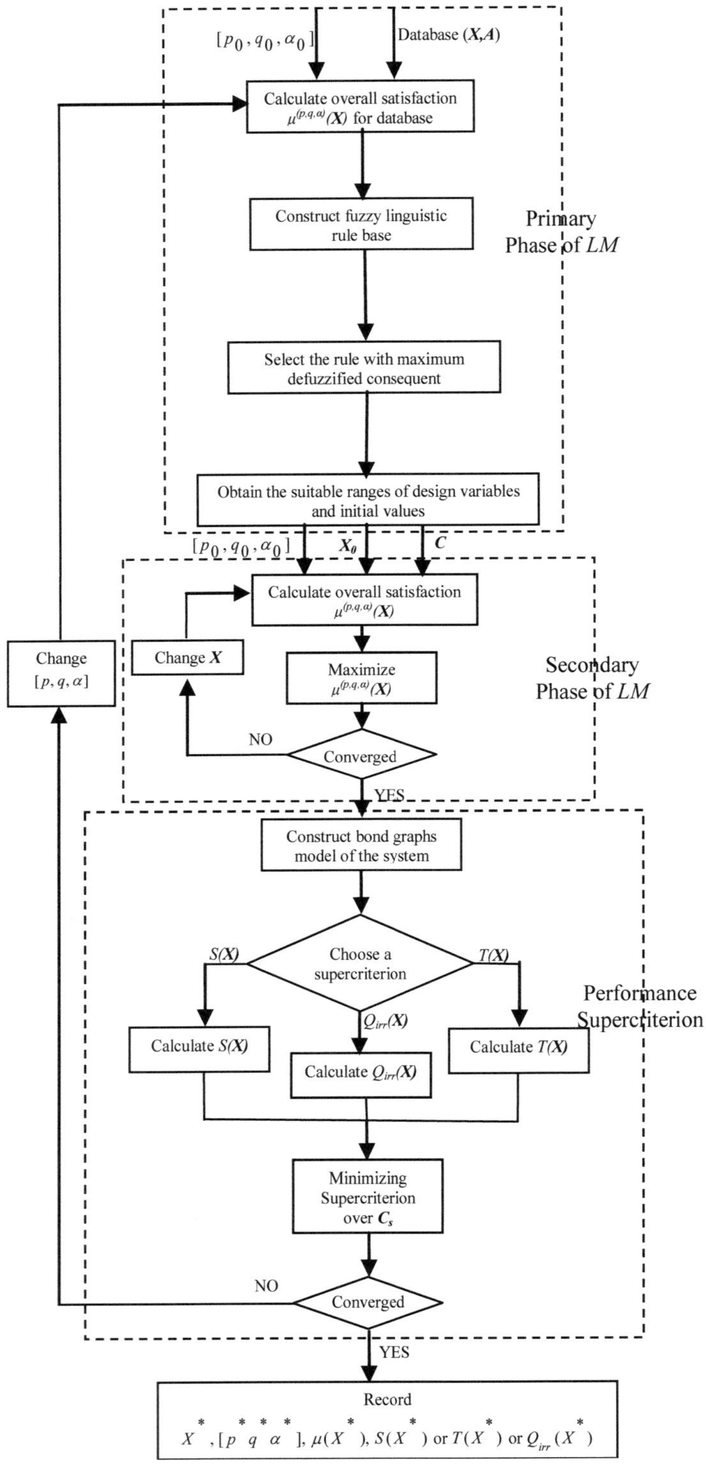

Fig. 1. The flow chart of Linguistic Mechatronics

$$t_{eq}(\boldsymbol{X}) = Inf\{t_0 : \forall t > t_0 \; \dot{Q}_{irr}(t,\boldsymbol{X}) < \varepsilon\} . \tag{34}$$

Consequently, the best design is attained in the set of optimally satisfactory solutions,

$$Q_{irr}(t_{eq}(\boldsymbol{X}^*)) = \min_{\boldsymbol{X}_s \in C_s} Q_{irr}(t_{eq}(\boldsymbol{X}_s); p,q,\alpha) . \tag{35}$$

3) Agility Criterion

Alternatively, for systems where response time is a crucial factor the rate of energy transmission through the system, or *agility*, can be used for defining the performance supercriterion. Thus, the supercriterion would be to minimize the time that the system needs to reach a steady state as the result of a unit step change of all input parameters at time zero. A system reaches the steady state when the rate of its *internal dynamic energy*, *K*, becomes zero. Internal dynamic energy is equivalent to the kinetic energy of masses in mechanical systems or the energy stored in inductors in electrical systems. Masses and inductors resist the change of velocity and current, respectively. In terms of bond graph modeling, both velocity and current are considered as *flow*. Consequently, internal dynamic energy is defined as the energy stored in the elements of system that inherently resist the change of *flow*. Therefore, Given a unit step change of input variables, the response time, denoted by $T(\boldsymbol{X})$, is the time instant after which the rate of change of internal dynamic energy, $\dot{K}$, remains below a small threshold, δ.

$$T(\boldsymbol{X}) = Inf\{t_0 : \forall t > t_0 \; \dot{K}(t,\boldsymbol{X}) < \delta\} . \tag{36}$$

As a design supercriterion, when the response time reaches its minimum value with respect to attitude parameters the best design is attained in $\boldsymbol{C}_s$.

$$T(\boldsymbol{X}^*) = \min_{\boldsymbol{X}_s \in C_s} T(\boldsymbol{X}_s; p,q,\alpha) . \tag{37}$$

The complete flowchart of *LM* is presented in Fig. 1.

3. Robotic Hardware-in-the-loop Simulation Platform

The increasing importance of several factors has led to an increase in the use of HIL simulation as a tool for system design, testing, and training. These factors are listed in (Maclay, 1997) as: reducing development time, exhaustive testing requirements for safety critical applications, unacceptably high cost of failure, and reduced costs of the hardware necessary to run the simulation. By using physical hardware as part of a computer simulation, it is possible to reduce the complexity of the simulation and incorporate factors that would otherwise be difficult or impossible to model. Therefore, HIL simulations can play an effective role in systems concurrent engineering. The HIL simulations have been successfully applied in many areas, including aerospace (Leitner, 1996), automotive (Hanselman, 1996), controls (Linjama et al., 2000), manufacturing (Stoeppler et al., 2005), and naval and defense (Ballard et al., 2002). They have proven as a useful design tool that reduces development time and costs (Stoeppler et al.; 2005; Hu, 2005). With the ever improving performance of today's computers it is possible to build HIL simulation without specialized and costly hardware (Stoeppler et al., 2005).

In the field of robotics, HIL simulation is receiving growing interest from researchers, and has been applied from a number of different perspectives. These approaches include: *robot-in-the-loop* simulations, such as the platform used for the task verification of the special-purpose dexterous manipulator at the Canadian Space Agency (Piedboeuf et al., 1999) or the use of both real and simulated mobile robots interacting with a virtual environment (Hu, 2005); *controller-in-the-loop* simulations, where a real control system interacts with a computer model of the robot (Cyril et al., 2000); and *joint-in-the-loop* simulations, which use a computer model to compute the dynamic loads seen at each joint and then emulate those loads on the real actuators (Temeltas et al., 2002). Each of these approaches applies the HIL concept slightly differently, but all have produced positive results. In a recent work (Martin & Emami, 2008), a modular and generic Robotic HIL Simulation (RHILS) platform was designed and developed for the industrial manipulators, and its performance was verified using the *CRS-CataLyst-5* manipulator from Thermo Fisher Scientific Inc. (Thermo, 2007). The RHILS platform was used in this work as the second constituent of robotic concurrent engineering, next to Linguistic Mechatronics. The architecture of the RHILS platform is illustrated in Fig. 2, and an overview of its modules is presented below:

3.1 RHILS Architecture

The RHILS platform architecture allows for simultaneous design and testing of both the joint hardware and control system of a robot manipulator. The architecture is designed to be adequately generic so that it can be applied to any serial-link robot manipulator system, and focuses on modularity and extensibility in order to facilitate concurrent engineering of a wide range of manipulators. This section presents a detailed breakdown of the main blocks of the architecture.

The architecture is separated into four subsystems: (a) the *User Interface*, (b) the *Computer Simulation*, (c) *Hardware Emulation*, and (d) the *Control System*, which are described below with reference to Fig. 2. These subsystems are further partitioned into two major categories: RHILS Platform components (indicated with a white background), and Test System components (indicated with a grey background). The RHILS Platform components are generic and should remain largely consistent over multiple applications, while the Test System components are part of the system being designed and/or tested on the platform. Depending on how much of the system is implemented in hardware versus how much is simulated it is possible to tailor the setup to all phases of the design cycle, and the architecture is designed to make adjusting this ratio as easy as possible.

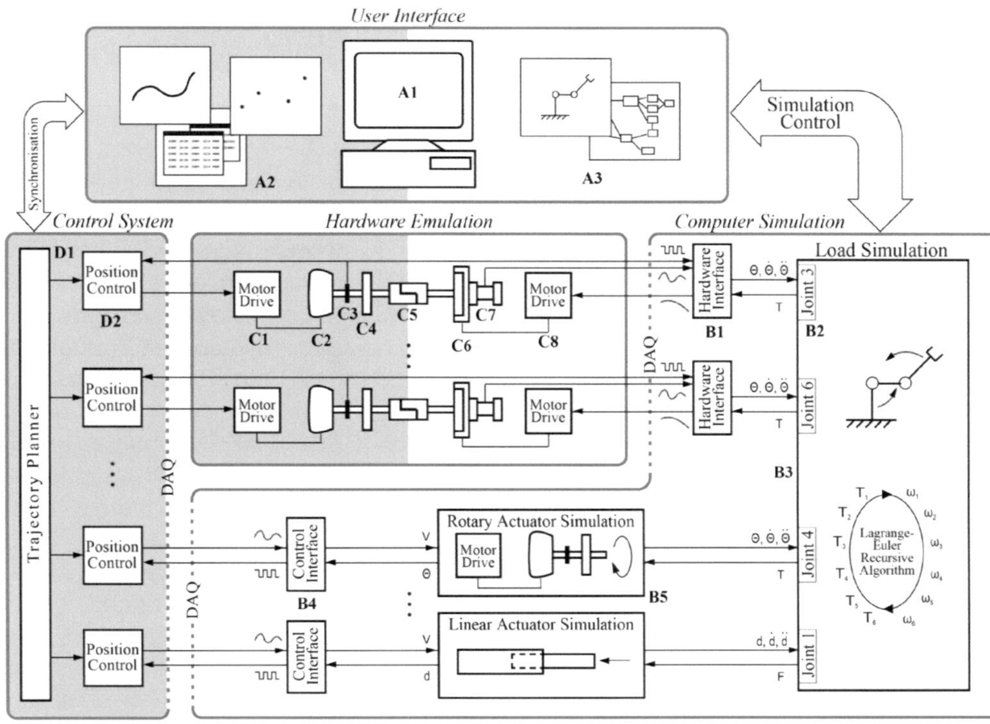

A1 User interface host computer
A2 Control system user interface and trajectory setup
A3 Simulation user interface and scheduler
B1 Motor interface block, converts between actual hardware signals and the standardized form used in the simulation
B2 Joint assignment for the module
B3 Inverse dynamics simulation
B4 Control interface block, converts between actual control signals and the standardized form used with simulated actuators
B5 Simulated model of an actuator, for cases where the hardware module is unavailable, impractical, or unnecessary
C1 Drive electronics for Test Motor
C2 Test Motor
C3 Differential rotary encoder
C4 Harmonic drive transmission
C5 Detachable coupling to allow test hardware to be swapped in and out
C6 Load Motor
C7 Reaction torque transducer, for closed loop control and data acquisition
C8 Drive electronics for Load Motor
D1 Trajectory planner
D2 Position controller

A gray background indicates that section is part of the system being designed and tested using the RHIL platform

Fig. 2. RHILS Platform Architecture

A. User Interface Block

This block contains the most overlap between the RHILS Platform and the Test System. Because it is necessary to synchronize initial conditions before starting a simulation, this block acts as an intermediary between the custom control system and the generic simulation. On the RHILS Platform side robot configurations and parameters are chosen, as well as specifying any external conditions, for example zero-gravity or end-effector payloads, that will be used during a simulation. For the Test System side any configurable control parameters are set in the control system, such as the planned trajectories and feedback loop gains. Finally, the duration of the simulation and the type of data logging to be performed are selected.

B. Computer Simulation Block

The *Computer Simulation* performs three primary roles. Its first and most obvious task, represented by the *Load Simulation* block, is to run the inverse dynamics computations based on the instantaneous position, velocity, and acceleration of each joint, and solve for the dynamic load applied to each joint actuator. Due to the recursive algorithm used for computing the inverse dynamics (Li & Sankar, 1992) on the dedicated kernel, it is possible to specify any reasonable number of joints in any configuration and still attain the computational efficiency necessary to run the simulation in real-time. The second task is to convert the hardware signals read in and sent out through a data acquisition board into the standardized format used by the load simulation, which is shown by the *Hardware Interface* blocks. These hardware interface blocks play a key role in the modularity of the architecture since they allow different hardware to be used without significant changes to the simulation. The third task of the *Computer Simulation* is to simulate any joints that do not have a corresponding hardware module. In some situations it may be desirable to have one or more joint actuators without a hardware component, for example when the hardware is unavailable, too costly, or simply unnecessary. Then the computer simulation must model the joint and interface directly with the control system, shown in the *Actuator Simulation* and *Control Interface* blocks. This third task makes it possible to utilize the RHILS platform at early stages of the design as well as making it more cost effective to set up tests if only one section of the manipulator is under study.

C. Hardware Emulation Block

The *Hardware Emulation* system consists of separate modules for each joint, and each module interfaces with both the *Control System* and the *Computer Simulation*. These modules are further separated into two parts: a *Test Module*, the joint actuator that is being designed/tested, and a *Load Module*, the load-emulating device that mimics the dynamic loads that would be seen in a real system. The *Test Module* includes not only the real actuator, but also the transmission system, position/speed sensors, and motor drive that would be used in the real manipulator, all of which can lead to significant inaccuracies in a pure computer-based simulation. The *Test Module* interfaces directly with the *Control System*, which controls the motor as if it were part of a physical robot. The *Load Module* is coupled to the output of the transmission system, ideally without the use of a secondary transmission that may introduce unwanted uncertainty in the load emulation mechanism. For the range required by most applications, it was found that *torque motors* can supply the necessary torque directly and have other desirable features including consistent torque at low speeds, low inertia, and proper heat dissipation characteristics. The *Load Module* is controlled through a feedback loop that follows the torque calculated by the *Computer Simulation* block. This torque represents the arm dynamics that must be reflected on each joint actuator to have a genuine simulation of the real system. To emulate the dynamic torque accurately closed-loop control is needed, which requires that the torque generated by the *Load Module* be identified. This is done through a unique installation of the torque sensor as a cantilever support for the torque motor (Martin & Emami, 2008).

D. Control System Block

This block can range from running in software on a standard PC to running on dedicated custom hardware depending on the nature and requirements of the application. It is possible to use the real control system for the robot, since as far as the control system is concerned it is connected to the real actuators in a physical robot. This has significant benefits over running a simulated or modified version of the control system: in many applications intense testing of the final control system is required, which can now begin before the final hardware is complete without building expensive prototypes. On the other hand, when the control system is not the focus of the design the flexibility of this architecture allows any simple controller to be quickly implemented and used.

4. LM-RHILS Based Concurrent Engineering of Robot Manipulators

In this section, the *LM* methodology along with the RHILS platform are implemented for building a framework to concurrently design kinematic, dynamic and control parameters of robot manipulators. This framework includes various phases of *LM*, and the *RHILS* is used to evaluate the design attributes and performance supercriterion.

4.1 Architecture

The architecture of the concurrent design framework consists of two parallel workstations, namely *Host* and *Target*, and physical components of a robot manipulator, i.e., three physical joint modules and a controller unit. For each joint module a load emulator is employed to apply simulated dynamic loads during the real-time execution. The collection of load emulators, joint modules and control system is called *Hardware Emulation* block. The entire design architecture and the real physical joint modules are shown in Fig. 3. Although the concurrent engineering framework discussed here is generic and can be applied to any robot manipulator, the *CRS CataLyst-5* manipulator is used in the following implementations for further illustration.

A. Host Workstation

The Host computer is the link between the system and the engineer(s). All design preferences and options are set in this block, where the main code that governs the design process is executed. The preferences are reflected in the satisfactions defined on the design attributes, and the simulation options include initial configuration, the predefined end-effector trajectories, gravity, payload, and the simulation time. This block communicates with the controller to load control gains through an FTP connection, and sends the command signals to the trajectory planner using Python® software. It also loads the kinematic and dynamic parameters and inverse dynamic model of a design candidate to the Target workstation via a TCP/IP connection, and gathers position and torque data that are saved on the Target PC using MATLAB® xPC Target® toolbox. The data are processed and the design attributes are calculated by the Host computer, and considering the design availabilities, the satisfactions are assigned to the design variables and attributes. According to the *LM* methodology, the overall satisfaction of the design candidate is calculated and it is maximized using the MATLAB® optimization toolbox. The optimization of the performance supercriterion is also carried out on the Host computer.

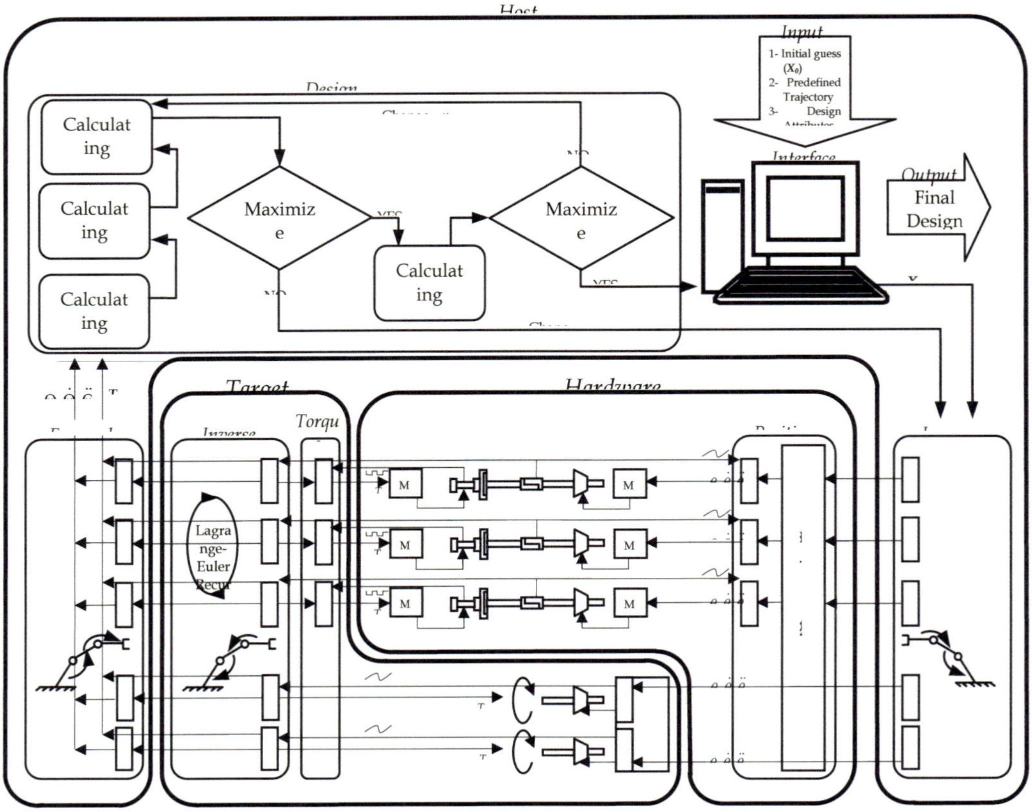

Fig. 3. The LM-RHILS concurrent design architecture

B. Target Workstation

This block is a barebones PC running the xPC Target® real time kernel. On this workstation a servo torque controller for the load emulators and an inverse dynamics model of the manipulator, built in Simulink® and compiled through Real-Time Workshop®, are executed. In the dynamics model, torque signals are calculated based on the kinematics and dynamics of the candidate manipulator and the joints position, velocity and acceleration. The Target computer contains several interface boards to communicate with the joint modules and load emulators. Furthermore, to gather data from the hardware components a data acquisition board and an RS232 port are utilized

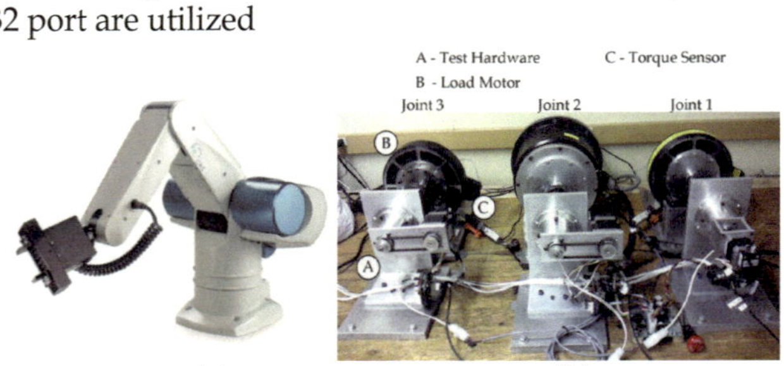

(a) (b)

Fig. 4. (a) CRS CataLyst-5 robot, (b) RHILS platform

C. Hardware Emulation
All physical pieces that remain unchanged in the design process form the Hardware Emulation block. Industrial manipulators often have 5 or 6 degrees of freedom (d.o.f.). The first three joints are often used to position the end-effector and the last joints help the wrist change its orientation. Since the first three links are more massive, more force or torque is applied on the corresponding joints, and they play a crucial role in the serial link manipulator performance. Hence, in the design architecture, the first three joint modules of *CRS CataLyst-5* are physically included as a part of the *RHILS* platform, and the rest of the joints are virtually modeled on the Target computer. The corresponding load emulators are also coupled to the joints and the *CRS DM* Master Controller unit is used to control the joint positions. Each joint module consists of a stepper motor, an encoder mounted on the motor shaft, a harmonic drive as a transmission mechanism, and the driver unit. The module interfaces with both the controller and Target workstation in order to receive control signals via motor driver and send joint position to the Target workstation.
The load emulators are coupled directly to the joint shafts to apply the computed loads. These torque signals represent the arm's dynamics and weight and payload effects that must be reflected on each joint actuator to have a genuine simulation of the real system. Since the applied torque should be followed accurately, a servo torque controller is designed and calibrated for each load emulator module. A reaction torque sensor is also installed between the load emulator case (stator) and its mounting fixture to measure the feedback signal. Thus, the load emulator module sends and receives the command and feedback torque signals to and from the Target PC where the torque controller is located (Martin & Emami, 2008).
The controller unit includes a trajectory planner and a typical feedback/feedforward controller for each physical joint module. The trajectory planner generates instantaneous desired position signals with a frequency of 1 KHz based on the input of the controller. Joint trajectories are divided into three sections: first, accelerating to the maximum speed with the nominal acceleration of the joint module, second, constant speed motion and finally, decelerating to the final position with the nominal acceleration.

4.2 Manipulator Concurrent Design Process
In this section, the design architecture is employed to concurrently redesign kinematic, dynamic and control parameters of *CRS-CataLyst-5*. This industrial manipulator consists of 5 rotary joints, three of which are included in the *RHILS* platform. Fig. 4 shows the *CRS-CataLyst-5 manipulator* next to its *RHILS* platform.
In general, the *LM* design framework can be divided into five steps: **a)** decision about design variables and attributes, **b)** assignment of satisfactions, **c)** the primary phase, **d)** the secondary phase, and **e)** the performance supercriterion. However, in this case study, since the existing design is modified and the process can be safely started from the current configuration, the primary phase is not required.

A. Design Variables and Attributes
The kinematic characteristics of a manipulator can be represented by the standard Denavit-Hartenberg convention. Therefore, length (l_i), offset (d_i) and twist (α_i) are considered as kinematic design variables of the ith link. In order to take into account dynamic parameters of the robot, each link is considered as an L-shaped circular cylinder along the link length

and offset. The radius of such cylinder (r_i), as a design variable, specifies dynamic parameters of the ith link knowing the link density. The *CRS DM* Master Controller unit generates control signals for each joint consisting of proportional (P_i) and integral (I_i) gains along with gains for feedback velocity ($Kv_{fb,i}$) and acceleration ($Ka_{fb,i}$) and also feedforward velocity ($Kv_{ff,i}$) and acceleration ($Ka_{ff,i}$). Consequently, the design problem deals with $10 \times ndof$ design variables, where *ndof* is the number of degrees of freedom, to identify the most desirable kinematic, dynamic, and control configuration of the manipulator. In the case of *CRS CataLyst-5*, since the last two joints are small at the tip of the manipulator with much less moments of inertia than that of the other joints, their control gains are not considered in the design. Consequently, the design problem deals with thirty-eight design variables in total.

In *LM*, design attributes are divided into *must* and *wish* attributes. The following *must* design attributes are considered:

Design availabilities: Each design variable has an acceptable range of values, considering its physical nature and manufacturing constraints. They are taken into account by the following inequality expression.

$$X_j^{\min} \le X_j \le X_j^{\max} \quad (j = 1,\ldots,n); \tag{38}$$

where $X_j^{\min}$ and $X_j^{\max}$ are the minimum and maximum values for X_j, respectively.

Joint constraint: Since real joint modules are used in the design process, the motor constraints are considered automatically; however, the joints displacements are restricted due to the shape and location of links. This constraint is checked at k^{th} working point for the ith joint angle (θ_i^k) by means of an inequality.

Torque constraint: Each joint module can handle a maximum amount of torque ($\tau_i^{\max}$), usually corresponding to the stall torque of the ith joint motor. Therefore,

$$_{\max}\left|\tau_i\right|^k \le \tau_i^{\max} \quad (i = 1,\ldots, ndof\,; k = 1,\ldots, N); \tag{39}$$

where $_{\max}\left|\tau_i\right|^k$ is the ith joint maximum absolute value of the torque between k^{th} and $(k\text{-}1)^{th}$ working points.

Maximum reachability: The farthest point that the manipulator can reach is the maximum reachability of the robot (R) and because of environmental constraints it should not exceed a certain number (R^{max}).

The main mission of a robot is reflected in the *wish* attributes. In this research, the following *wish* attributes are deemed as the design objectives.

End-effector error: The typical ultimate task for a robot manipulator is to follow predefined trajectories. Therefore, the measured error at the working points is an appropriate *wish* attribute to minimize. If Δ_{tk} and δ_{tk} are the maximum permitted errors for the end-effector position and orientation, respectively, at the kth working point of the tth trajectory, then the end-effector error can be defined as:

$$E = \frac{1}{NT}\sum_{t=1}^{T}\sum_{k=1}^{N}\left(\frac{\sqrt{\Delta x_{tk}^{2} + \Delta y_{tk}^{2} + \Delta z_{tk}^{2}}}{\Delta_{tk}} + \frac{\sqrt{\delta x_{tk}^{2} + \delta y_{tk}^{2} + \delta z_{tk}^{2}}}{\delta_{tk}}\right) \quad (40)$$

where Δx_{tk}, Δy_{tk} and Δz_{tk} are the position errors in *x*, *y* and *z* directions, δx_{tk}, δy_{tk} and δz_{tk} are the orientation errors about *x*, *y* and *z* axes at the kth working point of the tth trajectory, and *T* is the number of trajectories. Note that orientation errors are assumed to be sufficiently small so that the overall orientation error can be considered as a vector. Also, for the 5 d.o.f. *CataLyst-5* manipulator only yaw and roll angles of the end-effector were considered. A maximum of 1*mm* for the translational error and 6° for the orientation error are assigned for this design.
Manipulability: The manipulability index is used for checking the manipulator singularity at the working points. This measure can be expressed as (Bi et al., 1997):

$$M = \frac{1}{N}\sum_{k=1}^{N} cond(J_k^{\,0}) ; \quad (41)$$

where $cond(J_k^{\,0})$ is the condition number of Jacobian matrix with respect to the base frame at k^{th} working point. At the singular points the manipulability index approaches infinity and its minimum value is one. Therefore, this *wish* attribute is satisfied when manipulability index is close enough to one.
Structural length index: A desirable manipulator is the one with a smaller Structural length index,

$$Q_L = \left(\sum_{i=1}^{ndof}(l_i + d_i)\right) / \sqrt[3]{V} ; \quad (42)$$

where *V* is the workspace volume that can be numerically calculated based on a method detailed in (Ceccarelli et al., 2006).
Total required torque: The total required torque at the kth working point, expressed in (43), can be considered as another *wish* attribute that should be minimized.

$$\tau_T^{\,k} = \sum_{i=1}^{ndof}\left|\tau_i^{\,k}\right| ; \quad (43)$$

where $\tau_i^{\,k}$ is the torque of joint *i* at the kth working point.

B. Satisfactions Assignment
Satisfactions are defined as fuzzy membership functions over the range of values that design variables and attributes can obtain. The availability constraints and *must* attributes often satisfy inequalities, while *wish* attributes should be as satisfactory as possible. Since *LM* methodology employs fuzzy set theory, by redefining the notions of inequality and optimization, their restricted binary behaviour can be turned into a flexible and fuzzy one. This brings subjective aspects of design into the scope; in addition, simplifies the design

process. One of the popular fuzzy membership functions is the trapezoidal membership function. This function possesses four parameters, i.e., four corners of the trapezoid that the designer should decide about to specify the range in which the satisfaction is one and the slopes of the sides. This decision is made considering the design requirements and the designer's preferences. In other words, the trapezoidal parameters reflect how conservative or aggressive the designer is in interpreting the design attributes. The trapezoids, which are used in this case study, are depicted in Fig. 5. The first and last points of a *must* satisfaction mapping are the minimum and maximum values of the corresponding inequality, respectively. The middle points are picked in a manner that the definition of the inequality is neither too fuzzy nor too crisp, and it obeys the design requirements. For a *wish* satisfaction mapping, the last point is the maximum allowed value of the attribute (for an attribute approaching a minimum), and as it decreases the corresponding satisfaction approaches to one. The middle point is selected based on designer's consensus of the notion of minimum. All minimum and maximum values of design variables and attributes are listed in Table I. Note that since this design problem starts with an existing manipulator configuration and the simulation platform is sufficiently accurate, strict parameters are chosen for defining *wish* satisfactions. This indicates smaller middle ranges and, hence, less steep trapezoid sides.

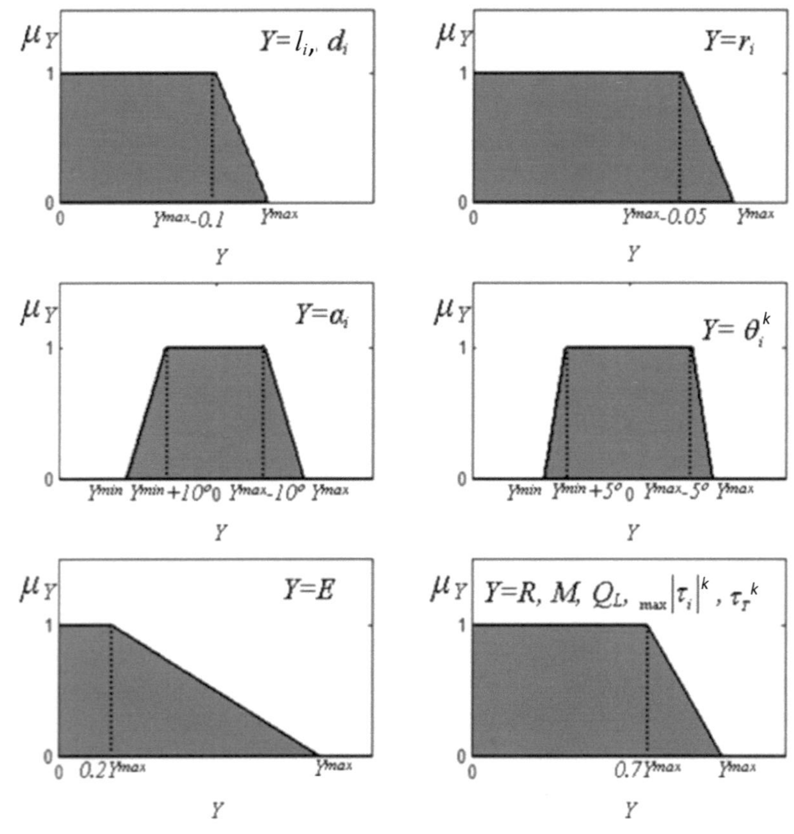

Fig. 5. Satisfactions on design variables and attributes

C. Secondary Phase

To calculate the overall satisfaction, design attributes are determined utilizing the *RHILS* platform that simulates the candidate configuration while it follows a predefined pick-and-place trajectory. In this procedure, first the Denavit-Hartenberg table and dynamic parameters of the design candidate are determined based on the kinematic parameters and the links radii. They are loaded onto the Target workstation as the parameters of the inverse dynamic model of the manipulator. The control gains are also loaded on the controller. On the Host computer an inverse kinematic code is executed to transform the end-effector trajectory to the joint trajectories. The corresponding command signals are sent to the controller from the Host workstation using Python® software and simultaneously, while the real joint modules are moving the joint torques calculated in the Target PC are applied on them by means of the load emulators. Subsequently, the position and torque signals are saved on the Target workstation for further computations. On the Host PC, the design availability, maximum reachability, manipulability and structural length index attributes are calculated using the kinematic parameters. And the joint restriction, torque restriction and total torque required design attributes are determined based on the saved position and torque signals. In addition, a forward kinematic code is executed to compute the actual end-effector position at the working points in order to evaluate the end-effector error. Finally, the corresponding satisfactions are identified and aggregated using the attitude parameters.

The secondary phase searches for the design variables that maximize the overall design satisfaction. A function in the optimization toolbox of MATLAB®, called *fminsearch*, has been employed to perform this single-objective maximization. This function uses a derivative-free search algorithm based on the simplex method that is suitable for handling discontinuity, sharp corners and noise in the objective function, which is the case in this problem. This real-time process takes almost one minute for evaluating each configuration.

i	1	2	3	4	5
$r_i(mm)$	[0,200]	[0,200]	[0,200]	[0,200]	[0,200]
$l_i(mm)$	[0,500]	[0,500]	[0,500]	[0,500]	[0,500]
$d_i(mm)$	[0,500]	[0,500]	[0,500]	[0,500]	[0,500]
$\alpha_i(o)$	[-180,180]	[-180,180]	[-180,180]	[-180,180]	[-180,180]
$\theta_i(o)$	[-180,180]	[-110,0]	[-90.6,35]	[-110,110]	[-180,180]
$_{\max}\lvert\tau_i\rvert(N.m)$	[0,13.8]	[0,13.8]	[0,13.8]	[0,4.8]	[0,2.4]
$R(m)$			[0,0.87]		
E			[0,2]		
M			[1,24]		
Q_L			[0,1.6]		
$\tau_T(N.m)$			[0,12.5]		
Control Gains			$(-\infty,+\infty)$		

Table 1. - Design Variables and Attributes and their Range

D. Performance Supercriterion

By altering the designer's attitude parameters (*p, q* and *a*) the secondary phase generates a set of optimally satisfactory solutions for design. The physical performance of the system should also be checked against an objective *supercriterion*, which is selected to be the total energy consumption at the joints, in order to adjust the designer's attitude.

$$Energy(\boldsymbol{X}_S;p,q,\alpha)=\sum_{i=1}^{ndof}\int_{\theta_i^1}^{\theta_i^N}\left|\tau_i d\theta_i\right|; \tag{44}$$

where θ_i^k is the ith joint angle at the kth working point and τ_i is the torque at the ith joint. Ultimately, by minimizing this criterion over optimally satisfactory solutions set ($\boldsymbol{C}_S$), the best design ($\boldsymbol{X}^*$) is achieved.

$$Energy(\boldsymbol{X}^*)=\min_{X_S\in C_S}(Energy(\boldsymbol{X}_S;p,q,\alpha)). \tag{45}$$

4.3 Some Results and Discussions

The *CRS CataLyst-5* manipulator was redesigned according to the LM-RHILS based concurrent methodology, and the results are shown in Table II. With respect to the manipulator dynamic parameters, the mass of link 3 was reduced by 17.5% as a result of decreasing the link radius and length by 10% and 0.7%, respectively. In addition, all other kinematic and dynamic parameters have been modified slightly, which resulted in enhancing the manipulator performance in terms of the error in the end-effector trajectory, manipulator reachability, workspace and manipulability, and total energy consumption. For example the radius of the first and second links has been changed by almost 0.1% and 0.7%, respectively. The length of link 2 and the offset of link 1 have also been altered by 0.1% and 0.4%, respectively. On the other hand, twist angles have remained almost unchanged. Therefore, in terms of dynamic and kinematic design, the third link has been modified considerably.

In addition, since the controller of the existing manipulator was tuned prior to the redesign process, the control gains have made only slight modifications by an average of 0.8%. Even these small changes in the control parameters significantly affected the end-effector error, *E*, which observed in the results. The error in the end-effector trajectory after the redesign process is approximately 78 times less than its initial value. An increase in the level of satisfaction for all other *wish* attributes can be observed from Table II, as well. Therefore, based on the designer's preferences, all the considered attributes have been enhanced. The total *must* satisfaction has improved, which indicates that the new system is far from its performance limits, and hence the new design is more reliable.

The design candidates obtained from the *LM* secondary phase were optimized against an objective *supercriterion,* which is the total consumed energy, through altering attitude parameters. Ultimately, the configuration with the minimum energy consumption was picked as the final design. The energy consumption was improved by 10%. By looking at the variation of designer's attitude parameters during the design process, one realizes that the initial designer's attitude in aggregating *must* satisfactions was appropriate. That is, the value of *p* did not change through the attitude adjustment. However, in aggregating *wish* satisfactions the designer was originally too conservative. Therefore, *q* was decreased by

50% and α was increased by 140%, approximately, through the attitude adjustment. This implies that instead of focusing on the worst *wish* attribute, the designer should equally stress all *wish* design attributes in order to improve the system energy consumption. Overall, the results show that the original designers of the manipulator (prior to the redesign process) could have been more aggressive (optimistic) in the design of *CRS CataLyst-5*.

5. Conclusion

Concurrent engineering is a promising paradigm for the analysis and synthesis of complex, multidisciplinary systems, such as robot manipulators. It brings *synergy* as a direct consequence of utilizing design knowledge from all participating disciplines, while interacting with each other, and offering equal opportunities to them to contribute to each state of design simultaneously. The advantage, however, does not come at no cost; one must deal with highly-complicated mechatronic system models, and handle optimizations with a large set of multidisciplinary objective and constraint functions and a great number of design variables. The compromise seems to be either to simplify the system model to reduce dimensions of the design space, or to give up the transparency of the design process and appeal to parallel computing algorithms. This chapter discussed an alternative methodology that does not imply any of the above compromises. The new methodology makes the system model computations efficient without compromising design transparency, because it uses the physical system components in the simulation loop, next to the computational model of those modules that need to be designed. The robotic hardware-in-the-loop simulation platform enables the designer to take into account some complex phenomena that are difficult to model, yet execute the entire simulation in real-time. Using hardware components in concurrence with the computational model of the modules that are to be designed results in an effective platform for rapid design alterations. Moreover, the new methodology alleviates the optimization complexities of concurrent design, because it employs Linguistic Mechatronics that not only transforms the multi-objective constrained optimization problem into a single-objective unconstrained formulation, but also formalizes subjective notions and brings the linguistic aspects of communication into the design process.

	$r_i(mm)$					$l_i(mm)$				
	i=1	i=2	i=3	i=4	i=5	i=1	i=2	i=3	i=4	i=5
Initial	65.6	27.7	24.1	10.0	10.0	0.0	254.0	254.0	0.0	0.0
Final	65.7	28.0	21.8	10.0	10.0	0.0	253.6	255.9	0.0	0.0
	$d_i(mm)$					a_i(□)				
	i=1	i=2	i=3	i=4	i=5	i=1	i=2	i=3	i=4	i=5
Initial	254.0	0.0	0.0	0.0	0.0	-90.0	0.0	0.0	-90.0	0.0
Final	255.0	0.0	0.0	0.0	0.0	-90.8	0.0	0.0	-90.7	0.0
	P_i			I_i			$Kv_{fb,i}$			$[p,q,a]$
	i=1	i=2	i=3	i=1	i=2	i=3	i=1	i=2	i=3	
Initial	18.32	20.00	12.00	0.073	0.050	0.100	40.7	40.0	20.0	[10,1.5,0.5]
Final	18.46	20.16	12.10	0.074	0.050	0.101	41.0	40.3	20.2	[10,0.7,1.2]
	$Ka_{fb,i}$			$Kv_{ff,i}$			$Ka_{ff,i}$			*Energy (J)*
	i=1	i=2	i=3	i=1	i=2	i=3	i=1	i=2	i=3	

Initial	43.4	100.0	80.0	59.0	40.0	30.0	3473.0	100.0	120.0	6.2549
Final	43.8	100.8	80.6	59.5	40.3	30.2	3483.6	100.8	120.9	5.6307

	Wish Design Attributes								
	E	M	Q_L	$\tau_T^k\,(N.m)$					
				k=1	k=2	k=3	k=4	k=5	k=6
Initial	1.4787	20.7223	1.3091	9.3557	10.2754	9.3561	9.3561	10.2172	10.2172
Final	0.0189	19.4921	1.3025	8.3071	9.1391	8.3071	8.3071	9.1394	8.3071

	Wish Satisfactions									Overall Satisfaction
	μ_E	μ_M	μ_{Q_L}	$\mu_{\tau_T^k}$						
				k=1	k=2	k=3	k=4	k=5	k=6	μ
Initial	0.000	0.606	0.455	0.838	0.593	0.838	0.838	0.609	0.609	0.250
Final	1.000	0.620	0.626	1.000	0.896	1.000	1.000	0.896	1.000	0.607

Table 2. – Results of Concurrent Design

The new methodology of concurrent engineering was used to redesign the kinematic, dynamic, and control parameters of an industrial manipulator, namely *CRS CataLyst-5*, whose joint modules had been installed in the *RHILS* platform. Despite the fact that the existing manipulator design had been well developed, the new design enhanced the system performance (end-effector trajectory error, manipulator reachability, workspace and manipulability, and total energy consumption) by changing the current manipulator configuration.

6. References

Ballard, B. L., Elwell Jr., R. E., Gettier, R. C., Horan, F. P., Krummenoehl, A. F. and Schepleng, D. B. (2002). Simulation Approaches for Supporting Tactical System Development, *John Hopkins APL Technical Digest* (*Applied Physics Laboratory*), Vol. 23, No. 2-3, pp. 311-324.

Bi, Z. M., Li, Y. F. and Zhang, W. J (1997). A New Method For Dimensional Synthesis of Robotic Manipulators, *5th National Applied Mechanisms and Robotics Conference*, Cincinnati.

Bi, Z. M. and Zhang, W. J. (2001). Concurrent Optimal Design of Modular Robotic Configuration, *Journal of Robotic Systems*, Vol. 18, No. 2.

Borutzky, W. (2006). Bond Graph Modeling and Simulation of Mechatronic Systems: An Introduction into the Methodology, *Proceeding 20th European Conference on Modeling and Simulation (ECMS)*.

Breedveld, P. C. (2004). Port-based Modeling of Mechatronic Systems, *Mathematics and Computers in Simulation*, Vol. 66, pp. 99-127.

Castano, A., Behar, A. and Will, P. M. (2002). The Conro modules for reconfigurable robots, *IEEE/ASME Transactions on Mechatronics*, Vol. 7, No. 4, pp. 403-409.

Ceccarelli, M., Carbone, G. and Ottaviano, E. (2005). An Optimization Problem Approach for Designing Both Serial And Parallel Manipulators, *The International Symposium on Multibody Systems and Mechatronics Proceedings of MUSME*, Brazil.

Chhabra, R. and Emami, M.R. (2009). Design Criteria in Mechatronics, submitted to *IEEE/ASME Transactions on Mechatronics*, September 2009.

Chirikjian, G. S. (1994). Kinematics of a Metamorphic System, *Proceedings of the 1994 IEEE International Conference on Robotics and Automation*, pp. 449-455.

Coello, C.A. (1999). A Comprehensive Survey of Evolutionary Based Multiobjective Optimization Techniques, *Knowledge and Information Systems An International Journal*, 1(3): 269–308.

Cyril, X., Jaar, G. and St-Pierre, J. (2000). Advanced Space Robotics Simulation for Training and Operations, *AIAA Modeling and Simulation Technologies Conference*, pp. 1-6, Denver, USA.

Dhingra, A. K., Rao, S. S. and Miura, H. (1990). Multiobjective Decision Making in a Fuzzy Environment with Application to Helicopter Design, *AIAA Journal*, Vol. 28, No. 4.

Dhingra, A. K. and Rao, S. S. (1995). A Cooperative Fuzzy Game Theoretic Approach to Multiple Objective Design Optimization, *European Journal of Operational Research*, Vol. 83, pp. 547-567.

Emami, M. R. (1997). *Systematic Methodology of Fuzzy-logic Modeling and Control and Application to Robotics*, Ph.D. Dissertation, Department of Mechanical and Industrial Engineering, University of Toronto, Canada.

Emami, M. R., Turksen, I.B., Goldenberg, A.A. (1998). Development of a systematic methodology of fuzzy-logic modeling, *IEEE Trans. Fuzzy Systems*, Vol. 6, No. 3, pp. 346-361.

Emami, M. R., Türksen, I. B. and Goldenberg, A. A. (1999). A Unified Parameterized Formulation of Reasoning in Fuzzy Modeling and Control, *Fuzzy Sets and Systems*, Vol. 108, pp. 59-81.

Hanselman, H. (1996). Hardware-in-the-loop Simulation Testing and its Integration into a CACSD Toolset, *IEEE International Symposium on Computer-Aided Control System Design*, pp. 15-18.

Hewit, J. (1996). Mechatronics design - the key to performance enhancement, *Robotics and Autonomous Systems*, No. 19, pp. 135-142.

Hu, X. (2005). Applying Robot-in-the-loop Simulation to Mobile Robot Systems, *12th International Conference on Advanced Robotics (ICAR 2005)*.

Kurapati, A., Azarm, S. and Wu, J. (2000). Constraint Handling in Multiobjective Genetic Algorithms, In: *Proceedings of 8th AIAA/NASA/USAF/ISSMO Symposium on Multidisciplinary Analysis and Optimization*, Long Beach, CA, Paper No. AIAA-2000-4893.

Leitner, J. (1996). Space Technology Transition Using Hardware in the Loop Simulation, *Proceedings of the 1996 Aerospace Applications Conference*, Vol. 2 , pp. 303-311.

Li, C.-J. and Sankar, T. S. (1992). Fast inverse dynamics computation in real-time robot control, *Mechanism & Machine Theory*, Vol. 27, No. 6, pp. 741-750.

Linjama, M., Virvalo, T., Gustafsson, J., Lintula, J., Aaltonen, V. and Kivikoski, M. (2000). Hardware-in-the-loop Environment for Servo System Controller Design, Tuning, and Testing, *Microprocessors and Microsystems*, Vol. 24, No. 1, pp. 13-21.

Maclay, D. (1997). Simulation gets into the loop, *IEE Review*, Vol. 43, No. 3, pp. 109-112.

Martin, A. and Emami, M. R. (2008). Design and Simulation of Robot Manipulators using a Modular Hardware-in-the-loop Platform, in M. Ceccarelli (ed.) *Robot Manipulators: Programming, Design, and Control*, I-Tech Education and Publishing, Vienna, Austria, pp. 347-372.

Murata, S., Yoshida, E., Tomita, K., Kurokawa, H., Kamimura, A. and Kokaji, S. (2000). Hardware Design of Modular Robotic System, *Proceedings of the 2000 IEEE International Conference on Intelligent Robots and Systems,* pp. 2210-2217.

Otto, K. N. and Antonsson, E. K. (1995). Imprecision in Engineering Design, *ASME Journal of Mechanical Design,* Vol. 117(B), pp. 25-32.

Paredis, C.J.J. (1996). *An Agent-based Approach to the Design of Rapidly Deployable Fault Tolerant Manipulators*, PhD Thesis, Department of Electrical and Computer Engineering, Carnegie Mellon University, Pittsburgh, USA.

Paynter, H. M. (1961). Analysis and Design of Engineering Systems, M.I.T. Press, Cambridge, Massachusetts, USA.

Piedboeuf, J.-C., Carufel, J. de, Aghili, F. and Dupuis, E. (1999). Task Verification Facility for the Canadian Special Purpose Dexterous Manipulator, *Proceedings of the 1999 IEEE International Conference on Robotics and Automation*, pp. 1077–1083, USA.

Ramachandran, S. and Chen, I.-M. (2000). Distributed Agent Based Design of Modular Reconfigurable Robots, *Proceeding of the 5th International Conference on Computer Integrated Manufacturing*, Singapore, pp. 447–458.

Rus, D. and McGray, C. (1998). Self-Reconfigurable Modular As 3-D Metamorphic Robots, *Proceedings of the 1998 IEEE International Conference on Intelligent Robots and Systems,* pp.837-842.

Rus, D. and Vona, M. (2000). A Physical Implementation of Self-Reconfigurable Crystalline Robot, *Proceedings of the 2000 IEEE International Conference on Robotics and Automation,* pp. 1726-1733.

Stoeppler, G., Menzel, T. and Douglas, S. (2005). Hardware-in-the-loop simulation of machine tools and manufacturing systems, *Computing & Control Engineering Journal*, Vol. 16, No. 1, pp. 10-15.

Temeltas, H.; Gokasan, M., Bogosyan, S. and Kilic, A. (2002). Hardware in the Loop Simulation of Robot Manipulators through Internet in Mechatronics Education, *The 28th Annual Conference of the IEEE Industrial Electronics Society*, Vol. 4, pp. 2617-2622, Sevilla, Spain.

Thermo Fisher Scientific Inc. (2007). CRS CataLyst-5 Robot System, Website: http://www.thermo.com/com/cda/product/detail/0,1055,21388,00.html.

Yager, R. R. and Filev, D. P. (1994). Essentials of Fuzzy Modeling and Control, *New York: John Wiley and Sons.*

Yoshida, E., Kokaji, S., Murata, S., Kurokawa H. and Tomita, K. (1999). Miniaturised Self-Reconfigurable System Using Shape Memory Alloy, *Proceedings of the 1999 IEEE International Conference on Intelligent Robots and Systems,* pp. 1579-1585.

13

Desktop Cartesian-Type Robot with Abilities of Compliant Motion and Stick-Slip Motion

Fusaomi Nagata, Shintaro Tani and Takanori Mizobuchi
Tokyo University of Science, Yamaguchi
Japan
Tetsuo Hase and Zenku Haga
R&D Center, Meiho Co. Ltd., Fukuoka
Japan
Keigo Watanabe
Okayama University, Okayama
Japan

1. Introduction

In manufacturing process of small lens molds, 3D CAD/CAM systems and high precision NC (numerically controlled) machining centers are used generally, and these advanced systems have drastically rationalized the design and manufacturing process. For example, recently, an ultra precision multi-axis control machining system has been developed for spherical micro-lens array molds. As a result, the mold with a spherical micro lens array has been effectively shaped with high accuracy (Oba et al., 2008).
In case of an LED lens mold as shown in Fig. 1, however, the finishing process after the machining process has been hardly automated yet, because the LED lens mold has plural small concave areas to be finished, in which each diameter is 4 mm.

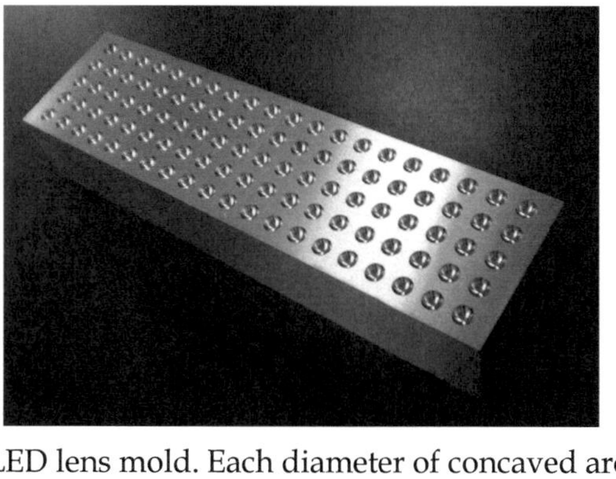

Fig. 1. Example of a target LED lens mold. Each diameter of concaved area is 4 mm.

That means the target mold is not axis-symmetric, so that it is difficult for conventional effective polishing systems (e.g., Kuriyagawa et al., 2003; Suzuki et al., 2006), which are good at dealing with axis-symmetric workpieces, to be applied to the LED lens mold well. Accordingly, such axis-asymmetric LED lens molds are finished by skilled workers in almost cases. The skilled workers usually finish a mold by using a small wood stick tool with diamond lapping paste, while qualitatively checking the concave surface through a microscope.

In this paper, a new desktop Cartesian-type robot, which has abilities of compliant motion and stick-slip motion, is first presented for finishing small metallic molds with curved surface. The Cartesian-type robot is also called the orthogonal-type robot. The robot consists of three single-axis robots with a high position resolution of 1 μm. A thin wood stick tool is attached to the tip of the z-axis. The tool tip has a small ball-end shape. The control system is composed of a force feedback loop, position feedback loop and position feedforward loop. The force feedback loop controls the polishing force consisting of tool's contact force and kinetic friction forces. The position feedback loop controls the position in spiral direction, i.e., z-direction. The position feedforward loop leads the tool tip along a desired trajectory called cutter location data (CL data). The CL data are generated from the main-processor of a CAM system. The proposed Cartesian-type robot has realized a compliant motion required for the surface following control along a spiral path.

In order to improve the finishing performance, a small stick-slip motion control strategy is further added to the control system. The small stick-slip motion is orthogonally generated to the direction of the tool moving direction. Generally, the stick-slip motion is an undesirable phenomenon and should be eliminated in precision machineries (Bilkay & Anlagan, 2004; Mei et al., 2004). However, the proposed Cartesian-type robot employs a small stick-slip motion to improve the finishing quality. The effectiveness of the robot was examined through an actual finishing test of an LED lens mold with a diameter of 4 mm. It was observed that the undesirable small cusps could be removed uniformly. Further, it was confirmed from the result that the proposed Cartesian-type robot with the abilities of compliant motion and stick-slip motion has a superior performance to achieve a higher quality surface like a mirror finishing.

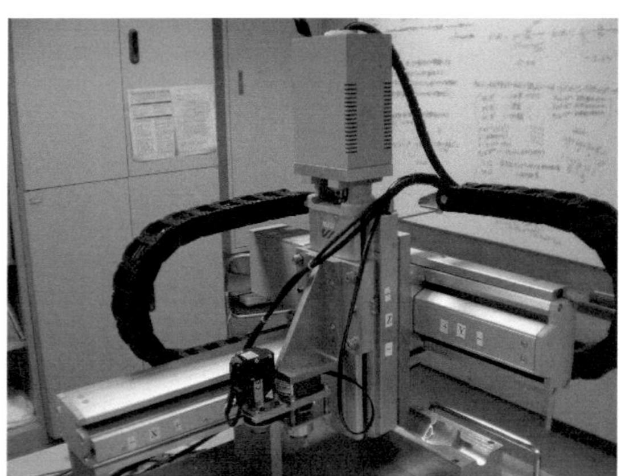

Fig. 2. Developed desktop Cartesian-type robot with abilities of compliant motion and stick-slip motion.

2. Desktop Cartesian-Type Robot

Figure 2 shows the developed desktop Cartesian-type robot consisting of three single-axis robots with position resolution of 1 μm. The size of the robot is 850×645×700 mm. The single-axis robot is a position control device ISPA with high-precision resolution provided by IAI Corp., which is comprised of a base, linear guide, ball-screw, AC servo motor. The effective strokes in x-, y- and z-directions are 400, 300 and 100 mm, respectively. The tool axis is designed to be parallel to the z-axis of the robot. A thin wood stick tool is fixed to the tip through a compact force sensor with 3 degree-of-freedom. To regulate the rotation, a servo spindle is located parallel to the tool axis. The hardware block diagram of the robot is shown in Fig. 3. For example, if one pulse is given to x-, y- and z-axis robots, then each single-axis robot can simultaneously move 1 μm, respectively. Figure 4 shows an example of the static relation between the position and contact force in case of using a wood stick tool, in which ■ and ● show values in press and unpress motions, respectively.

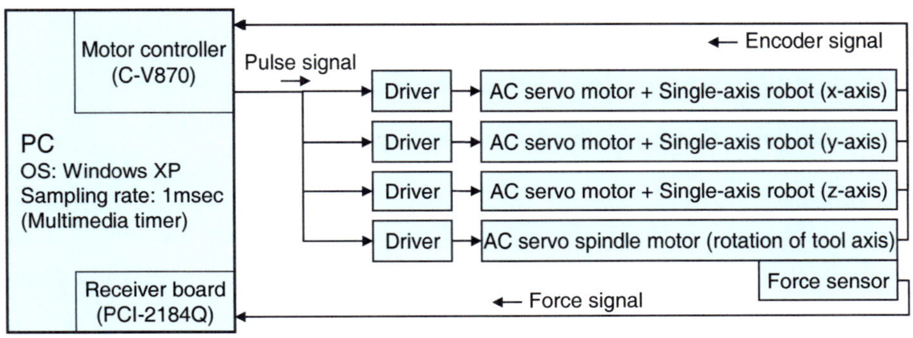

Fig. 3. Hardware block diagram of the proposed Cartesian-type robot composed of three single-axis robots with position resolution of 1 μm.

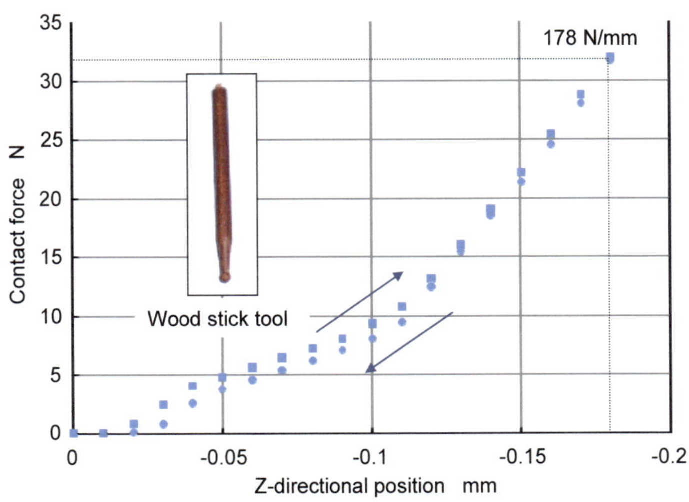

Fig. 4. An example of the static relation between position and contact force in case of using a wood stick tool, in which squares and circles show press and unpress motions, respectively.

The quantity of the position is the z-directional component at the tip of the wood stick tool. The quantity of force is yielded by contacting the tool tip with a workpiece and is measured by the force sensor. The experiment was conducted while giving the relative position 0.01 mm in press motion and –0.01 mm in unpress motion. In the experiment, the tool tip approached to an aluminum workpiece with a low speed, and after touching the workpiece, i.e., after detecting a small contact force, the tool tip was pressed to the workpiece with every 0.01 mm. The graph drawn with ■ in Fig. 4 shows the relation the position and contact force. The force is about 32 N when the position of the tool tip is –0.18 mm, so that the effective stiffness within the range can be estimated about 178 N/mm. After the press motion, the tool tip was away from the workpiece once, and returned to the position again where 32 N had been obtained. After that, the tool tip was unpressed every 0.01 mm. The graph drawn with ● in Fig. 4 shows the relation of the position and contact force of this case. It is observed that the undesirable backlash is largely decreased compared with an articulated-type industrial robot (Nagata et al, 2008). It is expected that the force resolution about 0.178 N can be performed due to the position resolution of 1 μm.

3. Compliant Motion of Wood Stick Abrasive Tool

3.1 Position/Force Control with Weak Coupling

The basic finishing strategy is conducted along a continuous spiral path while performing stable polishing force control. In this section, the control system incorporated in the Cartesian-type robot is explained. The tool tip is controlled by the translational velocity ${}^W\boldsymbol{v}(k) = \left[{}^W v_x(k)\ {}^W v_y(k)\ {}^W v_z(k)\right]^T$, which is composed of three velocities as given by

$$ {}^W\boldsymbol{v}(k) = {}^W\boldsymbol{v}_t(k) + {}^W\boldsymbol{v}_n(k) + {}^W\boldsymbol{v}_p(k) \tag{1} $$

where k denotes the discrete time; superscript ^{W} denotes the work coordinate system. Note that the control system realizes 1 msec sampling time by using the Windows multimedia timer. It is assumed that the polishing force is the resultant force of the contact force and kinetic friction forces. The kinetic friction forces are generated by Coulomb friction and viscous friction. The polishing force can be obtained as the resultant force of x-, y- and z-directional force sensor measurements. Figure 5 shows the proposed CAD/CAM-based position/force controller with weak coupling (Nagata et al, 2007). First of all, ${}^W\boldsymbol{v}_t(k)$ is the manipulated variable generated from the feedforward control law based on cutter location data called the CL data. The CL data consist of sequential position and orientation components. ${}^W\boldsymbol{v}_t(k)$ is the tangent velocity and written by

$$ {}^W\boldsymbol{v}_t(k) = v_{tangent}(k)\frac{{}^W t(k)}{\left\|{}^W t(k)\right\|} \tag{2} $$

where $v_{tangent}(k)$ is a velocity norm. ${}^W t(k)$ is the tangent vector calculated by using the position compenents of two adjacent steps in the CL data. Also, ${}^W\boldsymbol{v}_n(k)$ is the manipulated variable generated from the force feedback control law. ${}^{W}\boldsymbol{v}_n(k)$ is the normal velocity and written by

$$ {}^{W}\boldsymbol{v}_n(k) = v_{normal}(k)\ {}^{W}\boldsymbol{o}_d(k) \tag{3} $$

where ${}^{W}\boldsymbol{o}_d(k)$ is the normalized normal direction vector calculated by using the orientation compenents in CL data. The scalar $v_{normal}(k)$ representing the normal velocity is the output of the impedance model following force control law (Nagata et al, 2007) given by

$$ v_{normal}(k) = v_{normal}(k-1)\, e^{-\frac{Bd}{Md}\Delta t} + \left(e^{-\frac{Bd}{Md}\Delta t} - 1 \right) \frac{K_f}{B_d} E_f(k) \tag{4} $$

where K_f is the force feedback gain, and impedance parameters M_d and B_d are the desired mass and damping coefficients, respectively. Δt is the sampling time. Also, $E_f(k)$ is the error between the desired polishing force F_d and the norm of force vector ${}^{S}\boldsymbol{F}(k) \in \Re^3$ measure by the force sensor, which is given by

$$ E_f(k) = F_d - \left\| {}^{S}\boldsymbol{F}(k) \right\| \tag{5} $$

where superscript ^{S} represents the sensor coordinate system. Further, ${}^{W}\boldsymbol{v}_p(k)$ is the manipulated variable yielded by a position feedback control law based on a simple PI action, which is given by

$$ {}^{W}\boldsymbol{v}_p(k) = \boldsymbol{S}_p \left(\boldsymbol{K}_p \boldsymbol{E}_p(k) + \boldsymbol{K}_i \sum_{n=1}^{k} \boldsymbol{E}_p(n) \right) \tag{6} $$

where the switch matrix $\boldsymbol{S}_p = \mathrm{diag}(S_{px}, S_{py}, S_{pz})$ makes the weak coupling control to the force control active or inactive in each direction; $\boldsymbol{E}_p(k) = {}^{W}\boldsymbol{x}_d(k) - {}^{W}\boldsymbol{x}(k)$ is the position error. The desired position ${}^{W}\boldsymbol{x}_d(k)$ is calculated by using the position components in CL data. ${}^{W}\boldsymbol{x}(k)$ is the current position measured by the encoders of servo motor. $\boldsymbol{K}_p = \mathrm{diag}\,(K_{px}, K_{py}, K_{pz})$ and $\boldsymbol{K}_i = \mathrm{diag}\,(K_{ix}, K_{iy}, K_{iz})$ are proportional and integral gains for the position feedback control. Due to the weak coupling control, it is simultaneously realized that stable polishing force control and profiling control along a fine spiral path.

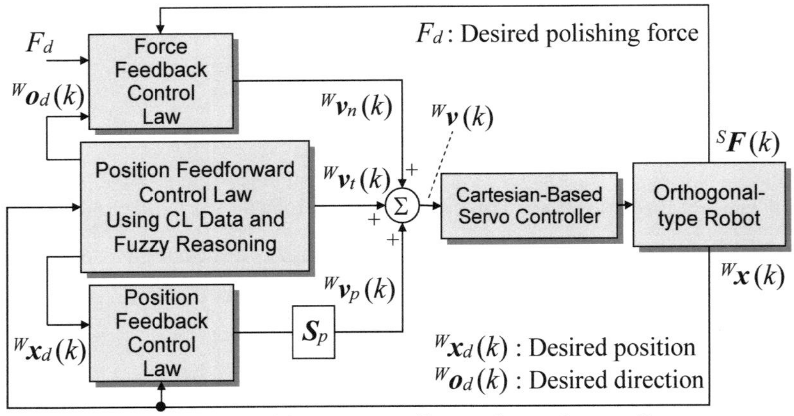

Fig. 5. CAD/CAM-based position/force controller with weak coupling.

3.2 Tuning of Desired Damping

Next, a tuning method of the desired damping is proposed by using the effective stiffness of the Cartesian-type robot. When the polishing force is controlled, the characteristics of force control system can be varied according to the combination of impedance parameters such as desired mass M_d and damping B_d. In order to increase the force control stability the desired damping, which has much influence on force control stability, should be tuned suitably. In this section, a tuning method of the desired damping is proposed based on the critical damping condition with the effective stiffness of the Cartesian-type robot. Eq. (4) is derived from the following impedance model.

$$M_d(\ddot{x}-\ddot{x}_d)+B_d(\dot{x}-\dot{x}_d)=K_f(F-F_d) \tag{7}$$

where $\ddot{x}$, $\dot{x}$ and F are the acceleration, velocity and force scalars in the direction of force control, respectively. $\ddot{x}_d$, $\dot{x}_d$ and F_d are the desired acceleration, velocity and force scalars, respectively. When the force control is active, $\ddot{x}_d$ and $\dot{x}_d$ are set to zero. It is assumed that F is the external force given by the environment and is modelled as

$$F=-B_m\dot{x}-K_m x \tag{8}$$

where B_m and K_m are the viscosity and stiffness coefficients of the environmnet, respectively. Eqs. (7) and (8) lead to the following second order lag system.

$$\ddot{x}+\frac{B_d+K_f B_m}{M_d}\dot{x}+\frac{K_f K_m}{M_d}x=0 \tag{9}$$

The characteristics equation of Eq. (9) is written by

$$s^2 + \frac{B_d + K_f B_m}{M_d} s + \frac{K_f K_m}{M_d} = 0 \tag{10}$$

In this case, the damping coefficient ζ and natural frequency ω_n are given by

$$\zeta = \frac{B_d + K_f B_m}{2\sqrt{M_d K_f K_m}}, \quad \omega_n = \sqrt{\frac{K_f K_m}{M_d}} \tag{11}$$

Further, solving Eq. (10) for B_d using the critical damping condition, the following simple condition is obtained.

$$B_d = 2\sqrt{M_d K_f K_m} - K_f B_m \tag{12}$$

In profiling control experiment, the base value for the desired damping is calculated with Eq. (12). The desired damping should be fine-tuned around the base value according to the actual system.

3.3 Finishing Experiment without Stick-Slip Motion Control

In this subsection, the proposed Cartesian-type robot is applied to a finishing experiment of an LED lens mold. Figure 6 shows the model of an LED lens mold designed by the 3D CAD/CAM Pro/ENGINEER. The model is designed based on an edge of the mold profile. First of all, an inner edge is drawn in 2D. Then, z-axis is defined as shown in Fig. 6. The z-axis is called the spiral direction. After that, an inner surface can be created by revolving the edge around the z-axis. Figure 7 shows the image of a spiral path generated from the main-processor of the CAM, which is used in the finishing experiment. The spiral path has sequential position and orientation components. The control parameters and finishing conditions tuned for the experiment are tabulated in Table 1.

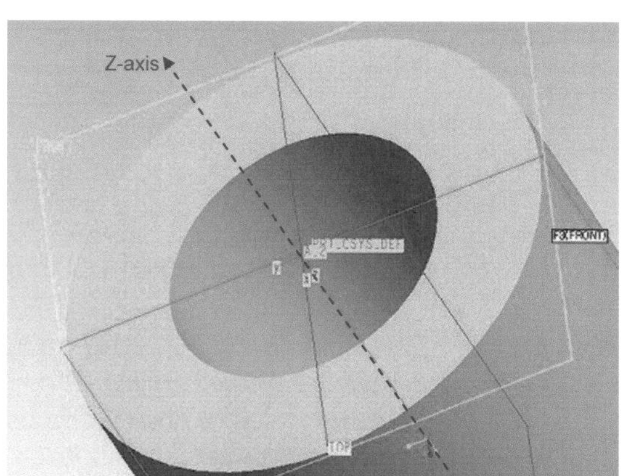

Fig. 6. 3D model of an LED lens mold designed by 3D CAD/CAM Pro/Engineer.

Conditions	Values
Desired trajectory along curved surface	Spiral path
Pitch of spiral path	0.01 mm
Radius of ball-end abrasive tool R	1 mm
Grain size of diamond lapping paste	4 μm
Desired polishing force F_d	5 N
Tool revolution per minute	400 rpm
Tangent directional velocity $\|\|{}^W\boldsymbol{v}_t\|\|$	0.1 mm/s
Desired mass coefficient M_d	0.01 N·s^2/mm
Desired damping coefficient B_d	0.183 N·s/mm
Force feedback gain K_f	0.01
Diagonal elements of switch matrix $\boldsymbol{S}_p$	$0, 0, 1$
Position feedback gain K_{px}, K_{py}, K_{pz}	$0, 0, 0.001$
Integral control gain K_{ix}, K_{iy}, K_{iz}	$0, 0, 0.00001$
Sampling time Δt	1 ms

Table 1. Control parameters and finishing conditions tuned for an experiment.

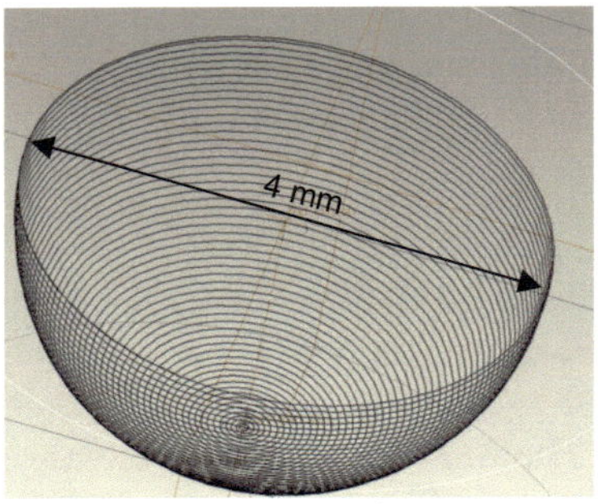

Fig. 7. Image of spiral path generated by using the main-processor of the CAM, which is used for the desired trajectory of the wood stick tool.

Fig. 8. Dexterous finishing scene by using the wood stick tool and diamond lapping paste.

Figure 8 shows the finishing scene of the LED lens mold, where a special oil including the diamond lapping paste is poured. The diameter of the concaved area is 4 mm. In this case, a small ball-end tool lathed from a wood stick is used, whose tip diameter is 1 mm. Actually, the tool tip moves along an inner path compared to the CL data used as a desired trajectory. The reason is that the position of the tool tip is corrected by the force feedback loop according to the diameter of the tool tip.

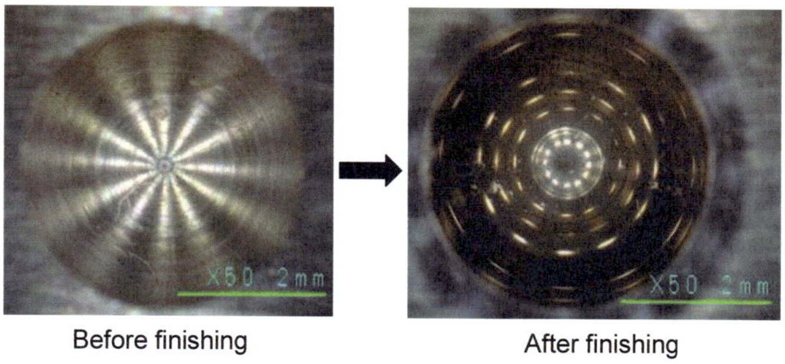

Fig. 9. Finished surfaces before and after the finishing experiment.

Figure 9 shows the surfaces before and after the finishing process. It is observed that the concaved surface area has a good quality like a mirror surface reflecting the room lights. Although small oily spots are observed, they can be cleaned easily.

4. Stick-Slip Motion of Wood Stick Abrasive Tool

4.1 Stick-Slip Motion Control

In this section, the effectiveness of the tool's stick-slip motion is evaluated to improve the surface quality. Generally, the stick-slip motion is an undesirable phenomenon and should be eliminated in various precise machine tools. However, the proposed Cartesian-type robot employs a small stick-slip motion not only to partially improve the finishing quality but also to skillfully emphasize the polishing energy. Figure 10 shows the images of stick-slip motion seen like small vibrations along a straight path and a curved path. The stick-slip motion is given along curved surface and also to orthogonal directions to tool's tangent velocity ${}^{W}\boldsymbol{v}_t(k)$. Here, how to generate small stick-slip motion vectors is explained in detail by using Fig. 11. In Fig. 11, point O is the origin in work coordinate system, where the tool tip initially contacts to the workpiece. Point P is the current contact point.

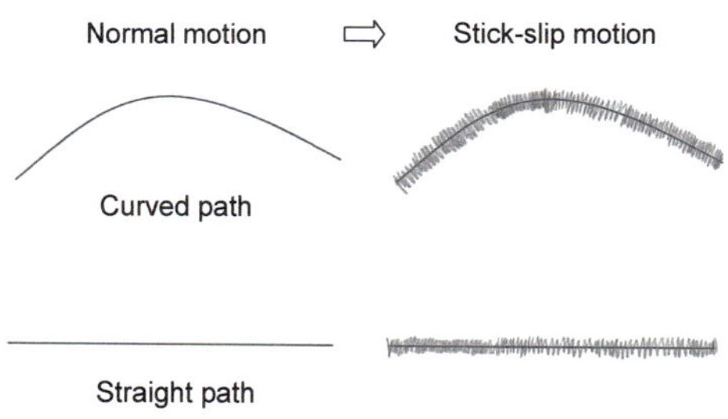

Fig. 10. Image of the small stick-slip motion seen like small vibrations for an abrasive tool.

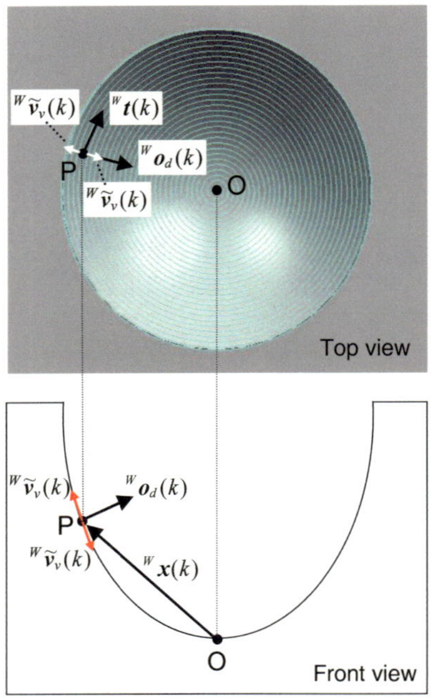

Fig. 11. Theoretical idea to calculate the stick-slip motion vector ${}^W\boldsymbol{v}_v(k)$, where the directions of ${}^W\boldsymbol{v}_t(k)$ and ${}^W\boldsymbol{v}_n(k)$ are the same ones of ${}^W\boldsymbol{t}(k)$ and ${}^W\boldsymbol{o}_d(k)$, respectively ${}^W\boldsymbol{x}(k)$ is the position vector given by ${}^W\boldsymbol{x}(k)=\left[{}^W x(k)\ {}^W y(k)\ {}^W z(k)\right]^T$ viewed from point O; ${}^W\boldsymbol{o}_d(k)$ is the normalized normal vector at the point P written by ${}^W\boldsymbol{o}_d(k)=\left[{}^W o_{dx}(k)\ {}^W o_{dy}(k)\ {}^W o_{dz}(k)\right]^T$ and is calculated with the orientation components in CL data; ${}^W\boldsymbol{t}(k)=\left[{}^W t_x(k)\ {}^W t_y(k)\ {}^W t_z(k)\right]^T$ is the tangent vector at the point P. Here, it is assumed that ${}^W\boldsymbol{v}_v(k)=\left[{}^W v_{vx}(k)\ {}^W v_{vy}(k)\ {}^W v_{vz}(k)\right]^T$ is a small stick-slip vector to be considered in this section.

In this example, the tool approaches to the workpiece with a low speed and follow the spiral path after contacting the point O. Because ${}^W\boldsymbol{v}_v(k)$ is perpendicular to ${}^W\boldsymbol{o}_d(k)$, the following relation is obtained.

$$ {}^W v_{vx}(k)\,{}^W o_{dx}(k)+{}^W v_{vy}(k)\,{}^W o_{dy}(k)+{}^W v_{vz}(k)\,{}^W o_{dz}(k)=0 \tag{13}$$

Also, ${}^W\boldsymbol{v}_v(k)$ and ${}^W\boldsymbol{t}(k)$ are orthogonal each other, so that

$$ {}^W v_{vx}(k)\,{}^W t_x(k)+{}^W v_{vy}(k)\,{}^W t_y(k)+{}^W v_{vz}(k)\,{}^W t_z(k)=0 \tag{14}$$

Further, ${}^W\boldsymbol{v}_v(k)$ is located in a plane which includes both ${}^W\boldsymbol{o}_d(k)$ and ${}^W\boldsymbol{x}(k)$, so that the components of ${}^W\boldsymbol{v}_v(k)$ are represented by

$$ {}^{W}v_{vx}(k) = i\,{}^{W}o_{dx}(k) + j\,{}^{W}x(k) \tag{15} $$

$$ {}^{W}v_{vy}(k) = i\,{}^{W}o_{dy}(k) + j\,{}^{W}y(k) \tag{16} $$

$$ {}^{W}v_{vz}(k) = i\,{}^{W}o_{dz}(k) + j\,{}^{W}z(k) \tag{17} $$

where i and j are real numbers. By solving Eqs. (13), (14), (15), (16) and (17), ${}^{W}v_{vx}(k)$, ${}^{W}v_{vy}(k)$ and ${}^{W}v_{vz}(k)$ can be obtained. Here, however, a simpler calculation is used. First of all, substituting Eqs. (15), (16) and (17) into Eq. (13) leads to

$$ i\left(\left({}^{W}o_{dx}(k)\right)^2 + \left({}^{W}o_{dy}(k)\right)^2 + \left({}^{W}o_{dz}(k)\right)^2\right) + j\left({}^{W}o_{dx}(k)\,{}^{W}x(k) + {}^{W}o_{dy}(k)\,{}^{W}y(k) + {}^{W}o_{dz}(k)\,{}^{W}z(k)\right) = 0 \tag{18} $$

$\left\| {}^{W}\boldsymbol{o}_d(k) \right\| = 1$ further leads to

$$ i = -j\left({}^{W}o_{dx}(k)\,{}^{W}x(k) + {}^{W}o_{dy}(k)\,{}^{W}y(k) + {}^{W}o_{dz}(k)\,{}^{W}z(k)\right) \tag{19} $$

Accordingly, by giving Eq. (19) into Eqs. (15), (16) and (17), the following equations are obtained.

$$ {}^{W}v_{vx}(k) = j\left({}^{W}x(k) - \left[{}^{W}o_{dx}(k)\,{}^{W}x(k) + {}^{W}o_{dy}(k)\,{}^{W}y(k) + {}^{W}o_{dz}(k)\,{}^{W}z(k)\right]{}^{W}o_{dx}(k)\right) \tag{20} $$

$$ {}^{W}v_{vy}(k) = j\left({}^{W}y(k) - \left[{}^{W}o_{dx}(k)\,{}^{W}x(k) + {}^{W}o_{dy}(k)\,{}^{W}y(k) + {}^{W}o_{dz}(k)\,{}^{W}z(k)\right]{}^{W}o_{dy}(k)\right) \tag{21} $$

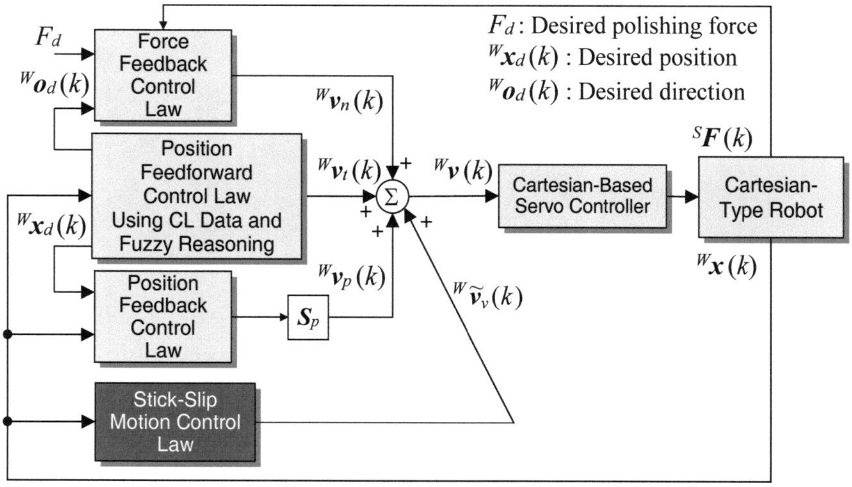

Fig. 12. Proposed CAD/CAM-based position/force controller with stick-slip motion.

$$ {}^{W}v_{vz}(k)= j\left({}^{W}z(k)-\left[{}^{W}o_{dx}(k)\,{}^{W}x(k)+{}^{W}o_{dy}(k)\,{}^{W}y(k)+{}^{W}o_{dz}(k)\,{}^{W}z(k)\right]{}^{W}o_{dz}(k)\right) \qquad (22) $$

Because both ${}^{W}\boldsymbol{o}_d(k)$ and ${}^{W}\boldsymbol{x}(k)$ are known, ${}^{W}\boldsymbol{v}_v(k)$ can be normalized as ${}^{W}\boldsymbol{v}_v(k)/\|{}^{W}\boldsymbol{v}_v(k)\|$. Further, by using a scalar K_v and a sign $\mathrm{SIGN}(k)$, the stick-slip motion vector is finally obtained as

$$ {}^{W}\tilde{\boldsymbol{v}}_v(k)=\mathrm{SIGN}(k)\,K_v\frac{{}^{W}\boldsymbol{v}_v(k)}{\|{}^{W}\boldsymbol{v}_v(k)\|} \qquad (23) $$

where $\mathrm{SIGN}(k)$ is given by

$$ \mathrm{SIGN}(k)=\begin{cases} 1 & \text{if } k=\text{odd number} \\ -1 & \text{otherwise} \end{cases} \qquad (24) $$

${}^{W}\tilde{\boldsymbol{v}}_v(k)$ is a velocity vector to yield another polishing energy, and which is given to the tool tip alternately changing the direction every sampling period. Figure 12 shows the block diagram of the controller with the stick-slip motion control method. As can be seen from Eqs. (2) and (3), the directions of ${}^{W}\boldsymbol{v}_t(k)$ and ${}^{W}\boldsymbol{v}_n(k)$ are the same ones of ${}^{W}\boldsymbol{t}(k)$ and ${}^{W}\boldsymbol{o}_d(k)$, respectively. Also, ${}^{W}\boldsymbol{v}_p(k)$ is generated in the direction of z-axis called the spiral direction as shown in Fig. 6.

4.2 Experiment of Stick-Slip Motion Control

Next, the effectiveness of the stick-slip motion control is examined through an actual finishing test. In the conventional finishing method shown in Fig. 5, the proposed stick-slip motion was not applied. Although the concaved area after finishing shown in Fig. 9 may be seen as a high quality surface, uneven lines are observed as shown in Fig. 13, in which small cusps still remain.

Fig. 13. Large scale photo of the LED lens mold, where undesirable small cusps still remain on the surface.

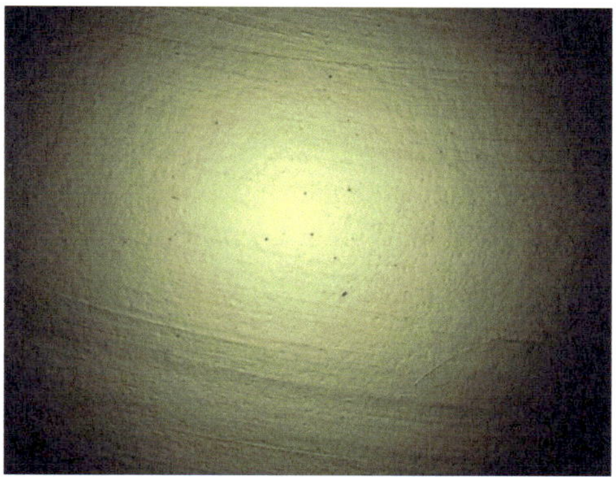

Fig. 14. Large scale photo of the LED lens mold after the finishing process by partially using the proposed stick-slip motion control.

Figure 14 shows the large scale photo of the LED lens mold after the finishing process by partially using the proposed stick-slip motion control. It is observed that the undesirable cusps can be removed uniformly. It has been confirmed from the result that the proposed finishing strategy by using the stick-slip motion control has a promising effectiveness to achieve a higher quality surface.

5. Conclusions and Future Works

The final goal of this study is the development of a novel Cartesian-type robot with compliance controllability that can be applied to from the cusp mark removing process to the finishing process for mirror-like surface of LED lens molds. In this chapter, a desktop Cartesian-type robot with 3 degree-of-freedom was first designed by combining three single-axis robots with a high position resolution of 1 μm. The position resolution and force resolution, and effective stiffness were examined through a simple contact experiment, so that the force resolution of 0.178 N was obtained due to the position resolution of 1 μm. Next, a hybrid position/force controller with compliance controllability was proposed for the finishing task of LED lens molds, in which position control, force control or their weak coupling control can be selected according to each finishing strategy. A systematic tuning method of the desired damping was also considered. The desired damping was calculated from the critically damped condition using the static relation between the position and force. Further, a stick-slip motion control for a wood stick tool was developed to finely improve the finishing quality. The proposed desktop Cartesian-type robot using these peripheral techniques was applied to a finishing experiment of an LED lens mold, so that the high performance and promise were successfully confirmed. In future work, we plan to consider other potential applications using the Cartesian-type robot.

6. References

Oba, T.; Kawai, T.; Takino, H. & Takeuchi, Y. (2008). Ultraprecision multi-axis control machining of spherical micro-lens array mold. *Proceedings of JSPE Semestrial Meeting*, B67, Tohoku University, September 2008.

Kuriyagawa, T.; Saeki, M. & Syoji, K. (2002). Electrorheological fluid-assisted ultra-precision polishing for small three-dimensional parts. *Precision engineering*, Vol. 26, No. 4, pp. 370–380.

Suzuki, H.; Moriwaki, T.; Okino, T. & Ando Y. (2006). Development of ultrasonic vibration assisted polishing machine for micro aspheric die and mold. *CIRP Annals - Manufacturing Technology*, Vol. 55, No. 1, pp. 385–388.

Bilkay, O. & Anlagan, O. (2004). Computer simulation of stick-slip motion in machine tool slideways. *Tribology International*, Vol. 37, No. 4, pp. 347–351.

Mei, X.; Tsutsumi, M.; Tao, T. & Sun, N. (2004). Study on the compensation of error by stick-slip for high-precision table. *International Journal of Machine tools & Manufacture*, Vol. 44, No. 5, pp. 503–510.

Nagata, F.; Hase, T.; Haga, Z.; Omoto, M. & Watanabe, K. (2007). CAD/CAM-based position/force controller for a mold polishing robot. *Mechatronics*, Vol. 17, No. 4–5, pp. 207-216.

Nagata, F.; Hase, T.; Haga, Z.; Omoto, M. & Watanabe, K. (2008). CAD/CAM-based Position/Force Control for a Ball-end Abrasive Tool and Its Application to an Industrial Robot. *Journal of Advanced Mechanical Design, Systems, and Manufacturing*, Vol. 2, No. 4, pp. 742–752.

14

Kinematic calibration of articulated arm coordinate measuring machines and robot arms using passive and active self-centering probes and multipose optimization algorithm based in point and length constrains

Jorge Santolaria and Juan José Aguilar
Universidad de Zaragoza
Spain

1. Introduction

The kinematic modeling of articulated arm coordinate measuring machines (AACMM) has inherited both the previous developments in the field of robot arms and manipulators, and their calibration and parameter identification techniques, given the similarity of their mechanical characteristics. Traditional approaches to the problem of kinematic parameters identification in both cases use an objective function in terms of quadratic sum of errors of measurement or positioning, formulated as the euclidean distance between the points materialised by a gauge or measuring instrument, and the points obtained through the kinematic model. By capturing data at various positions in the workspace, those approaches follow a resolution scheme that involves indirect optimization, such as the Moore-Penrose pseudoinverse or other methods for solving systems of linear equations to obtain the set of kinematic model parameters which minimize the error in the positions considered.
This chapter first presents a review of developments and state of the art concerning the kinematic modeling of robot manipulators and AACMM, as well as aspects to consider regarding the influence of the model chosen on the subsequent parameters identification procedure. Secondly an optimization algorithm based on an objective function that considers terms related to the accuracy and repeatability is shown. This algorithm follows a pure optimization scheme from data obtained through probing several spheres of a ball-bar gauge placed at several positions in the working range for both systems. In addition to the distance errors from the nominal coordinates of the gauge, it is possible to optimize the repeatability from the captured pose values for the same sphere in several arm orientations. To capture data, a passive self-centering probe and an active self-centering probe are used to directly probe the center of each sphere for a large number of arm orientations, and also to analyze the effect of probing force in the identification process and the generalization of the error results for any arm position and orientation. Experimental results of the capture and identification technique are presented with both probes linked to a Faro AACMM, as well as

the comparison of these results with those obtained applying traditional identification techniques. Finally a description of the technique for a Kuka robot with the active self-centering probe is shown, and also the integration of both mathematical models to obtain the sphere center coordinates in the robot global reference frame from readings expressed on the reference system of the self-cetering probe.

2. Evolution of kinematic modelling for robot calibration

Since the development of the well-known model proposed by Denavit and Hartenberg (D-H) (Denavit & Hartenberg, 1955), multiple variants of kinematic models based on homogenous transformations have been considered. These later approaches solve problems of indetermination derived from the use of D-H model in robot calibration procedures. In (Roth et al., 1987), three levels of calibration for robots are established: joint level, kinematic model level and dynamic model level. The work presented in this chapter is focused on the second level of calibration. Thereby, the possible influences in the final measurement error of an AACMM will be studied to determine the mathematical model parameters. Those influences can be both geometrical -due to the kinematic model- and non-geometrical -due to other factors as assembly, damage in joints or transmission errors-. The approach of models which try to separate geometrical and non-geometrical errors has been a constant in literature on robot arms, without any apparent success from the point of view of a generalizable model regarding both robots and AACMMs. Based on the mathematical separation of both influences through the consideration of additional parameters in the model, the determination of a set of parameters related to each type of modeled error is carried out following optimization techniques without maintaining a direct relation between the real physical parameters and those obtained by optimization. Thus, these parameters cannot be considered more than an adjustment to the captured data which minimizes the positioning error in the case of robots but which do not have any numerical restriction to maintain the relation between results and real physical parameters, not being justifiable in many cases additional parameters that cause redundancy and complicate unnecessarily the mathematical model. In (Goswami & Bosnik, 1993) the relationship between the mathematical results and the physical reality, along with the influence of the redundancy in model parameters on a calibration method, are studied. The detection of the error sources is carried out by the way the model is approached. However, traditionally, the second calibration level only studies the identification of geometric parameters, whereas the non-geometrical and temperature influences are considered in a different way from model to model.
Regarding the D-H model itself, Hayati and Mirmirani (Hayati, 1983; Hayati & Mirmirani, 1985) outlined the discontinuity and non-proportionality of the model proposed by D-H for robotic arms with parallel or nearly parallel joint axes. In these cases the elimination of a parameter and the inclusion of a new rotation parameter are proposed. Many other authors have tried to solve the problem of the model proposed by D-H. In (Hsu & Everett, 1985) a modification of the D-H parameterization is proposed, adding a fourth joint parameter which describes with a variable the reference frame position in the direction of the previous frame axis. Subsequently, new approximations considering the general equation of rigid body spatial movement (Mooring & Tang, 1984) to model transformations between joints appeared, solving the indetermination of the D-H model but introducing redundant

parameters. However, it is desirable, and indeed essential for many researchers, that a calibration model is free of redundancies in order to identify each error with a parameter, to establish reliable correction methods and even to improve the machine design. This assumption cannot be generalized, since to detect an error and to find its relation to a particular source depends on many other factors. Although (Everett et al., 1987) and (Everett & Suryohadiprojo, 1988) describe the maximum number of independent parameters necessary in order to define the geometry of an arm depending on the type of joints, redundancy may exist. In (Goswami et al., 1993), a series of considerations regarding the minimum number of parameters to describe the kinematics of an arm can be found, along with the need for redundancy in certain robot configurations in order to maintain the physical-mathematical link in model parameters.
The specific inclusion of non-geometrical errors effect on a robot calibration process by mean of model parameters or series approximation of them appears in (Mooring, 1983) and (Whitney et al., 1984), beyond considering the arm errors in terms of variation of geometrical parameters. In (Chen & Chao, 1986) there is an approximation to the non-geometrical errors model based on the mathematical model proposed by Sheth and Uicker (Sheth & Uicker, 1971), which, like the previous ones, allows maintaining the nominal positions of the joints reference frames, since it uses different frames to model the geometrical and non-geometrical errors. In this way, the definition of the nominal model does not require any particular modeling procedure, since the errors are modeled by way of new matrices and transformations to the nominal systems. Subsequently, the inclusion of intermediate error matrices has been given extensive consideration in the biographical resources on robots to model non-geometrical errors. In (Drouet et al., 2002), an efficient method to consider any kind of errors is shown, including an intermediate matrix which relates joint nominal frame position with a second joint frame in its real position.
Stone and others (Stone et al., 1986) proposed a model, known as S-model, which adds two parameters to the basic D-H notation in order to allow variable positioning of the reference frames. Again, a solution to the indetermination problem is obtained, resulting in a complete but non-proportional model. Hollerbach and Wampler (Hollerbach & Wampler, 1996) later proposed a model based on a mix of the D-H parameters and the Hayati notation , using model parameters depending on the type of joint in order to avoid indeterminations. In the search for a model which fulfils the three basic properties to be used for calibration (Everett et al., 1987), Zhuang et al. (Zhuang & Roth, 1992) proposed the CPC model. Based on the idea that an incomplete model can be made complete by adding parameters, from the D-H model, they considered two more parameters to model the arbitrary position of the frames. Finally, models which consider the physical reference position of the encoder zero and which determine the relation between this position and the initial nominal position of the model (Mooring & Tang, 1984; Mooring, 1983; Park & Brockett, 1994) should be considered as the last main group of kinematic models.
Apart from the modeling techniques presented, all of them with later approximations in bibliography, and those which combine parameters associated to geometrical and non-geometrical error (Caenen & Angue, 1990; Vincze et al., 1999)– given the difficulty of modeling separately -, one of the current modeling trends involves an approximation to the final error based on the simplest kinematic models (Alici & Shirinzadeh, 2005). In this way, they do not try to explicitly model each parameter based on a parametric model which includes the error. So far, the models reviewed increase in mathematical complexity when

modeling each supposedly identifiable error by way of parameters not directly measurable. Once the value of these parameters has been established by optimization, it is possible to correct the final position of the end of the arm through an error model using off-line correction tools. Using different linear and non-linear regression mathematical tools, such as polynomials, Bezier curves, Fourier series, wavelets, neural networks, etc..., it is possible to make an approximation to the arm error in line with the error characteristics observed in different positions of the work space. The non-parametric nature of these tools makes more difficult to link the geometrical and non-geometrical parameters of a model with the physical reality of the arm. Most applications which use regression as an approximation to errors try to combine traditional modeling and regression techniques to allow the error to be linked in some way to measurable arm parameters.

In this way, despite not fulfilling the conditions of proportionality and equivalence (Everett et al., 1987), the D-H model avoids redundancies and perfectly describes the kinematics of a measurement arm, not presenting indeterminations in practically all AACMMs on the market, whose dual joints define consecutive perpendicular axes. For this reason, in the presented work the AACMM has been modeled by way of D-H when dealing with the identification procedure, although it is easily generalizable to any kinematic model.

3. AACMM Kinematic model

AACMMs make up a special group within coordinate measurement due to their special characteristics and differences with regards to traditional coordinate measuring machines (CMMs). While the CMMs, whether they are bridge, gantry or horizontal arm type, have a cartesian configuration which allows the measurement of the physical displacement of each of the three linear axes, in the AACMMs the measured point is a result of a series of coordinate transformations, depending on the model. AACMMs adopt the kinematic structure and model of robotic arms, and, similarly, are made up of a series of straight sections linked by rotary joints which provide them with the dof necessary to reach the required measurement positions. The differences compared to robotic arms are great, both in terms of accuracy and functionality, paying special attention to the accuracy of the sensors, materials used and dynamic conditions. Moreover, another important difference compared to CMMs and robotic arms is their manual and portable operation, instead of having machine axes automatically controlled.

The growing use of the AACMMs has been accompanied by an absence of standards on verification and calibration procedures, both from the point of view of the user and of the manufacturer. Traditionally, each AACMM manufacturer has adopted its own evaluation procedures. Firstly, it is necessary to determine the value of the parameters of the kinematic model of the arm. To this end, each manufacturer uses its own methods depending on the model and parameters implemented in each arm. Both the mathematical model considered and the method used to identify parameters constitutes restricted information which is not available to the final user. These evaluation methods are based on the procedures set out by the predecessors of the three main standards for performance evaluation in current CMMs, UNE-EN ISO 10360, ASME B89.4.1 and VDI/VDE 2617. However, the special characteristics of the AACMMs require different verification procedures. While for a CMM an X, Y, Z point clearly defines a position of the three machine axes, for an AACMM the possible positions of its elements to achieve a fixed point defined in the measurement volume are practically infinite, according with its inverse kinematics. Moreover, for CMMs, besides the general evaluation methods, it is possible to carry out evaluation tests in order to extract the positioning errors to finally implement correction models (Trapet & Wäldele, 1991). The CMM evaluation tests achieve a high level of maintenance of the physical-mathematical relations between the error model parameters and the error physically committed by the machine. This is not the case of AACMMs, in which the application of these models does not make sense, given the difficulty of directly relating the error committed with the model parameters, which are obtained by using optimization procedures.

ASME B89.4.22-2004 and VDI/VDE 2617-9:2008 are nowadays the only standards existing in the field of AACMM verification. They recommend the methods which should be followed for reliable performance evaluation of the measurement arms. These documents deal with complete standardization in this field. They are focused on contact measuring with active and passive probes, covering the common measurement arm applications. In order to uniformize and eliminate ambiguity in evaluation methods for AACMMs, these standards fulfil their function, without making any indications with regards to parameter identification methods, calibration or correction.

The static calibration of an AACMM establish a parametric model of its kinematic behavior in order to determine, numerically, the relationship between the joint variables and the probe position for any arm posture. A direct kinematic model takes the form of (1).

$$y = f(\theta_i, q) \tag{1}$$

with i=1,...,n for an arm with n rotating joints. This model calculates the position and orientation of the AACMM probe y,according to the value of the joint variables θ_i and to the equations of the model defined in f, which depend on the parameters vector q. This parameters vector contains the geometric parameters of the model, which must be optimized in order to obtain the lowest possible measurement error. Depending on the chosen kinematic model, the way the equations are obtained in f changes, along with the number of geometric parameters necessary to be included in q. The D-H basic model uses four parameters (d_i, a_i, θ_i and α_i) to model the transformation of coordinates between successive reference systems. The homogeneous transformation matrix between frame i and i-1 of equation (2) depends on those four parameters. Calculating successive transformations of coordinates, by pre-multiplying the transformation matrix between a frame and the previous one, it is possible to obtain the global transformation matrix of the arm, which obtains coordinates of the center of the probe sphere with respect to the base of the AACMM. In this manner, considering 0 as the global fixed reference system of the base and reference frame 6 moving with the rotation of the last joint, the desired homogeneous transformation can be obtained by way of equation (3). Apart from the four joint parameters indicated, it is necessary to consider an extra joint parameter θ_{0i} which will define the relationship between the encoder physical reference mark position of each joint and the initial position considered in the definition of the kinematic model.

$$^{i-1}A_i = T_{z,d}T_{z,\theta}T_{x,a}T_{x,\alpha} = \begin{bmatrix} \cos\theta_i & -\cos\alpha_i\sin\theta_i & \sin\alpha_i\sin\theta_i & a_i\cos\theta_i \\ \sin\theta_i & \cos\alpha_i\cos\theta_i & -\sin\alpha_i\cos\theta_i & a_i\sin\theta_i \\ 0 & \sin\alpha_i & \cos\alpha_i & d_i \\ 0 & 0 & 0 & 1 \end{bmatrix} \quad (2)$$

$$^{0}T_6 = {}^{0}A_1\,{}^{1}A_2\,{}^{2}A_3\,{}^{3}A_4\,{}^{4}A_5\,{}^{5}A_6 \qquad \overline{X}_{AACMM} = {}^{0}T_6\,\overline{X}_{Probe} \quad (3)$$

Thus, the joint variable θ_i of the model is related to the rotation reading provided by the encoder through equation (4), where θ_{0i} must also be identified from its nominal value defined for the initial position.

$$\theta_i = \theta_{iEnc} - \theta_{0i} \quad (4)$$

The use of a reference system defined in the robotic hand responds generally to the need to have a frame which follows the characteristic directions of the tool, which will be subsequently controlled. In this case, since the aim is to obtain the sphere center coordinates of a static probe, the use of a reference system attached to the probe is omitted, assuming the link between the probe center and frame 6 to be a translation expressed by the coordinates X_{Probe}, Y_{Probe}, Z_{Probe} of the sphere center respect to this frame. With this, the model has 4 joint parameters plus the three coordinates of the probe sphere center in frame 6, what makes a total of 27 parameters for the 6 dof AACMM considered. The initial values taken for the arm model parameters are shown in Fig. 1.

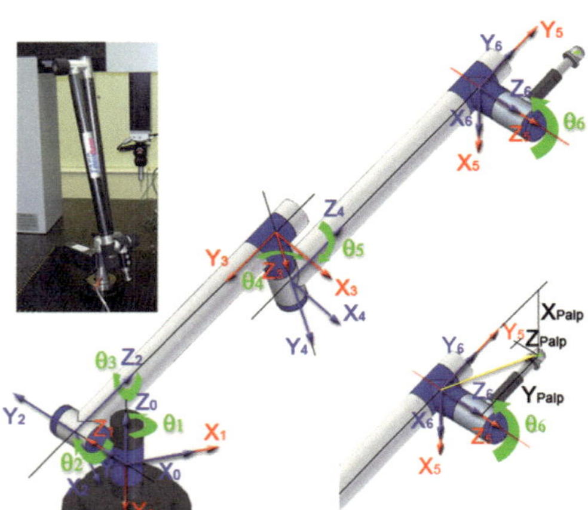

Joint	a_i (mm)	α_i (°)	d_i (mm)	θ_{0i} (°)
1	0.0	-90.0	0.0	0.0
2	0.0	90.0	47.8	15.0
3	0.0	-90.0	645.83	-90.0
4	0.0	90.0	54.5	0.0
5	0.0	90.0	615.7	90.0
6	0.0	0.0	0.0	0.0
	X_{Probe} (mm)	Y_{Probe} (mm)	Z_{Probe} (mm)	
	0.0	138.5	54.5	

Fig. 1. Model definition posture of FARO AACMM with D-H convention. Initial values of parameters.

4. Parameter identification

The AACMM used in the present work is a 6 dof Sterling series FARO arm with a typical 2-2-2 configuration and a-b-c-d-e-f deg rotation, with a nominal value of 2σ = ± 0.102 mm obtained in a single-point articulation performance test of the arm manufacturer, without specifying the number of positions of the kinematic seat and points captured. Simple evaluation tests have been carried out to evaluate the current state of the measurement arm using the manufacturer capture software and model, in order to know the expected initial values for accuracy and repeatability with the arm in the current situation. Using a spherical-point passive probe of 6 mm in diameter, 14 spheres of a ball-bar gauge have been measured in a single gauge position, obtaining a mean deviation of the lengths measured with regards to the gauge lengths of 0.854 mm. On the other hand, a single-point articulation performance test has been carried out using a conical seat to capture 100 points in different AACMM orientations for a single kinematic mount position. The result obtained in this test was 0.347 mm for 2σ, with respect to the mean point of the captured data. Without the need for further initial evaluation tests or complete tests in accordance with ASME B89.4.22-2004, it can clearly be appreciated that the AACMM is out of calibration with the manufacturer parameters and model. Hence, it will be taken as the initial situation when dealing with the proposed calibration procedure.

4.1 Data capture setup and evaluation

Once the mathematical model is defined, the next step involves the capture of nominal coordinates in the workspace of the AACMM. All the calibration procedures, both for robotic arms and AACMMs, establish a system which acquires coordinates or nominal distances in the workspace, in order to capture points which allow the error to be evaluated and minimized.

Current identification procedures for robotic arms are based on a small set of positions where the coordinates of the robot hand are captured discretely. In (Alici & Shirinzadeh, 2005), based on the results obtained in (Driels & Pathre, 1990) and (Borm & Menq, 1991), 85 discrete robot positions were selected to cover the entire work range and extrapolate the results obtained to any arm position. The selection of capture positions to identify kinematic parameters must be preceded by tests which characterize the influence of each joint on the final error, to finally choose positions in accordance with this influence. For this reason, the application of a number of specific positions is not generalizable from one robotic arm to another, since the errors committed by each robot will depend on their configuration and assembly defects. In (Chunhe et al., 2000), the method described to identify the geometrical parameters takes 30 capture positions, without any apparent justification. It is common to consider a very different number and type of robot positions in the bibliography, constituting a very important step of the procedure for the extrapolation of the results. The referenced procedures successfully carry out their task, and hence both the capture of points and the subsequent test positions are captured discretely in robot configurations which are very similar to each other and also similar to the common robot work positions. This means that the positioning accuracy of the robot is improved globally in common work configurations. Since the optimization of the kinematic parameters is a least squares method, the adjustment of the parameters which minimize the error in the robot identification positions will mean small errors occur in positions similar to those of capture, but will cause bigger errors in very different positions.

In this work, a continuous data capture method has been developed. This technique allows the massive capture of arm positions corresponding to several points of the workspace. To this end, a ball-bar gauge of 1.5 m long was placed in 7 positions within the workspace of the arm in order to cover the maximum number of possible AACMM positions, to subsequently extrapolate the results obtained throughout the volume. Fig. 2 shows the considered positions for the bar in a quadrant of the workspace. The ball-bar comprises a carbon fiber profile and 15 ceramic spheres of 22 mm in diameter, reaching calibrated distances between the centers with an uncertainty, in accordance with its calibration certificate, of (1+0.001L)µm, with L in mm. The ball-bar profile is made of a carbon fiber layer having a balanced pair of carbon fiber plies embedded in a resin matrix, with a nominal coefficient of thermal expansion (CTE) between $\pm 0.5 \times 10^{-6}$ K^{-1}. The position of the fibers in the profile allows compensating this coefficient, obtaining a mean CTE near zero.

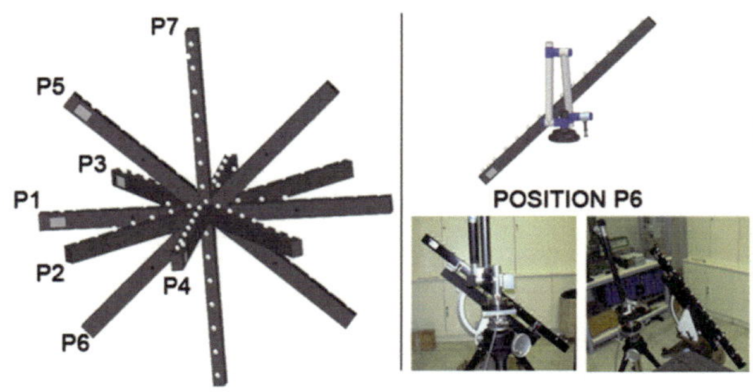

Pos. No.	Affected quadrants	Inclination	Distance B1	Direction
P1	Q1-Q2-Q4	Horizontal	Medium	Tangential
P2	Q1-Q4	Horizontal	Near	Tangential
P3	Q1-Q2	Horizontal	Far	Tangential
P4	Q1	Horizontal	Near	Radial
P5	Q1-Q2-Q4	+45°	Medium	Tangential
P6	Q1-Q2-Q4	-45°	Medium	Tangential
P7	Q1	Vertical	Medium	Tangential

Fig. 2. Ball bar positions for each quadrant. Sample of position P6.

The capture of data both for calibration and for verification of the arms is usually performed by way of discrete contact probing of surface points of the gauge in order to obtain the center of the spheres from several surface measurements. This means that the time required for the capture of positions is high, and then, identification is generally carried out with a relatively low number of arm positions. In the present work, two specific probes, capable of directly probing the center of the spheres of the gauge without having to probe surface points, were designed. As seen in Fig. 3, one of the probes comprises three tungsten carbide spheres of 6 mm in diameter, laid out at 120° on the end of the probe. Since the ceramic spheres of the gauge have a diameter of 22 mm, it is necessary to establish the geometrical relationships in order to ensure the proper contact of the three spheres and the stability of this contact. In general, in order to maintain this stability, it is recommended a contact between the spheres of the kinematic mount and the sphere to fit between them at 45° with respect to the plane formed by the centers of the mount spheres. Thereby, the centering of the probe direction with regards to the sphere center is ensured, making this direction cross it (Fig. 3) for any orientation of the probe. Thus, in this case, it is possible to define a probe with zero probe sphere radius and with the distance from the position of the housing to the center of the probed sphere of 22 mm as length, allowing direct probing of the sphere center when the three spheres of the probe and the sphere of the gauge are in contact.
On the other hand, we have reproduced the process of data capture for all positions of the gauge with a self-centering active probe. This probe, specifically designed for probing spheres, is composed of three styli positioned to form a trihedron with their probing directions. Each individual stylus has been designed by using a linear way together with a LED+PSD sensor combination to measure its displacement. From the readings of the displacement of the three styli and its mathematical model, the probe is able to get the center of the probed sphere in its reference system. So, it is necessary to link the reference system

with the last reference system of the kinematic chain of the AACMM, to express the center of the probed sphere in the global reference system. The method followed to determine the relationship between the two frames is described in the last section of this chapter, since this probe is particularly suitable for parameter identification procedures in robots.

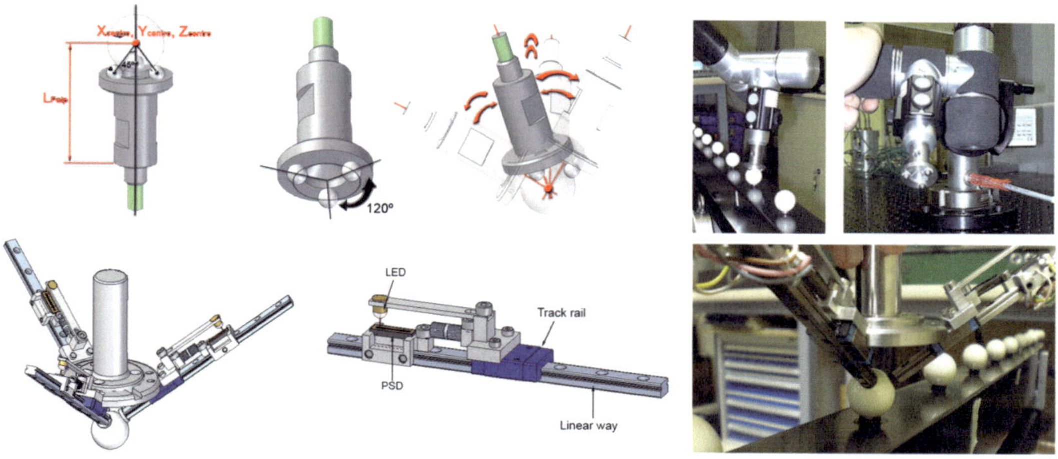

Fig. 3. Pasive and active self centering probes used in the capture of AACMM positions for parameter identification

Besides characterizing and optimizing the behavior of the arm with regards to error in distances, its capacity to repeat measurements of a same point is also tested. Hence, an automatic arm position capture software has been developed to probe each considered sphere of the gauge and to replicate the arm behavior in the single-point articulation performance test, but in this case, to include the positions captured in the optimization from the point of view of this repeatability. The rotation angle values of the arm joints for each position, reached in the continuous probing of each sphere, are stored to obtain the coordinates of the measured point with respect to the global reference system for any set of parameters considered. In this way, it is possible to capture the maximum possible number of arm positions, thus covering a large number of arm configurations for each sphere considered. Fig. 4 shows the capture scheme followed. As a general rule, the indicated trajectories will be followed for a probed sphere. Moreover, positions causing maximum variation of the arm joints in all the possible directions at the start, end and midpoint of each trajectory will be searched. The capture will be continuous and we will try to capture data in symmetrical trajectories in the sphere, in order to minimize the effect of probing force on the gauge. Thereby, around 400 rotation angle combinations θ_{iEnc} (i=1,…,6) have been captured for the joints to cover the positions of the arm probing the center of the measured sphere.
With this configuration, 4 spheres of the gauge in each of the 7 positions considered for each of the quadrants of the arm work volume were probed. The measuring of a sphere center with the self centering probe from different arm orientations should result in the same point measured.

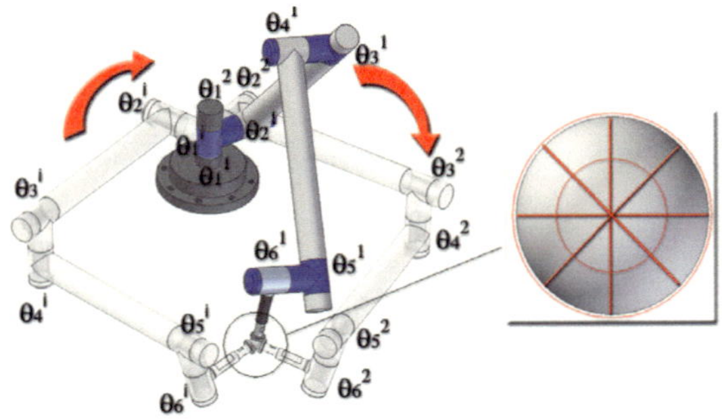

Fig. 4. Data capture procedure and capture trajectories. The readings from each of the 6 joint encoders are stored continuously for all capture AACMM positions.

The unsuitable value of the kinematic parameters of the model will be shown by way of a probing error. This error produces different coordinates obtained for the same measured point in different arm orientations. In this manner, by probing four spheres of each position of the gauge with an approximate average of 400 arm positions per sphere for the passive self centering probe (250 for the active self centering probe), a series of 400 XYZ coordinates measured for each sphere center will be obtained. The deviations, initially due to the value of the parameters of the model between these 400 points in each sphere, will be used to characterize and optimize the arm point repeatability. In addition, in each gauge location 6 nominal distances between the four probed spheres are reached (Fig. 5a). The nominal distances of the gauge will be compared to the distances measured by the arm. Since an average of 400/250 centers per sphere are captured, the mean point of the set of points captured will be taken as the center of the sphere measured, in order to determine the distances between spheres probed by the arm (Fig. 5b). Thereby, a method for the subsequent combined optimization of the AACMM error in distances and point repeatability is defined.

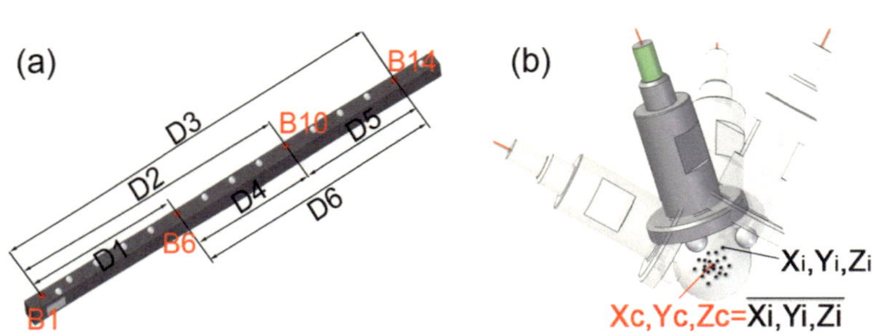

Fig. 5. Nominal parameters used in identification: (a) distances between spheres centers and (b) center considered to evaluate distances between spheres measured and point repeatability.

In order to analyze the metrological characteristics of the AACMM for a specific set of parameters, both the error in distances of the arm and the dispersion of the points captured for each probed sphere center will be studied. As can be seen in Fig. 5, the parameters to evaluate are the six distances between the centers of the four spheres probed by bar location and the standard deviation of the points captured for each of the spheres probed. The 3D distance between pairs of spheres, based on the mean points calculated in each of them, is shown in equation (5).

$$D_{i_{jk}} = \sqrt{\left(\overline{X}_{ij} - \overline{X}_{ik}\right)^2 + \left(\overline{Y}_{ij} - \overline{Y}_{ik}\right)^2 + \left(\overline{Z}_{ij} - \overline{Z}_{ik}\right)^2} \quad (5)$$

in which $D_{i_{jk}}$ represents the Euclidean distance between sphere j and sphere k of the gauge i location, with coordinates corresponding to the mean of the points captured for sphere j and sphere k according to equation (6).

$$\overline{X}_{ij} = \frac{\sum_{m=1}^{n_{ij}} X(m)_{ij}}{n_{ij}} \quad (6)$$

In equation (6), n_{ij} is the number of angle combinations captured for sphere j in identification position i of the gauge, analogously for the coordinates Y and Z. In this manner, considering D_{0jk} as the nominal distance between spheres j and k obtained in the gauge calibration table, it is possible to calculate the error in distance between spheres j and k in location i in accordance with equation (7).

$$E_{i_{jk}} = \sqrt{\left(D_{i_{jk}} - D_{0jk}\right)^2} \quad (7)$$

Since there are 4 spheres per gauge location, a total of 6 distances in each one are calculated, bringing a total of 42 distances. Considering in the previous equations i=1,…,7, and j and k covering each one of the spheres of the gauge for each position (1, 6, 10 and 14), eliminating the terms in which j=k comes about, and considering that $D_{i_{jk}} = D_{i_{kj}}$, it is possible to evaluate the errors in distances obtained for all the gauge locations. Moreover, as a second quality value of a set of parameters the maximum standard deviation of the points captured in each of the measured spheres is chosen.

$$2\sigma_{Xij} = 2\sqrt{\frac{\sum_{m=1}^{n_{ij}} \left(X(m)_{ij} - \overline{X}_{ij}\right)^2}{n_{ij} - 1}} \quad (8)$$

In equation (8), σ_{Xij} represents the standard deviation in coordinate X of the points obtained for sphere j of position i. Analogously for the coordinates Y and Z. Following the diagram presented in Fig. 6 we obtain the values of all the possible standard deviations and distances for the data captured, considering a given set of parameters. The quality indicators for the self centering passive probe of the initial set of parameters considered in Fig. 1 are shown in Fig. 6. In the first column is shown the maximum error in distances for all the positions of the gauge, the position of identification in which it is produced and the distance at which the maximum error has been obtained. Likewise, the minimum error and its location, and the mean value of all the errors in distance observed are shown. In the case of standard deviation, Fig. 6 also includes the coordinate in which the value has been obtained, since both parameters are calculated separately for the three point coordinates. As can be seen, the values obtained for the initial set of parameters are large, as was expected given the initial lack of adjustment of the AACMM kinematic parameters.

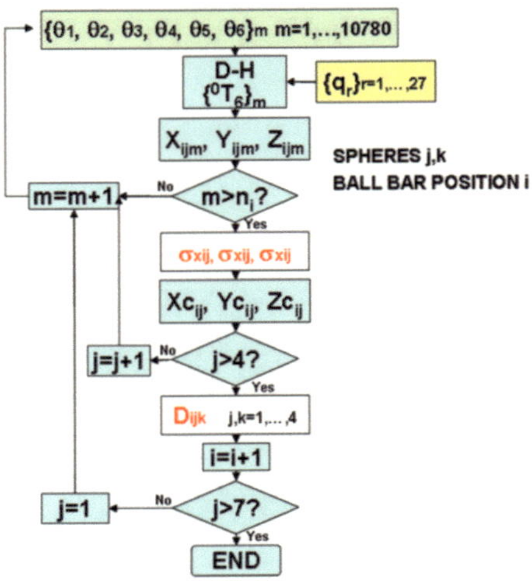

Distance Error (mm)		2σ by Sphere (mm)	
Max.	49,427	Max.	28,227
Causing Pos.	POS2	Causing Pos.	POS 4
Causing Dist.	D2	Causing Sph.	B10
Min.	0,276	Causing Coord.	Y
Causing Pos.	POS2	Min.	2,580
Causing Dist.	D5	Causing Pos.	POS6
Med.	22,429	Causing Sph.	B10
		Causing Coord.	X
		Med.	11,689

Fig. 6. Evaluation of a set of parameters q in identification positions. Results for data captured with the self-centering passive probe and initial set of paramenters.

4.2 Non-linear least squares identification

Kovac and Klein present in (Kovac & Klein, 2002) an identification method based on nominal data obtained with the gauge developed in (Kovac & Frank, 2001). This method uses an objective function as used in robots, along with commercial software to identify kinematic parameters, without focusing the study on the particularities of the measurement arms. In (Furutani et al., 2004), Furutani et al. describe an identification procedure for measurement arms and make an approximation to the problem of determination of AACMM uncertainty. This study is centered on the type of gauge to be used according to the arm configuration and analyses the minimum number of necessary measurement positions for identification, as well as the possible gauge configurations to be used. Again, this work does not specify the procedure to obtain the parameters of the model, nor the type of model implemented, and does not show experimental results for the method proposed. In (Ye et al., 2002), Ye et al. develop a simple parameters identification procedure based on arm positions captured for a specific point of the space. In (Lin et al., 2006), Lin et al. perform an error propagation analysis from the definition of several error geometrical parameters. This study shows the influence of the error parameters defined by its authors in their model and, even though it is not generalized to the geometrical errors propagation from the parameters identification, it shows an effective method to elaborate a software-based error correction procedure.

As indicated in section 3, the kinematic model implemented in the measurement arm can be described, for any arm position, by way of equation (9), based on the formulation of direct kinematic problem.

$$p = f\left(a_i, \alpha_i, d_i, \theta_{0i}, X_{Probe}, Y_{Probe}, Z_{Probe}, \theta_{iEnc}\right) \quad i = 1,\ldots,6 \tag{9}$$

in which $p=[X\ Y\ Z\ 1]^T$ are the coordinates of the point measured with respect to the arm global reference frame at the base, corresponding to the value of the geometrical parameters and to the joints rotation angles in the current arm position. There are many alternatives when dealing with an optimization procedure, although the most widely used in the field of robot arms and AACMMs are the formulations based on least squares fitting. Given the non-linear nature of the arm kinematic model, it is not possible to obtain an analytical solution to the problem of parameter identification. Therefore, it is necessary to use non-linear optimization iterative procedures. In this way, for the mathematical formulation of the optimization method it is common to define the objective function to minimize in terms of square error components. Based on the nominal coordinates reached by the gauge and those corresponding to the points measured, we can obtain the arm measurement error as the Euclidean distance between both points, as shown in equation (5), although applied to the difference between the measured point and the nominal point. Since the identification procedure both in robots and in AACMMs is based on the capture of discrete positions within the workspace, all the reviewed optimization procedures use equation (10) as basic objective function to minimize.

$$\begin{aligned} \phi &= \sum_{i=1}^{m} [\Delta p]^T [\Delta p] \\ [\Delta p]_i &= [\delta x \quad \delta y \quad \delta z]_i^T = \left(p - p_0\right)_i \end{aligned} \tag{10}$$

Equation (10) quantifies the error in distances between the nominal point and the point reached for all the positions captured, formulated as the quadratic sum. There are variants of this expression in those cases in which a capture procedure based on nominal distances is proposed, where the measurement error of the arm is obtained in accordance with equation (7) for each distance considered, to obtain the errors in all the distances and with the target function being the quadratic sum.

In this work, in order to choose the objective function to be minimized, consideration has been given to the error in distances presented in equation (7) for the 42 distances measured. Therefore, it is possible to evaluate all the combinations of six values of joint angles captured for each set of kinematic parameters, and to obtain the centers as the mean value of the coordinates corresponding to each sphere as shown in equation (6). Finally, we evaluate all the distances in each iteration of the optimization procedure. The objective function can be formulated as the quadratic sum of all the errors in distances calculated by way of equation (7). Hence an objective function similar to those commonly chosen in robot and AACMMs parameter identification is obtained.

Given the arm positions capture setup used, and the fact that point repeatability in any arm probe orientation is a very important parameter in order to characterize the metrological behavior, unlike traditional expressions, our objective function in equation (11) includes both the errors in distance and the deviation of the points measured in each sphere showing the influence of the volumetric accuracy and point repeatability, minimizing simultaneously the errors corresponding to both parameters.

$$\phi=\sum_{i=1}^{r}\sum_{j,k=1}^{s}\left[\left(D_{i_{jk}}-D_{0jk}\right)^{2}+\left(2\sigma_{Xij}\right)^{2}+\left(2\sigma_{Yij}\right)^{2}+\left(2\sigma_{Zij}\right)^{2}\right] \tag{11}$$

In the objective function proposed, with the capture setup described, r=7 positions of the ball bar and s=4 spheres (1, 6, 10 and 14) per bar position. Again, in equation (11) it is necessary to consider the elimination of the terms in which j=k, in order to avoid the inclusion of null terms or considering as duplicate the influence of the error on distances, taking into account that $D_{i_{jk}}=D_{i_{kj}}$. The first term of equation (11) corresponds to the error in distances in position i of the gauge between sphere j and sphere k, whereas the other terms refer to twice the standard deviation in each of the three coordinates for sphere j in position i of the gauge. Finally, again by mathematical formulation of the optimization problem, it is necessary to consider the sum of all the square errors calculated. With the objective function of equation (11), 126 quadratic error terms will be obtained to calculate the final value of the objective function after each optimization algorithm stage. This value will show the influence of the kinematic parameters as well as of the joint variables through the calculation of the points coordinates corresponding to the arm positions captured in both cases, active and passive probe.

The Levenberg-Marquardt (L-M) method (Levenberg, 1944; Marquardt, 1963) has been chosen as optimization algorithm for parameter identification, given its proven efficiency in robot parameter identification procedures (Goswami et al., 1993; Alici & Shirinzadeh, 2005). The selection of a specific optimization procedure implies to avoid the influence of the mathematical method itself with regards to the data captured on the result. One of the most suitable methods to solve this problem is the L-M algorithm. Table 1 shows the AACMM kinematic model parameters finally identified, based on the initial values and for the objective function of equation (11) and the arm positions considered with the passive self-centering probe. Also, the error values obtained for the identified set of parameters for the passive self centering probe are shown in Table 1.Results of distance errors between centers have been obtained for each of the 6 distances materialized in each of the 7 ball bar positions for the two probes considered. Measured distances for each sphere in the 7 different positions were compared with the distances obtained with the ball bar gauge thus obtaining the error in distance (Fig. 7a), as well as the differences between the distance errors of the active and the passive self centering probes in all 42 positions that were considered (Fig. 7b). In Fig. 7b, a positive difference represents a smaller error in the active probe and in that case this probe is considered better than the passive one. In the case of positions 3, 4 and 7, three spheres were not measured, so a value of zero was assigned in the graphs. From Fig. 7a, we can observe that on average, the error made by the self-centering active probe was less than the one corresponding to the self-centering passive probe; the errors obtained with the active probe, when greater than those corresponding to the passive probe, can be associated to AACMM as it approaches its workspace frontier.

Joint	a_i (mm)	α_i (°)	d_i (mm)	θ_{0i} (°)
1	0.036962	-90.052249	-0.000002	-0.126434
2	0.102485	90.044751	47.891183	14.942165
3	0.097868	-90.020699	645.780523	-88.99688
4	-0.133079	90.068899	54.240741	-3.636896
5	0.057606	90.011014	615.242600	89.770488
6	0.367275	-0.522698	0.150712	-0.878373
	X_{Probe} (mm)	Y_{Probe} (mm)	Z_{Probe} (mm)	
	0.367276	139.450887	54.657060	

Distance Error (mm)		2σ by Sphere (mm)	
Max.	**0.144**	**Max.**	**0.249**
Causing Pos.	POS2	Causing Pos.	POS 1
Causing Dist.	D1	Causing Sph.	B1
Min.	0.006	Causing Coord.	Z
Causing Pos.	POS1	Min.	0.035
Causing Dist.	D2	Causing Pos.	POS4
Med.	**0.066**	Causing Sph.	B6
		Causing Coord.	Y
		Med.	**0.104**

Table 1. Identified values for the model parameters by L-M algorithm and quality indicators for these parameters over 7 ball bar locations with equation (11) as objective function. Data from passive probe.

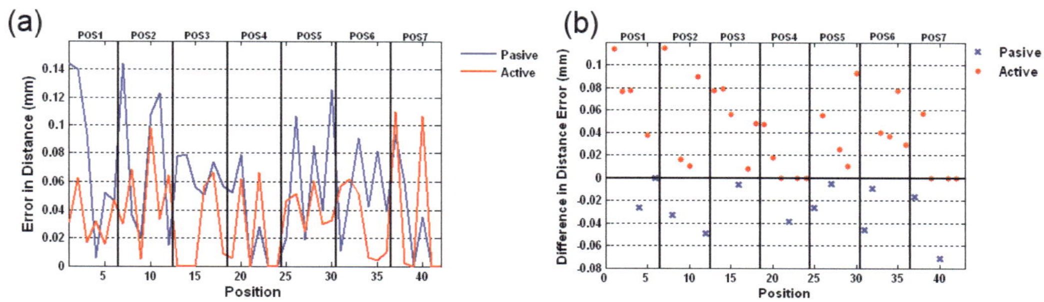

Fig. 7. Comparison between passive and active self-centering probes with the identified parameters in each case over the identification data: (a) Error in distance of the centers measured, (b) Difference in distance errors.

The repeatability error values for all measured points are shown in Fig. 8a and 8b, for the self-centering active probe and self-centering passive probe respectively. These values represent the errors made in *X*, *Y* and *Z* coordinates of each one of the approximately 10000 points obtained with each probe, corresponding to the 7 positions of the ball-bar gauge with regards to the mean obtained for each sphere. The repeatability error value for each coordinate as a function of the 6 joint rotation angles is given by equation (12). This information can also be used to obtain empirical error correction functions as a function of the angles (Santolaria et al., 2008).

$$\begin{aligned}
\varepsilon_{Xijk}(\theta_1,\theta_2,\theta_3,\theta_4,\theta_5,\theta_6) &= \overline{X}_{ij} - X_{ij} \\
\varepsilon_{Yijk}(\theta_1,\theta_2,\theta_3,\theta_4,\theta_5,\theta_6) &= \overline{Y}_{ij} - Y_{ij} \\
\varepsilon_{Zijk}(\theta_1,\theta_2,\theta_3,\theta_4,\theta_5,\theta_6) &= \overline{Z}_{ij} - Z_{ij}
\end{aligned} \tag{12}$$

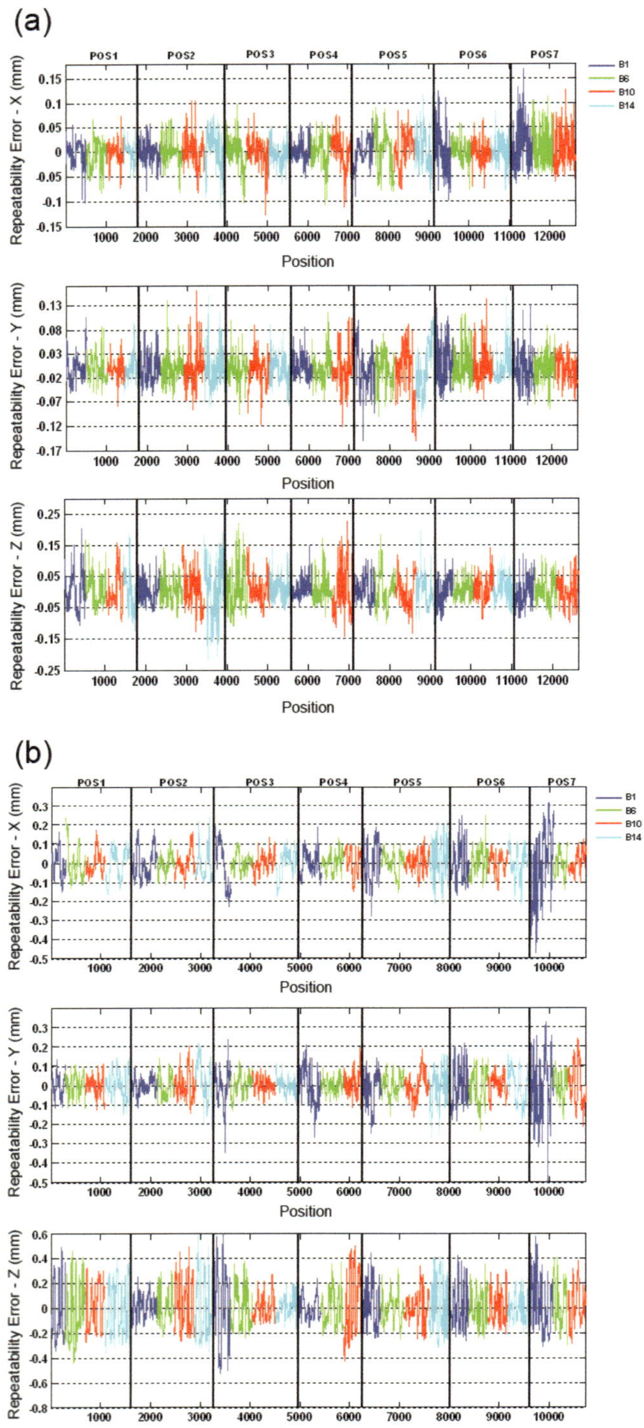

Fig. 8. Point repeatability errors for the optimal sets of model parameters over identification AACMM positions: (a) Active probe, (b) Passive probe.

It can be observed that the error made by the self-center active probe is a lot smaller than the error made by the self-center passive probe and that in both graphs the error shows an increment in the Z coordinate. This behavior in the Z coordinate, could be explained by the fact that, unlike what happens in the X and Y coordinates, there is no self-compensation effect in the gauge deformation due to the probing force in this coordinate.

In Fig. 9 we can observe the standard deviation corresponding to the 7 different positions in X, Y and Z for both types of probes. As expected, the standard deviation in the self-centering active probe is smaller than the one obtained with the self-centering passive probe, except as mentioned earlier, in the positions were spheres were not measured and a value of zero was assigned in the graph.

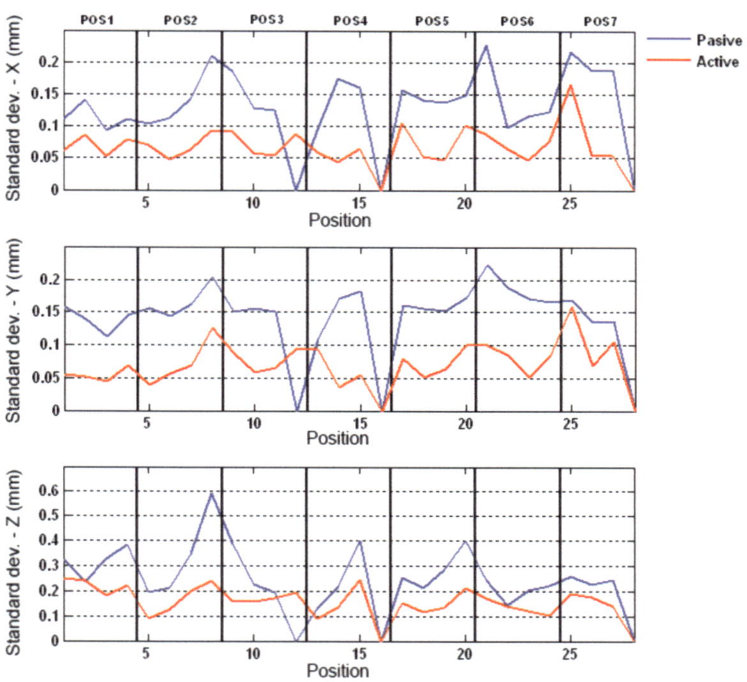

Fig. 9. Standard deviation of the center of the spheres probed.

In order to study the influence of the inclusion of the standard deviation on the objective function, we have complete optimizations taking as function only the terms corresponding to the error in distances for the 10,780 positions captured with the passive probe, as would correspond to a common objective function for parameter identification of robots.

$$\phi = \sum_{i=1}^{r} \sum_{j,k=1}^{s} \left[\left(D_{i_{jk}} - D_{0\,jk} \right)^2 \right] \tag{13}$$

Compared to the maximum and mean error obtained in Table 1, using equation (13) as objective function, a maximum error of 15 µm was obtained and a mean error of 5 µm for the same arm positions. However, for the parameters identified with the objective function of equation (13), the maximum value obtained for 2σ is 1.8932 mm compared to 0.249 mm obtained using equation (11), and the mean value is 1.009 mm. As can be seen in the results, an optimization equivalent to those commonly found in robots produces excellent results for errors in distance but inadequate results for range and standard deviation. Hence, to obtain a set of parameters which allows the arm to be repeatable in a point for any measurement orientation and not only in the orientation captured for optimization, it is necessary to consider the range or the standard deviation in objective function. There may exist cases of robot arms in which an optimization scheme without considering repeatability evaluation parameters is useful for work positions and orientations similar to those used in identification. However, in general for robots and always in the case of AACMM parameter identification, regarding the standard deviation results, the traditional objective functions should be completed with repeatability evaluation parameters, obtaining kinematical parameters that makes more reliable the generalization to the measurement volume of the error values obtained.

5. Generalization tests with the identified sets of parameters

The generalization of an identified set of parameters to the rest of the measurement volume involves the obtaining of deviation and error values smaller than the maximums obtained for the identification process for any arm position. For this reason, the use of at least one test position different to the identification positions is recommended. Thereby, the maximum error for the identification positions, in those cases in which a lower number of gauge or arm positions have been taken, has proven to be better than that finally considered as optimum. However, in these conditions, the evaluation of the identified parameters on positions not considered before has resulted in worse values than those obtained in identification with consideration of all the positions captured. For this reason, the use of all the positions captured as representative of the arm measurement volume was the option taken. Thus, a sufficiently representative set is obtained in order to absorb all the influences on the final error and to obtain a set of kinematic parameters which make the error obtained in identification be realistic and truly the maximum for the arm for any position in the measurement volume. As is shown in Table1, a maximum error of 144 μm and a mean error of 66 μm are obtained for all the measurement volume of quadrant 1 with the passive probe. This can be compared to maximum error in distances (0.854 mm) and to mean error (0.262 mm) obtained in one single evaluation position in the initial situation. In normal operation of the arm - probing discrete points of the center of the sphere probe - the error obtained with the identified set of parameters for the passive probe will be normally around the mean value of 66 μm, producing the maximum error in certain specific arm positions.

Once the optimization process is complete, as the final stage of the presented parameter identification procedure, it is necessary to evaluate the behavior of the arm with the optimum set of parameters on arm positions different to those used during identification. The more similar the evaluation positions subsequent to those used in identification, the better the results. Hence, it is necessary to find different measurement arm positions to evaluate the level of fulfillment of the error values obtained in other measurement volume positions.

In this case, as test bar location subsequent to identification was chosen in the upper part of quadrant 1. Based on the same orientation of position P1, the bar was rotated approximately 25° both horizontally and vertically. For this ball bar location, angle combinations corresponding to the arm positions probing the centers of the 14 gauge spheres were captured for both probes. In this way around 6.000 arm positions were captured for the test position for each probe, which is a reliable check of the measurement arm error on positions not used. Table 2 shows the error values obtained for the 14 test position spheres.

(a)

Distance Error (mm)		2σ by Sphere (mm)	
Max.	**0,128**	**Max.**	**0,202**
Causing Pos.	POS_TEST	Causing Pos.	POS_TEST
Causing Dist.	D1-10	Causing Sph.	B14
Min.	0,012	Causing Coord.	Z
Causing Pos.	POS_TEST	Min.	0,027
Causing Dist.	D1-2	Causing Pos.	POS_TEST
Med.	0,068	Causing Sph.	B3
		Causing Coord.	X
		Med. X	**0,054**
		Med. Y	**0,069**
		Med. Z	**0,141**

(b)

Distance Error (mm)		2σ by Sphere (mm)	
Max.	**0,083**	**Max.**	**0,167**
Causing Pos.	POS_TEST	Causing Pos.	POS_TEST
Causing Dist.	D1-12	Causing Sph.	B13
Min.	0,007	Causing Coord.	Z
Causing Pos.	POS_TEST	Min.	0,014
Causing Dist.	D1-2	Causing Pos.	POS_TEST
Med.	**0,024**	Causing Sph.	B2
		Causing Coord.	X
		Med. X	**0,046**
		Med. Y	**0,053**
		Med. Z	**0,112**

Table 2. Quality indicators for the identified sets of model parameters over 14 spheres of ball bar test location: (a) Passive probe, (b) Active probe.

As can be seen in the results obtained, the mean error is of the same order as in the identification positions and the maximum is below the maximum obtained in that case for both probes. It should be considered that the maximum values of standard deviation are obtained in the end spheres of the gauge, in more forced positions of the measurement arm. Given that we check the error values in one single test position of the gauge, better results in the arm behavior could be expected. However, it should be remembered that, for this ball bar location, over 6.000 arm positions are evaluated, both from the point of view of point repeatability and error in distances based on the calculation of the mean point probed for each sphere. For this reason, as the conclusion of the evaluation test, the importance of the data captured should be again emphasized. A high number of arm positions, different to those chosen for identification, should be searched in the way recommended in normalized evaluation test, in order to conclude with the acceptance of the identified model parameters. In this case, the number of arm positions considered for evaluation is high compared to those used in identification, obtaining values below the maximum error, meaning the arm behavior is verified in accordance with these maximum errors within the volume considered.

6. Application to kinematic calibration of robot arms with active self-centering probes

This section describes the application of the identification method presented to robot arms. Due to its automatic movement, it is not appropriate in this case to probe the spheres of the gauge with a self-centering passive probe. Influences of probing force or incorrect position of the robot's hand are removed by using a self-centering active probe (Fig. 10).

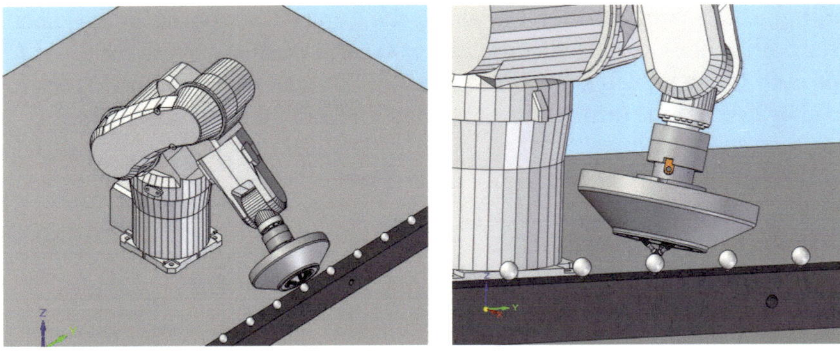

Fig. 10. Self-centering active probe in a robot arm.

Both the data capture procedure and the identification are the same as the ones presented in section four for AACMMs, so it is necessary to capture points of several spheres of the gauge at various gauge positions distributed within the workspace of the robot. This makes it necessary manual probing of the first and last sphere in each gauge position, to know its center coordinates in robot reference system. Once these coordinates are known, it is possible to automatically generate the measuring program for a gauge position. Thus, from the nominal positions of the spheres of the gauge expressed in robot reference system, it is possible to generate the probing trajectories of each sphere through the inverse kinematics model (Fig. 11). As a result, the inverse model will provide the position and orientation of the robot's hand. At each point of the trajectory, by inverse kinematics, it should be captured the maximum possible robot positions for this position and orientation of the hand. This will capture all the possible influences of the joints on the position and orientation at each probing point.

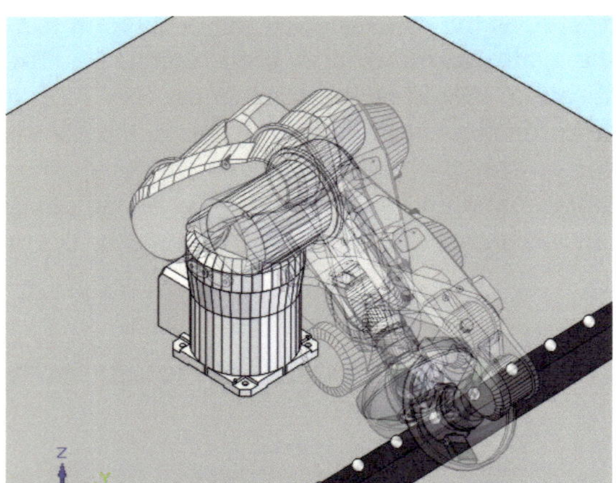

Fig. 11. Several probing poses obtained by inverse kinematics for a gauge sphere.

After probing the selected spheres of all the positions of the gauge, we will have information related to both volumetric accuracy and point repeatability. So, with the objective function of equation (11) and the described procedure it is possible to identify the parameters of the kinematic model of the robot. This procedure will lead to a set of parameters that will improve the accuracy of the robot throughout its workspace, considering also its ability to reach a point from many different postures, unlike the procedures that identify parameters only in some specific working positions of the robot.

6.1 Linking the mathematical model of the probe with the kinematic model of the robot

As discussed above, the self-centering active probe obtains in its reference system the coordinates of the probed sphere center from the readings of displacement of its three styli. Therefore it is necessary to obtain the homogeneous transformation matrix that relates the coordinates of a sphere expressed in the probe reference system with the coordinates of the same sphere in the last reference system of the robot arm, once mounted the probe. This will provide the coordinates of the sphere center in the global reference system of the robot. This can be achieved following several methods, all of them based on least squares.
This section presents the method for obtaining this homogeneous matrix from the probing of a single sphere. This self-calibration method will allow obtaining the matrix that relates the two reference systems without knowing the coordinates of the probed sphere in robot reference system. Assuming a robot with six joints, the equation (14) obtains the coordinates of the center of a sphere in robot reference system from the coordinates of the sphere expressed in probe reference system.

$$\begin{bmatrix} X \\ Y \\ Z \\ 1 \end{bmatrix}_{0\_ROBOT} = {}^{0}T_{6\_i} \cdot P \cdot \begin{bmatrix} X \\ Y \\ Z \\ 1 \end{bmatrix}_{PROBE\_i} \tag{14}$$

where $[X\ Y\ Z\ 1]^T_{PROBE\_i}$ are the coordinates of the probed sphere expressed in the probe reference system; P is the 4x4 matrix that relates the probe frame with the last joint frame of the robot, constant for any position; ${}^{0}T_{6\_i}$ is the 4x4 robot matrix in the probing pose i, that relates coordinates in the last joint frame of the robot with coordinates expressed in the base frame; and $[X\ Y\ Z\ 1]^T_{0\_ROBOT}$ are the coordinates of the probed sphere center in robot base frame, invariants for any position and orientation of the robot.
In equation (14), both robot matrix and the coordinates of the points probed expressed in the probe frame are known for each probing posture, while P matrix and the coordinates of the sphere center expressed in robot reference system are unknown. Thus, it is possible to propose a least squares resolution for the overdetermined system of equation (15).

$$Aq = b \tag{15}$$

where

$$A = \begin{bmatrix} t_{11}X_{P\_i} & t_{11}Y_{P\_i} & t_{11}Z_{P\_i} & t_{11} & t_{12}X_{P\_i} & t_{12}Y_{P\_i} & t_{12}Z_{P\_i} & t_{12} & t_{13}X_{P\_i} & t_{13}Y_{P\_i} & t_{13}Z_{P\_i} & t_{13} & -1 & 0 & 0 \\ t_{21}X_{P\_i} & t_{21}Y_{P\_i} & t_{21}Z_{P\_i} & t_{21} & t_{22}X_{P\_i} & t_{22}Y_{P\_i} & t_{22}Z_{P\_i} & t_{22} & t_{23}X_{P\_i} & t_{23}Y_{P\_i} & t_{23}Z_{P\_i} & t_{23} & 0 & -1 & 0 \\ t_{31}X_{P\_i} & t_{31}Y_{P\_i} & t_{31}Z_{P\_i} & t_{31} & t_{32}X_{P\_i} & t_{32}Y_{P\_i} & t_{32}Z_{P\_i} & t_{32} & t_{33}X_{P\_i} & t_{33}Y_{P\_i} & t_{33}Z_{P\_i} & t_{33} & 0 & 0 & -1 \\ \vdots & \vdots & \vdots & \vdots & \vdots & \vdots & \vdots & \vdots & \vdots & \vdots & \vdots & \vdots & \vdots & \vdots & \vdots \\ t_{11}X_{P\_n} & t_{11}Y_{P\_n} & t_{11}Z_{P\_n} & t_{11} & t_{12}X_{P\_n} & t_{12}Y_{P\_n} & t_{12}Z_{P\_n} & t_{12} & t_{13}X_{P\_n} & t_{13}Y_{P\_n} & t_{13}Z_{P\_n} & t_{13} & -1 & 0 & 0 \\ t_{21}X_{P\_n} & t_{21}Y_{P\_n} & t_{21}Z_{P\_n} & t_{21} & t_{22}X_{P\_n} & t_{22}Y_{P\_n} & t_{22}Z_{P\_n} & t_{22} & t_{23}X_{P\_n} & t_{23}Y_{P\_n} & t_{23}Z_{P\_n} & t_{23} & 0 & -1 & 0 \\ t_{31}X_{P\_n} & t_{31}Y_{P\_n} & t_{31}Z_{P\_n} & t_{31} & t_{32}X_{P\_n} & t_{32}Y_{P\_n} & t_{32}Z_{P\_n} & t_{32} & t_{33}X_{P\_n} & t_{33}Y_{P\_n} & t_{33}Z_{P\_n} & t_{33} & 0 & 0 & -1 \end{bmatrix} \tag{16}$$

$$b = \begin{bmatrix} -t_{14} \\ -t_{24} \\ -t_{34} \\ \vdots \\ -t_{14} \\ -t_{24} \\ -t_{34} \end{bmatrix} \tag{17}$$

As a result of the resolution of the previous system, the vector $q = \begin{bmatrix} p_{11} & p_{12} & p_{13} & p_{14} & p_{21} & p_{22} & p_{23} & p_{24} & p_{31} & p_{32} & p_{33} & p_{34} & X_R & Y_R & Z_R \end{bmatrix}$ is obtained. This vector contains the searched terms of the P matrix and also the coordinates of the sphere center in robot reference system. It is possible to follow the same resolution strategy but probing more than one sphere and introducing the corresponding nominal distance constrains between gauge spheres, leading to a more accurate solution in fewer iterations.

7. Conclusions

In this chapter, a comparison between two different probing systems applied to capturing data for parameter identification and verification of AACMM is presented. Besides the probing systems traditionally used in the verification of AACMM, self-centering probing systems with kinematic coupling configuration and self-center active probing systems have also been used for the presented method. Such probing systems are very suitable for use in verification procedures and capturing data for parameter identification, because they drastically reduce the capture time and the required number of positions of the gauge as compared to the usual standard and manufacturer methods. These systems are also very suitable for their capacity of capturing multiple positions of the AACMM for a single gauge position, so that the accuracy results obtained after a procedure of identification or verification are more generalizable than those obtained with the traditional probing systems.

The effect of auto compensation of the gauge deformation has been shown by properly defining the trajectories of capture or the direction of probing during the process of capturing data. Moreover, it has been demonstrated that the smallest influence of the probing force is obtained in the case of the self-centering active probe, this being the most adequate system in tasks of verification or capturing data for the identification of kinematic parameters if no configuration or application restrictions are imposed, specially for robot arms.

8. References

Alici, G. & Shirinzadeh, B. (2005). A systematic technique to estimate positioning errors for robot accuracy improvement using laser interferometry based sensing. *Mechanism and Machine Theory*, 40, 879-906

Borm, J.H. & Menq, C.H. (1991). Determination of optimal measurement configurations for robot calibration based on observability measure. *International Journal of Robotics Research*, 10(1), 51–63

Caenen, J.L. & Angue, J.C. (1990). Identification of geometric and non geometric parameters of robots. *Proceedings of the IEEE International Conference on Robotics and Automation*, 2, pp. 1032-1037

Chen, J. & Chao, L.M. (1986). Positioning error analysis for robot manipulator with all rotary joints. *IEEE International Conference On Robotics And Automation*, 2, pp.1011-1016

Chunhe, G.; Jingxia, Y. & Jun, N. (2000). Nongeometric error identification and compensation for robotic system by inverse calibration. *International Journal of Machine Tools and Manufacture*, 40, 2119-2137

Denavit, J. & Hartenberg, R.S. (1955). A kinematic notation for lower-pair mechanisms based on matrices. *Journal of Applied Mechanics, Transactions of the ASME*, 77, 215-221

Driels, M.R. & Pathre, U.S. (1990). Significance of observation strategy on the design of robot calibration experiments. *Journal of Robotic Systems*, 7(2), 197–223

Drouet, P.H.; Dubowsky, S.; Zeghloul, S. & Mavroidis, C. (2002). Compensation of geometric and elastic errors in large manipulators with an application to a high accuracy medical system. *Robotica*, 20(3), 341-352

Everett, L.J.; Driels, M. & Mooring, B.W. (1987). Kinematic modelling for robot calibration. *IEEE International Conference on Robotics and Automation*, 1, pp. 183-189

Everett, L.J. & Suryohadiprojo, A.H. (1988). A study of kinematic models for forward calibration of manipulators. *IEEE International Conference of Robotics and Automation*, pp. 798-800

Furutani, R.; Shimojima, K. & Takamasu, K. (2004). Parameter calibration for non-cartesian CMM. *VDI Berichte*, 1860, 317-326

Goswami, A. & Bosnik, J.R. (1993). On a relationship between the physical features of robotic manipulators and the kinematic parameters produced by numerical calibration. *Journal of Mechanical Design, Transactions of the ASME*, 115(4), 892-900

Goswami, A.; Quaid, A. & Peshkin, M. (1993). Identifying robot parameters using partial pose information. *IEEE Control Systems Magazine*, 13(5), 6–14

Hayati, S. (1983). Robot arm geometric link parameter estimation. Proceedings of the IEEE Conference on Decision and Control, 3, pp. 1477-1483

Hayati, S. & Mirmirani, M. (1985). Improving the absolute positioning accuracy of robot manipulators. *Journal of Robotic Systems*, 2(4), 397-413

Hollerbach, J.M. & Wampler, C.W. (1996). The calibration index and taxonomy for robot kinematic calibration methods. *International Journal of Robotics Research*, 15(6), 573-591

Hsu, T.W. & Everett, L.J. (1985). Identification of the kinematic parameters of a robot manipulator for positional accuracy improvement. *Computers in Engineering, Proceedings of the International Computers in Engineering Conference and exhibition*, 1, pp. 263-267

Kovac, I. & Frank, A. (2001). Testing and calibration of coordinate measuring arms. *Precision Engineering*, 25(2), 90-99

Kovac, I. & Klein, A. (2002). Apparatus and a procedure to calibrate coordinate measuring arms. *Journal of Mechanical Engineering*, 48(1), 17-32

Levenberg, K. (1944). A method for the solution of certain non-linear problems in least squares. *Quarterly of Applied Mathematics-Notes*, 2(2), 164-168

Lin, S.W.; Wang, P.P.; Fei, Y.T. & Chen, C.K. (2006). Simulation of the errors transfer in an articulation-type coordinate measuring machine. *International Journal of Advanced Manufacturing Technology*, 30, 879-886

Marquardt, D.W. (1963). An algorithm for least-squares estimation of nonlinear parameters. *Journal of the Society for Industrial and Applied Mathematics*, 11(2), 431-441

Mooring, B.W. (1983). The effect of joint axis misalignment on robot positioning accuracy. *Computers in Engineering, Proceedings of the International Computers in Engineering Conference and Exhibit*, 2, pp. 151-155

Mooring, B.W. & Tang, G.R. (1984). An improved method for identifying the kinematic parameters in a six axis robots. *Computers in Engineering, Proceedings of the International Computers in Engineering Conference and Exhibit*, 1, pp. 79-84

Park, F.C. & Brockett, R.W. (1994). Kinematic dexterity of robotic mechanisms. *International Journal of Robotics Research*, 13 (1), 1-15

Roth, Z.S.; Mooring, B.W. & Ravani, B. (1987). An overview of robot calibration. *IEEE Journal of Robotics and Automation*, 3(5), 377-385

Santolaria, J.; Aguilar, J.J.; Yagüe, J.A. & Pastor, J. (2008). Kinematic parameter estimation technique for calibration and repeatability improvement of articulated arm coordinate measuring machines. *Precision Engineering*, 32, 251-268

Sheth, P.N. & Uicket, J.J. (1971). A generalized symbolic notation of mechanism. *ASME Journal of engineering for industrial*, 102-112

Stone, H.W.; Sanderson, A.C. & Neumann, C.F. (1986). Arm signature identification. *IEEE International Conference On Robotics And Automation*, 1, pp. 41-48

Trapet, E. & Wäldele, F. (1991). A reference object based method to determine the parametric error components of coordinate measuring machines and machine tools. *Measurement*, 9(1), 17-22

Vincze, M.; Spiess, S.; Parotidis, M. & Götz, M. (1999). Automatic Generation of Non Redundant and Complete Models for Geometric and Non Geometric Errors of Robots. *International Journal of Modelling and Simulation*, 19(3), 236-243

Whitney, D.E.; Lozinski, C.A. & Rourke, J.M. (1984). Industrial robot calibration method and results, *Computers in Engineering, Proceedings of the International Computers in Engineering Conference and Exhibit*, 1, pp. 92-100

Ye, D.; Che, R.S. & Huang, Q.C. (2002). Calibration for kinematics parameters of articulated CMM, *Proceedings of the Second International Symposium on Instrumentation Science and Technology*, 3, pp. 145-149

Zhuang, H. & Roth, Z.S. (1992). Robot calibration using the CPC error model. *Journal of Robotics and Computer Integrated Manufacturing*, 9(3), 227-237

15

Two Cooperating Manipulators with Fractional Controllers

N. M. Fonseca Ferreira
Department of Electrical Engineering, Institute of Engineering of Coimbra
Rua Pedro Nunes, 3031-601 Coimbra, Portugal
nunomig@isec.pt

J. A. Tenreiro Machado
Department of Electrical Engineering, Institute of Engineering of Porto
Rua Dr. António Bernardino de Almeida, 4200-072 Porto, Portugal
jtm@isep.ipp.pt

József K. Tar
Budapest Tech, John von Neumann, Faculty of Informatics
Inst. of Intelligent Engineering Systems
Bécsiút 96/B, H-1034, Budapest, Hungary
tar.jozsef@nik.bmf.hu

Abstract
This paper analyzes the dynamic performance of two cooperative robot manipulators. It is studied the implementation of fractional-order algorithms in the position/force control of two cooperating robotic manipulators holding an object. The simulations reveal that fractional algorithms lead to performances superior to classical integer-order controllers.

1. Introduction

Two robots carrying a common object are a logical alternative for the case in which a single robot is not able to handle the load. The choice of a robotic mechanism depends on the task or the type of work to be performed and, consequently, is determined by the position of the robots and by their dimensions and structure. In general, the selection is done through experience and intuition; nevertheless, it is important to measure the manipulation capability of the robotic system (Y. C. Tsai & A.H Soni., 1981) that can be useful in the robot operation. In this perspective it was proposed the concept of kinematic manipulability measure (T. Yoshikawa, 1985) and its generalization to dynamical manipulability (H. Asada, 1983) or, alternatively, the statistical evaluation of manipulation (J. A. Tenreiro. Machado & A. M. Galhano, 1997). Other related aspects such as the coordination of two robots handling

objects, collision avoidance and free path planning have been also investigated (Y. Nakamura, K. Nagai, T. Yoshikawa, 1989) but they still require further study.
With two cooperative robots the resulting interaction forces have to be accommodated and consequently, in addition to position feedback, force control is also required to accomplish adequate performances (T. J. Tarn, A. K. Bejczy, P. K., 1996) and (N. M. Fonseca Ferreira, J. A. Tenreiro Machado, 2000) and (A. K. Bejczy and T. Jonhg Tarn, 2000). There are two basic methods for force control, namely the hybrid position/force and the impedance schemes. The first method (M. H. Raibert and J. J. Craig, 1981) separates the task into two orthogonal sub-spaces corresponding to the force and the position controlled variables. Once established the subspace decomposition two independent controllers are designed. The second method (N. Hogan, 1985) requires the definition of the arm mechanical impedance. The impedance accommodates the interaction forces that can be controlled to obtain an adequate response. Others authors (Kumar, Manish; Garg, Devendra 2005, Ahin Yildirim, 2005, Jufeng Peng, Srinivas Akella, 2005) present advance methodologies to optimize the control of two cooperating robots using the neural network architecture and learning mechanism to train this architecture online. This paper analyzes the manipulation and the payload capability of two arm systems and we study the position/force control of two cooperative manipulators, using fractional-order (*FO*) algorithms (J. A. Tenreiro Machado, 1997) and (N. M. Fonseca Ferreira & J. A. Tenreiro Machado 2003, 2004 and 2005).
Bearing these facts in mind this article is organized as follows. Section two presents the controller architecture for the position/force control of two robotic arms. Based on these concepts, section three develops several simulations for the statistical analysis and the performance evaluation of *FO* and classical *PID* controllers, for robots having several types of dynamic phenomena at the joints. Finally, section four outlines the main conclusions.

2. Control of Two Arms

The dynamics of a robot with n links interacting with the environment is modelled as:

$$\boldsymbol{\tau} = \mathbf{H(q)\ddot{q}} + \mathbf{C(q,\dot{q})} + \mathbf{G(q)} - \mathbf{J}^{\mathrm{T}}\mathbf{(q)F} \tag{1}$$

where $\boldsymbol{\tau}$ is the $n \times 1$ vector of actuator torques, $\mathbf{q}$ is the $n \times 1$ vector of joint coordinates, $\mathbf{H(q)}$ is the $n \times n$ inertia matrix, $\mathbf{C(q,\dot{q})}$ is the $n \times 1$ vector of centrifugal/Coriolis terms and $\mathbf{G(q)}$ is the $n \times 1$ vector of gravitational effects. The $n \times m$ matrix $\mathbf{J}^{\mathrm{T}}\mathbf{(q)}$ is the transpose of the Jacobian of the robot and $\mathbf{F}$ is the $m \times 1$ vector of the force that the (m-dimensional) environment exerts in the gripper.
We consider two robots with identical dimensions (Fig. 1). The contact of the robot gripper with the load is modelled through a linear system with a mass M, a damping B and a stiffness K (Fig. 2). The numerical values adopted for the *RR* (where *R* denote rotational joints) robots and the object are $m_1 = m_2 = 1.0$ kg, $l_1 = l_2 = l_b = l_0 = 1.0$ m, $\alpha_0 = 0$ deg, $B_1 = B_2 = 1$ Nsm^{-1} and $K_1 = K_2 = 10^4$ Nm^{-1}.

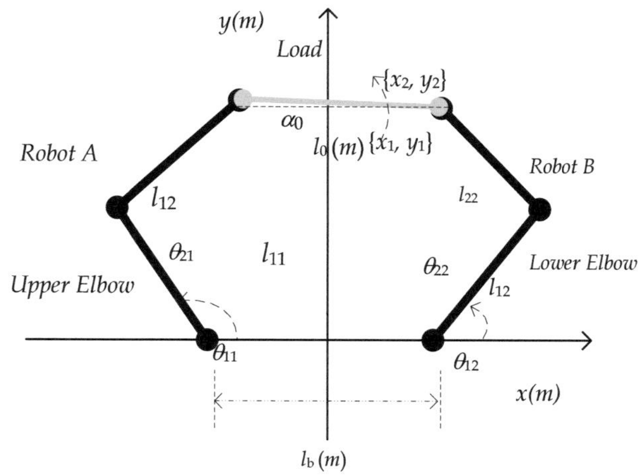

Fig. 1. Two *RR* robots working cooperation for the manipulation of an object with length l_0 and orientation α_0.

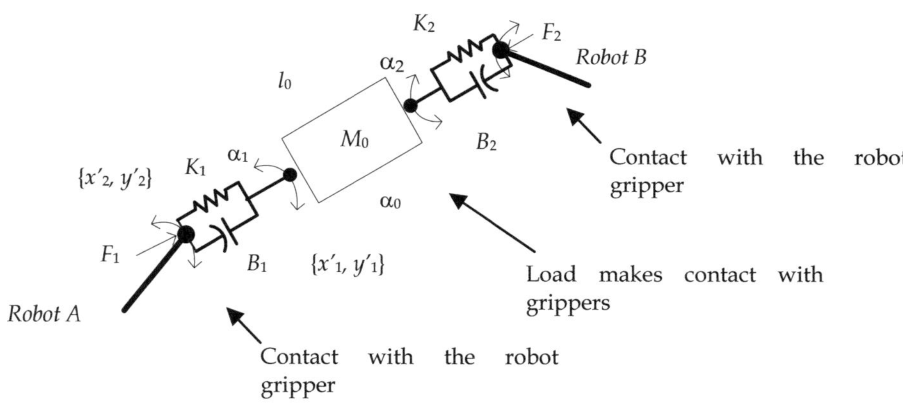

Fig. 2. The contact between the robot gripper and the object.

The controller architecture (Fig. 3), is inspired on the impedance and compliance schemes. Therefore, we establish a cascade of force and position algorithms as internal an external feedback loops, respectively, where x_d and F_d are the payload desired position coordinates and contact forces.

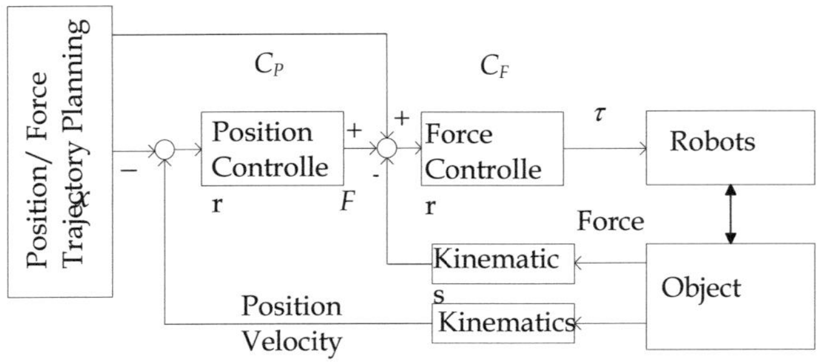

Fig. 3. The position/force cascade controller.

In the position and force control loops we consider *FO* controllers of the type $C(s) = K_p + K_\alpha s^\alpha$, $-1 < \alpha < 1$, that are approximated by 4th order discrete-time Pade expressions (a_i, b_i, $\Re$, $k = 4$):

$$C(z) \approx K_p + K_\alpha \frac{\sum_{k=0}^{4} a_k z^{4-k}}{\sum_{k=0}^{4} b_k z^{4-k}} \tag{2}$$

We compare the response with the classical *PD–PI* algorithms therefore, in the position and force loops we consider, respectively.

$$\begin{gathered} C(s) = K_p + K_d s \\ C(z) = K_p + K_d(1 - z^{-1}) \end{gathered} \tag{3}$$

$$\begin{gathered} C(s) = K_p + K_i \frac{1}{s} \\ C(z) = K_p + K_i \frac{z}{z-1} \end{gathered} \tag{4}$$

Both algorithms were tuned by trial and error, having in mind getting a similar performance in the two cases (Tables 1 and 2).

i	K_p	K_α	α	K_p	K_α	α
1	0.1259	$1.555\ 10^{-3}$	$\frac{1}{2}$	10.59	$2.0\ 10^{-3}$	$-\frac{1}{5}$
2	0.1259	$1.555\ 10^{-3}$	$\frac{1}{2}$	10.59	$2.0\ 10^{-3}$	$-\frac{1}{5}$
	(a) Position controller			(b) Force controller		

Table 1. The parameters of the position and force *FO* controllers.

i	K_p	K_d	K_p	K_i
1	$250\ 10^2$	$2.5\ 10^1$	$5.0\ 10^1$	$10.0\ 10^1$
2	$250\ 10^2$	$2.5\ 10^1$	$5.0\ 10^1$	$10.0\ 10^1$
	(a) Position controller		(b) Force controller	

Table 2. The parameters of the position and force *PD–PI* controllers.

3. Analysis of the system performance

In order to study the system dynamics we apply a small amplitude rectangular pulse δy_d at the position reference and we analyze the system response.

The simulations adopt a controller sampling frequency f_c = 10 kHz, contact forces of the grippers $\{Fx_j, Fy_j\} \equiv \{0.5, 5\}$ Nm, a operating point of the center of the object A $\equiv \{x, y\} \equiv \{0, 1\}$ and a load orientation of $\alpha = 0^o$.

In a first phase we consider robots with ideal transmissions at the joints. Figure 4 depicts the time response of robot A under the action of the *FO* and PD–*PI* algorithms.

In a second phase (figure 5) we analyze the response of robots with dynamic backlash at the joints. For the ith joint (i = 1, 2), with gear clearance h_i, the backlash reveals impact phenomena between the inertias, which obey the principle of conservation of momentum and the Newton law:

$$\dot{q}'_i = \frac{\dot{q}_i(J_{ii} - \varepsilon J_{im}) + \dot{q}_{im} J_{im}(1+\varepsilon)}{J_{ii} + J_{im}} \tag{5}$$

$$\dot{q}'_{im} = \frac{\dot{q}_i J_i (1+\varepsilon) + \dot{q}_{im}(J_{im} - \varepsilon J_{ii})}{J_{ii} + J_{im}} \tag{6}$$

where $0 \le \varepsilon \le 1$ is a constant that defines the type of impact ($\varepsilon = 0$ inelastic impact, $\varepsilon = 1$ elastic impact) and $\dot{q}_i$ and $\dot{q}_{im}$ ($\dot{q}'_i$ and $\dot{q}'_{im}$) are the velocities of the ith joint and motor before (after) the collision, respectively. The parameter J_{ii} (J_{im}) stands for the link (motor) inertias of joint i. The numerical values adopted are $h_i = 1.8\ 10^{-4}$ rad and $\varepsilon_i = 0.8$ (i = 1, 2).

In a third phase (figure 6) we study the *RR* robot with compliant joints. For this case the dynamic model corresponds to model (1) augmented by the equations:

$$\boldsymbol{\tau} = \mathbf{J_m}\ddot{\mathbf{q}}_\mathbf{m} + \mathbf{B_m}\dot{\mathbf{q}}_\mathbf{m} + \mathbf{K_m}(\mathbf{q_m} - \mathbf{q}) \tag{7}$$

$$\mathbf{K_m}(\mathbf{q_m} - \mathbf{q}) = \mathbf{J}(\mathbf{q})\ddot{\mathbf{q}} + \mathbf{C}(\mathbf{q},\dot{\mathbf{q}}) + \mathbf{G}(\mathbf{q}) \tag{8}$$

where $\mathbf{J_m}$, $\mathbf{B_m}$ and $\mathbf{K_m}$ are the $n \times n$ diagonal matrices of the motor and transmission inertias, damping and stiffness, respectively. In the simulations we adopt $\boldsymbol{K}_{mi} = 2\ 10^6$ Nm rad^{-1} and $\boldsymbol{B}_{mi} = 10^4$ Nms rad^{-1} (i = 1, 2).

The time response characteristics (Tables 3 and 4), namely the percent overshoot *PO%*, the steady-state error e_{ss}, the peak time T_p and the settling time T_s reveal that, if we consider similar performances for robots with ideal transmissions at the joints, the *FO* is superior to the *PD–PI* algorithms at the cases of robots with joint dynamic phenomena. I conclusion the *FO* have good dynamic response for position and force perturbations.

N_o	*C(s)*	*PO%*	$e_{ss}[mm]$	$T_p[s]$	$T_s[s]$
1	*PD–PI*	43.0	$5.0\ 10^{-3}$	$2.1\ 10^{-2}$	$15.0\ 10^{-2}$
	FO	39.0	$0.9\ 10^{-3}$	$3.6\ 10^{-2}$	$15.0\ 10^{-2}$
2	*PD–PI*	0.2	$2.7\ 10^{-3}$	$12.2\ 10^{-2}$	$12.0\ 10^{-2}$
	FO	0.3	$3.5\ 10^{-3}$	$10.0\ 10^{-2}$	$10.0\ 10^{-2}$
3	*PD–PI*	0.3	$64.0\ 10^{-2}$	$16.0\ 10^{-2}$	$16.0\ 10^{-2}$
	FO	0.3	$50\ 10^{-3}$	$8.0\ 10^{-2}$	$8.0\ 10^{-2}$

Table 3. Time response characteristics for a pulse δy_d at the robot A position reference.

N_o	*C(s)*	*PO%*	$e_{ss}[mm]$	$T_p[s]$	$T_s[s]$
1	*PD–PI*	400.0	$9.8\ 10^{-1}$	$11.0\ 10^{-2}$	$2.0\ 10^{-1}$
	FO	115.0	$77.0\ 10^{-3}$	$25.0\ 10^{-2}$	$2.0\ 10^{-1}$
2	*PD–PI*	400.0	$9.8\ 10^{-1}$	$15.0\ 10^{-2}$	$10.0\ 10^{-1}$
	FO	100.0	$77.0\ 10^{-3}$	$10.0\ 10^{-2}$	$4.0\ 10^{-1}$
3	*PD–PI*	100.0	$9.8\ 10^{-1}$	$15.0\ 10^{-2}$	$10.0\ 10^{-1}$
	FO	100.0	$77.0\ 10^{-3}$	$10.0\ 10^{-2}$	$4.0\ 10^{-1}$

Table 4. Time response characteristics for a pulse δF_d at the robot A force reference.

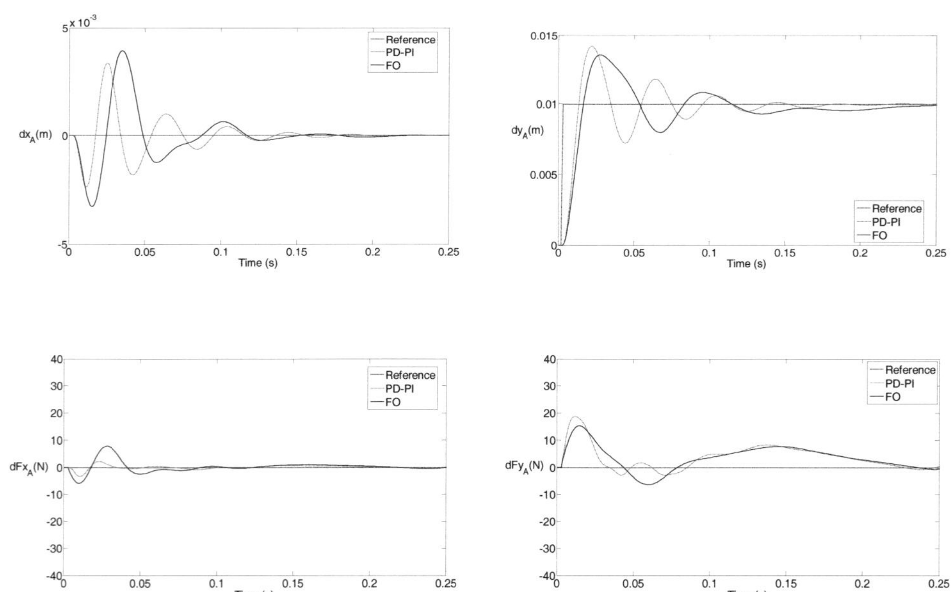

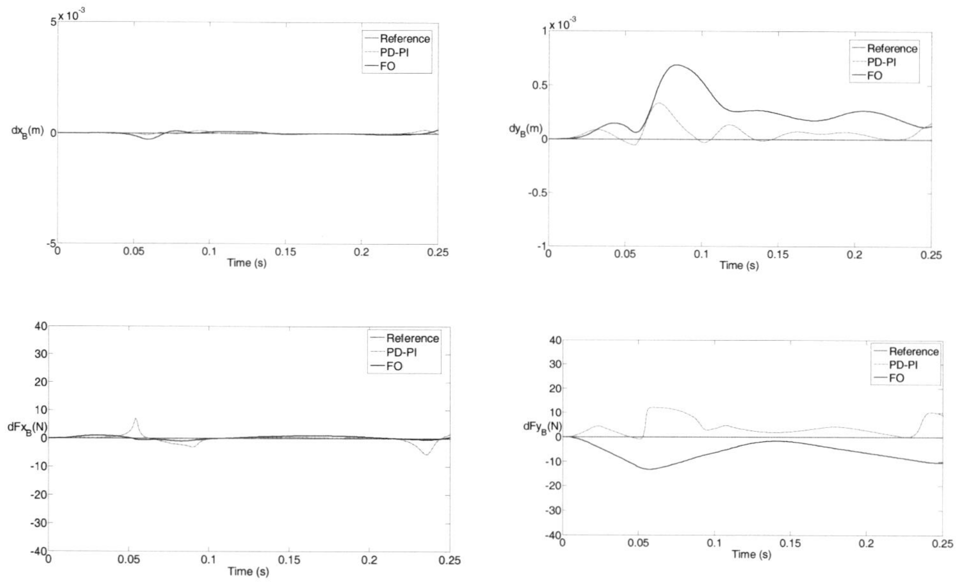

Fig. 4. Time response for robots A and B with ideal joints under the action of the *FO* and the *PD–PI* algorithms for a pulse perturbation at the robot A position reference $\delta y_d = 10^{-3}$ m and a payload with $M = 1$ kg, $B_i = 10$ Nsm^{-1} and $K_i = 10^3$ Nm^{-1}.

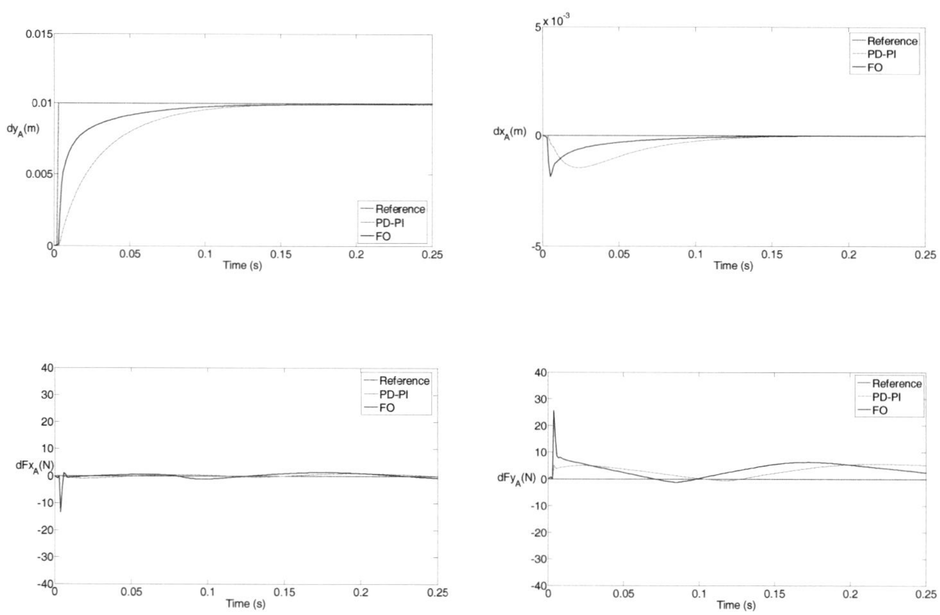

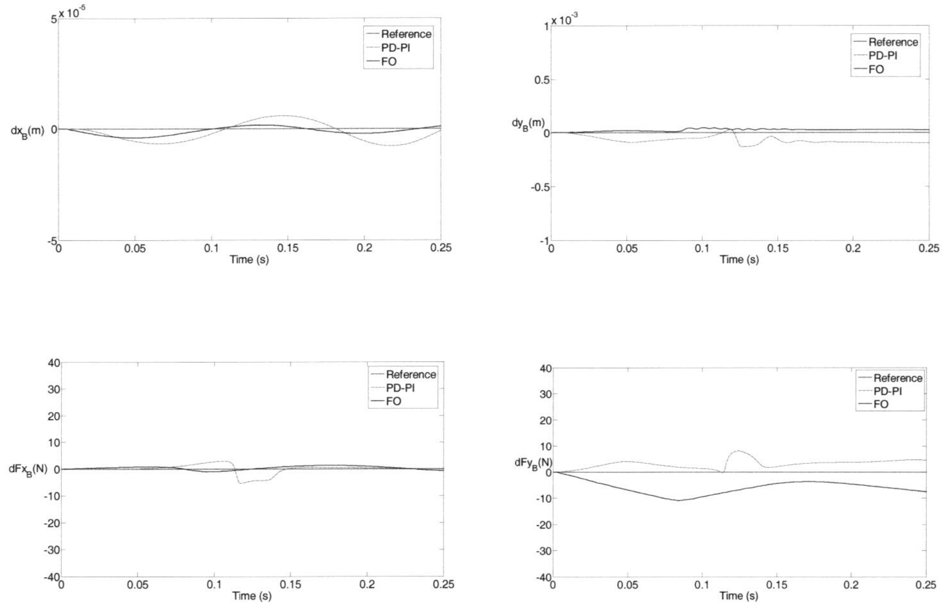

Fig. 5. Time response for robots A and B with joints having backlash under the action of the *FO* and the *PD–PI* algorithms for a pulse perturbation at the robot A position reference $\delta y_d = 0.1$ m and a payload with $M = 1$ kg, $B_i = 10$ Nsm^{-1} and $K_i = 10^3$ Nm^{-1}.

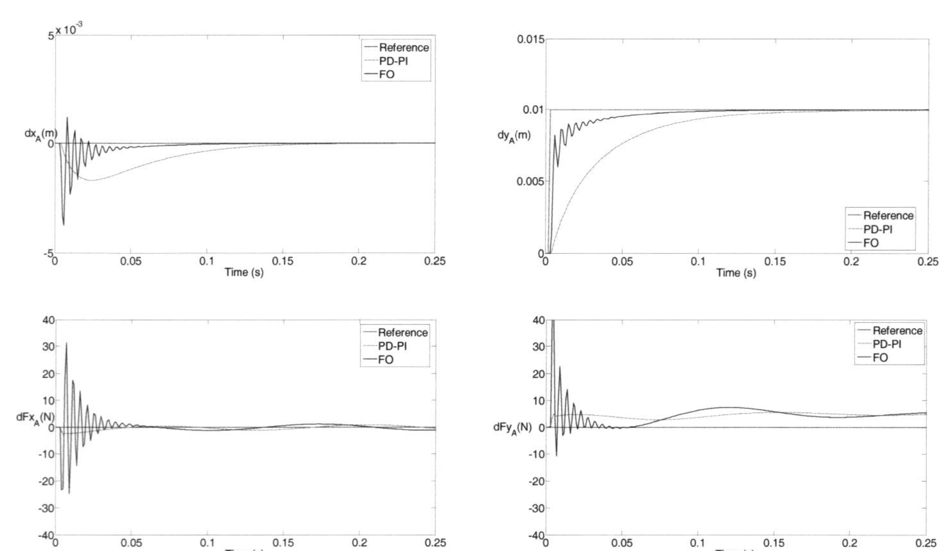

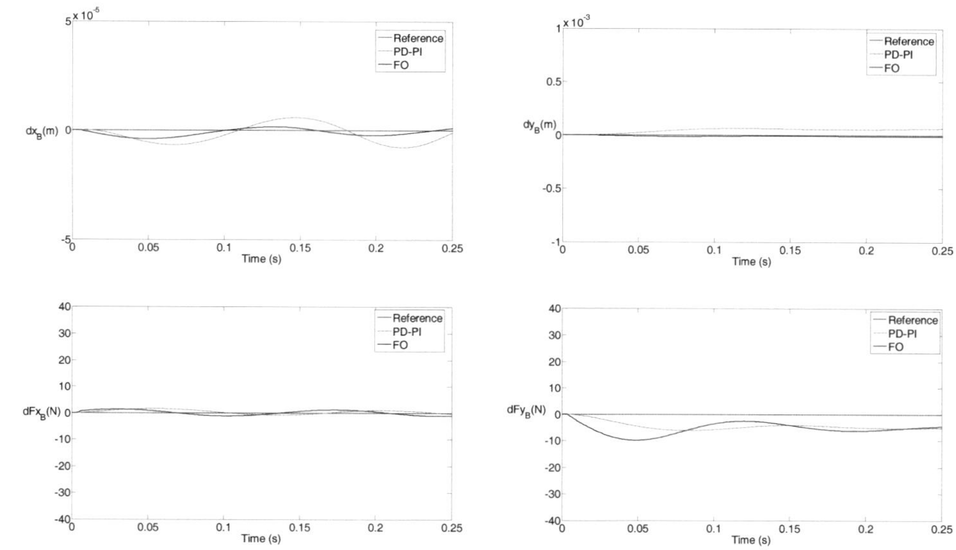

Fig. 6. Time response for robots A and B with joints having flexibility under the action of the *FO* and the *PD–PI* algorithms for a pulse perturbation at the robot A position reference δy_d = 0.1 m and a payload with M = 1 kg, B_i = 10 Nsm^{-1} and K_i = 10^3 Nm^{-1}.

4. Conclusion

This paper studied the position/force control of two robots working in cooperation using fractional and integer order control algorithms. The system time response was analyzed for manipulators having several types of dynamical phenomena at the joints. The transient response of the system shows the superior performance of the FO controller.

5. References

Y. C. Tsai & A.H Soni. (1981) "Accessible Region and Synthesis of Robot Arms". *ASME Journal of Dynamic Systems, Measurement and Control.* Vol. 103, pp. 803-811, 1981.

T. Yoshikawa, (1985) "Manipulability of Robotic Mechanisms". *Int. J. Robotics Research.* Vol. 4, pp. 3-9.

H. Asada, (1983) "A Geometrical Representation of Manipulator Dynamics and its Application to Arm Design". *ASME Journal of Dynamic Systems, Measurement and Control.* Vol. 5, pp. 131-142.

J. A. Tenreiro. Machado & A. M. Galhano, (1997) "A Statistical and Harmonic Model for Robot Manipulators". *In: Proc. of IEEE Int. Conf. on Robotics and Automation,* Albuquerque, New Mexico, USA, pp. 231-242.

J. A. Tenreiro Machado, (1997) "Analysis and Design of Fractional-Order Digital Control Systems". *Journal Systems Analysis, Modelling and Simulation.* Vol. 27, No 1, pp. 107-122.

Y. Nakamura, K. Nagai, T. Yoshikawa, (1989) "Dynamics and Stability in Coordination of Multiple Robotic Mechanisms". *International Journal of Robotics Research.* Vol. 8, pp. 44-61.

T. J. Tarn, A. K. Bejczy, P. K., (1996) "Analysis of the Dynamic Ability of Two Robot Arms in Object Handling". *Advanced Robotics*. Vol. 10, No 10, pp. 301-315.

N. M. Fonseca Ferreira, J. A. Tenreiro Machado, (2000) "Manipulability Analysis of Two-Arm Robotic Systems". *In: Proc. of IEEE International Conference on Intelligent Engineering Systems*. Vol. 1, No 1, pp. 101-109.

A. K. Bejczy and T. Jonhg Tarn, (2000) "Redundancy in Robotics Connected Robots Arms as Redundant Systems". *In: Proc. of IEEE International Conference on Intelligent Engineering Systems*, Vol. 1, No 1, pp. 310-319.

M. H. Raibert and J. J. Craig, (1981) "Hybrid Position/Force Control of Manipulators". *ASME Journal of Dynamic Systems, Measurement and Control.* Vol. 1, pp. 226-233.

N. Hogan, (1985) "Impedance control: An Approach to Manipulation, Parts I-Theory, II-Implementation, III- Applications". *ASME J. of Dynamic Systems, Measurement and Control*. Vol. 107, No 1 pp. 1-24,.

Kumar, Manish; Garg, Devendra P.(2005), "Neuro-fuzzy control applied to multiple cooperating robots", Industrial Robot: An *International Journal*, Volume 32, Number 3, pp. 234-239.

Ahin Yildirim, (2005) "A Proposed Hybrid Recurrent Neural Control System for Two Co-operating Robots", *Journal of Intelligent and Robotic Systems*, v.42 n.1, p.95-111.

Jufeng Peng, Srinivas Akella, (2005) "Coordinating Multiple Robots with Kinodynamic Constraints Along Specified Paths" *The International Journal of Robotics Research* Vol. 24, No. 4, pp. 295-310.

N. M. Fonseca Ferreira & J. A. Tenreiro Machado, (2003) "Fractional-Order Hybrid Control of Robotic Manipulator". *In: Proc. of 10th IEEE Int. Conf. on Advanced Robotics*, Coimbra, Portugal. Vol. 1, No 1, pp. 221-229.

N. M. Fonseca Ferreira, J. A. Tenreiro Machado, J. Boaventura Cunha, "Fractional - Order Position/Force Robot Control". *In: Proc. of 2nd IEEE Int. Conference on Computational Cybernetics*, Vienna, Austria. Vol. 1, No 1, pp. 126-133, 2004.

N. M. Fonseca Ferreira & J. A. Tenreiro Machado, (2005) "Fractional-Order Position/Force Robot Control". *Journal of Advanced Computational Intelligence and Intelligent Informatics*. Vol. 9, No 4, pp. 379-386.

16

MFR (Multi-purpose Field Robot) based on Human-robot Cooperative Manipulation for Handling Building Materials

Seungyeol Lee
Technical University Munich
Germany

1. Introduction

A field robot is defined as one that executes tasks while moving around in a dynamic environment where structures, operators and equipments are constantly changing. The basic elements of a field robot consist of a mobile platform for executing a particular operation in a dynamic environment, sensors and intelligence technology to recognize and cope with barriers in the routes in which they move, and a manipulator for executing a desired operation in place of a human (Lee et al., 1997; Wong & Spetsakis, 2000). Field robots have been designed specifically for a particular environment and used in various industries such as agriculture, construction, engineering, space exploration and deep sea diving, due to the inherent dangers and costs associated with these fields (Hollingum, 1999; Kangari, 1991; LeMaster et al., 2003; Whitcomb, 2000).

To date, the development of field robots has focused on the basic elements, plan for a specific work or a single task. This planning leads to not only the inefficient use of time and resources but also to limited utility. To solve this problem, this paper suggests a 'MFR (Multi-purpose Field Robot)' as shown in Fig. 1. At the end of the 70's, Shimizu Co. developed 'MTV (Multi-purpose Travelling Vehicle for concrete slabs)' which could transport and guide various robotic working modules. From the viewpoint of operational characteristics, the MTV can be thought of as a construction robot designed to perform automatic grinding and cleaning of concrete surfaces. On the contrary, the MFR can be considered as a field robot designed specifically for a particular environment and used in various industries such as agriculture, construction, engineering, space exploration and deep sea diving, due to the inherent dangers and costs associated with these fields (Han, 2005).

A MFR can best be conceived in two parts: a 'basic system' consisting of a manipulator and a mobile platform, and an 'additional module' which includes sensor and intelligence technology to execute particular operations in various areas such as construction, national defense and rescue by changing this additional module. Also, continuous system maintenance and improvement is simplified due to the modularization. In this paper, the development of constituent technology for MFR will be discussed to realize the robotization of building materials handling at construction sites.

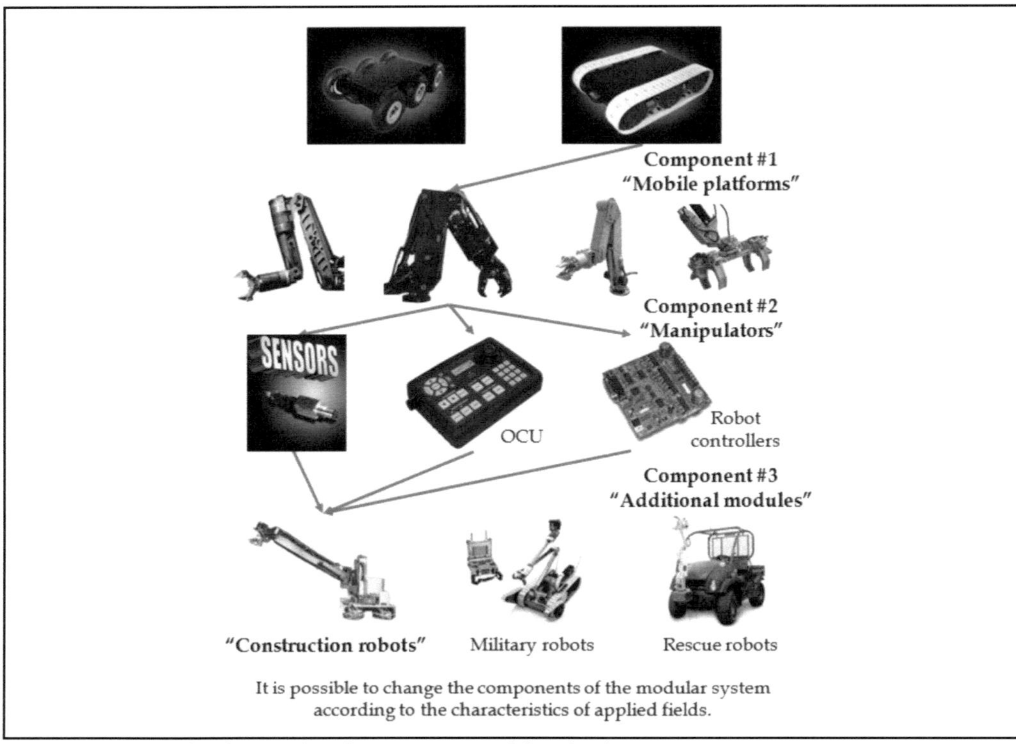

Fig. 1. Framework of MFR (Multi-purpose Field Robot)

Recent research has found that a lack of skilled manpower in the construction industry is rapidly becoming a serious problem. For example, it is estimated that there will be a shortage of about 423,000 skilled laborers by 2010 in case of Korea. This problem of a shrinking workforce, coupled with an aging society, leads to higher wages, a drop in construction quality, project delays, increased costs and the increased likelihood of accidents occurring at construction sites. One of the solutions suggested to solve these problems is robotization and automation in construction. The issue of applying 'Automation System and Robotics in Construction' has been raises as a result of the need for improvement in the safety, productivity, quality and working environment (Roozbeh, 1985; Warszawski, 1985; Albus, 1986; bernold, 1987; Skibniewski, 1988; Wen et al., 1991; Cusack, 1994; Poppy, 1994, Kochan, 2000). Sequentially, operation with automation systems and robots are widely employed at construction sites (Isao et al., 1996; Gambao et al., 2000; Choi et al., 2005; Skibniewski, 1989; Santos et al., 2003; Skibniewski & Wooldridge, 1992; Masatoshi et al., 1996). Generally, almost half of construction work is said to be material handling. Materials and equipment used for construction are heavy and bulky for humans. Handling heavy materials has been, for the most part, eliminated for outside work by cranes and other various lifting equipment. Such equipment, however, is not available for precise work. To address curtain-walls handling needs for precise work, especially, 'ASCI (Automation

System for Curtain-wall Installation)' has been successfully developed and applied as shown in Fig. 2.

Through the case studies on constructions, to which ASCI was applied, however, we could find some factors to be improved. Unlike the automation lines of the general manufacturing industry, construction sites rarely shows repeated operational patterns use to its unstructured processes. That is, construction robots are defined as field robots that execute tasks while operating in a dynamic environment where structures, operators, and equipment are constantly changing. To date, a guidance or remote-controlled system is the natural way to implement construction robot manipulators. However, a remote-controlled system has to solve some problems according to upper working conditions. Firstly, it is difficult to cope with malfunctions immediately when unexpected situation is occurred during construction works. Secondly, there is difficult to gain environmental information for operator's suitable commands.

Fig. 2. An ASCI (Automation System for Curtain-wall Installation)

One of the solutions to address these problems is the technology of 'Human-Robot Cooperative Manipulation (Fukuda et. al., 1991 a; Fukuda et. al., 1991 b)' as shown in Fig. 3. Studies on the human-robot cooperation have been ceaselessly performed so far. In 1960s, the Department of Defense developed the 'Suit of Armor', which enhanced the capability of soldiers in carrying heavy materials (Miller, 1968). In 1962, the Cornell Aeronautical Lab. conducted a study on the 'Master-Slave System', which allowed a man to walk with heavy materials (Miller, 1968). The study on the Master-Slave System led up to the 'Hardiman' of GE that existed from 1966 to 1971 (Mosher, 1967). Afterward, Kazerooni suggested the 'Extender' that, unlike the Master-Slave System, delivered the operational force and information to a robot at the same time by the contact force (Kazerooni, 1989 a; Kazerooni, 1989 b; Kazerooni & Mahoney, 1991). The Extender presented unique features of modeling through using each impedance of the three factors: human arm, extender and environment, and the controller for the operator's force and robot's force. Kosuge suggested a control algorithm for the man-robot cooperation, using maneuverability and amplification factor (Kosuge, 1993). To implement the human-robot cooperation in restricted environments, the

impedance control method, which was proposed by Hogan, has been used as a basic force control method (Hogan, 1985).
To implement this technology, the additional module of MFR that will be combined to the basic system will be explained in two parts, first the hardware and then software. The HRI (Human-Robot Interface) device will be introduced to realize to human-robot cooperative manipulation in the hardware part. We also design a robot control algorithm, for handling building materials in cooperation between an operator and a robot. Especially, considerations on interactions among operator, robot and environment are applied to design of the robot controller. We examined the influences, which the parameters of the impedance model gave to performance of the cooperation system, through an experimental system. An installation method for the specified building material appropriate to the developed system is also suggested and a mock installation is carried out.

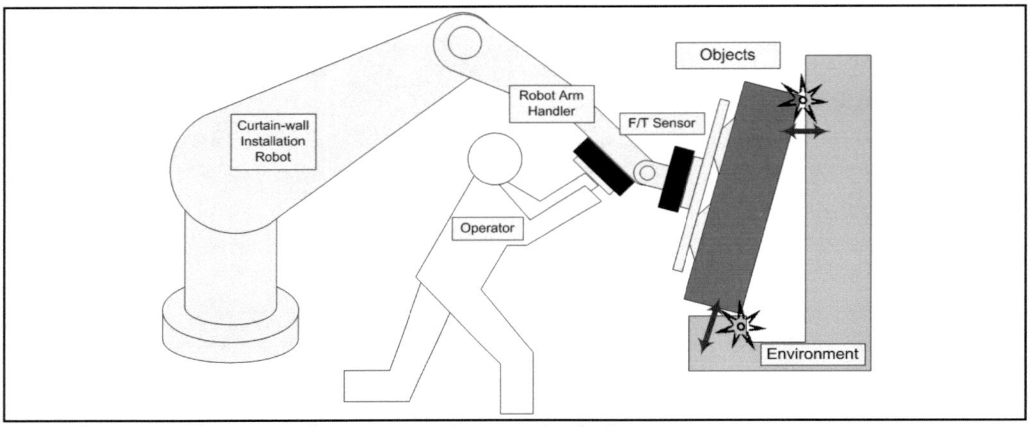

Fig. 3. Concept of human-robot cooperative manipulation

2. Modelling of human-robot cooperative system

Building materials handling works can be divided into 'the environment-contacting cases' and 'non-environment-contacting cases'. During contact with an environment, it acts as a dynamic constraint and affects an operator. These constraining conditions are usually avoidable through controlling actions, but some of phenomena can be considered as 'the virtual dynamic behaviors' against external forces from the environments including the operator. The mechanical relationship, between an external force and the motion toward the external force, is defined as 'the impedance', and the desired dynamic behavior of virtual system is defined as 'the target dynamics'. The operational force is measured by 'the operational force sensor' that is mounted on the last link of a manipulator, and the contact force from an environment is measured by 'the experimental force sensor' that is located between the last link and the end effector. While the case, an operator handles a building material on an obstacle-free place, is defined as 'the unconstrained condition', the case, an operator performs work under interactions with environments, is defined as the operation in 'the constrained condition'.

2.1 Unconstrained condition

In Fig. 4, the force(torque), measured by the operational force sensor which is generated by the operator, is $F_h(T_h)$, and the impedance parameters, that are related to a desired dynamic behavior, are $M_{pt}(M_{ot})$ and $B_{pt}(B_{ot})$ ($n \times n$ positive definite diagonal inertia and damping matrices) respectively. Here, the desired dynamic behavior of a robot can be given, with the input $F_{h,}(T_h)$, by an impedance equation (1). The subscript 'p' stands for the position and 'o' stands for the orientation, and λ means the power assist ratio of an operator.

The K (stiffness matrices) parameter, having the property of a spring, was excluded as it disturbed the operation to move a building material to a desired position with the operational force. Ultimately, adjusting each of the impedance parameters equals to adjusting the dynamic behavior of a virtual system. The dynamic behavior, generated from the impedance equation (1) when an operator applies force to a virtual system, is used as a reference that a robot system should follow to move a building material.

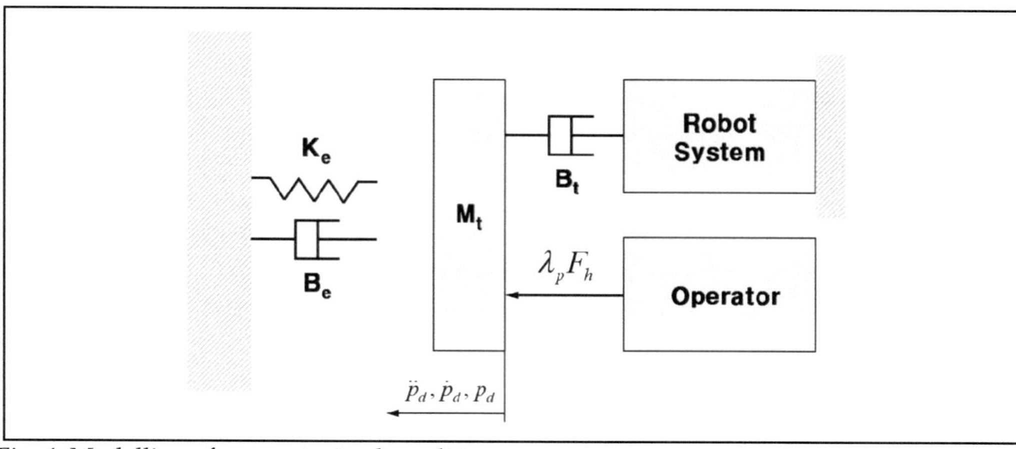

Fig. 4. Modelling of unconstrained condition

$$
\begin{aligned}
& M_{pt}\ddot{p}_d + B_{pt}\dot{p}_d = \lambda_p F_h \\
& M_{ot}\ddot{\varphi}_d + B_{ot}\dot{\varphi}_d = \lambda_o T^T(\varphi_d)T_h \\
& where,\ \ \varphi = \begin{bmatrix} \alpha & \beta & \gamma \end{bmatrix}^T \\
& \qquad T = \begin{bmatrix} 0 & -s\alpha & c\alpha s\beta \\ 0 & c\alpha & s\alpha s\beta \\ 1 & 0 & c\beta \end{bmatrix}
\end{aligned} \tag{1}
$$

2.2 Constrained condition

In Fig. 5, the force (torque) that was measured by the operational force sensor is $F_h(T_h)$, and the force (torque) that was measured by the experimental force sensor is $F_e(T_e)$. With the input values of $F_h(T_h)$ and $F_e(T_e)$, unlike an operation in the unconstrained condition, the desired dynamic behavior of a robot can be described by an impedance equation (2). For the

same reason with case of the unconstrained condition, the K (stiffness matrices), having the property of a spring, was excluded.
In case interactions with an environment occur (a constraint condition), the end effector should endow with a behavior, considering the compliance. In this regard, we defined the relationship between the contact force (torque) and the dynamic behavior (position, velocity and acceleration) error of the end effector, through the generalized active impedance, as in (3). Thus, the end effector can have linear and dependant impedance characteristics to the translation part, for which the contact force F_e was considered, and the rotation part, for which the equivalent contact moment $T^T T_e$ was considered. In the (3), $M_{pe}(M_{oe})$, $B_{pe}(B_{oe})$, $K_{pe}(K_{oe})$ are the environmental impedance parameters that determine a dynamic behavior of the end effector for interactions with an environment.

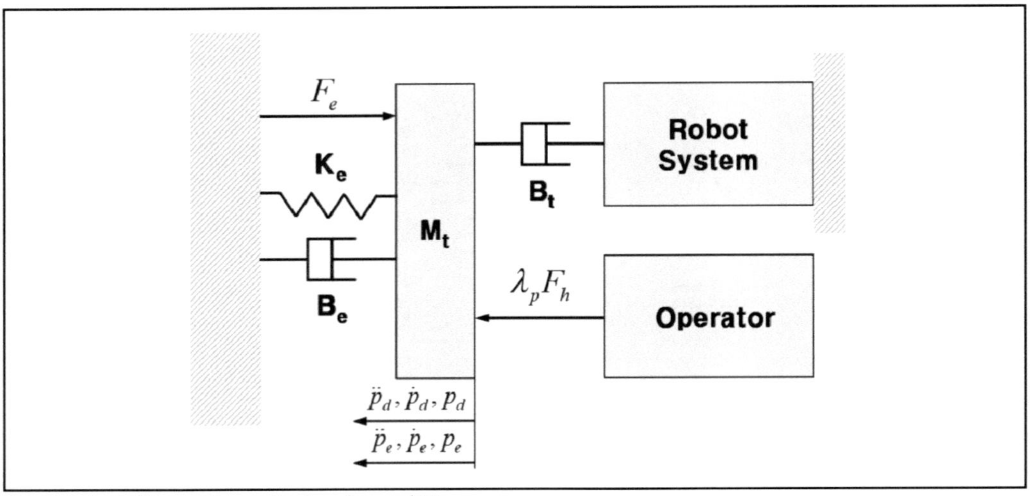

Fig. 5. Modelling of constrained condition

$$M_{pt}\ddot{p}_d + B_{pt}\dot{p}_d = \lambda_p F_h - F_e$$
$$M_{ot}\ddot{\varphi}_d + B_{ot}\dot{\varphi}_d = T^T(\varphi_d)(\lambda_o T_h - T_e) \tag{2}$$

$$M_{pe}\Delta\ddot{p}_{de} + B_{pe}\Delta\dot{p}_{de} + K_{pe}\Delta p_{de} = F_e$$
$$M_{oe}\Delta\ddot{\varphi}_{de} + B_{oe}\Delta\dot{\varphi}_{de} + K_{oe}\Delta\varphi_{de} = T^T(\varphi_e)T_e \tag{3}$$
$$where, \quad \Delta p_{de} = p_d - p_e$$

The dynamic behavior is determined, as shown in (3), by the environmental contact force and impedance characteristics. The dynamic behavior error indicates the difference between the desired dynamic behavior and the actual dynamic behavior of the end effector. That is to say, it explains that a robot system cannot practically follow a desired dynamic behavior, but indicates a level of compliance with an environment.

3. Control strategy of human-robot cooperative manipulation

3.1 Adjustment of impedance parameters

We categorized the building materials handling work through the human-robot cooperative manipulation into the environment-contacting case and the non-environment-contacting case. From the viewpoint of operational characteristics, the former case can be thought as 'the press fit operation' under interactions with other building materials or obstacles, which requires relatively higher stability. On the contrary, the latter case can be considered as the operation of moving building materials promptly to an installation site, which requires relatively higher mobility. In a human-robot cooperative system, we can make the virtual system have specific impedance characteristics through the modeling of human-robot cooperative system. To get the high stability, the virtual system may have a damping characteristic. However, too much damping decreases the mobility of the system. In this strategy, the impedance parameters of the virtual system are adjusted corresponding to the operational characteristics of the building materials handling work.

In the impedance equation (2), the impedance parameters $M_{pt}(M_{ot})$ and $B_{pt}(B_{ot})$ are switched to proper values when an operator requires stability or mobility according to the operational characteristics of the work process. The appropriate parameter values are determined through enough simulations with an experimental system. Also, each of the impedance parameters should be adjusted by stage according to the choice of an operator. This adjusting way is also applied exactly to adjustment of the power assist ratio (λ) of an operator.

3.2 Inner motion control loop

The selection of suitable impedance parameters that guarantee a satisfactory compliant behavior during the interaction may turn out to be inadequate to ensure accurate tracking of the desired position and orientation trajectory when the robot moves in unconstraint condition. A solution to this drawback can be devised by separating the motion control action from the impedance control action as follows. The motion control action is purposefully made stiffness so as enhance disturbance rejection but, rather than ensuring tracking of a reference position and orientation, it shall ensure tracking of a reference position and orientation resulting from the impedance control action. In other words, the desired position and orientation together with the measured contact force and moment are input to the impedance equation which, via a suitable integration, generates the position and orientation to be used as a reference for the motion control action.

In order to realize the above solution, it is worth introducing a reference frame other than the desired frame specified by a desired position vector p_d and a desired rotation matrix R_d. This frame is referred to as the compliant frame, and is specified by a position vector p_c and a rotation matrix R_c. In this way, the inverse dynamics motion control strategy can be still adopted as long as the actual end effector position p_e and orientation R_e is taken to coincide with p_c and R_c in lieu of p_d and R_d, respectively. Accordingly, the actual end effector linear velocity $\dot{p}_e$ and angular velocity ω_e are taken to coincide with $\dot{p}_c$ and ω_c, respectively.

A block diagram of the resulting scheme is sketched in Figure 16 and reveals the presence of an inner motion control loop with respect to the outer impedance control loop. In view (2), the impedance equation is chosen so as to enforce an equivalent mass-damper-spring behavior for the position displacement when the end effector exerts a force (torque) $F_e\,(T_e)$ on the environment, i.e.

$$\begin{aligned} &M_{pe}\Delta\ddot{p}_{dc} + B_{pe}\Delta\dot{p}_{dc} + K_{pe}\Delta p_{dc} = F_e \\ &M_{oe}\Delta\ddot{\varphi}_{dc} + B_{oe}\Delta\dot{\varphi}_{dc} + K_{oe}\Delta\varphi_{dc} = T^T(\varphi_e)T_e \\ &where,\quad \Delta p_{dc} = p_d - p_c \end{aligned} \tag{4}$$

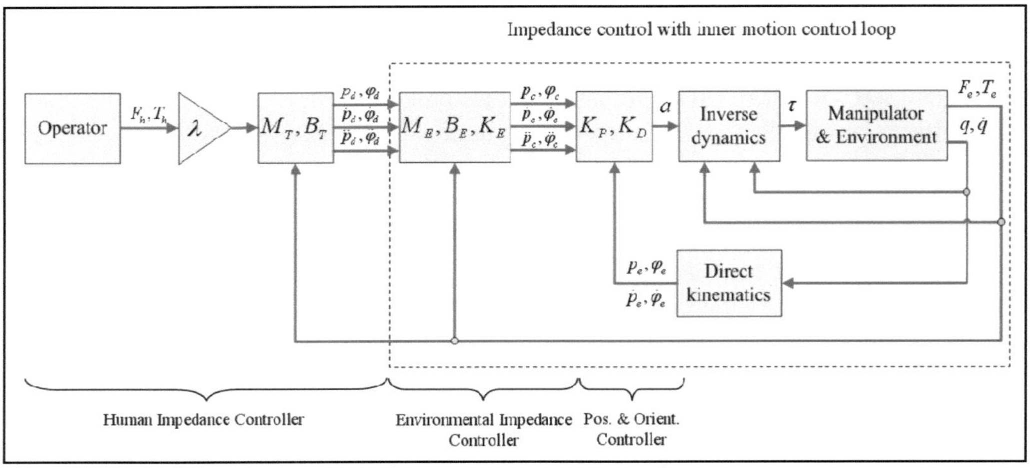

Fig. 6. Block diagram for human-robot cooperative manipulation

With reference to the scheme in Fig. 6, the impedance control generates the reference position for the inner motion control. Therefore, in order to allow the implementation of the complete control scheme, the acceleration shall be designed to track the position and the velocity of the compliant frame, i.e.

$$\begin{aligned} &a_p = \ddot{p}_c + K_{Dp}\Delta\dot{p}_{ce} + K_{Pp}\Delta p_{ce} \\ &a_o = T(\varphi_e)(\ddot{\varphi}_{ce} + K_{Do}\Delta\dot{\varphi}_{ce} + K_{Po}\Delta\varphi_{ce}) + \dot{T}(\varphi_e,\dot{\varphi}_e)\dot{\varphi}_e \\ &where,\quad \Delta p_{ce} = p_c - p_e \end{aligned} \tag{5}$$

Notice that p_c and its associated derivatives can be computed by forward integration of the impedance equation (4) with input $F_e\,(T_e)$ available from the force/torque sensor.

3.3 Experiments and results

As shown in Fig. 7, we mounted two sensors in the 2DOF manipulator, moving in the x and y axes. One receives operational signals from an operator, and the other, positioned between

the end effector and the 15 kg weighing object, can detect the contact force from the environment. With the signals that are received by the two sensors, the control signals, the manipulator should follow, are generated from the robot controller. The manipulator is controlled using an impedance control with inner motion loop method based on the robot force control. It assumes that the manipulator follows a commanded force derived by equation (6) (a Lagrange formulation). The sampling time for the force analysis and controlling the manipulator is settled as 1 msec. A mount string was used for the environmental system. The stiffness for an actual environment can be adjusted through replacement of the spring.
The experimental methods for the human-robot cooperation-work can be categorized into four staged.

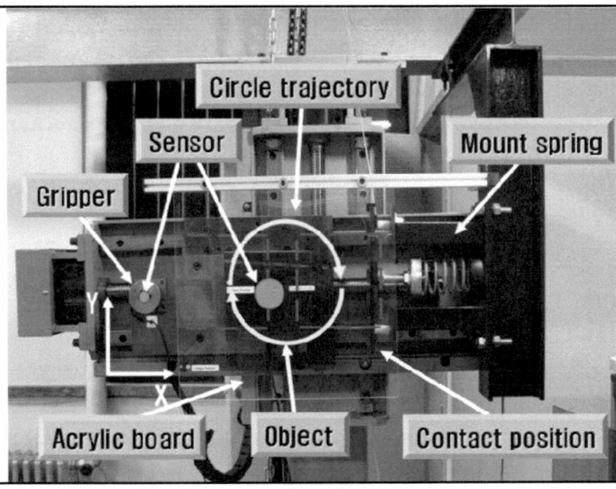

Fig. 7. Experimental system (2-DOF manipulator)

$$\begin{bmatrix} F_1 \\ F_2 \end{bmatrix} = \begin{bmatrix} m_1 + m_2 & 0 \\ 0 & m_2 \end{bmatrix} \begin{bmatrix} \ddot{d}_1 \\ \ddot{d}_2 \end{bmatrix} + \begin{bmatrix} m_1 g + m_2 g \\ 0 \end{bmatrix} + \begin{bmatrix} F_{ey} \\ F_{ez} \end{bmatrix} \tag{6}$$

1) An indicator (a needle), mounted on an object, automatically moves to the start position.
2) The operator applies force to the gripper (HRI device), so that the indicator follows a circle trajectory that is described on an acrylic board.
3) Based on the operational force, the robot follows the circle trajectory through the impedance control in the unconstraint condition.
4) The robot contacts a mount spring (environmental system) while following the circle trajectory. The contact force, generated at this time, enables the impedance control, and the robot finishes a building material handling work in compliance with an environment.

The experimental contents are as follows; Firstly, the influences of each parameter are to be observed for adjustment of the impedance parameters. Secondly, the performance of the suggested impedance control with inner motion control loop is to be evaluated to reduce the position following error for operation of a robot in an unconstrained condition. Thirdly, the influences of F_h and F_e, according to change of the power assist ratio (λ) of an operator, are to be studied. Finally, the changes of F_h and F_e, according to the changes of the actual environmental stiffness, are to be investigated (Lee et al., 2007).

3.3.1 Influence of impedance parameters

The factors are given as the impedance parameters $M_{pt}(M_{ot})$ and $B_{pt}(B_{ot})$ in (2), and the levels are given as 1, 3, and 10. The force, used for the two experiments, is 5 N and applied for around 10 sec. The K value is to be set to 1 N/m for convergence of graphs. As a result of experiments, $M_{pt}(M_{ot})$ is related to an operation that requires mobility in building materials handling processes. That is to say, this operation does not require relatively higher stability, but requires prompt movement of an object to a desired position with small operational force. As the $M_{pt}(M_{ot})$ value rises, an object can be moved to a long-distance place with big operational force. However, B_{pt} is related to an operation that requires stability in building materials handling processes. That is to say, this case does not require relatively higher mobility, but requires a precise and stable operation. As $B_{pt}(B_{ot})$ value rises, the distance of movement by the same operational force gets shorter. As the mobility is reduced, the more demanding force may make an operator feel the minimum moving distance shorter.

3.3.2 Influence of inner motion control loop

Without the operational force, the constant force, which is input to the controller, makes the robot follow an already-programmed circle trajectory. It can be recognized that the path tracking accuracy is rather poor during execution of the whole tasks if the stiffness parameter is small. The small stiffness parameter also causes reduction of the contact force in the constraint condition. These results occur due to a larger end effector position error in operation. To solve these problems, we proposed the impedance control with an inner motion control loop. The inner motion loop gains in (5) have been set as K_{Dp}=1.5I and K_{Pp}=15I. According to the experiment, the robot operation without the inner motion control loop shows inferior desired-position-following performance in an unconstraint condition. By using the inner motion control loop, the high following performance can be obtained.

3.3.3 Influence of power assist ratio

The power assist ratio (λ), suggested in (1) and (2), plays a role of controlling the scale of the force that is required by an operator for a human-robot cooperative manipulation. In the experiments, we studied changes of F_h and F_e when the power assist ratio (λ) was increased from 3 to 6. As λ increased to 6 from 3, F_h, required in case of no contact with the environment, was reduced by a half, and F_h for the environment-contacting case was also reduced by around a half. We can see that the force, required by an operator, gets smaller as λ increases, but there is no significant change in the force (F_e) that reflects in the contacting condition.

3.3.4 Influence of environmental stiffness parameter

The environmental stiffness (K_e) depends on the characteristics of materials, composing an environment. The purpose of this experiment, changing the environmental stiffness, lies on comparison of the contact force (F_e) that is felt by an operator according to operational conditions such as a case contact with an obstacle occurs or a case the press fit is required. It can be recognized that, as K_{pe} increases, the force, required by an operator, gets increased and the force (F_e), reflecting in the contact condition, gets increased, too.

4. The basic system of a MFR (Multi-purpose Field Robot) for handling building materials

The MFR system combines a basic system with an additional module for construction. Considering workspace and mobility, a 6DOF (degree-of-freedom) manipulator and a 3DOF mobile platform were suggested for use in the basic system. Moreover, it was possible to change the components of the basic system according to load specifications.

4.1 A multi (6)-DOF manipulator

Fig. 8 shows the 6DOF manipulator of a basic system. This robot is a special case manipulator where the centers of the last 3 axes meet in the center of the robot wrist. The kinematic analysis in such form of manipulator can be divided into 2 link chains (the first 3 link chains and then the other 3 link chains). Table 1 shows the specifications of the manipulator.

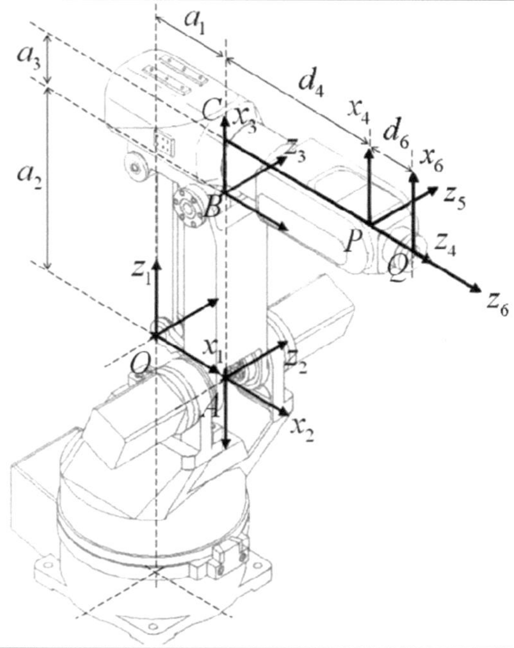

Fig. 8. A 6DOF manipulator (Samsung Electronics Co. ltd)

specification		value
Degree of Freedom		6
Weight Capacity		58.38 (*N*)
Arm Length(max)		858 (*mm*)
Velocity of End-effector		30 (°/*sec*)
Weight	Manipulator	588 (*N*)
	Controller	245 (*N*)

Table 1. Specifications of the 6DOF manipulator

The forward kinematics of the manipulator is defined by the question of solving for the position and direction of the end-effector according to each degree of the joints. That is, it is the problem of solving the position vector and rotational matrix of the end-effector. The kinematic analysis of the manipulator can be executed with any coordinate system but the most typical one is Denavit-Hartenberg Notation (noted as D-H notation below). The unknown kinematics values of a series manipulator can be solved by multiplying the homogeneous matrix defined in the following (7) and solved similar to (8).

$$H = R(\theta_i, z)T(d_i, z)T(a_i, x)R(\alpha_i, x)$$
$$= \begin{bmatrix} \cos\theta_i & -\sin\theta_i \cos\alpha_i & \sin\theta_i \sin\alpha_i & a_i \cos\alpha_i \\ \cos\theta_i & \cos\theta_i \cos\alpha_i & -\cos\theta_i \sin\alpha_i & a_i \sin\alpha_i \\ 0 & \sin\alpha_i & \cos\alpha_i & d_i \\ 0 & 0 & 0 & 1 \end{bmatrix} \tag{7}$$

$${}_0^nH = {}_0^1H \cdot {}_1^2H_i \cdots {}_{n-1}^{\;\;n}H_i = \begin{bmatrix} {}_0^nR & {}^0P_n \\ 0 & 1 \end{bmatrix}$$
$${}_6^0H = \begin{bmatrix} u_x & v_x & w_x & q_x \\ u_y & v_y & w_y & q_y \\ u_z & v_z & w_z & q_z \\ 0 & 0 & 0 & 1 \end{bmatrix} \tag{8}$$

Table 2 shows the D-H parameters through the D-H notation using the coordinate system defined in Figure 8. The position of the end-effector can be calculated by substituting a given parameter into the D-H transformation matrix in (8).

i	α_i (°)	a_i (mm)	d_i (mm)	θ_i (°)
1	-90	a_1	0	θ_1
2	0	a_2	0	θ_2
3	-90	a_3	0	θ_3
4	90	0	d_4	θ_4
5	-90	0	0	θ_5
6	0	0	d_6	θ_6

$a_1 = 150mm,\ a_2 = 350mm,\ a_3 = 100mm,\ d_4 = 350mm,\ d_6 = 195mm$

Table 2. D-H parameters of the 6DOF manipulator

4.2 A mobile platform

A 6DOF manipulator was fitted to the top plate of the mobile platform. Thus, movement of the manipulator was possible according to the platform's DOF. Traveling on uneven surfaces or surfaces with barriers was made possible using caterpillar tread.

Table 3 shows the specifications of the suggested mobile platform. This mobile platform largely consisted of caterpillar tread, a top plate and a controller as shown in Fig. 9. The caterpillar tread was powered by 2 DC motors with a reduction gear. Also, perpendicular movement of the top plate was achieved through a hydraulic cylinder. Through such movement mechanisms, 3DOF movement of forward and backward (T_y), left and right rotation (R_z) and perpendicular movement of the top plate was realized with the central axis (Z) of the mobile platform as the base. It was possible to control movement through both wired and wireless controllers and traveling speed was controlled through an internal controller.

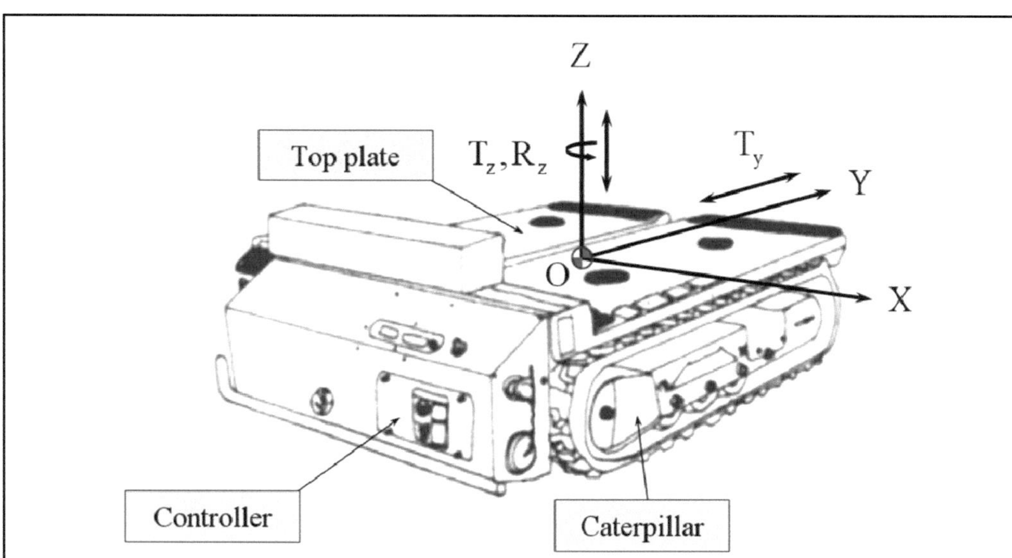

Fig. 9. A mobile platform (Kajima Mechatro Engineering Co.)

specification		value
Maximum load of carriage		9800 (N)
Weight		3920 (N)
Length		1,260 (mm)
Breadth		900 (mm)
Velocity	Maximum	2.5 (km/h)
	Minimum	0.6 (km/h)
Inclination of degree		20(°)
Power consumption		0.8 (kW)
Source of electricity		Charging battery

Table 3. Specifications of the mobile platform

5. The additional module of a MFR (Multi-purpose Field Robot) for handling building materials

An additional module, which is used for construction work along with various devices, was suggested for incorporating the MFR into construction work. This module consisted of hardware (HRI: Human-Robot Interface) and software (HRC control: Human-Robot Cooperative control).

5.1 A HRI (Human-Robot Interface)

First, the robot controller needs to be able to implement DOF for a mobile platform and a 6DOF manipulator.

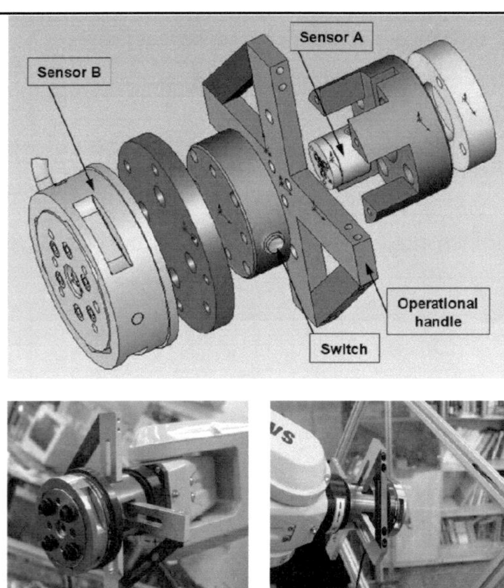

Fig. 10. The first robot controller (HRI device)

The first robot controller (HRI device) is shown in Fig. 10. As seen in this figure, if an operator puts external force containing an operation command on a handler of the robot controller, it is converted into a control signal to operate the robot with 'sensor A (6DOF force/torque sensor; ATI Industrial Automation, Inc.)'.
Here, if the robot comes in contact with an external object, information on the contact force is transmitted to the robot controller through 'sensor B (6DOF force/torque sensor)'. It is important to note that external force transmitted through sensor B and that transmitted to sensor A should operate separately from each other. In addition, the switch attached to the HRI device should be able to control the manipulator and mobile platform separately. That is, it plays a role of determining whether external force being inputted is a control signal for the manipulator or that for the mobile platform.
In MFR system, the operator can select between two communication methods: wired or wireless control. The wireless control system is used to carry materials long distances or to move a robot to places that are difficult for an operator to reach. The wired control system is used to install construction materials by cooperation or in an emergency.
For the wireless communication system, it is then possible to choose between the mobile platform control system and the manipulator control system. In other words, it is possible to control a mobile platform and a manipulator with one wireless controller (Fig. 11(a)). Each control signal is transmitted to the controller of a manipulator and a mobile platform through a main controller via a RF communication module and a converter.
For the wired communication system, it is again possible to choose between the cooperation-based control system and the emergency control system. Unlike the wireless communication system, the wired communication system uses a separate control unit. The cooperation-based control system operates through main controllers including industrial computers and sensors, and the first robot controller (HRI device) mentioned in Paragraph 5.1. The emergency control system can operate through the teach pendant of a manipulator and a mobile platform in emergency situations, as seen in Figure 11(b).

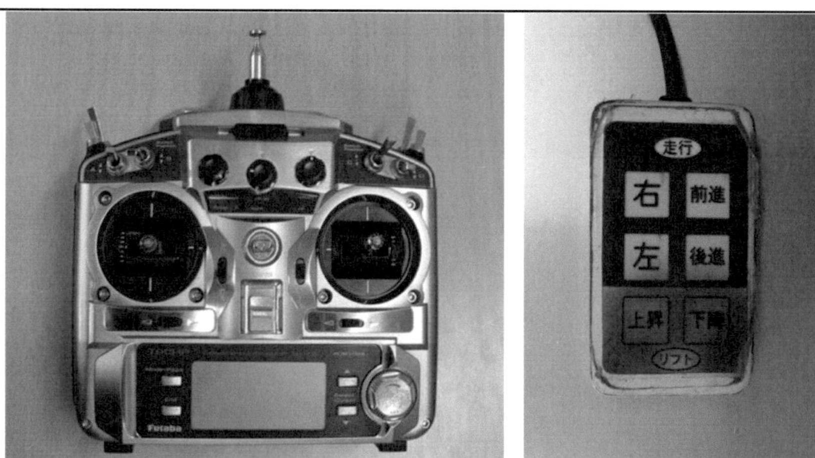

(a) Remote control system (b) Teach pendant

Fig. 11. The second robot controller (remote control system & teach pendant)

5.2 The software of the additional module

The software of the suggested additional module for construction works refers to a control algorithm, primarily necessary for handling building materials by human-robot cooperative manipulation. In this study, remote control, human-robot cooperation-based control, and emergency control are proposed as methods to control a MFR for handling building materials.

Fig. 12 shows an overall flow-chart for the suggested robot control methods. The operator can select between two communication methods: wired or wireless control. The wireless control system is used to carry materials long distances or to move a robot to places that are difficult for an operator to reach. The wired control system is used to handle building materials by cooperation or in an emergency. For the wireless communication system, it is then possible to choose between the mobile platform control system and the manipulator control system. In other words, it is possible to control a mobile platform and a manipulator with one wireless controller (Figure 11(a)). For the wired communication system, it is again possible to choose between the cooperation-based control system and the emergency control system. Unlike the wireless communication system, the wired communication system uses a separate control unit. The cooperation-based control system operates through main controllers including industrial computers and sensors, and the first robot controller (HRI device).

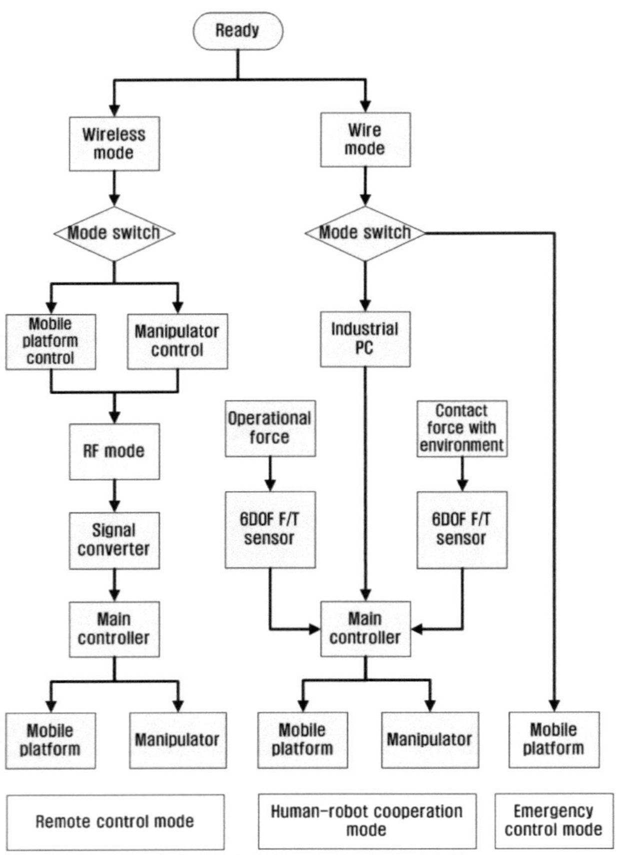

Fig. 12. Flow-chart for MFR control methods

The emergency control system can operate through the teach pendant of a manipulator and a mobile platform in emergency situations, as seen in Figure 11(b).
Handling of building materials by the cooperation-based control system can be largely divided as below.

1) Process of transporting materials to an installation position
2) Process of inserting them into the correct position or doing press pits, depending on the environment

In previous chapter 2, the former is defined as free space movement (motion under unconstrained conditions) and the latter as motion under constrained conditions. Free space motion needs rapid movement with relatively low precision while motion under constrained conditions needs precise motion with relatively low motion velocity. According to modeling of the interactions among the operator, robot and environment, we designed an impedance controller for the human-robot cooperation (Fig. 13). When an operator judges that the position (X) to which a robot carries materials fails to agree with the position (X_d) to which he or she wants to carry them, his or her force is transmitted to sensor A. In particular, external force (F_h) measured by sensor A can be used by operators from various age groups through the force augmentation ratio (α). That is, all people, regardless of muscular strength, can operate a robot by the force augmentation ratio. In terms of an operator's inputted force and the contact force (F_e) with environments inputted from sensor B, the target dynamics needed for operation are determined by the following equation (8) for impedance. Of the dynamics values, the deviation between the target position (X_d) and the present position (X) decreases as feedback is received through the encoder of a position/direction controller, resulting in 0. In other words, the current deviation is inputted into a servo controller, which causes a manipulator to pursue the target position value. In addition, it is possible to adapt the operation properties of a robot's motion characteristics by controlling the impedance parameters (M_t, B_t) in (9). Relatively rapid and precise motions can be implemented by controlling these parameters.

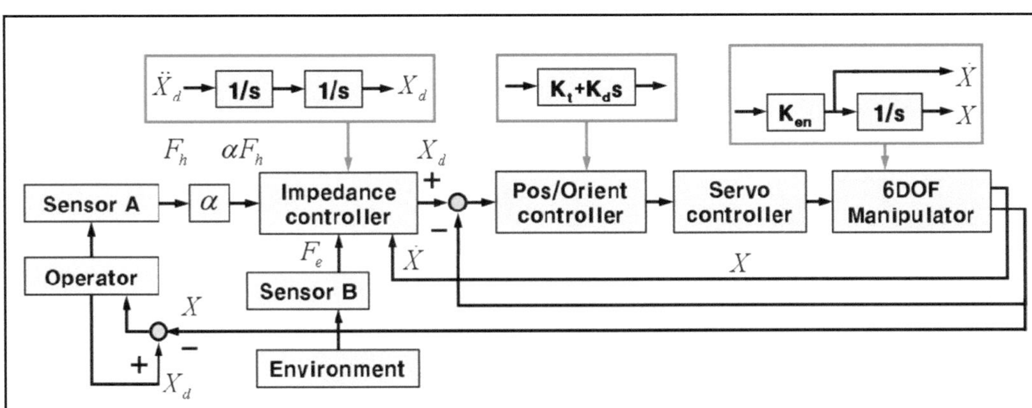

Fig. 13. Block diagram for human-robot manipulation of MFR

$$\ddot{X}_d = (M_t)^{-1}\{(\alpha F_h - F_e) - B_t \dot{X}_d\} \tag{9}$$

where, $\ddot{X}_d$: Acceleration related target dynamics

$\dot{X}_d$: Velocity related target dynamics

M_t : Inertia related impedance parameter in the virtual system

B_t : Damping related impedance parameter in the virtual system

6. Experiment with the prototype of a MFR

Fig. 14 shows the prototype of a MFR for handling building materials. In this figure, the basic system consists of a 6DOF manipulator and a mobile platform with caterpillar tread; the portion that excludes building material is an additional module (robot controller, vacuum suction device, F/T sensor and controller etc.) for handling building materials.

The development of a MFR applied to the construction area is not achieved by actual system production alone. Studies on system operation technology are also necessary for the developed system to be fully effective in operation. The construction material installation method suitable for a robot which was developed in this paper is shown in Fig. 15. Each process can be outlined as follows:

1) First, construction materials piled on the ground are fixed to a robot with an adsorption device. The type of loading for materials carried from the ground is determined by the most efficient adsorption posture within the operation range of a manipulator.
2) An operator rapidly moves the robot to an installation site through a wireless controller . Here, a mobile platform whose velocity can be controlled by the input of a control sig nal is principally used. The posture of construction materials is adjusted by the motion of a manipulator if necessary.
3) Construction materials carried to the vicinity of an installation position are installed thr ough interaction with materials already installed by an operator. That is, compliance oc curs upon contact, so that press pits for materials and systems are completed safely.
4) After the operation is completed, the robot is returned to the site of construction materi als loading through a wireless controller for the next operation.

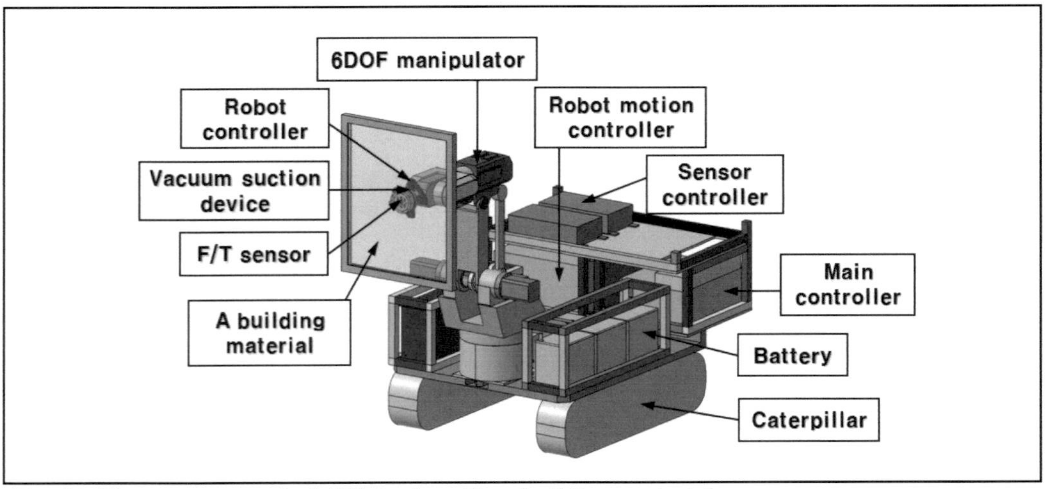

Fig. 14. The prototype of a MFR for handling building materials

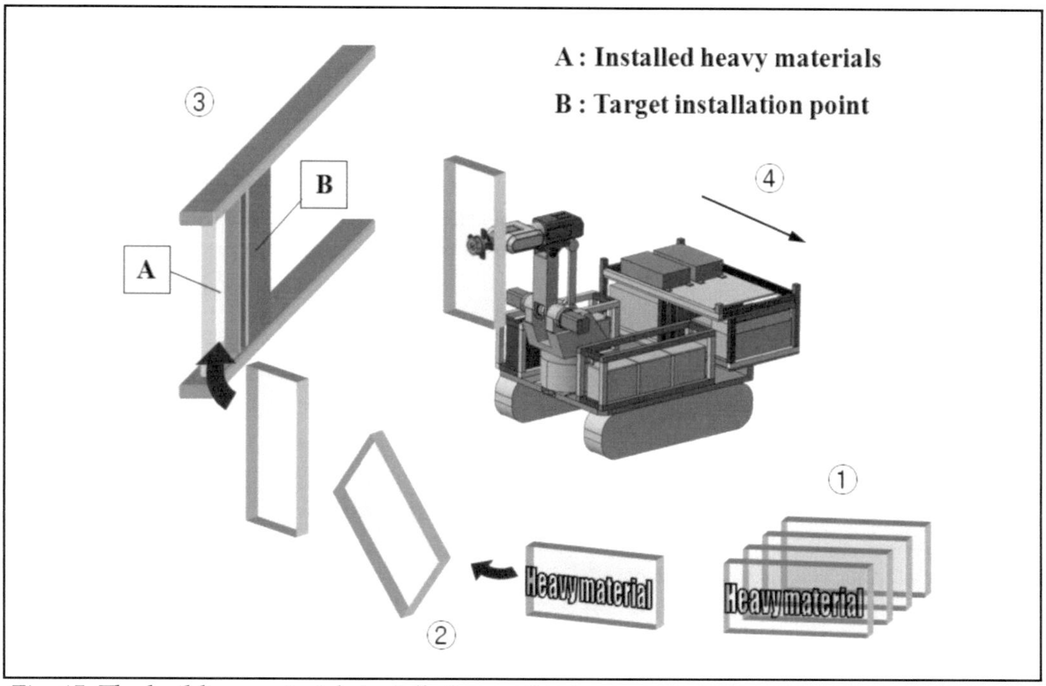

Fig. 15. The building materials installation method

A simulation for handling building materials is implemented to evaluate the performance of the prototype of MFR for construction works. The test is implemented indoors with an operation environment similar to that of an actual construction site. An experimental system to implement press pits after inserting building materials into the correct position was designed as in Fig. 16.

Fig. 16. An experimental system

Inserting building materials between the supporting board and the L-board is substituted for actual installation operation. As the gap is narrower than the thickness of building materials, they are moved horizontally and vertically with the supporting board connected to 'spring A' being pressed in order to complete the installation operation. If the supporting board is pressed, it means that compliance occurred; if the length of compression exceeds a certain range, the result is contact force which causes the robot to move in the opposite direction. In this experiment, building materials were limited to 60 N, considering the specifications of the manipulator, and manufactured into models of curtain wall or panel.
Fig. 17 shows a simulation for handling building materials through an experimental system. Once building materials are completely fixed to a robot through a vacuum suction device at a loading site, the robot is moved relatively rapidly to the vicinity of the installation position through a wireless controller. Precise positioning is performed by human-robot cooperative control. In handling building materials, an operator is encouraged to collect information on the operation in real time in order to cope with changing environments. Here, the speed or efficiency of operation is proportional to an operator's workmanship.
Fig. 18 shows the result of a mock-up test of a building material handling work using an experimental system. A comparison was made between the contact force (F_e) with environments and an operator's force (F_h), measured by sensors during the handling of building materials. F_h and F_e refer to the mean value of forces measured in the x, y, and z directions by a force/torque sensor during operation time T_h and T_e, respectively, as shown in the following (10).

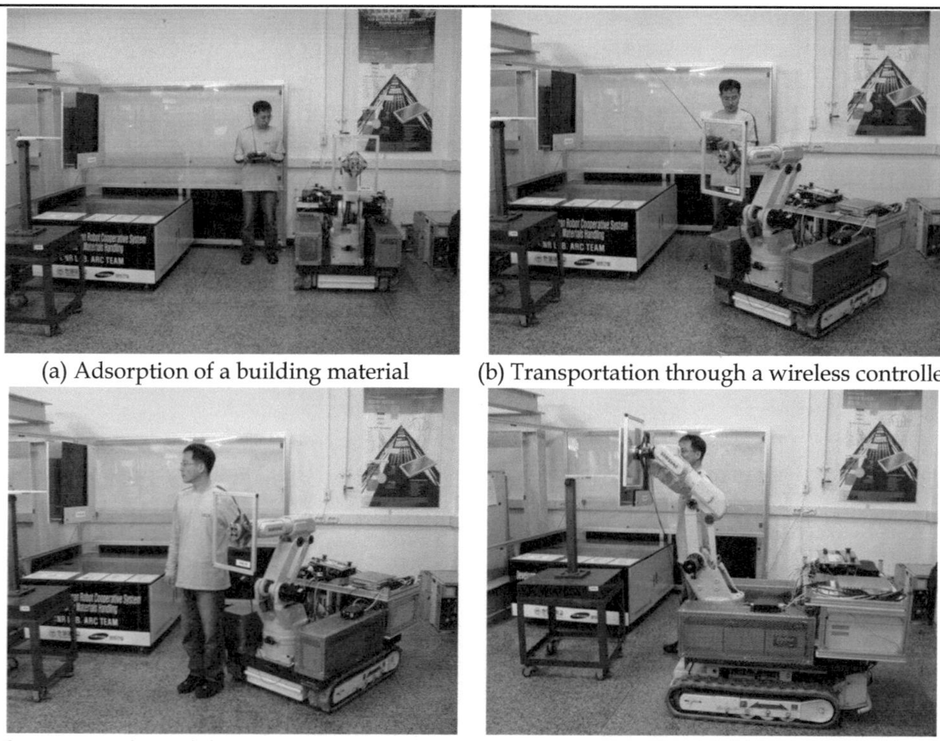

(a) Adsorption of a building material (b) Transportation through a wireless controller

(c) Positioning through human-robot cooperation (d) Installing (coupling) of a building material

Fig. 17. An experimental system

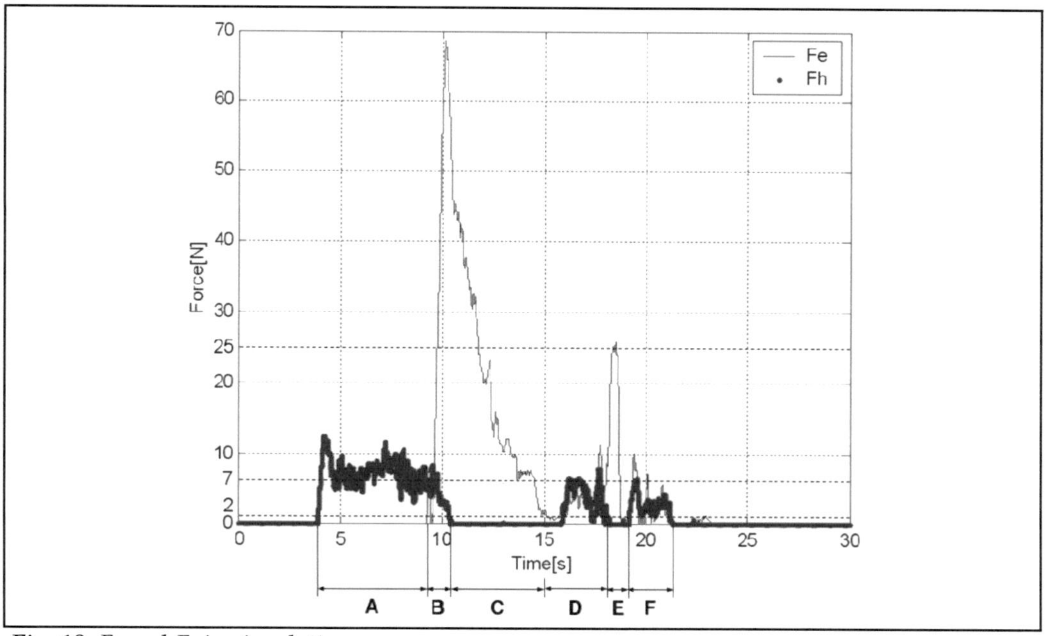

Fig. 18. F_e and F_h in simulation

$$F_h = \int_0^{T_h} \frac{\sqrt{F_{hx}^2 + F_{hy}^2 + F_{hz}^2}}{T_h} dt$$
$$F_e = \int_0^{T_e} \frac{\sqrt{F_{ex}^2 + F_{ey}^2 + F_{ez}^2}}{T_e} dt \qquad (10)$$

Each section can be described as follows:

1) Section A – A building material is carried to an installation position by the operators' fo rce (F_h). As seen in the graph, about 50 N of building materials are carried by about 70 N of external force supplied by an operator. The force augmentation ratio (a) is about 7, which is necessary to access the supporting board of the experimental system.
2) Section B - Contact with the environment (experimental system) begins to occur, genera ting a maximum of 70 N of contact force (F_e). Even at the moment of contact, the operat or's force is maintained to press the supporting board of the experimental system.
3) Section C - Not the operator's force but rather his or her torque is transmitted to improv e posture. About 2 N of compliance force is generated by the correlation between the ex ternal force provided and the impedance parameters of the experimental system. This value is used to press a spring connected to a supporting board into a position with a c ertain value.
4) Section D – A building material is carried horizontally to be inserted between the supp orting board and the L-board.
5) Section E – A building material is inserted; about 7 N of external force is provided by a n operator to make press pits, generating about 25 N of contact force.
6) Section F - Inserted horizontally, a building material is then inserted vertically.

A total of about 17 seconds is spent on the test, with an average 7 N or less required of an operator.

7. Conclusions and future works

The prototype of MFR for handling building materials presented in this study combines a manipulator and a mobile platform standardized in modular form to compose its basic system. Also, the hardware and software necessary for each area of application were composed of additional modules and combined with the robot's basic system. The suggested MFR can execute particular operations in various areas such as construction, national defense and rescue by changing these additional modules. One of the advantages of the proposed MFR can be handled building materials through human-robot cooperation. For this cooperation, the robot controller (HRI device) and end-effector (vacuum suction device) are combined in the basic system. Also, human-robot cooperative control is done through target dynamics modeling of human, robot, environment and control of impedance and external force inputted from the power/torque sensor attached to the additional module. In addition, a wireless control and emergency control function were added through other extra equipment.
Applying the suggested MFR to construction works can be used as one of solutions to the problem of unbalanced supply of manpower, a problem raised in construction industry. Also, construction safety will be assured because when a construction material is implemented by press fit with a material already installed, compliance occurs within the elastic range of the material and it is installed without damaging either object.
The expected results applying to advantages and disadvantages of both existing curtain wall installation robot (ASCI) and MFR are compared with and analyzed in Table 4. As seen from the table, the proposed robot will be expected that it will be safer and more efficient than the existing one.

	ASCI	MFR
Control mode	Wire	Wire/Wireless/Human-robot cooperation
Number of workers	2	1
Working condition	Receive limited accurate information	Install materials intuitively
Compatibility	Be restricted in specific work	Be compatible in various work through a change of a basic system and additional modules
Safety	Damage to construction materials and robot system by malfunction	Protection construction materials and system through force reflection

Table 4. Comparison and analysis of the ASCI and MFR in a construction site

A manipulator and a mobile platform, the basic system of the MFR, are combined to suit various working conditions and construction materials as module type. Therefore it is possible to install a variety of construction materials in various construction sites. Actual size of MFR for construction works is developing through the experiment result executed in our laboratory.

To apply a MFR at real construction sites, we must execute additional work required for application. Firstly, according to analysis of job definition and working condition, it is deduced that the conceptual design of a construction robot for installing bulk building materials. Secondly, practical arts (including robotized construction process) for applying to real construction sites should be proposed. Finally, after field test at a real construction site, productivity and safety of the developed system are compared with the existing construction equipment.

In next study, we will apply a MFR to a real construction site to install bulk glass ceiling that is installed 15m above the ground. Also, in realizing the potential of the suggested MFR, additional modules which consider the abilities and specifications required by national defense and rescue operations will be developed in the future.

8. Acknowledgement

This work was supported by the Korea Research Foundation Grant funded by the Korean Government [KRF-2008-357-D00010].

9. References

Albus, James S. (1986). Trip Report: Japanese Progress in Robotics for Construction, *Rototics Magazine*, (Spring 1986)

Bernold, L.E. (1987). Automation and robotics in construction: A challenge and change for an industry in transition. *International Journal of Project Management: The Journal of the International Project Management Association*, Vol. 5, No. 3, page numbers (155–160), ISSN 0263-7863

Choi, H.S., Han, C.S., Lee, K.Y. & Lee, S.H. (2005). Development of hybrid robot for construction works with pneumatic actuator. *Automation in Construction*, Vol. 14, No. 4, (November 2004) page numbers (452–459), ISSN 0926-5805

Cusack, M. (1994). Automation and robotics the interdependence of design and construction systems. *Industrial Robot*, Vol. 21, No. 4, page numbers (10–14), ISSN 0143-991X

Fukuda, T., Fujisawa, Y., Arai, F., Muro, H., Hoshino, K., Miyazaki, K., Ohtsubo, K. & Uehara, K. (1991). A New Robotic Manipulator in Construction Based on Man-Robot Cooperation Work. *Proceedings of the 8th International Symposium on Automation and Robotics in Construction*, pp. 239-245, Stuttgart, Germany, June 1991, IAARC, Eindhoven

Fukuda, T., Fujisawa, Y., Kosuge, K., Arai, F., Muro, H., Hoshino, K., Miyazaki, K., Ohtsubo, K. & Uehara, K. (1991). Manipulator for Man-Robot Cooperation. *International Conference on Industrial Electronics, Control and Instrumentation*, pp. 996-1001, Kobe, Japan, October 1991, IEEE, CA

Gambao, E., Balaguer, C. & Gebhart, F. (2000). Robot assembly system for computer-integrated construction. *Automation in Construction*, Vol. 9, No. 5-6, (June 2000) page numbers (479–487), ISSN 0926-5805

Han, H. (2005). *Automated construction technologies: analyses and future development strategies,* Master's thesis of science in architecture studies at the Massachusetts Institute of Technology, MA

Hogan, N. (1985). Impedance control: an approach to manipulation, Part I-III. *ASME Journal of Dynamic Systems, Measurements and Control*. Vol. 107, No. 3, (September 1985) page numbers (1-24), ISSN 0022-0434

Hollingum, J. (1999). Robots in agriculture. *The Industrial Robot*, Vol. 26, No. 6, (1999) page numbers (438-445), ISSN 0143-991X

Isao, S., Hidetoshi, O., Nobuhiro, T. & Hideo, T. (1996). Development of automated exterior curtain wall installation system. *Proceedings of International Symposium on Automation and Robotics in Construction*, pp. 915-924, Tokyo, Japan, June 1996, IAARC, Eindhoven

Kangari, R. (1991). Advanced robotics in civil engineering and construction. *91 ICAR. Fifth International Conference on Advanced Robotics*, pp. 375-378, ISBN 0-7803-0078-5, Pisa, Italy, 19-22 Jun 1991, IEEE, CA

Kazerooni, H. (1989). Human/robot interaction via the transfer of power and information signals - part I & II: Dynamics and control analysis. *IEEE Proc. of IEEE International Conference on Robotics and Automation*, pp. 1632-1647, AZ, USA, May 1989, IEEE, CA

Kazerooni, H. & Mahoney, S.L. (1991). Dynamics and control of robotic systems worn by humans, *ASME Journal of Dynamic Systems, Measurement and Control*, Vol. 133, No. 3, (September 1991) page numbers (379-387), ISSN 0022-0434

Kochan, A. (2000). Robots for automating construction—An abundance of research. *Industrial Robot*, Vol. 27, No. 2, page numbers (111–113), ISSN 0143-991X

Kosuge K., Fujisawa, Y. & Fukuda, T. (1993). Mechanical system control with man-machine-environment interactions. *Proc. of IEEE International Conference on Robotics and Automation*, pp. 239-244, Atlanta, USA, May 1993, IEEE, CA

Lee, S.H. Adams, T.M. & Ryoo B.Y. (1997). A fuzzy navigation system for mobile construction robot. *Automation in Construction,* Vol. 6, No. 2, (May 1997) page numbers (97-107), ISSN 0926-5805

Lee, S.Y., Lee, K.Y, Lee, S.H, Kim, J.W. & Han, C.S. (2007). Human-Robot Cooperation Control for Installing Heavy Construction Materials. *Autonomous Robots*, Vol. 22, No. 3, (April 2007) page numbers (305-319), ISSN 0929-5593

LeMaster, E.A. & Rock, S.M. (2003). A local-area GPS pseudolitebased navigation system for mars rovers. *Autonomous Robots,* Vol. 14, No. 2-3, (March 2003) page numbers (209-224), ISSN 0929-5593

Miller, J.S. (1968). *The Myotron – A Servo-controlled exoskeleton for the measurement of muscular kinetics*. Cornell Aeronautical Laboratory Report VO-2401-E-1

Masatoshi, H., Yukio, H., Hisashi, M., Kinya, T., Sigeyuki, K., Kohtarou, M., Tomoyuki, T. & Takumi, O. (1996). Development of interior finishing unit assembly system with robot: WASCOR IV research project report. *Automation in Construction*, Vol. 5, No. 1, (1996) page numbers (31–38), ISSN 0926-5805

Mosher, R.S. (1967). *Handyman to Hardiman.* Automotive Engineering Congress. SME670088

Poppy, W. (1994). Driving force and status of automation and robotics in construction in Europe. *Automation in Construction*, Vol. 2, No. 4, page numbers (281–289), ISSN 0926-5805

Roozbeh K. (1985). Advanced Robotics in Civil Engineering and Construction. *Proc. of IEEE International Conference on Robotics and Automation*, pp. 375-378, Tokyo, Japan, September 1985, IEEE, CA

Santos, P.G., Estremera, J., Jimenez, M.A., Garcia, E. & Armada, M. (2003). Manipulators helps out with plaster panels in construction. *The Industrial Robot*, Vol. 30, No. 6, (2003) page numbers (508–514), ISSN 0143-991X

Skibniewski, M.J. (1988). *Robotics in civil engineering*, Van Nostrand–Reinhold, ISBN 0442319258, New York

Skibniewski, M.J. & Wooldridge, S.C. (1992). Robotic materials handling for automated building construction technology. *Automation in Construction*, Vol. 1, No. 3, (1992) page numbers (251– 266), ISSN 0926-5805

Warszawski, A. (1985). Economic implications of robotics in building. *Building and Environment*, Vol. 20, No. 2, page numbers (73~81), ISSN 0360-1323

Wen, X., Romano, V.F. & Rovetta, A. (1991). Remote control and robotics in construction engineering. 91 ICAR. Fifth International Conference on Advanced Robotics, ISBN 0-7803-0078-5, Pisa, Italy, 19-22 Jun 1991, IEEE, CA

Whitcomb, L.L. (2000). Underwater robotics: Out of the research laboratory and into the field. *Proceedings of IEEE International Conference on Robotics and Automation*, pp. 709-716, ISBN 0-7803-5886-4, San Francisco, USA, April 2000, IEEE, CA

Wong, B. & Spetsakis, M. (2000). Scene reconstruction and robot navigation using dynamic fields. *Autonomous Robots*, Vol. 8, No. 1, (January 2000) page numbers (71-86), ISSN 0929-5593

17

A Sensor Classification Strategy for Robotic Manipulators

Miguel F. M. Lima [1], J. A. Tenreiro Machado [2] and António Ferrolho [3]
[1,3] *Dept. of Electrical Engineering, School of Technology, Polytechnic Institute of Viseu, Portugal, {lima, antferrolho}@mail.estv.ipv.pt*
[2] *Dept. of Electrical Engineering, Institute of Engineering, Polytechnic Institute of Porto, Portugal, jtm@isep.ipp.pt*

1. Introduction

In practice the robotic manipulators present some degree of unwanted vibrations. The advent of lightweight arm manipulators, mainly in the aerospace industry, where weight is an important issue, leads to the problem of intense vibrations. On the other hand, robots interacting with the environment often generate impacts that propagate through the mechanical structure and produce also vibrations.

In order to analyze these phenomena a robot signal acquisition system was developed. The manipulator motion produces vibrations, either from the structural modes or from end-effector impacts. The instrumentation system acquires signals from several sensors that capture the joint positions, mass accelerations, forces and moments, and electrical currents in the motors. Afterwards, an analysis package, running off-line, reads the data recorded by the acquisition system and extracts the signal characteristics.

Due to the multiplicity of sensors, the data obtained can be redundant because the same type of information may be seen by two or more sensors. Because of the price of the sensors, this aspect can be considered in order to reduce the cost of the system. On the other hand, the placement of the sensors is an important issue in order to obtain the suitable signals of the vibration phenomenon. Moreover, the study of these issues can help in the design optimization of the acquisition system. In this line of thought a sensor classification scheme is presented.

Several authors have addressed the subject of the sensor classification scheme. White (White, 1987) presents a flexible and comprehensive categorizing scheme that is useful for describing and comparing sensors. The author organizes the sensors according to several aspects: measurands, technological aspects, detection means, conversion phenomena, sensor materials and fields of application. Michahelles and Schiele (Michahelles & Schiele, 2003) systematize the use of sensor technology. They identified several dimensions of sensing that represent the sensing goals for physical interaction. A conceptual framework is introduced that allows categorizing existing sensors and evaluates their utility in various applications. This framework not only guides application designers for choosing meaningful sensor

subsets, but also can inspire new systems and leads to the evaluation of existing applications.
Today's technology offers a wide variety of sensors. In order to use all the data from the diversity of sensors a framework of integration is needed. Sensor fusion, fuzzy logic, and neural networks are often mentioned when dealing with problem of combing information from several sensors to get a more general picture of a given situation. The study of data fusion has been receiving considerable attention (Esteban et al., 2005; Luo & Kay, 1990). A survey of the state of the art in sensor fusion for robotics can be found in (Hackett & Shah, 1990). Henderson and Shilcrat (Henderson & Shilcrat, 1984) introduced the concept of logic sensor that defines an abstract specification of the sensors to integrate in a multisensor system.
The recent developments of micro electro mechanical sensors (MEMS) with unwired communication capabilities allow a sensor network with interesting capacity. This technology was applied in several applications (Arampatzis & Manesis, 2005), including robotics. Cheekiralla and Engels (Cheekiralla & Engels, 2005) propose a classification of the unwired sensor networks according to its functionalities and properties.
This paper presents a development of a sensor classification scheme based on the frequency spectrum of the signals and on a statistical metrics.
Bearing these ideas in mind, this paper is organized as follows. Section 2 describes briefly the robotic system enhanced with the instrumentation setup. Section 3 presents the experimental results. Finally, section 4 draws the main conclusions and points out future work.

2. Experimental platform

The developed experimental platform has two main parts: the hardware and the software components (Lima et al., 2005). The hardware architecture is shown in Fig. 1. Essentially it is made up of a robot manipulator, a personal computer (PC), and an interface electronic system.
The interface box is inserted between the robot arm and the robot controller, in order to acquire the internal robot signals; nevertheless, the interface captures also external signals, such as those arising from accelerometers and force/torque sensors. The modules are made up of electronic cards specifically designed for this work. The function of the modules is to adapt the signals and to isolate galvanically the robot's electronic equipment from the rest of the hardware required by the experiments.
The software package runs in a Pentium 4, 3.0 GHz PC and, from the user's point of view, consists of two applications: the acquisition application and the analysis package. The acquisition application is a real time program for acquiring and recording the robot signals.
After the real time acquisition, the analysis package processes the data off-line in two phases, namely, pre-processing and processing. The preprocessing phase consists of the signal selection in time, and their synchronization and truncation. The processing stage implements several algorithms for signal processing such as the auto and cross correlation, and Fourier transform (FT).

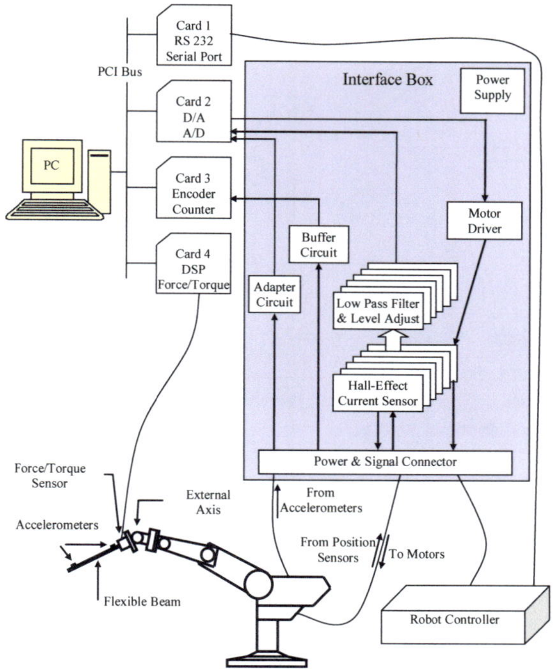

Fig. 1. Block diagram of the hardware architecture

3. Experimental results

According to the platform described in Section 2 a set of experiments is developed. Based on the signals captured from the robot this section presents several results obtained both in the time and frequency domains.

In the experiments a flexible link is used that consists of a long and round flexible steel rod clamped to the end-effector of the manipulator. In order to analyze the impact phenomena in different situations two types of beams are used. Their physical properties are shown in Table 1. The robot motion is programmed in a way such that the rods move against a rigid surface. Figure 2 depicts the robot with the flexible link and the impact surface.

During the motion of the manipulator the clamped rod is moved by the robot against a rigid surface. An impact occurs and several signals are recorded with a sampling frequency of f_s = 500 Hz. The signals come from several sensors, such as accelerometers, force and torque sensor, position encoders, and current sensors.

In order to have a wide set of signals captured during the impact of the rods against the vertical screen thirteen trajectories were defined. Those trajectories are based on several points selected systematically in the workspace of the robot, located on a virtual Cartesian coordinate system (see Fig. 3). This coordinate system is completely independent from that used on the measurement system. For each trajectory the motion of the robot begins in one of these points, moves against the surface and returns to the initial point. A parabolic profile was used for the trajectories.

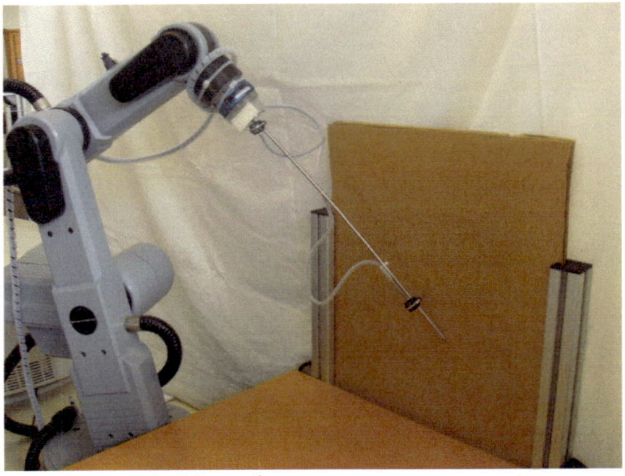

Fig. 2. Steel rod impact against a rigid surface

Characteristics	Thin rod	Gross rod
Material	Steel	Steel
Density [kg m^{-3}]	4.34×10^3	4.19×10^3
Mass [kg]	0.107	0.195
Length [m]	0.475	0.475
Diameter [m]	5.75×10^{-3}	7.9×10^{-3}

Table 1. Physical properties of the flexible beams

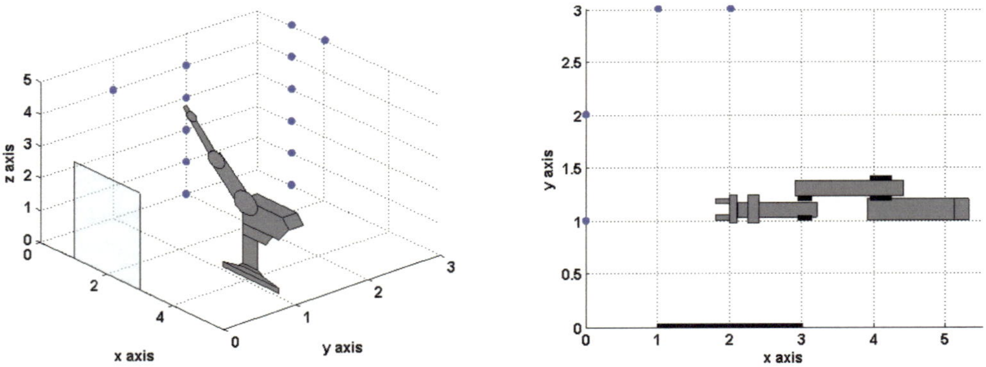

Fig. 3. Schematic representation 3D (left) and 2D (right) of the robot and the impact surface on the virtual cartesian coordinate system

3.1 Analysis in the time domain

Figures 4 to 7 depict some of the signals corresponding to the cases: (*i*) without impact, (*ii*) the impact of the rod on a gross screen and (*iii*) the impact of the rod on a thin screen using either the thin or the gross rod.

In this chapter only the most relevant signals are depicted, namely the forces and moments at the gripper sensor, the electrical currents of the robot's axes motors, and the rod accelerations. The signals present clearly a strong variation at the instant of the impact that occurs, approximately, at t = 3 s. Consequently, the effect of the impact forces (Fig. 4 left) and moments (Fig. 4 right) is reflected in the current required by the robot motors (Fig. 6). Moreover, as expected, the amplitudes of forces due to the gross screen (case ii) are higher than those corresponding to the thin screen (case iii). On the other hand, the forces with the gross rod (Fig. 4 right) are higher than those that occur with the thin rod (Fig. 4 left). The torques present also an identical behavior in terms of its amplitude variation for the tested conditions (see Fig. 5).

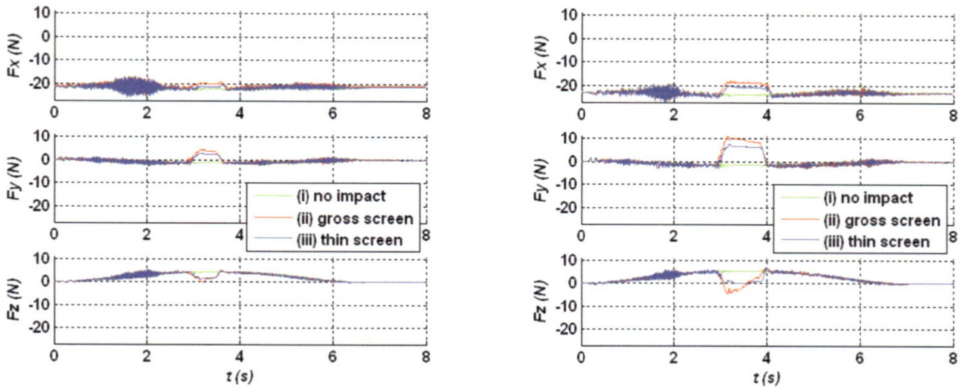

Fig. 4. Forces { F_x, F_y, F_z } at the gripper sensor: thin rod (left); gross rod (right)

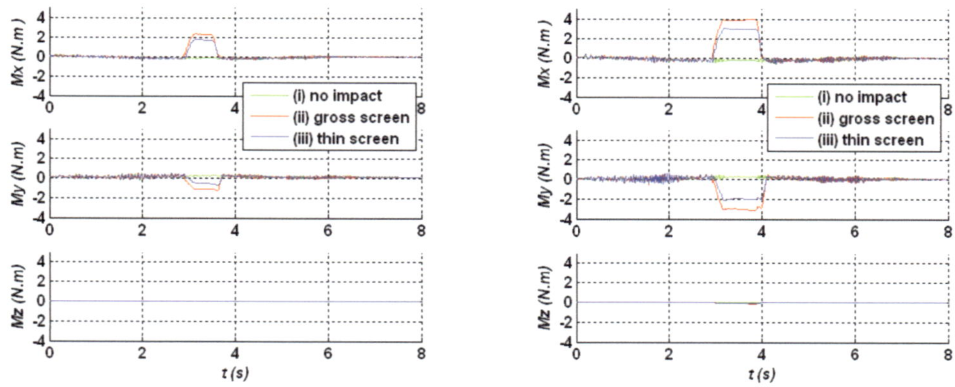

Fig. 5. Moments { M_x, M_y, M_z } at the gripper sensor: thin rod (left); gross rod (right)

Figure 7 presents the accelerations at the rod free-end (accelerometer 1), where the impact occurs, and at the rod clamped-end (accelerometer 2). The amplitudes of the accelerometers signals are higher near the rod impact side. Furthermore, the values of the accelerations obtained for the thin rod (Fig 7 left) are higher than those for the gross rod (Fig 7 right), because the thin rod is more flexible.

3.2 Analysis in the frequency domain

Figures 8 and 9 show, as examples, the amplitude of the Fast Fourier Transform (FFT), of two signals captured during the same impact trajectory. These figures illustrate the different behaviors of the spectrum, depending on the signal in study. All the signals of the trajectories set referred previously were studied, but only the most relevant are depicted.

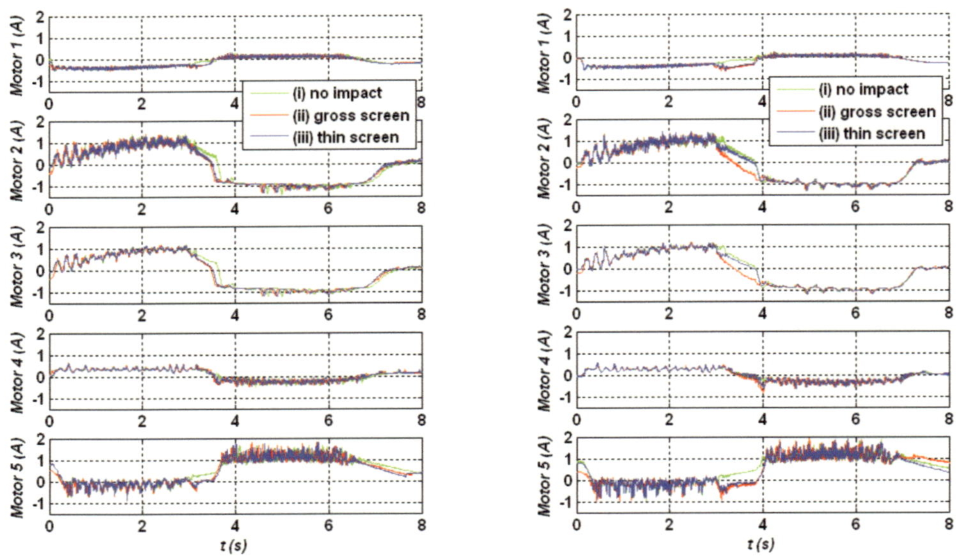

Fig. 6. Electrical currents { I_1, I_2, I_3, I_4, I_5 } of the robot's axes motors: thin rod (left); gross rod (right)

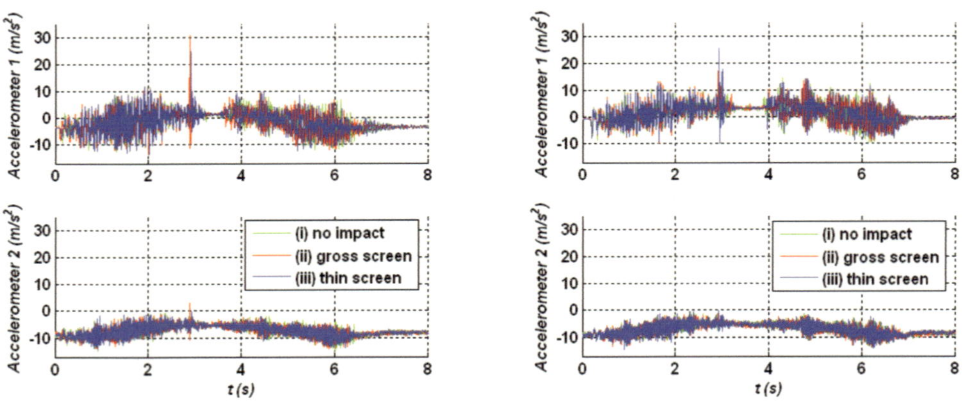

Fig. 7. Rod accelerations { A_1, A_2 }: thin rod (left); gross rod (right)

In order to examine the behavior of the FT signal, in a systematic way, a trendline was superimposed over the spectrum over, at least, one decade. The trendline is based on the power law approximation (Lima et al., 2006).

$$\left|\mathrm{F}\left\{f(t)\right\}\right| \approx c\omega^{m} \tag{1}$$

where F is the FT of the signal, $c \in \Re$ is a constant that depends on the amplitude, ω is the frequency, and $m \in \Re$ is the slope.

For each type of signal, the frequency interval was defined approximately in the middle range of the frequency content of the signal.

Figure 8 shows the FFT amplitude of the electrical current of the axis 3 motor that occurs in the case of impact with the thin rod. A trendline was calculated, and superimposed to the signal (case ii), with slope $m = -1.31$. The others current signals were studied, revealing also an identical behavior in terms of its spectrum spread, both under impact and no impact conditions, either for the thin rod or the gross rod. The spectrum was approximated by trendlines in a frequency range larger than one decade.

According to the robot manufacturer specifications (Robotec, 1996) the loop control of the robot has a cycle time of $t_c = 10$ ms. This fact is observed approximately at the fundamental ($f_c = 100$ Hz) and multiple harmonics in all spectra of motor currents.

The FFT amplitudes of the axes positions signals were studied (Lima et al., 2007), revealing also a behavior similar to the electrical current in terms of the spectrum spread for the tested conditions (impact, no impact, thin rod and gross rod).

Figure 9 shows the FFT amplitude of the F_z force (case *i*) due to the impact with the thin rod. This spectrum is not so well defined in a large frequency range. Nevertheless, the spectrum was approximated by a trendline in a frequency range of approximately one decade in order to get a systematic method of comparison. The trendline has a slope of $m = -0.13$.

The torques and accelerations signals were studied also for the distinct test conditions, namely: impact, no impact, thin rod and gross rod. Their FFT amplitudes revealed also an identical behavior in terms of its spectrum spread for the tested conditions.

Whereas the trendlines used for the electrical currents and position signals FT seem appropriate, the same technique used for the forces/moments and acceleration signals is questionable. However, in spite of this, trendlines were used for all FT signals in order to obtain comparable units. In fact, the purpose of this research is to establish a relationship between signals of the same system based on the spectrum behavior.

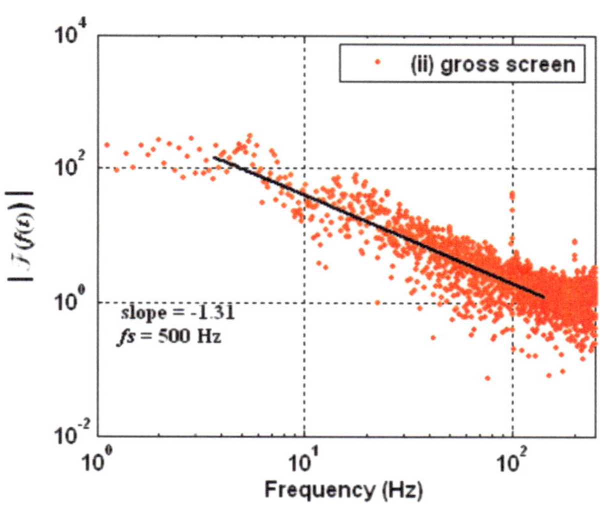

Fig. 8. Spectrum of the axis 3 motor current I_3 for the thin rod

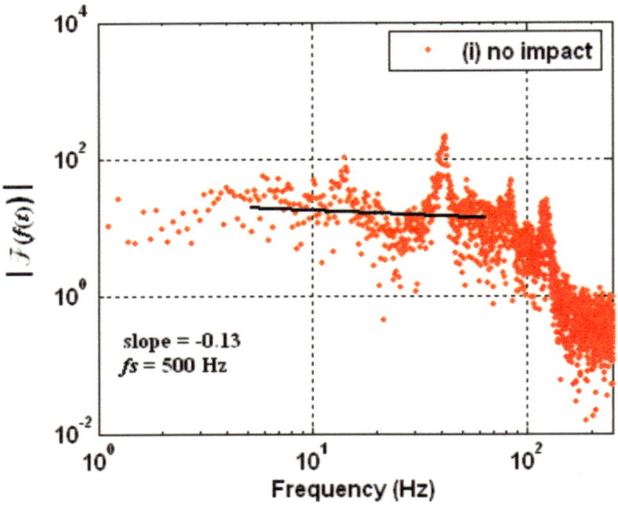

Fig. 9. Spectrum of the F_z force for the thin rod

3.3 Analysis of the spectrum trendlines slopes

Based on the several values of the spectrum trendlines slopes several statistics can be performed. During each trajectory of the robot eighteen signals were captured. For each trajectory there are three cases: (*i*) without impact, (*ii*) the impact of the rod on a gross screen, and (*iii*) the impact of the rod on a thin screen. As referred before, thirteen trajectories were defined. Additionally, the same trajectories were executed with the thin rod and with the gross rod. These samples lead to a population of 1404 slope values.

A box plot provides a visual summary of many important aspects of a data distribution. It indicates the median, upper and lower quartile, upper and lower adjacent values (whiskers), and the outlier individual points. Figure 10 shows a box plot of the spectrum trendlines slopes for the three cases of the thin rod impact, namely: (*i*) without impact, (*ii*) the impact of the rod on a gross screen, and (*iii*) the impact of the rod on a thin screen. Moreover, Fig. 11 depicts the respective interquartile range (IQR) versus the median. The IQR is obtained by subtracting the lower (first) quartile value from the upper (third) quartile value.

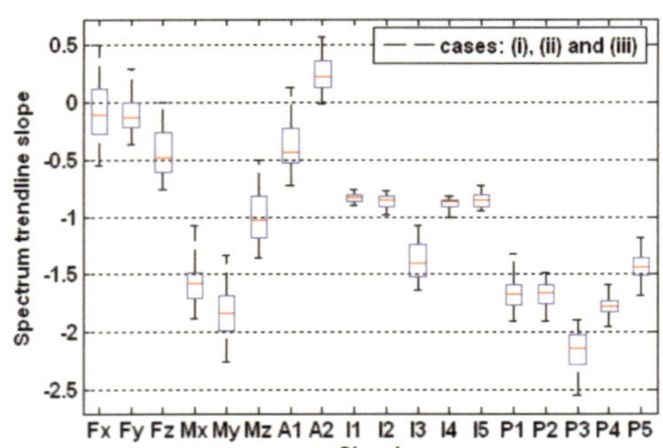

Fig. 10. Statistics of spectrum trendlines slopes for all the cases (*i, ii, iii*) using the thin rod

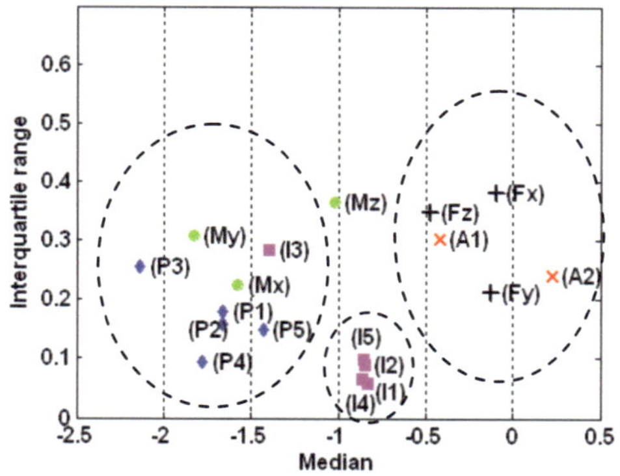

Fig. 11. IQR *versus* median for all the cases (*i, ii, iii*) using the thin rod

The IQR is a robust way of describing the dispersion of the data. From Fig. 11 three groups of signals can be defined. The ellipses depicted in the chart represent these groups. The forces $\{F_x, F_y, F_z\}$ and the accelerations $\{A_1, A_2\}$ signals are located close to each other. Positions $\{P_1, P_2, P_3, P_4, P_5\}$, moments $\{M_x, M_y\}$, and I_3 signals are located on the left side of the Fig. 11. Finally, the other electrical currents $\{I_1, I_2, I_4, I_5\}$ are situated in the middle of the chart and near each other. It rests the M_z signal that apparently is alone.

Figures 12 and 13 show the same statistic analysis described previously, but now for the gross rod. In Fig. 13 again three groups of signals can be defined. One groups the $\{F_x, F_y, F_z, A_1, A_2\}$ signals, and the second is formed of the $\{I_1, I_2, I_4, I_5\}$ signals. The third group consists of the $\{P_1, P_2, P_3, P_4, P_5, M_x, M_y, M_z, I_3\}$ signals. Comparing with the thin rod case, it can be seen that now the M_z signal joined the group of "torques and positions".

Finally, figures 14 to 15 depict the statistics of the overall spectrum trendlines slopes, considering the data for the thin and gross rods. Three groups are observed again: the group of "positions and torques", the group of "currents" and the group of "forces and accelerations". As can be seen the I_3 signal continues to remain in the same group of "positions and torques". A deeper insight into the nature of this feature must be envisaged to understand the behavior of the I_3 signal.

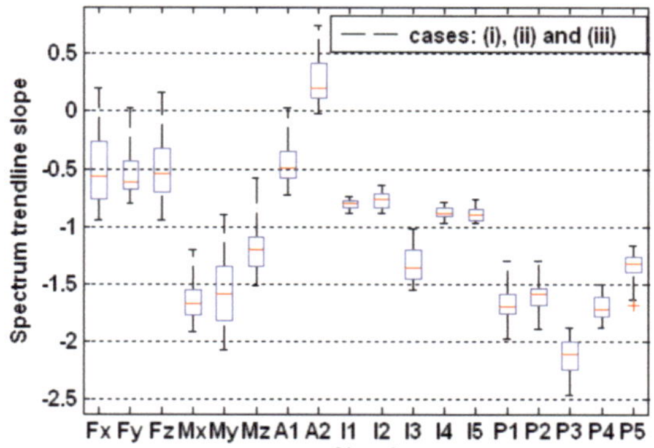

Fig. 12. Statistics of spectrum trendlines slopes for all the cases (*i, ii, iii*) using the gross rod

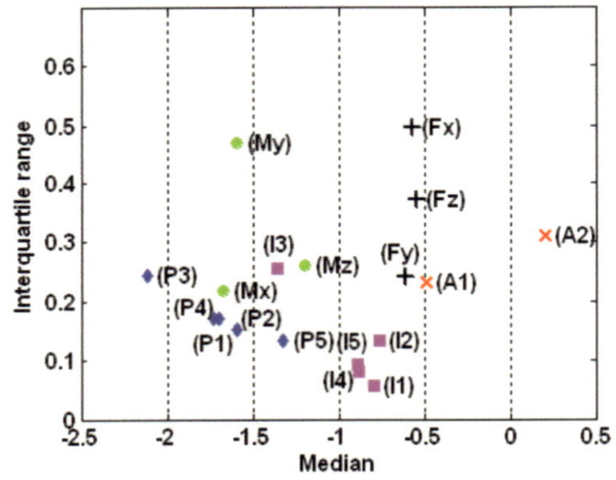

Fig. 13. IQR *versus* median for all the cases (*i, ii, iii*) using the gross rod

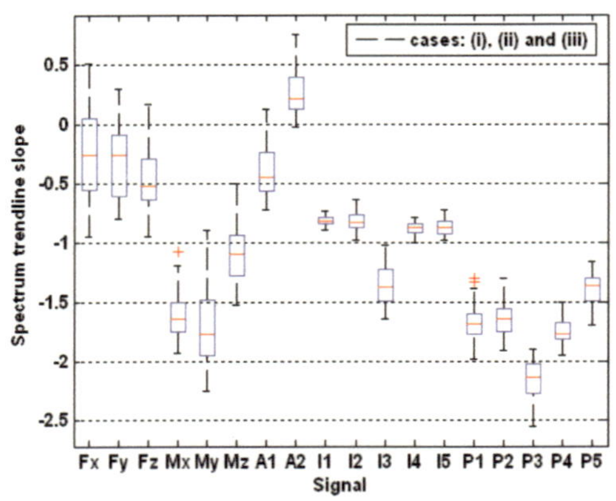

Fig. 14. Statistics of spectrum trendlines slopes for all the cases (*i, ii, iii*) using the thin and gross rods

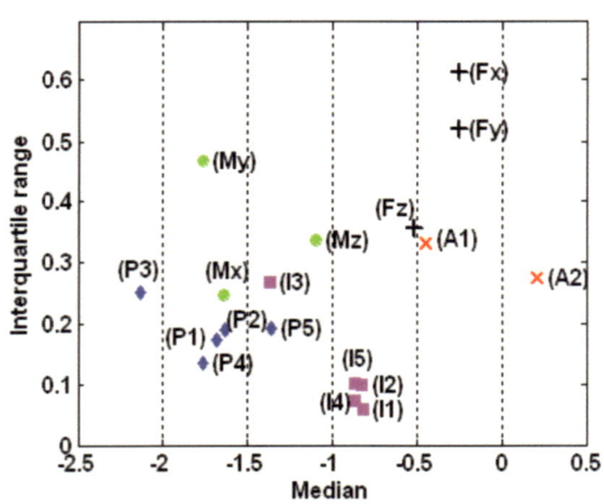

Fig. 15. IQR *versus* median for all the cases (*i, ii, iii*) using the thin and gross rods

3.4 Metrics in the time domain

Several indices can be used to evaluate the relashionship between the signal, including statistical, entropy and information theory approaches. These metrics are based on a bidimensional probability density function associated with the two signals $x_1(t)$ and $x_2(t)$ acquired in the same time interval and can be calculated according with the expression:

$$P(x_1,x_2)=\frac{\beta(x_1,x_2)}{\iint \beta(x_1,x_2)dx_1dx_2} \tag{2}$$

where β is the bidimensional histogram.

The marginal probability distributions of the signals $x_1(t)$ and $x_2(t)$ are denoted as $P(x_1)$ and $P(x_2)$, respectively. The expected values, $E(x_1)$ and $E(x_2)$, and the variances, $V(x_1)$ and $V(x_2)$, are then easily obtained.

The correlation coefficient R (Orfanidis, 1996) is a statistical index that provides a measurement of the similarity between two signals $x_1(t)$ and $x_2(t)$ and is define as

$$R(x_1,x_2)=\frac{E(x_1x_2)-E(x_1)E(x_2)}{\sqrt{V(x_1)V(x_2)}} \tag{3}$$

where $E(x_1x_2)$ is the joint expected value.

The mutual information (Shannon, 1948; Cover, 2006), or transinformation (Spataru, 1970) is the index that measures the dependence of two variables in the viewpoint of the information theory. The mutual information for the two signals $x_1(t)$ and $x_2(t)$ is:

$$I(x_1,x_2)=\log_2\frac{P(x_1,x_2)}{P(x_1)P(x_2)} \tag{4}$$

The average mutual information between the two signals is given by:

$$I_{av}(x_1,x_2)=\iint P(x_1,x_2)\log_2\frac{P(x_1,x_2)}{P(x_1)P(x_2)}dtdt \tag{5}$$

The entropy (Shannon, 1948) is a statistical measure of randomness. This index applied to the two signal $x_1(t)$ and $x_2(t)$ gives the join entropy (MacKay, 2003) between the two signal defined as:

$$H(x_1,x_2)=-\iint P(x_1,x_2)\log_2 P(x_1,x_2)dtdt \tag{6}$$

Figure 16 shows the squared correlation coefficient R^2 between the signals captured during the same impact trajectory, for an experiment in the case of (*i*) using the gross rod. The results obtain with R^2 are simetric relative to the diagonal formed by $R^2(x_i,x_j)$ for $i = j$, where the metric is maximum, as expected. To clearly visualize the results only one side is shown. The correlation between the same families of signals is higher than the correlation between different families. For example, the correlation between the currents and positions are low. The same occurs between the currents and the forces, moments and accelerations. It exists a strong correlation between the positions and the forces, moments and accelerations that depends, as expected, on the trajectory.

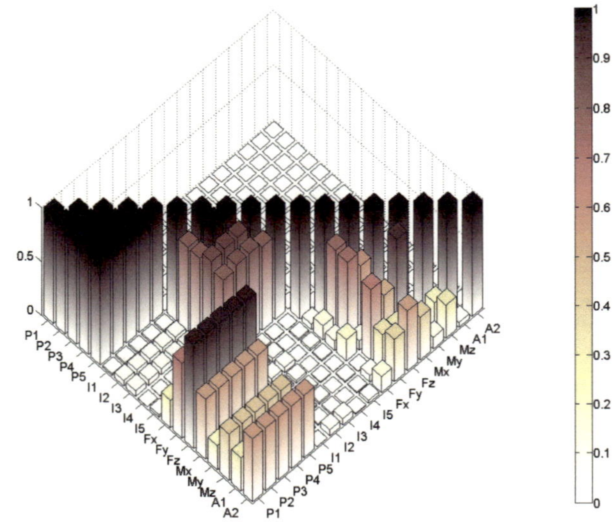

Fig. 16. Correlation between the signals for the case (*i*) using the gross rod

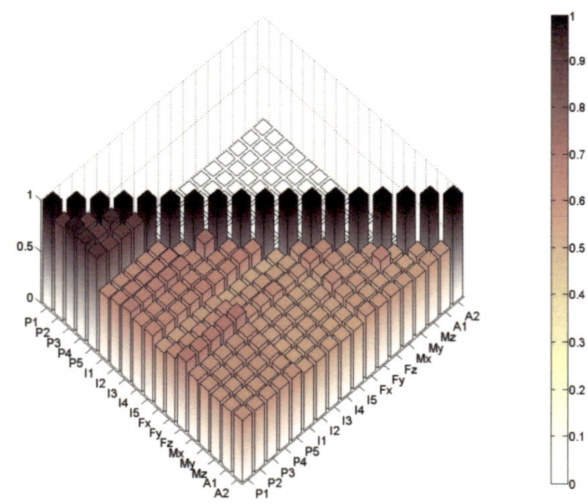

Fig. 17. Average mutual information between the signals for the case (*i*) using the gross rod

Figure 17 shows the average mutual information between the signals for the same experiment used previously for the correlation. Again the results obtain with $I_{av}(x_1, x_2)$ are simetric relative to the diagonal where the metric is maximum. Due to the same reason referred before only one side is shown. The values presented are normalized. The values of the index $I_{av}(x_1, x_2)$ between the positions are high.

Figure 18 shows a chart based on the entropy between the signals for the same experiment used previously for the other metrics. In fact, the values shown are proportional to the inverse of the index $H(x_1, x_2)$ due to the normalization used. Again, the values of this index between the positions are high.
The metrics shown in figures 16–18 were obtained for an experiment corresponding to one trajectory. In future this approach should be applied for all the thirteen trajectories referred before.

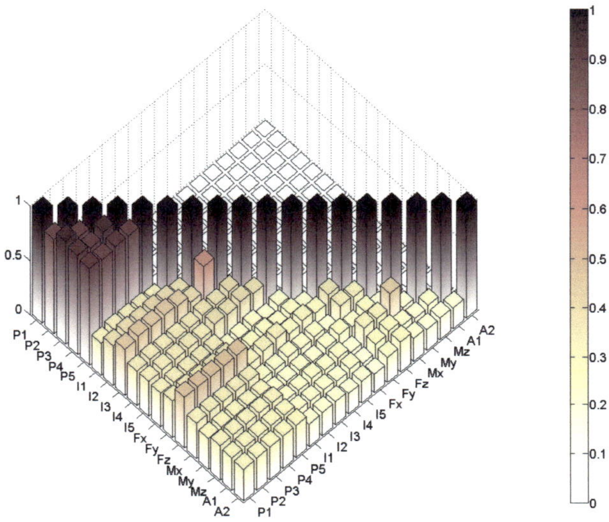

Fig. 18. Metric based on the entropy between the signals for the case (*i*) using the gross rod

4. Conclusion

In this paper an experimental study was conducted to investigate several robot signals. A new sensor classification strategy was proposed.
One of the adopted methodologies leads to arrange the robotic signals in terms of identical spectrum behavior, obtaining three groups of signals: the group of "positions and torques", the group of "currents" and the group of "forces and accelerations".
The other methodology is based on several metrics used to evaluate the relashionship between the signals in the time domain, namely the correlation, the average mutual information and the entropy. These indices revealed the hidden relationships between the signals.
The results merit a deeper investigation as they give rise to new valuable concepts towards instrument control applications. In this line of thought, in future, we plan to pursue several research directions to help us further understand the behavior of the signals.

5. Acknowledgment

The authors would like to acknowledge the GECAD unit.

6. References

Arampatzis, T. & S. Manesis (2005). A Survey of Applications of Wireless Sensors and Wireless Sensor Networks. *Proc. IEEE Int. Symp. on Intelligent Control*. pp. 719-724.

Cheekiralla, S. & W. Engels (2005). A functional taxonomy of wireless sensor network devices. *2nd International Conference on Broadband Networks IEEE*. - Vol. 2. - pp. 949-956. doi:10.1109/ICBN.2005.1589707.

Cover, T. & Thomas, J. (2006). *Elements of Information Theory*. John Wiley & Sons, ISBN-10 0-471-24195-4.

Esteban, J.; Starr, A.; Willetts, R.; Hannah, P. & Bryanston-Cross, P. (2005). A review of data fusion models and architectures: towards engineering guidelines. *Neural Computing & Applications*, Springer , London, 14(4):pp. 273–281.

Hackett, J. & Shah, M. (1990). Multi-sensor fusion: a perspective. *Proc. IEEE Int. Conf. on Robotics & Automation*, pages 1324–1330.

Henderson, T. & Shilcrat, E. (1984), E. Logical sensor systems. *J. of Robotic Systems*. - 2. - Vol. 1. - pp. 169-193.

Lima, M.; Machado, J. & Crisóstomo, M. (2005). Experimental Set-Up for Vibration and Impact Analysis in Robotics, *WSEAS Trans. on Systems*, Issue 5, vol. 4, May, pp. 569-576.

Lima, M.; Machado, J. & Crisóstomo, M. (2006). Windowed Fourier Transform of Experimental Robotic Signals with Fractional Behavior. *Proc. IEEE Int. Conf. on Computational Cybernetics*, August, Tallin, Estonia.

Lima, M.; Machado, J.; Crisóstomo, M. & Ferrolho, M. (2007). On the sensor classification scheme of robotic manipulators. *Int. Journal of Factory Automation, Robotics and Soft Computing - International Society for Advanced Research*, 3, pp.26–31, ISSN1828–6984.

Lima, M.; Machado, J. & Crisóstomo, M. (2007). Towards a sensor classification scheme of robotic manipulators, *CIRA- 7th IEEE International Symposium on Computational Intelligence in Robotics and Automation*, ISBN1424407907, Florida, USA, June,Jacksonville.

Luo, R. & Kay, M. (1990). A tutorial on multisensor integration and fusion. *IEEE 16th Annual Conf. of Industrial Electronics Society*, pp. 707–722, 1990.

MacKay, D. (2003). *Information Theory, Inference, and Learning Algorithms*. Cambridge University Press, August, ISBN 0521642981.

Michahelles, F. & Schiele, B. (2003). Sensing opportunities for physical interaction. *Workshop on Physical Interaction of Mobile HCI conference*, September, Udine, Italy

Orfanidis, S. (1996), *Optimum Signal Processing. An Introduction*. 2nd Edition, Prentice-Hall, Englewood Cliffs, NJ, 1996.

Robotec, E. (1996). *Scorbot ER VII, User's Manual*, Eshed Robotec, ISBN9652910333.

Shannon, C. (1948). A Mathematical Theory of Communication. *The Bell System Technical Journal*. July; October, Vol. 27, pp. 379-423; 623-656.

Spataru, Al. (1970). *Theorie de la Transmission de l'Information – Signaux et Bruits*. Editura technical, Bucarest, Roumanie.

White, R. M. (1987). A sensor classification scheme. *IEEE Trans. on Ultrasonics, Ferroelectrics and Frequency Control*, 34(2):124–126.

Passivity-based Visual Force Feedback Control for Eye-to-Hand Systems

Hiroyuki Kawai [1], Toshiyuki Murao [2] and Masayuki Fujita [3]
[1] *Kanazawa Institute of Technology, Ishikawa*
Japan
[2] *Advanced Institute of Industrial Technology, Tokyo*
Japan
[3] *Tokyo Institute of Technology, Tokyo*
Japan

1. Introduction

Robotics and intelligent machines need sensory information to behave autonomously in dynamical environments. Visual information is particularly suited to recognize unknown surroundings. Vision based control of robotic systems involves the fusion of robot kinematics, dynamics, and computer vision to control the motion of the robot in an efficient manner. The combination of mechanical control with visual information, so-called visual feedback control or visual servoing, is important when we consider a mechanical system working under dynamical environments (Chaumette & Hutchinson, 2008).
For the theoretically problem of three dimensional(3D) visual servo control based on the robot control theory, (Kelly et al., 2000) considered a simple image-based controller under the assumption that the objects' depths are known. (Chen et al., 2007) addressed the field-of-view problem for 3D dynamic visual feedback system using an image-space navigation function. In our previous works, we discussed the dynamic visual feedback control for 3D target tracking based on passivity (Fujita et al., 2007) (Murao et al., 2008). On the other hand, applications of visual feedback system are also increasing in many fields. For example, recent applications of visual feedback system include the autonomous injection of biological cells (Yu & Nelson, 2001), laparoscopic surgery (Omote et al., 1999) and others. Although visual information is necessary in order to recognize environments, only visual information is not enough to complete tasks in these applications. For example, not only visual information but also force information are needed to inject DNA to biological cells. Hence, integrating visual feedback control with force control is important for the modern robot.
(Xiao et al., 2000) developed sensor fusion scheme for controlling an end-effector to follow an unknown trajectory on a contact surface. (Baeten et al., 2003) addressed a hybrid control structure for the eye-in-hand vision and force control. Although many practical methods are reported with experimental results, rigorous results have hardly been obtained in terms of the nonlinear control aspects. For this problem, (Dean-Leon et al., 2006) has combined image-based visual feedback control with force control and discussed the stability of the

nonlinear system. The authors have proposed passivity based visual force feedback control law for force control with target tracking (Kawai et al., 2007). Although these control laws guarantee Lyapunov stability and are effective for the visual force feedback system, they are restricted to planar manipulators.

This chapter deals with 3D visual force feedback control for eye-to-hand systems as depicted in Fig. 1. In our proposed method, we can control not only the position but also the orientation of the robot hand with a contact force in the visual force feedback system. The main contribution of this chapter is to show that the 3D visual force feedback system has the passivity which allows us to prove stability in the sense of Lyapunov. Both the passivity of the manipulator dynamics and the passivity of the visual feedback system are preserved in the 3D visual force feedback system. Finally simulation results are shown to verify the stability of the proposed method.

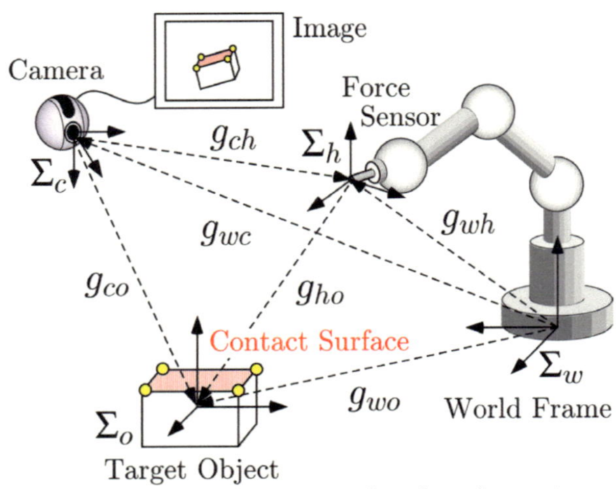

Fig. 1. Visual force feedback system with an eye-to-hand configuration

2. Visual Feedback System

This section mainly reviews our previous works (Fujita et al., 2007) (Murao et al., 2008) via the passivity based visual feedback control. Throughout this chapter, we use the notation $e^{\hat{\xi}\theta_{ab}} \in \Re^{3\times 3}$ to represent the change of the principle axes of a frame Σ_b relative to a frame Σ_a. $\xi_{ab} \in \Re^3$ specifies the direction of rotation and $\theta_{ab} \in \Re$ is the angle of rotation. For simplicity we use $\hat{\xi}\theta_{ab}$ to denote $\hat{\xi}_{ab}\theta_{ab}$. The notation '$\wedge$' (wedge) is the skew-symmetric operator such that $\hat{\xi}\theta = \xi \times \theta$ for the vector cross-product $\times$ and any vector $\theta \in \Re^3$. The notation '$\vee$' (vee) denotes the inverse operator to '$\wedge$', i.e., $so(3) \to \Re^3$. Recall that a skew-symmetric matrix corresponds to an axis of rotation (via the mapping $a \mapsto \hat{a}$). We use the 4×4 matrix

$$g_{ab} = \begin{bmatrix} e^{\hat{\xi}\theta_{ab}} & p_{ab} \\ 0 & 1 \end{bmatrix} \tag{1}$$

as the homogeneous representation of $g_{ab} = (p_{ab}, e^{\hat{\xi}\theta_{ab}}) \in SE(3)$ describing the configuration of a frame Σ_b relative to a frame Σ_a. The adjoint transformation associated with g_{ab} is denoted by $\mathrm{Ad}_{(g_{ab})}$ (Murray et al., 1994).

2.1 Basic Representation for Visual Feedback System

Visual feedback systems with an eye-to-hand configuration typically use four coordinate frames which consist of a world frame Σ_w, a target object frame Σ_o, a camera frame Σ_c and a hand (end-effector) frame Σ_h as in Fig. 1. Then, $g_{wh} = (p_{wh}, e^{\hat{\xi}\theta_{wh}}) \in SE(3)$, $g_{wc} = (p_{wc}, e^{\hat{\xi}\theta_{wc}}) \in SE(3)$ and $g_{wo} = (p_{wo}, e^{\hat{\xi}\theta_{wo}}) \in SE(3)$ denote the rigid body motion from Σ_w to Σ_h, from Σ_w to Σ_c and from Σ_w to Σ_o, respectively. Similarly, the relative rigid body motion from Σ_c to Σ_h, from Σ_c to Σ_o and from Σ_h to Σ_o can be represented by $g_{ch} = (p_{ch}, e^{\hat{\xi}\theta_{ch}}) \in SE(3)$, $g_{co} = (p_{co}, e^{\hat{\xi}\theta_{co}}) \in SE(3)$ and $g_{ho} = (p_{ho}, e^{\hat{\xi}\theta_{ho}}) \in SE(3)$, respectively, as shown in Fig. 1. The objective of the visual feedback control is to bring the actual relative rigid body motion g_{ho} to a given reference $g_d = (p_d, e^{\hat{\xi}\theta_d})$ which is constant in this chapter. In other words, our goal is to determine the motion of hand by using the visual information. The relative rigid body motion from Σ_c to Σ_o can be led by using the composition rule for rigid body transformations ((Murray et al., 1994), Chap. 2, pp. 37, eq. (2.24)) as follows:

$$g_{co} = g_{wc}^{-1} g_{wo} \tag{2}$$

The relative rigid body motion involves the velocity of each rigid body. To this aid, let us consider the velocity of a rigid body as described in (Murray et al., 1994). We define the body velocity of the camera relative to the world frame Σ_w as $V_{wc}^b = \begin{bmatrix} v_{wc}^T & \omega_{wc}^T \end{bmatrix}^T$, where v_{wc} and ω_{wc} represent the velocity of the origin and the angular velocity from Σ_w to Σ_c, respectively ((Murray et al., 1994), Chap. 2, eq. (2.55)).
Differentiating (2) with respect to time, the body velocity of the relative rigid body motion g_{co} can be written as follows (See (Fujita et al., 2007)):

$$V_{co}^b = -\mathrm{Ad}_{(g_{co}^{-1})} V_{wc}^b + V_{wo}^b \tag{3}$$

where V_{wo}^b is the body velocity of the target object relative to Σ_w. In the case of the eye-to-hand configuration, i.e. $V_{wc}^b = 0$, the model of the relative rigid body motion g_{co} can be rewritten as

$$V_{co}^b = V_{wo}^b \tag{4}$$

Roughly speaking, if both the camera and the target object move, then the relative rigid body motion g_{co} will be derived from the difference between the camera velocity V_{wc}^b and the target object velocity V_{wo}^b. Hence, the model of the relative rigid body motion from Σ_c to Σ_o equals the target object velocity V_{wo}^b.

2.2 Estimation Error System

The relative rigid body motion g_{co} can not be immediately obtained in the visual feedback system, because the target object velocity is unknown and furthermore can not be measured directly. Hence, we consider the estimation problem of the relative rigid body motion g_{co}. The visual feedback control task requires information of the relative rigid body motion g_{co}. Since the measurable information is only the image information $f(g_{co})$ in the visual feedback system, we consider a visual motion observer in order to estimate the relative rigid body motion g_{co} from the image information $f(g_{co})$.

Firstly, using the basic representation (4), we choose estimates $\bar{g}_{co}$ and $\bar{V}_{co}^b$ of the relative rigid body motion and velocity, respectively as

$$\bar{V}_{co}^b = u_e. \tag{5}$$

The new input u_e is to be determined in order to drive the estimated values $\bar{g}_{co}$ and $\bar{V}_{co}^b$ to their actual values.

In order to establish the estimation error system, we define the estimation error between the estimated value $\bar{g}_{co}$ and the actual relative rigid body motion g_{co} as

$$g_{ee} = \bar{g}_{co}^{-1} g_{co}. \tag{6}$$

Using the notation $e_R(e^{\hat{\xi}\theta})$, the vector of the estimation error is defined as $e_e := [p_{ee}^T \; e_R^T(e^{\hat{\xi}\theta_{ee}})]^T$. Note that $e_e = 0$ iff $p_{ee} = 0$ and $e_R(e^{\hat{\xi}\theta_{ee}}) = I_3$. Therefore, if the vector of the estimation error is equal to zero, then the estimated relative rigid body motion $\bar{g}_{co}$ equals the actual relative rigid body motion g_{co}.

Suppose the attitude estimation error θ_{ee} is small enough that we can let $e^{\hat{\xi}\theta_{ee}} \approx I + \mathrm{sk}(e^{\hat{\xi}\theta_{ee}})$. Therefore, using a first-order Taylor expansion approximation, the estimation error vector e_e can be obtained from image information $f(g_{co})$ and the estimated value of the relative rigid body motion $\bar{g}_{co}$ as follows (Fujita et al., 2007) :

$$e_e = J^{+}(\bar{g}_{co})(f - \bar{f}), \tag{7}$$

where $\bar{f}$ is the estimated value of image information. In the same way as the basic representation (4), the estimation error system can be represented by

$$V_{ee}^{b} = -\mathrm{Ad}_{(g_{ee}^{-1})} u_e + V_{wo}^{b}. \tag{8}$$

It should be noted that if the vector of the estimation error is equal to zero, then the estimated relative rigid body motion $\bar{g}_{co}$ equals the actual one g_{co}.

2.3 Control Error System

In this subsection, let us consider the dual of the estimation error system, which we call the control error system, in order to establish the visual feedback system. We assume that g_{wc} and g_{wh} can be obtained accurately by a prior calibration procedure, then the estimated value of g_{ho} is calculated as $\bar{g}_{ho} = g_{ch}^{-1} \bar{g}_{co}$ where $\bar{g}_{co}$ is the estimated value which discussed in the previous subsection. Here, we define the control error between the actual relative rigid body motion g_{ho} and desired one g_d as

$$g_{ec} = g_d^{-1} g_{ho}. \tag{9}$$

It should be noted that g_{ho} can not be measured directory. Similar to the definition of e_e, the vector of the control error is defined as $e_c := [p_{ec}^T \; e_R^T(e^{\hat{\xi}\theta_{ec}})]^T$.

Here we have to consider the way of deriving g_{ec} (9), because g_{ho} can not be measured directory. Using g_{ee}, the control error can be transformed as

$$g_{ec} = g_d^{-1} g_{ho} = g_d^{-1} \bar{g}_{ho}^{-1} \bar{g}_{ho} g_{ho} = g_d^{-1} \bar{g}_{ho}^{-1} g_{ee}. \tag{10}$$

In Equation (10), g_d and $\bar{g}_{ho}$ are available information. While the estimation error vector e_e can be obtained as Equation (7), the estimation error matrix g_{ee} cannot be directly obtained, because g_{ee} is defined using non-measurable value g_{co} as Equation (6). Therefore, we consider the way of deriving g_{ee} from e_e.

Because of the definition of the estimation error vector e_e, i.e., $e_e = [p_{ee}^T \; e_R^T(e^{\hat{\xi}\theta_{ee}})]^T$, the position estimation error p_{ee} can be derived directly from e_e. Under the condition $-\frac{\pi}{2} \le \theta_{ee} \le \frac{\pi}{2}$, $\xi\theta_{ee}$ can be derived as follows (Murao et al., 2008):

$$\xi\theta_{ee} = \frac{\sin^{-1}\left\| e_R(e^{\hat{\xi}\theta_{ee}}) \right\|}{\left\| e_R(e^{\hat{\xi}\theta_{ee}}) \right\|} e_R(e^{\hat{\xi}\theta_{ee}}) \tag{11}$$

Hence, g_{ee} can be derived from e_e through $\xi\theta_{ee}$.

The reference of the relative rigid body motion g_d is constant in this chapter, i.e., $\dot{g}_d = 0$, hence, $V_{ec}^b = V_{ho}^b$. Thus, the control error system can be represented as

$$V_{ec}^b = -\mathrm{Ad}_{(g_{ec}^{-1})}\mathrm{Ad}_{(g_d^{-1})}V_{wh}^b + V_{wo}^b. \tag{12}$$

This is dual to the estimation error system.

2.4 Passivity of Visual Feedback System

Combining (8) and (12), we construct the visual feedback system as follows:

$$\begin{bmatrix} V_{ec}^b \\ V_{ee}^b \end{bmatrix} = \begin{bmatrix} -\mathrm{Ad}_{(g_{ec}^{-1})} & 0 \\ 0 & -\mathrm{Ad}_{(g_{ee}^{-1})} \end{bmatrix} u_{ce} + \begin{bmatrix} I \\ I \end{bmatrix} V_{wo}^b \tag{13}$$

where $u_{ce} := [(\mathrm{Ad}_{(g_d^{-1})}V_{wh}^b)^T \ u_e^T]^T$ denotes the control input. For the design of the visual feedback system, it is assumed that the hand velocity V_{wh}^b can be directly chosen. Let us define the error vector of the visual feedback system as $e := [e_c^T \ e_e^T]^T$ which consists of the control error vector e_c and the estimation error vector e_e.

Next, we show an important relation between the input and the output of the visual feedback system.

Lemma 1 (Fujita et al., 2007) : If $V_{wo}^b = 0$, then the visual feedback system (13) satisfies

$$\int_0^T u_{ce}^T \nu_{ce} dt \geq -\beta_{ce}, \ \forall T > 0 \tag{14}$$

where ν_{ce} is defined as $\nu_{ce} := -e$ and β_{ce} is a positive scalar.

Using the following positive definite function, we can prove Lemma 1.

$$V_{ce} = E(g_{ec}) + E(g_{ee}) \tag{15}$$

where $E(g_{ab}) := \frac{1}{2}\|p_{ab}\|^2 + \phi(e^{\hat{\xi}\theta_{ab}})$ and $\phi(e^{\hat{\xi}\theta_{ab}}) := \frac{1}{2}\mathrm{tr}(I - e^{\hat{\xi}\theta_{ab}})$ is the error function of the rotation matrix (see, e.g., (Bullo & Lewis, 2004)).

The block diagram of the passivity of the visual feedback system is shown in Fig. 2. Let us take u_{ce} as the input and ν_{ce} as its output in Fig. 2. Thus, Lemma 1 implies that the visual feedback system (13) is *passive* from the input u_{ce} to the output ν_{ce} as in the definition in (Schaft, 2000).

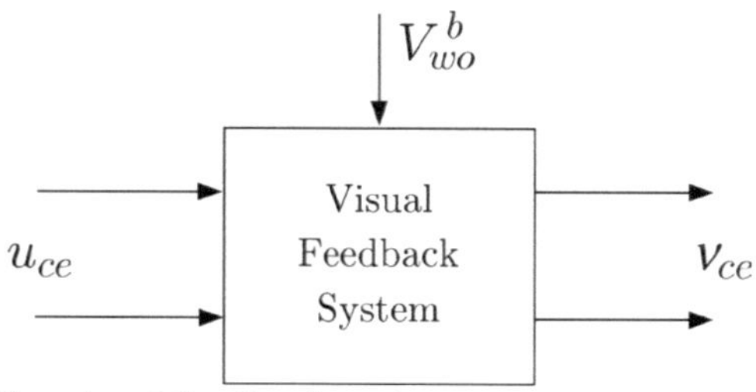

Fig. 2. Block diagram of the visual feedback system

3. Dynamic Visual Force Feedback Control

A force is important to complete tasks which need a contact with environments. This section considers the visual force feedback control with a contact force (we call the dynamic visual force feedback control) based on the passivity which is a main contribution in this chapter.

3.1 Dynamic Visual Force Feedback System

The dynamics of n-link rigid robot manipulators with the end-effector constraint can be written as follows (Liu et al., 1999)

$$M(q)\ddot{q}+C(q,\dot{q})\dot{q}+g(q)=\tau+J_{\varphi}(q)\lambda \tag{16}$$

where q, $\dot{q}$ and $\ddot{q}$ are the joint angle, velocity and acceleration, respectively, τ is the vector of the input torque. $M(q)\in\Re^{n\times n}$ is the manipulator inertia matrix, $C(q,\dot{q})\in\Re^{n\times n}$ is the Coriolis matrix and $g(q)\in\Re^{n}$ is the gravity vector. $\lambda\in\Re$ is the contact force, $J_{\varphi}(q)\in\Re^{n}$ is the normalized Jacobian of the kinematic constraint $\varphi(q)=0\in\Re$ and defined as follows.

$$J_{\varphi}^{T}(q)\dot{q}=0,\quad J_{\varphi}(q)=\left[\frac{\partial\varphi(q)}{\partial q}\right]^{T}\in\Re^{n} \tag{17}$$

Equation (16) possesses several important properties which will be used in the sequel. The manipulator dynamics (16) is passive from τ to $\dot{q}$, that is $\int_{0}^{T}\tau^{T}\dot{q}dt\geq-\beta_{m}$ where β_{m} is a positive scalar. Moreover, $\dot{M}(q)-2C(q,\dot{q})$ is skew-symmetric by defining $C(q,\dot{q})$ using the Christoffel symbols.

Now, we propose the control law for the manipulator as

$$\tau=M(q)\ddot{q}_{r}+C(q,\dot{q})\dot{q}_{r}+g(q)+J_{b}^{T}(q)\mathrm{Ad}_{(g_{d}^{-1})}^{T}e_{e}-J_{\varphi}\lambda_{d}+u_{s}+J_{\varphi}u_{F} \tag{18}$$

where

$$\dot{q}_{r}=Q_{\varphi}(q)\dot{q}_{d}+\alpha J_{\varphi}F_{e}. \tag{19}$$

α is positive constant and

$$F_e := \int_0^t (\lambda - \lambda_d) d\tau = \int_0^t e_\lambda \, d\tau \in \Re \tag{20}$$

where $e_\lambda := \lambda - \lambda_d$ is the force error. Then, the following relation holds with respect to about the force error

$$\dot{F}_e = e_\lambda . \tag{21}$$

Because we consider the single point contact in this chapter, the projection matrix $Q_\varphi(q)$ can be simply defined as

$$Q_\varphi(q) = I - J_\varphi(q) J_\varphi^T(q) \tag{22}$$

which arises on the tangent space at the contact surface $\varphi(q) = 0$ (Liu et al., 1999).
On the other hand, the body velocity of the hand V_{wh}^b is given by

$$V_{wh}^b = J_b(q)\dot{q} = J_b(q) Q_\varphi \dot{q} \tag{23}$$

where $J_b(q)$ is the body manipulator Jacobian (Murray et al., 1994). Moreover, we define the desired body velocity of the hand $u_h = [v_{uh}^T \ \omega_{uh}^T]^T$ which will be obtained from the visual feedback system. Then, u_h can be represented as $u_h = J_b(q) Q_\varphi \dot{q}_d$ by using the projection matrix $Q_\varphi(q)$.
We define the error vector with respect to the joint velocity of the manipulator dynamics as

$$s := \dot{q} - \dot{q}_r \in \Re^n \tag{24}$$

Here, we know that the following relation holds (Liu et al., 1999)

$$J_\varphi^T s = -\alpha F_e . \tag{25}$$

Using (13)(16) and (18), the visual force feedback system with manipulator dynamics (we call the dynamic visual force feedback system) can be derived as follows:

$$\begin{bmatrix} \dot{s} \\ \dot{F}_e \\ V_{ec}^b \\ V_{ee}^b \end{bmatrix} = \begin{bmatrix} -M^{-1}\left(Cs - J_b^T \mathrm{Ad}_{(g_d^{-1})}^T e_c - J_\varphi e_\lambda\right) \\ e_\lambda \\ -\mathrm{Ad}_{(g_{ho}^{-1})} J_b s \\ 0 \end{bmatrix} + \begin{bmatrix} M^{-1} & M^{-1}J_\varphi & 0 & 0 \\ 0 & 0 & 0 & 0 \\ 0 & 0 & -\mathrm{Ad}_{(g_{ec}^{-1})} & 0 \\ 0 & 0 & 0 & -\mathrm{Ad}_{(g_{ee}^{-1})} \end{bmatrix} u + \begin{bmatrix} 0 \\ 0 \\ I \\ I \end{bmatrix} w \tag{26}$$

where

$$x := \begin{bmatrix} s \\ F_e \\ e_c \\ e_e \end{bmatrix} \quad u := \begin{bmatrix} u_s \\ u_F \\ \mathrm{Ad}_{(g_d^{-1})} u_h \\ u_e \end{bmatrix} \quad w := V_{wo}^b,$$

are defined as the state, the input and the disturbance of the dynamic visual force feedback system, respectively. Here, we formulate the manipulator control problem as follows:
Control problem : For the dynamic visual force feedback system with the eye-to-hand configuration described by (26), design a control input u such that

$$\lim_{t\to\infty} s = 0, \quad \lim_{t\to\infty} F_e = 0, \quad \lim_{t\to\infty} e_c = 0 \quad \textit{and} \quad \lim_{t\to\infty} e_e = 0.$$

3.2 Passivity of Dynamic Visual Force Feedback System

Before constructing the dynamic visual force feedback control law, we derive an important lemma.

Lemma 2 : If $w = 0$, then the dynamic visual force feedback system (26) satisfies

$$\int_0^T u^T \nu \, dt \geq -\beta, \ \forall T > 0 \tag{27}$$

where $\nu := Nx$, $N := \mathrm{diag}\{I_n, -\alpha, -I_6, -I_6\}$ and β is a positive scalar.

Proof: Consider the following positive definite function

$$V = \frac{1}{2} s^T M s + \frac{1}{2} \alpha F_e^2 + E(g_{ec}) + E(g_{ee}). \tag{28}$$

Differentiating (28) with respect to time yields

$$\begin{aligned} \dot{V} &= s^T M \dot{s} + \frac{1}{2} s^T \dot{M} s + \alpha F_e \dot{F}_e \\ &\quad + p_{ec}^T e^{\hat{\xi}\theta_{ec}} e^{-\hat{\xi}\theta_{ec}} \dot{p}_{ec} + e_R^T (e^{\hat{\xi}\theta_{ec}}) e^{\hat{\xi}\theta_{ec}} \omega_{ec} + p_{ee}^T e^{\hat{\xi}\theta_{ee}} e^{-\hat{\xi}\theta_{ee}} \dot{p}_{ee} + e_R^T (e^{\hat{\xi}\theta_{ee}}) e^{\hat{\xi}\theta_{ee}} \omega_{ee} \\ &= s^T \begin{bmatrix} M & 0 & 0 & 0 \\ 0 & \alpha & 0 & 0 \\ 0 & 0 & \mathrm{Ad}_{(e^{\hat{\xi}\theta_{ec}})} & 0 \\ 0 & 0 & 0 & \mathrm{Ad}_{(e^{\hat{\xi}\theta_{ee}})} \end{bmatrix} \begin{bmatrix} \dot{s} \\ \dot{F}_e \\ V_{ec}^b \\ V_{ee}^b \end{bmatrix} + \frac{1}{2} s^T \dot{M} s. \end{aligned} \tag{29}$$

Observing that the skew-symmetry of the matrices $\hat{p}_{ec}$ and $\hat{p}_{ee}$, i.e., $p_{ec}^T \hat{p}_{ec} e^{-\hat{\xi}\theta_d} \omega_{ud} = -p_{ec}^T (e^{-\hat{\xi}\theta_d} \omega_{ud})^\wedge p_{ec} = 0$, $p_{ee}^T \hat{p}_{ee} \omega_{ue} = -p_{ee}^T \hat{\omega}_{ue} p_{ee} = 0$, the above equation along the trajectories of the system (26) can be transformed into

$$\dot{V} = -s^T C\dot{s} + s^T s_b^T \mathrm{Ad}^T_{(g_d^{-1})} e_c + s^T J_\varphi e_\lambda + \alpha F_e e_\lambda - e_c^T \mathrm{Ad}_{(g_{ho}^{-1})} J_b s + \frac{1}{2} s^T \dot{M} s$$

$$+ s^T \begin{bmatrix} M & 0 & 0 & 0 \\ 0 & \alpha & 0 & 0 \\ 0 & 0 & \mathrm{Ad}_{(e^{\hat{\xi}\theta_{ec}})} & 0 \\ 0 & 0 & 0 & \mathrm{Ad}_{(e^{\hat{\xi}\theta_{ee}})} \end{bmatrix} \begin{bmatrix} M^{-1} & M^{-1} J_\varphi & 0 & 0 \\ 0 & 0 & 0 & 0 \\ 0 & 0 & -\mathrm{Ad}_{(g_{ec}^{-1})} & 0 \\ 0 & 0 & 0 & -\mathrm{Ad}_{(g_{ee}^{-1})} \end{bmatrix} + \frac{1}{2} s^T \dot{M} s.$$

$$= \frac{1}{2} s^T (\dot{M} - 2C) s + e_\lambda^T (J_\varphi^T s + \alpha F_e) + x^T N^T u = x^T N^T u. \tag{30}$$

Integrating (30) from 0 to T, we obtain

$$\int_0^T u^T \nu \, dt V = V(T) - V(0) \geq -V(0) = -\beta \tag{31}$$

where β is a positive scalar that only depends on the initial states of s, F_e, e_c and e_e.
The block diagram of the passivity of the 3D dynamic visual force feedback system is shown in Fig. 3.

Remark 1 : The visual feedback system (13) satisfies the passivity property as described in (14). It is well known that the manipulator dynamics (16) also has the passivity. In Lemma 2, the inequality (27) says that the dynamic visual force feedback system (26) is *passive* from the input $u = [u_s^T \; u_F^T \; (Ad_{(g_d^{-1})} u_h)^T \; u_e^T]^T$ to the output $\nu = [s^T \; -\alpha F_e^T \; -e_c^T \; -e_e^T]^T$ as shown in Fig. 3.

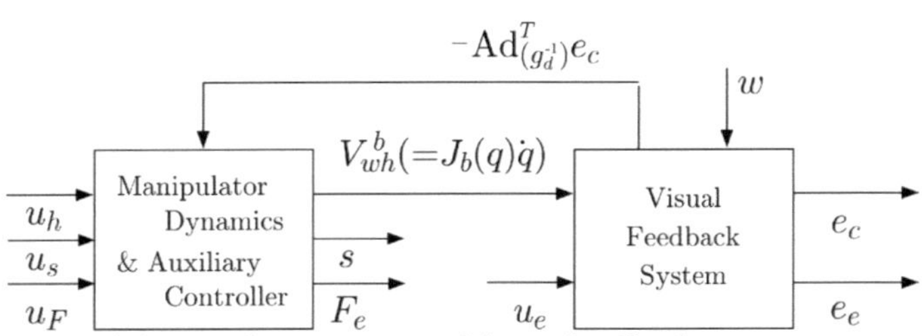

Fig. 3. Block diagram of the 3D dynamic visual force feedback system

3.3 Passivity-based Dynamic Visual Force Feedback Control

We now propose the following control input for the interconnected system:

$$u = -K\nu = -KNx \tag{32}$$
$$K := \mathrm{diag}\ \{K_s, k_F, K_c, K_e\} \in \Re^{n+13}$$

where $K_s := \mathrm{diag}\{k_{s1}, \cdots, k_{sn}\}$, k_F, $K_c := \mathrm{diag}\{k_{c1}, \cdots, k_{c6}\}$ and $K_e := \mathrm{diag}\{k_{e1}, \cdots, k_{e6}\}$ denote the positive gain matrices.

Theorem 1 : If w = 0, then the equilibrium point x = 0 for the closed-loop system (26) and (32)is asymptotic stable.

Proof: In the proof of Lemma 2, we have already derived that the time derivative of V along the trajectory of the system (26) is formulated as (30). Using the control input (32), (30) can be transformed into

$$\dot{V} = -x^T N^T K N x. \tag{33}$$

This completes the proof.
Theorem 1 shows the stability via Lyapunov method for the dynamic visual force feedback system. It is interesting to note that stability analysis is based on the passivity as described in (27). Our proposed method is valid for the 3D dynamic visual force feedback system, while previous works (Dean-Leon et al., 2006) (Kawai et al., 2007) consider the 2D dynamic visual force feedback control. Hence, we can control not only the position but also the orientation of the robot hand with a contact force in the visual force feedback system.

4. Simulation Results

The simulation results on 3DOF planar manipulator as depicted in Fig. 4 are shown in order to understand our proposed method simply, though it is valid for 3D dynamic visual force feedback systems.

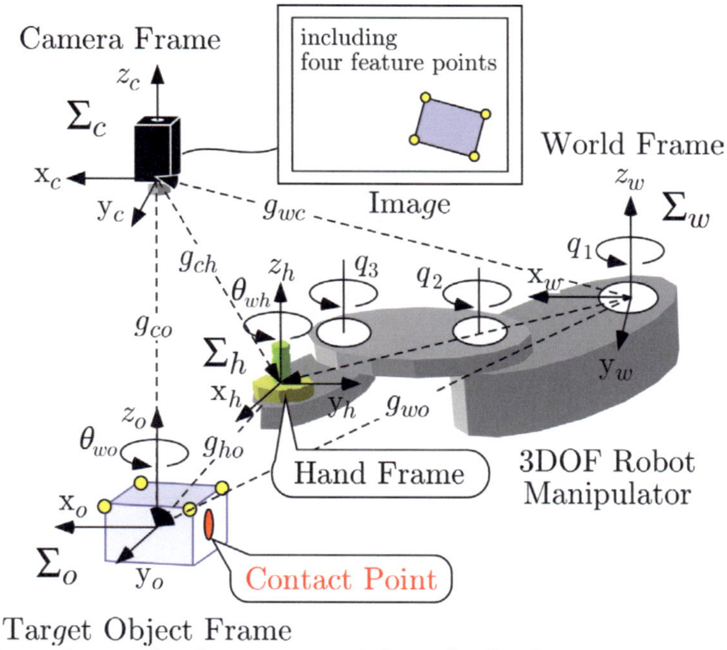

Fig. 4. Coordinate frames for dynamic visual force feedback system with three degree of freedom manipulator

We present results for the stability analysis with a static target object. The simulation is carried out with the conditions $p_{wo} = [0.47 \quad 0.05 \quad 0]^T$ [m], $\xi\theta_{wo} = [0 \quad 0 \quad 0]^T$ [rad],

$p_{wc} = [0.47 \quad 0.05 \quad 1]^T$ [m], $\xi\theta_{wc} = [0 \quad 0 \quad 0]^T$ [rad]. The lengths of the three links of the manipulator are $l_1 = 0.2$ [m], $l_2 = 0.2$ [m] and $l_3 = 0.1$ [m], respectively. The initial angles of the manipulator is set as $q_1(0) = -\pi/2$ [rad], $q_3(0) = \pi/2$ [rad] and $q_2(0) = \pi/2$ [rad]. In other words, the initial relative rigid body motion is $p_{wo} = [0.15 \quad -0.27 \quad 0]^T$ [m], $\xi\theta_{ho} = [0 \quad 0 \quad -\pi/2]^T$ [rad]. The desired force λ_d and the desired relative rigid body motion $g_d = (p_d, e^{\hat{\xi}\theta_d})$ are given by $\lambda_d = 5$ [N], $p_d = [0.03 \quad 0 \quad 1]^T$ [m] and $\xi\theta_d = [0 \quad 0 \quad 0]^T$ [rad] in this simulation. The initial errors of force and vision are calculated as $\lambda_e = -5$ [N], $p_{ec} = [0.12 \quad -0.27 \quad 0]^T$ [m], $\xi\theta_{ec} = [0 \quad 0 \quad -\pi/2]^T$ [rad], $p_{ee} = [0.53 \quad 0.95 \quad 0]^T$ [m] and $\xi\theta_{ee} = [0 \quad 0 \quad \pi/4]^T$ [rad], respectively. The controller parameters for Equation (32) were empirically selected as $K_s = \mathrm{diag}\{10,10,10\}$, $k_F = 25$, $\alpha = 1$, $K_c = \mathrm{diag}\{40,40,20,20,20,40\}$ and $K_e = 50I_6$. The simulation results are shown in Figs. 5-7.

Figs. 5-7 illustrate the control error e_c, the estimatoin error e_e, and the contact force λ, respectively. In Figs. 5 and 6, we focus on the errors of the translations of x and y and the rotation of z, because the errors of the translation of z and the rotations of x and y are zeros ideally on the 3DOF planar manipulator. The control error e_c and the estimation error e_e tended to zero, thus we can confirm that the relative rigid body motion g_{ho} coincided with the desired one g_d by using image information. In Fig. 7, the pulse siglnal means the contact transition at around 0.2 [s]. The contact force λ tended to 5 [N], i.e., converged to the desired one λ_d. From these figures, the asymptotic stability can be also confirmed.

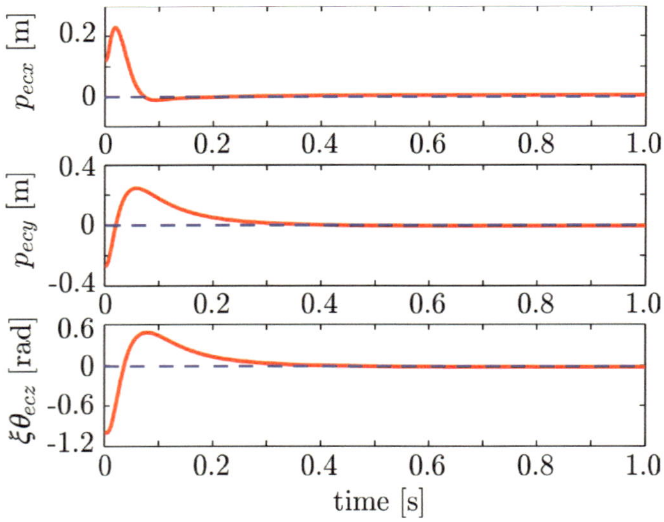

Fig. 5. The control error e_c which defined the error between the relative rigid body motion of the robot hand g_{ho} and desired one g_d. Initial control errors are $p_{ecx} = 0.12$ [m], $p_{ecy} = -0.27$ [m] and $\xi\theta_{ecz} = -\pi/2$ [rad], respectively.

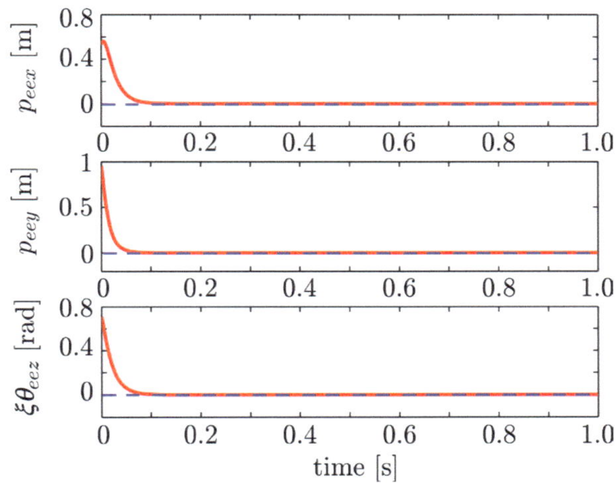

Fig. 6. The estimation error e_e which defined the error between the estimated value $\bar{g}_{co}$ and the actual relative rigid body motion g_{co}. Initial estimation errors are $p_{eex} = 0.53$ [m], $p_{eey} = -0.95$ [m] and $\xi\theta_{eez} = \pi / 4$ [rad], respectively.

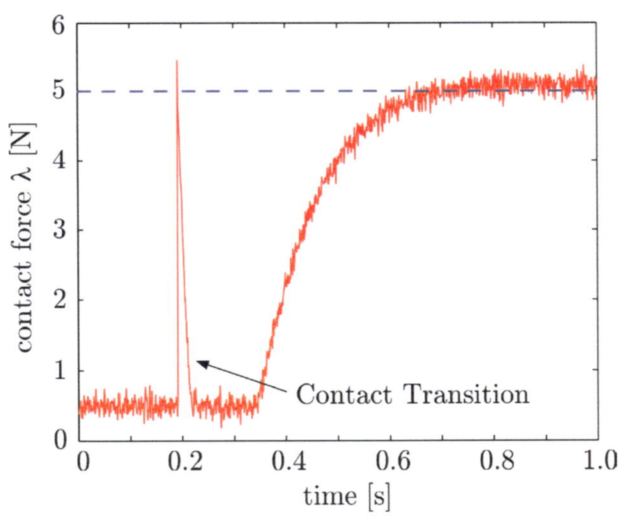

Fig. 7. The contact force trajectory λ

5. Conclusions

This chapter considers 3D visual force feedback control for eye-to-hand systems. In our approach, we can control not only the position but also the orientation of the robot hand with a contact force by using visual information. The proposed method can be regarded as an extension of the hybrid position/force control to the hybrid vision/force control. The main contribution of this chapter is to show that the visual force feedback system has the passivity which allows us to prove stability in the sense of Lyapunov. Both the passivity of the manipulator dynamics and the passivity of the visual feedback system are preserved in the visual force feedback system. Finally simulation results are shown to verify the stability of the proposed method.

6. References

Baeten, J.; Bruyninckx, H. & Schutter, J. D. (2003). Integrated Vision/Force Robotic Servoing in the Task Frame Formalism, *International Journal of Robotics Research*, Vol. 22, No. 10-11, pp. 941-954.

Bullo, F. & Lewis, A.D. (2004). *Geometric Control of Mechanical Systems*, Springer-Verlag.

Chaumette, F. & Hutchinson, S. (2008). Visual Servoing and Visual Tracking, In: *Springer Handbook of Robotics*, B. Siciliano and O. Khatib (Eds), pp. 563-583. Springer-Verlag.

Chen, J.; Dawson, D. M.; Dixon, W. E. & Chitrakaran, V. K. (2007). Navigation Function-based Visual Servo Control, *Automatica*, Vol. 43, No. 7, pp. 1165-1177.

Dean-Leon, E. C.; Parra-Vega, V. & Espinosa-Romero, A. (2006). Visual Servoing for Constrained Planar Robots Subject to Complex Friction, *IEEE/ASME Trans. on Mechatronics*, Vol. 11, No. 4, pp. 389-400.

Fujita, M.; Kawai, H. & Spong, M. (2007). Passivity-based Dynamic Visual Feedback Control for Three Dimensional Target Tracking: Stability and L_2-gain Performance Analysis, *IEEE Transactions on Control Systems Technology*, Vol. 15, No. 1, pp. 40-52.

Kawai, H.; Murao, T. & Fujita, M. (2007). Passivity-based Visual Force Feedback Control for Planar Manipulators with Eye-in-Hand Configuration, *Proc. of the 2007 IEEE Multi-conference on Systems and Control*, pp. 1480-1485.

Kelly, R.; Carelli, R.; Nasisi, O.; Kuchen, B. & Reyes, F. (2000). Stable Visual Servoing of Camera-in-Hand Robotic Systems, *IEEE Trans. Mechatronics*, Vol. 5, No. 1, pp. 39 - 48.

Liu, Y.; Kitagaki, K.; Ogasawara, T. & Arimoto, S. (1999). Model-based Adaptive Hybrid Control for Manupilators Under Multiple Geometric Constraints, *IEEE Trans. on Control Systems Technology*, Vol. 7, No. 1, pp. 97-109.

Murao, T.; Kawai, H. & Fujita, M. (2008). Predictive Visual Feedback Control with Eye-in/to-Hand Configuration via Stabilizing Receding Horizon Approach, *Proc. of the 17th IFAC World Congress on Automatic Control*, pp. 5341-5346.

Murray, R.; Li Z. & Sastry, S. S. (1994). *A Mathematical Introduction to Robotic Manipulation*, CRC Press,

Omote, K. et al., (1999). Self-Guided Robotic Camera Control for Laparoscopic Surgery Compared with Human Camera Control, *The American Journal of Surgery*, Vol. 177, No. 4, pp. 321-324.

van der Schaft, A. (2000). *L_2-Gain and Passivity Techniques in Nonlinear Control* (2nd ed.), Springer-Verlag,

Xiao, D.; Ghosh, B. K.; Xi, N. & Tarn, T. J. (2000). Sensor-based Hybrid Position/Force Control of a Robot Manipulator in an Uncalibrated Environment, *IEEE Transactions on Control Systems Technology*, Vol. 8, No. 4, pp. 635-645,

Yu, S. & Nelson, B. J. (2001). Autonomous Injection of Biological Cells Using Visual Servoing, In: *Experimental Robotics VII*, D. Rus and S. Singh (Eds), pp. 169-178, Springer-Verlag.

19

Kinematic Analysis of 3-UCR Parallel Robot Leg

Cheng Gang and Ge Shi-rong
College of Mechanical and Electrical Engineering Xuzhou
China University of Mining and Technology
China

1. Introduction

Mine rescue robot was developed to enter mines during emergencies, such as underground explosion, roof fall or water inundation, to locate possible escape routes for those trapped inside and determine whether it is safe for human to enter or not. Comparing with wheeled robots (Baker et al., 2004) and tracked robots (Tanaka et al., 2005; Wang et al., 2007), legged robots are flexible and effective to move on uneven surfaces and natural environments because of its adaptability to the geometry of the terrain and in principle support on very steep surfaces. Though thought as the promising systems, it is always challenging and complex to achieve robust locomotion of legged robots.

Most of the recent prototypes of legged robots, including Lauron III (Gabmann et al., 2005), RHex (Koditschek et al., 2004) and SILO4 (Santos et al., 2005), adopted the serial leg manipulator. Due to better system rigidity, rapid motion velocity, high nominal load to weight ratio and flexible end position-stance, the parallel manipulators are feasible to be the leg manipulators of mine rescue robots. But it is well known that the close chain often leads to difficulty in its mechanical design. Since Clavel and his Delta structures (Clavel, 1988) in the late 80's have reached extremely high performances, lower-mobility parallel manipulators have been under intensive study for over many years. Lower-mobility parallel manipulators have simpler mechanical structure, simpler control system, high speed performance, low manufacturing and operations cost (Kim, 2001). Therefore, they have been applied in some fields, including telescope applications (Carretero et al., 2000), flight simulation (Pouliot et al., 1996) and beam aiming applications (Dunlop & Jones, 1997). Among lower-mobility parallel manipulators, special attention has been paid to optimization and innovation of 3 degree-of-freedom (DOF) parallel manipulators. Some 3-DOF translational and spherical parallel manipulators were proposed respectively.

The topological structure of 3-UCR symmetrical parallel robot leg is described in Fig. 1(a). The parallel manipulator consists of a fixed base, a moving platform and three limbs with identical structure. In Fig. 1(a), O_O-$X_OY_OZ_O$ is the static coordinate system attached to the base, while $O_{O'}$-$X_{O'}Y_{O'}Z_{O'}$ is the moving coordinate system attached to the moving platform. The lengths of the equilateral triangle lines in the moving platform and the base, such as $L_{A'B'}$ and L_{AB}, are denoted as L_m and L_B, respectively. Each limb connects the moving platform to the base by a revolute joint (R) followed by cylindrical joint (C) and universal joint (U) in sequence. The lengths of limbs are given as $L_{II'}$ where I can be substituted by A, B, C.

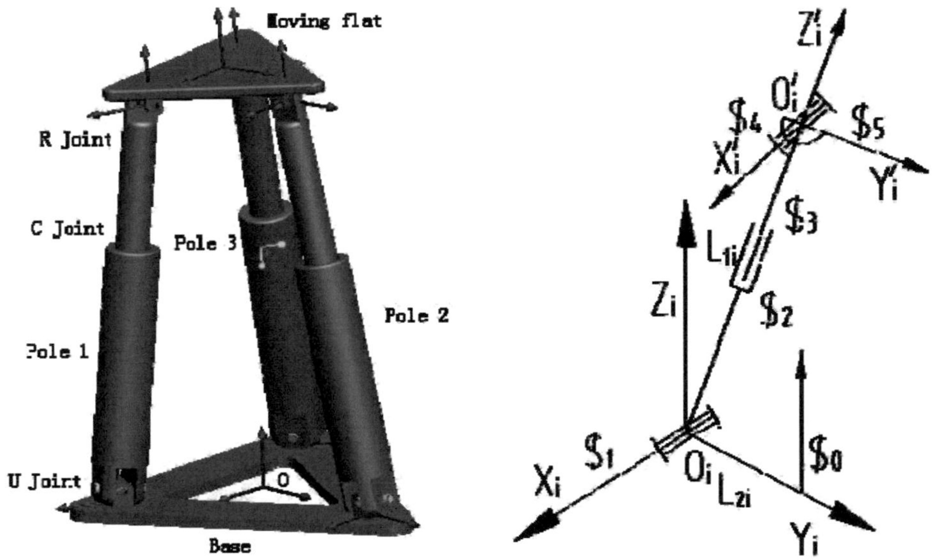

Fig. 1. 3-UCR spatial parallel manipulator: (a) 3-UCR parallel manipulator, (b) UCR topological limb

As shown in Fig. 1(a), the number of links is n=8 for 1 platform, 3 cylinders, 3 piston-rods, and 1 base; the number of joints is 9 for 3 revolute joints, 3 cylindrical joints, and 3 universal joints. Based on a revised Kutzbach-Grübler equation, the DOF of 3-UCR parallel robot leg is 3. Therefore, any screw of the manipulator consisted of three linearly independent principal screws and it is known as third-order screw system.

2. Instantaneous Kinematic Characteristics of 3-UCR Parallel Robot Leg

The performance of parallel robot leg largely depends on the characteristics of the end-effector, including the DOF number, workspace, singularity and dynamic performance, decided by the instantaneous kinematics. Because of the kinematic coupling of parallel robot leg caused by the interaction of limbs, it is complicated to describe the instantaneous kinematics of parallel robot leg. The reciprocal screw theory was introduced by Fang and Alon and the spatial screw restrictions were used to analyze the kinematic characteristics of moving platform (Fang & Tsai, 2004; Alon & Moshe, 2006). Sokolov made the similar study on the basis of differential geometrical method and he further developed the singularities of moving platform (Sokolov & Xirouchakis, 2006). For the purpose of the instantaneous kinematic analysis of 3-UCR parallel robot leg with 2R1T DOF, the principal screw theory in reciprocal screw theory can be proposed to analyze the robot leg. By contrast with the first-order influence coefficient matrix between base and moving platform, the principal screw model can be established. So the spatial screw restrictions and the corresponding kinematics of moving platform can be also obtained.

2.1 Principal Screw Theory

Screw theory is important to analyze the kinematic characteristics of manipulators. The principal screws intersect perpendicularly each other in screw system, and the number of principal screws equals to the order of screw system. The principal screws describe the instantaneous independent motion of rigid body, and any screw in screw system is the linear combination of principal screws. Therefore, the principal screws are an important tool in the analysis of kinematic characteristics. In order to further understand the kinematic nature and all possible motions of the manipulators at any given instant, the principal screws of the manipulators should be identified from screw system. Compared with the principal screws in second-order screw system, the solving process of principal screws in third-order screw system is more complicated and there are three principal screw pitches h_1, h_2 and h_3. All screw pitch values exist in the scale between two extreme values denoted as h_1 and h_3. Ball deduced the principal screws of third-order system by means of mapping geometry method (Ball, 1900) and Fang analyzed for the same purpose based on conic section degenerating theory (Fang & Huang, 1998).

The study (Hunt, 1978) indicated that the quadric surface would degenerate into principal screw axis 1or 3 when $h=h_1$ or $h=h_3$. The quadric surface would degenerate into a pair of intersecting planes and the corresponding intersecting line is principal screw axis 2 when $h=h_2$. The middle symmetry axis in hyperboloid of one sheet is regarded as principal screw axis 1, and the semi-major axis of middle ellipse is principal screw axis 2 when $h_1>h>h_2$. The mid-symmetry axis in hyperboloid of one sheet is principal screw axis 3, and the semi-major axis of middle ellipse in the plane decided by axis 1 and axis 2 is still axis 2 when $h_2>h>h_3$. The second equation would degenerate if any pitch value in the system equals to any of h_1, h_2 and h_3. For example, the planar quadratic equation would degenerate into two virtual lines crossing vertically at one point when $h=h_1$ or $h=h_3$, and the equation would degenerate into two solid lines when $h=h_2$.

Based on the screw theory (Hunt, 1978), a screw can be written as

$$\$ = (S, S_0 + hS), \tag{1}$$

where S, S_0 denote the direction and position of a line in space respectively and h is the pitch of the screw. h=0 and $h=\infty$ correspond to pure rotation and pure translation of a rigid body, and the screw will have the form (S, S_0) and (0, hS) respectively.

2.2 Principal Screw Model

From the theory as above mentioned, the screw pitch of rotation joint and translation joint respectively is 0 and ∞. For the purpose of analyzing DOF and kinematics characteristics of 3-UCR parallel robot leg, the revolute and cylindrical characteristics of moving parts should be obtained. Therefore, it is necessary to put emphasis on the analysis of principal screw pitch.

In order to solve the principal screw pitch, the first-order influence coefficient matrix G and G' need to be analyzed firstly. The first-order influence coefficient matrix G_i^0 (i=1, 2, 3) of limbs can be deduced by means of imaginary mechanism principle as

$$\left[G_i^0\right] = \left[A_i^0\right]\left[\$_0^{(i)} \quad \$_1^{(i)} \quad \$_2^{(i)} \quad \$_3^{(i)} \quad \$_4^{(i)} \quad \$_5^{(i)}\right], \tag{2}$$

where A_i^0 is the transform matrix between the limb coordinate system and the static coordinate system. So the equation about the angular velocity ω of the end-effector in moving platform, the linear velocity v_p of the chosen reference point and the input velocity vector $\dot{\xi}^{(i)}$ of every limb can be shown as

$$[\omega \quad v_P]^T = \left[G_i^0\right]\left[\dot{\xi}_0^{(i)} \quad \dot{\xi}_1^{(i)} \quad \dot{\xi}_2^{(i)} \quad \dot{\xi}_3^{(i)} \quad \dot{\xi}_4^{(i)} \quad \dot{\xi}_5^{(i)}\right]. \tag{3}$$

If the matrix $\left[G_i^0\right]$ is non-singular, there would be $\left[G_0^i\right] = \left[G_i^0\right]^{-1}$. If the velocities of cylindrical joint 1 and 3 can be chosen as the input velocities in third-order screw system, taking and combining the first row and third row corresponding to $\xi_0^{(i)}$ and $\xi_2^{(i)}$ from the matrix $\left[G_0^i\right]$ of every limbs in Eq. (3) could get the following equation as

$$\left[G_H^q\right] = \left[\left[G_0^1\right]_3 \quad \left[G_0^2\right]_3 \quad \left[G_0^3\right]_3 \quad \left[G_0^1\right]_1 \quad \left[G_0^2\right]_1 \quad \left[G_0^3\right]_1\right]^T \in R^{6\times 6}. \tag{4}$$

The manipulators with different DOF need different number of input parameters, except for redundant driving, so the forms of first-order influence matrix G and G′ are decided by the number of DOF. Since 3-UCR manipulator has three DOF, it is necessary to have three inputs. Therefore, the corresponding relation of inputs and outputs can be established as

$$[\omega \quad v_P]^T = \left[G_L^H\right]\left[\dot{L}_1 \quad \dot{L}_2 \quad \dot{L}_3\right] = [G' \quad G]\left[\dot{L}_1 \quad \dot{L}_2 \quad \dot{L}_3\right], \tag{5}$$

where $\dot{L}_1$, $\dot{L}_2$, $\dot{L}_3$ are the known input velocities of limbs and G_L^H formed by taking the first three columns of $\left[G_H^q\right]^{-1}$ is a 6×3 matrix.

Fang analyzed the screw characteristics of instantaneous kinematics of robot, and the pitch and the axis of screw can be solved as

$$h = \frac{\{\mu \quad \sigma \quad 1\}^T [G']^T [G] \{\mu \quad \sigma \quad 1\}}{\{\mu \quad \sigma \quad 1\}^T [G']^T [G'] \{\mu \quad \sigma \quad 1\}}, \tag{6}$$

$$[r][G'\{\mu \quad \sigma \quad 1\}] = ([G] - h[G'])\{\mu \quad \sigma \quad 1\}, \tag{7}$$

where $(\mu \quad \sigma) = \left(\dot{L}_1/\dot{L}_3 \quad \dot{L}_2/\dot{L}_3\right)$.

It is obviously that any pair of variables (μ, σ) corresponds to a screw in space. Therefore, the screw in third-order system can be represented by using any point (μ, σ). Eliminating (x, y, z) from Eq. (6), we obtain

$$a_{11}\mu^2 + 2a_{12}\mu\sigma + a_{22}\sigma^2 + 2a_{13}\mu + 2a_{23}\sigma + a_{33} = 0, \tag{8}$$

where 3×3 matrix a_{ij} (i, j=1, 2, 3) is the function about screw pitch of moving platform and elements of matrix [G] and [G'].

If the quadric curve expressed by Eq. (8) degenerates, that is σ becomes the linear function of μ, the following condition must be satisfied,

$$D = \begin{vmatrix} a_{11} & a_{12} & a_{13} \\ a_{21} & a_{22} & a_{23} \\ a_{31} & a_{32} & a_{33} \end{vmatrix} = 0 . \tag{9}$$

Expending Eq. (9), we obtain

$$c_1 h^3 + c_2 h^2 + c_3 h + c_4 = 0 , \tag{10}$$

where the coefficients c_i (i=1, 2, 3, 4) are formed by the elements in matrix [G] and [G']. Three possible roots corresponding to the pitches of three principal screws can be obtained by solving Eq. (10), and the instantaneous characteristics of the manipulator can be gotten by analyzing them.

2.3 Screw Analysis of 3-UCR Parallel Robot Leg

In order to solve first-order influence coefficient matrixes, the limb screw system should be constructed as shown in Fig. 1(b) by means of imaginary mechanism principal firstly. An imaginary link and an imaginary revolute pair denoted by a screw with zero pitch are added to every limb. Moreover, there are five unit-DOF kinematic pairs in every limb of 3-UCR parallel robot leg. So every limb of this manipulator has six unit pairs. For the purpose of keeping equivalent kinematic effect between the imaginary manipulator and the real one, let the velocity amplitude of the imaginary unit screw $\$_0$ of every limb be zero, and $\$_0$ is linearly independent with the other five real screws of the primary limb. Therefore, the plücker coordinate of the imaginary screw $\$_0$ in the static coordinate system is

$$\$_0 = \begin{pmatrix} 0 & 0 & 1 & l_2 & 0 & 0 \end{pmatrix}, \tag{11}$$

where l_2 is the length from the origin to the centers of three imaginary pairs which is shown in Fig. 1(b). And the plücker coordinates of three original limbs in base system are

$$\left.\begin{aligned} \$_1 &= \begin{pmatrix} 0 & 1 & 0 & 0 & 0 & 0 \end{pmatrix}, \\ \$_2 &= \begin{pmatrix} 0 & 0 & 0 & -\sin(\alpha_i) & 0 & \cos(\alpha_i) \end{pmatrix}, \\ \$_3 &= \begin{pmatrix} -\sin(\alpha_i) & 0 & \cos(\alpha_i) & 0 & 0 & 0 \end{pmatrix}, \\ \$_4 &= \begin{pmatrix} 0 & 1 & 0 & l_{1i}\cos(\alpha_i) & 0 & l_{1i}\sin(\alpha_i) \end{pmatrix}, \\ \$_5 &= \begin{pmatrix} \cos(\alpha_i) & 0 & \sin(\alpha_i) & 0 & -l_{1i} & 0 \end{pmatrix}, \end{aligned}\right\} \tag{12}$$

where l_{1i} is the instantaneous length of every limb and α_i is the angel between limb l_i and Z_i in the limb coordinate system, $\alpha_i = \arcsin((r - r')/l_i)$. The first-order influence matrix G_i^0 of every limb can be obtained by Eq. (11) and Eq. (12). By selecting the imaginary pair and the prismatic motion in cylindrical joint as the initiative inputs, the influent matrix G_H^q can be obtained as

$$G_H^q = \begin{bmatrix} 0 & -c\alpha_1 & 0 & -s\alpha_1 & 0 & c\alpha_1 \\ \dfrac{\sqrt{3}rc\alpha_2}{2} & \dfrac{1}{2rc\alpha_2} & 0 & \dfrac{s\alpha_2}{2} & \dfrac{-\sqrt{3}s\alpha_2}{2} & c\alpha_2 \\ \dfrac{\sqrt{3}rc\alpha_3}{2} & \dfrac{1}{2rc\alpha_3} & 0 & \dfrac{s\alpha_3}{2} & \dfrac{\sqrt{3}s\alpha_3}{2} & c\alpha_3 \\ \dfrac{l_{11}c\alpha_1}{l_2+l_{11}s\alpha_1} & 0 & \dfrac{r+l_{11}s\alpha_1}{l_2+l_{11}s\alpha_1} & 0 & \dfrac{1}{(l_2+l_{11}s\alpha_1)} & 0 \\ \dfrac{-l_{12}c\alpha_2}{2(l_2+l_{12}s\alpha_2)} & \dfrac{\sqrt{3}l_{12}c\alpha_2}{2(l_2+l_{12}s\alpha_2)} & \dfrac{r+l_{12}s\alpha_2}{l_2+l_{12}s\alpha_2} & \dfrac{-\sqrt{3}}{2(l_2+l_{12}s\alpha_2)} & \dfrac{-1}{2(l_2+l_{12}s\alpha_2)} & 0 \\ \dfrac{-l_{13}c\alpha_3}{2(l_2+l_{13}s\alpha_3)} & \dfrac{-\sqrt{3}l_{13}c\alpha_3}{2(l_2+l_{13}s\alpha_3)} & \dfrac{r+l_{13}s\alpha_3}{l_2+l_{13}s\alpha_3} & \dfrac{\sqrt{3}}{2(l_2+l_{13}s\alpha_3)} & \dfrac{-1}{2(l_2+l_{13}s\alpha_3)} & 0 \end{bmatrix}. \qquad (13)$$

The first-order influence coefficient matrixes G and G' of 3-UCR parallel robot leg can be solved by the above principal screw model, and the pitches of the general screw can be gotten by Eq. (6) and Eq. (10). Analyzing the pitches shows that 3-UCR parallel robot leg with different position and orientation has only 3 DOF including 2R1T. In the following part, the principal screw when the base is parallel to the moving platform, as an example of numerical simulation, would be analyzed.

2.4 Analysis of principal screw when the base is parallel to the moving platform

According to the topological structure of UCR limb, the parameters need to be taken in this configuration. They are l_{1i}=180*mm*, l_2=35*mm*, r=45*mm*, and r'=35*mm*. By substituting those parameters into the above principal screw model, we obtain the influent matrix G_H^q as

$$G_H^q = \begin{bmatrix} 0 & -4.9938 & 0 & -4.9979\times10^{-2} & 0 & 9.9875\times10^{-1} \\ 4.3248 & 2.4969 & -7.4742\times10^{-17} & 2.4989\times10^{-2} & -4.3283\times10^{-2} & 9.9875\times10^{-1} \\ -4.3248 & 2.4969 & 7.4742\times10^{-17} & 2.4989\times10^{-2} & 4.3283\times10^{-2} & 9.9875\times10^{-1} \\ 3.9953 & 0 & 1.2 & 0 & 2.0002\times10^{-2} & 0 \\ -1.9976 & 3.46 & 1.2 & -1.7322\times10^{-2} & -0.01 & -6.9389\times10^{-18} \\ -1.9976 & -3.46 & 1.2 & 1.7322\times10^{-2} & -0.01 & 6.9389\times10^{-18} \end{bmatrix}. \qquad (14)$$

By analyzing the previous equation, the first-order influence coefficient matrixes G and G' are written as

$$G = \begin{bmatrix} -2.2224 & 1.1112 & 1.1112 \\ 4.4409\times10^{-16} & -1.9246 & 1.9246 \\ 3.3375\times10^{-1} & 3.3375\times10^{-1} & 3.3375\times10^{-1} \end{bmatrix}, \qquad (15)$$

$$G' = \begin{bmatrix} -1.7347\times10^{-18} & 9.635\times10^{-3} & -9.635\times10^{-3} \\ -1.1126\times10^{-2} & 5.5629\times10^{-3} & 5.5629\times10^{-3} \\ 3.7565\times10^{-18} & -2.3217\times10^{-16} & 2.2927\times10^{-16} \end{bmatrix}. \quad (16)$$

Substituting Eq. (14) and Eq. (15) into Eq. (6) gets the following equation as

$$9.0939\times10^{-44}h^3 + 6.097\times10^{-26}h^2 + 1.8077\times10^{-36}h - 1.1126\times10^{-49} = 0. \quad (17)$$

By solving Eq. (16), the pitches of the instantaneous 3-UCR parallel robot leg are obtained as

$$h = \begin{bmatrix} -6.7045\times10^{17} & -2.971\times10^{-9} & 6.1421\times10^{-12} \end{bmatrix}. \quad (18)$$

From the Eq. (18), it is shown that two possible roots of h are close to zero and one possible root of h can be seen as infinite in spite of any inputs. Based on the definition of screw theory, we can see that the screw is pure rotation when the pitch is zero and pure translation when the pitch is close to infinite. Therefore, the 3-UCR parallel robot leg has two instantaneous rotations and one instantaneous translation when the base is parallel to the moving platform.

3. Inverse Kinematic Analysis of 3-UCR Parallel Robot Leg

For better motion control, it is essential to analyze the kinematics of the parallel robot leg. Kim and Park considerably simplified the kinematic algorithm of 3-RS parallel manipulator based on Sylvester's elimination method (Kim & Park, 2001). Kindermann and Cruse proposed the mean of multiple computations to solve the kinematics of manipulators of nearly arbitrary configuration and validated the method by the calculation of a hexapod walking system (Kindermann & Cruse, 2002). Sokolov introduced subtly some novel geometrical parameters and established the inverse kinematic model about a 3-RPS parallel manipulator (Sokolov & Xirouchakis, 2005). Through the double semi-ellipses approximate distribution model, Wang obtained the corrected inverse kinematic solution of the variable geometry parallel manipulator (Wang & Yang, 2005). It is necessary to choose the descriptions of the attitude motion of rigid bodies including direction-cosine, Euler angles, quaternion and Rodrigues parameters. But direction-cosine method needs nine parameters and six constraint equations and it is difficult to be solved. Quaternion method has four parameters which is more than the least number of parameters required to describe the orientation of a rotating rigid body that is three. Though described by three parameters, Euler angle method has singularities. Such this case is Rodirgues parameters can be used. Moreover, Rodrigues parameters stand for trigonometric functions in the kinematic model and improve the ability of real-time control. But when calculating by Rodirgues parameters, eigenaxis rotations greater than 180° cannot be allowed because of the corresponding singularities. The kinematic characteristics of the parallel robot leg decide that its motions are within the range of angle limits. So Rodrigues parameter method is adopted to describe the parallel robot leg.

3.1 RODRIGUES PARAMETERS

In 1840, Rodrigues published a paper on the transformation groups, and Rodrigues parameters that integrate the direction cosines of a rotation axis with the tangent of half the rotation angle were presented with three quantities. The angles of the rotations appear as half-angles which occurred for the first time in the study of rotations. The half-angles are an essential feature of the parameterization of rotations and are the measure of pure rotation for the most elegant representation of rotations in kinematics (Dai, 2006). The Rodrigues parameters were further taken by Cayley to comprise a skew symmetric matrix which then formed Cayley's formula (Cayley, 1875) for a rotation matrix (Altmann, 1986).

Cayley-Rodrigues parameters can be used to eliminate the constraints associated with the Euler parameters, and further these parameters reduce the number of coordinates that describe the rigid body orientation from four to three. This fact can be established by defining so-called Cayley-Rodrigues parameters as follows:

$$\Phi_i = \lambda_i / \lambda_0 \qquad (i\text{=}1, 2, 3), \tag{19}$$

where λ_i (i=0, 1, 2, 3) are defined as the Euler parameters.

The Cayley-Rodrigues parameters λ_i (i=1, 2, 3) are also components of the Gibbes which defined as

$$\Phi_i = \hat{p}_i \tan(\theta/2) \quad (i\text{=}1, 2, 3), \tag{20}$$

where $\hat{p}_i$ (i=1, 2, 3) are the components of principal vector of rotation $\hat{p}$ referred to the body axes.

Considering Φ_i (i=1, 2, 3) as the projection of connected coordinates composes the Rodrigues vector denoted as $\mathbf{\Phi}$. So the direction-cosine matrix with Rodrigues parameters can be written as

$$\mathbf{A} = \frac{1}{1+\Phi_1^2+\Phi_2^2+\Phi_3^2}\begin{bmatrix} 1+\Phi_1^2-\Phi_2^2-\Phi_3^2 & 2(\Phi_1\Phi_2-\Phi_3) & 2(\Phi_1\Phi_3+\Phi_2) \\ 2(\Phi_2\Phi_1+\Phi_3) & 1-\Phi_1^2+\Phi_2^2-\Phi_3^2 & 2(\Phi_2\Phi_3-\Phi_1) \\ 2(\Phi_3\Phi_1-\Phi_2) & 2(\Phi_3\Phi_2+\Phi_1) & 1-\Phi_1^2-\Phi_2^2+\Phi_3^2 \end{bmatrix}. \tag{21}$$

3.2 Inverse kinematics of moving platform

As shown in the topological figure of 3-UCR parallel robot leg, $O_0I'O_0'$ is considered as the closed-loop kinematic chain, and the corresponding vector of the chain can be written as

$$\overrightarrow{O_0I'} = \mathbf{A}\overrightarrow{O_0'I'} + \overrightarrow{O_0O_0'}\,, \tag{22}$$

where O_0I', $O_0'I'$ and O_0O_0' correspond to the coordinates of the vector O_oI', $O_o'I'$ and O_oO_o', respectively.

Due to the geometrical characteristics of the parallel manipulator, the kinematic spaces of limbs are limited in three planes that are defined, respectively, as x=0 in limb AA', $x=-\sqrt{3}y$ in limb BB' and $x=\sqrt{3}y$ in limb CC'. So substituting Eq. (21) into Eq. (22) gives the following equations:

$$\Phi_3 = 0, \tag{23}$$

$$x_{O'} = -2L_m \Phi_1 \Phi_2 / \sqrt{3}\lambda_0^2, \tag{24}$$

$$y_{O'} = L_m \left(\Phi_2^2 - \Phi_1^2\right) / \sqrt{3}\lambda_0^2. \tag{25}$$

Suppose that the velocity of origin on the moving coordinate system is denoted as $V_{O'}$ and the corresponding sub-velocities along coordinate axes are denoted as $V_{O'x}$, $V_{O'y}$, and $V_{O'z}$. According to the corresponding differential equations of the above three equations with respect to time, the sub-velocities can be written as

$$V_{O'x} = 2L_m \left(\Phi_2 \left(1 - \Phi_1^2 + \Phi_2^2\right)\dot{\Phi}_1 + \Phi_1 \left(1 + \Phi_1^2 - \Phi_2^2\right)\dot{\Phi}_2\right) / \sqrt{3}\lambda_0^4, \tag{26}$$

$$V_{O'y} = 2L_m \left(-\Phi_1 \left(1 + 2\Phi_2^2\right)\dot{\Phi}_1 + \Phi_2 \left(1 + 2\Phi_1^2\right)\dot{\Phi}_2\right) / \sqrt{3}\lambda_0^4, \tag{27}$$

$$V_{O'z} = \dot{z}, \tag{28}$$

where λ_0^2 equals to $1 + \Phi_1^2 + \Phi_2^2$ and $\dot{\Phi}_1$, $\dot{\Phi}_2$ represent the change velocities of the corresponding Rodrigues parameters, respectively.

Suppose that the angular velocity of origin on the moving coordinate system is denoted as $\boldsymbol{\omega}_{O'}$. Analyzing the geometrical characteristics of the parallel manipulator with 3-UCR limbs gives the following equation as

$$\mathbf{V}_{I'} = \mathbf{V}_{O'} + \boldsymbol{\omega}_{O'} \times \overrightarrow{O'I'}, \tag{29}$$

where $\mathbf{V}_{I'}$ (I'=A', B', C') represents the velocity of the connectors on the moving platform relative to the static coordinate system and the vector between O' and I' is denoted as $\overrightarrow{O'I'}$. Due to the restrictions of revolute joints on the base, the limb coordinates are located in three planes perpendicular to the axes of revolute joints. The normalization of the three axes can be obtained as $\mathbf{e}_A = (-1\ 0\ 0)^T$, $\mathbf{e}_B = \left(1/2\ \sqrt{3}/2\ 0\right)^T$ and $\mathbf{e}_C = \left(1/2\ -\sqrt{3}/2\ 0\right)^T$, so one can have

$$\mathbf{V}_{I'} \mathbf{e}_I = 0. \tag{30}$$

Analyzing the above equation gives the angular velocities of centroid on the moving platform as follows:

$$\omega_{O'x} = \frac{\left(s_{22}s_{33} - s_{32}s_{23}\right)s_{41} + \left(s_{31}s_{23} - s_{33}s_{21}\right)s_{42} + \left(s_{32}s_{21} - s_{31}s_{22}\right)s_{43}}{s_{12}\left(s_{33}s_{21} - s_{31}s_{23}\right) + s_{13}\left(s_{31}s_{22} - s_{32}s_{21}\right)}, \tag{31}$$

$$\omega_{O'y} = \frac{\left(s_{13}s_{32} - s_{12}s_{33}\right)s_{41} - s_{13}s_{31}s_{42} + s_{12}s_{31}s_{43}}{s_{12}\left(s_{33}s_{21} - s_{31}s_{23}\right) + s_{13}\left(s_{31}s_{22} - s_{32}s_{21}\right)}, \tag{32}$$

$$\omega_{O'z} = \frac{\left(s_{12}s_{23} - s_{13}s_{22}\right)s_{41} + s_{13}s_{21}s_{42} - s_{21}s_{12}s_{43}}{s_{12}\left(s_{33}s_{21} - s_{31}s_{23}\right) + s_{13}\left(s_{31}s_{22} - s_{32}s_{21}\right)}, \tag{33}$$

where the variables in the above three equations can be expressed analytically.

According to the differential forms of sub-velocities of origin on the moving coordinate system with respect to time, the corresponding linear accelerations can be obtained as

$$a_{O'x} = \dot{V}_{O'x}\,, \tag{34}$$

$$a_{O'y} = \dot{V}_{O'y}\,, \tag{35}$$

$$a_{O'z} = \dot{V}_{O'z}\,. \tag{36}$$

Similarly, the differential forms of angular velocities of origin on the moving coordinate system with respect to time give the corresponding angular acceleration as

$$\zeta_{O'x} = \dot{\omega}_{O'x}\,, \tag{37}$$

$$\zeta_{O'y} = \dot{\omega}_{O'y}\,, \tag{38}$$

$$\zeta_{O'z} = \dot{\omega}_{O'z}\,. \tag{39}$$

3.3 Inverse kinematics of limbs

The coordinates of the connectors in the moving platform being reference to the static coordinate system can be obtained by substituting Eq. (23), Eq. (24) and Eq. (25) into Eq. (22). So the lengths of limbs described by Rodrigues parameters can be gotten as

$$L_{AA'}{}^2 = (0)^2 + \left(\frac{\sqrt{3}L_m}{3\lambda_0^2}\left(1-2\Phi_1^2+2\Phi_2^2\right) - \frac{\sqrt{3}}{3}\eta L_m\right)^2 + \left(\frac{2\sqrt{3}L_m\Phi_1}{3\lambda_0^2} + z_{O'}\right)^2, \tag{40}$$

$$L_{BB'}{}^2 = \left(\frac{L_m}{\lambda_0^2}\left(\frac{\left(1+\Phi_1^2-\Phi_2^2\right)}{2} - \sqrt{3}\Phi_1\Phi_2\right) - \frac{\eta L_m}{2}\right)^2 + \left(\frac{L_m}{\lambda_0^2}\left(\Phi_1\Phi_2 - \frac{\sqrt{3}\left(1+\Phi_1^2-\Phi_2^2\right)}{6}\right) + \frac{\sqrt{3}}{6}\eta L_m\right)^2 + \left(\frac{L_m}{\lambda_0^2}\left(-\Phi_2 - \frac{\Phi_1}{\sqrt{3}} + \frac{z_{O'}\lambda_0^2}{L_m}\right)\right)^2, \tag{41}$$

$$L_{CC'}{}^2 = \left(\frac{L_m}{\lambda_0^2}\left(\frac{-\left(1+\Phi_1^2-\Phi_2^2\right)}{2} - \sqrt{3}\Phi_1\Phi_2\right) + \frac{\eta L_m}{2}\right)^2 + \left(\frac{L_m}{\lambda_0^2}\left(-\Phi_1\Phi_2 - \frac{\sqrt{3}\left(1+\Phi_1^2-\Phi_2^2\right)}{6}\right) + \frac{\sqrt{3}}{6}\eta L_m\right)^2 + \left(\frac{L_m}{\lambda_0^2}\left(\Phi_2 - \frac{\Phi_1}{\sqrt{3}} + \frac{z_{O'}\lambda_0^2}{L_m}\right)\right)^2. \tag{42}$$

The parallel manipulator includes three limbs denoted as *AA'*, *BB'* and *CC'*. The following part shows the kinematic calculation of limb *AA'* firstly.
The coordinates of the connector *A'* can be expressed as

$$x_{A'} = 0\,, \tag{43}$$

$$y_{A'} = \frac{\sqrt{3}L_m}{3\lambda_0^2}\left(1-2\Phi_1^2+2\Phi_2^2\right), \tag{44}$$

$$z_{A'} = \frac{2\sqrt{3}L_m\Phi_1}{3\lambda_0^2} + z_{O'}\,. \tag{45}$$

Differentiating the above three equations with respect to time gets the sub-velocities of the connector *A'* as follows:

$$v_{A'x} = 0\,, \tag{46}$$

$$v_{A'y} = \frac{2L_m\left(-\Phi_1\left(3+4\Phi_2^2\right)\dot{\Phi}_1 + \Phi_2\left(1+4\Phi_1^2\right)\dot{\Phi}_2\right)}{\sqrt{3}\lambda_0^4}\,, \tag{47}$$

$$v_{A'z} = \frac{2L_m\left(\left(1-\Phi_1^2+\Phi_2^2\right)\dot{\Phi}_1 - 2\Phi_1\Phi_2\dot{\Phi}_2\right)}{\sqrt{3}\lambda_0^4} + \dot{z}\,. \tag{48}$$

The driving velocity of limb AA' obtained by differentiating its length with Rodrigues parameters can be shown as

$$\mathbf{v}_{AA'} = \frac{d\left(L_{AA'}\right)}{dt} = \frac{d\left(L_{AA'}^2\right)}{2L_{AA'}dt}\,. \tag{49}$$

According to the motion of limb AA' and the geometrical characteristics of this parallel manipulator, one can have

$$\mathbf{v}_{A'} = \mathbf{v}_{AA'}\mathbf{e}_{AA'} + \boldsymbol{\omega}_{AA'} \times \left(L_{AA'}\mathbf{e}_{AA'}\right). \tag{50}$$

By dot-multiplying both sides of the above equation with $\mathbf{e}_{AA'} \times \left(L_{AA'}\mathbf{e}_{AA'}\right)$, Eq. (50) can be simplified as

$$\boldsymbol{\omega}_{AA'} = \mathbf{v}_{A'}\left(L_{AA'}\mathbf{e}_{AA'}\right)/L_{AA'}^2\,, \tag{51}$$

where $\mathbf{e}_{AA'}$ represents the unit vector of $\overrightarrow{AA'}$, and $\boldsymbol{\omega}_{AA'}$ is the angular velocity of limb AA'. Moreover, $\mathbf{v}_{A'} = \left[v_{A'x}\ v_{A'y}\ v_{A'z}\right]^T$.

Differentiating the velocities of the connector A' with respect to time gets the linear accelerations of this connector relative to the static coordinate system, three sub-accelerations of which along different axes of the static coordinate system can be written as

$$a_{A'x} = 0\,, \tag{52}$$

$$a_{A'y} = \frac{2L_m\left(-4\left(\Phi_1\dot{\Phi}_1+\Phi_2\dot{\Phi}_2\right)\left(-\Phi_1\left(3+4\Phi_2^2\right)\dot{\Phi}_1+\Phi_2\left(1+4\Phi_1^2\right)\dot{\Phi}_2\right)\right)}{\sqrt{3}\lambda_0^6} + \frac{2L_m\left(-\left(3+4\Phi_2^2\right)\dot{\Phi}_1^2+\left(1+4\Phi_1^2\right)\dot{\Phi}_2^2-\Phi_1\left(3+4\Phi_2^2\right)\ddot{\Phi}_1+\Phi_2\left(1+4\Phi_1^2\right)\ddot{\Phi}_2\right)}{\sqrt{3}\lambda_0^4}\,, \tag{53}$$

$$a_{A'z} = \frac{8L_m\left(\Phi_1\dot{\Phi}_1+\Phi_2\dot{\Phi}_2\right)\left(-\left(1-\Phi_1^2+\Phi_2^2\right)\dot{\Phi}_1+2\Phi_1\Phi_2\dot{\Phi}_2\right)}{\sqrt{3}\lambda_0^6} + \frac{2L_m\left(-2\Phi_1\left(\dot{\Phi}_1^2+\dot{\Phi}_2^2\right)+\left(1-\Phi_1^2+\Phi_2^2\right)\ddot{\Phi}_1-2\Phi_1\Phi_2\ddot{\Phi}_2\right)}{\sqrt{3}\lambda_0^4} + \ddot{z}\,. \tag{54}$$

The angular acceleration can be obtained by differentiating the angular velocity of limb AA' with respect to time as follows:

$$\zeta_{AA'} = \left(a_{AA'}\left(L_{AA'}^2\right) - \mathbf{v}_{AA'}\left(L_{AA'}^2\right)'\right)\left(L_{AA'}\mathbf{e}_{AA'}\right)\Big/L_{AA'}^4\,. \tag{55}$$

Similarly, the velocities and the accelerations of the other limbs can also be gotten by the corresponding calculations.

4. Forward Kinematic Analysis of 3-UCR Parallel Robot Leg

The forward position-stance analysis of the parallel manipulators is the basis of structure synthesis, kinematic analysis and dynamic optimization, and many researchers had paid more attention to it gradually (Ruggiu, 2008; Kim & Park, 2001; Jaime et al., 2006; Lu et al., 2008). However, the forward position-stance analysis of the parallel manipulator is more difficult than the inverse position-stance analysis because it is essential to solve the multivariate nonlinear equations.

4.1 Analytical model

The constraint equations of the parallel robot leg can be obtained by the geometrical characteristics, and the variables in the equations can be eliminated by the successive elimination method. Then the constraint equations are changed into the unary polynomial equation.

Three joints, denoted as A', B', and C', can be described by another method by the angle between driving limbs and moving platform. The above angles are assumed as α_A, α_B, and α_C respectively. And the joints in the moving platform can be expressed as

$$(x_{A'},y_{A'},z_{A'})=\left(0,\frac{\sqrt{3}}{3}L_m-L_{AA'}\cos(\alpha_A),L_{AA'}\sin(\alpha_A)\right), \tag{56}$$

$$(x_{B'},y_{B'},z_{B'})=\left(-\frac{1}{2}L_m+\frac{\sqrt{3}}{2}L_{BB'}\cos(\alpha_B),-\frac{\sqrt{3}}{6}L_m+\frac{1}{2}L_{BB'}\cos(\alpha_B),L_{BB'}\sin(\alpha_B)\right), \tag{57}$$

$$(x_{C'},y_{C'},z_{C'})=\left(\frac{1}{2}L_m-\frac{\sqrt{3}}{2}L_{CC'}\cos(\alpha_C),-\frac{\sqrt{3}}{6}L_m+\frac{1}{2}L_{CC'}\cos(\alpha_C),L_{CC'}\sin(\alpha_C)\right). \tag{58}$$

Three joints in the moving platform are symmetrical and the distances between two joints of them are denoted as L_m. So the equations can be gotten as follows:

$$(x_{B'}-x_{A'})^2+(y_{B'}-y_{A'})^2+(z_{B'}-z_{A'})^2=L_m^{\ 2}, \tag{59}$$

$$(x_{C'}-x_{A'})^2+(y_{C'}-y_{A'})^2+(z_{C'}-z_{A'})^2=L_m^{\ 2}, \tag{60}$$

$$(x_{C'}-x_{B'})^2+(y_{C'}-y_{B'})^2+(z_{C'}-z_{B'})^2=L_m^{\ 2}. \tag{61}$$

Substituting the coordinates of three joints into the above equation gives

$$L_{AA'}^{\ 2}+L_{BB'}\left(-\sqrt{3}L_m\cos(\alpha_B)+L_{BB'}\right)+L_{AA'}\left(-\sqrt{3}L_m\cos(\alpha_A)+\left(\cos(\alpha_A)\cos(\alpha_B)-2\sin(\alpha_A)\sin(\alpha_B)\right)L_{BB'}\right)=0, \tag{62}$$

$$L_{AA'}^{\ 2}+L_{CC'}\left(-\sqrt{3}L_m\cos(\alpha_C)+L_{CC'}\right)+L_{AA'}\left(-\sqrt{3}L_m\cos(\alpha_A)+\left(\cos(\alpha_A)\cos(\alpha_C)-2\sin(\alpha_A)\sin(\alpha_C)\right)L_{CC'}\right)=0, \tag{63}$$

$$L_{BB'}^{\ 2}+L_{CC'}\left(-\sqrt{3}L_m\cos(\alpha_C)+L_{CC'}\right)+L_{BB'}\left(-\sqrt{3}L_m\cos(\alpha_B)+\left(\cos(\alpha_B)\cos(\alpha_C)-2\sin(\alpha_B)\sin(\alpha_C)\right)L_{CC'}\right)=0. \tag{64}$$

Substituting the universal trigonometric functions into the above transcendental equations and supposing $\sin(\alpha_i) = 2x_i/(1+x_i^2)$, $\cos(\alpha_i) = (1-x_i^2)/(1+x_i^2)$ where $x_i = \tan(\alpha_i/2)$ (*i*=*A*, *B*, *C*), can simplify the above equations as follows:

$$G_1 x_2^2 + H_1 x_2 + J_1 = 0\,, \tag{65}$$

$$G_2 x_3^2 + H_2 x_3 + J_2 = 0\,, \tag{66}$$

$$G_3 x_3^2 + H_3 x_3 + J_3 = 0\,, \tag{67}$$

where $G_1 = -\sqrt{3}L_m L_{AA'} + L_{AA'}^2 + \sqrt{3}L_m L_{BB'} - L_{AA'}L_{BB'} + L_{BB'}^2 + \sqrt{3}L_m L_{AA'}x_1^2 + L_{AA'}^2 x_1^2 + \sqrt{3}L_m L_{BB'}x_1^2 + L_{AA'}L_{BB'}x_1^2 + L_{BB'}^2 x_1^2$,

$G_2 = -\sqrt{3}L_m L_{AA'} + L_{AA'}^2 + \sqrt{3}L_m L_{CC'} - L_{AA'}L_{CC'} + L_{CC'}^2 + \sqrt{3}L_m L_{AA'}x_1^2 + L_{AA'}^2 x_1^2 + \sqrt{3}L_m L_{CC'}x_1^2 + L_{AA'}L_{CC'}x_1^2 + L_{CC'}^2 x_1^2$,

$G_3 = -\sqrt{3}L_m L_{BB'} + L_{BB'}^2 + \sqrt{3}L_m L_{CC'} - L_{BB'}L_{CC'} + L_{CC'}^2 + \sqrt{3}L_m L_{BB'}x_2^2 + L_{BB'}^2 x_2^2 + \sqrt{3}L_m L_{CC'}x_2^2 + L_{BB'}L_{CC'}x_2^2 + L_{CC'}^2 x_2^2$,

$J_1 = -\sqrt{3}L_m L_{AA'} + L_{AA'}^2 - \sqrt{3}L_m L_{BB'} + L_{AA'}L_{BB'} + L_{BB'}^2 + \sqrt{3}L_m L_{AA'}x_1^2 + L_{AA'}^2 x_1^2 - \sqrt{3}L_m L_{BB'}x_1^2 - L_{AA'}L_{BB'}x_1^2 + L_{BB'}^2 x_1^2$,

$J_2 = -\sqrt{3}L_m L_{AA'} + L_{AA'}^2 - \sqrt{3}L_m L_{CC'} + L_{AA'}L_{CC'} + L_{CC'}^2 + \sqrt{3}L_m L_{AA'}x_1^2 + L_{AA'}^2 x_1^2 - \sqrt{3}L_m L_{CC'}x_1^2 - L_{AA'}L_{CC'}x_1^2 + L_{CC'}^2 x_1^2$,

$J_3 = -\sqrt{3}L_m L_{BB'} + L_{BB'}^2 - \sqrt{3}L_m L_{CC'} + L_{BB'}L_{CC'} + L_{CC'}^2 + \sqrt{3}L_m L_{BB'}x_2^2 + L_{BB'}^2 x_2^2 - \sqrt{3}L_m L_{CC'}x_2^2 - L_{BB'}L_{CC'}x_2^2 + L_{CC'}^2 x_2^2$,

$H_1 = -2L_{AA'}L_{BB'}x_1$, $H_2 = -2L_{AA'}L_{CC'}x_1$, $H_3 = -2L_{BB'}L_{CC'}x_2$.

The above equations can be simplified into the polynomial with sixteen degrees having one variable, and the main process can be described as the following steps.
At first, simplifying the two equations about x_3 gives

$$\begin{bmatrix} G_2H_3 - G_3H_2 & G_2J_3 - J_2G_3 \\ J_2G_3 - G_2J_3 & J_2H_3 - H_2J_3 \end{bmatrix} \begin{bmatrix} x_3 \\ 1 \end{bmatrix} = \begin{bmatrix} 0 \\ 0 \end{bmatrix}. \tag{68}$$

By analyzing the geometrical characteristics of the parallel robot leg, it is obvious that $x_3 \neq 0$. So the following equation can be gotten as

$$(G_2H_3 - G_3H_2)(J_2H_3 - H_2J_3) - (G_2J_3 - J_2G_3)^2 = 0\,. \tag{69}$$

Then simplifying x_2 in the above equation gives

$$Q_4x_2^4 + Q_3x_2^3 + Q_2x_2^2 + Q_1x_2 + Q_0 = 0\,, \tag{70}$$

where Q_i (*i*=0, 1, 2, 3, 4) is the polynomial about x_1 having not more than four degrees. Because of $G_1x_2^2 + H_1x_2 + J_1 = 0$, combining the above two equations can get the following equation as

$$\begin{bmatrix} H_1Q_4 - G_1Q_3 & J_1Q_4 - G_1Q_2 & -G_1Q_1 & -G_1Q_0 \\ G_1Q_2 - J_1Q_4 & G_1Q_1 + H_1Q_2 - J_1Q_3 & G_1Q_0 + H_1Q_1 & H_1Q_0 \\ G_1 & H_1 & J_1 & 0 \\ 0 & G_1 & H_1 & J_1 \end{bmatrix} = 0\,. \tag{71}$$

The above equation is the polynomial about x_1 having sixteen degrees, and the corresponding solutions have sixteen groups. Putting the angle values of a_i (i=1, 2, 3) into the coordinates joints in the moving platform gives the forward position-stance solutions of three spheral joints. Because three spheral joint coordinates do not exist in the same line, the plane decided by the spheral joints can be solved. Moreover, the coordinates of any points in the moving platform can also be gotten. So knowing the exact values of the inputs, denoted as $L_{AA'}$, $L_{BB'}$, $L_{CC'}$, can get the corresponding values of the outputs, denoted as Φ_1, Φ_2, $Z_{O'}$. And the forward position-stance model of the parallel robot leg with the analytical form has been established.

4.2 Numerical model

Though all position-stance solutions of the robot leg can be gotten by the analytical model, the elimination process is complicated and sometimes it is not necessary to get all of them in practice. In the given workspace, the only one forward position-stance solution of the structure is available. So the numerical solutions can be easier to be calculated and it becomes the feasible method to analyze the forward kinematics.

4.2.1 Iterative algorithm

Bracketing methods such as the bisection method and the false position method of finding roots of a nonlinear equation require bracketing of the root by two guesses. These methods are always convergent since they are based on reducing the interval between the two guesses to zero on the root. In the Newton-Raphson method, only one initial guess of the root is needed to get the iterative process started to find the root of an equation. This method is based on the principle that if the initial guess of the root of f(x)=0 is at x_i , then if one draws the tangent to the curve at f(x_i), the point x_{i+1} where the tangent crosses the x-axis is an improved estimate of the root. So the Newton-Raphson method is applied as the iterative algorithm.

The iterative steps of numerical model of the parallel robot leg can be written as follows.

At first, the iterative function is defined as

$$\mathrm{g}(\Phi_1,\Phi_2,Z_{O'}) = \mathbf{L}_{II'}(\Phi_1,\Phi_2,Z_{O'}) - \mathbf{L}_{II'M}\,, \tag{72}$$

where $\mathbf{L}_{II'}(\Phi_1,\Phi_2,Z_{O'}) = \left[AA'(\Phi_1,\Phi_2,Z_{O'})\ BB'(\Phi_1,\Phi_2,Z_{O'})\ CC'(\Phi_1,\Phi_2,Z_{O'})\right]^T$ and $\mathbf{L}_{II'M}$ is the measured values of three driving limbs. Substituting the iterated values of three outputs into the inverse kinematic model of the parallel robot leg can obtain the theoretical values of three driving limbs.

Based on the Newton-Raphson method, supposing Q_K as $(\Phi_{1K},\Phi_{2K},Z_{O'K})$ gets the following equation as

$$Q_{K+1} = Q_K - \frac{g(Q_K)}{g'(Q_K)}, \tag{73}$$

where g′(Q_K) can be replaced by the Jacobi matrix of the robot leg. That can be expressed as

$$Q_{K+1} = Q_K - \frac{\left(\mathbf{L}_{II'}(\Phi_1, \Phi_2, Z_{O'}) - \mathbf{L}_{II'M}\right)}{J_E}. \tag{74}$$

The tolerances of the driving limbs are defined as $\mathbf{L}_{II'□}$. If the iterative terminational condition could be reached, the corresponding outputs about Q_{K+1} can be calculated by the above equation. Corresponding to the preceding inputs, the values of three outputs are the forward kinematic solutions of 3-UCR parallel robot leg.

4.2.2 Numerical simulation

In order to validate the iterative process of forward kinematics, the initial structure parameters of 3-UCR parallel robot leg need to be defined and put into the Matlab program written by the preceding steps. Then the output values of 3-UCR parallel robot leg can be obtained after several iterative circles.

Firstly, the distances between the joints in the moving platform, denoted as L_m, are initialized as $50\sqrt{3}$ mm, and the circumcircle radius of the equilateral triangle formed by three spheral joints is set as 50mm. The distances between the rotational joints, denoted as L_B, are $68\sqrt{3}$ mm, and the corresponding circumcircle radius of the equilateral triangle is 68mm.

For the purpose of getting the target values of the outputs, it is necessary to assume the position-stance outputs as $[\Phi_1 \Phi_2\ z_{O'}]$ = [-0.2 0.5 320] in advance. By the relations among three spheral joints and the outputs, the position-stance output values caused by the other related DOF can be obtained as $[\Phi_3\ x_{O'}\ y_{O'}]$ = [0 7.7519 8.1395].

Substituting the output values into the inverse kinematic model gives $[L_{AA'}\ L_{BB'}\ L_{CC'}]$ = [304.7719 295.1586 364.8734]. The units in the above matrix are millimeter, and the above input values of three driving limbs are assumed as the measured values by the displacement sensors.

The choice of the initial values in the course of calculation is important, especially the parallel manipulators, because the number of the forward position-stance solutions of the parallel manipulators is more than the number of the serial manipulators. If the errors of initial values are enough large, the other group of forward solutions would be gotten. So the initial search values of the outputs are set as Q_0 = $[-0.22\ 0.48\ 318]^T$.

By calculating circularly the iterative parameter, denoted as $\{Q_K\}_{K=0}^{\infty}$, and defining the terminational tolerance value as $\mathbf{L}_{II'□}$=0.0001, the accurate values of the outputs can be obtained when the calculated tolerance is less than the terminational tolerance. According to the above parameter choice, the output values in the different iterative steps have been solved and the corresponding values are written in Table 1.

	Φ_1	Φ_2	$z_{O'}$
0	-0.22	0.48	318
1	-0.196089	0.505499	320.46100
2	-0.200635	0.499836	319.90179
3	-0.199858	0.499836	320.01285
4	-0.200032	0.499996	319.99716
5	-0.199993	0.500000	320.00063
6	-0.200002	0.500000	319.99992

Table 1. Numerical solution of the outputs parameters of forward kinematics

The data in Table 1 have been calculated by the taking or rejecting way, and the values of the last two iterative steps meet with the iterative terminational condition. Calculation of the six cycles shows that the Newton-Raphson method can search the exact forward position-stance solutions rapidly. However, for the reason of the choice of initializations and limitation of iteration step, it is necessary to pay attention to solution precision and algorithm stability. So we need take the following measures during the calculation process of forward kinematics.

At first, if the function equals to zero, the program would have faults. So we need to judge the value of $f'(Q)$ and eliminate the condition. Then the slope value of $f'(Q)$ is so little that $\{Q_K\}$ converges to another group of the solutions. So we should define the initial values accurately. Finally, if the items of $\{Q_K\}$ tend to repetition, the calculation process would run into limitless cycles. So the maximum steps should be chosen to improve the validity of the program.

5. Conclusions

Based on principal screw theory and imaginary manipulator method, the kinematic characteristics of 3-UCR spatial parallel robot leg with three DOF were analyzed. According to the topologic structure of limbs, the screw coordinate system was obtained and the kinematics of limbs was studied. By the relation of the matrices of influence coefficient between limbs and moving platform, the kinematic model with the screw coordinates was established. It shows that the matrices of influence coefficient is only dependent on the inputs and kinematic parameters and the method analyzing instantaneous motion is fit for others kinds of lower-mobility parallel manipulators. The instantaneous pitches of the principal screws gained decide that the kind of manipulator has always three DOF including one translation and two rotations. By the numerical simulation when the moving platform is parallel to the base, the pitch analysis of principal screws validates the kinematic characters of 3-UCR parallel robot leg.

A new method to describe the position-stance of 3-UCR parallel robot leg was proposed based on the Rodrigues theory. Comparing with others methods, the kinematic model with Rodrigues parameters has the advantages including least computational parameters, no trigonometric function calculation and convenient real-time control. The model of the inverse kinematics was established and the inverse solutions of the position-stance were obtained by analyzing the topologic structure of 3-UCR parallel robot leg. By analyzing the vectors of the manipulator, the velocity and acceleration models of moving platform, limbs

and end-effector were deduced. According to the designed kinematic track, it is convenient to control accurately 3-UCR parallel robot leg by the inverse kinematic model.
According to the topologic system of 3-UCR parallel robot leg, the geometrical constraints are obtained. And the forward kinematic model with analytical expressions can be established by eliminating the unknown terms. It is shown that the analytical solutions of the forward kinematic model have 16 groups. In order to decrease the number of solutions and get the exact position-stance of 3-UCR parallel robot leg, the Newton-Raphson method was used to search the best numerical solutions by the judgment of the terminal value. The corresponding numerical simulation proved that the exact forward solution can be found rapidly by the iterative steps. Moreover, aiming at improving the numerical precision, some measures on the choice of initial value and iterative step had been put forward. The forward kinematic model provides the basis of the perfect control of 3-UCR parallel robot leg.

6. Acknowledgment

Financial support for this work, provided by the National Natural Science Foundation of China (Grant No. 50905180, 60808017) and the youth foundation and Qihang Project of China University of Mining and Technology, are gratefully acknowledged.

7. References

Alon, W. & Moshe, S. (2006). Screw theory tools for the synthesis of the geometry of a parallel robot for a given instantaneous task. *Mechanism and Machine Theory*, 41: 656-670.

Altmann, S. L. (1986). *Rotations, quaternions, and double groups*, Clarendon Press, Oxford, England.

Baker, C.; Morris, A. & Ferguson, D. (2004). A campaign in autonomous mine mapping, *IEEE International Conference on Robotics and Automation*, New Orleans, pp. 2004-2009.

Ball, R. S. (1900). *A treatise on the theory of screws* . Cambridge University Press, Cambridge.

Carretero, J. A.; Podhorodeski, R. P. & Nahon, M. A. ; et al. (2000). Kinematic analysis and optimization of a new three degree-of-freedom spatial parallel manipulator, *Journal of Mechanical Design*, 122(1): 17-24.

Cayley, A. (1875). On three-bar motion, *Proceedings of the London Mathematical Society V II*, pp. 136-166.

Clavel R. (1988). DELTA, A Fast Robot with Parallel Geometry. *The 18th International Symposium on Industrial Robot*, Lausanne, pp. 91-100.

Dai, J. (2006). An historical review of the theoretical development of rigid body displacements from Rodrigues parameters to the finite twist. *Mechanism and Machine Theory*, 41: 41-52.

Dunlop, G. R. & Jones, T. P. (1997). Position analysis of a 3-DOF parallel manipulator, *Mechanism and Machine Theory*, 32(8): 903-920.

Fang, Y. F. & Huang Z. (1998). Analytical identification of the principal screws of the third order screw system. *Mechanism and Machine Theory*, 33(7): 987-992.

Fang, Y. F. & Tsai, L. W. (2004). Structure synthesis of a class of 3-DOF rotational parallel manipulators. *IEEE Transactions on Robotics and Automation*, 1: 117-121.

Gabmann, B.; Zacharias, F. & Zollner, J. (2005). Location of walking robots, *IEEE International Conference on Robotics and Automation*, Barcelona, Spain, pp. 1471-1476.

Hunt, K. H. (1978). *Kinematic geometry of mechanisms.* Cambridge University Press, Cambridge.

Jaime, G. A.; Jose, M. R. & Gursel, A. (2006). Kinematics and singularity analysis of a 4-dof parallel manipulator using screw theory. *Mechanism and Machine Theory*, 41: 1048-1061.

Kim, J. (2001). Design and analysis of an overactuated parallel mechanism for rapid machining, *IEEE Trans. Robotics & Automation*, 17(4): 423-434.

Kim, J. & Park, F. (2001). Direct kinematic analysis of 3-RS parallel mechanisms. *Mechanism and Machine theory*, 36: 1121-1134.

Kindermann, H. & Cruse, H. (2002). MMC – a new numerical approach to the kinematics of complex manipulators. *Mechanism and Machine Theory*, 37: 375-394.

Koditschek, D.; Full, R. & Buehler, M. (2004). Mechanical aspects of legged locomotion control, *Arthropod Structure & Development*, 33: 251-272.

Lu, Y.; Shi, Y. & Hu, B. (2008). Kinematic analysis of two novel 3UPU I and 3UPU II PKMs. *Robotics and Autonomous Systems*, 56: 296-305.

Pouliot, N. A.; Nahon, M. A. & Gosselin, C. M. (1996). Analysis and comparison of the motion simulation capabilities of three degree-of-freedom flight simulators, *Proceedings of AIAA Flight Simulation Technologies Conference*, San Diego, PP. 37-47.

Ruggiu, M. (2008). Kinematics analysis of the CUR translational manipulator. *Mechanism and Machine Theory*, 43: 1087-1098.

Santos, P.; Estremera, J. & Garcia, E. (2005). Including joint torques and power consumption in the stability margin of walking robots, *Autonomous Robots*, 18: 43-57.

Sokolov, A. & Xirouchakis, P. (2005). Kinematics of a 3-DOF parallel manipulator with an R-P-S joint structure. *Robotica*, 23: 207-217.

Sokolov, A. & Xirouchakis, P. (2006). Singularity analysis of a 3-DOF parallel manipulator with R-P-S joint structure. *Robotica*, 24: 131-142.

Tanaka, J.; Suzumori, K. & Takata, M. (2005). A mobile Jack robot for rescue operation, *IEEE International Workshop on Safety, Security and Rescue Robotics*, Kobe Japan, pp. 99-104.

Wang, W.; Du, Z. & Sun, L. (2007). Obstacle performance analysis of mine research robot based on terramechanics, *IEEE International Conference on Mechatronics and Automation*, Harbin, China, pp. 1382-1387.

Wang, Y. & Wang, Y. (2005). Inverse kinematics of variable geometry parallel manipulator. *Mechanism and Machine Theory*, 40: 141-155.

20

Digital Control of Free Floating Space Robot Manipulators Using Transpose of Generalized Jacobian Matrix

Shinichi Sagara[1] and Yuichiro Taira[2]
[1]*Kyushu Institute of Technology*
[2]*National Fisheries University*
Japan

1. Introduction

Space robots having manipulators are expected to work in future space missions (Xu & Kanade, 1993). Since it is difficult to supply fuel to the robots equipped with rocket motors during manipulation, control methods for free-floating space robots consisting of a base and a manipulator have been proposed (Dubowsky & Papadopulos, 1993; Masutani et al., 1989a;b; Sagara et al., 1998a;b; Shin et al., 1995; Umetani & Yoshida, 1989; Yamamoto et al., 1995). Most of them use the inverse of the Generalized Jacobian Matrix (GJM) which is a coefficient matrix between the velocity of the end-effector of the manipulator and the manipulator's joint velocity (Umetani & Yoshida, 1989). Therefore, in a case that the robot manipulator gets into a singular configuration, the inverse of the GJM does not exist and the manipulator is out of control. For this problem, a continuous-time control method using the transpose of the GJM has been proposed for manipulators equipped with joint torque controllers (Masutani et al., 1989a;b).

In practical systems digital computers are utilized for controllers. So, we have proposed a discrete-time control method using the transpose of the GJM (Taira et al, 2001). The control method using the transpose of the GJM uses position and orientation errors between the desired and actual values of the end-effector. Namely, the control method belongs to a class of constant value control such as PID control. Therefore, the value of the errors depends on the desired linear and angular velocities of the end-effector based on the desired trajectory. To obtain higher control performance we have proposed a digital trajectory tracking control method that has variable feedback gains depending on the desired linear and angular velocities of the end-effector (Sagara & Taira, 2007). Moreover, we have also proposed the control method for manipulators with velocity type joint controllers (Sagara & Taira, 2008b).

In addition, it is considered that many tasks will be achieved by cooperative motions of several space robots in future space missions. We have studied control problems for realizing cooperative manipulations, and reported that a system consisting of space robots with manipulators and a floating object can be treated as a kind of distributed system (Katoh et al., 1997; Sagara et al., 1998b). Using the distributed system representation, each robot consisting of the distributed system can be designed by the control system individually, and we have reported a cooperative manipulation of a floating object by some space robots with the control methods using the transpose of the GJM (Sagara & Taira, 2008a; 2009).

In this chapter, our proposed control methods for space robot manipulators using the transpose of the GJM are described and the computer simulations are performed. First, for manipulators equipped with joint torque controllers we explain about a basic control method using constant feedback gains with the proof of stability. Next, to obtain higher control performance, we introduce a trajectory tracking control method with variable feedback gains for both torque and velocity joint inputs. Moreover, we address a cooperative manipulation of a floating object by some space robots with the control methods using the transpose of the GJM. Simulations where manipulators get into a singular configuration are also performed for the cooperative manipulation.

2. Modelling of space robot

We consider a free-floating space robot manipulator, as shown in Figure 1. It has an uncontrolled base and an n-DOF manipulator with revolute joints. Let link 0 denote a base main body, link i ($i = 1, \cdots, n$) the i-th link of the manipulator and joint j a joint connecting link $(i-1)$ and link i. The target of the end-effector of the manipulator is stationary in an inertial coordinate frame.

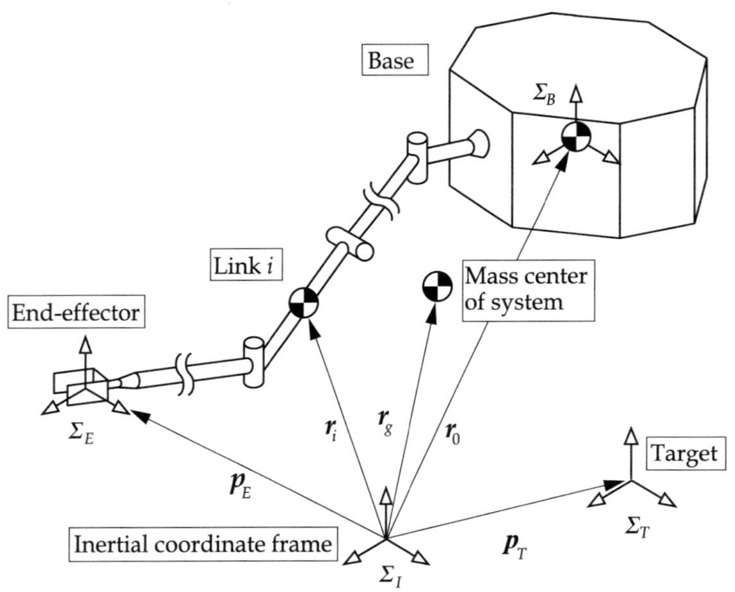

Fig. 1. Model of a space robot manipulator with an uncontrolled base

Assumptions and symbols used in this chapter are defined as follows:

Assumptions

A1) All elements of the space robot are rigid.

A2) The robot system is standing still at an initial state, i.e., the initial linear momentum and angular momentum of the space robots are zero.

A3) No external force acts on the robot system.

A4) Positions and attitude angles of robots and an object in inertial coordinate frame can be measured.

Symbols

Σ_I : inertial coordinate frame

Σ_B : base coordinate frame

Σ_E : end-effector coordinate frame

Σ_T : target coordinate frame

Σ_i : link i coordinate frame

$\boldsymbol{p}_E$: position vector of origin of Σ_E

$\boldsymbol{p}_T$: position vector of origin of Σ_T

$\boldsymbol{p}_i$: position vector of joint i

$\boldsymbol{r}_0$: position vector of center of mass of base

$\boldsymbol{r}_i$: position vector of center of mass of link i

$\boldsymbol{r}_g$: position vector of center of mass of system

$\boldsymbol{v}_0$: linear velocity vector of origin of Σ_B

$\boldsymbol{v}_E$: linear velocity vector of origin of Σ_E

$\boldsymbol{\omega}_0$: angular velocity vector of origin of Σ_B

$\boldsymbol{\omega}_E$: angular velocity vector of origin of Σ_E

$\boldsymbol{k}_i$: unit vector indicating joint axis direction of joint i

$\boldsymbol{\phi}$: joint angle vector

$\boldsymbol{\tau}$: joint torque vector

m_0 : mass of base

m_i : mass of link i

$\boldsymbol{I}_0$: inertia tensor of base

$\boldsymbol{I}_i$: inertia tensor of link i

$\boldsymbol{E}$: identity matrix

$\tilde{\{\cdot\}}$: Tilde operator stands for a cross product such that $\tilde{\boldsymbol{r}}\boldsymbol{a} = \boldsymbol{r} \times \boldsymbol{a}$

Note that all vectors and inertia tensors are defined with respect to the inertial reference frame.

2.1 kinematics and dynamics in continuous-time domain

In this subsection, kinematics and dynamics of the robot shown in Figure 1 are briefly described. The derivation of the following equations are founded in reference (Umetani & Yoshida, 1989).

First, a kinematic equation of the robot is described as

$$\boldsymbol{\nu}(t) = \begin{bmatrix} \boldsymbol{v}_E(t) \\ \boldsymbol{\omega}_E(t) \end{bmatrix} = \boldsymbol{J}_s \begin{bmatrix} \boldsymbol{v}_0(t) \\ \boldsymbol{\omega}_0(t) \end{bmatrix} + \boldsymbol{J}_m \dot{\boldsymbol{\phi}}(t) \tag{1}$$

where

$$\boldsymbol{J}_s = \begin{bmatrix} \boldsymbol{E} & \tilde{\boldsymbol{r}}_0 - \tilde{\boldsymbol{p}}_E \\ \boldsymbol{0} & \boldsymbol{E} \end{bmatrix}$$

and

$$J_m = \begin{bmatrix} \tilde{k}_1(p_E - p_1) & \tilde{k}_2(p_E - p_2) & \cdots & \tilde{k}_n(p_E - p_n) \\ k_1 & k_2 & \cdots & k_1 \end{bmatrix}$$

are Jacobian matrices for a robot base motion dependent part and a manipulator dependent part, respectively.
Next, with an assumption that total momentum of the system is held to zero, a linear and an angular momentum of the system, P and L, are described as follows:

$$\begin{bmatrix} P \\ L \end{bmatrix} = \begin{bmatrix} wE & w(\tilde{r}_g - \tilde{r}_0)^T \\ w\tilde{r}_g & I_\omega \end{bmatrix} \begin{bmatrix} v_0(t) \\ \omega_0(t) \end{bmatrix} + \begin{bmatrix} J_{Tw} \\ I_\phi \end{bmatrix} \dot{\phi}(t) = \mathbf{0} \tag{2}$$

where

$$I_\omega = \sum_{i=1}^{n} \{ I_i - m_i \tilde{r}_i (\tilde{r}_i - \tilde{r}_0) \} + I_0,$$

$$I_\phi = \sum_{i=1}^{n} (I_i J_{R_i} + m_i \tilde{r}_i J_{T_i}),$$

$$J_{Tw} = \sum_{i=1}^{n} m_i J_{T_i},$$

$$J_{T_i} = \left[\tilde{k}_1(r_i - p_1),\ \tilde{k}_2(r_i - p_2),\ \cdots,\ \tilde{k}_i(r_i - p_i),\ \mathbf{0},\ \cdots,\ \mathbf{0} \right],$$

$$J_{R_i} = \left[k_1,\ k_2,\ \cdots,\ k_i,\ \mathbf{0},\ \cdots,\ \mathbf{0} \right].$$

Furthermore, an equation of motion of the robot is

$$H\ddot{\phi}(t) + C\dot{\phi}(t) = \tau(t) \tag{3}$$

where

$$H = H_\phi - \begin{bmatrix} J_{Tw}^T & H_{\omega\phi}^T \end{bmatrix} \begin{bmatrix} wE & w(\tilde{r}_g - \tilde{r}_0)^T \\ w(\tilde{r}_g - \tilde{r}_0) & H_\omega \end{bmatrix}^{-1} \begin{bmatrix} J_{Tw} \\ H_{\omega\phi} \end{bmatrix},$$

$$C = \frac{1}{2}\dot{H} + S,$$

$$S\dot{\phi} = \frac{1}{2}\dot{H}\dot{\phi} - \frac{\partial}{\partial \phi} \left\{ \frac{1}{2} \dot{\phi}^T H \dot{\phi} \right\},$$

$$H_\omega = \sum_{i=1}^{n} \left\{ I_i + m_i (\tilde{r}_i - \tilde{r}_0)^T (\tilde{r}_i - \tilde{r}_0) \right\} + I_0,$$

$$H_{\omega\phi} = \sum_{i=1}^{n} \left\{ I_i J_{R_i} + m_i (\tilde{r}_i - \tilde{r}_0) J_{T_i} \right\},$$

$$H_\phi = \sum_{i=1}^{n} \left(J_{R_i}^T I_i J_{R_i} + m_i J_{T_i}^T J_{T_i} \right).$$

From Equations (1) and (2), the following relationship between the end-effector velocity and the joint velocity can be obtained:

$$\boldsymbol{\nu}(t) = \begin{bmatrix} \boldsymbol{v}_E(t) \\ \boldsymbol{\omega}_E(t) \end{bmatrix} = \boldsymbol{J}\dot{\boldsymbol{\phi}}(t) = \begin{bmatrix} \boldsymbol{J}_L \\ \boldsymbol{J}_A \end{bmatrix} \dot{\boldsymbol{\phi}}(t) \tag{4}$$

where

$$\boldsymbol{J} = \bar{\boldsymbol{J}}_m - \bar{\boldsymbol{J}}_s \boldsymbol{I}_s^{-1} \boldsymbol{I}_m,$$

$$\boldsymbol{I}_s = \sum_{i=0}^{n} \{\boldsymbol{I}_i - m_i(\tilde{\boldsymbol{r}}_i - \tilde{\boldsymbol{r}}_g)(\tilde{\boldsymbol{r}}_i - \tilde{\boldsymbol{r}}_g)\},$$

$$\boldsymbol{I}_m = \boldsymbol{I}_\phi - \tilde{\boldsymbol{r}}_g \boldsymbol{J}_{Tw},$$

$$\bar{\boldsymbol{J}}_m = \boldsymbol{J}_m - \boldsymbol{J}_s \begin{bmatrix} \frac{1}{w} \boldsymbol{J}_{Tw} \\ \boldsymbol{0} \end{bmatrix},$$

$$\bar{\boldsymbol{J}}_s = \boldsymbol{J}_s \begin{bmatrix} \tilde{\boldsymbol{r}}_g - \tilde{\boldsymbol{r}}_0 \\ \boldsymbol{E} \end{bmatrix}.$$

$\boldsymbol{J}$ is the GJM that is an extended and generalized form of manipulator Jacobian implying the reaction dynamics of free-floating systems. $\boldsymbol{J}_L$ and $\boldsymbol{J}_A$ are the GJMs of the linear and angular velocities, respectively.

2.2 Discrete-time representation

To design digital control systems, discrete-time representation of Equations (4) and (3) are given.
Discretizing Equation (4) by a sampling period T, the following discrete-time relationship between the end-effector velocity and the joint velocity can be obtained:

$$\boldsymbol{v}_E(k) = \boldsymbol{J}_L(k)\dot{\boldsymbol{\phi}}(k), \tag{5}$$

$$\boldsymbol{\omega}_E(k) = \boldsymbol{J}_A(k)\dot{\boldsymbol{\phi}}(k). \tag{6}$$

Note that the discrete-time kT is abbreviated to k.
Similarly, the equation of motion (3) is descretized applying $\ddot{\boldsymbol{\phi}}(k)$ to the backward Euler approximation:

$$\dot{\boldsymbol{\phi}}(k) = \dot{\boldsymbol{\phi}}(k-1) - T\boldsymbol{H}^{-1}(k)\left\{\boldsymbol{C}(k)\dot{\boldsymbol{\phi}}(k) - \boldsymbol{\tau}(k)\right\} \tag{7}$$

where

$$\boldsymbol{C}(k) = \frac{1}{2}\dot{\boldsymbol{H}}(k) + \boldsymbol{S}(k),$$

$$\boldsymbol{S}(k)\dot{\boldsymbol{\phi}}(k) = \frac{1}{2}\dot{\boldsymbol{H}}(k)\dot{\boldsymbol{\phi}}(k) - \frac{\partial}{\partial \boldsymbol{\phi}(k)}\left\{\frac{1}{2}\dot{\boldsymbol{\phi}}^T(k)\boldsymbol{H}(k)\dot{\boldsymbol{\phi}}(k)\right\}.$$

2.3 Position and orientation errors

In this subsection, a position and an orientation errors, which are used as feedback signals in the control methods described below, are defined (Taira et al, 2001).

2.3.1 Position error

The discrete-time position error between the target and the end-effector is defined as

$$\boldsymbol{e}_L(k) = \boldsymbol{p}_T - \boldsymbol{p}_E(k). \tag{8}$$

Using Equations (5) and (8), and the backward Euler approximation of $\boldsymbol{v}_E$, i.e.,

$$\boldsymbol{v}_E(k) = \frac{1}{T}\{\boldsymbol{p}_E(k) - \boldsymbol{p}_E(k-1)\},$$

the following difference equation of the position error is obtained:

$$\begin{aligned}\boldsymbol{e}_L(k) &= \boldsymbol{e}_L(k-1) - T\boldsymbol{v}_E(k) \\ &= \boldsymbol{e}_L(k-1) - T\boldsymbol{J}_L(k)\dot{\boldsymbol{\phi}}(k).\end{aligned} \tag{9}$$

2.3.2 Orientation error

In order to represent orientation of the end-effector and the target, we introduce the following rotation matrices:

$$\boldsymbol{A}_E(t) = \begin{bmatrix}\boldsymbol{n}_E(t) & \boldsymbol{s}_E(t) & \boldsymbol{a}_E(t)\end{bmatrix}, \tag{10}$$

$$\boldsymbol{A}_T = \begin{bmatrix}\boldsymbol{n}_T & \boldsymbol{s}_T & \boldsymbol{a}_T\end{bmatrix} \tag{11}$$

where $\boldsymbol{n}_*$, $\boldsymbol{s}_*$ and $\boldsymbol{a}_*$ $(* = E,\ T)$ are unit vectors along the axes of Σ_* with respect to Σ_I, and

$$\left.\begin{aligned}\dot{\boldsymbol{n}}_E(t) = \tilde{\boldsymbol{\omega}}_E(t)\boldsymbol{n}_E(t) = -\tilde{\boldsymbol{n}}_E(t)\boldsymbol{\omega}_E(t) \\ \dot{\boldsymbol{s}}_E(t) = \tilde{\boldsymbol{\omega}}_E(t)\boldsymbol{s}_E(t) = -\tilde{\boldsymbol{s}}_E(t)\boldsymbol{\omega}_E(t) \\ \dot{\boldsymbol{a}}_E(t) = \tilde{\boldsymbol{\omega}}_E(t)\boldsymbol{a}_E(t) = -\tilde{\boldsymbol{a}}_E(t)\boldsymbol{\omega}_E(t)\end{aligned}\right\} \tag{12}$$

is satisfied.

The continuous-time orientation error between the target and the end-effector is defined as

$$\boldsymbol{E}_A(t) = \begin{bmatrix}\boldsymbol{n}_T - \boldsymbol{n}_E(t) \\ \boldsymbol{s}_T - \boldsymbol{s}_E(t) \\ \boldsymbol{a}_T - \boldsymbol{a}_E(t)\end{bmatrix}. \tag{13}$$

Using Equations (6), (12) and (13), we have

$$\dot{\boldsymbol{E}}_A(t) = -\begin{bmatrix}\dot{\boldsymbol{n}}_E(t) \\ \dot{\boldsymbol{s}}_E(t) \\ \dot{\boldsymbol{a}}_E(t)\end{bmatrix} = \boldsymbol{E}_X(t)\boldsymbol{J}_A(t)\dot{\boldsymbol{\phi}} \tag{14}$$

where

$$\boldsymbol{E}_X(t) = \begin{bmatrix}\tilde{\boldsymbol{n}}_E(t) \\ \tilde{\boldsymbol{s}}_E(t) \\ \tilde{\boldsymbol{a}}_E(t)\end{bmatrix}.$$

Discretizing Equation (14) by the sampling period T, and the backward Euler approximation of $\dot{\boldsymbol{E}}_A$, the following difference equation of the orientation error is obtained:

$$\boldsymbol{E}_A(k) = \boldsymbol{E}_A(k-1) + T\boldsymbol{E}_X(k)\boldsymbol{J}_A(k)\dot{\boldsymbol{\phi}}(k). \tag{15}$$

Furthermore, the discrete-time orientation error used in the control method described below is defined as

$$\boldsymbol{e}_A(k) = \frac{1}{2}\left\{\tilde{\boldsymbol{n}}_E(k)\boldsymbol{n}_T + \tilde{\boldsymbol{s}}_E(k)\boldsymbol{s}_T + \tilde{\boldsymbol{a}}_E(k)\boldsymbol{a}_T\right\}. \tag{16}$$

It should be noted that a relationship between $\boldsymbol{E}_A$ and $\boldsymbol{e}_A$ is given by the following equation:

$$\boldsymbol{e}_A(k) = -\frac{1}{2}\boldsymbol{E}_X{}^T(k)\boldsymbol{E}_A(k). \tag{17}$$

Hence, if a condition

$$\boldsymbol{n}_E{}^T(k)\boldsymbol{n}_T + \boldsymbol{s}_E{}^T(k)\boldsymbol{s}_T + \boldsymbol{a}_E{}^T(k)\boldsymbol{a}_T > -1$$

is satisfied, $\boldsymbol{E}_A(k) = \boldsymbol{0}$ is equivalent to $\boldsymbol{e}_A(k) = \boldsymbol{0}$ (Masutani et al., 1989b).

3. Basic control method using transpose of GJM

In this section, we address a basic digital control method using the transpose of the GJM (Taira et al, 2001). The control method guarantees the stability of system in discrete-time domain by using Lyapunov's direct method for difference equations.
A joint torque command is given by

$$\boldsymbol{\tau}_d(k) = \boldsymbol{J}_L{}^T(k)\left\{k_L(k)\boldsymbol{e}_L(k) - \boldsymbol{K}_L(k)\boldsymbol{v}_E(k)\right\} + \boldsymbol{J}_A{}^T(k)\left\{k_A(k)\boldsymbol{e}_A(k) - \boldsymbol{K}_A(k)\boldsymbol{\omega}_E(k)\right\} \tag{18}$$

where, $k_\dagger$ ($\dagger = L,\ A$) is a positive scalar gain and $\boldsymbol{K}_\dagger$ is a symmetric and positive definite gain matrix.
The following theorem establishes the stability properties of the Equation (18).

Theorem 1. *Assume that the joint torque of the manipulator is identical to the joint torque command given by Equation (18), i.e.,* $\boldsymbol{\tau}(k) \approx \boldsymbol{\tau}_d(k)$, *and number of joints is* $n = 6$. *In the system represented by Equations (7), (9), (15) and (18), the equilibrium state*

$$\boldsymbol{r}_E(k) = \boldsymbol{r}_T,\ \boldsymbol{n}_E(k) = \boldsymbol{n}_T,\ \boldsymbol{s}_E(k) = \boldsymbol{s}_T,\ \boldsymbol{a}_E(k) = \boldsymbol{a}_T,\ \dot{\boldsymbol{\phi}}(k) = \boldsymbol{0} \tag{19}$$

is asymptotically stable, if the following conditions hold during manipulation:

$$rank\left[\boldsymbol{J}_L{}^T(k) \quad \boldsymbol{J}_A{}^T(k)\right] = 6 \quad (full\ rank), \tag{20}$$

$$\boldsymbol{n}_E{}^T(k)\boldsymbol{n}_T + \boldsymbol{s}_E{}^T(k)\boldsymbol{s}_T + \boldsymbol{a}_E{}^T(k)\boldsymbol{a}_T > -1. \tag{21}$$

Proof.
We choose

$$W(k) = W_1(k) + W_2(k) + W_3(k) \tag{22}$$

as a candidate for a Lyapunov function which is positive definite, and its first difference $\Delta W(k)$ is defined as

$$\Delta W(k) = W(k) - W(k-1) = \sum_{i=1}^{3} \Delta W_i(k) \tag{23}$$

where

$$W_1(k) = k_p \boldsymbol{e}_L{}^T(k)\boldsymbol{e}_L(k), \tag{24}$$

$$W_2(k) = \frac{1}{2}k_A \boldsymbol{E}_A{}^T(k)\boldsymbol{E}_A(k), \tag{25}$$

$$W_3(k) = \dot{\boldsymbol{\phi}}^T(k)\boldsymbol{H}(k)\dot{\boldsymbol{\phi}}(k), \tag{26}$$

$$\Delta W_i(k) = W_i(k) - W_i(k-1)\ (i = 1,\ 2,\ 3). \tag{27}$$

From Equation (9), $\Delta W_1(k)$ is represented as follows:

$$\Delta W_1(k) = -2k_L T\dot{\boldsymbol{\phi}}^T(k)\boldsymbol{J}_L{}^T(k)\boldsymbol{e}_L(k) - k_L T^2\dot{\boldsymbol{\phi}}^T(k)\boldsymbol{J}_L{}^T(k)\boldsymbol{J}_L(k)\dot{\boldsymbol{\phi}}(k). \tag{28}$$

Similarly, using Equations (15) and (17) and the property that $\boldsymbol{E}_X{}^T(k)\boldsymbol{E}_X(k) = 2\boldsymbol{E}$, $\Delta W_2(k)$ can be rewritten as

$$\begin{aligned}\Delta W_2(k) &= k_A T\dot{\boldsymbol{\phi}}^T(k)\boldsymbol{J}_A{}^T(k)\boldsymbol{E}_X{}^T(k)\boldsymbol{E}_A(k) - \frac{1}{2}k_A T^2\dot{\boldsymbol{\phi}}^T(k)\boldsymbol{J}_A{}^T(k)\boldsymbol{E}_X{}^T(k)\boldsymbol{E}_X(k)\boldsymbol{J}_A(k)\dot{\boldsymbol{\phi}}(k)\\ &= -2k_A T\dot{\boldsymbol{\phi}}^T(k)\boldsymbol{e}_A(k) - k_A T^2\dot{\boldsymbol{\phi}}^T(k)\boldsymbol{J}_A{}^T(k)\boldsymbol{J}_A(k)\dot{\boldsymbol{\phi}}(k).\end{aligned} \tag{29}$$

For $\Delta W_3(k)$, assuming $\boldsymbol{H}(k) \approx \boldsymbol{H}(k-1)$ for one sampling period, i.e., $\dot{\boldsymbol{H}}(k) \approx \boldsymbol{0}$, and using Equation (7) and the property

$$\dot{\boldsymbol{\phi}}^T(k)\boldsymbol{S}(k)\dot{\boldsymbol{\phi}}(k) = \boldsymbol{0},\ ^{\forall}k$$

we have

$$\Delta W_3(k) = 2T\dot{\boldsymbol{\phi}}^T(k)\boldsymbol{\tau}(k) - T^2\{\boldsymbol{S}(k)\dot{\boldsymbol{\phi}}(k) - \boldsymbol{\tau}(k)\}^T\boldsymbol{H}^{-1}(k)\{\boldsymbol{S}(k)\dot{\boldsymbol{\phi}}(k) - \boldsymbol{\tau}(k)\} \tag{30}$$

Substituting $\boldsymbol{\tau}(k) \approx \boldsymbol{\tau}_d(k)$ and Equations (5), (6) and (18) into Equation (30) yields

$$\begin{aligned}\Delta W_3(k) =& 2k_L T\dot{\boldsymbol{\phi}}^T(k)\boldsymbol{J}_L{}^T(k)\boldsymbol{e}_L(k) + 2k_A T\dot{\boldsymbol{\phi}}^T(k)\boldsymbol{J}_A{}^T(k)\boldsymbol{e}_L(k)\\ &- 2T\dot{\boldsymbol{\phi}}^T(k)\boldsymbol{J}_L{}^T(k)\boldsymbol{K}_A\boldsymbol{v}_E(k) - 2T\dot{\boldsymbol{\phi}}^T(k)\boldsymbol{J}_A{}^T(k)\boldsymbol{K}_A\boldsymbol{\omega}_E(k)\\ &- T^2\{\boldsymbol{S}(k)\dot{\boldsymbol{\phi}}(k) - \boldsymbol{\tau}_d(k)\}^T\boldsymbol{H}^{-1}(k)\{\boldsymbol{S}(k)\dot{\boldsymbol{\phi}}(k) - \boldsymbol{\tau}_d(k)\}\\ =& 2k_L T\dot{\boldsymbol{\phi}}^T(k)\boldsymbol{J}_L{}^T(k)\boldsymbol{e}_L(k) + 2k_A T\dot{\boldsymbol{\phi}}^T(k)\boldsymbol{J}_A{}^T(k)\boldsymbol{e}_L(k)\\ &- 2T\dot{\boldsymbol{\phi}}^T(k)\boldsymbol{J}_L{}^T(k)\boldsymbol{K}_A\boldsymbol{J}_L(k)\boldsymbol{\phi}(k) - 2T\dot{\boldsymbol{\phi}}^T(k)\boldsymbol{J}_A{}^T(k)\boldsymbol{K}_A\boldsymbol{J}_A(k)\boldsymbol{\phi}(k)\\ &- T^2\{\boldsymbol{S}(k)\dot{\boldsymbol{\phi}}(k) - \boldsymbol{\tau}_d(k)\}^T\boldsymbol{H}^{-1}(k)\{\boldsymbol{S}(k)\dot{\boldsymbol{\phi}}(k) - \boldsymbol{\tau}_d(k)\}.\end{aligned} \tag{31}$$

From Equations (28), (29) and (31), Equation (23) can be rewritten as

$$\begin{aligned}\Delta W(k) =& -k_L T^2 \dot{\phi}^T(k) \boldsymbol{J}_L{}^T(k) \boldsymbol{J}_L(k) \dot{\phi}(k) - k_A T^2 \dot{\phi}^T(k) \boldsymbol{J}_A{}^T(k) \boldsymbol{J}_A(k) \dot{\phi}(k) \\ & -2T\dot{\phi}^T(k) \boldsymbol{J}_L{}^T(k) \boldsymbol{K}_A \boldsymbol{J}_L(k) \phi(k) - 2T\dot{\phi}^T(k) \boldsymbol{J}_A{}^T(k) \boldsymbol{K}_A \boldsymbol{J}_A(k) \phi(k) \\ & -T^2\{\boldsymbol{S}(k)\dot{\phi}(k) - \boldsymbol{\tau}_d(k)\}^T \boldsymbol{H}^{-1}(k)\{\boldsymbol{S}(k)\dot{\phi}(k) - \boldsymbol{\tau}_d(k)\} \\ =& -2T\dot{\phi}^T(k) \bar{\boldsymbol{K}}(k) \dot{\phi}(k) - T^2 \dot{\phi}^T(k) \bar{\boldsymbol{Q}}(k) \dot{\phi}(k) \\ & -T^2\{\boldsymbol{S}(k)\dot{\phi}(k) - \boldsymbol{\tau}_d(k)\}^T \boldsymbol{H}^{-1}(k)\{\boldsymbol{S}(k)\dot{\phi}(k) - \boldsymbol{\tau}_d(k)\} \\ \leq& 0 \end{aligned} \tag{32}$$

where

$$\bar{\boldsymbol{K}}(k) = \begin{bmatrix} \boldsymbol{J}_L{}^T(k) & \boldsymbol{J}_A{}^T(k) \end{bmatrix} \begin{bmatrix} \boldsymbol{K}_L & \boldsymbol{0} \\ \boldsymbol{0} & \boldsymbol{K}_A \end{bmatrix} \begin{bmatrix} \boldsymbol{J}_L(k) \\ \boldsymbol{J}_A(k) \end{bmatrix},$$
$$\bar{\boldsymbol{Q}}(k) = \begin{bmatrix} \boldsymbol{J}_L{}^T(k) & \boldsymbol{J}_A{}^T(k) \end{bmatrix} \begin{bmatrix} k_L\boldsymbol{E} & \boldsymbol{0} \\ \boldsymbol{0} & k_A\boldsymbol{E} \end{bmatrix} \begin{bmatrix} \boldsymbol{J}_L(k) \\ \boldsymbol{J}_A(k) \end{bmatrix}$$

and $\bar{\boldsymbol{K}}(k)$ and $\bar{\boldsymbol{Q}}(k)$ are positive definite matrices.
In Equation (32), $\Delta W(k) = 0$ is satisfied if and only if $\dot{\phi}(k) = \boldsymbol{0}$ and $\boldsymbol{\tau}_d(k) = \boldsymbol{0}$. In addition, from Equations (5) and (6), the condition of $\dot{\phi}(k) = \boldsymbol{0}$ holds $\boldsymbol{v}_E(k) = \boldsymbol{0}$ and $\boldsymbol{\omega}_E(k) = \boldsymbol{0}$. Then, in the case of $\dot{\phi}(k) = \boldsymbol{0}$ and $\boldsymbol{\tau}_d(k) = \boldsymbol{0}$, Equation (18) becomes

$$\begin{bmatrix} \boldsymbol{J}_L{}^T(k) & \boldsymbol{J}_A{}^T(k) \end{bmatrix} \begin{bmatrix} k_L\boldsymbol{e}_L(k) \\ k_A\boldsymbol{e}_A(k) \end{bmatrix} = \boldsymbol{0} \tag{33}$$

and this equation is equivalent to $\boldsymbol{e}_L(k) = \boldsymbol{0}$ and $\boldsymbol{e}_A(k) = \boldsymbol{0}$ under the condition (20). Furthermore, $\boldsymbol{e}_A(k) = \boldsymbol{0}$ is equivalent to $\boldsymbol{E}_{OI} = \boldsymbol{0}$ under the condition (21).
Therefore, since $W(k) = 0$ and $\Delta W(k) = 0$ are satisfied if and only if the state is (19), and the other state keeps $W(k) > 0$ and $\Delta W(k) < 0$, $W(k)$ is a Lyapunov function and the equilibrium state (19) is is asymptotically stable. ∎
In order to validate the effectiveness of the control law (18), computer simulations were performed. Figure 2 shows the simulation model which has a 6-DOF manipulator, whose physical parameters are shown in Table 1.
To avoid excessive inputs and to improve the end-effector path from the initial point to the target, the following conditions are used:

$$\bar{\boldsymbol{e}}_L(k) = \begin{cases} \boldsymbol{e}_L(k) & (||\boldsymbol{e}_L(k)|| \leq e_{L_{\max}}) \\ \dfrac{\boldsymbol{e}_L(k)}{||\boldsymbol{e}_L(k)||} e_{L_{\max}} & (||\boldsymbol{e}_L(k)|| > e_{L_{\max}}) \end{cases}, \tag{34}$$

$$\bar{\boldsymbol{e}}_A(k) = \begin{cases} \boldsymbol{e}_A(k) & (||\boldsymbol{E}_A(k)|| \leq \sqrt{2} e_{A_{\max}}) \\ \dfrac{\sqrt{2}\boldsymbol{e}_A(k)}{||\boldsymbol{E}_A(k)||} e_{A_{\max}} & (||\boldsymbol{E}_A(k)|| > \sqrt{2} e_{A_{\max}}) \end{cases}, \tag{35}$$

Simulation was carried out under the following condition. The sampling period is $T = 0.01$[s] and t the controller parameters are $k_L = 150$, $k_A = 150$, $\boldsymbol{K}_L = \mathrm{diag}[300\ 150\ 600]$, $\boldsymbol{K}_A = \mathrm{diag}[10\ 10\ 50]$, $e_{L_{\max}} = 0.2$ and $e_{A_{\max}} = 0.2$.

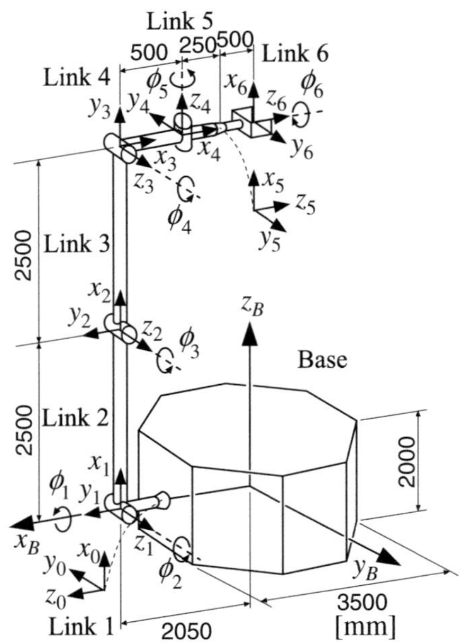

Fig. 2. Simulation model of 6-DOF space robot

	Length m	Mass kg	Moment of inertia kg·m^2		
			I_{ix}	I_{iy}	I_{iz}
Base	3.5	2000	1933	1936	2537
Link 1	0.25	5	0.16	0.06	0.17
Link 2	2.5	50	0.79	35.0	35.4
Link 3	2.5	50	0.79	35.0	35.4
Link 4	0.5	10	0.27	0.45	0.27
Link 5	0.25	5	0.04	0.06	0.06
Link 6	0.5	5	0.07	0.23	0.26

Table 1. Physical parameters of 6-DOF space robot

Simulation result is shown in Figure 3. From this figure, we can see that the position and orientation of the end-effector are well controlled nevertheless the base is moved.

4. Trajectory tracking control

The basic control law (18) uses position and orientation errors between the desired and actual values of the end-effector of the manipulator. Namely, the basic control method using Equation (18) belong to a class of constant value control such as PID control. So, the basic control method can adopt a desired trajectory of the end-effector for practical purposes. The value of errors, however, depends on the desired linear and angular velocities of the end-effector based on the desired trajectory. In this section, we address digital trajectory tracking control methods using the transpose of the GJM (Sagara & Taira, 2007; 2008b).

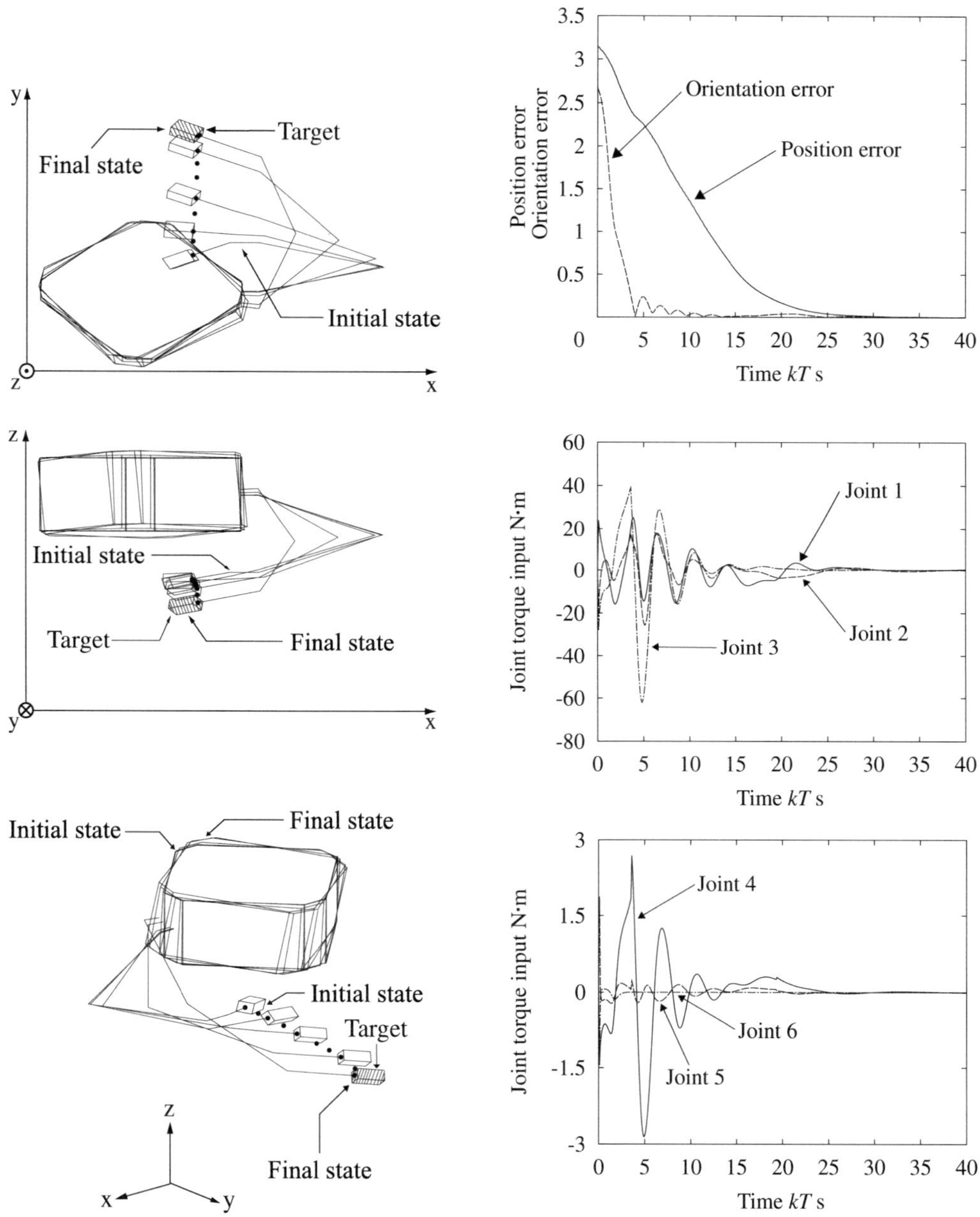

(a) Motion (b) Time history

Fig. 3. Simulation result of 6-DOF robot using basic control law (18)

4.1 Tracking control

To apply the basic control law (18) to tracking control, the following equations are utilized instead of Equations (8) and (17):

$$\bar{e}_L(k) = p_T(k) - p_E(k), \quad (36)$$

$$\bar{e}_A(k) = -\frac{1}{2} E_X{}^T(k) \bar{E}_A(k) \quad (37)$$

where

$$\bar{E}_A(k) = \begin{bmatrix} n_T(k) - n_E(k) \\ s_T(k) - s_E(k) \\ a_T(k) - a_E(k) \end{bmatrix}.$$

To examine the performance of the tracking control using Equation (18), three types of simulations were performed using a horizontal planar 3-DOF robot shown in Figure 4 with object.

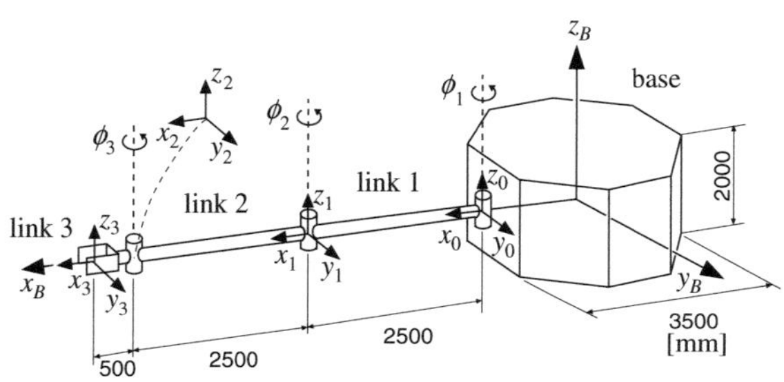

	Length[m]	Mass[kg]	Moment of inertia[kg·m^2]
Base	3.5	2000	3587.9
Link 1	2.5	50	26.2
Link 2	2.5	50	26.2
Link 3	0.5	5	0.23
Object	4.0	100	200.0

Fig. 4. Simulation model of 3-DOF space robot

Simulations were carried out under the following condition. A point of interest of the object moves along a straight path from the initial position to the target position, and the object angle is set to the initial value. The sampling period is $T = 0.01$s and the feedback gains are set for the following cases.

Case 1: This is the basic case. The feedback gains are $k_L = k_A = 50000$, $K_L = \text{diag}\{5000, 5000\}$ and $K_A = 5000$.

Case 2: Position and orientation feedback gains, k_L and k_A, are set to larger values than those in Case 1. The gains are $k_L = k_A = 100000$, $K_L = \text{diag}\{5000, 5000\}$ and $K_A = 5000$.

Case 3: Linear and angular velocity feedback gains K_L and K_A are set to smaller values than those in Case 1. The gains are $k_L = k_A = 50000$, $K_L = \text{diag}\{3000, 3000\}$ and $K_A = 3000$.

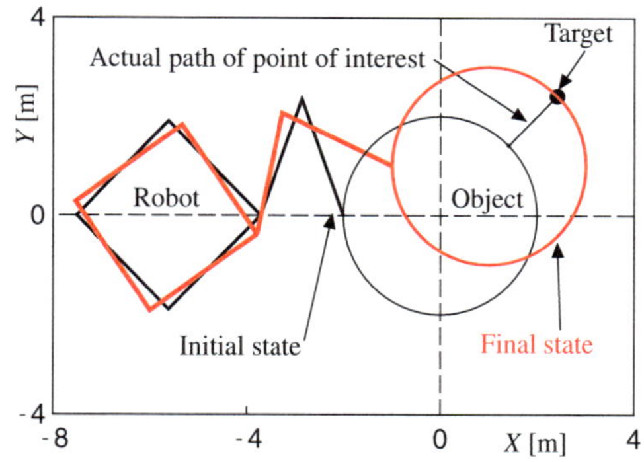

(a) Motion (Case 1)

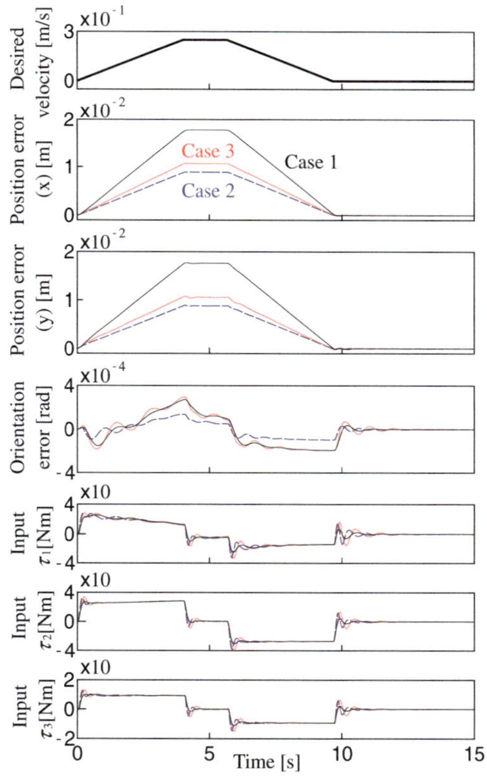

(b) Time history

Fig. 5. Simulation result of using basic control law (18)

Figure 5(a) shows the motion of the robot in Case 1. From Figure 5(a), it can be seen that tracking control using the control law (18) is valid. Figure 5(b) shows the simulation results in all cases. This figure consists of the norm of the desired linear velocity of the object, the position errors of x and y directions, the orientation error, and the control inputs of joint 1, 2 and 3. Furthermore, Figure 6 shows the time-history of the position and orientation errors after the object reaches the vicinity of the target. From Figure 5(b), the tracking errors in Case 2 and 3 are smaller than the errors in Case 1, while the desired velocity is transitional. From Figure 6, however, the convergent performance of the target becomes slightly worse.

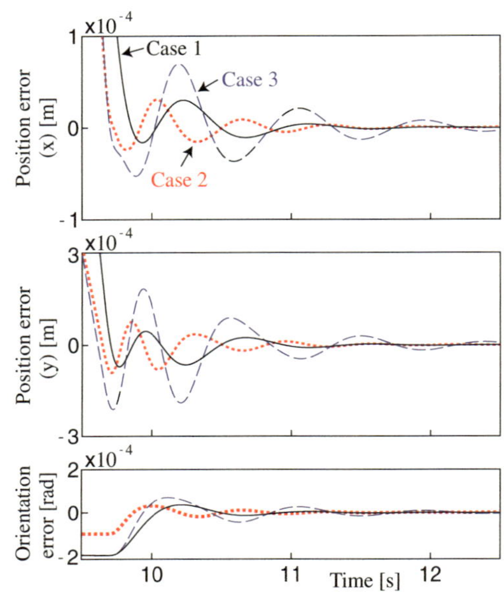

Fig. 6. Position and orientation errors after the object reaches the vicinity of the target

So, to get good control performance, we propose the modified control law (Sagara & Taira, 2007):

$$\boldsymbol{\tau}_d(k) = \boldsymbol{J}_L{}^T(k)\left\{\hat{k}_L(k)\hat{\boldsymbol{e}}_L(k) - \hat{\boldsymbol{K}}_L(k)\boldsymbol{v}_E(k)\right\} + \boldsymbol{J}_A{}^T(k)\left\{\hat{k}_A(k)\hat{\boldsymbol{e}}_A(k) - \hat{\boldsymbol{K}}_A(k)\boldsymbol{\omega}_E(k)\right\} \quad (38)$$

where

$$\hat{k}_\dagger(k) = k_\dagger\{1 + \alpha_\dagger \nu_\dagger(k)\}, \quad \hat{\boldsymbol{K}}_\dagger(k) = \boldsymbol{K}_\dagger\{1 - \beta_\dagger \nu_\dagger(k)\} \quad (\dagger = L,\ A),$$

$$\nu_L(k) = \frac{||\boldsymbol{v}_{\mathrm{int}_d}(k)||}{v_{d_{\max}}}, \quad \nu_A(k) = \frac{||\boldsymbol{\omega}_{\mathrm{int}_d}(k)||}{\omega_{d_{\max}}},$$

and $\boldsymbol{v}_{E_d}(k)$ and $\boldsymbol{\omega}_{E_d}(k)$ are the desired velocities of $\boldsymbol{v}_E(k)$ and $\boldsymbol{\omega}_E(k)$, respectively, $v_{d_{\max}}$ and $\omega_{d_{\max}}$ are the maximum values of the norm of $\boldsymbol{v}_{E_d}(k)$ and $\boldsymbol{\omega}_{E_d}(k)$, α_* ($\alpha_* \geq 0$) and β_* ($0 \leq \beta_* \leq 1$) ($* = L,\ A$) are setting parameters.
To verify the validity of the tracking control law (38), a simulation was performed. The values of the feedback gains, k_L, k_A, $\boldsymbol{K}_L$ and $\boldsymbol{K}_A$, for this simulation and for Case 1 are the same. The setting parameters are $\alpha_A = \alpha_L = 0.8$ and $\beta_A = \beta_L = 0.3$.

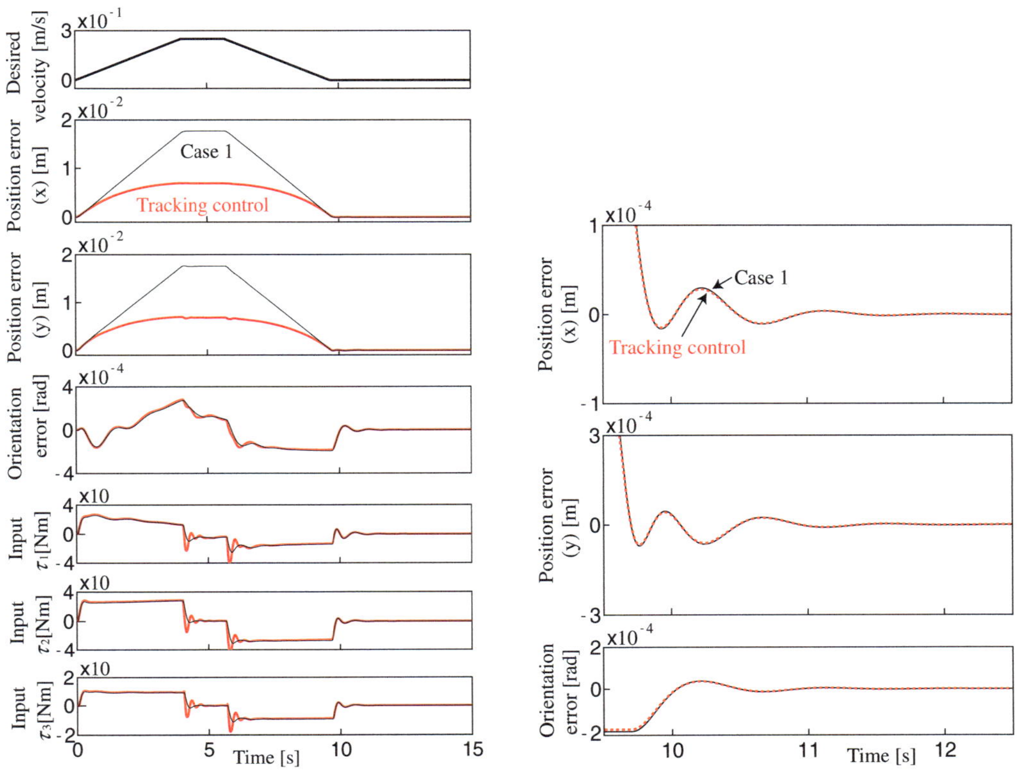

Fig. 7. Simulation result using tracking control law (38)

Figure 7 shows simulation result using the tracking control law (38). From Figure 7, it can be seen that good control performance can be achieved using the tracking control law.

4.2 Tracking control for joint velocity controller

The tracking control law (38) can be utilized for manipulators equipped with joint torque controller. It is considered that joint velocity controllers are also used for space robot manipulators. In this subsection, we address a tracking control method for robot manipulators equipped with joint velocity controller (Sagara & Taira, 2008b).
Figure 8(a) shows the time-history of the joint torque and velocity of the simulation result in case of Figure 7. And Figure 8(b) shows the relation between the actual joint input torque and the joint angular velocity. From Figure 8(b) we can see that the value of angular velocity is varying with the constant torque input during the sampling interval T. In other words, if the control inputs vary roughly for manipulators with joint velocity controllers, the joint controllers give large torques to the robot.
For manipulators with joint velocity controllers the tracking control law (38) cannot be applied directly. To obtain similar control performance to the case of the joint torque controllers, we use the following discrete-time dynamic equation of the robot.

$$\dot{\phi}(k_1) = \dot{\phi}(k_1 - 1) - T_1 \boldsymbol{H}^{-1}(k_1) \{\boldsymbol{c}(k_1) - \boldsymbol{\tau}(k_1)\} \tag{39}$$

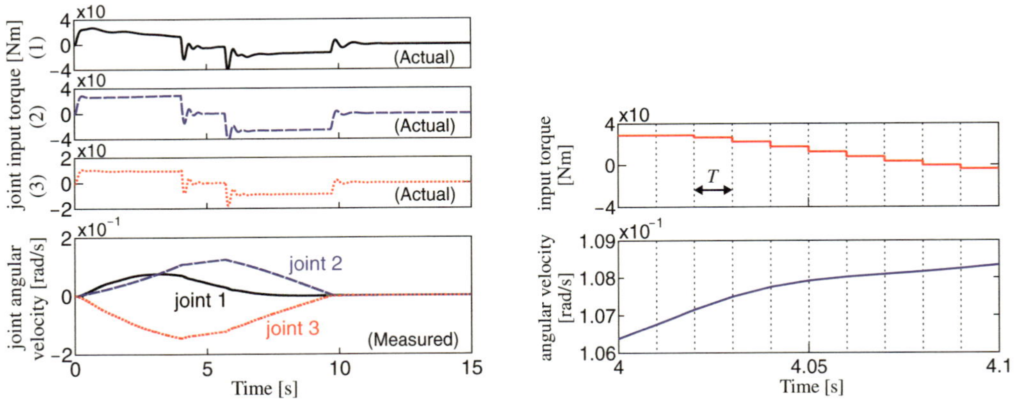

(a) All joints

(b) Relation between torque and velocity of joint 2

Fig. 8. Joint torque and velocity in case of tracking control for joint torque controller

where the sampling period T_1 ($T = nT_1$, n is positive integer) is used to Equation (7) instead of T, and $\boldsymbol{c}(k_1) = \boldsymbol{C}(k_1)\dot{\boldsymbol{\phi}}(k_1)$. Here, we assume that $\boldsymbol{\phi}$ is constant during the sampling interval T and $\boldsymbol{c}(k_1) \approx \boldsymbol{c}(k_1 - 1)$ during the sampling interval T_1. Then for Equation (39) the actual joint velocity control input $\dot{\boldsymbol{\phi}}_d(k_1)$ is determined as

$$\dot{\boldsymbol{\phi}}_d(k_1) = \dot{\boldsymbol{\phi}}(k_1 - 1) - T_1\boldsymbol{H}^{-1}(k)\left\{\boldsymbol{c}(k_1 - 1) - \boldsymbol{\tau}(k)\right\} \tag{40}$$

To verify the validity of the control law (38) with Equation (40) simulations were performed. The condition is the same to the torque input case and the sampling period for Eq. (40) is $T_1 = 0.001$s ($n = 10$).

Simulation result is shown in Figure 9. Furthermore, Figure 10 shows the difference of the joint angular velocity between the case of torque input control and velocity input control. From these figures, the both control performances are similar and good control performance can be achieved using the control law (38) with Equation (40).

5. Cooperative manipulation of object

It is considered that many tasks will be achieved by cooperative motions of some space robots in future space missions. We have studied on control problems for realizing cooperative manipulations and reported that a system consisting of some space robots with manipulators and a floating object can be treated as a kind of distributed system (Katoh et al., 1997; Sagara et al., 1998b). Using the distributed system representation each robot constituting the distributed system can be designed the control system individually.

In this section, the tracking control method using the transpose of the GJM is applied to cooperative manipulations of a floating object by some space robots (Sagara & Taira, 2008a; 2009).

5.1 Robot system model

In this subsection, we consider a space robot system consisting of M robots with manipulators and a floating object shown in Figure 11. The h-th robot ($h = 1, \cdots, M$) is consisting of an uncontrolled base and n_h-DOF manipulator with revolute joints.

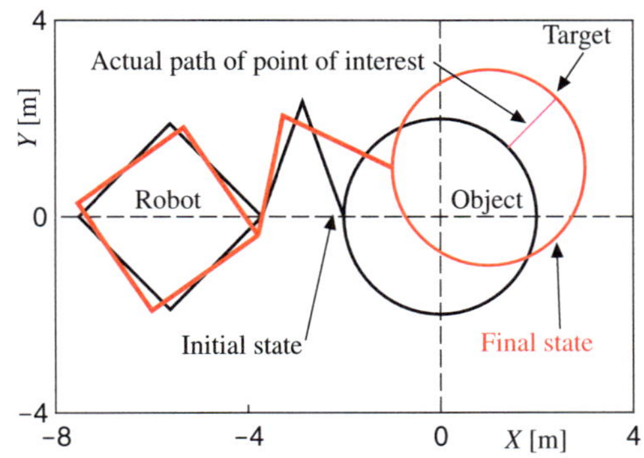

(a) Motion

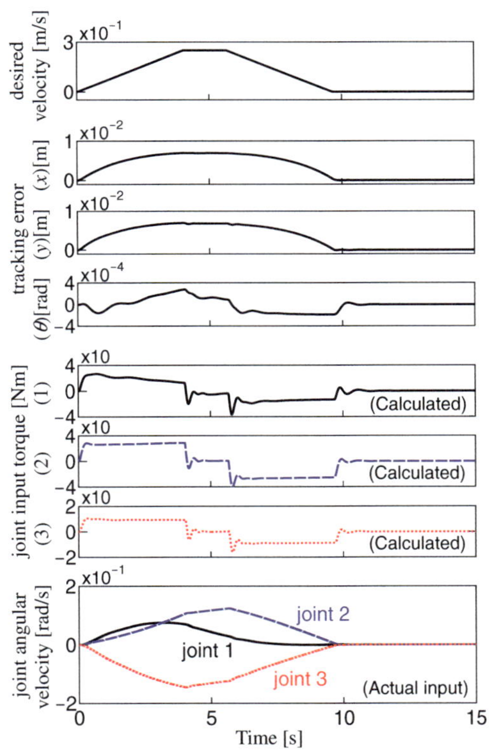

(b) Time history

Fig. 9. Simulation result using tracking control for joint velocity controller

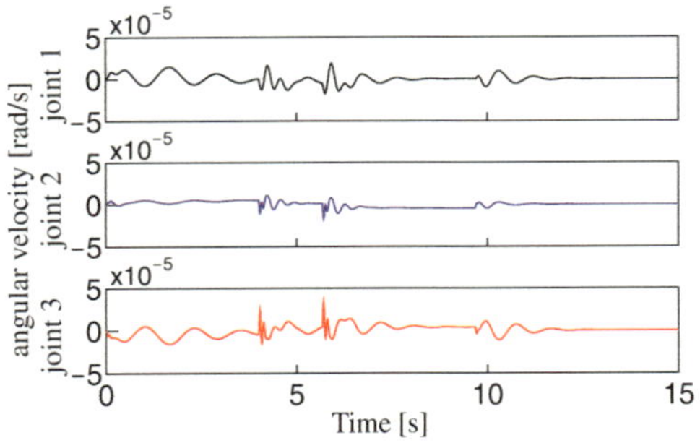

Fig. 10. Error comparison between torque controller and velocity controller

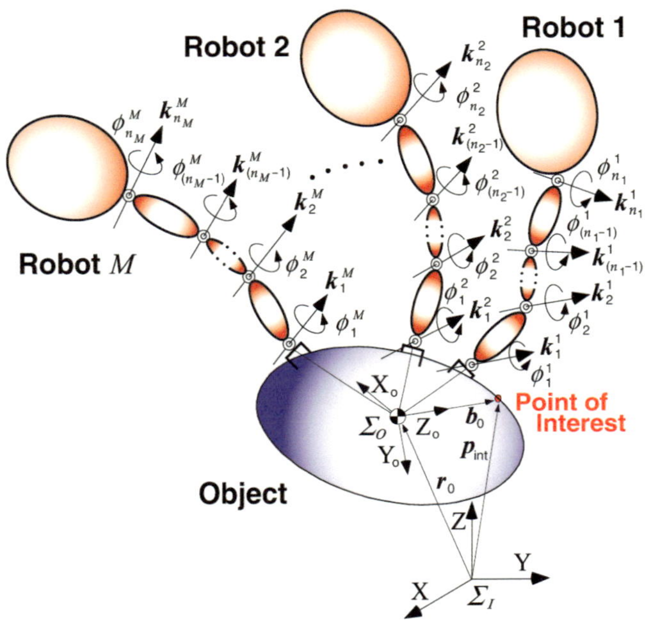

Fig. 11. Model of space robot system

The robot system shown in Figure 11 can be understood as one robot with M manipulators by regarding the object as a robot body, and M robot arms and robot bodies as M manipulators. The kinematic formulation of such space system has been derived by (Yoshida et al., 1991a;b). So, several symbols defined in Section 2 are changed. These symbols used in this section are redefined as follows:

$\boldsymbol{r}_0$: position vector of center of mass of base

$\boldsymbol{p}_E$: position vector of center of object

$\boldsymbol{p}_{\text{int}}$: position vector of point of interest on object

$\boldsymbol{p}_T$: position vector of point of interest

$\boldsymbol{v}_0$: linear velocity vector of center of mass of object

$\boldsymbol{v}_{\text{int}}$: linear velocity vector of point of interest

$\boldsymbol{\omega}_0$: angular velocity vector of center of mass of object

$\boldsymbol{\omega}_{\mathbf{int}}$: angular velocity vector of point of interest

i^h: number of link or joint i of robot h

$\boldsymbol{p}_i^h$: position vector of joint i^h

$\boldsymbol{r}_i^h$: position vector of mass center of link i^h

$\boldsymbol{k}_i^h$: unit vector indicating joint axis direction of joint i^h

$\boldsymbol{r}_g^h$: position vector of mass center of robot h

ϕ_i^h: relative angle of joint i^h

m_0: mass of object

m_i^h: mass of link i^h

$\boldsymbol{I}_0$: inertia tensor of object

$\boldsymbol{I}_i^h$: inertia tensor of link i^h

The relation obtained from the geometrical relationships of the robot system, and the conservation laws of linear momentum and angular momentum as follows:

$$\boldsymbol{\nu}_{\text{int}} = \begin{bmatrix} \boldsymbol{v}_{\text{int}} \\ \boldsymbol{\omega}_{\text{int}} \end{bmatrix} = \boldsymbol{J}_s^c \begin{bmatrix} \boldsymbol{v}_0 \\ \boldsymbol{\omega}_0 \end{bmatrix}, \quad \boldsymbol{H}_s^c \begin{bmatrix} \boldsymbol{v}_0 \\ \boldsymbol{\omega}_0 \end{bmatrix} + \boldsymbol{H}_m^c \dot{\boldsymbol{\phi}} = \mathbf{0} \tag{41}$$

where

$$\boldsymbol{J}_s^c = \begin{bmatrix} \boldsymbol{E} & \tilde{\boldsymbol{r}}_0 - \tilde{\boldsymbol{p}}_{\text{int}} \\ \mathbf{0} & \boldsymbol{E} \end{bmatrix}, \qquad \boldsymbol{H}_s^c = \begin{bmatrix} w\boldsymbol{E} & w(\tilde{\boldsymbol{r}}_0 - \tilde{\boldsymbol{r}}_g) \\ w\tilde{\boldsymbol{r}}_g & \boldsymbol{I}_w^c \end{bmatrix}, \qquad \boldsymbol{H}_m^c = \begin{bmatrix} \boldsymbol{J}_{T_w}^c \\ \boldsymbol{I}_\phi^c \end{bmatrix},$$

$$\boldsymbol{\phi}^c = \left[(\boldsymbol{\phi}^1)^T, (\boldsymbol{\phi}^2)^T, \cdots, (\boldsymbol{\phi}^M)^T \right]^T,$$

$$\boldsymbol{I}_w^c = \sum_{h=1}^{M} \boldsymbol{I}_w^h + \boldsymbol{I}_o, \qquad \boldsymbol{J}_{T_w}^c = \sum_{h=1}^{M} \boldsymbol{J}_{T_w}^h, \qquad \boldsymbol{I}_\phi^c = \sum_{h=1}^{M} \boldsymbol{I}_\phi^h,$$

$$\boldsymbol{I}_w^h = \sum_{i=1}^{n_h} \{ \boldsymbol{I}_i^h - m_i^h \tilde{\boldsymbol{r}}_i^h (\tilde{\boldsymbol{r}}_i^h - \tilde{\boldsymbol{r}}_0^h) \}, \qquad \boldsymbol{J}_{T_w}^h = \sum_{i=1}^{n_h} m_i^h \boldsymbol{J}_{T_i}^h, \qquad \boldsymbol{I}_\phi^h = \sum_{i=1}^{n_h} (\boldsymbol{I}_i^h \boldsymbol{J}_{R_i}^h + m_i^h \tilde{\boldsymbol{r}}_i^h \boldsymbol{J}_{T_i}^h),$$

$$\boldsymbol{J}_{T_i}^h = \begin{bmatrix} \boldsymbol{O}_a & \bar{\boldsymbol{J}}_{T_i}^h & \boldsymbol{O}_b \end{bmatrix}, \qquad \boldsymbol{J}_{R_i}^h = \begin{bmatrix} \boldsymbol{O}_a & \bar{\boldsymbol{J}}_{R_i}^h & \boldsymbol{O}_b \end{bmatrix},$$

$$\bar{\boldsymbol{J}}_{T_i}^h = \left[\tilde{\boldsymbol{k}}_1^h (\boldsymbol{r}_i^h - \boldsymbol{p}_1^h), \cdots, \tilde{\boldsymbol{k}}_i^h (\boldsymbol{r}_i^h - \boldsymbol{p}_i^h), \mathbf{0}, \cdots, \mathbf{0} \right], \qquad \bar{\boldsymbol{J}}_{R_i}^h = \left[\boldsymbol{k}_1^h, \cdots, \boldsymbol{k}_i^h, \mathbf{0}, \cdots, \mathbf{0} \right],$$

and $\boldsymbol{O}_a \in \boldsymbol{R}^{3 \times n_a}$ ($n_a = \sum_{i=1}^{h-1} n_i$) and $\boldsymbol{O}_b \in \boldsymbol{R}^{3 \times n_b}$ ($n_b = \sum_{i=h+1}^{M} n_i$) are zero matrices.
Form Equation (41), the relation between velocity $\boldsymbol{\nu}_{\text{int}}$ of the object and joint angular velocity $\dot{\boldsymbol{\phi}}^c$ of the manipulator can be derived as

$$\boldsymbol{\nu}_{\text{int}} = \boldsymbol{J}^c \dot{\boldsymbol{\phi}}^c \tag{42}$$

where $\boldsymbol{J}^c = -\boldsymbol{J}_s^c (\boldsymbol{H}_s^c)^{-1} \boldsymbol{H}_m^c$ is a GJM of the system shown in Figure 11.

5.2 System partition

For the system shown in Figure 11 control systems can be easily constructed by using Equation (42). However, if the number of robots is changed, Equation (42) must be recalculated. Furthermore, if the number of robots becomes increased, a large amount of calculation for the system is necessary. To solve the problems described above, This total robot system is regarded as a distributed system.

By examining parameters and variables included in the matrix $\boldsymbol{H}_s^c$ and vector $\boldsymbol{H}_m^c \dot{\boldsymbol{\phi}}^c$ in Equation (41), the matrix and vector can be rewritten as

$$\boldsymbol{H}_s^c = \boldsymbol{H}_s^0 + \sum_{h=1}^{M} \boldsymbol{H}_s^h, \ \ \boldsymbol{H}_m^c \dot{\boldsymbol{\phi}}^c = \sum_{h=1}^{M} \boldsymbol{H}_m^h \dot{\boldsymbol{\phi}}^h \tag{43}$$

where

$$\boldsymbol{H}_s^0 = \begin{bmatrix} m_0 \boldsymbol{E} & \boldsymbol{0} \\ m_0 \tilde{\boldsymbol{r}}_0 & \boldsymbol{I}_0 \end{bmatrix}, \qquad \boldsymbol{H}_s^h = \begin{bmatrix} m^h \boldsymbol{E} & m^h (\tilde{\boldsymbol{r}}_0^h - \tilde{\boldsymbol{r}}_g^h) \\ m^h \tilde{\boldsymbol{r}}_g^h & \boldsymbol{I}_w^h \end{bmatrix}, \qquad \boldsymbol{H}_m^h = \begin{bmatrix} \bar{\boldsymbol{J}}_{T_w}^h \\ \bar{\boldsymbol{I}}_\phi^h \end{bmatrix},$$

$$m^h = \sum_{i=1}^{n_h} m_i^h, \qquad \bar{\boldsymbol{J}}_{T_w}^h = \sum_{i=1}^{n_h} m_i^h \bar{\boldsymbol{J}}_{T_i}^h, \qquad \bar{\boldsymbol{I}}_\phi^h = \sum_{i=1}^{n_h} \left(\boldsymbol{I}_i^h \bar{\boldsymbol{J}}_{R_i}^h + m_i^h \tilde{\boldsymbol{r}}_i^h \bar{\boldsymbol{J}}_{T_i}^h \right).$$

$\boldsymbol{H}_s^h$ and $\boldsymbol{H}_m^h$ are matrices including parameters the h-th robot only, and $\boldsymbol{H}_s^0$ is a matrix including parameters of the object only.

Equations (41) and (43) make the following relation:

$$\left(\boldsymbol{H}_s^0 + \sum_{h=1}^{M} \boldsymbol{H}_s^h \right) (\boldsymbol{J}_s^c)^{-1} \boldsymbol{\nu}_{\text{int}} + \sum_{h=1}^{M} \boldsymbol{H}_m^h \dot{\boldsymbol{\phi}}^h = \boldsymbol{0}. \tag{44}$$

It is clear that the following set of equations is one of solutions of Equation (44), when a constant and diagonal matrix $\boldsymbol{A}_h$ is introduced.

$$\bar{\boldsymbol{H}}_s^h (\boldsymbol{J}_s^c)^{-1} \boldsymbol{\nu}_{\text{int}} + \boldsymbol{H}_m^h \dot{\boldsymbol{\phi}}^h = \boldsymbol{0} \quad (h = 1, \cdots, M) \tag{45}$$

where

$$\bar{\boldsymbol{H}}_s^h = \boldsymbol{H}_s^h + \boldsymbol{A}_h \boldsymbol{H}_s^0, \qquad \sum_{h=1}^{M} \boldsymbol{A}_h = \boldsymbol{E}.$$

Then, the following relation can be derived from Equations (45).

$$\boldsymbol{\nu}_{\text{int}} = -\boldsymbol{J}_s^c \left(\bar{\boldsymbol{H}}_s^h \right)^{-1} \boldsymbol{H}_m^h \dot{\boldsymbol{\phi}}^h \quad (h = 1, \cdots, M). \tag{46}$$

Therefore, for each robot of the system the control system can be designed individually.

5.3 Simulation

To examine the performance of the control methods using the transpose of the GJM, simulations are performed by using three of the horizontal planar 3-DOF robots shown in Figure 12 and an object. Note that the physical parameters of the robots shown in Figure 12 and 4 are the same, except for numbering the links and the joints.

All simulations were carried out under the following condition. A point of interest on the object moves along a straight path from the initial position to the target position and the object

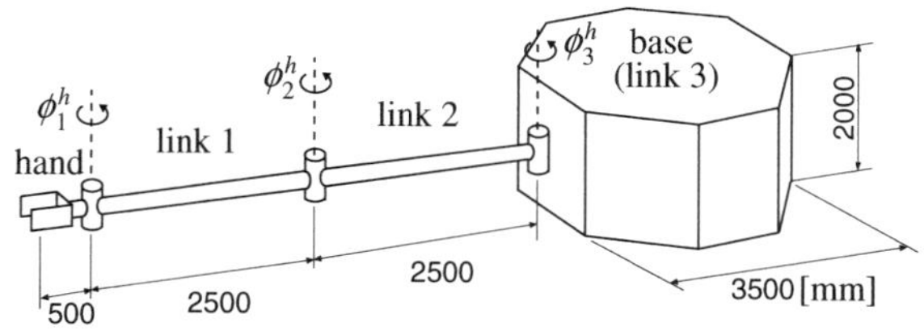

Fig. 12. Simulation model of 3-DOF space robot

angle is set up the initial value. The sampling period is $T = 0.01$s and the coefficient matrices are $\boldsymbol{A}_1 = \boldsymbol{A}_2 = 0.33\boldsymbol{E}$ and $\boldsymbol{A}_3 = 0.34\boldsymbol{E}$.

5.3.1 Robustness for singular configuration

Our proposed control methods using the transpose of the GJM can be utilized to the case of a singular configuration of manipulator. The robustness for the singular configuration is demonstrated by simulations using the basic control law (18).

Simulations were carried out under the following condition. The feedback gains are $k_L = k_A = 3 \times 10^4$, $\boldsymbol{K}_L = \text{diag}\{2 \times 10^3,\ 2 \times 10^3\}$ and $\boldsymbol{K}_A = 2 \times 10^3$. Furthermore, one of the robots breaks down 2.5s after simulations are started, and the physical parameters of the floating object are shown in Table 2.

	Diameter [m]	Mass [kg]	Moment of inertia [kg·m^2]
Object	3.0	100	106.2

Table 2. Physical parameters of floating object

Figure 13 shows the simulation result. From this figure, we can see that the determinant of GJM of Robot 1 changes from positive values to negative values affected by the breakdown of Robot 3, and the floating object can be manipulated by the robot system in spite of the singular configuration of Robot 1. Therefore, our proposed control methods using the transpose of the GJM have the robustness for the singular configuration of manipulator.

5.3.2 Tracking control for joint torque controller

Next, the tracking control law (38) is applied to the robot system (Sagara & Taira, 2008a).

Simulations were carried out under the following condition. The feedback gains are $k_L = k_A = 2 \times 10^5$, $\boldsymbol{K}_L = \text{diag}\{2 \times 10^4,\ 2 \times 10^4\}$ and $\boldsymbol{K}_A = 2 \times 10^4$. The setting parameters are $\alpha_\dagger = 2$ and $\beta_\dagger = 0.2$ ($\dagger = L,\ A$). The physical parameters of the floating object are shown in Table 3.

	Diameter [m]	Mass [kg]	Moment of inertia [kg·m^2]
Object	4.0	1200	2400.0

Table 3. Physical parameters of floating object for tracking control method

Figure 14 shows the simulation result. Figure 14(b) also shows the case of the basic control, i. e., $\alpha_\dagger = \beta_\dagger = 0$. From Figure 14, it can be seen that good control performance can be achieved using the tracking control law (38).

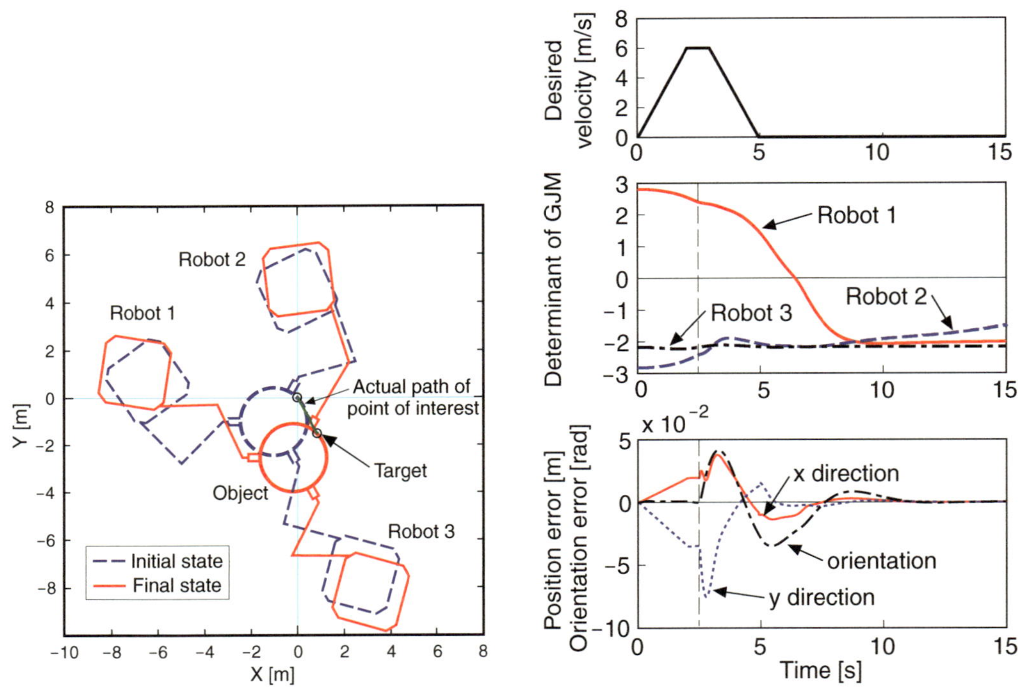

(a) Motion (b) Time history

Fig. 13. Simulation result of robot system using basic control law (18)

5.3.3 Tracking control for joint velocity controller

From the simulation result shown in Figure 9, the tracking control law (38) with discrete-time dynamic equation (40) can be applied to single manipulators equipped with joint velocity controllers. For the robot system shown in Figure 11, the dynamics of each robot is affected by other robots. Thus, the external force affected by other robots must be considered (Sagara & Taira, 2009).

Based on the dynamic equation (39), the dynamic equation of the h-th robot shown in Figure 11 is obtained:

$$\boldsymbol{\chi}^h(k_1) = \boldsymbol{\chi}^h(k_1 - 1) - T_1 \boldsymbol{H}^{h^{-1}}(k_1)\left\{\boldsymbol{c}^h(k_1) - \boldsymbol{u}^h(k_1)\right\}. \tag{47}$$

where

$$\boldsymbol{\chi} = \begin{bmatrix} \boldsymbol{\eta}_0 \\ \dot{\boldsymbol{\phi}}^h \end{bmatrix}, \quad \boldsymbol{u} = \begin{bmatrix} \boldsymbol{f}^h \\ \boldsymbol{\tau}^h \end{bmatrix},$$

and $\boldsymbol{H}^h$ is the symmetric and positive definite inertia matrix, $\boldsymbol{c}^h$ is the vector of Coliolis and centrifugal forces, $\boldsymbol{\eta}_0 = [\boldsymbol{v}_0{}^T, \ \boldsymbol{\omega}_0{}^T]^T$ is the velocity of the mass center of object, $\boldsymbol{f}^h$ is the external force affected by other robots.

For Eq. (47) the actual joint velocity control input $\dot{\boldsymbol{\phi}}_d^h(k_1)$ is determined as

$$\boldsymbol{\chi}_d^h(k_1) = \boldsymbol{\chi}^h(k_1 - 1) - T_1 \boldsymbol{H}^{h^{-1}}(k)\left\{\boldsymbol{c}^h(k_1 - 1) - \boldsymbol{u}_d^h(k_1)\right\} \tag{48}$$

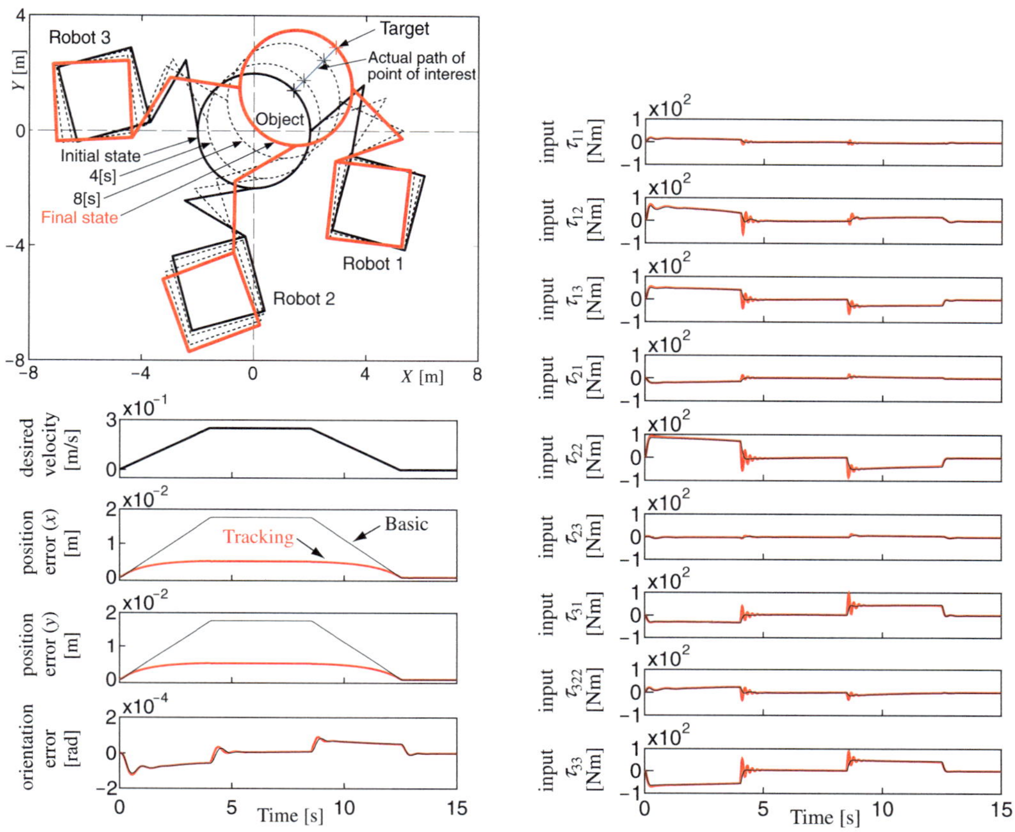

Fig. 14. Simulation result of robot system using tracking control law (38) (torque input)

where

$$\boldsymbol{\chi}_d(k_1) = \begin{bmatrix} \boldsymbol{\eta}^h_{0_d}(k_1) \\ \dot{\boldsymbol{\phi}}^h_d(k_1) \end{bmatrix}, \quad \boldsymbol{u}_d(k_1) = \begin{bmatrix} \boldsymbol{f}^h(k_1) \\ \boldsymbol{\tau}^h_d(k) \end{bmatrix}.$$

Since the value of the external force $\boldsymbol{f}_h$ affected by other robots cannot obtained directly, a disturbance observer in discrete time (Godler et al., 2002) is used to estimate $\boldsymbol{f}_h$.
The equation of motion of the floating object with respect to the h-th robot is

$$\boldsymbol{H}^h_0(k_1)\dot{\boldsymbol{\eta}}_0(k_1) = \boldsymbol{f}^h(k_1) + \bar{\boldsymbol{H}}^h_0\dot{\boldsymbol{\eta}}^h_{0_d}(k_1) \tag{49}$$

where $\boldsymbol{H}^h_0(k_1)$ is the inertia matrix of the floating object and

$$\bar{\boldsymbol{H}}^h_0 = \boldsymbol{A}_h \begin{bmatrix} m_0\boldsymbol{E} & \boldsymbol{0} \\ \boldsymbol{0} & \boldsymbol{I}_0 \end{bmatrix}$$

is the nominal model of $\boldsymbol{H}^h_0(k_1)$.
For Equation (49) the estimated value of $\boldsymbol{f}_h$, $\hat{\boldsymbol{f}}_h$, can be obtained from the disturbance observer as shown in Figure 15. In this figure, (a), (b) and (c) show the basic configuration in

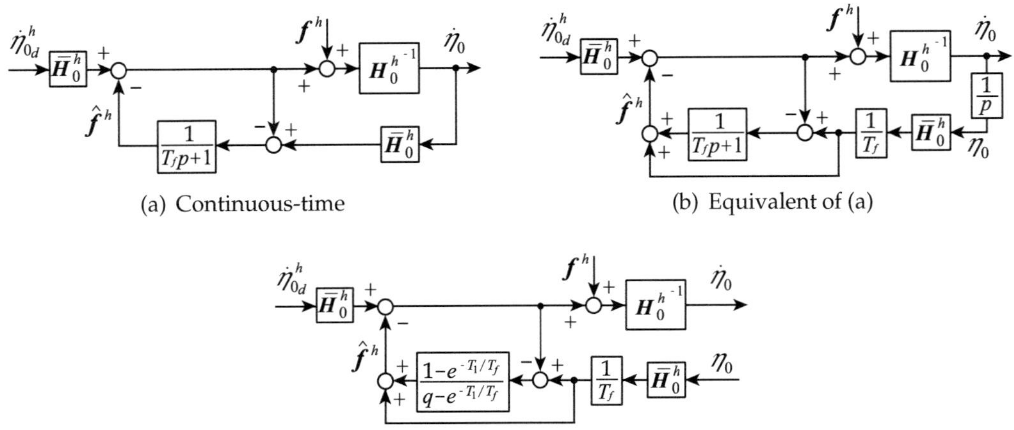

(a) Continuous-time (b) Equivalent of (a)

(c) Discrete-time

Fig. 15. Disturbance observer

continuous-time, the equivalent transformation of (a) and the discrete-time version, respectively, T_f is a time constant of a low-pass filter, p and q are the differential and shift operators. From Figure 15(c), the estimated force can be obtained as follows:

$$\begin{aligned}\hat{\boldsymbol{f}}^h(k_1) =& \hat{\boldsymbol{f}}^h(k_1-1) + \left(1 - e^{-T_1/T_f}\right)\bar{\boldsymbol{H}}_0^h \dot{\boldsymbol{\eta}}_{0_d}(k_1-1) \\ &+ \frac{1}{T_f}\bar{\boldsymbol{H}}_0^h\left\{\boldsymbol{\eta}_0(k_1) + \left(1 - 2e^{-T_1/T_f}\right)\boldsymbol{\eta}_0(k_1-1)\right\}. \end{aligned} \quad (50)$$

Simulations were performed to validate the effectiveness of the control method described above. The simulation condition for the joint velocity controllers is the same as those for the joint torque controllers. In addition, the time constant for the low-pass filter is $T_f = 1$s. Figure 16 shows the simulation result. From Figure 16(a), the object is successfully moved by three robots. And from Figure 16(b), it can be seen that good control performance can be achieved.

6. Conclusion

In this chapter, our proposed control methods for space robot manipulators using transpose of GJM were described and the computer simulations were performed. For manipulators equipped with joint torque controllers we explained about a basic control method using constant feedback gains with the proof of stability. To obtain higher control performance, we addressed trajectory tracking control methods with variable feedback gains for both torque and velocity joint inputs. Moreover, we addressed a cooperative manipulation of a floating object by some space robots with the control methods using transpose of GJM. For the cooperative manipulation, simulations where manipulators get into a singular configuration were also performed.

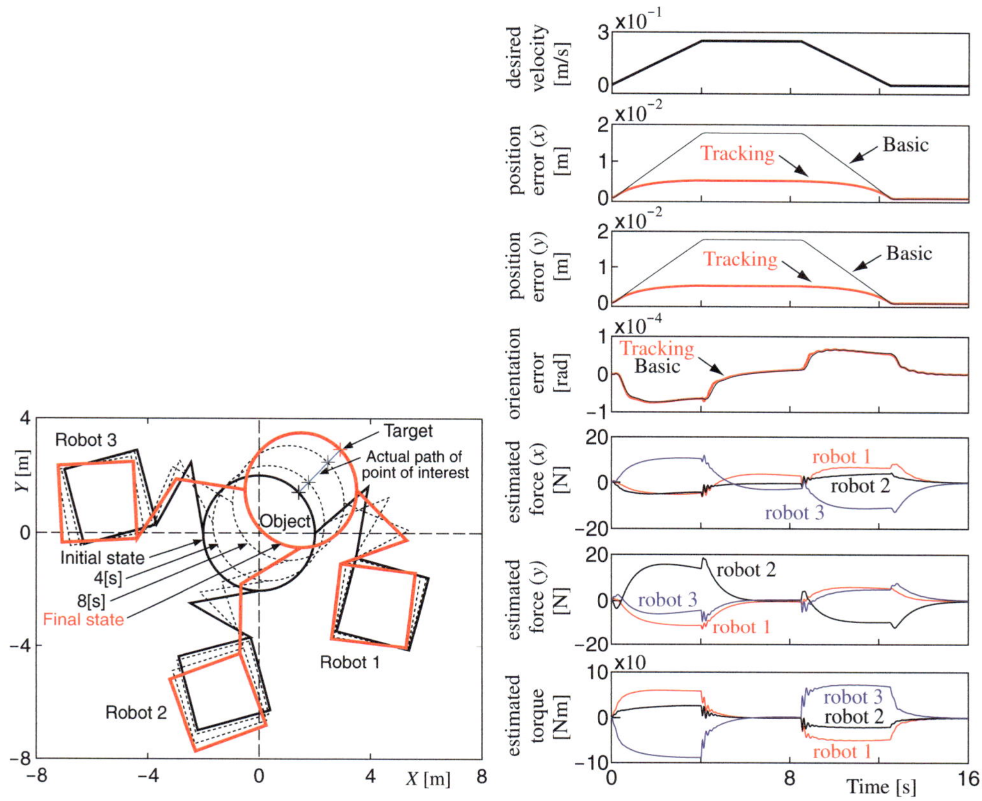

(a) Motion (b) Time history

Fig. 16. Simulation result of the robot system equipped with joint velocity controllers

7. References

Dubowsky, S. & Papadorpulos, E. (1993). The Kinematics, Dynamics, and Control of Free-Flying and Free-Floating Space Robotic Systems, *IEEE Transactions on Robotics and Automation*, Vol. 9, No. 5, pp. 531–543

Godler, I., Honda, H. & Ohnishi, K. (2002). Design Guidelines for Disturbance Observer's Filter in Discrete Time, *Proceedings of the 7th International Workshop on Advanced Motion Control*, Maribor, pp. 390 – 395

Katoh, R., Nakatsuka, K., Sagara, S., Kobayshi, J. & Ohkawa, F. (1997). Manipulation of a Floating Object by Two Space Manipulators, *Proceedings of 23rd IEEE Industrial Electronics Society International Conference*, pp. 1270–1275

Masutani, Y., Miyazaki, F. & Arimoto, S. (1989a). Sensory Feedback Control for Space Manipulators, *Proceedings 1989 IEEE International Conference on Robotics and Automation*, pp. 1346 – 1351

Masutani, Y., Miyazaki, F. & Arimoto, S. (1989b). Modeling and Sensory Feedback Control for Space Manipulators, *Proceedings of the NASA Conference on Space Telerobotics*, pp. 287 – 296

Sagara, S., Taira, Y., Katoh, R., Yamashita, T., Ohkawa, F. & Suehiro, T. (1998a). Digital Adaptive Control of Space Robot Manipulator Having Input Constraints, *Proceedings of the 1998 IEEE/RSJ International Conference on Intelligent Robots and Systems*, pp. 1449–1454

Sagara, S., Hideura, M., Katoh, R., Yamashita, T., Kobayashi, J. & Ohkawa, F. (1998b). Adaptive RMRC for Cooperative Manipulation of a Floating Object by Two Free-Based Space Robots, *Proceedings of the 1998 IEEE/RSJ International Conference on Intelligent Robots and Systems*, pp. 1467–1472

Sagara, S. & Taira, Y. (2007). Digital Tracking Control of Space Robots Using a Transpose of the Generalized Jacobian Matrix, *Artificial Life and Robotics*, Vol. 11, pp. 82–86

Sagara, S. & Taira, Y. (2008a). Cooperative Manipulation of a Floating Object by Some Space Robots: Application of a Tracking Control Method Using the Transpose of the Generalized Jacobian Matrix, *Artificial Life and Robotics*, Vol. 12, pp. 138–141

Sagara, S. & Taira, Y. (2008b). Digital Control of Space Robot Manipulators with Velocity Type Joint Controller Using the Transpose of the Generalized Jacobian Matrix, *Artificial Life and Robotics*, Vol. 13, pp. 355–358

Sagara, S. & Taira, Y. (2009). Cooperative Manipulation of a Floating Object by Some Space Robots with Joint Velocity Controllers: Application of a Tracking Control Method Using the Transpose of the Generalized Jacobian Matrix, *Artificial Life and Robotics*, Vol. 14, pp. 392–396

Shin, J. H., Jeong, I. K., Lee, J. J. & Ham, W. (1995). Adaptive Robust Control for Free-Flying Space Robots Using Norm-Bounded Property of Uncertainty, *Proceedings of the 1995 IEEE/RSJ International Conference on Intelligent Robots and Systems*, pp. 59–64

Taira, Y., Sagara, S. & Katoh, R. (2001). Digital Control of a Space Robot Manipulators Using Transpose of the Generalized Jacobian Matrix (in Japanese), *Transactions of The Japan Society of Mechanical Engineers*, Series C, Vol. 67, No. 663, pp. 3540–3547

Umetani, Y. & Yoshida, K. (1989). Resolved Motion Rate Control of Space Manipulators with Generalized Jacobian Matrix, *IEEE Transactions on Robotics and Automation*, Vol. 5, No. 3, pp. 303 – 314

Xu, Y. & Kanade, T. (Ed.) (1993). *Space Robotics: Dynamics and Control*, Kluwer Academic Publishers, 0-7923-9265-5, Massachusetts

Yamamoto, T., Kobayashi, J., Katoh, R. & Ohkawa, F. (1995). Digital Adaptive Control of a Manipulator Mounted on Free-Flying Robot in Space, *Proceedings Ninth World Congress on the Theory of Machines and Mechanism*, pp. 2151–2155

Yoshida, K., Kurazume, R & Umetani, Y. (1991a). Dual Arm Coordination in Space Free-Flying Robot, *Proceedings of the 1991 IEEE International Conference on Robotics and Automation*, pp. 2516–2521

Yoshida, K., Kurazume, R & Umetani, Y. (1991b). Coordinated control of multiple manipulators in space robots (in Japanese), *Journal of the Robotics Society of Japan*, Vol. 9, pp. 718–726

21

Kinematics, Singularity and Dexterity Analysis of Planar Parallel Manipulators Based on DH Method

Serdar Kucuk
Kocaeli University
Turkey

1. Introduction

Parallel manipulators have separate serial kinematic chains that are linked to the ground and the moving platform at the same time. They have some potential advantages over serial robot manipulators such as high accuracy, greater load capacity, high mechanical rigidity, high velocity and acceleration (Kang et al., 2001; Kang & Mills, 2001). Planar Parallel Manipulators (PPMs), performing two translations along the x and y axes, and rotation through an angle of ϕ around the z axis are a special group among the parallel robot manipulators. They have potential advantages for microminiaturization (Hubbard et al., 2001) and pick-and-place operations (Heerah et al., 2003). However, due to the complexity of the closed-loop chain mechanism, the kinematics analysis of parallel manipulators is more difficult than their serial counterparts. Therefore selection of an efficient mathematical model is very important for simplifying the complexity of the kinematics problems in parallel robots. In this book chapter, the forward and inverse kinematics problems of PPMs are solved based on D-H method (Denavit & Hartenberg, 1955) which is a common kinematic modelling convention using 4x4 homogenous transformation matrices. The easy physical interpretation of the robot mechanisms is the main advantage of this method. The Forward kinematics problem calculates the position and orientation of the end-effector if the set of joint angles are known. The inverse kinematics problem solves for the joint angles when the position and orientation of the end effector is given. In contrast to serial manipulators, the forward kinematics problem is much more difficult than the inverse kinematic problem for parallel manipulators. Afterwards very practical definitions are provided for Jacobian matrix and workspace determination of PPMs which are required for singularity and dexterity analyses. Rest of this book chapter is composed of the following sections. Some fundamental definitions about D-H method as a kinematics modelling convention, Jacobian matrix, condition number, global dexterity index, singularity and workspace determination are presented in Section 2. A two-degree-of-freedom (2-dof) PPM and a 3-dof Fully Planar Parallel Manipulator (FPPM) are given as examples to illustrate the methodology in the following Section. FPPMs are composed of a moving platform linked to the ground by three independent limbs including one active (actuated) joint in each (Merlet, 1996). The conclusions with final remarks are presented in the last section.

2. Background

2.1 Denavit-Hartenberg convention

The D-H convention (Denavit & Hartenberg, 1955) based on 4x4 homogenous transformation matrices is commonly used kinematic modelling convention for the robotics community. The easy physical interpretation of the robot mechanisms is the main advantage of this method. A spatial transformation between two consecutive links can be described by a set of parameters, namely α_{j-1}, a_{j-1}, θ_j and d_j. Using these parameters, the general form of the transformation matrix for adjacent coordinate frames, j-1 and j is obtained as

$$ {}^{j-1}_{j}T = \begin{bmatrix} \cos\theta_j & -\sin\theta_j & 0 & a_{j-1} \\ \sin\theta_j \cos\alpha_{j-1} & \cos\theta_j \cos\alpha_{j-1} & -\sin\alpha_{j-1} & -\sin\alpha_{j-1} d_j \\ \sin\theta_j \sin\alpha_{j-1} & \cos\theta_j \sin\alpha_{j-1} & \cos\alpha_{j-1} & \cos\alpha_{j-1} d_j \\ 0 & 0 & 0 & 1 \end{bmatrix} \tag{1} $$

The transformation of the link n coordinate frame into the base coordinate frame of the robot manipulator is given by

$$ {}^{j-1}_{n}T = {}^{j-1}_{j}T \; {}^{j}_{j+1}T \; {}^{j+1}_{j+2}T \cdots {}^{n-1}_{n}T \tag{2} $$

where j=1, 2, 3, ..., n.

2.2 Jacobian Matrix, Condition Number and Global Dexterity Index

Dexterity is an essential topic for design and conceptual control of robotic manipulators. It can be described as the ability to perform infinitesimal movement in the arbitrary directions as easily as possible in the workspace of the robotic manipulators. It is based on the condition number (κ) of the manipulator Jacobian matrix (Gosselin & Angeles, 1991).

$$ \kappa = \|J\| \|J^{-1}\| \tag{3} $$

where J illustrates the Jacobian matrix and $\| \cdot \|$ denotes the Frobenius norm of the matrix. κ varies between 1 to ∞. In general, $\eta=1/\kappa$ is used for limiting the dexterity between 0 and 1. Thus dexterity of a manipulator can be easily measured. It is well known that as $\eta=1$ represents a perfect isotropic dexterity, $\eta=0$ illustrates singular configuration. Then the kinematic relations for the planar parallel manipulators can be expressed as

$$ f(\mathbf{x},\mathbf{q})=0 \tag{4} $$

where f is the function of $\mathbf{x}=(x,y,\phi)^T$ and $\mathbf{q}=(q_1, q_2,$ and $q_n)^T$, and 0 is n dimensional zero vector. The parameters x, y and ϕ are the position and orientation of the end-effector in terms of base frame. Additionally, q_1, q_2 and q_n represent the actuated joint variables. The following term is obtained differentiating the equation 4 with respect to the time.

$$A\dot{\mathbf{x}}+B\dot{\mathbf{q}}=0 \tag{5}$$

where $\dot{\mathbf{x}}$ and $\dot{\mathbf{q}}$ are the time derivatives of $\mathbf{x}$ and $\mathbf{q}$, respectively. A and B are two separate Jacobian matrices. The overall Jacobian matrix for a parallel manipulator can be obtained as

$$J=-B^{-1}A \tag{6}$$

Since the dexterity demonstrates the local property of a mechanism, designers requires a global one. Gosselin and Angeles (1991) introduced Global Dexterity Index (GDI) in order to measure the global property of the manipulators. GDI defined as

$$GDI=\frac{A}{B} \tag{7}$$

where B represents the area of the workspace and A is

$$A=\iiint_{\phi\, y\, x}\left(\frac{1}{\kappa}\right)dxdyd\phi \tag{8}$$

2.3 Singularity

Three types of singularities exist for planar parallel manipulators (Tsai, 1999; Merlet, 2000). They occur when either determinant of the matrices A or B or both are zero. When the determinant of A and B goes to zero, the direct and inverse kinematics singularities occur, respectively. Direct kinematics singularity occurs inside the workspace of the manipulator. Inverse kinematics singularity occurs at the boundaries of the manipulator's workspace where any limb is completely stretched-out or folded-back.

2.4 Workspace Determination

Workspace determination is a very significant step in analyzing robot manipulators. Its determination is necessary for conceptual design and trajectory planning. Many researches (Agrawal, 1990; Gosselin, 1990; Merlet, 1995; Merlet et al., 1998) focused on the workspace determination of parallel manipulators. Merlet, Gosselin and Mouly (1998) outlined some workspace definitions for PPMs such as constant orientation workspace, reachable workspace and dexterous workspace. The constant orientation workspace (considered also in this book chapter) is the area that can be reached by the end-effector with constant orientation. In general, geometrical (Bonev & Ryu, 2001; Gosselin & Guillot, 1991) and numerical methods (Kim et al., 1997) are used for workspace determination of parallel manipulators. Computation of workspace based on geometrical method is difficult and may not be obtained due to the complexity of the closed-loop chain mechanisms (Fichter, 1986).

Alternatively, numerical methods based on the discretization of Cartesian space (Kim et al., 1997) in general, are also used in this study due to their simplicity according to the geometrical methods. In discretization method, firstly a Cartesian area is discretized into "PxQ" points along the X and Y axes. Secondly each pose in this discretized Cartesian area is examined whether belonging to the workspace of the manipulator.

3. Examples

3.1. The 2-dof RPR planar parallel manipulator

The 2-dof RPR (Revolute-Prismatic-Revolute) planar parallel manipulator given by Figure 1 has two limbs B_iM_i, where i=1, 2, a moving platform M_1M_2 and a fixed base B_1B_2. The lengths of the actuated prismatic joints vary between $d_{i(min)}$ and $d_{i(max)}$. The end-effector is attached to the pint P on the moving platform. The position $P(p_x, p_y)$ and orientation (ϕ) of the end-effector is defined with respect to the {XYZ} coordinate system.

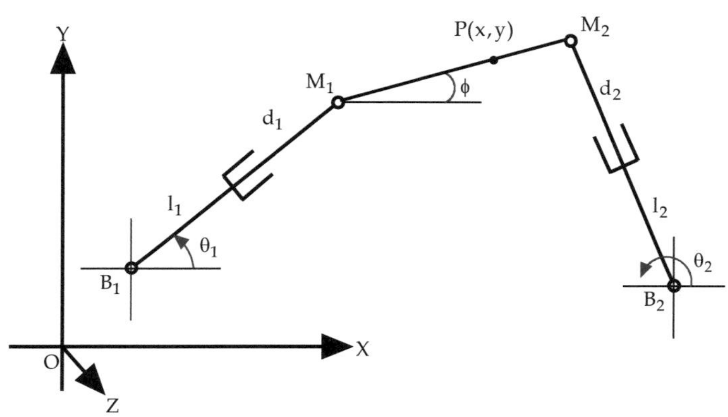

Fig. 1. The planar 2-dof RPR parallel manipulator.

The following equation can be written using the geometric identities on Figure 1.

$$OB_i + B_iM_i = OP + PM_i \tag{9}$$

where i=1, 2.

In order to solve **the inverse kinematics problem**, the relation in equation 9 is adapted to the manipulator in Figure 1. Thus the ${}^{M_i}_{O_i}T^1$ and ${}^{M_i}_{O_i}T^2$ homogeneous transformation matrices can be obtained as

$$
{}^{M_i}_{O_i}T^1 = \begin{bmatrix} 1 & 0 & 0 & OB_{x_i} \\ 0 & 1 & 0 & OB_{y_i} \\ 0 & 0 & 1 & 0 \\ 0 & 0 & 0 & 1 \end{bmatrix} \begin{bmatrix} \cos\theta_i & -\sin\theta_i & 0 & 0 \\ \sin\theta_i & \cos\theta_i & 0 & 0 \\ 0 & 0 & 1 & 0 \\ 0 & 0 & 0 & 1 \end{bmatrix} \begin{bmatrix} 1 & 0 & 0 & 0 \\ 0 & 0 & 1 & l_i + d_i \\ 0 & -1 & 0 & 0 \\ 0 & 0 & 0 & 1 \end{bmatrix}
$$

$$
= \begin{bmatrix} \cos\theta_i & 0 & \sin\theta_i & OB_{x_i} - \sin\theta_i (d_i + l_i) \\ \sin\theta_i & 0 & -\cos\theta_i & OB_{y_i} + \cos\theta_i (d_i + l_i) \\ 0 & 1 & 0 & 0 \\ 0 & 0 & 0 & 1 \end{bmatrix} \tag{10}
$$

$$
{}^{M_i}_{O_i}T^2 = \begin{bmatrix} 1 & 0 & 0 & p_x \\ 0 & 1 & 0 & p_y \\ 0 & 0 & 1 & 0 \\ 0 & 0 & 0 & 1 \end{bmatrix} \begin{bmatrix} \cos\gamma_i & -\sin\gamma_i & 0 & 0 \\ \sin\gamma_i & \cos\gamma_i & 0 & 0 \\ 0 & 0 & 1 & 0 \\ 0 & 0 & 0 & 1 \end{bmatrix} \begin{bmatrix} 1 & 0 & 0 & PM_i \\ 0 & 1 & 0 & 0 \\ 0 & 0 & 1 & 0 \\ 0 & 0 & 0 & 1 \end{bmatrix}
$$

$$
= \begin{bmatrix} -\cos\phi & \sin\phi & 0 & p_x + PM_i \cos\gamma_i \\ -\sin\phi & -\cos\phi & 0 & p_y + PM_i \sin\gamma_i \\ 0 & 0 & 1 & 0 \\ 0 & 0 & 0 & 1 \end{bmatrix} \tag{11}
$$

where $\gamma_1 = \pi + \phi$ and $\gamma_2 = \phi$. Since the position vectors of ${}^{M_i}_{O_i}T^1$ and ${}^{M_i}_{O_i}T^2$ matrices are equal, one can write the following equations easily.

$$
\begin{bmatrix} -\sin\theta_i (d_i + l_i) \\ \cos\theta_i (d_i + l_i) \end{bmatrix} = \begin{bmatrix} p_x + PM_i \cos\gamma_i - OB_{x_i} \\ p_y + PM_i \sin\gamma_i - OB_{y_i} \end{bmatrix} \tag{12}
$$

Summing the squares of the both sides in the equation 12, we obtain, after simplification,

$$
\begin{aligned} Q_i = {} & PM_i^2 + p_x^2 + p_y^2 + OB_{x_i}^2 + OB_{y_i}^2 - 2p_x OB_{x_i} - 2p_y OB_{y_i} \\ & + 2\cos\gamma_i PM_i (p_x - OB_{x_i}) + 2\sin\gamma_i PM_i (p_y - OB_{y_i}) - (d_i + l_i)^2 = 0 \end{aligned} \tag{13}
$$

The active joint variables d_i can be found using equation 13 as follows.

$$
d_i = \sqrt{\begin{aligned} & PM_i^2 + p_x^2 + p_y^2 + OB_{x_1}^2 + OB_{y_1}^2 - 2p_x OB_{x_i} - 2p_y OB_{y_i} \\ & \qquad + 2\cos\gamma_i PM_i (p_x - OB_{x_i}) + 2\sin\gamma_i PM_i (p_y - OB_{y_i}) \end{aligned}} - l_i \tag{14}
$$

Once the active joint variables d_1 and d_2 are solved, the passive joint variable θ_1 and θ_2 can be found from the equation 12 by back substitution.

$$\theta_i = a\tan 2\left(\frac{-p_x - PM_i \cos\gamma_i + OB_{x_i}}{(d_i + l_i)}, \frac{p_y + PM_i \sin\gamma_i - OB_{y_i}}{(d_i + l_i)} \right) \tag{15}$$

The Jacobian matrix of the 2-dof RPR is computed using the equation 13 as follows.

$$A\dot{x} + B\dot{q} = 0$$

$$\begin{bmatrix} \frac{\partial Q_1}{\partial p_x} & \frac{\partial Q_1}{\partial p_y} \\ \frac{\partial Q_2}{\partial p_x} & \frac{\partial Q_2}{\partial p_y} \end{bmatrix} \begin{bmatrix} \dot{p}_x \\ \dot{p}_y \end{bmatrix} + \begin{bmatrix} \frac{\partial Q_1}{\partial d_1} & \frac{\partial Q_1}{\partial d_2} \\ \frac{\partial Q_2}{\partial d_1} & \frac{\partial Q_2}{\partial d_2} \end{bmatrix} \begin{bmatrix} \dot{d}_1 \\ \dot{d}_2 \end{bmatrix} = 0$$

$$\begin{bmatrix} p_x - OB_{x_1} + \cos\gamma_1 PM_1 & p_y - OB_{y_1} + \sin\gamma_1 PM_1 \\ p_x - OB_{x_2} + \cos\gamma_2 PM_2 & p_y - OB_{y_2} + \sin\gamma_2 PM_2 \end{bmatrix} \begin{bmatrix} \dot{p}_x \\ \dot{p}_y \end{bmatrix} + \begin{bmatrix} -d_1 - l_1 & 0 \\ 0 & -d_2 - l_2 \end{bmatrix} \begin{bmatrix} \dot{d}_1 \\ \dot{d}_2 \end{bmatrix} = 0 \tag{16}$$

The overall Jacobian matrix is obtained as

$$J = \begin{bmatrix} -\frac{p_x - OB_{x_1} + \cos\gamma_1 PM_1}{-d_1 - l_1} & -\frac{p_y - OB_{y_1} + \sin\gamma_1 PM_1}{-d_1 - l_1} \\ -\frac{p_x - OB_{x_2} + \cos\gamma_2 PM_2}{-d_2 - l_2} & -\frac{p_y - OB_{y_2} + \sin\gamma_2 PM_2}{-d_2 - l_2} \end{bmatrix} \tag{17}$$

***Numerical example for the inverse kinematics*:** The dimensions of the planar 2-dof RPR parallel manipulator are given as $l_1=l_2=12$, $PM_1=PM_2=3$, $OB_{x1}=0$, $OB_{y1}=0$, $OB_{x2}=15$, $OB_{y2}=0$, $p_x=11$, $p_y=20$ and $\phi=30^o$. According to the data above, the active joint variables d_1 and d_2 can be found as 8.3185 and 9.5457, respectively. The passive joint variables θ_1 and θ_2 are found as -24.4256 and 3.7307, respectively by back substitution.

Numerical example for the Jacobian matrix and condition number: Using the dimensions of the inverse kinematics example above, the Jacobian matrix and condition number are found as

$$J = \begin{bmatrix} 0.4135 & 0.9105 \\ -0.0651 & 0.9979 \end{bmatrix} \text{ and } \kappa = 2.1192 \tag{18}$$

Numerical example for the workspace determination: The dimensions of the manipulator are given as $l_1=l_2=10$, $PM_1=PM_2=2$, $OB_{x1}=0$, $OB_{y1}=0$, $OB_{x2}=20$ and $OB_{y2}=0$ and. The lengths of the actuated prismatic joints vary between $d_i(min)=0$ and $d_i(max)=7$, where i=1,2. The workspace is obtained as 102.3888 for $\phi=0^o$ orientation. The limits of the discretized Cartesian area are denoted by the red rectangle shown in Figure 2a. The black and green areas show reachable and non-reachable workspaces of the manipulator, respectively.

Numerical example for the dexterity and singularity analysis: The Figure 2b shows the inverse dexterity of the manipulator using the same dimensions given by example for workspace determination. The GDI of the manipulator are found as 0.0012. As seen in Figure 2b, since the inverse of the condition numbers colored with red areas are so close to the unity, there isn't any singular point occurred inside reachable workspace. The red and blue areas show reachable and non-reachable workspaces of the manipulator, respectively.

3.2. The 3-dof RPR planar parallel manipulator

The 3-dof RPR planar parallel manipulator shown in Figure 3 has three limbs B_iM_i, where i=1, 2, 3. P denotes the point where the end-effector is located at the moving platform which is chosen as equilateral triangle. The angle ϕ represents the orientation of the end-effector. If a line AB passing through the point *P* is drawn parallel to the M_1M_2, the angles PM_1M_2 (σ_1) and PM_2M_1 (σ_2) are equal to the angles APM_1 and M_2PB, respectively. The angles between the lines BP and PM_1, M_2P and PB, BP and PM_3 are denoted as λ_1, λ_2 and λ_3, respectively.

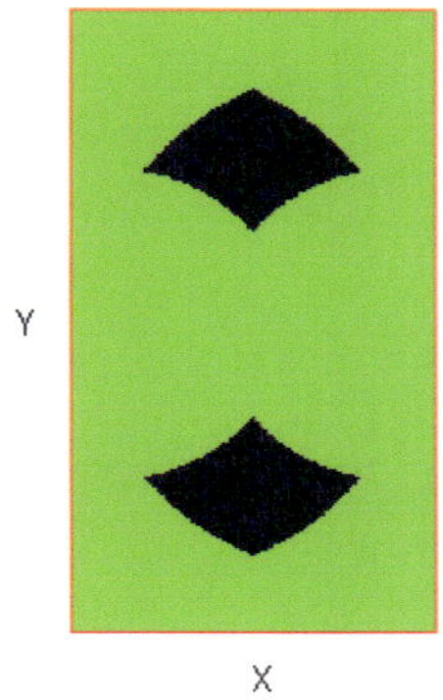

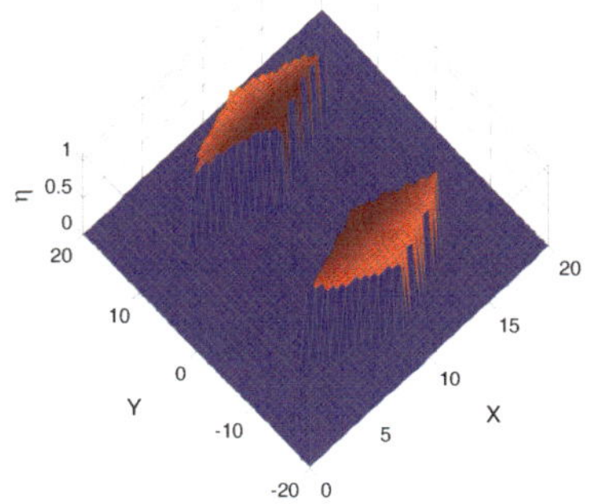

Fig. 2. a) Workspace, b) dexterity graph, for planar 2-dof RPR parallel manipulator.

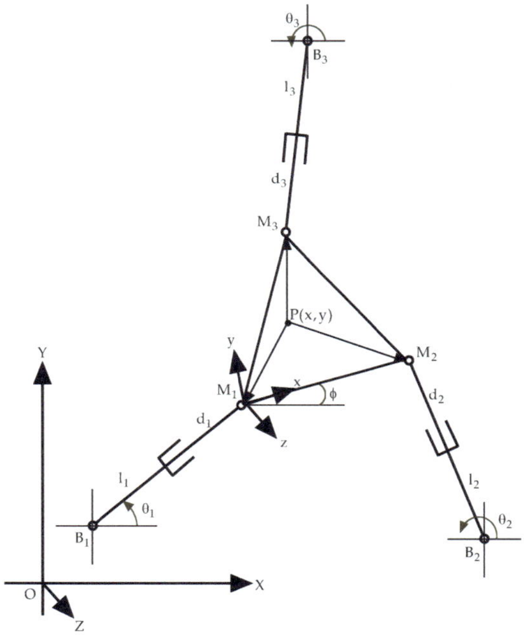

Fig. 3. The planar 3-dof RPR parallel manipulator.

If the relation in equation 9 is adapted to the manipulator in Figure 3, the ${}^{M_i}_{O_i}T^1$ and ${}^{M_i}_{O_i}T^2$ homogeneous transformation matrices can be obtained as

$$
{}^{M_i}_{O_i}T^1 = \begin{bmatrix} 1 & 0 & 0 & OB_{x_i} \\ 0 & 1 & 0 & OB_{y_i} \\ 0 & 0 & 1 & 0 \\ 0 & 0 & 0 & 1 \end{bmatrix} \begin{bmatrix} \cos\theta_i & -\sin\theta_i & 0 & 0 \\ \sin\theta_i & \cos\theta_i & 0 & 0 \\ 0 & 0 & 1 & 0 \\ 0 & 0 & 0 & 1 \end{bmatrix} \begin{bmatrix} 1 & 0 & 0 & 0 \\ 0 & 0 & 1 & l_i + d_i \\ 0 & -1 & 0 & 0 \\ 0 & 0 & 0 & 1 \end{bmatrix}
$$

$$
= \begin{bmatrix} \cos\theta_i & 0 & \sin\theta_i & OB_{x_i} - \sin\theta_i (d_i + l_i) \\ \sin\theta_i & 0 & -\cos\theta_i & OB_{y_i} + \cos\theta_i (d_i + l_i) \\ 0 & 1 & 0 & 0 \\ 0 & 0 & 0 & 1 \end{bmatrix} \tag{19}
$$

where i=1, 2, 3.

$$
{}^{M_i}_{O_i}T^2 = \begin{bmatrix} 1 & 0 & 0 & p_x \\ 0 & 1 & 0 & p_y \\ 0 & 0 & 1 & 0 \\ 0 & 0 & 0 & 1 \end{bmatrix} \begin{bmatrix} \cos(\lambda_i + \phi) & -\sin(\lambda_i + \phi) & 0 & 0 \\ \sin(\lambda_i + \phi) & \cos(\lambda_i + \phi) & 0 & 0 \\ 0 & 0 & 1 & 0 \\ 0 & 0 & 0 & 1 \end{bmatrix} \begin{bmatrix} 1 & 0 & 0 & PM_i \\ 0 & 1 & 0 & 0 \\ 0 & 0 & 1 & 0 \\ 0 & 0 & 0 & 1 \end{bmatrix}
$$

$$
= \begin{bmatrix} \cos(\lambda_i + \phi) & -\sin(\lambda_i + \phi) & 0 & p_x + PM_i \cos(\lambda_i + \phi) \\ \sin(\lambda_i + \phi) & \cos(\lambda_i + \phi) & 0 & p_y + PM_i \sin(\lambda_i + \phi) \\ 0 & 0 & 1 & 0 \\ 0 & 0 & 0 & 1 \end{bmatrix} \tag{20}
$$

where $\lambda_1 = \pi + \sigma_1$ and $\lambda_2 = -\sigma_2$. Since the position vectors of ${}^{M_i}_{O_i}T^1$ and ${}^{M_i}_{O_i}T^2$ matrices are equal, one can write easily the following equations.

$$
\begin{bmatrix} -\sin\theta_i (d_i + l_i) \\ \cos\theta_i (d_i + l_i) \end{bmatrix} = \begin{bmatrix} p_x + \upsilon_{x_i} \cos\phi - \upsilon_{y_i} \sin\phi - OB_{x_i} \\ p_y + \upsilon_{x_i} \sin\phi + \upsilon_{y_i} \cos\phi - OB_{y_i} \end{bmatrix} \tag{21}
$$

where $\upsilon_{x_i} = PM_i \cos\lambda_i$ and $\upsilon_{y_i} = PM_i \sin\lambda_i$. Summing the squares of the both sides in the equation 21, we obtain, after simplification,

$$
\begin{aligned}
Q_i = {} & p_x^2 + p_y^2 + OB_{x_i}^2 + OB_{y_i}^2 + \upsilon_{x_i}^2 + \upsilon_{y_i}^2 - 2p_x OB_{x_i} - 2p_y OB_{y_i} \\
& + 2\cos\phi (p_x \upsilon_{x_i} + p_y \upsilon_{y_i} - OB_{x_i} \upsilon_{x_i} - OB_{y_i} \upsilon_{y_i}) \\
& + 2\sin\phi (p_y \upsilon_{x_i} - p_x \upsilon_{y_i} - OB_{y_i} \upsilon_{x_i} + OB_{x_i} \upsilon_{y_i}) - (d_i + l_i)^2 = 0
\end{aligned} \tag{22}
$$

The inverse kinematics problem is solved using the equation 22 as follows.

$$d_i = \sqrt{\begin{array}{l} p_x^2 + p_y^2 + OB_{x_i}^2 + OB_{y_i}^2 + \upsilon_{x_i}^2 + \upsilon_{y_i}^2 - 2p_x OB_{x_i} - 2p_y OB_{y_i} \\ \quad + 2\cos\phi(p_x \upsilon_{x_i} + p_y \upsilon_{y_i} - OB_{x_i}\upsilon_{x_i} - OB_{y_i}\upsilon_{y_i}) \\ \quad\quad + 2\sin\phi(p_y \upsilon_{x_i} - p_x \upsilon_{y_i} - OB_{y_i}\upsilon_{x_i} + OB_{x_i}\upsilon_{y_i}) \end{array}} - l_i \quad (23)$$

Once the d_1, d_2 and d_3 are obtained, the angles θ_1, θ_2 and θ_3 can be determined from the equation 21 by back substitution.

$$\theta_i = a\tan 2\left(-\frac{p_x + \upsilon_{x_i}\cos\phi - \upsilon_{y_i}\sin\phi - OB_{x_i}}{(d_i + l_i)}, \frac{p_y + \upsilon_{x_i}\sin\phi + \upsilon_{y_i}\cos\phi - OB_{y_i}}{(d_i + l_i)} \right) \quad (24)$$

The forward kinematics problem is found rearranging the equation 22 as follows.

$$p_x^2 + p_y^2 + a_i p_x + b_i p_y + c_i = 0 \qquad (i=1,2,3) \quad (25)$$

where $a_i = 2(\cos\phi\upsilon_{x_i} - OB_{x_i} - \sin\phi\upsilon_{y_i})$, $b_i = 2(\cos\phi\upsilon_{y_i} + \sin\phi\upsilon_{x_i} - OB_{y_i})$ and $c_i = OB_{x_i}^2 + OB_{y_i}^2 + \upsilon_{x_i}^2 + \upsilon_{y_i}^2 - 2\cos\phi(OB_{x_i}\upsilon_{x_i} + OB_{y_i}\upsilon_{y_i}) + 2\sin\phi(OB_{x_i}\upsilon_{y_i} - OB_{y_i}\upsilon_{x_i}) - (d_i + l_i)^2$

One can obtain the following new system of equations using equation 25.

$$\begin{array}{l} (a_1 - a_2)p_x + (b_1 - b_2)p_y + (c_1 - c_2) = 0 \\ (a_1 - a_3)p_x + (b_1 - b_3)p_y + (c_1 - c_3) = 0 \\ (a_2 - a_3)p_x + (b_2 - b_3)p_y + (c_2 - c_3) = 0 \end{array} \quad (26)$$

One can write the following equations using the equation 26.

$$\begin{array}{l} a_{12}p_x + b_{12}p_y = c_{12} \\ a_{13}p_x + b_{13}p_y = c_{13} \end{array} \quad (27)$$

where $a_{12} = a_1 - a_2$, $b_{12} = b_1 - b_2$, $c_{12} = c_2 - c_1$, $a_{13} = a_1 - a_3$, $b_{13} = b_1 - b_3$ and $c_{13} = c_3 - c_1$. The p_x and p_y can be obtained applying the elimination method to the equation 27 as

$$p_x = \frac{b_{12}c_{13} - b_{13}c_{12}}{(b_{12}a_{13} - b_{13}a_{12})} \quad \text{and} \quad p_y = \frac{a_{13}c_{12} - a_{12}c_{13}}{(a_{13}b_{12} - a_{12}b_{13})} \quad (28)$$

The orientation angle (ϕ) can be found using i=3 in equation 25.

$$p_x^2 + p_y^2 + a_3 p_x + b_3 p_y + c_3 = 0 \tag{29}$$

If p_x and p_y in equation 28 are substituted in equation 29, one can obtain after simplification

$$\chi^2 + \xi^2 + a_3 \chi \vartheta + b_3 \xi \vartheta + c_3 \vartheta^2 = 0 \tag{30}$$

where $\vartheta = (b_{12}a_{13} - b_{13}a_{12})$, $\chi = b_{12}c_{13} - b_{13}c_{12}$ and $\xi = a_{13}c_{12} - a_{12}c_{13}$. The equation 30 can be converted into the eight-degree polynomial using $\sin\phi = \frac{2t}{1+t^2}$ and $\cos\phi = \frac{1-t^2}{1+t^2}$.
The roots of this polynomial are the answer of the forward kinematics problem. Once ϕ is determined, p_x and p_y can be found easily substituting the angle ϕ in equation 28. After having position (p_x, p_y) and orientation (ϕ), the passive joint angles can be found using the equation 21 by back substitution

$$\theta_i = A\tan 2\left(-\frac{p_x + \upsilon_{x_i}\cos\phi - \upsilon_{y_i}\sin\phi - OB_{x_i}}{(d_i + l_i)}, \frac{p_y + \upsilon_{x_i}\sin\phi + \upsilon_{y_i}\cos\phi - OB_{y_i}}{(d_i + l_i)} \right) \tag{31}$$

The Jacobian matrix of the 3-dof PPM shown in Figure 3 is computed as follows.

$$\begin{bmatrix} \frac{\partial Q_1}{\partial p_x} & \frac{\partial Q_1}{\partial p_y} & \frac{\partial Q_1}{\partial \phi} \\ \frac{\partial Q_2}{\partial p_x} & \frac{\partial Q_2}{\partial p_y} & \frac{\partial Q_2}{\partial p\phi} \\ \frac{\partial Q_3}{\partial p_x} & \frac{\partial Q_3}{\partial p_y} & \frac{\partial Q_3}{\partial p\phi} \end{bmatrix} \begin{bmatrix} \dot{p}_x \\ \dot{p}_y \\ \dot{\phi} \end{bmatrix} + \begin{bmatrix} \frac{\partial Q_1}{\partial d_1} & \frac{\partial Q_1}{\partial d_2} & \frac{\partial Q_1}{\partial d_3} \\ \frac{\partial Q_2}{\partial d_1} & \frac{\partial Q_2}{\partial d_2} & \frac{\partial Q_2}{\partial d_3} \\ \frac{\partial Q_3}{\partial d_1} & \frac{\partial Q_3}{\partial d_2} & \frac{\partial Q_3}{\partial d_3} \end{bmatrix} \begin{bmatrix} \dot{d}_1 \\ \dot{d}_2 \\ \dot{d}_3 \end{bmatrix} = 0$$
$$\begin{bmatrix} \delta_1 & \gamma_1 & \rho_1 \\ \delta_2 & \gamma_2 & \rho_2 \\ \delta_3 & \gamma_3 & \rho_3 \end{bmatrix} \begin{bmatrix} \dot{p}_x \\ \dot{p}_y \\ \dot{\phi} \end{bmatrix} + \begin{bmatrix} -d_1 - l_1 & 0 & 0 \\ 0 & -d_2 - l_2 & 0 \\ 0 & 0 & -d_3 - l_3 \end{bmatrix} \begin{bmatrix} \dot{d}_1 \\ \dot{d}_2 \\ \dot{d}_3 \end{bmatrix} = 0 \tag{32}$$

where

$$\delta_i = p_x - OB_{x_i} + \cos\phi\upsilon_{x_i} - \sin\phi\upsilon_{y_i} \tag{33}$$

$$\gamma_i = p_y - OB_{y_i} + \cos\phi\upsilon_{y_i} + \sin\phi\upsilon_{x_i} \tag{34}$$

$$\begin{aligned} \rho_i = &-\sin\phi(p_x\upsilon_{x_i} + p_y\upsilon_{y_i} - OB_{x_i}\upsilon_{x_i} - OB_{y_i}\upsilon_{y_i}) \\ &+ \cos\phi(p_y\upsilon_{x_i} - p_x\upsilon_{y_i} - OB_{y_i}\upsilon_{x_i} + OB_{x_i}\upsilon_{y_i}) \end{aligned} \tag{35}$$

The overall Jacobian matrix is obtained as

$$J=\begin{bmatrix} \dfrac{\delta_1}{d_1+l_1} & \dfrac{\gamma_1}{d_1+l_1} & \dfrac{\rho_1}{d_1+l_1} \\ \dfrac{\delta_2}{d_2+l_2} & \dfrac{\gamma_2}{d_2+l_2} & \dfrac{\rho_2}{d_2+l_2} \\ \dfrac{\delta_3}{d_3+l_3} & \dfrac{\gamma_3}{d_3+l_3} & \dfrac{\rho_3}{d_3+l_3} \end{bmatrix} \tag{36}$$

Numerical example for the inverse kinematics: The dimensions of the planar 3-dof RPR parallel manipulator are given as l_1=10, l_2=10, l_3=10, OB_{x1}=0, OB_{y1}=0, OB_{x2}=20, OB_{y2}=0, OB_{x3}=10, OB_{y3}=45, M_1M_2=15. The position of the point P in terms of the coordinate frame {xyz} attached to the M_1 corner of the moving platform is chosen as (4, 5). The angles λ_1, λ_2 and λ_3 are computed as 231.3402, -24.4440 and 66.3453, where $\sigma_1 = 51.3402$ and $\sigma_2 = 24.4440$, respectively. Moreover, the lengths PM_1, PM_2 and PM3 are determined as 6.4031, 12.0830 and 8.7233, respectively. If the position and orientation in terms of the base frame {XYZ} are given as p_x=12, p_y=18 and ϕ=30°, the active joint variables d_1 d_2 and d_3 can be found as 6.0617, 9.5881 and 8.3594, respectively. The passive joint variables θ_1, θ_2 and θ_3 are found as -43.4006, -11.8615 and -176.7655, respectively by back substitution.

Numerical example for the forward kinematics: The dimensions of manipulator are given as l_1=11, l_2=11, l_3=11, OB_{x1}=0, OB_{y1}=0, OB_{x2}=10, OB_{y2}=0, OB_{x3}=5, OB_{y3}=32, M_1M_2=10. The position of the point P in terms of the coordinate frame {xyz} is chosen as (5, 2.8868). The angles λ_1, λ_2 and λ_3 are computed as 210.0004, -30.0004 and 90, where $\sigma_1 = 30$ and $\sigma_2 = 30.0004$, respectively. Moreover, the lengths PM_1, PM_2 and PM_3 are determined as 8.4853, 7.8102 and 3.5616, respectively. If the active and passive joint variables are given as d_1=8, d_2=7.9999, d_3=7.9999 and θ_1=-52.1046, θ_2=-52.105, θ_3=-127.8937, the position and orientation of end-effector in terms of base coordinate frame {XYZ} are found as p_x=19,9936 p_y=14.5569 and ϕ=0.000932°, respectively.

Numerical example for the Jacobian matrix and condition number: Using the dimensions of numerical example for the inverse kinematics above, the Jacobian matrix and condition number are found as

$$J=\begin{bmatrix} 0.6871 & 0.7266 & 3.6489 \\ 0.2055 & 0.9786 & 11.5290 \\ 0.0564 & -0.9984 & 0.4734 \end{bmatrix} \text{ and } \kappa = 8.1351 \tag{37}$$

Numerical example for the workspace determination: The dimensions of the manipulator are given as l_1=l_2=l_3=8.5, OB_{x1}=0, OB_{y1}=0, OB_{x2}=17, OB_{y2}=0, OB_{x3}=9, OB_{y3}=40, M_1M_2=16. The lengths of the actuated prismatic joints vary between d_i(min)=0 and d_i(max)=8.5, where i=1,2,3. If the position of the point P in terms of the coordinate frame {xyz} is chosen as (7, 7), the angles λ_1, λ_2 and λ_3 are computed as 225, -37.8750 and 81.7020, where $\sigma_1 = 45$ and

$\sigma_2 = 37.8750$, respectively. Moreover, the lengths PM_1, PM_2 and PM_3 are determined as 9.8995, 11.4018 and 6.9289, respectively. The workspace is obtained as 110.6180 for $\phi=0^o$ orientation. The red rectangle given by Figure 4 illustrates the limits of the discretized Cartesian area. Additionally, the black region shows the reachable workspaces of the manipulator.

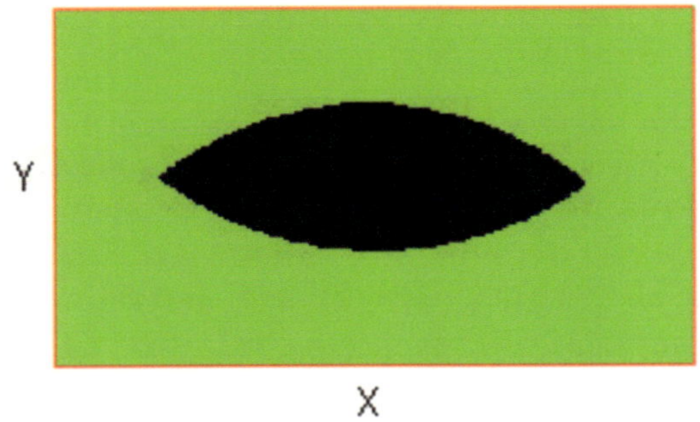

Fig. 4. The workspace of the planar 3-dof RPR parallel manipulator.

Numerical example for the dexterity and singularity analysis: The Figure 5 shows the inverse dexterity of the manipulator using the same dimensions given by the example for workspace determination. The GDI of the manipulator is found as 0.00018179. As can be illustrated in Figure 5, as the inverse of the condition numbers increase the manipulator accomplish better gross motion capability like the regions represented with red color. At the same time, the diagonal line divides the workspaces of the manipulator into two regions, illustrates the singular points. The dark blue areas illustrate the non-reachable workspaces of the manipulator.

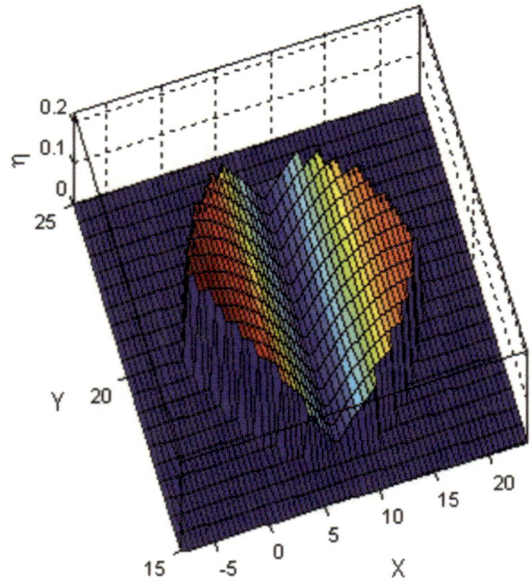

Fig. 5. The inverse dexterity graph for planar 3-dof RPR parallel manipulator.

4. Conclusion

In this book chapter, the forward and inverse kinematics problems of planar parallel manipulators are obtained using Denavit & Hartenberg (1955) kinematic modelling convention. Afterwards some fundamental definitions about Jacobian matrix, condition number, global dexterity index, singularity and workspace determination are provided for performing the analyses of a two-degree-of-freedom (2-dof) PPM and a 3-dof Fully Planar Parallel Manipulator.

5. References

Agrawal, S. K. (1990). Workspace boundaries in-parallel manipulator systems. *Int. J. Robotics Automat*, Vol., 6(3), pp. 281-290

Bonev, I. A. & Ryu, J. (2001). A geometrical method for computing the constant-orientation workspace of 6-PRRS parallel manipulators. *Mechanism and Machine Theory*, Vol., 36 (1), pp. 1–13

Denavit, J. & Hartenberg, R. S., (1955). A kinematic notation for lower-pair mechanisms based on matrices. *Journal of Applied Mechanics*, Vol., 22, 1955, pp. 215–221

Fichter, E. F. (1986). A Stewart platform-based manipulator: General theory and practical consideration. *Journal of Robotics Research*, pp. 157-182

Gosselin C., (1990). Determination of the workspace of 6-d.o.f parallel manipulators. *ASME Journal of Mechanical Design*, Vol., 112, 1990, pp. 331-336

Gosselin, C. & Angeles, J. (1991). A global performance index for the kinematic optimization of robotic manipulators. *Journal of Mech Design*, Vol., 113, 1991, pp. 220-226

Gosselin, C. M. & Guillot, M. (1991). The synthesis of manipulators with prescribed workspace. *Journal of Mechanical Design*, Vol., 113, pp. 451–455

Heerah, I.; Benhabib, B.; Kang, B. & Mills, K. J. (2003). Architecture selection and singularity analysis of a three-degree-of freedom planar parallel manipulator. *Journal of Intelligent and Robotic Systems*, Vol., 37, 2003, pp. 355-374

Hubbard, T.; Kujath, M. R. & Fetting, H. (2001). MicroJoints, Actuators, Grippers, and Mechanisms, *CCToMM Symposium on Mechanisms, Machines and Mechatronics*, Montreal, Canada

Kang, B.; Chu, J. & Mills, J. K. (2001). Design of high speed planar parallel manipulator and multiple simultaneous specification control, *Proceedings of IEEE International Conference on Robotics and Automation*, pp. 2723-2728, South Korea

Kang, B. & Mills, J. K. (2001). Dynamic modeling and vibration control of high speed planar parallel manipulator, *In Proceedings of IEEE/RJS International Conference on Intelligent Robots and Systems*, pp. 1287-1292, Hawaii

Kim, D. I.; Chung, W. K. & Youm, Y. (1997). Geometrical approach for the workspace of 6-DOF parallel manipulators, *Proceedings of IEEE International Conference on Robotics and Automation*, pp. 2986-2991, Albuquerque, New Mexico

Merlet, J. P. (1995). Determination of the orientation workspace of parallel manipulators. *Journal of Intelligent and Robotic Systems*, Vol., 13, pp. 143-160

Merlet, J. P. (1996). Direct kinematics of planar parallel manipulators, *Proceedings of IEEE International Conference on Robotics and Automation*, Vol., 4, pp. 3744-3749

Merlet, J. P.; Gosselin, C. M. & Mouly, N. (1998). Workspaces of planar parallel manipulators. *Mechanism and Machine Theory*, Vol., 33, pp. 7–20

Merlet, J. P. (2000). *Parallel robots*, Kluwer Academic Publishers

Tsai, L. W. (1999). *Robot analysis: The mechanics of serial and parallel manipulators*, A Wiley-Interscience Publication

22

Robot Manipulator Probabilistic Workspace Applied to Robotic Assistance

Fernando A. Auat Cheein, Fernando di Sciascio,
Juan Marcos Toibero and Ricardo Carelli
Instituto de Automatica, National University of San Juan
Argentina

1. Introduction

The probabilistic modelling of a robot manipulator workspace when combined with a Human-Machine Interface (HMI) allows the extraction and learning of the spatial preferences of the user. Furthermore, the knowledge of the most accessed zones of the robot's workspace permits the bounding of the time needed for the robot to reach a given position at its workspace.

From its early beginning, the use of robot manipulators within the robotic assistance field was concerned to emulate an orthopaedic arm (Fukuda et al, 2003; Zecca et al, 2002; Lopez et al, 2009). Therefore, the robot manipulator was considered as the final actuator of the assistive system where its main goal was to imitate the movements of an arm. Depending on the user/patient capabilities, the robot manipulator was commanded by either electro-myographic or electro-encephalic signals (Ferreira et al, 2006a; 2006b; 2008). A robotic device controlled by a Muscle-Computer Interface (MCI) can be found in (Artemiadis & Kyriakopoulos, 2006; Lopez et al, 2007; Millan et al, 2004; Ferreira et al, 2006b; Lopez et al, 2009; Ferreira et al, 2008). In these works, the electro-miographic signal is acquired, processed, classified and converted to motion commands. The system is closed by a bio-feedback loop. When used with a robot manipulator, a MCI is usually connected to a set of muscles that the patient is able to move at its own will. The number of channels used by the interface increases as increases the number of the degrees of freedom (DOF) of the robot (Yatsenko et al, 2007; Lopez et al, 2009; Ferreira et al, 2008). Thus, for a single 2DOF robot manipulator are necessary three different muscles: two to govern each DOF and a third to set a sign (if the manipulator is moving to the left or to the right), for a direct control of the robot manipulator.

For robotic devices controlled by Brain-Computer Interfaces (BCI's), the situation is analogous to the MCI; the number of signals or patterns to be extracted from the EEG (electro-encephalogram) increases as increases with the number of DOF's -or actions- to be performed by the robot. Although most applications of BCI's are oriented to govern a mobile robot -because of its direct analogy with a motorized wheelchair (Bastos Filho et al, 2007a; Bastos Filho, 2007; Ferreira et al, 2008; Bastos Filho et al, 2007b)- some works have been published showing the implementation of a BCI to control the movements of a robot

manipulator (Oskoei & Hu, 2007; Farry et al, 1996; Auat Cheein et al, 2007a; Auat Cheein et al, 2007b; Auat Cheein et al, 2008).
The way that the patient interacts with the interface is determined by the objective of the HMI. Thus, for a HMI connected to a robot, the proper motion of the automata serves as a bio-feedback channel to the user (Wolpaw et al, 2002 ; Ferreira et al, 2008). A more suitable way to interact with the user is by means of scan modes. The scan modes have been widely used in the recent years for assistance technology. It is not restricted only to BCI's or MCI's but to most of HMI.
The scan mode applied to a HMI's in robotic assistance is mainly composed by a series of icons; each icon represents an action or a place attached to the robotic task; the patient, using its biological signals, displace the scan over the several icons; once an icon is selected -also using the biological signals- the robot performs the task associated with the selected icon. This kind of interface decreases the effort of the user when generating the biological signals, relegating the control of the interface to the control of the scan. Moreover, some HMI include an automatic scan; thus the user only has to accept -or cancel if needed- an icon within the scan. Figure 1 shows a general architecture of a HMI with a scan mode incorporated.

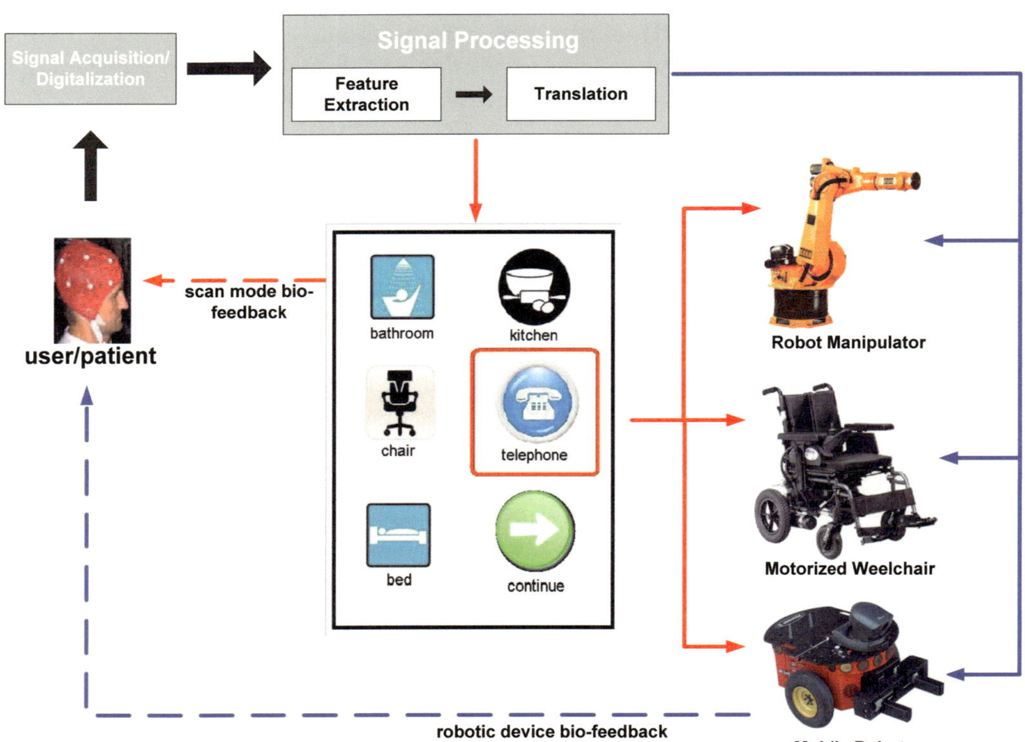

Fig. 1. General architecture of a HMI. The dashed and solid blue-lines correspond to a HMI that control the robotic device in a direct way; the solid and dashed-red lines refer to a HMI with a scan mode incorporated on it.

Figure 1 shows the general architecture of a HMI. When the HMI has a scan mode, the bio-feedback loop is closed by both the robot's movements and the scan mode itself. As was stated before, the scan mode reduces the work of the HMI's user, although it restricts the interface to a number of actions associated with the icons within the scan.
In this work we show the implementation of a robot manipulator workspace scan mode applied to a HMI. Although we have used a BCI, any other HMI can be adapted to the system presented in this work. The robotic device used is a SCARA robot manipulator (Sciavicco & Siciliano, 2000). The robot's workspace is divided into cells. Each cell has a probability value associated with it. The scan mode is run according the probability information of the cell, decreasing in that way the time needed for the robot to reach the desire cell. Once a cell is chosen by the user, the robot manipulator reaches the center of mass of the cell using a kinematic controller.
Three types of scan modes are shown in this work: an uniform scan mode and two based on the probabilistic information of the workspace. The three scan modes are tested by Monte Carlo experiments. The probability-based scan modes learn the preferences of the patient by the measure of the probability of a cell in being selected by the user. The probability value of each cell at the robot's workspace is dynamically maintained according to the patient cell's selection. Considering that the robot's workspace behaves as a probability density function, the Kullback-Liebler (relative entropy) measure is shown in order to establish a measure of divergence between the different scan modes.

2. Overview of the System

The main objective of the HMI is that the user/patient be capable of generating a set of commands by means of his/her biological signals in order to control the robotic device used. In this work, the HMI used is a BCI but is not restricted to it. The BCI is controlled by event related potentials (ERP) acquired from the occipital lobe of the patient. Such potentials are the event related synchronization/desynchronization (ERS/ERD). The ERS/ERD potentials are measured in the alpha band (8–13 Hz) of the EEG from the occipital lobe. The BCI conform an alphabet of commands from the ERS and ERD which are then sent to a Finite State Machine (FSM). The FSM contains actions or desired behaviors of the robot to be performed if the user generates the appropriate command. The BCI was tested in a population of 25 cognitive normal volunteers and the results can be seen in (Ferreira et al, 2008).
The robot manipulator is the actuator device that completes the HMI system. The robot used in this work is a SCARA type, Bosch SR-800 (see Fig. 2). In the absence of perturbations and without considering gravity effects, the dynamic of a rigid robot manipulator with n joints can be expressed as (1).

$$M(q)\ddot{q} + C(q,\dot{q})\dot{q} + f(\dot{q}) = \tau \tag{1}$$

In (1), q is an $n{\times}1$-array of joint angular positions; τ is an $n{\times}1$-array of applied torques; $M(q)$ is the $n{\times}n$ positive definite inertia matrix; $C(q,\dot{q})$ is an $n{\times}n$ matrix of centripetal and *Coriolis* effect and $f(\dot{q})$ is an $n{\times}1$-array of friction moments. More information about the robot's parameters can be found in (Bastos Filho et al, 2006 ; Ferreira et al, 2006c).

Considering that the objective of the HMI is to position the robot manipulator at a specific cell of its workspace, a kinematic controller was used (Kelly et al, 2004).

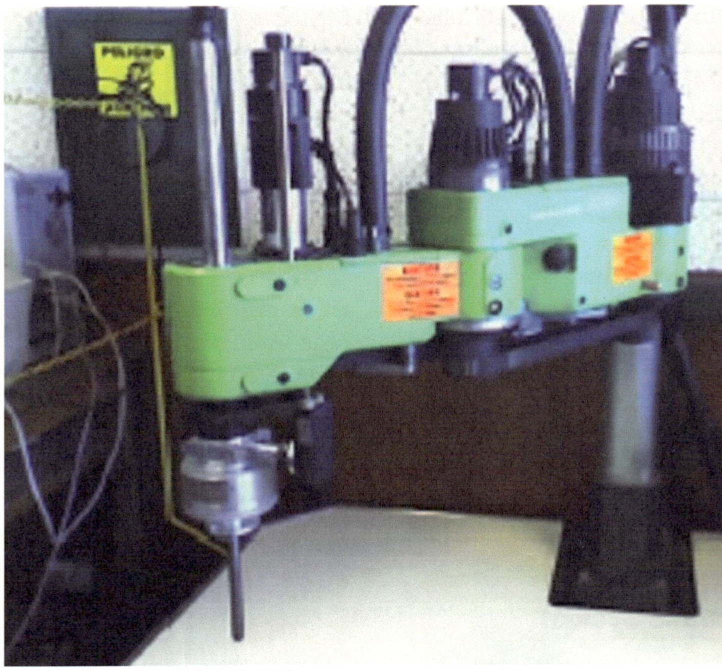

Fig. 2. Robot manipulator Bosch SR-800.

3. Sequential Scan Mode

The scan mode developed in this work is an automatic one. The workspace of the robot manipulator is divided into cells. As a first approach lets considerer that the cells do not have a probability value associated with them. Thus, the probability of a cell to be chosen by the patient is the same for all the cells at the robot's workspace. Moreover, the probability of being scanned is also the same. Figure 3.a shows the cells' distribution of the robot's workspace. The number of cells can be increased if the application requires it. Instead of scanning cell by cell, the robot's workspace is divided into three main zones. Then, the scan mode is first made zone by zone until one zone is selected by the user (see Fig. 3.b). Once a zone is selected, the scan is made row by row of the selected zone. If after three row-by-row scans, no row is selected, the scan returns to be executed zone-by-zone. On the other hand, if a row is selected, then the scan passes to be cell-by-cell until a cell is selected (see Fig. 3.c). The scan returns to be row-by-row if no cell is selected after three runs. Once a cell is selected, the robot is positioned at the center of mass of the cell and the scan mode begins again zone-by-zone.

The three zones division of the robot manipulator's workspace is based on the fact that a normal limbed patient grasps the elements of the environment from the right, center or left to his/her body.

The final interface is presented in Fig. 4.

a)

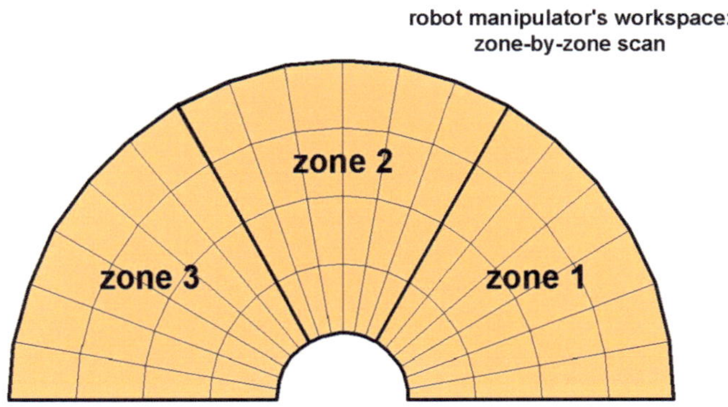

b)

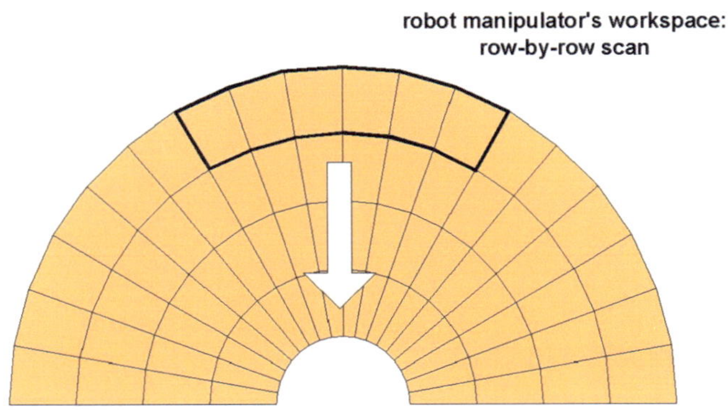

c)

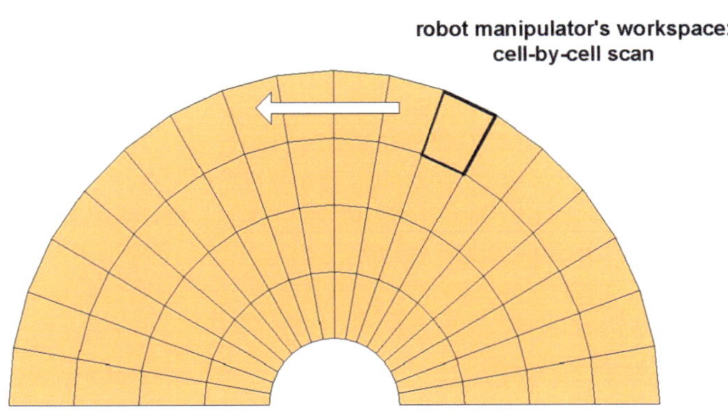

Fig. 3. Sequential scan mode. a) The scan is first executed zone-by-zone until one zone is selected by the user; b) then the scan is executed row-by-row of the selected zone; c) the scan is run cell-by-cell of the selected row.

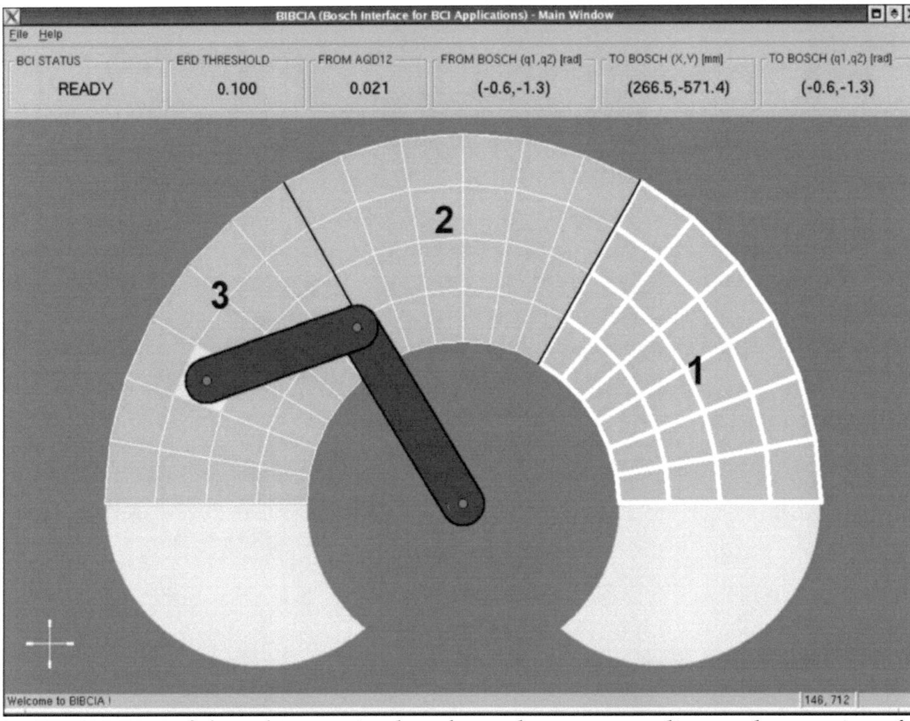

Fig. 4. Visualization of the robot manipulator's workspace according to the user interface.

4. Probability-based Scan Mode

Two probability-based scan modes have been implemented in this work. A first approach is based on maintaining the workspace zones with a fixed number of cells. On the other hand, the second approach has a variable number of cells assigned to each zone depending on their probability values. The probability value in both approaches represents the probability of accessing a cell.

4.1 First Approach: Fixed zones division

The first approach of the probability-based scan mode works as follows:

- The workspace resolution is set to 72 cells (4 rows and 18 columns) and can be easely changed depending on the application. The worspace behaves as a *pmd* (probability mass distribution) and can be interpreted as a *4×18*-matrix (where $c_{i,j}$ is the cell at the *i*-row and *j*-column).
- When the HMI is loaded by the first time, the robot's workspace has a uniform distribution over their cells ($p(c_{i,j}) = \frac{1}{72}$, con $i = 1..4$ and $j = 1..18$).
- Each zone (see Fig. 4) has a probability value associated with it, which is the sum of the probability values contained in each zone. Thus, for example:

$p(zone_1) = \sum_{i=1,j=1}^{i=4,j=6} p(c_{i,j})$. Also, each row's probability equals the sum of the probability value of the cells contained in that row. This is so because the robot's workspace is considered as a *pmd*, then the sum of all probability values of the cells sum the unity.

- The division of the zones remain unchanged with the evolution of the probability values of the cells.
- The scan starts at the zone with the highest probability value. If such a zone does not exists, the scan starts at the rightest zone (i.e., zone 1 in Fig. 4). The scan highlights one zone at a time and waits 2 seconds (Ferreira et al, 2008) until passing to the next zone. The scan is cyclic, which means that if the scan has passed from zone 2 to zone 3, it then goes to zone 1 instead of returning to zone 2.
- Once the patient selects a zone -using his/her ERS/ERD potentials- the scan is made row-by-row in the selected zone. As before, the scan starts at the row with the highest probability value and runs decreasing to the row with the lowest probability value of the same zone. The scan is also cyclic. Once a row is selected, the scan is run cell-by-cell of the selected row only, starting from the cell with the highest probability value.
- Once a cell is selected, the robot manipulator is driven to the center of mass of such a cell and the probability value of it is updated. The update procedure is performed by the recursive *Bayes* rule. Once a probablity value of a cell is updated, the probability values of all cells at the robot's workspace are also updated. The update procedure is shown in the following sections.
- When the scan is on the row-by-row mode or cell-by-cell mode, after three runs with no selection, the scan returns to its previous mode.
- Finally, in every mode, the scan waits 2 seconds before changing the highlighting.

4.2 Second Approach: Dynamic zone division

The second approach works under the same criteria of the first approach: the workspace is divided into three zones, although in this case, the number of cells that belong to each zone is determined by their probability values. The scan mode can be summarized as follows.

- When the scan starts the first time, the robot's workspace is considered as a uniform probability mass distribution; each cell has the same probability value. The scan mode up to this point has the same execution than the first approach. Experimentally was determined that after 20 runs (i.e., twenty cells were reached by the robot manipulator) the dynamic zone's division is able to be performed.
- Let a and b be the higher and lower probability cells' values respectively after the first 20 runs. Then, the workspace is divided into three zones according to these values. Table 1 shows how the division is made. The number of cells belonging to each zone is now governed by their probability values and is not a fixed number - of 24 cells- as it is in the first approach.
- Then, each zone is divided into three sub-zones under the same criteria of the last paragraph: each one of these sub-zones contains a set of probability weighted cells according to Table 1 but divided with respect to the zone instead the workspace.

- The scan mode proceeds as follows:
    - First, the zone with the highest probability value at the workspace is highlighted. If after two seconds this zone is not selected by the user of the HMI, the second probability zone is highlighted. If it is not selected, the highlight passes to the third and last zone -which has the lowest probability value associated with it-. The scan is cyclic -as it is the first approach- and the probability of each zone equals the sum of the probability values of the cells contained within them.
    - When a zone is selected, the scan highlights the sub-zone with the highest probability value inside the previous zone. Thus, all cells within this sub-zone are highlighted. Also, the cells may not be consecutive. If the user does not select this sub-zone, the scan passes to the next one. The scan mode is also cyclic.
    - Once a sub-zone is selected, the scan highlights each cell of the current sub-zone. In this case, the scan starts at the cell with the highest probability value associated with it and finishes at the cell with the lowest probability value. If all cells of the sub-zone have the same probability value, the scan starts randomly.
    - At every stage, the scan is cyclic and remains highlighting during two seconds before changing. If no selection is performed by the user after three runs, the system returns to the previous scan mode.
    - If a cell is selected, the probability value of the cell and of the rest of the cells of the workspace is updated according to the recursive *Bayes* rule.
    - The dimensions of the zones and sub-zones are re-calculated after a cell is selected.

a_t	Highest probability cell value at time *t*
b_t	Lowest probability cell value at time *t*
$zone1 = \left\{ \forall c_{i,j}, i = 1..4, j = 1..18 : b_t + \frac{2}{3}(a_t - b_t) < p(c_{i,j} \mid W) \le a_t \right\}$	Zone 1: is the set of all cells whose probabilities are the highest of the workspace (W)
$zone2 = \left\{ \forall c_{i,j}, i = 1..4, j = 1..18 : b_t + \frac{(a_t - b_t)}{3} < p(c_{i,j} \mid W) \le b_t + \frac{2}{3}(a_t - b_t) \right\}$	Zone 2: is the set of all cells whose probabilities are of middle range
$zone3 = \left\{ \forall c_{i,j}, i = 1..4, j = 1..18 : b_t \le p(c_{i,j} \mid W) \le b_t + \frac{(a_t - b_t)}{3} \right\}$	Zone 3: is the set of all cells with the lowest probability values.

Table 1. Workspace zones definitions.

As it can be seen, the number of cells that belong to a sub-zone or zone is variable. Then, the organization of the zones at the robot's workspace is dynamic. This allows the improving of the scan mode in order to access in a priority way to the most probably cells.

4.3 Recursive *Bayes* Rule

The probability update of each cell at the robot's workspace is based on the recursive *Bayes* rule (Fishman, 1996 ; Papoulis, 1980). Once the patient directs the HMI to access a specific cell, the probability value of that cell changes according to (2).
Let $c_{i,j}$ be a cell at the robot's workspace and G a zone to which that cell belongs (e.g., *zone 1*). Thus, the updating algorithm is given by,

$$p_t(c_{i,j}\,|\,G)=\frac{p_t(G\,|\,c_{i,j})p_{t-1}(c_{i,j}\,|\,G)}{p_t(G\,|\,c_{i,j})p_{t-1}(c_{i,j}\,|\,G)+p_t(G\,|\,c_{i,j})p_{t-1}(c_{i,j}\,|\,G)} \tag{2}$$

In (2), $p_t(c_{i,j}\,|\,G)$ is the current probability of the cell $c_{i,j}$ given the zone G; $p_{t-1}(c_{i,j}\,|\,G)$ is the previous probability value of $c_{i,j}$. Though (2) is mainly used in very simple applications (Auat Cheein et al, 2008; Ferreira et al, 2006c) it fits as an updating rule for the purpose of this work. The rest of the cells at the robot's workspace are also updated using (2), but with $p_t(c_{i,j}\,|\,G)$ -the probability of $c_{i,j}$ not been selected with respect to the zone- instead of $p_t(c_{i,j}\,|\,G)$ in the appropriate places of (2).

In order to make sense to the use of the recursive *Bayes* rule, an initial probability value must be given to all cells at the workspace. Thus, it is initialized as a uniform *pmd*. Figure 4 shows the evolution of a cell's probability when it is successively accessed by the patient. As the probability of this cell tends to 1, the probability value of the non-accessed cells tends to zero according to (2).

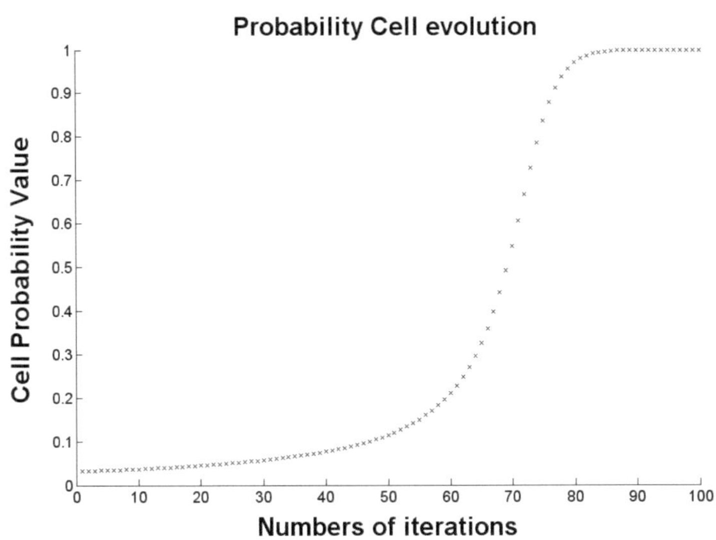

Fig. 4. Probability cell evolution when it is constantly accessed (Auat Cheein et al, 2008).

Considering that the scan mode of this approach is based on the probability weight of the cells of the robot's workspace, if a cell, e.g., has the highest probability value of the *zone 2* at time t, and it is successively accessed, then at the instant $t + k$ that cell might abandon the *zone 2* -because of it probability value increases- and enter to *zone 1*. Nevertheless, having into account that zone 1 is composed by those cells with the highest probability values of the entire robot's workspace, then the cell that has recently enter this zone, might not be one of the first to be scanned. This situation will be shown in the experimental results.

5. Experimental Results

This section shows the time consuming differences of the three scan modes shown in this work. In order to do this, a Monte Carlo experiment was designed (Auat Cheein et al, 2008; Fishman, 1996). The time analysis establishes a performance comparison between the three scan modes. In addition, the final distribution of the robot's workspace for a right-handed user after 100 trials is also shown. This distribution is shown for both probability scan modes -the accessed cells are the same-. The Kullback-Liebler relative entropy is calculated between both probability-based workspaces in order to establish a divergence value.

5.1 Monte Carlo Experiment

In this work, the robot's workspace is divided into 72 cells (see Fig. 3). It also can be considered as a *4×18-matrix*. Then, a cell's position can be defined by the number of row and column of the *4×18-matrix*. Considering the number of row and column as a random variable, the Monte Carlo experiment was designed as follows:

- A uniform random source generates two random variables: x and y.
- The random variable x is mapped into the rows of the robot's workspace (m).
- The random variable y is mapped into the columns of the robot's workspace (n).
- The pair $<m,n>$ determines the position of a cell at the robot's workspace. This cell is the one that will be accessed by the scan modes. The three scan modes start.
- The time needed by the patient to reach a cell is recorded for each scan mode.
- Once the robot reaches the proposed cell, a new cell is generated and the process starts all over.

5.2 Mapping Functions

Let f_x be a row mapping function of the form,

$$\begin{aligned} f_x &: A \to B \\ x &\to m \end{aligned} \tag{3}$$

where $A = \{x \sim U(0,1)\}$ and $B = \{m : m \in \{1,2,3,4\} \subset N\}$, with N the set of positive integers. Let also f_y be a column mapping function of the form,

$$\begin{aligned} f_y &: A \to C \\ y &\to n \end{aligned} \tag{4}$$

where A is again $A = \{x \sim U(0,1)\}$ and $C = \{n : n \in \{1,2,...,18\} \subset N\}$. Equations (3) and (4) show the domain and range of the mapping functions. The mapping is made according to the following statements (Auat Cheein et al, 2008).

- Let $p(c_{i,j})$ the probability value of a cell located at the *i-row* and *j-column* of the robot's workspace.
- Let x be an outcome of the uniform random source f_x.
  - If $0 \le x \le \sum_{i=1, j \in C} p(c_{i,j})$, then $f_x(x) = i = 1$. This means that the value of $x \in A$ should be lower than the sum of all probability cell's values over row 1 for $f_x(x)$ be equal to 1.
  - If $\sum_{i=1, j \in C} p(c_{i,j}) \le x < \sum_{i=2, j \in C} p(c_{i,j})$, then $f_x(x) = i = 2$. This means that $x \in A$ should be greater or equal to the sum of all probability cell's values over row 1 and lower than the sum of all probability cell's values over row 2 for $f_x(x) = i = 2$.
  - The process continues up to the last row whose expression is: if $\sum_{i=3, j \in C} p(c_{i,j}) \le x \le \sum_{i=4, j \in C} p(c_{i,j})$, then $f_x(x) = i = 4$.
- For the mapping over rows, the procedure is the same. However, in this case, the sum of probability values is made over the columns.
- Each time a cell is reached by the scan, the mapping functions (f_x and f_y) vary. It is so because these functions depend on the probability values of the cells at the robot's workspace. Thus, the most accessed cells will have more probability in being mapped. This allows us to emulate, e.g., a right-handed user.

5.3 Monte Carlo Simulation Results

The objective of the Monte Carlo experiments was to test the performance of both scanning methods: probability-based and sequential ones. The performance is measured in function of the time needed to access a given position. This position is generated by the uniform random source shown in (3) and (4). After 500 trials the mean time needed to access a random position by the first approach of the probability-based scan was of 16.8 seconds. For the second approach of the probability-based scan the mean time needed was of 20.4 and for the sequential scan was of 19.8 seconds. The three results are in the same order but the probability based first approach of the scan mode requires less time.

Lets consider now only the right side of the workspace. This situation is suitable for a right-arm amputee patient. The mean time of access for all points belonging to the workspace right side is of 8.4 seconds under the second approach of the probability-based scan instead of 11.3 seconds corresponding to the first approach of the probability-based scan mode. Under sequential scan, the mean time is of 14.8 seconds. The probability-based scan mode second approach is 43% faster than the sequential scan for cells over the right side of the workspace whereas the first probability-based approach is 23.7% faster. Details of these results can be seen in (Auat Cheein et al, 2008). The accessing time values can be interpreted as the patient needs an overall time of 8.4 seconds to access any cell of the right side of the robot's workspace using the second scan mode approach instead of 11.3 seconds of the first probability-based approach. If the whole workspace is to be used, then the first approach is more convenient.

The Figures 5 and 6 show the mapping function for the probability-based scans at the end of the Monte Carlo simulation. The mapping functions are dependent on the probability values of the cells.

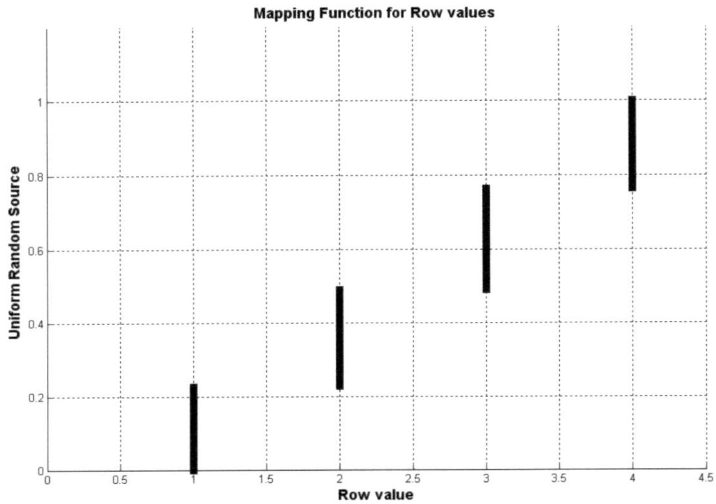

Fig. 5. Row mapping function of the Monte Carlo experiment -see Eq. (3)-.

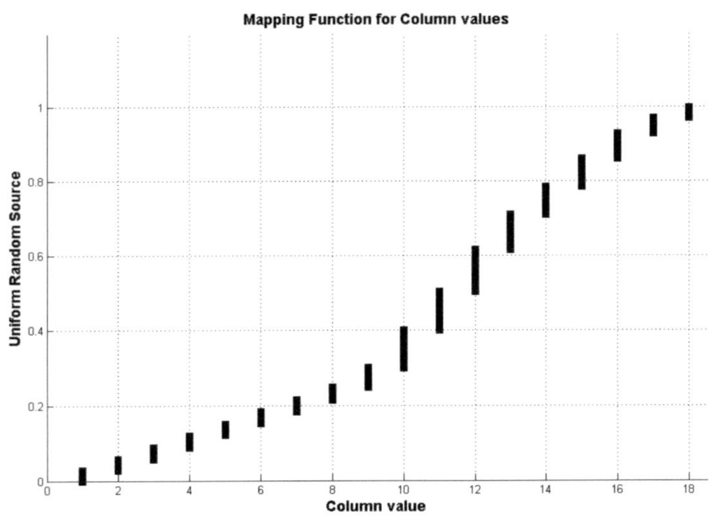

Fig. 6. Column mapping function of the Monte Carlo experiment -see Eq. (4)-.

In *section 4.3* was stated that a cell may change of zones as its probability grows or decrease during the scan mode of the second probability-based approach. Figure 7 shows this situation. A specific cell was successively accessed (see Fig. 4). The cell was initially in *zone 3* (the group with the lowest probability value). During the successive calls of the cell, its probability was growing and the cell passed from *zone 3* to *zone 2* and then to *zone 1*.

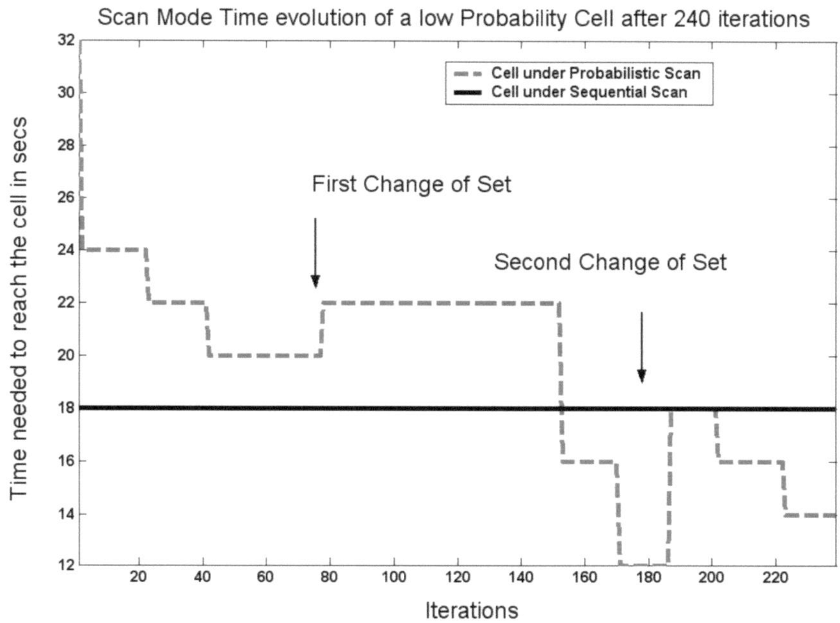

Fig. 7. Evolution of the accessing time of a cell during successive calls.

Figure 7 shows that the time of access of the cell by the sequential scan mode remains constant whereas varies for the second approach of the probability-based scan mode. When a cell changes to a greater probability zone, it abandons its scan priority. Thus, if it was previously in *zone 3* with the highest probability value in that zone (it is the first cell in being scanned inside that zone), when it passes to *zone 2* it will be one of the cells with the lowest probability value of that zone. This situation explains why the time of access of a cell grows every time a cell change its zone (as it is shown in Fig. 7). Nevertheless, the final time of access is lower than the initial value.

Figure 8 shows the final zone's division of the robot's workspace at the end of the Monte Carlo experiment for the two probability scan mode approaches. Figure 8.a shows the probability workspace distribution for the first probability based scan mode approach; the *zone 2* represents the most accessed zone. This situation means that, for a right-handed user, every location at the front of the hand will be the most reached. Figure 8.b shows the probability workspace distribution of the second probabilistic scan mode approach. The *zone 3* -which contains the cells with the lowest probability according to Table 1- is limited by a solid-black line; *zone 2* is formed by those cells with a solid-blue line at their limits. The *zone 1* -the set of cells with the highest probability value according to Table 1- remains unbounded in Fig. 8.b. The probability value of the cells and zones in Fig. 8 is represented by the gray-bar at the top of each figure.

a)

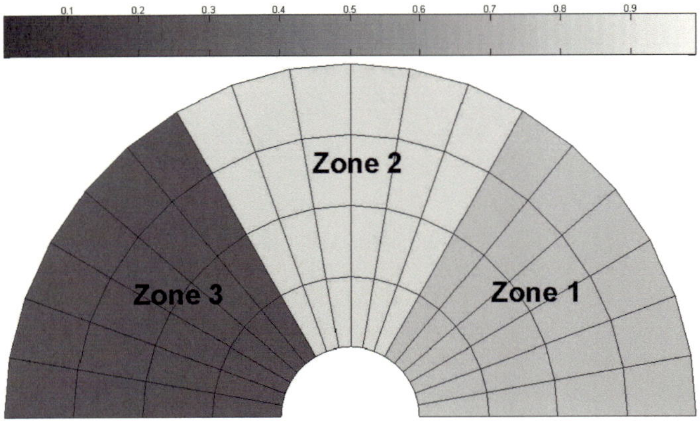

b)

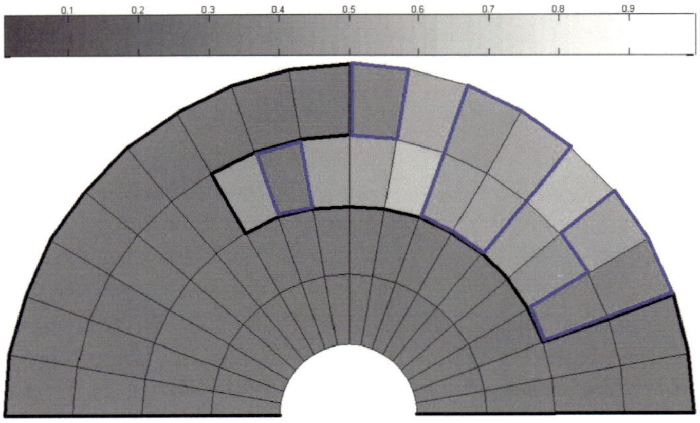

Fig. 8. Robot's workspace zones' division after the Monte Carlo experiment. a) Zones probability distribution according to the first approach of the probability scan mode. b) Cells probability distribution according to the second approach of the probability-based scan mode; cells with the lowest probability value -*zone 1*- are limited by the solid-black line; cells of the zone 2 are bounded by a solid-blue line; cells of the *zone 1* -with the highest probability value- remain unbounded.

5.4 Right-handed Workspace Distribution

Figure 8 has shown the zones divisions for the two probability-based scan modes for the Monte Carlo experiment. Figure 9, on the other hand, shows a real workspace distribution for a right-handed user. For the experiment, the BCI connected to the HMI was used. Figure 9 shows the cell's probabilities after 100 trials. As it can be seen, the right side of the workspace is the most accessed one.

By inspection, it is possible to see that both workspaces' *pmds* shown in Fig. 8 look very similar. Considering that the *pmd* represents the learning of the HMI with respect to the patient's needs, the Kullback-Liebler relative entropy is calculated between both *pmds* of Fig. 9 in order to have a divergence value of the workspaces probability mass distributions.

a)

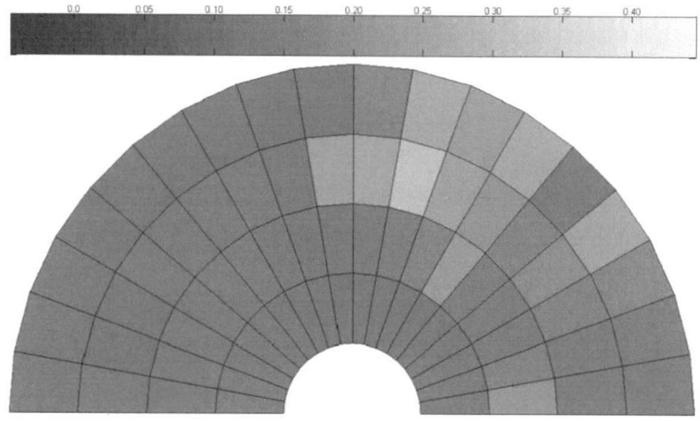

b)

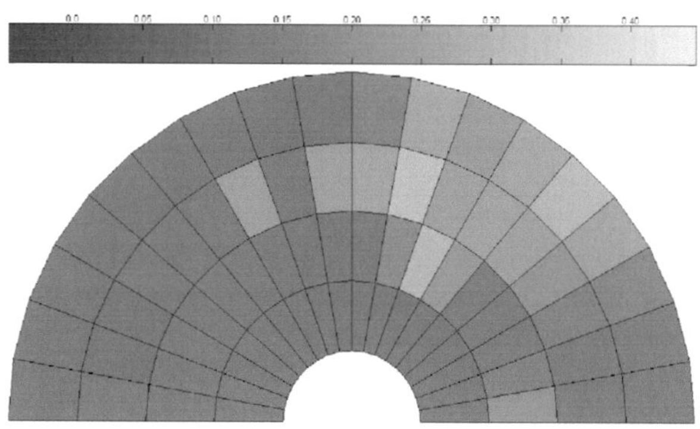

Fig. 9. Cells probability values for the two probability scan modes approaches after 100 trials. a) Final workspace's *pmd* for the first approach of the probability-based scan mode. b) Final workspace's *pmd* for the second approach of the probability-based scan mode. The difference between both *pmds* lies on the fact that the recursive *Bayes* rule is calculated based on the zone's probability information and not the workspace's probability (which is the unity).

5.5 Kullback-Liebler divergence

The Kullback-Liebler relative entropy is a pseudometric (Cover & Thomas, 2006) that gives a measure of the divergence between two probability density functions and it is presented in (5) -for the continuum case- and (6) -for the discrete case-. The reference functions in (5) and (6) are $g(.)$ -density function- and $q(.)$ -mass function- respectively.

$$D_{f\|g} = \int_{-\infty}^{\infty} f(x)\log(\frac{f(x)}{g(x)})dx \quad (5)$$

$$D_{p\|q} = \sum_{x=-\infty}^{\infty} p(x)\log(\frac{p(x)}{q(x)}) \quad (6)$$

Considering that the set of cells is the same for both probability based scan modes approaches, equation (6) is directly applied to the *pmds* of the robot manipulator. Thus, for the experiment shown in Fig. 9, the divergence between the first workspace probability based scan mode approach and the second workspace probability-based scan mode approach is of 0.0832. Thus, it is possible to say that both probability scan modes approaches lead to a very similar workspace's *pmds* -both learn the user's preferences in the same way-.

6. Conclusion

This work has shown the implementation of three scan modes of a robot manipulator workspace governed by a Human Machine Interface. The first scan approach was a sequential one with no consideration of preferences of the user/patient of the HMI. The other two scan approaches were based on the probability information attached to every cell at the robot's workspace. The probability value was updated by means of the recursive *Bayes* rule. The whole workspace was seen as a single probability mass distribution. Thus, each cell's probability value reflects the times that cell was accessed by the user and it governs the scan (the probability of been accessed). The cells with the highest probability value will be scanned first.

A time analysis between the three methods presented here showed that the probability-based scans improves the access time of the most accessed cells. Thus, if the HMI is going to be used for the first time (the robot manipulator workspace is an uniform *pmd*), the first approach of the probability-based scan mode has shown a better performance than the other approaches. On the other hand, once the preference of the patient (right/center/left side of the workspace) is well defined, the second approach of the probability based scan mode has shown lesser time requirements.

Although the system was primary designed to be implemented via a Brain Computer Interface, it could be used with any HMI with the same Finite State Machine implemented on it.

Experimental results showed that the time needed to access a specific position at the workspace is decreased each time that position is reached. This is due to the fact that the recursive *Bayes* algorithm implemented updates the probability value of the selected position after it is reached by the robot. A decrement of the time of access implies less effort by the user/patient of the HMI in reaching the objective.

In this work, a right-handed workspace distribution case was presented. This case showed that all cells to the right of the workspace have the higher probability value and the lower time needed to be accessed.

The robot workspace resolution was set to 72 cells in this work, although it could be adjusted according to the HMI destination. In addition, the interface can be extended to operate in 3D.

The probabilistic workspace configuration learns the user's preferences and uses this knowledge in the scan mode. It pays special attention to those cells with the highest probability value thus minimizing the time needed to access them.

7. References

Artemiadis, P.K. & Kyriakopoulos, K.J. (2006). EMG-based teleoperation of a robot arm in planar catching movements using ARMAX model and trajectory monitoring techniques. *IEEE Int. Conf. on Robotics and Automation*, pp. 3244-3249, ISBN: 0-7803-9505-0, Orlando, US, may 2006.

Auat Cheein, F.; di Sciasio, F.A.; Bastos Filho, T. & Carelli, R. (2007a). Probabilistic Scan Mode of a Robot Manipulator Workspace using EEG signals. *Journal of Physics: Conference Series*, Vol. 90, ISSN 1742-6588.

Auat Cheein, F.; di Sciasio, F.A.; Bastos Filho, T. & Carelli, R. (2008). Towards a Probabilistic Manipulator Robot´s Workspace Governed by a BCI. In: A. Fred, J. Filipe, and H. Gamboa (Eds.): *BIOSTEC 2008, CCIS 25. (Org.). Communications in Computer and Information Science.* Berlim: Springer-Verlag, 2008, 1, p. 73-84. ISBN: 1865-0929.

Auat Cheein, F.; Ferreira, A.; Bastos Filho, T.; di Sciasio, F.A. & Carelli, R. (2007b). Probabilistic Scan Mode of a Robot Manipulator Workspace using EEG signals. Part II. *Journal of Physics: Conference Series*, Vol. 90, ISSN 1742-6588.

Bastos Filho, T.F. (2007). Equipment Design and Development Directed for Human Rehabilitation. *Workshop on Robotics Applications and Automatic Control ROBCON'07*, 1, 1-78, ISSN:

Bastos Filho, T.F.; Ferreira, A.; Celeste, W.C.; Auat Cheeín, F. & Sarcinelli Filho, M. (2006). Human-Machine Interfaces Based on EMG and EEG Applied to Robotic Systems. *Proc. of the 2nd Int. Workshop on Biosignal Processing and Classification - Biosignals and Sensing for Human Computer Interface / ICINCO (Int. Conf. on Informatics in Control, Automation and Robotics), 2006*, Setúbal, 1, pp. 116-125.

Bastos Filho, T.F.; Ferreira, A.; Silva, R.L.; Celeste, W.C. & Sarcinelli Filho, M. (2007a). Human-Machine Interface Based on EMG and EEG Signals Applied to a Robotic Wheelchair. *Journal of Physics: Conference Series*, Vol. 90, ISSN 1742-6588.

Bastos Filho, T.F.; Ferreira, A.; Silva, R.L.; Celeste, W.C. & Sarcinelli Filho, M. (2007b). Human-Machine Interface Based on EMG and EEG Signals Applied to a Robotic Wheelchair. *Journal of Physics: Conference Series*, Vol. 90, ISSN 1742-6588.

Cover & Thomas. (2006). *Elements of Information Theory 2nd Ed.*, Wiley, 2006.

Farry, K. ; Walker, I. & Baraniuk, R.G. (1996). Myoelectric teleoperation of a complex robotic hand. *Proc. IEEE Int. Conf. Robotics and Automation*, pp. 775-788, ISBN :, conference location, month and year, publisher, city.

Ferreira, A. ; Auat Cheein F.A. ; Postigo J. ; Bastos Filho T.F. ; Carelli, R. & Sarcineli Filho M. (2006b). Teleoperação de um Manipulador para uma Cadeira de Rodas Usando Sinais de EEG IV Congresso Ibero-Americano Sobre Tecnologias de Apoio a Portadores de Deficiência. *Anais do IV Congresso Ibero-Americano Sobre Tecnologias de Apoio a Portadores de Deficiência*, 1, may 23-27, 2006.

Ferreira, A. ; Bastos Filho, T.F. ; Sarcinelli Filho, M. ; Auat Cheein, F.A. ; Postigo, J. & Carelli, R. (2006a). Teleoperation of an Industrial Manipulator Through a TCP/IP Channel Using EEG Signals. *Proc. of ISIE 2006, Int. Symp. on Industrial Electronics*, Montreal, vol. 1, pp. 3066–3071.

Ferreira, A.; Bastos Filho, T.F.; Sarcinelli Filho, M.; Auat Cheeín, F.; Postigo, J. & Carelli, R. (2006c). Teleoperation of an Industrial Manipulator Through a TCP/IP Channel Using EEG Signals. *Proc. Of the Int. Symposium on Industrial Electronics ISIE2006*, Montreal, 1, pp. 3066-3071.

Ferreira, A.; Celeste, W.C.; Auat Cheeín, F.; Bastos Filho, T.; Sarcinelli Filho, M. & Carelli, R. (2008). Human-machine interfaces based on EMG and EEG applied to robotic systems. Journal of NeuroEngineering and Rehabilitation, 5, 1-15.

Fishman, G.S. (1996). *Monte Carlo Concepts, Algorithms, and Applications*. New York, Springer-Verlag.

Fukuda, O. ; Tsuji, T. ; Kaneko, M. & Otsuka, A. (2003). A Human-Assisting Manipulator Teleoperated by EMG Signals and Arm Motions. *IEEE Trans. on Robotics and Automation*, 19, 210-222, ISSN: 1042-296X.

Kelly, R.; Carelli, R. & Soria, C. (2004). Sobre control cinemático de impedancia en robots industriales. *VI Congreso Mexicano de Robótica*, ISSN: 0035 0516, Torreón, México.

Lopez, N.M. ; di Sciascio, F.A. ; Soria, C.M. & Valentinuzzi M.E. (2009). Robust EMG sensing system based on data fusion for myoelectric control of a robotic arm. *BioMedical Engineering OnLine 2009*, pįgs. 8:5. ISSN 1475-925X, doi:10.1186/1475-925X-8-5.

Lopez, N.M. ; Soria, C.M. ; Orosco, E.C. ; di Sciascio, F.A. & Valentinuzzi, M.E. (2007). Two-Dimensional Myoelectric Control of a Robotic Arm for Upper Limb Amputees. *Journal of Physics: Conference Series*, Vol. 90, ISSN 1742-6588.

Millan, J. ; Renkens, F. ; Mouriño, J. & Gerstne, W. (2004). Non-invasive brain-actuated control of a mobile robot. *Proc. of the 18th Int. Joint Conf. on Artificial Intelligence*, ISBN: 0-127-05661-0, Acapulco, Mexico.

Oskoei, M. & Hu, H. (2007). Myoelectric control systems. A survey. *Biomedical Signal Processing and Control*, Vol.2, pp. 275–294, ISSN: 1746-8094.

Papoulis, A. (1980). *Probabilidad, variables alatorias y procesos estocásticos*. Eunibar, Barcelona.

Sciavicco, L. & Siciliano, B. (2000). *Modelling and Control of Robot Manipulators 2nd Edition*, Advanced Textbooks in Control and Signal Processing Series. ISBN: 1852332212, Edited by Springer-Verlag, London, UK.

Wolpaw, J.R. ; Birbaumer, N. ; McFarland, D.J. ; Pfurtscheller, G. & Vaughan, T.M. (2002). Braincomputer interfaces for communication and control. *Clin. Neurophysiol*, Vol.113, Nº6, 767–791, ISSN: 1537-1603.

Yatsenko, D. ; McDonnall, D. & Guillory, S. (2007). Simultaneous, Proportional, Multi-axis Prosthesis Control using Multichannel Surface EMG. *Proc. 29th Annual Int. Conf. of the IEEE EMBS*, ISSN: 1557-170X, pp.6134-6141, August 23-26, 2007.

Zecca, M.; Micera, S.; Carrozza, M.C. & Dario, P. (2002). Control of Multifunctional Prosthetic Hands by Processing the Electromyographic Signal. *Critical Reviews in Biomedical Engineering*, Vol.30, pp.459-485, ISSN: 0278-940X.

23

On the Design of Human-Safe Robot Manipulators

Vincent Duchaine, Nicolas Lauzier
and Clément Gosselin
Department of Mechanical Engineering
Université Laval
Canada

1. Introduction

Bringing robot manipulators in the same environment as humans seems a natural evolution in the path towards more advanced robotics. This upcoming co-existence will offer a tremendous potential to improve many industrial applications such as manufacturing and assembly. In this paradigm, an efficient synergy between human and robot can be obtained by combining the human's reasoning ability and adaptability in unstructured environments with the inexhaustible strength of robots.

The current generation of commercially available robot manipulators is not designed to fit the specific needs required by this novel collaboration. Indeed, control algorithms that enable an intuitive and efficient interaction between humans and robots are still missing to industrial robots. At an even more fundamental level, the way they are currently designed presents significant risks in the proximity of humans. Many studies have investigated this last aspect to demonstrate the potential danger of a robot Zinn, Khatib, Roth & Salisbury (2004a) and to understand and provide metrics to characterize to the level of this threat Haddadin, Albu-Schaffer, Frommberger & Hirzinger (2008); Haddadin, Albu-Schaffer & Hirzinger (2008); Yamada et al. (1997). The next step for robot designers should focus on increasing human safety to an acceptable level according to the conclusion of these studies.

The aim of this chapter is to present how, at the conceptual level, robot manipulators should be mechanically designed to be harmless for humans. Both established and novel concepts will be reviewed to provide actual guidelines to the robot designer. Serial elastic actuators (SAE), distributed macro-mini (DM2) and variable stiffness joints will be reviewed whereas more emphasis will be placed on force limiting devices (FLD), robot soft covering and a method for efficiently coupling robot joint actuators for reducing their potential of transferring energy to the surrounding environment.

2. Review

To build safe and dependable robots, engineers and researchers are using three different strategies:

1. to develop algorithms that use vision systems or proximity sensors to *anticipate* and *avoid* potentially harmful contacts Ebert et al. (2005); Lu et al. (2005);

2. to *detect* a collision by monitoring joint torques or a robot skin and quickly *react* to maintain the contact forces under a certain levelDe Luca et al. (2006); Duchaine, Lauzier, Baril, Lacasse & Gosselin (2009);
3. to *design* robots that are *intrinsically safe, i.e.,* that are physically unable to hurt a personChoi et al. (2008); Kim et al. (2007); Park et al. (2009; 2008); Sardellitti et al. (2007); Tonietti et al. (2005); Zinn, Khatib & Roth (2004).

It is clear that the *avoidance, reaction* and *design* strategies can be combined together to create safer and more dependable robots. However, the first two options alone cannot fully guarantee human safety. This can be explained by considering that a robot intended to interact physically with a person will require the ability to distinguish between desirable and undesirable contacts (or *good* and *bad* contacts). This can be done either by disabling safety sensors on the robot parts intended to interact or by running an algorithm that will decide if the upcoming contacts are desirable or not. In either case, safety is compromised either by unprotecting certain parts of the manipulator or by giving the robot some sort of 'judgment capability' which, even in the case of the human, is condemned to occasionally be wrong. Furthermore, the *avoidance* and *reaction* strategies rely on electronic components that can fail. Finally, one could argue that an operator would feel insecure working with a powerful machine with his safety guaranteed only by an algorithm. It can thus be concluded that the only way to obtain *safe* and *dependable* robots is to use the *design* strategy, which leads to the development of robots that are *intrinsically safe*.

2.1 Series Elastic Actuators (SEA)

To create *intrinsically safe* robots, the usual approach is to make them compliant. Indeed, compliance reduces the peak force attained during a collision. By extending the duration of the contact, it also allows the controller to sense it and react to reduce potential damages, under certain constraints (*i.e.*, reaction time). One early technique Pratt & Williamson (1995) to create compliant robots consists in placing the actuators of a serial robot at its base and linking them to the joints with an elastic transmission. However, the resulting *Series Elastic Actuators* (SEA) also limit the precision and stiffness of the robot. Moreover, as stated by the authors of Pratt & Williamson (1995), compliant joints can store potential energy. It can be argued that this energy could be harmful if released in an uncontrolled manner. Thus, a compromise must be achieved between safety and performance. The following sections present publications that propose solutions to increase safety while maintaining precision as much as possible.

2.2 Active Compliance

Active compliance Hogan (1987); Salisbury (1980) is a technique in which a regular robot is controlled to present a compliant interface at its effector. In a certain way, this technique is the ancestor of admittance control: efforts are measured at the effector and processed to command a displacement equal to the contact force divided by a virtual spring stiffness. Thus, the robot behaves like a spring around its trajectory and the compliance can be adapted online to match variable safety requirements. Unfortunatly, the response time of traditional actuators is longer than what is required to accommodate high frequency forces applied during collisions. Consequently, during a collision, the robot does not have a compliant behaviour and thus this technique is not suitable for the design of safe robots.

2.3 Programmable Passive Compliance

Programmable Passive Compliance Choi et al. (2008); Kim et al. (2007); Tonietti et al. (2005); Wolf & Hirzinger (2008) consists in using a compliant joint for each axis of the robot and a second set of actuators that allow the adjustment of each joint stiffness. This is obtained either by using two antagonistic actuators or by having a second actuator that adjusts the stiffness via a mechanism. This technique allows high stiffness (precision) at low velocity in addition to low stiffness (safety) at high velocity, *i.e.* when the manipulator is usually dangerous. This gives the controller the ability to continuously adjust the compromise between safety and performance. However, these characteristics are obtained by adding weight and complexity to the manipulator. Also, for the mechanisms currently proposed in the literature, the ratio between the largest and lowest stiffnesses is not sufficient to obtain high precision — by manufacturing standards — at low velocity, when collisions are less dangerous.

2.4 Distributed Macro-Mini Actuation (DM^2)

Distributed Macro-Mini Actuation (DM^2) Zinn, Khatib & Roth (2004), developed at Standford University, consists in actuating each joint with two actuators in parallel. The *macro* actuator is powerful but has a limited bandwidth. It is located at the base (to reduce inertia) and actuates the joint via an elastic cable transmission unable to transmit high frequency forces, characteristic of a collision. The *mini* actuator is directly located at the joint and has a large bandwidth. However, its size prevents the transmission of high torques, which makes the robot safer during collisions. The result is a combination of a large actuator that supplies large static torques, such as the ones induced by the robot's weight, and a smaller one that compensates for high frequency perturbations. In practice, this technique seems difficult to implement because it adds complexity to the manipulator's design, especially by doubling the number of required actuators. Recent developments Sardellitti et al. (2007) use two antagonistic pneumatic muscles as the *macro* actuator.

2.5 Nonlinear Passive Compliance

It has recently been proposed Park et al. (2009; 2008) to place on each joint a mechanism whose compliance varies by purely mechanical means. It is composed of two disks linked by a force transmitting pin and two mechanisms, each comprising two bars, one slider and one preloaded spring. In a normal situation, the force in the spring prevents the mechanism from moving. When the transmitted torque exceeds the threshold, the initial spring force is overcome and the mechanism starts moving in the slider. As the mechanism moves, the transmitted torque is reduced by the linkage geometry, even if the spring force is increased. The result is a rigid mechanism that becomes highly compliant when the transmitted torque exceeds a threshold that depends on the design parameters. Thus, the mechanism acts like a torque limiter (or a clutch) until the slider hits the end stop.

This is an interesting solution since by placing such torque limiter in series with each actuator, the resulting manipulator will be rigid unless external forces applied on it exceed a certain threshold, in which case it will become compliant and safer. This technique allows the design of robots that are stiff and accurate under normal conditions, but safer when collisions occur. Moreover, this principle is realized mechanically, which means that the reliability of this safety system does not depend on electronic components. Also, the mechanism is simple, compact and light. For all these reasons, nonlinear passive compliance is a promising approach.

However, this method also has some disadvantages. First, by adding a torque limiter on each joint of a serial robot, the force threshold will depend on the configuration of the manipulator.

This is because the relation between external forces and articular torques is determined by the Jacobian matrix of the manipulator, which is a function of the manipulator's pose. The threshold will also depend on the contact location and on the force orientation, which is not desirable since it means that the safety level will vary throughout the robot's external surface. Moreover, a manipulator in a singular configuration could theoretically apply infinite forces in certain directions that would not induce any articular torque. These issues arise from the articular architecture of the safety mechanism and consequently a mechanism using torque limiters in a Cartesian architecture would circumvent them and offer the same safety level, regardless of the manipulator's configuration.

Some Cartesian safety devices already exist. One of them Park et al. (2007) was developed by the same researchers as the previously mentioned torque limiter. In this case, the slider-spring mechanism is packaged in the robot's last member, allowing the end-effector to become compliant if a collision induces a large moment relative to the mechanism. The other one is a commercial product *Collision Sensor for Robotic Safety - Robotic Crash Sensors from ATI* (n.d.) that is mainly used to protect a tool if an unexpected contact occurs. These devices, however, possess disadvantages similar to those of the articular mechanisms. Since they are sensitive to external moments, the force threshold depends on the contact orientation and location on the end-effector. Therefore, they are reliable only when collisions occur at a pre-determined location, which is not the case in general for large end-effectors in an uncontrolled environment.

3. Cartesian Force Limiting Devices

A technique that combines torque limiters with parallel mechanisms to create Cartesian force limiting devices (CFLD) was recently proposed in Lauzier et al. (2009). The device behaves like a structure unless the external forces exceed a certain threshold, leading to the activation of one or more degree of freedom. If the parallel mechanism performs pure translation motions, replacing the actuators with torque limiters results in a CFLD that is sensitive to forces – not moments – and thus the threshold is independent from the location of the force application point. Cartesian force limiting devices are particularly well suited for ceiling-suspended robots because their end-effector is the only part on which a collision can occur. During a collision, the motion of the end-effector can thus be decoupled from the rest of the robot if a CFLD is placed between them.

Fig. 1 shows an example of a simple 1-DOF CFLD mounted between a suspended robot and its end-effector. The mechanism is a parallelogram linkage in which one revolute joint was replaced with a torque limiter. Under normal conditions, the torque limiter prevents the mechanism from moving and thus the end-effector is fixed rigidly to the robot. However, if a collision occurs, the couple passing through the torque limiter becomes too high and the mechanism is allowed to move, as shown in the middle and right pictures. This practically "disconnects" the end-effector from the robot and thus the person involved in the collision is only subjected to the inertia of the end-effector, which can be significantly lower than the inertia of the whole robot. For the mechanism to be effective in increasing the safety level, the collision has to be detected and the robot must stop before the mechanism reaches the end of its travel. The collision can be detected with a limit switch placed on one of the links and an emergency stop signal can be sent directly to brakes without passing through the controller, thus improving the reliability of the system by reducing the risks of electronic components failure. Once the robot is stopped, gravity tends to naturally return the mechanism to its original position. One important advantage of the parallelogram architecture is that the couple passing through the torque limiter only depends on the magnitude of the horizontal force applied on the end-effector and *is not affected by the height of the point of application of the force*. This implies that the same force level will cause the activation of the safety mechanism whether the collision occurs at the head or at the knee of the person. This is an important advantage since a collision can occur anywhere on the end-effector.

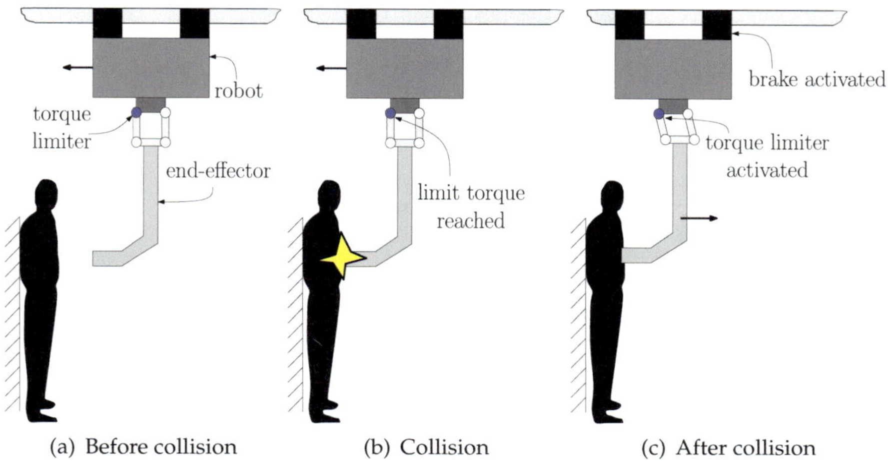

Fig. 1. Example of a 1-DOF Cartesian FLD using torque limiters(©2009 IEEE).

In Lauzier et al. (2009), this simple 1-DOF architecture was extended to a 2-DOF mechanism that reacts to collisions occuring in the whole horizontal plane. It is also possible to extend it to 3-DOF, thus covering all possible collisions occuring on the end-effector. For example, the Delta architecture can be used to create a 3-DOF CFLD by replacing the actuators with torque limiters, as shown in Fig. 2. However, since the mechanism is sensitive to vertical forces, the end-effector's weight (plus the carried load) will induce articular torques that will limit the robot's ability to apply forces before reaching its threshold. To circumvent this problem, the device has to be statically balanced.

Experiments were performed using a reduced-scale prototype of a 2-DOF CFLD to evaluate the behaviour of such a device during a collision. Fig. 3 shows the experimental setup and the contact force over time for a collision occuring at a low velocity. On the graph, it is possible to see that the contact force is slowly increasing until it reaches the preset threshold, after which it drops sharply to a level corresponding to the friction force until the motion is stopped. This shows that for collisions occuring at a low velocity, the maximum contact force is approximately limited to the preset threshold. For higher velocities, the inertia of the end-effector and the stiffness of the contact interface must be taken into account. More detailed results are presented in Lauzier et al. (2009).

3.1 Safety Improvements

It is important to evaluate the safety improvements and the limitations of the proposed approach which consists in placing a mechanism on the robot that can disconnect the end-effector if the forces applied on it are excessive.

Firstly, in the case of mechanisms performing horizontal motion only, the load to be carried by the robot is not limited. This is also the case for the 3-DOF architectures if gravity is compen-

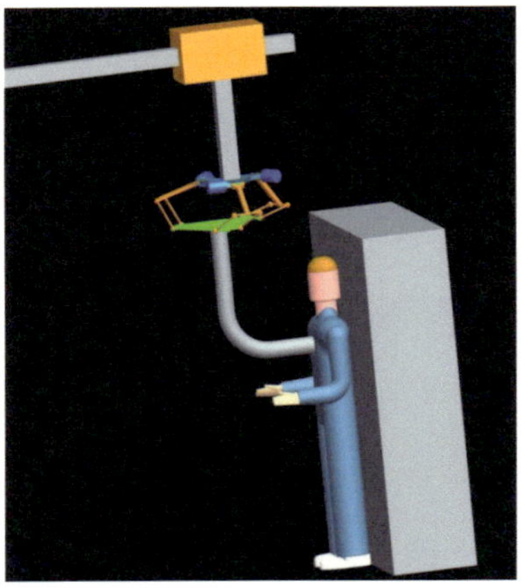

Fig. 2. Example of a collision between a person and a suspended robot with a 3-DOF CFLD using Delta architecture.

sated for. However, accelerations of the robot will induce inertial forces that can activate the torque limiters of the mechanism. Thus, for a given load, accelerations must be limited to a certain level to prevent the end-effector from being disconnected without collisions. This is a potential disadvantage because a robot usually aims at maximizing accelerations/decelerations.
Secondly, there is always a maximum velocity that can be imposed to a robot that will ensure that blunt, unconstrained collisions will be safe. This "safe" velocity is usually very low for heavy robots. However, if during a collision the end-effector is disconnected from the robot, the effective inertia to which the person is subjected is then greatly reduced. Therefore, it can be assumed that using this type of mechanism will allow to increase the maximum velocity of a robot moving around people. This maximum velocity should be evaluated using a collision model that considers all collision parameters, including the way the robot reacts when the collision is detected (braking force, delay before the brakes are applied, etc.).
Thirdly, as explained in Haddadin, Albu-Schaffer, Frommberger & Hirzinger (2008), collisions in which a human body is clamped to a wall by a robot can be very dangerous. With the mechanism described in this paper, however, the maximum clamping force that the robot can apply in quasistatic conditions is determined by the limit torques of the limiters. As the velocity increases, the safety is still improved because the inertia crushing the person's body against the wall is reduced. Again, the maximum velocity should be calculated using an appropriate model to ensure safety.
Also, because the mechanism is unable to store elastic potential energy (as opposed to compliant robots), it will not make the robot continue pushing on the person's body after the collision has taken place. This is an advantage since it will help the person to push the robot away after the collision.

Advantages over Existing Devices

Some robots designed for pHRI, such as the Kuka KR3-SI Haddadin, Albu-Schaffer & Hirzinger (2008), incorporate a flexible flange with breakaway function that links the tool to the manipulator. This device triggers an emergency stop when the contact force at the tool control point exceeds a certain threshold. Although this type of device behaves similarly on many aspects to the devices described in this section, it differs on certain points. Firstly, it lim-

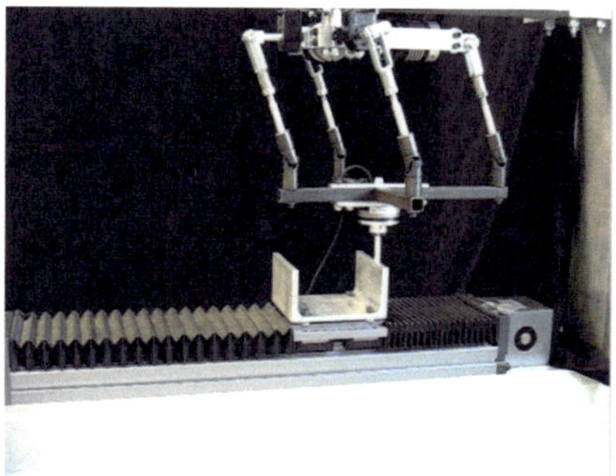

(a) Experimental setup(©2009 IEEE).

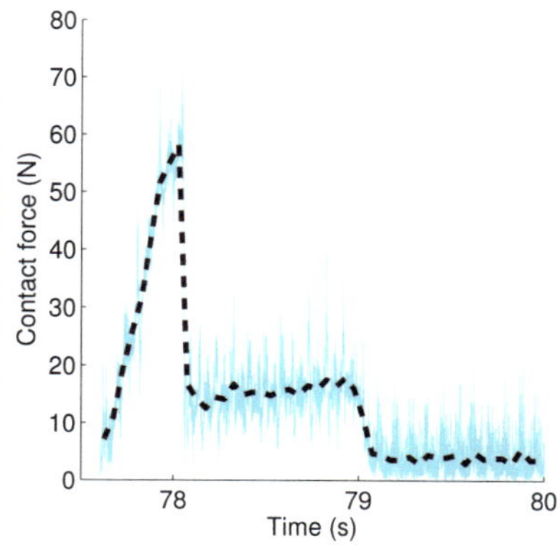

(b) Force over time for a low velocity collision(©2009 IEEE).

Fig. 3. Experimental collision between a 2-DOF CFLD prototype mounted on a structure and a linear actuator.

its the moment – not the force – that can be transmitted by the manipulator to the end-effector. This means that the threshold depends on the location of the collision point, as opposed to the proposed Cartesian FLDs. The latter behaviour is preferable since a collision can generally occur anywhere on the end-effector. Also, the proposed mechanism has a large achievable displacement compared to the existing device. This is an advantage since it yields the space required by a heavy ceiling mounted manipulator to stop without crushing the person involved in the collision.

4. Robot Soft Covering

The key idea behind compliant joints is to reduce the peak force attained during a collision. Covering a robot with a soft material can provide a very similar feature by absorbing directly the collision energy. However, since this safety element is an external cover, therefore isolated from the internal forces given by the robot dynamics, this approach does not suffer from the same drawback as compliant joints. Indeed, in this case the compliance has no effect on the robot end-effector stiffness and thus no tradeoff has to be made between the ideal compliance required for safety and minimum wanted robot stiffness. This characteristic gives to the robot designer a total freedom in the choice of the compliance.
This approach does not only have advantages. The thickness of the covering material required for attaining a good level of safety could be relatively large as mentioned in Zinn, Khatib, Roth & Salisbury (2004b). This extra material could significantly increased the weight of the robot with a negative impact on its dynamics performance . Fig. 4 shows some collision tests that have been presented in Duchaine, Lauzier, Lacasse, Baril & Gosselin (2009). From these curves it is possible to observe that with the 3 mm thick sample tested, even if the soft cover has a measurable impact in reducing the collision peak forces, this reduction is not enough to make the robot safe. The data of collision peak force for the case where the soft covering can provide sensing to detect contact is also provided. In this case the reduction is drastic and the robot is easily in the safe zone. This latter approach that combines soft covering and active sensing is often referred to robot skin. This concept is a promising solution that could

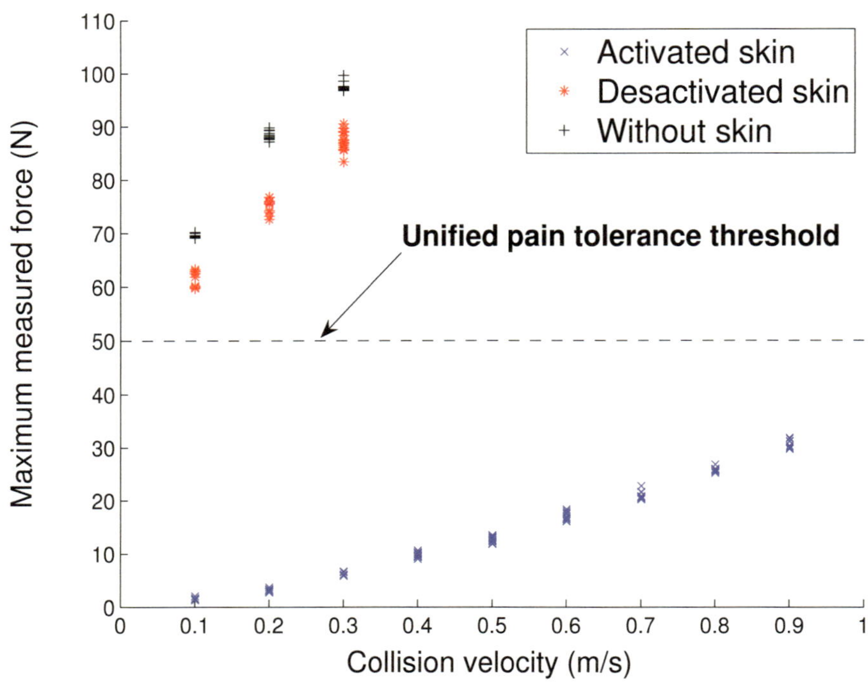

Fig. 4. Maximum force measured for different collision velocities. (©2009 IEEE)

become a *must have* feature on the next generation of cooperative robots. However, so far there is no commercially available robot skin on the market and there is still many research challenges associated to this topic such as manufacturing, wire routing and post processing of the signals. An interesting review of the state of the art in this area and an overview of the remaining challenges is presented in Cutkosky et al. (2008).

5. Efficient Joint Actuation Coupling

Conventional serial manipulators typically have one motor attached at each joint, thus leading to a direct relationship between the vectors of actuation torque (τ_m) and the joint torque (τ). In the design of such robots, each actuator is chosen so that it has the capability to deliver the maximum torque required at its associated joint. One way to improve safety in pHRI is by coupling the actuation of some of these joints. Indeed, actuation coupling for some very specific robot architectures could lead to a significant reduction of the maximum force that the robot can apply to a human being in the case of a malfunction or an unwanted contact. The capability of a robot to apply forces to its environment is maximized when the robot is in an equilibrium configuration, i.e., when no joint torque is required to maintain its pose. In such a configuration and for a short period of time, all the available torque can be directly used to apply forces to the surrounding environment. Reducing this maximum torque while maintaining the same static capability could greatly enhance the overall safety of a robot without affecting its performances.

i	θ_i	α_i	a_i	b_i
1	θ_1	$\frac{\pi}{2}$	0	l_1
2	θ_2	0	l_2	0

Table 1. HD parameters of the robot of figure 5.

For a general manipulator with a direct relationship between joint torque and actuator torque as described above, the maximum torque at each joint can happen independently from the others, therefore, one has:

$$\max(\tau_i + \tau_j) = \max(\tau_i) + \max(\tau_j), \quad \forall i, j \tag{1}$$

where τ_i is the ith joint torque. In this case, coupling the actuation of the joints will not lead to a significant improvement since instead of requiring two actuators with a maximum torque capability of τ we will end up with one more powerful actuator that can supply a torque of 2τ. However, for some specific serial architectures, it can be observed that eq. (1) becomes:

$$\max(\tau_i + \tau_j) < \max(\tau_i) + \max(\tau_j). \tag{2}$$

This equation means that, for this architecture, the maximum torque at each joint is not independent from the others and that the maxima cannot happen simultaneously. This kind of architecture can lead to what we call *efficient joint actuation coupling*. One good example of such an architecture is the human arm, where the largest muscle, the biceps, is involved in the upperarm supination as well as in the forearm extension.
The intent of this section is not to show how to mechanically achieve efficient joint *actuation* coupling but to demonstrate the potential contribution of this concept to safe pHRI and to illustrate how the joints to be coupled can be selected in order to maximize the benefits. Indeed, among all the possible joint arrangements in a serial robot, very few combinations will lead to a beneficial actuation coupling. Therefore, some design guidelines to achieve such coupling will be given by finding the corresponding constraints on the Denavit-Hartenberg (D-H) parameters.

5.1 Constraints on the Denavit-Hartenberg Parameters to Achieve Efficient Joint actuation Coupling

In this section, conditions on the Denavit-Hartenberg (D-H) parameters are derived that lead to a serial arrangement providing an efficient coupling of two of its joints. The conditions obtained are sufficient to satisfy eq. 2 but they may not be necessary. In other words, there is no proof yet that these conditions are the only possible combinations of D-H parameters that satisfy eq. 2.

5.1.1 Two-Degree-of-Freedom Serial Architecture

One simple architecture that leads to an efficient coupling is a two-dof serial combination of revolute joints with the D-H parameters given in Table (1). In this table, θ_i is used to denote the ith joint variable, α_i is the twist, a_i is the length and b_i the offset. A schematic representation of the corresponding robot is given in fig. 5. By observation of the figure, it is possible to observe that the maximum static torque at each joint cannot occur simultaneously for this serial arrangement. Using the expression of the corresponding joint torques helps understanding the reasons behind this behaviour.

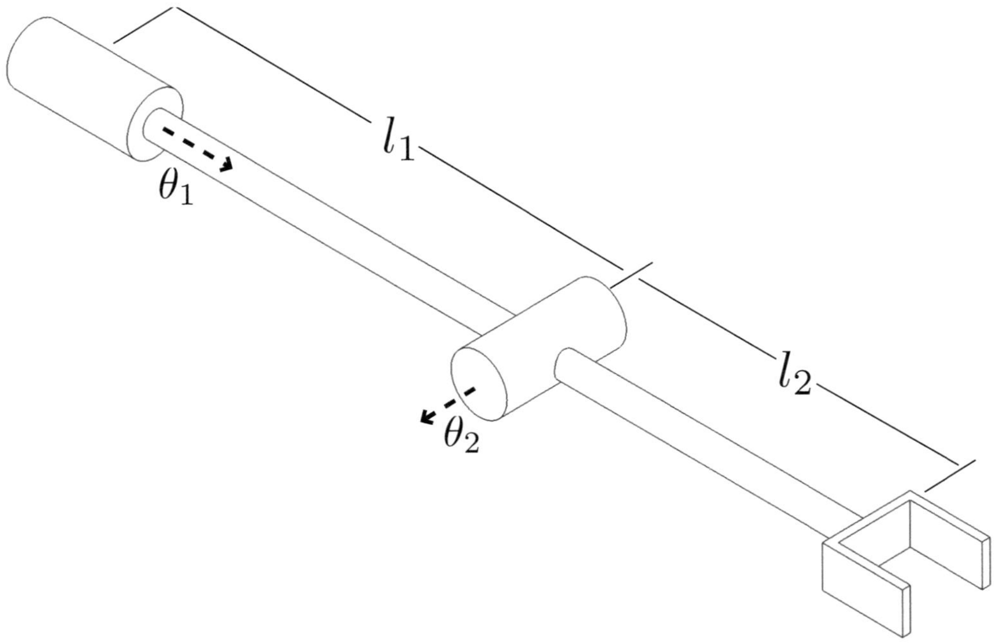

Fig. 5. Two *dof* serial architecture.

As suggested in the ANSI/RIA R15.06-1999 Standard for robot safety in factories, the maximum acceleration and velocity of a robot is typically low in the context of pHRI. Therefore, The maximum torque needed at each joint can be roughly estimated from the static forces, i.e., from the effect of gravity on the robot and its payload. The mathematical expressions for the torque induced by gravity at each joint of the robot of fig. 5 are given by:

$$\tau_1 = \frac{\partial V}{\partial \theta_1} = \frac{m_2 g l_2}{2} \cos\theta_1 \cos\theta_2 \tag{3}$$

$$\tau_2 = \frac{\partial V}{\partial \theta_2} = \frac{m_2 g l_2}{2} \sin\theta_1 \sin\theta_2, \tag{4}$$

where V is the gravitational potential energy given by

$$V = V(\boldsymbol{\theta}) = \sum_{i=1}^{n} m_i g h_i, \tag{5}$$

in which m_i is the mass of the ith member, g the gravitational acceleration, h_i the elevation of the centre of mass of the ith member measured from a fixed reference and n is the total number of links. In order to verify if this architecture can lead to efficient coupling by satisfying eq. (2), the sum of the joint torques is computed, namely:

$$\tau_1 + \tau_2 = \frac{m_2 g l_2}{2} \cos\theta_1 \cos\theta_2 + \frac{m_2 g l_2}{2} \sin\theta_1 \sin\theta_2. \tag{6}$$

Using the following trigonometric identity:

$$\cos(a \pm b) = \cos a \cos b \mp \sin a \sin b, \tag{7}$$

eq. (6) can be reduced to:

$$\tau_1 + \tau_2 = \frac{m_2 g l_2}{2} \cos(\theta_1 - \theta_2). \tag{8}$$

Therefore, since:

$$\max(\cos a) = \max(\sin a) = \max(\cos(\mathrm{a} + \mathrm{b} + ...)) = 1, \tag{9}$$

we obtain:

$$\max(\tau_1 + \tau_2) = \max(\tau_1) = \max(\tau_2) = \frac{m_2 g l_2}{2}. \tag{10}$$

This result is the minimum possible value for eq. 2, which means that it could be possible, with appropriate coupling, to drive these two joints with only one of the two motors. The key to this reduction lies in the fact that the sine and cosine expressions for the individual joint torques, when added together, can be combined into another cosine function by virtue of the trigonometric identity of eq. (7).

5.1.2 Generalization

The architecture described above is one possible example of application of efficient joint actuation coupling. However, it is important to generalize the results in order to determine all the possible serial architectures that can lead to efficient joint actuation coupling. One way to proceed is by finding the constraints on the DH parameters that allow the satisfaction of eq. (2).

For a 2-dof architecture, the expression of the joint static torques for a general value of the DH-parameters can be written as:

$$\tau_1 = \frac{1}{2} a_2 \cos\theta_1 \cos\theta_2 - \frac{1}{2} a_2 \cos\alpha_1 \sin\theta_1 \sin\theta_2 + \frac{1}{2} b_2 \sin\alpha_1 \sin\theta 1 \tag{11}$$

$$\tau_2 = -\frac{1}{2} a_2 \sin\theta_1 \sin\theta_2 + \frac{1}{2} a_2 \cos\alpha_1 \cos\theta_1 \cos\theta_2. \tag{12}$$

As demonstrated above, the trigonometric indentity of eq. (7) is the key that led to eq. (10). In order to make it possible for the sum of eqs. (11) and (12) to be manipulated using this trigonometric identity, the following constraints need to be introduced:

$$\cos\alpha_1 = b_2 \sin\alpha_1 = 0 \tag{13}$$

and

$$a_2 \neq 0 \tag{14}$$

This imposes the following constraints on the DH-parameters:

$$\alpha_1 = (2n+1)\frac{\pi}{2} \tag{15}$$

$$b_2 = 0 \tag{16}$$

where n is any integer.

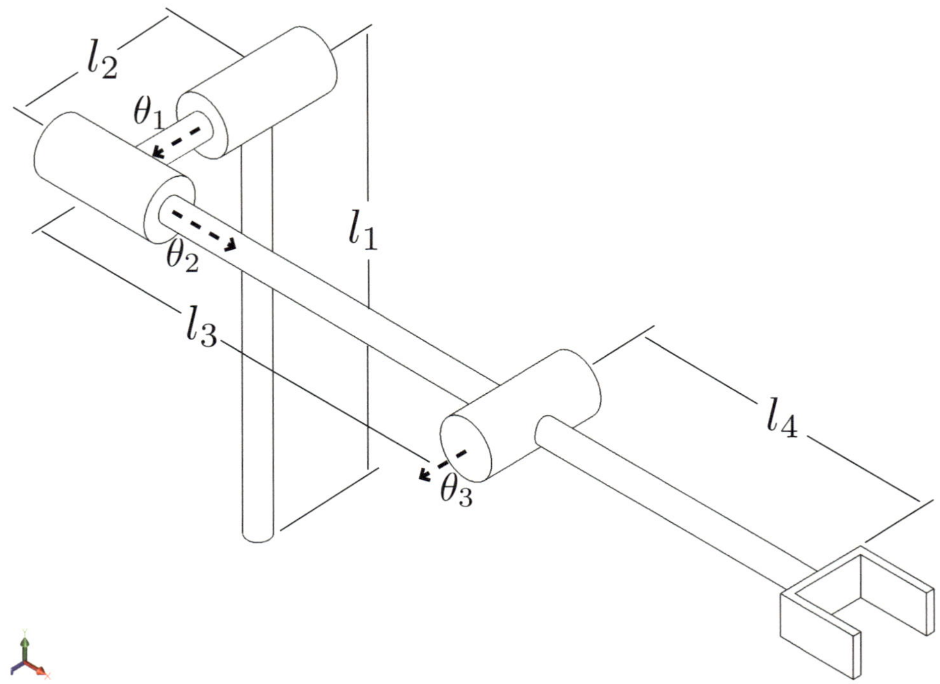

Fig. 6. Three dof serial architecture.

i	θ_i	α_i	a_i	b_i
1	θ_1	$\frac{\pi}{2}$	0	l_1
2	θ_2	$\frac{\pi}{2}$	0	l_2
3	θ_3	0	l_3	0

Table 2. HD parameters of the robot of figure 6.

5.1.3 Three-Degree-of-Freedom Serial Architecture

The previous example is rather trivial since the first member of the robot is fixed relative to the direction of gravity. In order to obtain a more realistic situation, a three-dof architecture is now considered. Table (2) provides the HD-parameters of the chosen architecture, which is illustrated schematically in fig. (6). The possible coupling of the last two joints is investigated. Computing the static forces from the potential energy as in eq. (5), the sum of the gravity torques of the last two joints of this serial architecture can be written as:

$$\tau_2 + \tau_3 = \frac{m_2 g l_2}{2} \left(\sin\theta_1 \sin\theta_2 \cos\theta_3 + \sin\theta_1 \cos\theta_2 \sin\theta_3 + \cos\theta_1 \cos\theta_3 \right). \tag{17}$$

The trigonometric identity of eq. (8) is now used, together with the following identity:

$$\sin(a \pm b) = \sin a \cos b \pm \cos a \sin b \tag{18}$$

and eq. (17) can then be reduced to

$$\tau_2 + \tau_3 = \frac{m_2 g l_2}{2} \left(\sin\theta_1 \sin(\theta_2 + \theta_3) + \cos\theta_1 \cos\theta_3 \right). \tag{19}$$

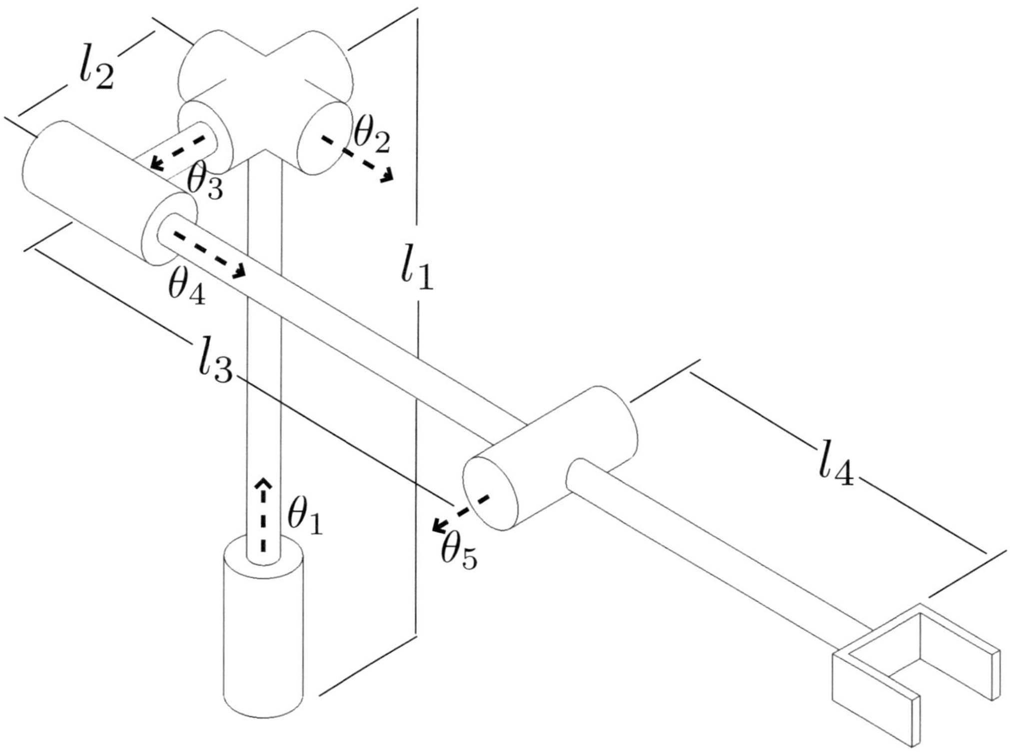

Fig. 7. Five dof serial architecture.

The maximum value of this expression can be obtained as:

$$\max(\tau_2 + \tau_3) = \frac{m_2 g l_2}{2} \max\left(\sin\theta_1 \sin(\theta_2 + \theta_3) + \cos\theta_1 \cos\theta_3\right) \quad (20)$$

$$= \frac{m_2 g l_2}{2} \max\left(\sin`_1 + \cos`_1\right) \quad (21)$$

$$= \frac{\sqrt{2}}{2} m_2 g l_2 \quad (22)$$

and therefore

$$\max(\tau_2 + \tau_3) = \sqrt{2}\max(\tau_2) = \sqrt{2}\max(\tau_3). \quad (23)$$

In this case, the gravity torque cannot be perfectly combined and it will not be possible to drive both joints with a motor that would have been selected for driving only one of the joints. However eq. (23) still satisfies eq. (2), meaning that combining the motion of both joints will require significantly less torque than driving both separately.

If the above exercise is repeated with three orghogonal revolute joints prior to the last two members, then the latter can have any possible orientation with respect to gravity. One then obtains:

$$\tau_4 + \tau_5 = \frac{m_5 g l_5}{2}(\sin\theta_2\cos\theta_3\sin(\theta_4+\theta_5) + \cos\theta_2\cos(\theta_4+\theta_5) - \sin\theta_2\sin\theta_3\cos\theta_5) \quad (24)$$

and the maximum for the sum of the torques is again given as

$$\max(\tau_4+\tau_5) = \frac{\sqrt{2}}{2}m_5 g l_5 = \sqrt{2}\max(\tau_4) = \sqrt{2}\max(\tau_5). \quad (25)$$

Figure (7) provides a schematic illustration of such an architecture. Since this architecture allows all possible orientations of the member of length l_3, it is possible to make a generalization of the results. Therefore, if the HD parameters associated with the last two dofs of the manipulator satisfy the constraints given by eqs. (15) and (16), no matter what will be the prior serial arrangement, coupling the actuation of the last two dofs will result in a significant reduction of the maximal torque compared to separate actuation.
If the designer wants to add other dofs after the member of length l_4 of the architecture presented in figure (7), eq. (25) will no longer be true. However, if all a_i and b_i for $i > 5$ in the HD parameters are kept as small as possible relative to the length of l_4, or if the centre of mass of these extra dofs are close to the end of l_4 or if the extra links are light, the gain can still be significant. The human arm is a good example of this situation, with is maximum reach mainly given by the upper arm(l_3) and the forearm (l_4).

6. Conclusion

Human-robot interaction is the next logical step in the evolution of robotics. However, the challenge of bringing robots in our environment is not simply about increasing their capabilities and their functionalities. Even before that, robots need to be built in a way that they cannot hurt human beings. In this chapter, we have reviewed several concepts that have been proposed in the recent years in order to address this particular challenge. The popular idea of compliant joints was exposed from Series Elastic Actuators (SEA) to the distributed Macro-Mini concept (DM2). A special emphasis was placed on the recent concept of Force Limiting Device (FLD), which we believe circumvents some of the drawbacks associated with compliant joints. We have also presented the concept of external compliance via soft covering of the robot. Finally, a new concept of efficient joint actuation coupling was proposed to reduce the capability of a robot to transfer energy to its environment while maintaining the same dynamic performances.

7. References

Choi, J., Park, S., Lee, W. & Kang, S. (2008). Design of a robot joint with variable stiffness, *Robotics and Automation, 2008. ICRA 2008. Proceedings of the 2008 IEEE International Conference on*, pp. 1760–1765.

Collision Sensor for Robotic Safety - Robotic Crash Sensors from ATI (n.d.). `http://www.ati-ia.com/products/collision_sensor/collision_sensor_articles.aspx`.

Cutkosky, M. R., Howe, R. D. & Provancher, W. R. (2008). Force and tactile sensors, *Springer Handbook of Robotics* pp. 455–476.

De Luca, A., Albu-Schäffer, A., Haddadin, S. & Hirzinger, G. (2006). Collision detection and safe reaction with the dlr-iii lightweight manipulator arm, *Intelligent Robots and Systems, 2006 IEEE/RSJ International Conference on*, pp. 1623–1630.
URL: *http://dx.doi.org/10.1109/IROS.2006.282053*

Duchaine, V., Lauzier, N., Baril, M., Lacasse, M.-A. & Gosselin, C. (2009). A flexible robot skin for safe physical human robot interaction, *Robotics and Automation, 2009. ICRA 2009. Proceedings of the 2009 IEEE International Conference on*, pp. 3676–3681.

Duchaine, V., Lauzier, N., Lacasse, M., Baril, M. & Gosselin, C. (2009). A Flexible Robot Skin for Safe Physical Human Robot Interaction, *IEEE International Conference on Robotics and Automation, 2009. ICRA 2009*, pp. 3606–3611.

Ebert, D., Komuro, T., Namiki, A. & Ishikawa, M. (2005). Safe human-robot-coexistence: emergency-stop using a high-speed vision-chip, *Intelligent Robots and Systems, 2005. (IROS 2005). 2005 IEEE/RSJ International Conference on*, pp. 2923–2928.
URL: *http://dx.doi.org/10.1109/IROS.2005.1545242*

Haddadin, S., Albu-Schaffer, A., Frommberger, M. & Hirzinger, G. (2008). The role of the robot mass and velocity in physical human-robot interaction-Part II: Constrained blunt impacts, *IEEE International Conference on Robotics and Automation, 2008. ICRA 2008*, pp. 1339–1345.

Haddadin, S., Albu-Schaffer, A. & Hirzinger, G. (2008). The role of the robot mass and velocity in physical human-robot interaction-Part I: Non-constrained blunt impacts, *IEEE International Conference on Robotics and Automation, 2008. ICRA 2008*, pp. 1331–1338.

Hogan, N. (1987). Stable execution of contact tasks using impedance control, *Robotics and Automation. Proceedings. 1987 IEEE International Conference on*, Vol. 4, pp. 1047–1054.
URL: *http://ieeexplore.ieee.org/xpls/abs_all.jsp?arnumber=1087854*

Kim, B.-S., Park, J.-J. & Song, J.-B. (2007). Double actuator unit with planetary gear train for a safe manipulator, *Robotics and Automation, 2007 IEEE International Conference on*, pp. 1146–1151.
URL: *http://dx.doi.org/10.1109/ROBOT.2007.363139*

Lauzier, N., Grenier, M., & Gosselin, C. (2009). 2 dof cartesian force limiting device for safe physical human-robot interaction, *Robotics and Automation, 2009. ICRA 2009. Proceedings of the 2009 IEEE International Conference on*, pp. 3676–3681.

Lu, Y., Zeng, L. & Bone, G. M. (2005). Multisensor system for safer human-robot interaction, *Robotics and Automation, 2005. ICRA 2005. Proceedings of the 2005 IEEE International Conference on*, pp. 1767–1772.
URL: *http://ieeexplore.ieee.org/xpls/abs_all.jsp?arnumber=1570369*

Park, J.-J., Kim, B.-S., Song, J.-B. & Kim, H.-S. (2007). Safe link mechanism based on nonlinear stiffness for collision safety, *Mechanism and Machine Theory* **In Press, Corrected Proof**.
URL: *http://dx.doi.org/10.1016/j.mechmachtheory.2007.10.004*

Park, J.-J., Kim, H.-S. & Song, J.-B. (2009). Safe robot arm with safe joint mechanism using nonlinear spring system for collision safety, *Robotics and Automation, 2009. ICRA 2009. Proceedings of the 2009 IEEE International Conference on*, pp. 3371–3376.

Park, J.-J., Song, J.-B. & Kim, H.-S. (2008). Safe joint mechanism based on passive compliance for collision safety, *Recent Progress in Robotics: Viable Robotic Service to Human* pp. 49–61.
URL: *http://dx.doi.org/10.1007/978-3-540-76729-9_5*

Pratt, G. & Williamson, M. M. (1995). Series Elastic Actuators, *Proc. IEEE/RSJ Int. Conf. on Intelligent Robots and Systems (IROS'95)*, Vol. 1, pp. 399–406.

Salisbury, J. (1980). Active stiffness control of a manipulator in cartesian coordinates, *Proc. of the IEEE Conference on Decision and Control*.

Sardellitti, I., Park, J., Shin, D. & Khatib, O. (2007). Air muscle controller design in the distributed macro-mini (dm2) actuation approach, *Intelligent Robots and Systems, 2007. IROS 2007. IEEE/RSJ International Conference on*, pp. 1822–1827.
URL: *http://dx.doi.org/10.1109/IROS.2007.4399073*

Tonietti, G., Schiavi, R. & Bicchi, A. (2005). Design and control of a variable stiffness actuator for safe and fast physical human/robot interaction, *Robotics and Automation, 2005. ICRA 2005. Proceedings of the 2005 IEEE International Conference on*, pp. 526–531.
URL: *http://ieeexplore.ieee.org/xpls/abs_all.jsp?arnumber=1570172*

Wolf, S. & Hirzinger, G. (2008). A new variable stiffness design: Matching requirements of the next robot generation, *Robotics and Automation, 2008. ICRA 2008. Proceedings of the 2008 IEEE International Conference on*, pp. 1741–1746.

Yamada, Y., Hirasawa, Y., Huang, S., Umetani, Y. & Suita, K. (1997). Human-robot contact in the safeguarding space, *Mechatronics, IEEE/ASME Transactions on* **2**(4): 230–236.

Zinn, M., Khatib, O. & Roth, B. (2004). A new actuation approach for human friendly robot design, *Robotics and Automation, 2004. Proceedings. ICRA '04. 2004 IEEE International Conference on*, Vol. 1, pp. 249–254 Vol.1.
URL: *http://dx.doi.org/10.1109/ROBOT.2004.1307159*

Zinn, M., Khatib, O., Roth, B. & Salisbury, J. (2004a). Playing it safe [human-friendly robots], *IEEE Robotics & Automation Magazine* **11**(2): 12–21.

Zinn, M., Khatib, O., Roth, B. & Salisbury, J. (2004b). Playing it safe [human-friendly robots], *IEEE Robotics & Automation Magazine* **11**(2): 12–21.

24

Vibration Based Control for Flexible Link Manipulator

Tamer Mansour, Atsushi Konno
and Masaru Uchiyama
Tohoku University
Japan

1. Introduction

Robotic manipulators are widely used to help in dangerous, monotonous, and tedious jobs. Most of the existing robotic manipulators are designed and built in a manner to maximize stiffness in an attempt to minimize the vibration of the end-effectors. This high stiffness is achieved by using heavy material and bulky design. Hence, the existing heavy rigid manipulators are shown to be inefficient in terms of power consumption or speed with respect to the operating payload. Also, the operation of high precision robots is severely limited by their dynamic deflection, which persists for a period of time after a move is completed. The settling time required for this residual vibration delays subsequent operations, thus conflicting with the demand of increased productivity. These conflicting requirements between high speed and high accuracy have rendered the robotic assembly task a challenging research problem. In addition, many industrial manipulators face the problem of arm vibrations during high-speed motion. In order to improve industrial productivity, it is required to reduce the weight of the arms and/or to increase their speed of operation. For these purposes, it is very desirable to build flexible robotic manipulators. Compared to the conventional heavy and bulky robots, flexible link manipulators have the potential advantage of lower cost, larger work volume, higher operational speed, greater payload-to-manipulator weight ratio, smaller actuators, lower energy consumption, better manoeuvrability, better transportability and safer operation due to reduced inertia. However, the major drawback of these robots is the inaccuracy of the end effectors due to low stiffness. Due to the importance and usefulness of these robots, researchers are nowadays engaged in the investigation of control of flexible manipulator. The issue of tip position control for flexible link manipulator has gained a lot of attention due to the great benefits, which can be achieved by changing the traditional rigid robots with flexible ones. Then, by measuring the elastic deformations of the link and using a more sophisticated control algorithm, the endpoint of the robot can be controlled with a relatively high degree of precision with minimal vibration. Using the vibration signal that is from the motion of the flexible links robot is one of the important methods used in controlling the tip position of the single-link arms. Compared with the common methods for controlling the base of the flexible arm the vibration feedback can improve the use of the flexible-link robot systems.

The control of a flexible robot arm has attracted many researchers either to design advanced and intelligent controllers or to use smart actuators in order to achieve a high positioning accuracy at the end of the arm. An initial experiment on the control of a single-link flexible robot moving in a plane was done by (Cannon & Schmitz, 1984). After then many researches have been done in all topics related to the control of flexible robot arms. Some researches focus on the modelling of the flexible arm such as (Zhu & Mote, 1997); (Kariz & Heppler, 2000). (Ge et al., 1997); (Ge et al., 1998) uses the finite element method to model flexible arms. The use of smart material and piezoelectric actuators in suppressing the vibration for a flexible robot has been investigated by (Tawfeic et al., 1997). (Lee et al., 1988) proposed PDS (proportional-derivative-strain) control for vibration suppression of multi-flexible-link manipulators and analyzed the Lyapunov stability of the PDS control while (Matsuno & Hayashi, 2000) applied the PDS control to a cooperative task of two one-link flexible arms. They aimed to accomplish the desired grasping force for a common rigid object and the vibration absorption of the flexible arms. Control policy had attracted (Menq & Xia, 1993) to investigate the use of classical control for the single-link flexible arm. The optimal control of the flexible link is highlighted by (Rai & Asada, 1995) while (Etxebarria et al., 2005) have proposed a robust control scheme for flexible link robotic manipulators.
The motivation for this research is to find a simple controller, which can be able to achieve final accurate tip position for the flexible arm and at the same time reduce resulting vibration. The use of the deflection signal or its derivatives in the feedback is one of the effective methods used in controlling the vibration of the tip position. A modified PID (MPID) control that uses the rate of change of the tip deflection is investigated in this chapter.
In this chapter, a Modified PID control (MPID) is proposed to control flexible link manipulator. The MPID control depends mainly on vibration feedback to improve the response of the flexible arm without the massive need of measurements. First, we give a brief introduction about the experimental set-up then the dynamic model of the system is driven. A detailed of the controller design is shown and the analysis of this controller is highlighted. The stability of the system is checked with the proposed controller. A case study for a single link flexible manipulator is chosen to verify the proposed controller. Simulation results are exposed for the system using the MPID to suppress the vibration. Finally, the experimental results for the response of the flexible manipulator are shown. A concluding summary is ending the chapter.
Unlike other research (Ge et al., 1998), the effect of static deformation is taken into consideration when evaluating the effect of the vibration on the control signal. As this control signal will drive the flexible manipulator, residual strain due to material defect and/or static deformation may lead to inaccurate movement. In addition to that, an experimental verification has been done in parallel with a simulation study to evaluate the performance of the MPID control. Using the rate of deflection at the tip of flexible manipulator as an indication for the vibration of the tip can remove successfully the effect of static deformation that may appears in the generated control signal.
The main contribution point with this controller is the usage of the rate of deflection at the tip as an indication of the vibration. The controller succeeds to remove the quasi-static component in the strain instead of using high-pass filter, which is used in general. However, a high-pass filter may bring a phase shift, which may cause the instability. The MPID

controller uses rate of deflection; therefore, neither quasi-static strains due to gravity nor residual strains in the material bring a problem.

2. Experimental setup

In this section, the details of the experimental flexible robot system are presented. As shown in Fig. 1, the flexible robot consists of a motor/actuator, an arm of length l and an end-point payload M_t. All the three elements are related to each other through the shear force and the bending moment. The motor torque T drives the whole system. The flexural rigidity EI and the mass per unit length ρ of the arm are assumed uniform along the length of the arm. A motor armature and gearbox are described by an equivalent moment of inertia I_h at the hub. A payload M_t is mounted at the tip of the arm. The variable $\delta(x, t)$ is the deflection of the arm at a point located a distance x from the hub, measured relative to the non-deformed position of the arm. θ is the rotary angle of the arm from its reference position. The geometry of the single-link flexible robot is shown in Fig. 1.

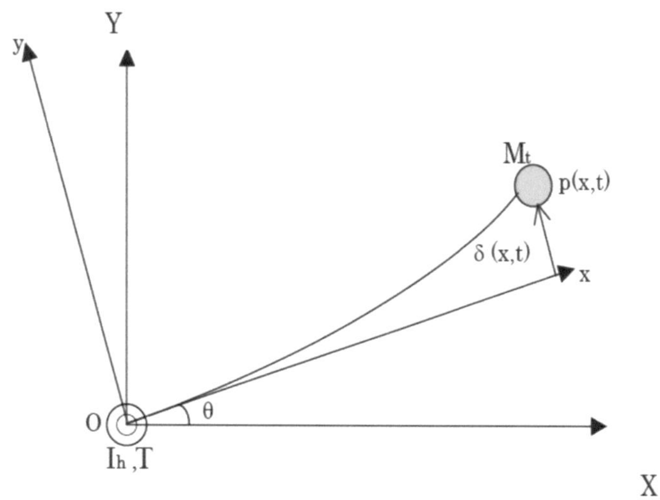

Fig. 1. Flexible arm system.

The experimental apparatus shown in Fig. 2 consists of a flexible arm, an actuator and sensors for tip and hub. An aluminum thin plate is used for the arm. The payload at the tip of the arm is detachable. The base end of the arm is clamped onto the hub that is driven by a DC permanent magnet servomotor, which is controlled by a PWM servo amplifier through a reduction gearbox. A tachometer is used to measure the rotary velocity of the joint. The flexible arm can freely bend in the horizontal plane but not in the vertical plane nor in torsion in order to eliminate the gravitational effects. Strain gauges are used to measure the strain at the base of the arm, which is an indication for the deflection of the tip. A measuring circuit with an amplifier is applied to get the value of the deflection. A/D converter is used to convert the analog signals into digital signals through an interface card. The hub position is measured using a rotary encoder. A digital controller is used through PC computer. The measurement instruments used for measuring the joint angle, joint velocity and the deflection is shown in Fig. 2.

Fig. 2. The experimental setup.

A program written in C language is used for the interfacing and controlling processes. In addition, a digital low pass filter based on Hamming window is used to eliminate the noise from the deflection-measured signal. The physical parameters of the system are shown in Table 1.

Parameter	Values
$l \times b \times h$ (Arm dim.)	0.5 ×0.003 × 0.05 m
ρ (Uniform mass/unit length)	0.403 kg/m
EI (Flexural rigidity)	7.85 Nm^2
M_t (Tip payload)	0.0, 0.25, 0.5 kg
K_1 (Motor amplifier gain)	4.8 V/V
K_2 (Motor torque const.)	0.11 Nm/A
K_3 (Back E.M.F const.)	0.117 V/rpm
G (Gear ratio)	80
L (Inductance)	1.4 Mh
R (Armature resistance)	0.4 .
b (Viscous friction coeff.)	0.003 Nm/(rad/s)
J (Inertia for the motor)	3.48×10^4 kgm^2

Table 1. Physical parameters of the system.

The mathematical equations, which represent the motor dynamics and the reduction gearbox, are expressed as:

$$v_a(t) = K_1 u(t)\,, \tag{1}$$

where $u(t)$ is the control signal generated from the controller and $v_a(t)$ is the armature voltage.
As the speed of a armature-controlled dc servo motor is controlled by the armature voltage $v_a(t)$ which is the output from the amplifier. The differential equation for the armature circuit is

$$v_a(t) = Ri_a(t) + L\frac{di_a(t)}{dt} + v_b(t)\,, \tag{2}$$

where $i_a(t)$ is the armature current and $v_b(t)$ is the back EMF voltage.
For a constant flux, the back EMF voltage $v_b(t)$ is directly proportional to the angular velocity $d\theta/dt$, or

$$v_b(t) = K_3\frac{d\theta(t)}{dt}\,. \tag{3}$$

The equations for torque equilibrium are

$$T_m(t) = J_o\frac{d^2\theta(t)}{dt^2} + b_o\frac{d\theta(t)}{dt}\,, \tag{4}$$

$$T_m(t) = K_2 i_a(t)\,, \tag{5}$$

where $T_m(t)$ is the output torque from the motor and J_o, b_o are the inertia and viscous friction of the combination of the motor, load, and gear referred to the motor shaft respectively.

3. Dynamic modelling

In this section, the mathematical model of the flexible link manipulator is driven in order to be used in the simulation program. First, we construct a simple block diagram to explain the complete system. The block diagram, which represents the system of the single-link flexible robot, is illustrated in Fig. 3. As shown previously in section 2 the mathematical equation of the actuator is driven starting from the output signal of the controller $u(t)$ to the output from the motor. Equation (6) gives the relation between the motor torque and arm torque as follows:

$$T_{arm}(t) = GT_m(t)\,, \tag{6}$$

where $T_m(t)$ is the motor torque and $T_{arm}(t)$ is the arm torque .

Three measurements are available on the experiment, the hub rotational angle $\theta(t)$ is measured using the rotary encoder, the tip deflection $\delta(l, t)$ is calculated from the strain at the base of the arm assuming the first vibration mode shape, and the velocity of the hub

$d\theta/dt$ is measured by the tachometer.

From the analysis of the single-link flexible arm in the experimental work, a continuous clamped-free beam approximates the flexible link. The flexible arm shown in Fig. 1 is rotating in the horizontal plane and the effect of gravity is not taken into consideration. Frame *O-XY* is the fixed base frame and frame *O-xy* is the local frame rotating with the hub. The deflection $\delta(x, t)$ is assumed to be small compared to the length of the arm. Let $p(x, t)$ represents the tangential position of a point on the flexible arm and with respect to the frame *O-XY*. From the assumption of the deflection of the flexible arm, the equation that describes the position is given by:

$$p(x,t) = x\theta(t) + \delta(x,t), \tag{7}$$

where $p(x, t)$ is the position of a point at distance x from the base of the arm at any time and $\delta(x, t)$ is the distance from the local rotating frame *O-xy* to the arm for a point at distance x from the base of the arm at any time.

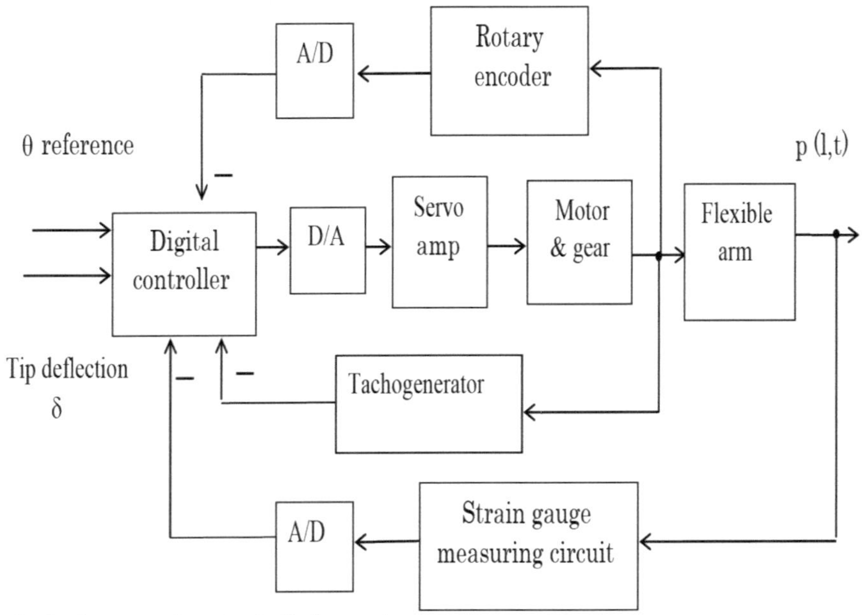

Fig. 3. Block diagram for single-link manipulator.

The flexible arm is modelled as Euler-Bernoulli beam under the assumption of simple beam theory, which is valid when the ratio between beam's length and its height is relatively large (>10) and if the beam does become too wrinkled because of flexure. Also, it is assumed that the beam has a uniform cross-sectional area and constant characteristics. If the flexible beam treated as a simple cantilever as shown in Fig. 4. The deflection at the free end of the beam can be estimated as

$$\delta(l,t) = \frac{Fl^3}{3EI}, \tag{8}$$

where F is the force at the free end and EI is the uniform flexural rigidity of the beam. Then, the Euler-Bernoulli equation for the link is given as follows :

$$EI\frac{\partial^4 p(x,t)}{\partial x^4}+\rho\frac{\partial^2 p(x,t)}{\partial t^2}=0\cdot \tag{9}$$

Substituting equation (7) into (9), the following equation is obtained:

$$EI\frac{\partial^4 \delta(x,t)}{\partial x^4}+\rho\frac{\partial^2 \delta(x,t)}{\partial t^2}=-\rho x\ddot{\theta}(t) \tag{10}$$

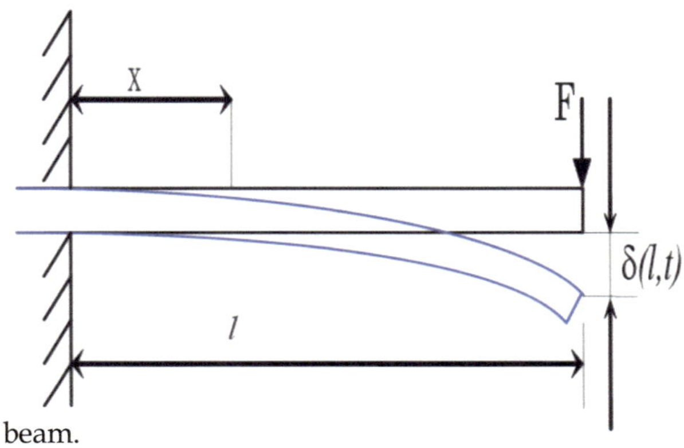

Fig. 4. Cantilever beam.

As the flexible arm is clamped at its base, both the deflection and the slope of the deflection curve must be zero at the clamped end (Meirovitch, 1967). Those conditions are represented by equations (11) and (12) respectively. Equation (13) presents the bending moment at the free end that is equal zero. Finally, if we make force balance at the tip of the flexible manipulator we can get equation (14). The boundary conditions can be summarized as follow:

$$\delta(x,t)\big|_{x=0}=0, \tag{11}$$

$$\frac{\partial\delta(x,t)}{\partial x}\bigg|_{x=0}=0, \tag{12}$$

$$EI\frac{\partial^2\delta(x,t)}{\partial x^2}\bigg|_{x=l}=0, \tag{13}$$

$$EI\frac{\partial^3\delta(x,t)}{\partial x^3}\bigg|_{x=l}=M_t\left[x\ddot{\theta}(t)+\frac{\partial^2\delta(x,t)}{\partial t^2}\right]_{x=l}, \tag{14}$$

where l is the length of the arm. Using the Lagrangian equations

$$T=\frac{d}{dt}\left(\frac{\partial L}{\partial\theta}\right)-\frac{\partial L}{\partial\theta}, \tag{15}$$

$$0=\frac{d}{dt}\left(\frac{\partial L}{\partial\delta}\right)-\frac{\partial L}{\partial\delta}\cdot \tag{16}$$

With the assumptions that the mass is only concentrated at the tip of the arm (i.e. neglect the weight of the link) and the deflection is small, the dynamic equations which describes the system can be written as

$$\left(I_h + \frac{1}{3}\rho l^3 + M_t l^2\right)\ddot{\theta}(t) + M_t l \ddot{\delta}(l,t) = T_{arm}(t), \tag{17}$$

$$M_t l \ddot{\theta}(t) + M_t l \ddot{\delta}(l,t) + \frac{3EI}{l^3}\delta(l,t) = 0. \tag{18}$$

4. Controller Design

The control of the single flexible link SFL robot has created a great deal of interest in the control theory field. It can be argued that it has become a benchmark problem for comparing the performance of newly developed control strategies. The reason for this is the inherent difficulty in controlling such a system. This is caused by several factors. First, this is mathematically an infinite-dimensional problem. This will make it very difficult to implement some control strategies, Controllers generally need to be finite-order in order to be implementable (with exception of time delays). Also, the internal damping in the beam is extremely difficult to model accurately, resulting in a plant with parametric uncertainties. Finally, if the tip deflection is chosen as the output, then the transfer function of the plant is nonminimum phase (i.e., it contains unstable zeros). This will make it very difficult to implement some control strategies which are commonly used for conventional rigid link robots. Not only that but the inherent non-minimum phase behavior of the flexible manipulator system makes it very difficult to achieve high level performance and robustness simultaneously. For the methods of collocating the sensors and actuators at the joint of a flexible manipulator, for example, the joint PD control, which is considered the most widely used controller for industrial robot applications, only a certain degree of robustness of the system can be guaranteed. As studied before (Spector & Flashner , 1990) and (Luo , 1993) the robustness of collocated controllers comes directly from the energy dissipating configuration of the resulting system. However, the performance of the flexible system with only a collocated controller, for example, the joint PD controller is often not very satisfactory because the elastic modes of the flexible beam are seriously excited and not effectively suppressed. Due to this reasons, numerous kinds of control techniques have been investigated as shown in section 1 to improve the performance of flexible systems. In general, the desired tip regulation performance of a flexible manipulator can be described as:

1- The joint motion converges to the final position fast.
2- The elastic vibrations are effectively suppressed.

Obviously there is a trade-off between the two requirements so the successive control try to achieve both of them together.

4.1 Controller analysis

The input for the flexible link system is a step input with a reference angle θ_{ref} with no deflection at the tip. Thus, the equivalent effect at the tip position, which is denoted herein as the effective input is ($l\theta_{ref}$ + zero deflection at the tip). The output of the system is the tip position, which is defined by rigid arm motion plus tip deflection. The error in the tip position can be defined as (effective input - output). Therefore, the following relation gives the error in the tip position of the flexible arm:

$$\begin{aligned} e(t) &= l\,[\theta_{ref} - \theta(t)] - \delta(l,t), \\ &= e_j(t) - \delta(l,t), \end{aligned} \tag{19}$$

where $e(t)$ is the total error in the tip. It is indicated from equation (19) that the error includes two components. The first component $e_j(t)$ is the tangential position error due to the joint motion and it equals to $l(\theta_{ref}\text{-}\theta(t))$ which is identical with the rigid arm error. The second one is much more important and is due to the flexibility of the arm and equals $\delta(l, t)$. These two error components are coupled to each other. On the other hand, a single controller is used to develop a single control signal $u(t)$ which drives a single actuator in the arm system. The drive torque $T(t)$ is proportional to the control signal $u(t)$ as expressed by

$$T\,(t) = K_1 K_2 G u(t), \tag{20}$$

where K_1, K_2 and G are presented in Table 1.
Thus, the current flexible arm control problem described by the two error components coupled to each other and having only one control command to actuate the joint actuator, is rather complicated and difficult to be solved by traditional controller strategies.
One of the best ways to overcome the problem of inaccuracy in the tip position of the flexible manipulator is to add a vibration feedback from the tip to the controller which control the base joint. Many researchers had used this algorithm like (Lee et al., 1988). They proposed PDS (proportional-derivative-strain) control, which is composed of a conventional PD control and feedback of strain detected at the root of link. Also (Matsuno & Hayashi, 2000), as they proposed the PDS control for a cooperative two one-link flexible arm. Other trails is done by (Ge et al., 1997); (Ge et al., 1998) to enhance the control of the flexible manipulator by using non-linear feedback controller based on the feedback of the vibration signal to the controller.
The Modified PID controller replaces the classical integral term of a PID control with a vibration feedback term to affect the effect flexible modes of the beam in the generated control signal. The MPID controller is formed as follows (Mansour et al., 2008):

$$u(t) = u_{bias} + K_{jp}\; e_j(t) + K_{jd}\; \dot{e}_j(t) + K_{vc}\; g(t)\; \mathrm{sgn}(\,\dot{e}_j(t))\; \int_0^t \left|\dot{e}_j(\tau)\right|\, g(\tau) d\tau, \tag{21}$$

where u_{bias} is the bias or null value.

K_{jp}, K_{jd} are the joint proportional and joint derivative gains respectively.

K_{vc} is the vibration control gain.

$g(t)$ is the vibration variable used in the controller.

The value of u_{bias} is the compensated control signal needed for the motor to overcome friction losses without causing any motion to the arm. The sign of this value depends on the

direction of motion, which means that if the arm motion is in the clockwise direction then the value of ubias is equal to (u_{hold}), and if the motion of the arm is reversed then the value of u_{bias} will be ($-u_{hold}$). The value of u_{bias} is evaluated as given in terms of the torque from the motor or voltage to the servo amplifier (Mansour et al., 2008).
The signum function (sgn) is defined as

$$\operatorname{sgn} \dot{e}(t) = \begin{cases} -1 & \dot{e}(t) < 0 \\ 0 & \dot{e}(t) = 0 \\ 1 & \dot{e}(t) > 0 \end{cases} \tag{22}$$

The value of $e_j(t)$ is defined in equation (19). The vibration variable $g(t)$ such as $\frac{\partial^2 \delta(0,t)}{\partial x^2}, \frac{\partial \delta(l,t)}{\partial t}, \ldots$etc.
One of the contributions of this research is the utilizing of rate of deflection signal as an indication of the vibration of the tip to enhance the response of the flexible manipulator. In this research the rate of change of the deflection at the tip $\dot{\delta}(l,t)$ is chosen as the vibration variable g(t), while (Ge et al., 1998) used $\delta''(0,t)$ for g(t). The use of $\dot{\delta}(l,t)$ has an advantage over the use of $\delta''(0,t)$ when the flexible-links have quasi-static strains due to gravity or initial strains due to material problems, because $\dot{\delta}(l,t)$ is not affected by such static deformations. When $\delta''(0,t)$ is used for g(t), the static components in $\delta''(0,t)$ must be removed by some means. (Ge et al., 1998) did not consider the static deformations; however, such static deformations are generally seen in a real manipulator system.
The mathematical equation for the MPID when using the rate of deflection as the vibration feedback signal is given by:

$$u(t) = u_{bias} + K_{jp}\ e_j(t) + K_{jd}\ \dot{e}_j(t) + K_{vc}\ \dot{\delta}(t)\ \operatorname{sgn}(\ \dot{e}_j(t)) \int_0^t \left|\dot{e}_j(\tau)\right|\ \dot{\delta}(\tau) d\tau \tag{23}$$

First, we wish to show the steps for enhancement the classic PD control to reach the MPID. The most common way to enhance the response is to include the vibration of the flexible manipulator in the generated control signal as in (Matsuno & Hayashi, 2000). A joint PD controller, which is given by:

$$u(t) = K_{jp}\ e_j(t) + K_{jd}\ \dot{e}_j(t), \tag{24}$$

is compared with an enhancement for the controller by feeding back the deflection signal. The mathematical equation, which represents the controller, in this case is give by:

$$u(t) = K_{jp}\ e_j(t) + K_{jd}\ \dot{e}_j(t) + K_d \delta(l,t), \tag{25}$$

where K_d is the deflection gain.

The response of the flexible manipulator using those two controllers is shown in Fig. 5. As shown form the response that feeding the deflection had improved the defection of the response but on the same time, it creates an overshoot on the joint response.

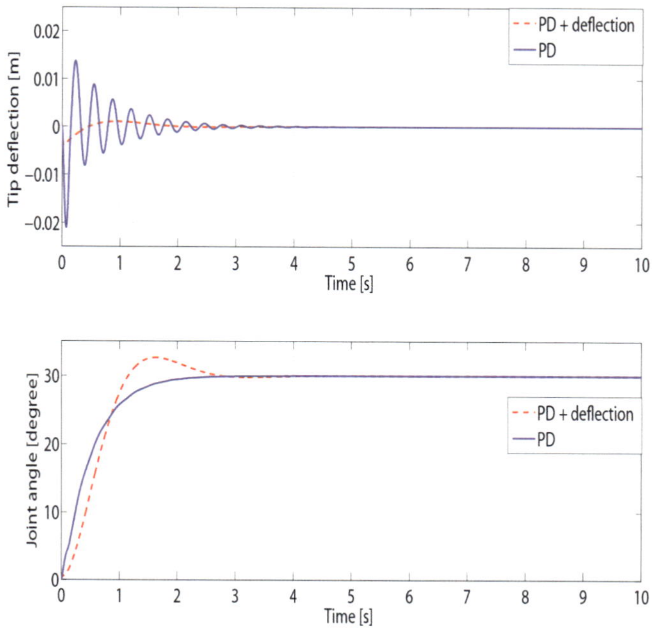

Fig. 5. Step response for the deflection and joint with reference angle 30^0 with 0.5 kg payload using PD and PD plus deflection.

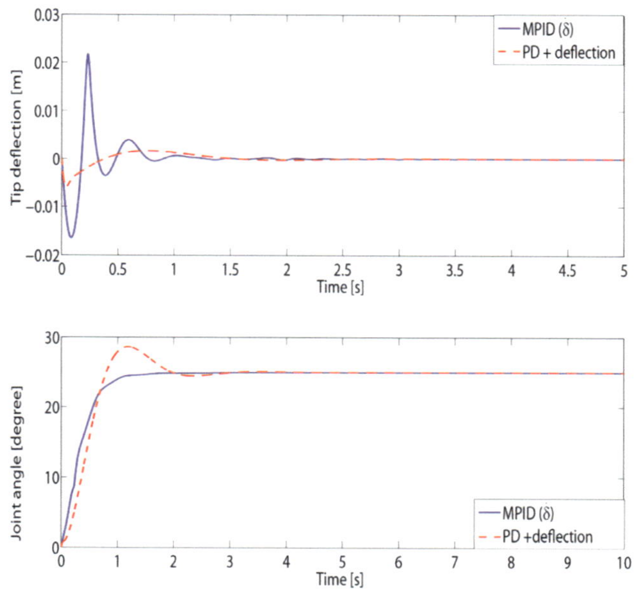

Fig. 6. Step response for the deflection and joint with reference angle 25^{0}with 0.5 kg payload.

The next step that we modify the effect of the vibration feedback and use it is an integral form as given by equation (23). The response of the flexible arm corresponding to 25^0 step input is presented in Fig. 6. Two figures are drawn one for the base joint of the flexible arm and the other for the tip deflection. Two types of controller are tested to control the flexible arm through the joint. First controller is a simple PD controller for the joint plus a proportional gain for the deflection of the tip and the second one is the MPID control.
The response for the first controller is represented with the dotted line while the response using the MPID is plotted using continuous line. The MPID control given by equation (23) uses the rate of deflection $\dot{\delta}(l,t)$ as a vibration feed back signal.

To compare between the behaviour of the classic PD controller and the proposed MPID controller Fig. 7 is drawn. In this figure both the PD controller and the MPID is used to control the joint of the flexible arm. The continuous lines represent the tip deflection and the joint angle when using the MPID controller while the dotted lines represent them when using PD control.

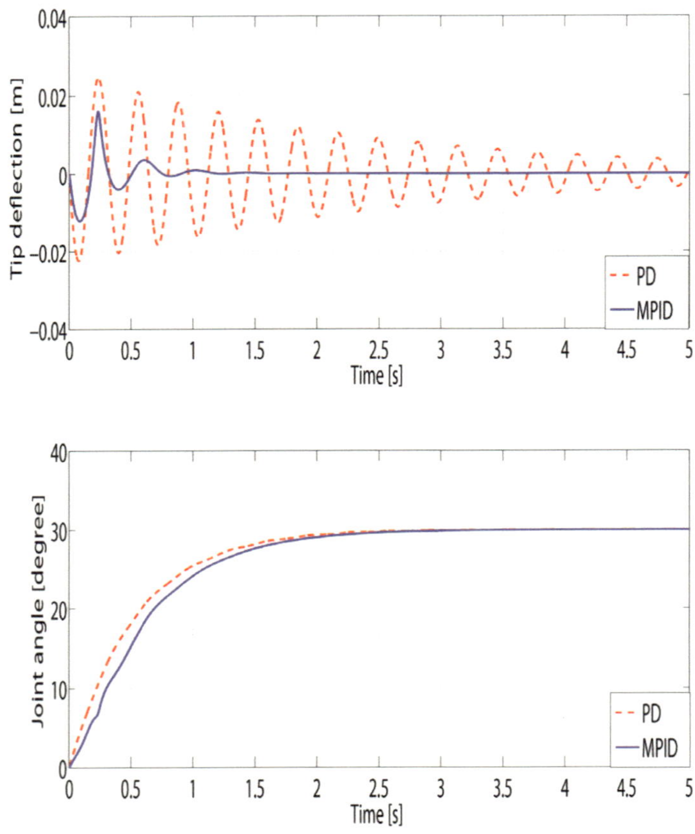

Fig. 7. Step response for the deflection and joint with reference angle $30^0$with 0.5 kg payload.

As it noticeable from Fig. 7 that the PD control can achieve a fast and accurate response for

the joint but on the same time it increased the oscillations on the tip while the MPID can achieve a damping for the tip deflection on approximate time for reaching the joint angle without causing overshoot for the response of the joint.
A simulation analysis for the single-link flexible manipulator system is presented using MATLAB software package. The mathematical equations used in building the simulation have been discussed in section 3. The aim of the simulation is to highlight the effect of adding the modified term, which contains the vibration feedback variable to the normal servo control for the joint. A simple joint PD controller and MPID controller are examined in the simulation. The MPID controller is compared with the traditional joint PD control to see the merits of using the rate of change of the tip deflection as the vibration variable in the feedback signal. The joint PD control is given by equation (24) while the MPID is designed using the rate of deflection at the tip of the flexible manipulator as the vibration variable g(t) as shown in equation (23).

4.2 Stability analysis

After the MPID control is analysed on subsection 4.1. The stability of the MPID controller around a stationary point $(\theta,\dot{\theta}) = (\theta_{ref},0)$ is analysed in this section. Note that $\dot{e}_j(t) = -\dot{\theta}(t)$ because θ_{ref} is constant.
Fundamental contribution to the stability theory for non-linear systems were made by the Russian mathematician Lyapunov where he investigated the non-linear differential equation

$$\frac{dx}{dt} = f(x), \quad f(0) = 0. \tag{26}$$

Since *f(x)* the equation has the solution *x(t)=0*. To guarantee that a solution exists and is unique, it is necessary to make some assumptions about f(x). A sufficient assumption is that *f(x)* is Lipschitz, that is

$$\|f(x) - f(y)\| \le L\,\|x - y\|, \quad L > 0, \tag{27}$$

in the neighborhood of the origin. Before we proceed in the stability prove two important definitions needs to be highlighted.

1- The solution $x(t) = 0$ to the differential equation (26) is called stable for given $\varepsilon > 0$ there exists a number $\zeta(\varepsilon) > 0$ such that all solutions with initial conditions

$$\|x(0)\| < \zeta, \tag{28}$$

have the property

$$\|x(t)\| < \varepsilon, \quad 0 \le t < \infty, \tag{29}$$

the solution is unstable if it is not stable. The solution is asymptotically stable if it is stable and ζ can be found such that all solutions with $\|x(0)\| < \zeta$ have the property that

$$\|x(0)\| \to 0 \quad \text{as} \quad t \to \infty \ . \tag{30}$$

2- A continuously differentiable function $V : R^n \to R$ is called positive definite in a region $U \subset R^n$ contains the origin if
   1- $V(0) = 0$
   2- $V(x) > 0, \; x \in U$ and $x \neq 0$,

and the function is called positive semi-definite if condition 2 is replaced by $V(x) \geq 0$.
As stated by the Lyapunov stability theorem, If there exists a function $V : R^n \to R$ that is positive definite such that its derivative along the solution of equation (26),

$$\frac{dV}{dt} = \frac{\partial V^T}{\partial t}\frac{dx}{dt} = \frac{\partial V^T}{\partial t} f(x) = -W(x) , \tag{31}$$

is negative semi-definite, then the solution $x(t) = 0$ to equation (26) is stable. If $\frac{dV}{dt}$ is negative definite, then the solution is also asymptotically stable. The function V is called a Lyapunov function for the system.

To check the stability of the MPID controller we start by forming the Lyapunov function $V(t)$. $V(t)$ is formed using the following relation

$$V(t) = K_E + P_E + \frac{1}{2} K_1 K_2 K_{jp} G \; e_j^2(t) + \frac{1}{2} K_1 K_2 K_{vc} G \left[\int_0^t \left| \dot{e}_j(\tau) \right| \dot{\delta}(\tau) d\tau \right]^2 , \tag{32}$$

where K_E is the total kinetic energy of the system and P_E the total potential energy of the system.
From the analysis of the flexible link manipulator system, the total Kinetic energy of the system can be calculated by

$$K_E = K_{Em} + K_{Eb} + K_{Ep} , \tag{33}$$

Where K_{Em}, K_{Eb}, K_{Ep} are the kinetic energy of the motor, beam and payload respectively. And

$$K_{Em} = \frac{1}{2} I_h \dot{\theta}^2(t) , \tag{34}$$

$$K_{Eb} = \frac{1}{2} \rho \int_0^t \dot{p}^2(x,t) dx , \tag{35}$$

$$K_{Ep} = \frac{1}{2} M_t \dot{p}^2(l,t) \ . \tag{36}$$

By substituting equations (34), (35), (36) into (33), the total kinetic energy of the system can be rewritten as

$$K_E = \frac{1}{2} I_h \dot{\theta}^2(t) + \frac{1}{2}\rho \int_0^l \dot{p}^2(x,t)dx + \frac{1}{2} M_t \dot{p}^2(l,t) \cdot \tag{37}$$

Consider that the beam only vibrates in horizontal direction, any effect of gravity are neglected such that the potential energy of the system is

$$P_E = \frac{1}{2} EI \int_0^l \left[\frac{\partial^2 p(x,t)}{\partial x^2} \right]^2 dx \cdot \tag{38}$$

Recalling equation (7), the total potential energy of the system can be rewritten as

$$P_E = \frac{1}{2} EI \int_0^l \left[\frac{\partial^2 \delta(x,t)}{\partial x^2} \right]^2 dx \cdot \tag{39}$$

Differentiating equation (32) and (37) with respect to time gives

$$\begin{aligned} \dot{V}(t) &= \dot{K}_E + \dot{P}_E + K_1 K_2 K_{jp} G \; e_j(t) \dot{e}_j(t) \\ &+ K_1 K_2 K_{vc} G \dot{e}_j(t) \operatorname{sgn}(\dot{e}_j(t)) \dot{\delta}(\tau) \int_0^t |\dot{e}_j(\tau)| \; \dot{\delta}(x,\tau) d\tau, \end{aligned} \tag{40}$$

$$\dot{K}_E = I_h \dot{\theta}(t)\ddot{\theta}(t) + \rho \int_0^l \dot{p}(x,t)\,\ddot{p}(x,t)dx + M_t \dot{p}(l,t)\,\ddot{p}(l,t) \cdot \tag{41}$$

From equation (7) the middle term of equation (41) can be written as

$$\begin{aligned} &\rho \int_0^l \left[x\dot{\theta}(t) + \dot{\delta}(x,t) \right]\left[x\ddot{\theta}(t) + \ddot{\delta}(x,t) \right] dx \\ &= \rho \int_0^l \left[x^2 \dot{\theta}(t)\ddot{\theta}(t) + x\dot{\theta}(t)\ddot{\delta}(x,t) + x\dot{\delta}(x,t)\ddot{\theta}(t) + \dot{\delta}(x,t)\ddot{\delta}(x,t) \right] dx \\ &= \frac{1}{3}\rho l^3 \dot{\theta}(t)\ddot{\theta}(t) + \rho \int_0^l \left[x\dot{\theta}(t)\ddot{\delta}(x,t) + \dot{\delta}(x,t)\left[x\ddot{\theta}(t) + \ddot{\delta}(x,t) \right] \right] dx, \end{aligned} \tag{42}$$

substituting equations (7) and (42) into equation (41) gives

$$\begin{aligned} \dot{K}_E &= I_h \dot{\theta}(t)\ddot{\theta}(t) + \frac{1}{3}\rho l^3 \dot{\theta}(t)\ddot{\theta}(t) + \rho \int_0^l \dot{\delta}(x,t)\left[x\ddot{\theta}(t) + \ddot{\delta}(x,t) \right] dx \\ &+ \rho \dot{\theta}(t) \int_0^i x\ddot{\delta}(x,t)dx + M_t l \dot{\theta}(t)\left[l\ddot{\theta}(t) + \ddot{\delta}(l,t) \right] + M_t \dot{\delta}(l,t)\left[l\ddot{\theta}(t) + \ddot{\delta}(l,t) \right] \cdot \end{aligned} \tag{43}$$

$$\dot{K}_E = \dot{\theta}(t)\left[I_h \ddot{\theta}(t) + M_t l \left[l\ddot{\theta}(t) + \ddot{\delta}(x,t) \right] + \frac{1}{3}\rho l^3 \ddot{\theta}(t) + \rho \int_0^l x\ddot{\delta}(x,t)\,dx \right]$$

$$+M_t\dot{\delta}(l,t)\left[l\ddot{\theta}(t)+\ddot{\delta}(l,t)\right]+\rho\int_0^l\dot{\delta}(x,t)\left[x\ddot{\theta}(t)+\ddot{\delta}(x,t)\right]dx \cdot \tag{44}$$

Substituting equation (17) into (44), we have

$$\dot{K}_E=T_{arm}(t)\dot{\theta}(t)+M_t\dot{\delta}(l,t)\left[l\ddot{\theta}(t)+\ddot{\delta}(l,t)\right]+\rho\int_0^l\dot{\delta}(x,t)\left[x\ddot{\theta}(t)+\ddot{\delta}(x,t)\right]dx \cdot \tag{45}$$

From equation (10), using integration by parts with the fourth boundary condition is it proved that

$$\dot{K}_E+\dot{P}_E=T_{arm}(t)\dot{\theta}(t)\,. \tag{46}$$

Substituting equation (46),(20) and (21) into equation (40) we get

$$\dot{V}(t)=-K_1K_2K_{jd}G\dot{\theta}^2(t)\ , \tag{47}$$

which is negative semi-definite as long as $K_{jd} \geq 0$ which means that the system is stable.

After showing the controller analysis and the stability analysis, some important points need to be highlighted.

- Include the deflection effect in the controller enable generating a control signal take into consideration the effect of the end effector vibration. The generated control signal have the ability to achieve accurate tip position without neither overshoot for the joint nor vibration at the tip.
- Only three measurements needed to apply this controller, the measurements are the base joint angle $\theta(t)$, base joint velocity $\dot{\theta}(t)$ and the rate of deflection $\dot{\delta}(l,t)$ unlike other types of controller which needs a full states measurements like (Cannon & Schmitz, 1984) and (Siciliano, 1988).
- The stability of the system is shown experimentally and theoretically when using the rate of deflection at the tip $\dot{\delta}(l,t)$ as the vibration signal in the controller. The stability is depend mainly of the joint derivative gain K_{jd} and will not be affected by the vibration feedback.

5. Case study

In this section, we test the proposed MPID control with the rate of change of deflection $\dot{\delta}(l,t)$ as a vibration signal to control a single link moving horizontally. The MPID represented by equation (23) is compared in simulation with a PI control as a classic control. The main function of the integral action in the PI is to make sure that the system output agrees with the set point in steady state. The equation representing the PI controller is

$$u(t) = u_{bias} + K_p\ e(t) + K_i \int_{t_0}^{t_f} e(\tau)\,d\tau \,. \tag{48}$$

where K_p, K_i are the proportional and integral feedback gains, respectively. The PI control is represented by equation (49)

$$u(t) = u_{bias} + K_{jp}\ e_j(t) + K_{ji} \int_{t_0}^{t_f} e_j(\tau)\,d\tau + K_{dp}\ \delta(t) + K_{di} \int_{t_0}^{t_f} \delta(\tau)\,d\tau \,, \tag{49}$$

where K_{jp}, K_{ji} are the joint proportional, joint integral gains while K_{dp}, K_{di} are the and deflection proportional, deflection integral gains respectively. As the tip deflection response is oscillatory, we set the deflection integral gain in equation (49) equal to zero to eliminate this problem. The mathematical equation representing the PI controller in this case is given by:

$$u(t) = u_{bias} + K_{jp}\ e_j(t) + K_{dp}\ \delta(t) + K_{ji} \int_{t_0}^{t_f} e_j(\tau)\,d\tau \,. \tag{50}$$

5.1 Simulation results

A simulation model using MATLAB-Simulink software is used to simulate the performance of the controller with different working conditions. As shown previously in section 3 the mathematical model of the flexible arm is used in the simulation.

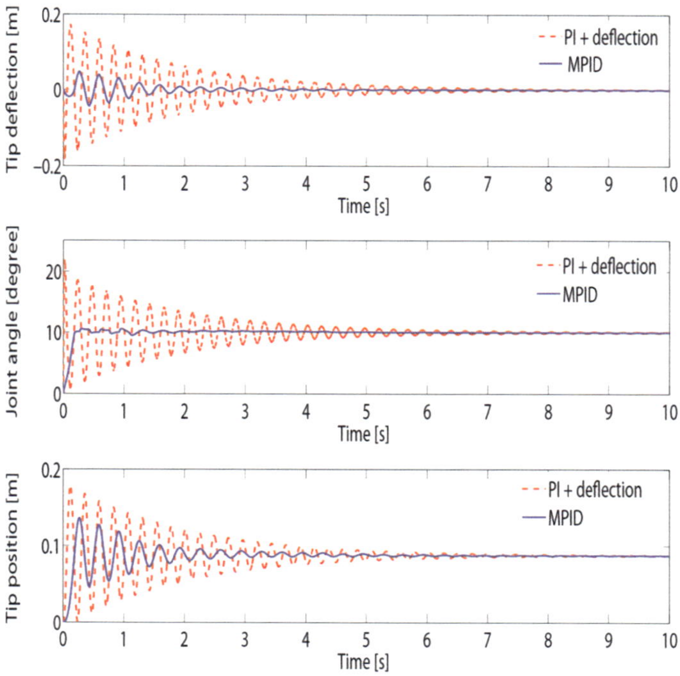

Fig. 8. Step response for the reference angle 10^0 with 0.5 kg payload (simulation).

The system does not model the friction of the motors so in the simulation we put the value of u_{bias} equals zero. As shown in Fig. 8 the dotted represents the response of the system when using the PI control plus the deflection feedback while the continuous line represents the response of the system when using the MPID given by equation (23).
It is clear that the MPID control can successfully suppress the vibration at the end effector of the flexible manipulator while it does not create an over shoot on the joint response.
After changing the tip payload and the input angle of the manipulator, the MPID control success to achieve a noticeable damping for the tip deflection of the flexible manipulator compared with the PI control as shown in Fig. 9. Compared with the MPID control based on rate of deflection at the tip as a vibration variable, the PI control can achieves an accurate joint angle at the steady state but it have an undesirable effect on the vibration of the end effector.

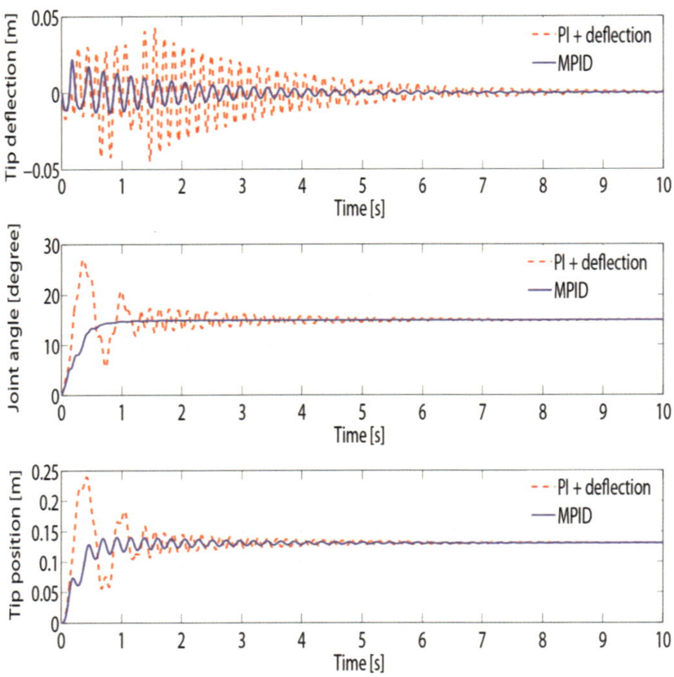

Fig. 9. Step response for the reference angle 15^0 with 0.25 kg payload (simulation).

Another set of simulation results is obtained by comparing the MPID with the PD joint control. Different payloads of 0.25 kg and 0.5 kg are tested in the simulation. A simulation result for the step input of 15^0 with tip payload 0.25 kg is shown in Fig. 10. The joint proportional gain K_{jp} and the joint differential gain K_{jd} for both PD and MPID control are set to be equal. The vibration control gain K_{vc} equals 744340 V.s^2/rad.m^2. The MPID succeeded to suppress the vibration in the tip of the flexible manipulator after 2 seconds as shown in Fig. 10(a). On the same time the joint angle reached its desired value.

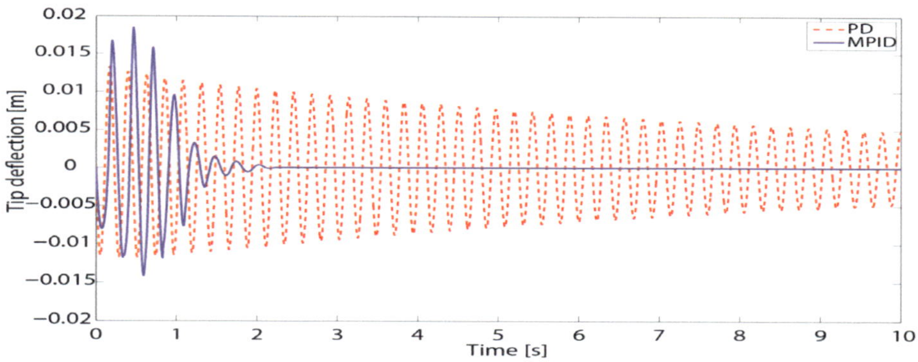

(a) Tip deflection with input 15^0.

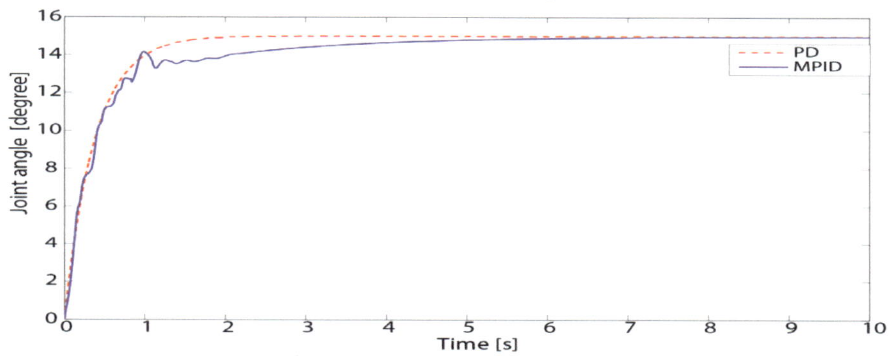

(b) Joint angle with input 15^0.

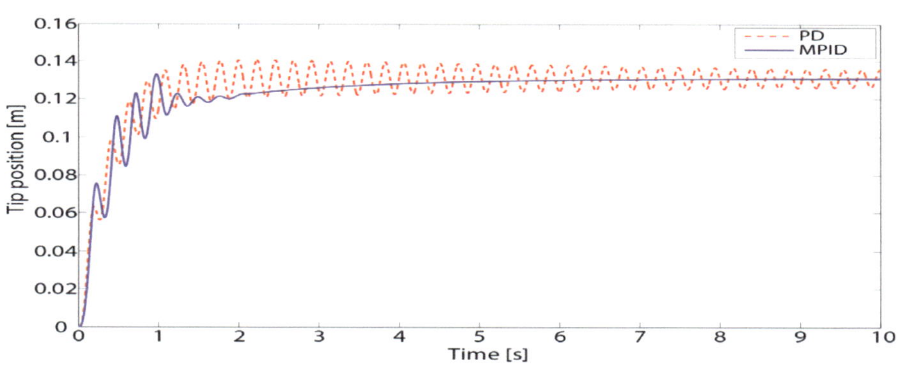

(c) Tip position with input 15^0.

Fig. 10. Step response for the reference angle 15^0 with 0.25 kg payload (simulation).

5.2 Experimental results

Since the performance of the new scheme is confirmed by simulation in the previous subsection, now it will be tested experimentally with PI controller as a classical controller. The experimental setup which had been highlighted before is used to verify the efficient of the MPID control. The MPID control given by equation (23) and PI control given by equation (50) are tested experimentally.

The experimental results of the tip position and the tip deflection with both PI and MPID controllers are shown for different payloads. The value of u_{bias} in equations (50) and (23) is determined experimentally. As a vibration variable g(t) in equation (21), the tip velocity is chosen in the experiments. The gains for the PI in is optimized using Ziegler- Nichols method while for the MPID it is first treated as PD controller to get the optimum gains then by trial and error get the values of K_{vc}. The response when using MPID controller is indicated with the continuous line, while the response with PI is indicated with the dashed line.

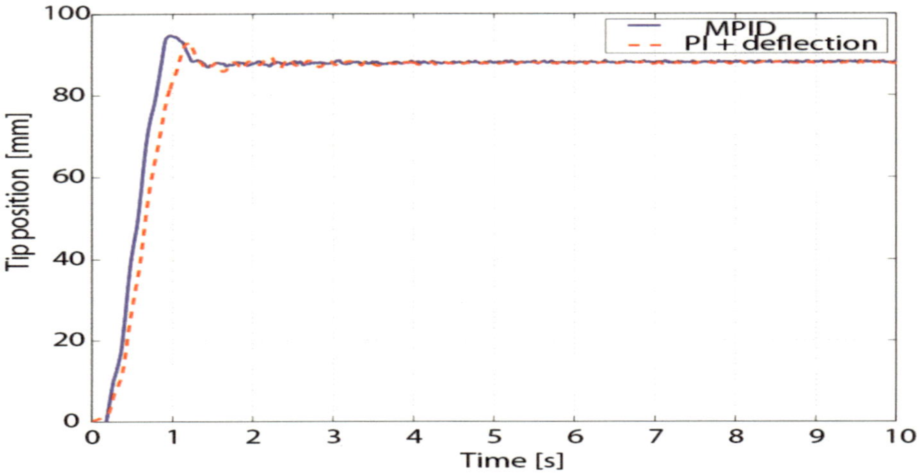

(a) Input 10^0

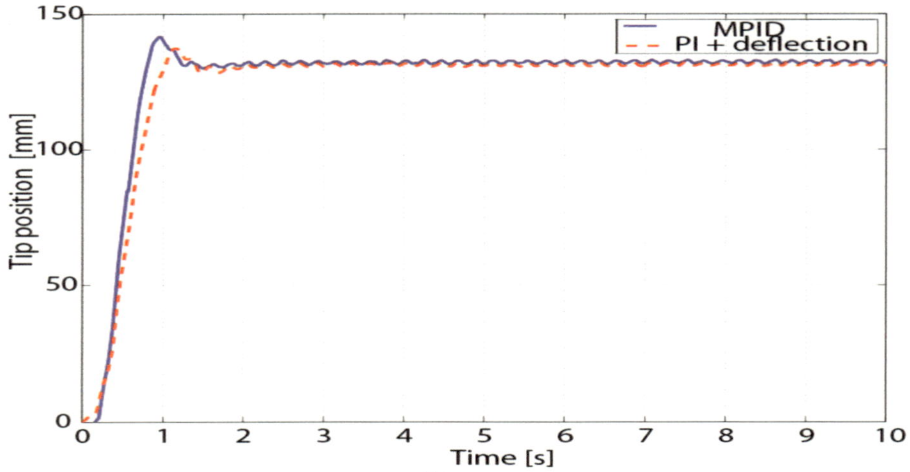

(b) Input 15^0

Fig. 11. Tip position with 0.25 kg payload for 10^0 and 15^0 step input (experimental).

First, a 0.25 kg tip payload is used, and tip position response with 10^0 step input for the joint angle shown in Fig. 11(a) and a step input response with 15^0 is shown in Fig. 11(b). Using the MPID the steady state error e_{ss} has a value of ±0.1 mm, while it reaches a value of ±1.3 mm when using PI controller for the same step input. It is noticed from the response that the MPID has a desirable response especially near the steady state.

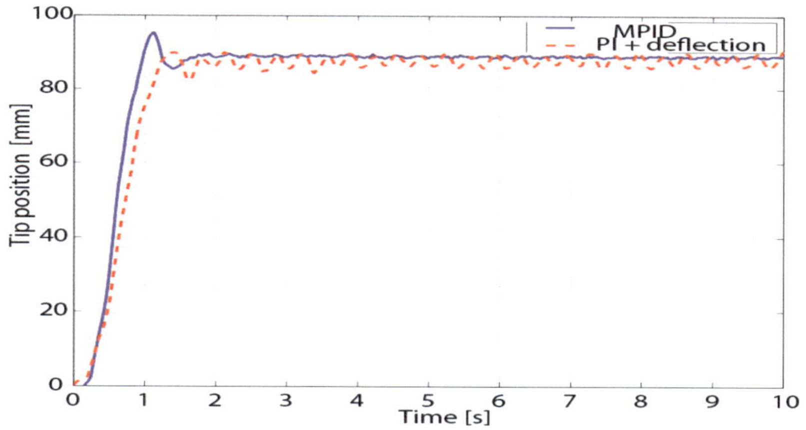

(a) Input 10^0

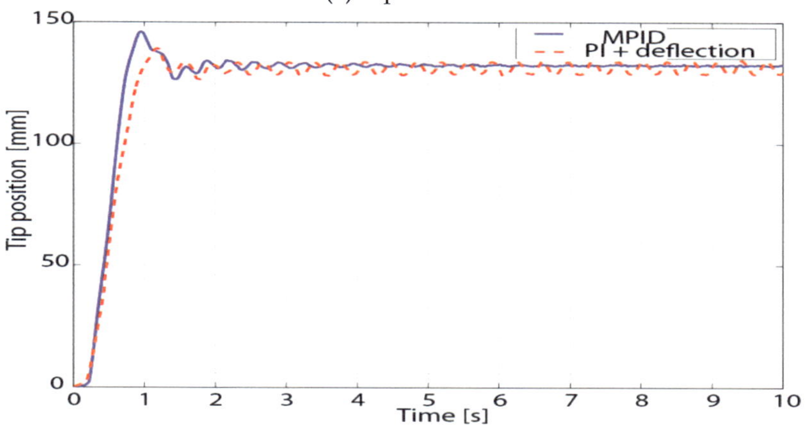

(b) Input 15^0

Fig. 12. Tip position with 0.5 kg payload for 10^0 and 15^0 step input (experimental).

After then the tip payload is increased to 0.5 kg and the tip response is recorded. In Fig. 12(a) and (b) the response of the single-link flexible arm is indicated. The same gains for both of the controllers, PI and MPID are used in the new experiment. In this case also the MPID gives a speedy rise time; t_r for the response of the tip position equals 0.95 s and e_{ss} ±0.2 mm while the PI shows rise time, t_r 1.23 s and steady state error ±2.0 mm.
To focus on the effect of the MPID controller on the response, the tip deflection with a 0.25 kg tip payload is shown in Fig. 13 (a) and also for the 0.5 kg tip payload appeared on Fig. 13 (b). It is well noticed that MPID controller could succeed to make remarkable vibration suppression for tip defection of the single-link flexible arm.

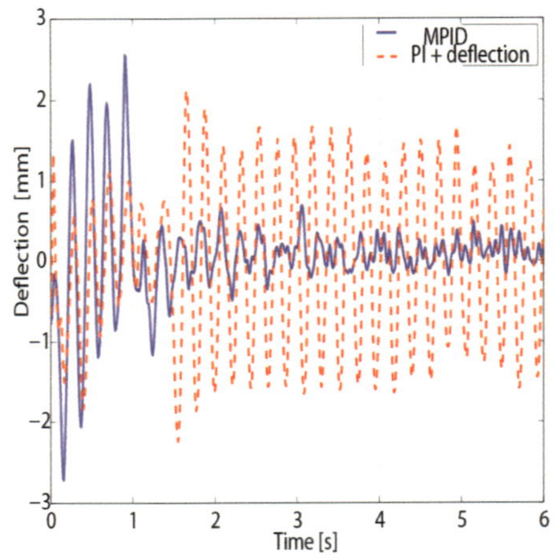

(a) payload 0.25 kg

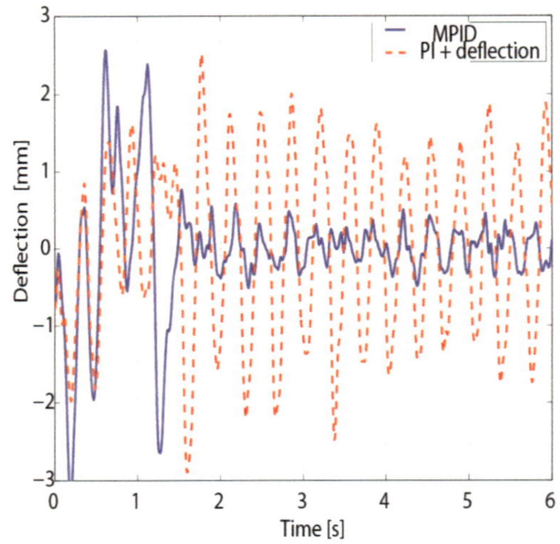

(b) payload 0.5 kg

Fig. 13. Tip deflection with different payload (experimental).

7. Conclusion

In this chapter, a Modified Proportional-Integral-Derivative (MPID) controller is utilized to solve the problem of achieving an accurate tip position of a flexible manipulator. The aim of the control of the flexible manipulator is to achieve the final position of the manipulator with the minimal vibration. The controller consists of a PD for the joint and an integral term including the vibration of the tip. As a result introducing vibration feedback, compared with traditional joint PD control allows explicit evaluation of vibration and, thus, provides explicit control effort in vibration suppression. The rate of change in the deflection of the tip is used as a vibration feedback signal in the MPID. In addition, the way the rate of deflection is used as a vibration feedback is different from other similar work. The advantage of the proposed MPID controller over the strain-based MPID controller is that the proposed controller does not receive a bad influence from the quasi-static strain or initial strain of the flexible links. The performance of the MPID controller is evaluated by simulation study. An experimental validation of the tip position control of a single-link flexible arm is carried out using the MPID. The implementation of the MPID showed that it is an easy controller to use. The MPID is compared with other standard controller in theoretical and experimental. The comparison shows good behave for the MPID. The settling time of the flexible manipulator using the MPID control is noticeably shorten compared with other control method. This will enable a fast task execution. The stability of the proposed controller is tested and it was proven that this controller can achieve stable operation for the flexible manipulator. The future work is aim to extend the use of MPID with the rate of deflection as vibration signal in the multiple link flexible manipulator

8. References

Cannon, R. H. & Schmitz, J. E. (1984). Initial experiments on the end-point control of a flexible one-link robot, *Int. J. of Robotics Research*, Vol. 3, No. 3, pp. 62–75, 0278-3649.

Etxebarria, A.; Sanz, A. & Lizarraga, I. (2005). Control of a Lightweight Flexible Robotic Arm Using Sliding Modes, *Int. J. of Advanced Robotic Systems,* Vol. 2, No. 2 , pp. 103- 110, 1729-8806

Ge, S. S.; Lee, T. H. & Zhu, G. (1997). A nonlinear feedback controller for a single link flexible manipulator based on a finite element model, *J. of Robotic Systems*, Vol 14, No. 3, pp. 165–178, 0741-2223.

Ge, S. S.; Lee, T. H. & Zhu, G. (1998). Asymptotically stable end-point regulation of a flexible SCARA/Cartesian robot, *IEEE/ASME Transactions on Mechatronics*, Vol. 3, No. 2, pp. 138–144, 1083-4435.

Kariz, Z. & Heppler, G. R. (2000). A Controller for an Impacted Single Flexible Link, *Journal of Vibration and Control*, Vol. 6, No. 3, pp 407-428, 1077-5463.

Lee, H. G.; Arimoto, S. & Miyazaki, F. (1988). Liapunov stability analysis for PDS control of flexible multi-link manipulators, *Proc. of IEEE Conf. of Decision and Control*, Austin, pp. 75–80.

Luo, Z. (1993). Direct strain feedback control of flexible robot arms: New theoretical and experimental results, *IEEE Trans. on Automatic Control*, Vol. 38, No. 11, pp. 1610–1622, 0018-9286.

Mansour, T.; Konno, A. & Uchiyama M. (2008). Modified PID Control of a Single- Link flexible Robot, *Advanced Robotics*, Vol. 22, No. 4, pp. 433-449, 0169-1864.

Matsuno, M. & Hayashi, A. (2000). PDS cooperative control of two one-link flexible arms, *Proc. of IEEE Int. Conf. on Robotics and Automation*, San Francisco, pp. 1490–1495.

Meirovitch, L. (1967). *Analytical Methods in Vibrations*, Macmillan Publishing Co., 0-02-3801409, NewYork

Menq, C. & Xia, J. Z. (1993). Experiments on the Tracking Control of A Flexible One-Link Manipulator, *Trans. of ASME, J. of Dynamic Systems, Measurement and Control*, Vol. 115, No. 2, pp. 306-308, 0022-0434.

Rai, S. & Asada, H. (1995). Integrated Structure/Control Design of High Speed Flexible Robots Based on Time Optimal Control, *Trans. of ASME, J. of Dynamic Systems, Measurement and Control*, Vol. 117, No. 4, pp. 503–512, 0022-0434.

Siciliano, B. & Book, W. J. (1988). A singular perturbation approach to control of lightweight flexible manipulators, *Int. J. of Robotics Research*, Vol. 7, No. 4 , pp. 79–90, 0278-3649.

Spector, V. A. & Flashner, H. (1990). Modelling and design implications of noncollocated control in flexible systems. *Trans. of ASME, J. of Dynamic Systems, Measurement and Control*, Vol. 112, No. 2, pp. 186–193, 0022-0434.

Tawfeic, S. R. ; Baz A.; Abo-Ismail A. A. & Azim, O. A. (1997). Vibration Control of a Flexible Arm with Active Constrained Layer Damping, *Journal of Low Frequency Noise, Vibration And Active Control*, Vol. 16, No. 4, pp. 271-287, 1461-3484.

Zhu, W. D. & Mote, C. D. (1997). Dynamic Modelling and Optimal Control of Rotating Euler-Bernoulli Beams, *Trans. of ASME, J. of Dynamic Systems, Measurement and Control*, Vol. 119, No. 4, pp. 802–808, 0022-0434.

Control of Robotic Systems with Flexible Components using Hermite Polynomial-Based Neural Networks

Gerasimos G. Rigatos
Unit of Industrial Automation
Industrial Systems Institute
26504, Rion Patras
Greece

1. Introduction

Flexible-link robots comprise an important class of systems that include lightweight arms for assembly, civil infrastructure, bridge/vehicle systems, military applications and large-scale space structures. Modelling and vibration control of flexible systems have received a great deal of attention in recent years (Kanoh, Tzafestas, et. al., 1986), (Rigatos, 2009), (Rigatos, 2006), (Aoustin, Fliess, et al.,1997). Conventional approaches to design a control system for a flexible-link robot often involve the development of a mathematical model describing the robot dynamics, and the application of analytical techniques to this model to derive an appropriate control law (Cetinkunt & Yu, 1991), (De Luca & Siciliano, 1993), (Arteaga & Siciliano, 2000). Usually, such a mathematical model consists of nonlinear partial differential equations, most of which are obtained using some approximation or simplification (Kanoh, Tzafestas, et al., 1986), (Rigatos, 2009). The inverse dynamics model-based control for flexible link robots is based on modal analysis, i.e. on the assumption that the deformation of the flexible link can be written as a finite series expansion containing the elementary vibration modes (Wang & Gao, 2004). However, this inverse-dynamics model-based control may result into unsatisfactory performance when an accurate model is unavailable, due to parameters uncertainty or truncation of high order vibration modes (Lewis, Jagannathan & Yesildirek, 1999).
In parallel to model-based control for flexible-link robots, model-free control methods have been studied (Rigatos, 2009), (Benosman & LeVey 2004). A number of research papers employ model-free approaches for the control of flexible-link robots based on fuzzy logic and neural networks. In (Tian & Collins, 2005) control of a flexible manipulator with the use of a neuro-fuzzy method is described, where the weighting factor of the fuzzy logic controller is adjusted by a dynamic recurrent identification network. The controller works without any prior knowledge about the manipulator's dynamics. Control of the end-effector's position of a flexible-link manipulator with the use of neural and fuzzy controllers has been presented in (Wai & Lee, 2004), (Subudhi & Morris, 2009), (Talebi, Khorasani, et. al, 1998), (Lin & Lewis, 2002), (Guterrez, Lewis & Lowe, 1998). In (Wai & Lee, 2004) an

intelligent optimal control for a nonlinear flexible robot arm driven by a permanent-magnet synchronous servo motor has been designed using a fuzzy neural network control approach. This consists of an optimal controller which minimizes a quadratic performance index and a fuzzy neural-network controller that learns the uncertain dynamics of the flexible manipulator. In (Talebi, Khorasani, et. al, 1998) a fuzzy controller has been developed for a three-link robot with two rigid links and one flexible fore-arm. This controller design is based on fuzzy Lyapunov synthesis where a Lyapunov candidate function has been chosen to derive the fuzzy rules. In (Subudhi & Morris, 2003) a neuro-fuzzy scheme has been proposed for position control of the end effector of a single-link flexible robot manipulator. The scale factors of the neuro-fuzzy controller are adapted on-line using a neural network which is trained with an improved back-propagation algorithm. In (Caswara & Ubenhauen, 2002) two different neuro-fuzzy feed-forward controllers have been proposed to compensate for the nonlinearities of a flexible manipulator. In (Renno, 2007) the dynamics of a flexible link has been modeled using modal analysis and then an inverse dynamics fuzzy controller has been employed to obtain tracking and deflection control. In (Shi & Trabia, 2006) a fuzzy logic controller has been applied to a flexible-link manipulator. In this distributed fuzzy logic controller the two velocity variables which have higher importance have been grouped together as the inputs to a velocity fuzzy controller while the two displacement variables which have lower importance degrees have been used as inputs to a displacement fuzzy logic controller. In (Hui, Fuchun & Zenghi, 2002) adaptive control for a flexible-link manipulator has been achieved using a neuro-fuzzy time-delay controller. In (Nguyen & Morris, 2007) a genetic algorithm has been used to improve the performance of a fuzzy controller designed to compensate for the links' flexibility and the joints' flexibility of a robotic manipulator.

In this paper, a neural controller using orthogonal wavelet basis functions is first proposed for the control of the flexible-link robot. The neural controller operates in parallel to a PD controller the gains of which are calculated assuming rigid link dynamics. Neural networks with wavelet basis functions, also known as 'wavelet networks', were first introduced in (Zhang & Benveniste, 1993) aiming at giving to feed-forward neural networks multi-resolution analysis features and at providing neural models with good approximation features while using a small number of tunable parameters. Wavelet neural networks can be classified into orthogonal and non-orthogonal. In orthogonal wavelet networks an orthonormal basis is generated, using the wavelet function. However, in order to create the orthonormal basis the wavelet function has to satisfy restrictions. The training of the orthonormal wavelet network is fast and its expansion is easy. On the other hand, the non-orthogonal wavelet network uses the so-called wavelet frame. The family of the wavelet functions that constitute a frame are such that the energy of the resulting wavelet coefficients lies within a certain bounded range of the energy of the original signal (Addison, 2002). Controllers based on Haar orthogonal wavelets have been used in vibration control problems (Karimi & Lohmann, 2006).

Next, a neural network with Gauss-Hermite polynomial basis functions is considered for the control of flexible-link manipulators. This neural model employs Gauss-Hermite basis functions which are localized both in space and frequency, as which, as wavelet basis functions, allow for better approximation of the multi-frequency characteristics of vibrating structures (Cannon & Slotine, 1995), (Krzyzak & Sasiadek, 1991), (Lin, 2006), (Sureshbabu & Farell, 1999). Gauss-Hermite basis functions have also some interesting properties

(Refregier, 2003), (Rigatos & Tzafestas, 2006): (i) they remain almost unchanged by the Fourier transform, which means that the weights of the associated neural network demonstrate the energy which is distributed to the various eigenmodes of the vibrating structure. This in turn enables to define thresholds for truncating the basis functions expansion and to design a neural controller with a small number of adaptable parameters, (ii) unlike wavelet basis functions the Gauss-Hermite basis functions have a clear physical meaning since they represent the solutions of differential equations describing stochastic oscillators and each neuron can be regarded as the frequency filter of the respective vibration eigenfrequency.

The structure of the chapter is as follows: In Section 2 the dynamic model of flexible-link robots is analyzed. In Section 3 a neural control scheme for flexible link robots is introduced. In Section 4 wavelet basis functions are proposed to implement the neural controller for the flexible-link manipulator. In Section 5 Hermite-polynomial basis functions are used to implement the neural controller which stabilizes the flexible-link robot dynamics. In Section 6 simulation experiments are presented. Finally in Section 7 concluding remarks are stated.

2. Dynamic model of flexible-link robots

A common approach in modelling of flexible-link robots is based on the Euler-Bernoulli model (Talebi, Khorasani, et. al, 1998) , (Wang & Gao, 2004). This model consists of nonlinear partial differential equations, which are obtained using some approximation or simplification. In case of a single-link flexible manipulator the basic variables of this model are $w(x,t)$ which is the deformation of the flexible link, and $\theta(t)$ which is the joint's angle.

$$E \cdot I \cdot w''''(x,t) + \rho \ddot{w}(x,t) + \rho x \ddot{\theta}(t) = 0 \tag{1}$$

$$I_t \ddot{\theta}(t) + \rho \int_0^L x \ddot{w}(x,t) dx = T(t) \tag{2}$$

In Eq. (1) and (2}), $w''''(x,t) = \frac{\partial^4 w(x,t)}{\partial x^4}$, $\ddot{w}(x,t) = \frac{\partial^2 w(x,t)}{\partial t^2}$, while I_t is the moment of inertia of a rigid link of length L, ρ denotes the uniform mass density and $E \cdot I$ is the uniform flexural rigidity with units $N \cdot m^2$. To calculate $w(x,t)$, instead of solving analytically the above partial differential equations, modal analysis can be used which assumes that $w(x,t)$ can be approximated by a weighted sum of orthogonal basis functions

$$w(x,t) = \sum_{i=1}^{n_e} \phi_i(x) v_i(t) \tag{3}$$

where index $i = [1,2,...,n_e]$ denotes the normal modes of vibration of the flexible link. Using modal analysis a dynamical model of finite-dimensions is derived for the flexible link robot. Without loss of generality assume a 2-link flexible robot (Fig. 1) and that only the first two vibration modes of each link are significant ($n_e = 2$). Σ_1 is a point on the first link with reference to which the deformation vector is measured. Similarly, Σ_2 is a point on the second link with reference to which the associated deformation vector is measured. In that case the dynamic model of the robot becomes (Wang & Gao, 2004), (Lewis, Jagannathan & Yelsidirek, 1999):

$$\begin{pmatrix} M_{11}(z) & M_{12}(z) \\ M_{21}(z) & M_{22}(z) \end{pmatrix} \cdot \begin{pmatrix} \ddot{\theta} \\ \ddot{v} \end{pmatrix} + \begin{pmatrix} F_1(z,\dot{z}) \\ F_2(z,\dot{z}) \end{pmatrix} + \begin{pmatrix} 0_{2\times 2} & 0_{2\times 4} \\ 0_{4\times 2} & D(z) \end{pmatrix} \cdot \begin{pmatrix} \dot{\theta} \\ \dot{v} \end{pmatrix} + \begin{pmatrix} 0_{2\times 2} & 0_{2\times 4} \\ 0_{4\times 2} & K(z) \end{pmatrix} \cdot \begin{pmatrix} \theta \\ v \end{pmatrix} = \begin{pmatrix} T(t) \\ 0_{4\times 1} \end{pmatrix} \quad (4)$$

where $z = [\theta \quad v]^T$, with $\theta = [\theta_1 \quad \theta_2]^T$, $v = [v_{11} \quad v_{12} \quad v_{21} \quad v_{22}]^T$ (vector of the vibration modes for links 1 and 2), and $\left[F_1(z,\dot{z}) \quad F_2(z,\dot{z}) \right]^T$ (centrifugal and Coriolis forces). The elements of the inertia matrix are: $M_{11} \in R^{2\times 2}$, $M_{12} \in R^{2\times 4}$, $M_{21} \in R^{4\times 2}$, $M_{22} \in R^{4\times 2}$. The damping and elasticity matrices of the aforementioned model are $D \in R^{4\times 4}$ and $K \in R^{4\times 4}$. Moreover the vector of the control torques is $T(t) = [T_1(t) \quad T_2(t)]^T$.

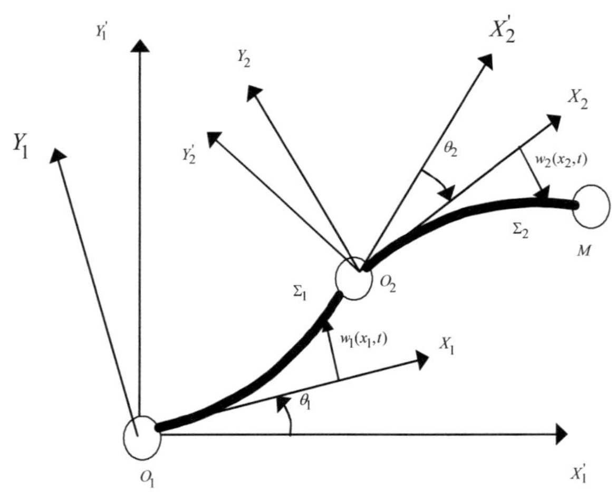

Fig. 1. A 2-DOF flexible-link robot

3. Neural control for flexible-link robots

3.1 Neural network-based control of flexible manipulators

Adaptive neural network control of robotic manipulators has been extensively studied (Lewis, Jagannathan & Yelsidirek, 1999), (Ge, Lee & Harris, 1998). Following (Tian, Wang & Mao, 2004) a method of neural adaptive control for flexible-link robots will be proposed.

Eq. (4) represents the dynamics of the flexible-link manipulator. It actually refers to a nonlinear transformation (mapping) from inputs (torques $T(t)$ generated by the motors) to outputs (motion of the joints). This nonlinear model can be written in the general form:

$$\ddot{\theta} = G(\theta, \dot{\theta}, v, \dot{v}, T(t)) \tag{5}$$

Consequently, the inverse dynamics of the flexible-link manipulator, is a relation that provides the torque that should be generated by the motors of the joints so as the joints angle, angular velocity and acceleration to take certain values. The inverse model of Eq. (5) is given by

$$T(t) = G^{-1}(\theta, \dot{\theta}, \ddot{\theta}, v, \dot{v}) \tag{6}$$

The dynamic model and its inverse are time dependent. If the inverse dynamic model of Eq. (6) can be explicitly calculated then a suitable control law for the flexible-link robot is available.

However, this model is not usually available and the system dynamics has to be adaptively identified. A neural network model can be used to effectively approximate the inverse dynamical model of Eq. (6). Variables θ, $\dot{\theta}$, $\ddot{\theta}$ can be measured while variables v, $\dot{v}$ are non-measurable. Thus, the inverse dynamics of the manipulator can be decomposed into n sub-models given in the following form:

$$T(t) = G^{-1}(\theta, \dot{\theta}, \ddot{\theta}) = \begin{pmatrix} g_1^{-1}(\theta, \dot{\theta}, \ddot{\theta}) \\ g_2^{-1}(\theta, \dot{\theta}, \ddot{\theta}) \\ \cdots \\ g_n^{-1}(\theta, \dot{\theta}, \ddot{\theta}) \end{pmatrix} \tag{7}$$

where each $g_i^{-1}, i = 1,2,\ldots,n$ defines the inverse dynamics of the corresponding joint, while n is the number of joints of the manipulator.

A neural network can be employed to approximate each sub-model g_i^{-1} of the flexible robot's inverse dynamics. Therefore, the inverse dynamics of the overall system can be represented by a neural network $N(\theta, \dot{\theta}, \ddot{\theta}, w)$ (Tian, Wang & Mao, 2004)

$$T(t) \cong N(\theta,\dot{\theta},\ddot{\theta},w) = \begin{pmatrix} N_1(\theta,\dot{\theta},\ddot{\theta},w_1) \\ N_2(\theta,\dot{\theta},\ddot{\theta},w_2) \\ \cdots \\ N_n(\theta,\dot{\theta},\ddot{\theta},w_n) \end{pmatrix} \tag{8}$$

where $N_i(\theta,\dot{\theta},\ddot{\theta},w_i)$, $i=1,2,...,n$ is the i-th neural network that approximates the i-th sub-model of robot's inverse dynamics and w_i is the associated weights vector.

Using a NN to model the dynamics of the flexible-link robot provides a mapping of the joint angles vector to the motor torques vector. The torque appearing at the output of the neural network can be combined with a PD-feedback controller to generate the overall control signal that will finally drive the motors (Tzafestas & Rigatos, 1998). Therefore, the control scheme can be given as

$$T(t) = N(\theta,\dot{\theta},\ddot{\theta},w) + K_p e + K_d \dot{e} \tag{9}$$

where $N(\theta,\dot{\theta},\ddot{\theta},w)$ is the neural network approximation of the actual inverse dynamics of the manipulators, while $K_p \in R^{n\times m}$ and $K_d \in R^{n\times m}$ are the diagonal gain matrices with entries K_p and K_d, respectively, denoting a servo feedback that is introduced to stabilize the system. In Eq. (9) $e=\theta-\theta_d$, and $\dot{e}=\dot{\theta}-\dot{\theta}_d$ denote the position and velocity error of the robot's joints, respectively, and $\theta_d \in R^2$ is the vector of the position and velocity set-points.

The architecture of NN-based control of flexible-link robots is depicted in Fig. 2. Making use of Eq. (7) and Eq. (8), the neural network controller that is described by Eq. (9) becomes

$$N(\theta,\dot{\theta},\ddot{\theta},w) + K_p e + K_d \dot{e} = G^{-1}(\theta,\dot{\theta},\ddot{\theta}) \tag{10}$$

or equivalently,

$$K_p e + K_d \dot{e} = G^{-1}(\theta,\dot{\theta},\ddot{\theta}) - N(\theta,\dot{\theta},\ddot{\theta},w) = \bar{N}(\theta,\dot{\theta},\ddot{\theta},w) \tag{11}$$

Eq. (11) represents a decoupled linear system, driven by the nonlinear vector function $\bar{N}(\theta,\dot{\theta},\ddot{\theta},w) \in R^n$. This function represents the error between the actual inverse dynamics $G^{-1}(\theta,\dot{\theta},\ddot{\theta})$ and its estimated model $N(\theta,\dot{\theta},\ddot{\theta},w)$ and can be written as

$$\bar{N}(\theta,\dot{\theta},\ddot{\theta},w)=\begin{pmatrix}\bar{N}_1(\theta,\dot{\theta},\ddot{\theta},w)\\ \cdots \\ \bar{N}_2(\theta,\dot{\theta},\ddot{\theta},w)\end{pmatrix}=\begin{pmatrix}g_1^{-1}(\theta,\dot{\theta},\ddot{\theta})-N_1(\theta,\dot{\theta},\ddot{\theta},w)\\ \cdots \\ g_n^{-1}(\theta,\dot{\theta},\ddot{\theta})-N_n(\theta,\dot{\theta},\ddot{\theta},w)\end{pmatrix} \tag{12}$$

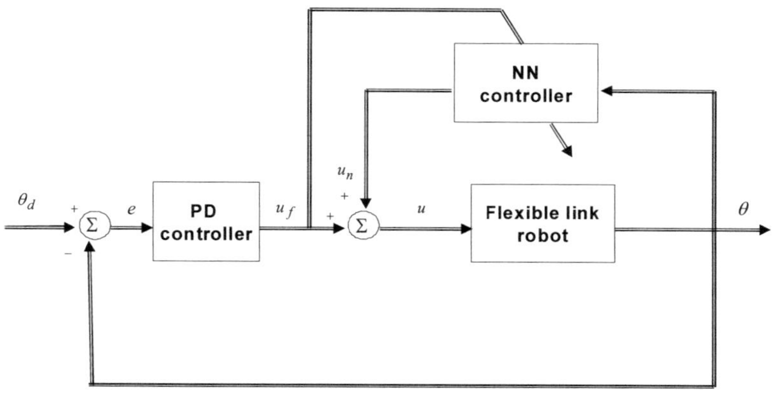

Fig. 2. NN-based control of the flexible-link robot

Instead of using one neural network to approximate the inverse dynamics of the flexible-link robot one can use a separate network for each joint of the manipulator. In that case and using Eq. (12), the error equation for the i-th joint of the manipulator becomes

$$K_p^i e_i + K_d^i \dot{e}_i = \bar{N}_i(\theta,\dot{\theta},\ddot{\theta},w),\ i=1,2,\ldots,n \tag{13}$$

where e_i and $\dot{e}_i$ denote the position and velocity errors of the i-th joint respectively, K_p^i and K_d^i are the proportional and derivative gains of the PD controller of the i-th joint, and $\bar{N}_i(\theta,\dot{\theta},\ddot{\theta},w)$ is the approximation error of the NN assigned to the i-th joint. The objective is to eliminate the approximation error, i.e. to succeed $\lim_{t\to\infty}\bar{N}_i(\theta,\dot{\theta},\ddot{\theta},w)=0$. In that case, a suitable selection of the control gains K_p^i and K_d^i results into

$$K_p^i e_i + K_d^i \dot{e}_i = 0 \Rightarrow \lim_{t\to\infty} e_i(t)=0, \quad \lim_{t\to\infty}\dot{e}_i(t)=0 \tag{14}$$

Eq. (14) denotes that, the convergence condition for the closed-loop system is to make the error surface $\varepsilon_i = K_p^i e_i + K_d^i \dot{e}_i$ approach zero. Then, a suitable selection of the control gains will result in the asymptotic convergence of $e(t)$ to 0

$$K_p^i e_i + K_d^i \dot{e}_i(t) = 0 \Rightarrow \dot{e}_i(t) = -\frac{K_p^i}{K_d^i} e_i(t) \Rightarrow e_i(t) = e_i(0) e^{-\frac{K_p^i}{K_d^i} t} \tag{15}$$

Thus a measure of the output error can be considered to be

$$\varepsilon_i = \bar{N}_i(\theta, \dot{\theta}, \ddot{\theta}, w) \tag{16}$$

which reflects the discrepancy between the actual inverse dynamics of the manipulator and its neural network approximation. To this end, the following cost function is defined for each joint

$$E_i(t) = \frac{1}{2}\varepsilon_i^2 \tag{17}$$

This cost function gives the squared distance of the error function $\varepsilon_i = K_p^i e_i + K_d^i \dot{e}_i$ from 0. The weights update algorithm is derived from the minimization of the cost function $E_i(t)$ over the weight space of the corresponding NN model.

3.2 Feed-forward neural networks for flexible-link robot control

The NN-based control for the flexible-link robots, which is depicted in Fig. 2, can be substantiated with the use of feed-forward neural networks (Rigatos, 2009). Feed-forward neural networks (FNN) serve as powerful computational tools, in a diversity of applications including function approximation, classification and pattern recognition. When equipped with procedures for learning from measurement data they can generate models of unknown systems. Feed-forward neural networks are the most popular neural architectures due to their structural flexibility, good representational capabilities, and availability of a large number of training algorithms.

The idea of function approximation with the use of feed-forward neural networks (FNN) comes from generalized Fourier series. It is known that any function $\psi(x)$ in a L^2 space can be expanded in a generalized Fourier series in a given orthonormal basis, i.e.

$$\psi(x) = \sum_{k=1}^{\infty} c_k \psi_k(x), \quad a \le x \le b \tag{18}$$

Truncation of the series yields in the sum

$$S_M(x) = \sum_{k=1}^{M} a_k \psi_k(x) \tag{19}$$

If the coefficients a_k are taken to be equal to the generalized Fourier coefficients, i.e. when $a_k = c_k = \int_a^b \psi(x)\psi_k(x)dx$, then Eq. (19) is a mean square optimal approximation of $\psi(x)$.

Unlike generalized Fourier series, in FNN the basis functions are not necessarily orthogonal. The hidden units in a FNN usually have the same activation functions and are often selected as sigmoidal functions or gaussians. A typical feed-forward neural network consists of n inputs x_i, $i = 1,2,...,n$, a hidden layer of m neurons with activation function $h: R \to R$ and a single output unit (see Fig. 3(a)). The FNN's output is given by

$$\psi(x) = \sum_{j=1}^{n} c_j h(\sum_{i=1}^{n} w_{ji} x_i + b_j) \tag{20}$$

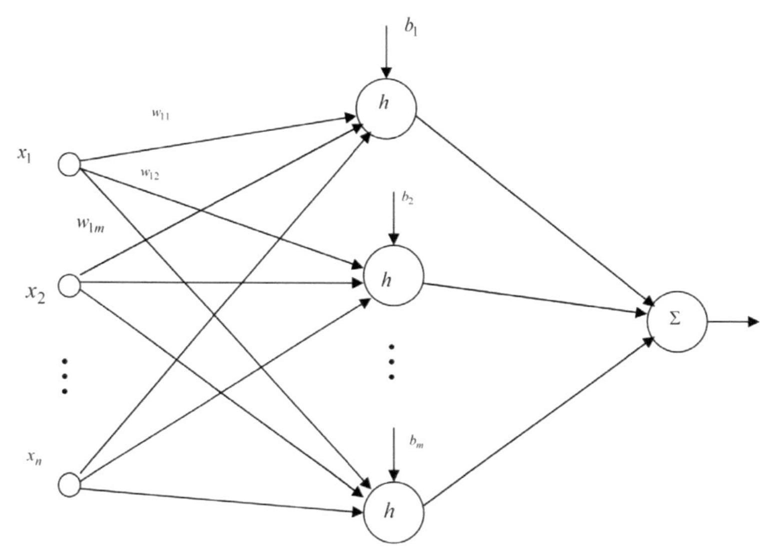

Fig. 3. (a) Feed-forward neural network

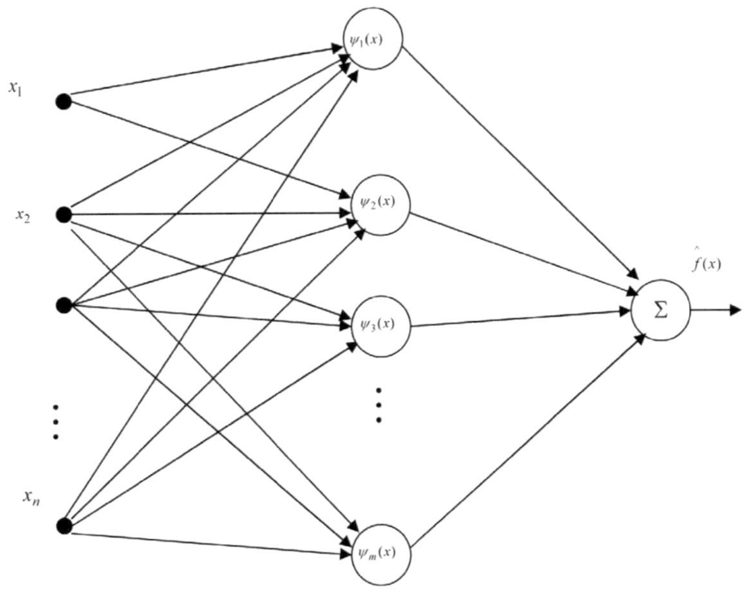

Fig. 3. (b) Neural network with Hermite basis functions

The root mean square error in the approximation of function $\psi(x)$ by the FNN is given by

$$E_{RMS} = \sqrt{\frac{1}{N}\sum_{k=1}^{N}(\psi(x^k) - \hat{\psi}(x^k))^2} \tag{21}$$

where $x^k = [x_1^k, x_2^k, ..., x_n^k]$ is the k-th input vector of the neural network. The activation function is usually a sigmoidal function $h(x) = \frac{1}{1+e^{-x}}$ while in the case of radial basis functions networks it is a Gaussian (Haykin, 1994). Several learning algorithms for neural networks have been studied. The objective of all these algorithms is to find numerical values for the network's weights so as to minimize the mean square error E_{RMS} of Eq. (21). The algorithms are usually based on first and second order gradient techniques. These algorithms belong to: i) batch-mode learning, where to perform parameters update the outputs of a large training set are accumulated and the mean square error is calculated (back-propagation algorithm, Gauss-Newton method, Levenberg-Marquardt method, etc.), ii) pattern-mode learning, in which training examples are run in cycles and the parameters update is carried out each time a new training pattern becomes available (Extended Kalman Filter algorithm).

Unlike conventional FNN with sigmoidal or Gaussian basis functions, Hermite polynomial-based FNN remain closer to Fourier series expansions by employing activation functions which satisfy the property of orthogonality (Zuo, Zhu and Cai, 2009). Other basis functions with the property of orthogonality are Hermite, Legendre, Chebyshev, and Volterra polynomials (Refregier, 2003), (Rigatos & Tzafestas, 2006), (Yang & Cheng, 1996).

4. Neural control using wavelet basis functions

4.1 Wavelet frames

First, as basis functions of the feed-forward neural network orthogonal wavelet functions will be considered. The continuous time wavelet is defined at scale a and b as

$$\psi_{a,b}(x) = \frac{1}{\sqrt{a}}\psi\left(\frac{x-b}{a}\right) \tag{22}$$

It will be shown that a continuous time signal $f(x)$ can be expressed as a series expansion of discrete wavelet basis functions. The discrete wavelet has the form (Addison, 2002)

$$\psi_{m,n}(x) = \frac{1}{\sqrt{a_0^m}}\psi\left(\frac{x-nb_0a_0^m}{a_0^m}\right) \tag{23}$$

The wavelet transform of a continuous signal $f(x)$ using discrete wavelets of the form of Eq. (23) is given by

$$T_{m,n} = \int_{-\infty}^{+\infty} f(x)\frac{1}{\sqrt{a_0^m}}\psi\left(\frac{x-nb_0a_0^m}{a_0^m}\right)dx \tag{24}$$

which can be also expressed as the inner product $T_{m,n} = <f, \psi_{m,n}>$. For the discrete wavelet transform, the values $T_{m,n}$ are known as wavelet coefficients. To determine how good the representation of a signal is in the wavelet space one can use the theory of wavelet frames. The family of wavelet functions that constitute a frame are such that the energy of the resulting wavelet coefficients lies within a certain bounded range of the energy of the original signal

$$AE \leq \sum_{m=-\infty}^{+\infty}\sum_{n=-\infty}^{+\infty}|T_{m,n}|^2 \leq BE \tag{25}$$

where $T_{m,n}$ are the discrete wavelet coefficients, A and B are the frame bounds, and E is the energy of the signal given by $E = \int_{-\infty}^{+\infty}|f(x)|^2dt = ||f(x)||^2$. The values of the frame bounds depend on the parameters a_0 and b_0 chosen for the analysis and the wavelet function used. If $A = B$ the frame is known as tight and has a simple reconstruction formula given by the infinite series

$$f(x) = \frac{1}{A} \sum_{m=-\infty}^{+\infty} \sum_{n=-\infty}^{+\infty} T_{m,n} \psi_{m,n}(x) \tag{26}$$

A tight frame with $A = B > 1$ is redundant, with A being a measure of the redundancy. When $A = B = 1$ the wavelet family defined by the frame forms an orthonormal basis. Even if $A \neq B$ a reconstruction formula of $f(x)$ can be obtained in the form:

$$f^{'}(x) = \frac{2}{A+B} \sum_{m=-\infty}^{+\infty} \sum_{n=-\infty}^{+\infty} T_{m,n} \psi_{m,n}(x) \tag{27}$$

where $f^{'}(x)$ is the reconstruction which differs from the original signal $f(x)$ by an error which depends on the values of the frame bounds. The error becomes acceptably small for practical purposes when the ratio B / A is near unity. The closer this ratio is to unity, the tighter the frame.

4.2 Dyadic grid scaling and orthonormal wavelet transforms

The dyadic grid is the simplest and most efficient discretization for practical purposes and is used for the construction of an orthonormal wavelet basis. Substituting $a_0 = 2$ and $b_0 = 1$ into Eq. (23) the dyadic grid wavelet can be written as

$$\psi_{m,n} = \frac{1}{\sqrt{2^m}} \psi(\frac{x - n2^m}{2^m}) \tag{28}$$

or more compactly

$$\psi_{m,n}(x) = 2^{-\frac{m}{2}} \psi(2^{-m} x - n) \tag{29}$$

Discrete dyadic grid wavelets are commonly chosen to be orthonormal. These wavelets are both orthogonal to each other and normalized to have unit energy. This is expressed as

$$\int_{-\infty}^{+\infty} \psi_{m,n}(x) \psi_{m^{'},n^{'}}(x) dx = \begin{cases} 1 \text{ if } m = m^{'} \text{ and } n = n^{'} \\ 0 \text{ otherwise} \end{cases} \tag{30}$$

Thus, the products of each wavelet with all others in the same dyadic system are zero. This also means that the information stored in a wavelet coefficient $T_{m,n}$ is not repeated elsewhere and allows for the complete regeneration of the original signal without redundancy. In addition to being orthogonal, orthonormal wavelets are normalized to have unit energy. This can be seen from Eq. (30), as using $m = m^{'}$ and $n = n^{'}$ the integral gives

the energy of the wavelet function equal to unity. Orthonormal wavelets have frame bounds $A = B = 1$ and the corresponding wavelet family is an orthonormal basis. An orthonormal basis has components which, in addition to being able to completely define the signal, are perpendicular to each other.
Using the dyadic grid wavelet of Eq. (28) the discrete wavelet transform is defined as

$$T_{m,n} = \int_{-\infty}^{+\infty} f(x)\psi_{m,n}(x)dx \tag{31}$$

By choosing an orthonormal wavelet basis $\psi_{m,n}(x)$ one can reconstruct the original signal $f(x)$ in terms of the wavelet coefficients T_{mn} using the inverse discrete wavelet transform:

$$f(x) = \sum_{m=-\infty}^{+\infty} \sum_{n=-\infty}^{+\infty} T_{m,n}\psi_{m,n}(x) \tag{32}$$

Moreover, the energy of the signal can be expressed as

$$\int_{-\infty}^{+\infty} |f(x)|^2 dx = \sum_{m=-\infty}^{+\infty} \sum_{n=-\infty}^{+\infty} |T_{m,n}|^2 \tag{33}$$

4.3 The scaling function and the multi-resolution representation

Orthonormal dyadic discrete wavelets are associated with 'scaling functions' and their dilation equations (Addison, 2002), (Mallat, 1999). The scaling function is associated with the smoothing of the signal and has the same form as the wavelet

$$\phi_{m,n}(x) = 2^{-m/2}\phi(2^{-m/2}x - n) \tag{34}$$

The scaling functions have the property

$$\int_{-\infty}^{+\infty} \phi_{0,0}(x)dx = 1 \tag{35}$$

where $\phi_{0,0}(x) = \phi(x)$ is sometimes referred as the mother scaling function or mother wavelet. The scaling function is orthogonal to translations of itself, but not to dilations of itself. The scaling function can be convolved with the signal to produce approximation coefficients as follows:

$$S_{m,n} = \int_{-\infty}^{+\infty} f(x)\phi_{m,n}(x)dx \tag{36}$$

One can represent a signal $f(x)$ using a combined series expansion using both the approximation coefficicents and the wavelet (detail) coefficients as follows:

$$f(x) = \sum_{n=-\infty}^{+\infty} S_{m_0,n}\phi_{m_0,n} + \sum_{m=-\infty}^{+\infty}\sum_{n=-\infty}^{+\infty} T_{m,n}\psi_{m,n}(x) \tag{37}$$

It can be seen from this equation that the original continuous signal is expressed as a combination of an approximation of itself, at arbitrary scale index m_0 added to a succession of signal details form scales m_0 down to negative infinity. The signal detail at scale m is defined as

$$d_m(x) = \sum_{n=-\infty}^{+\infty} T_{m,n}\psi_{m,n}(x) \tag{38}$$

and hence one can write Eq. (37)

$$f(x) = f_{m_0}(x) + \sum_{n=-\infty}^{m_0} d_m(x) \tag{39}$$

From this equation it can be shown that

$$f_{m-1}(x) = f_m(x) + d_m(x) \tag{40}$$

which shows that if one adds the signal detail at an arbitrary scale (index m) to the approximation at that scale he gets the signal approximation at an increased resolution (at a smaller scale index $m-1$). This is the so-called multi-resolution representation.

4.4 Examples of orthonormal wavelets

Wavelet functions can be further analyzed in terms of 'scaling functions'. The scaling equation (or dilation equation) describes the scaling function $\phi(t)$ in terms of contracted and shifted versions of itself as follows (Addison, 2002), (Mallat, 1999):

$$\phi(x) = \sum_k c_k \phi(2x-k) \tag{41}$$

where $\phi(2x-k)$ is a contracted version of $\phi(x)$ shifted along the $x-$axis by an integer step k and factored by an associated scaling coefficient c_k. The coefficients of the scaling equation should satisfy the condition

$$\sum_k c_k = 2 \tag{42}$$

$$\sum_k c_k c_{k+2k'} = \begin{cases} 2 \text{ if } k' = 0 \\ 0 \text{ otherwise} \end{cases} \tag{43}$$

This also shows that the sum of the squares of the scaling coefficients is equal to 2. The same coefficients are used in reverse with alternate signs to produce the associated wavelet equation

$$\psi(x) = \sum_k (-1)^k c_{1-k} \phi(2x-k) \tag{44}$$

This construction ensures that the wavelets and their corresponding scaling functions are orthogonal. For wavelets of compact support, which have a finite number of scaling coefficients N_k the following wavelet function is defined

$$\psi(x) = \sum_k (-1)^k c_{N_k-1-k} \phi(2x-k) \tag{45}$$

This ordering of scaling coefficients used in the wavelet equation allows for our wavelets and their corresponding scaling equations to have support over the same interval $[0, N_{k-1}]$. Often the reconfigured coefficients used for the wavelet function are written more compactly as

$$b_k = (-1)^k c_{N_k-1-k} \tag{46}$$

where the sum of all coefficients b_k is zero. Using this reordering of the coefficients Eq. (45) can be written as

$$\psi(x) = \sum_{k=0}^{N_k-1} b_k \phi(2x-k) \tag{47}$$

From the previous equations and examining the wavelet at scale index $m+1$ one can see that for arbitrary integer values of m the following holds

$$2^{-\frac{m+1}{2}} \varphi(\frac{x}{2^{m+1}} - n) = 2^{-m/2} 2^{-1/2} \sum_k c_k \varphi(\frac{2x}{2 \times 2^m} - 2nk) \tag{48}$$

which may be written more compactly as

$$\phi_{m+1,n}(x) = \frac{1}{\sqrt{2}} \sum_k c_k \phi_{m,2n+k}(x) \tag{49}$$

That is the scaling function at an arbitrary scale is composed of a sequence of shifted functions at the next smaller scale each factored by their respective scaling coefficients. Similarly, for the wavelet function one obtains

$$\psi_{m+1,n}(x) = \frac{1}{\sqrt{2}} \sum_k b_k \phi_{m,2n+k}(x) \tag{50}$$

4.5 The Haar wavelet

The Haar wavelet is the simplest example of an orthonormal wavelet (see Fig. 4 and Fig. 5). Its scaling equation contains only two nonzero scaling coefficients and is given by

$$\phi(x) = \phi(2x) + \phi(2x-1) \tag{51}$$

that is, its scaling coefficients are $c_0 = c_1 = 1$. These values can be obtained from Eq. (42) and Eq. (43). The solution of the Haar scaling equation is the single block pulse defined as

$$\phi(x) = \begin{cases} 1 & \text{if } 0 \le x < 1 \\ 0 & \text{otherwise} \end{cases} \tag{52}$$

Using this scaling function, the Haar wavelet equation is

$$\psi(x) = \phi(2x) - \phi(2x-1) \tag{53}$$

The Haar wavelet is finally found to be (see Fig. 4 and Fig. 5)

$$\psi(x) = \begin{cases} 1 & \text{if } 0 \le x < 1/2 \\ -1 & \text{if } 1/2 \le x < 1 \\ 0 & \text{elsewhere} \end{cases} \tag{54}$$

The mother wavelet for the Haar wavelet system $\psi(x) = \psi_{0,0}(x)$ is formed from two dilated unit block pulses sitting next to each other on the time axis, with one of them inverted. From the mother wavelet one can construct the Haar system of wavelets on a dyadic grid $\psi_{m,n}(x)$.

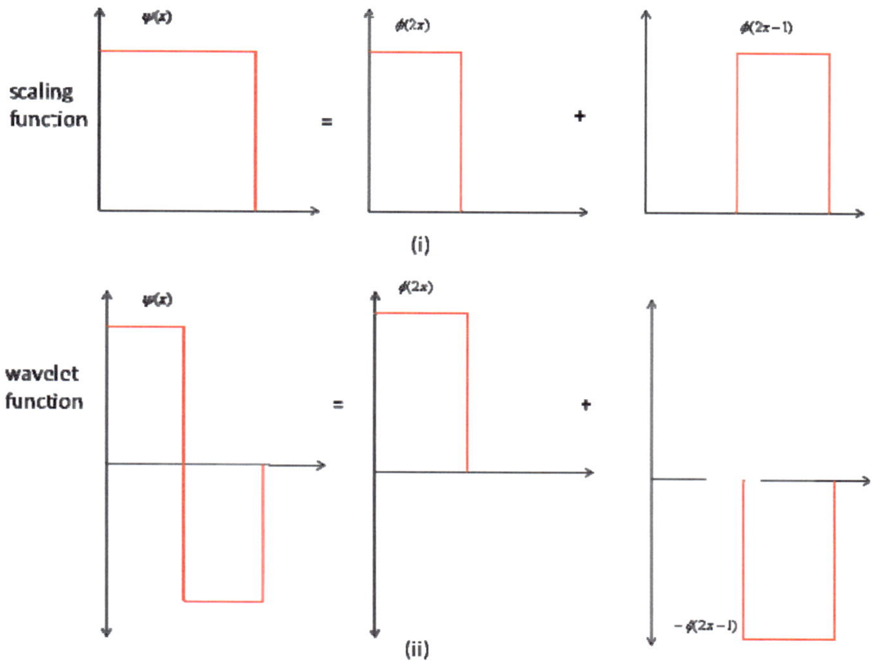

Fig. 4 (i) The Haar scaling function in terms of shifted and dilated versions of itself, (ii) The Haar wavelet in terms of shifted and dilated versions of the scaling function.

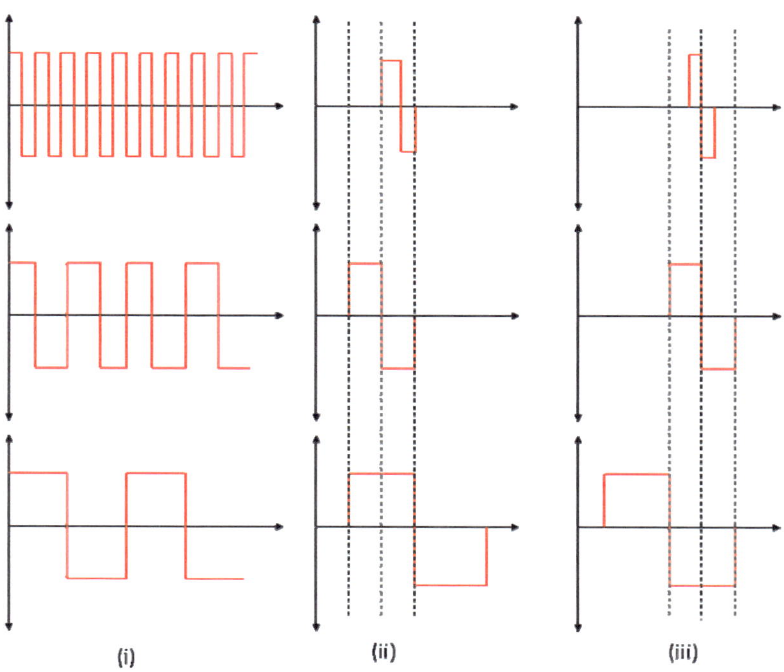

Fig. 5. (i) Three consecutive scales shown from the Haar wavelet family specified on a dyadic grid, e.g. from the bottom $\psi_{m,n}(x)$, $\psi_{m+1,n}(x)$, $\psi_{m+2,n}(x)$, (ii) Three Haar wavelets at three

consecutive scales on a dyadic grid, iii) Three Haar wavelets at different scales. In this case, the Haar wavelets are not defined on a dyadic grid and are hence not orthogonal to each other.

5. Neural Networks using Hermite activation functions

5.1 The Gauss-Hermite series expansion

Next, as orthogonal basis functions of the feed-forward neural network Hermite polynomials are considered. These are the spatial components $X_k(x)$ of the solution of Schrödinger's differential equation and describe a stochastic oscillation:

$$X_k(x) = H_k(x)e^{-\frac{x^2}{2}},\ k = 0,1,2,... \tag{55}$$

where $H_k(x)$ are the Hermite orthogonal functions (see Fig. 6(a) and Fig. 6(b)). The Hermite functions $H_k(x)$ are the eigenstates of the quantum harmonic oscillator. The general relation for the Hermite polynomials is

$$H_k(x) = (-1)^k e^{x^2} \frac{d^{(k)}}{dx^{(k)}} e^{-x^2} \tag{56}$$

According to Eq. (56) the first five Hermite polynomials are: $H_0(x) = 1$, $H_1(x) = 2x$, $H_2(x) = 4x^2 - 2$, $H_3(x) = 8x^3 - 12x$, $H_4(x) = 16x^4 - 48x^2 + 12$.

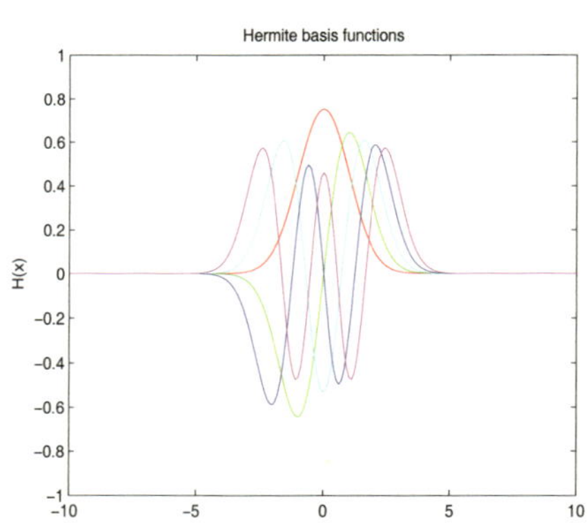

Fig. 6(a). First five one-dimensional Hermite polynomial basis functions

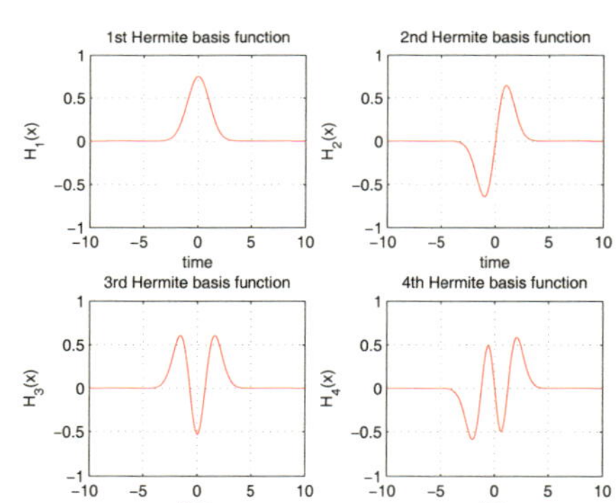

Fig. 6(b). Analytical represenation of the 1D-Hermite polynomial basis function

Hermite polynomials are orthogonal, i.e. it holds

$$\int_{-\infty}^{+\infty} e^{-x^2} H_m(x) H_k(x) dx = \begin{cases} 2^k k! \sqrt{\pi} & \text{if } m = k \\ 0 & \text{if } m \neq k \end{cases} \tag{57}$$

Using now, Eq. (57), the following basis functions can be defined (Refregier, 2003):

$$\psi_k(x) = [2^k \sqrt{\pi} k!]^{-\frac{1}{2}} H_k(x) e^{-\frac{x^2}{2}} \tag{58}$$

where $H_k(x)$ is the associated Hermite polynomial. From Eq. (57), the orthogonality of basis functions of Eq. (58) can be deduced, which means

$$\int_{-\infty}^{+\infty} \psi_m(x) \psi_k(x) dx = \begin{cases} 1 & \text{if } m = k \\ 0 & \text{if } m \neq k \end{cases} \tag{59}$$

Moreover, to succeed multi-resolution analysis Hermite basis functions of Eq. (58) are multiplied with the scale coefficient a. Thus the following basis functions are derived

$$\int_{-\infty}^{+\infty} \beta_m(x,a) \beta_k(x,a) dx = \begin{cases} 1 & \text{if } m = k \\ 0 & \text{if } m \neq k \end{cases} \tag{60}$$

which also satisfy orthogonality condition

$$\int_{-\infty}^{+\infty} \beta_m(x,a) \beta_k(x,a) dx = \begin{cases} 1 & \text{if } m = k \\ 0 & \text{if } m \neq k \end{cases} \tag{61}$$

Any function $f(x)$ $x \in R$ can be written as a weighted sum of the above orthogonal basis functions, i.e.

$$f(x) = \sum_{k=0}^{\infty} c_k \beta_k(x,a) \tag{62}$$

where coefficients c_k are calculated using the orthogonality condition

$$c_k = \int_{-\infty}^{+\infty} f(x) \beta_k(x,a) dx \tag{63}$$

Assuming now that instead of infinite terms in the expansion of Eq. (62), M terms are maintained, then an approximation of $f(x)$ is succeeded. The expansion of $f(x)$ using Eq. (62) is a Gauss-Hermite series. Eq. (62) is a form of Fourier expansion for $f(x)$. Eq. (62) can be considered as the Fourier transform of $f(x)$ subject only to a scale change. Indeed, the Fourier transform of $f(x)$ is given by

$$F(s)=\frac{1}{2\pi}\int_{-\infty}^{+\infty} f(x)e^{-jsx}dx \Rightarrow f(x)=\frac{1}{2\pi}\int_{-\infty}^{+\infty} F(s)e^{jsx}ds \quad (64)$$

The Fourier transform of the basis function $\psi_k(x)$ of Eq. (58) satisfies (Refregier, 2003)

$$\Psi_k(s)=j^n\psi_k(s) \quad (65)$$

while for the basis functions $b_k(x,a)$ using scale coefficient a it holds that

$$B_k(s,a)=j^n\beta_k(s,a^{-1}) \quad (66)$$

Therefore, it holds

$$f(x)=\sum_{k=0}^{\infty}c_k\beta_k(x,a)\overset{F}{\rightarrow} F(s)=\sum_{k=0}^{\infty}c_k j^n\beta_k(s,a^{-1}) \quad (67)$$

which means that the Fourier transform of Eq. (62) is the same as the initial function , subject only to a change of scale. The structure of a a feed-forward neural network with Hermite basis functions is depicted in Fig. 3(b).

5.2 Neural Networks using 2D Hermite activation functions

Two-dimensional feedforward neural networks can be constructed by taking products of the one dimensional basis functions $B_k(x,a)$ (Refregier, 2003). Thus, setting $x=[x_1,x_2]^T$ one can define the following basis functions

$$B_k(x,a)=\frac{1}{a}B_{k_1}(x_1,a)B_{k_2}(x_2,a) \quad (68)$$

These two dimensional basis functions are again orthonormal, i.e. it holds

$$\int d^2xB_n(x,a)B_m(x,a)=\delta_{n_1m_1}\delta_{n_2m_2} \quad (69)$$

The basis functions $B_k(x)$ are the eigenstates of the two dimensional harmonic oscillator and form a complete basis for integrable functions of two variables. A two dimensional function $f(x)$ can thus be written in the series expansion:

$$f(x)=\sum_{k_1,k_2}^{\infty}c_kB_k(x,a) \quad (70)$$

The choice of an appropriate scale coefficient a and maximum order k_{max} is of practical interest. The coefficients c_k are given by

$$c_k = \int dx^2 f(x) B_k(x,a) \tag{71}$$

Indicative basis functions $B_2(x,a)$, $B_6(x,a)$, $B_9(x,a)$, $B_{11}(x,a)$ and $B_{13}(x,a)$, $B_{15}(x,a)$ of a 2D feed-forward neural network with Hermite basis functions are depicted in Fig. 7, Fig. 8 and Fig. 9.

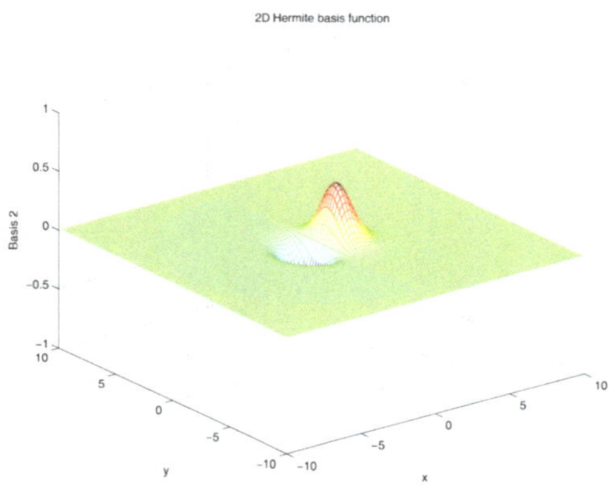

Fig. 7(a). 2D Hermite polynomial basis functions: basis function $B_2(x,a)$

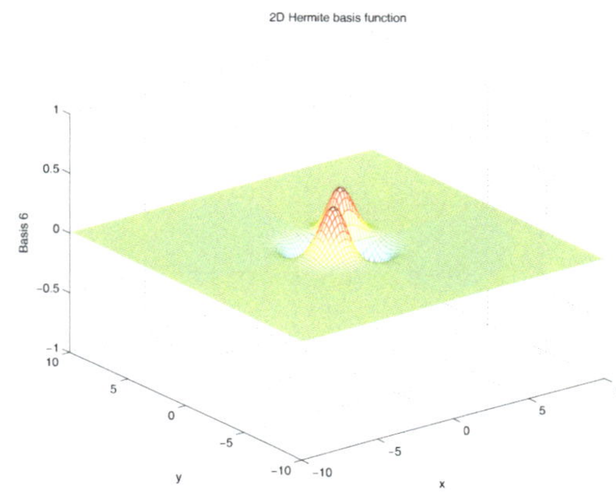

Fig. 7(b). 2D Hermite polynomial basis functions: $B_6(x,a)$

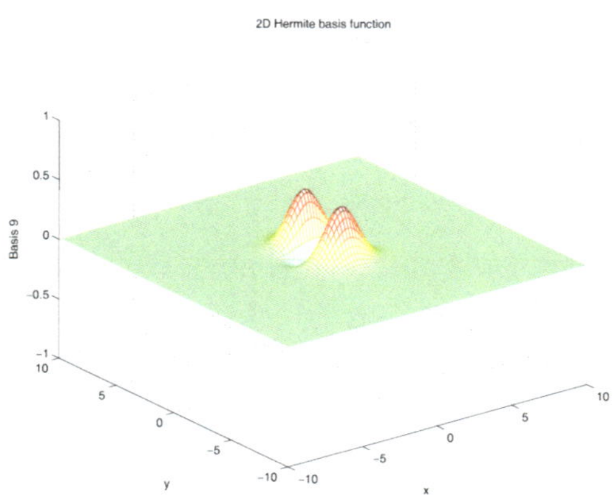

Fig. 8(a). 2D Hermite polynomial basis function $B_9(x,a)$

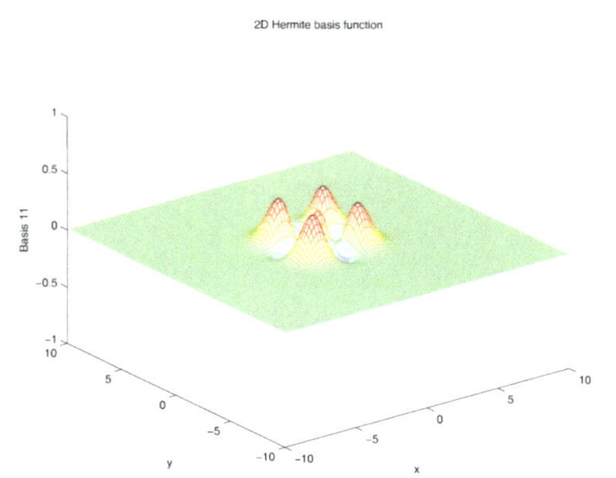

Fig. 8(b). 2D Hermite polynomial basis functions: $B_{11}(x,a)$

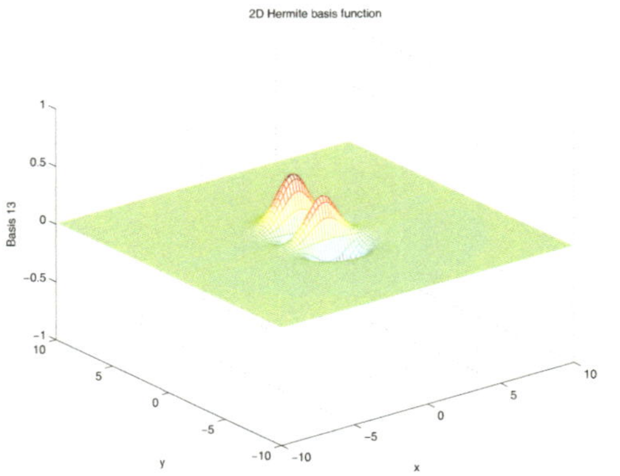

Fig. 9(a). 2D Hermite polynomial basis function $B_{13}(x,a)$

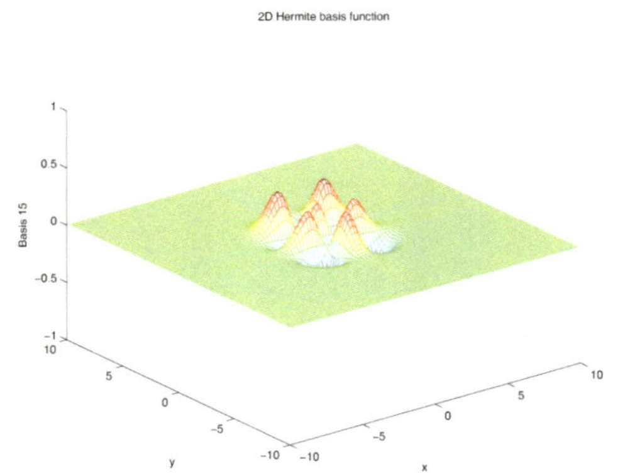

Fig. 9(b). 2D Hermite polynomial basis function $B_{15}(x,a)$

6. Simulation tests

The 2-DOF flexible link robot of Fig. 1 is considered. The robot is planar and consists of two flexible links of length $L_1 = 0.45m$ and $L_2 = 0.45m$, respectively. The dynamic model of the robot is given by Eq. (4). The elements of the inertia matrix M are:

$$M_{11} = \begin{pmatrix} 1 & 2 \\ 2 & 1 \end{pmatrix}, \quad M_{12} = M_{21}^T = \begin{pmatrix} 1 & 1 & 0.2 & 0.3 \\ 0.5 & 0.1 & 2 & 0.7 \end{pmatrix}, \quad M_{22} = \begin{pmatrix} 1 & 0 \\ 0 & 1 \end{pmatrix} \tag{72}$$

The damping matrix was taken to be $D = diag\{0.04, 0.08, 0.03, 0.06\}$ while the stiffness matrix was selected as $K = \{0.02, 0.04, 0.03, 0.06\}$. Initially only a PD controller is used. The selection of the gain matrices K_p and K_d determines the transient response of the closed loop system. The following controller gains have been considered: $K_p = diag\{0.2, 0.2\}$ and $K_d = diag\{0.1, 0.1\}$. The desirable joints positions are $\theta_1^d = 1rad$ and $\theta_2^d = 1.4rad$. Moreover, it is considered that an additive disturbance torque appears on each joint. The disturbance is given by $d_i(t) = 0.3\cos(t)$. The simulation diagram of Fig. 9(a) shows the evolution in time of the angles of the robot's joints θ_1 and θ_2, respectively, when only a PD controller is used in the loop and the flexibility of the link is not taken into account int the controller's design. In Fig. 9(b) the evolution in time of the vibration modes of the first link v_{11}, v_{12} and of the second link v_{21} and v_{22}, respectively, is presented. It can be seen that vibrations around the desirable joint positions cannot be eliminated.

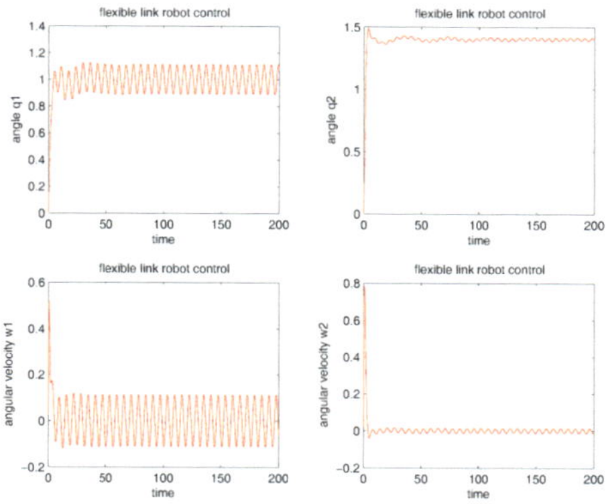

Fig. 9(a). (a) Diagrams of the angles θ_1 and θ_2, and of the angular velocities ω_1 and ω_2 of the joints of the flexible link manipulator when only a PD controller is used

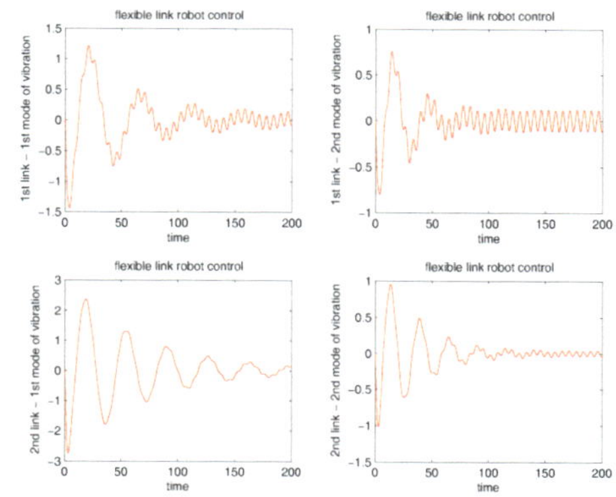

Fig. 9(b). Diagrams of the vibration modes the two flexible links of the robotic manipulator when only a PD controller is used

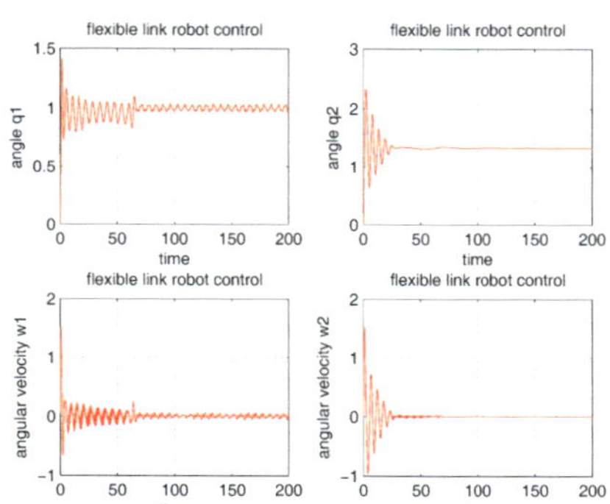

Fig. 10(a). Diagrams of the angles θ_1 and θ_2, and of the angular velocities ω_1 and ω_2 of the flexible-link robot when a wavelet-based NN controller is used to suppress vibrations.

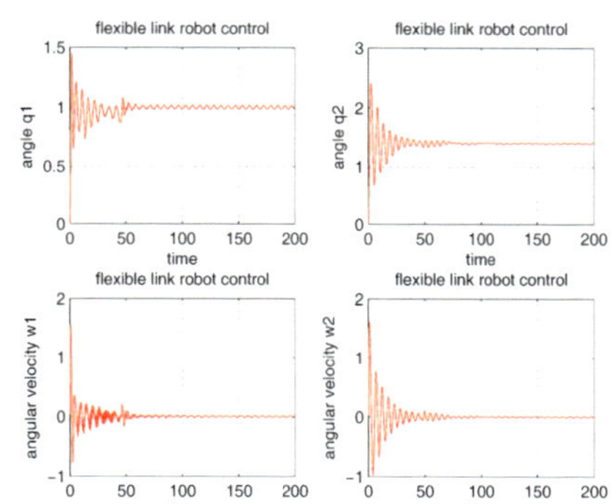

Fig. 10(b). Diagrams of the robot's joint angles θ_1 and θ_2, and of the angular velocities ω_1 and ω_2 of the flexible-link robot when a Hermite polynomial-based NN controller is used to suppress vibrations.

Next, a control loop with a NN which uses the Haar orthogonal wavelet functions of subsection 4.5 is considered. The neural controller is a single layer NN with wavelet basis functions, as shown in Fig. 3(b), and it is linear with respect to the output weights.
Fig. 10(a) presents the evolution in time of the joint angles of the robot when NN with wavelet basis functions are used for suppressing the vibrations of the flexible links. Fig. 10(b) shows the variation in time of the joint angles of the robot θ_1 and θ_2, respectively, when the neural controller uses Gauss-Hermite basis functions.
Finally, simulation diagrams are presented showing how the proposed neural controllers succeed the suppression of the vibration modes of the flexible links. Fig. 11 (a) shows the evolution in time of the vibration modes v_{11} and v_{12} of the first link, as well as of the

vibration modes v_{21} and v_{22} of the second link, when the neural controller uses wavelet basis functions. Similarly, Fig. 11 (b) shows the variation in time of the vibration modes v_{11}, v_{12} of the first link and v_{21}, v_{22} of the second link, respectively, when the neural controller uses Gauss-Hermite basis functions.

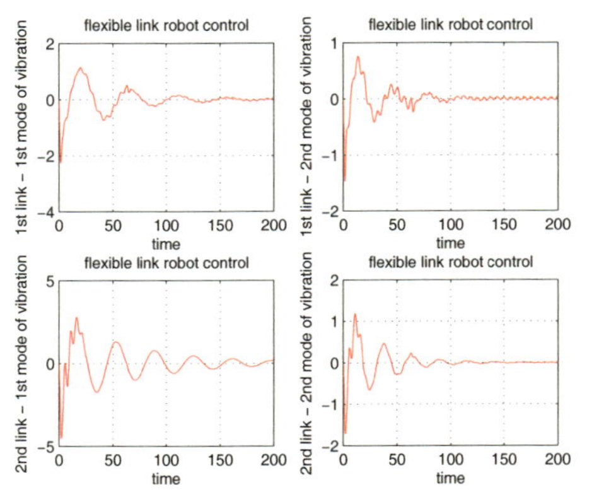

Fig. 11(a). Diagrams of the vibration modes of the two flexilbe links of the robotic manipulator, when a wavelet-based NN controller is used to suppress vibrations

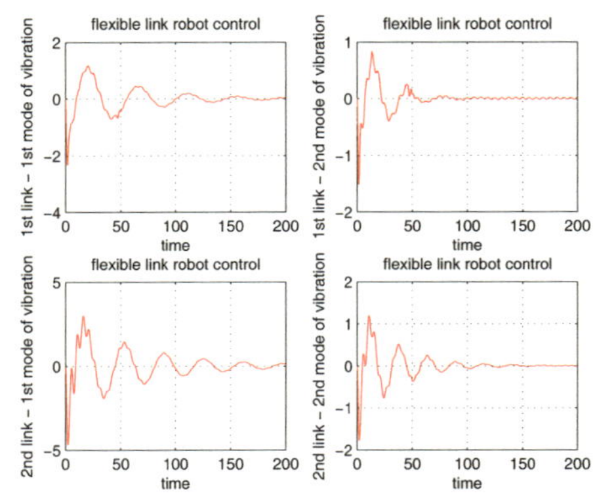

Fig. 11(b). Diagrams of the vibration modes of the two flexilbe links of the robotic manipulator, when a Hermite polynomial-based NN controller is used to suppress vibrations

From the simulation experiments it can be observed, that using a neural controller with basis functions which are localized both in space and frequency allows better approximation of the multi-frequency characteristics of the vibrating robot links. The angles of the robot's joints converge to the desirable set-points θ_1^d and θ_2^d , while fast and efficient suppression of the vibration modes v_{11}, v_{12} and v_{21}, v_{22} is also succeeded. Finally, in the above simulation experiments it was observed that the control signal (torque) generated by the neural controller with Hermite basis functions was smoother than the control signal of the neural controller which employs Haar wavelet basis functions.

Comparing the Gauss-Hermite basis functions to the Haar wavelet basis functions one can note the following: (i) both are basis functions which are local in space and spatial frequency. This allows better approximation of the multi-frequency characteristics of vibrating structures, such as the flexible-link robot, (ii) both satisfy the orthogonality property which helps to locally improve the accuracy of approximation of the unknown system dynamics. This means that the neural controller can be dynamically expanded by adding new basis functions which are orthogonal to the existing ones, while the coefficients of the new basis functions can be computed independently of the existing coefficients (iii) unlike wavelet basis functions, the Gauss-Hermite basis functions have a clear physical meaning, since they represent the solutions of differential equations of stochastic oscillators and each neuron can be regarded as the frequency filter of the respective vibration eigenfrequency, (iv) unlike the Haar wavelet basis functions, the Gauss-Hermite basis functions remain almost unchanged by the Fourier transform, which means that the weights of the associated neural network demonstrate the energy which is distributed in the various eigenmodes of the vibrating structure. This in turn allows to define thresholds for truncating the expansion and using neural controller with a small number of nodes and weight coefficients.

7. Conclusions

Neural networks with Gauss-Hermite polynomial basis functions have been proposed for the control of flexible-link manipulators. The Gauss-Hermite basis functions are local in both space and spatial frequency. Locality in spatial frequency allows the representation to be adaptively tuned to the variations of the local bandwidth of the system dynamics. Moreover,adding new basis functions helps to locally improve the accuracy of approximation of the unknown system dynamics. Since the new basis functions are orthogonal to the existing ones, the new NN weights can be computed independently of the existing weights. Comparing to Haar wavelet basis functions which are also orthogonal, the Gauss-Hermite basis functions have some interesting properties: (i) they remain almost unchanged by the Fourier transform, which means that the weights of the associated neural network demonstrate the energy which is distributed to the various eigenmodes of the vibrating structure, (ii) unlike wavelet basis functions the Gauss-Hermite basis functions have a clear physical meaning since they represent the solutions of differential equations of stochastic oscillators and each neuron can be regarded as the frequency filter of the respective vibration eigenfrequency.

The proposed neural controller operates in parallel to a PD controller the gains of which have been selected assuming rigid-link robot dynamics. The performance of the control scheme has been tested through simulation experiments. Comparison to a neural controller with Haar orthogonal wavelet functions has been provided. The simulation tests showed that using a neural controller with basis functions which are localized both in space and frequency allows for better approximation of the multi-frequency characteristics of the vibrating robot links.

8. References

Addison P.S (2002). *The illustrared wavelet transform handbook*, Institute of Physics Publishing.

Aoustin Y.; Fliess M., Mounier H., Rouchon P., & Rudoplh J. (1997). Theory and practice in the motion planning control of a flexible robot arm using Mikusinki operators, *Proc. of 4th Symposium on Robotics and Control*, pp. 287-293 Nantes.

Arteaga M.A. & Siciliano B. (2000). On Tracking Control of Flexible Robot Arms, *IEEE Transactions on Automatic Control*, Vol. 45, No. 3, pp. 520-527.

Benosman A. & Le Vey G (2004). Control of flexible manipulators: a survey, *Robotica* , Cambridge University Press, Vol. 22, No 5, pp. 533-545.

Bernard C.P. & Slotine J.J.E, (1997). Adaptive Control with Multiresolution *Bases, Proc. IEEE CDC 97, 36th IEEE Intl. Conference on Decision and Control, San Diego*, Califormia, USA.

Cannon M. & Slotine J.J.E. (1995). Space-frequency localized basis function networks for nonlinear system estimation and control, *Neurocomputing*, Elsevier, Vol. 9, No 3, pp. 293-342.

Caswara F.M. & Unbenhauen H. (2002). A neuro-fuzzy approach to the control of a flexible-link manipulator, *IEEE Transactions on Robotics and Automation*, Vol.18, No.6, pp. 932-944.

Cetinkunt S. & Yu W.L. (1991). Closed Loop Behavior of a Feedback Controlled Flexible Arm: A Comparative Study, *International Journal of Robotics Research*, Sage Publications, Vol. 10, No. 3, pp. 263-275.

Ge S.S.; Lee T.H. & Harris C.J. (1998). *Adaptive neural network control of robotic manipulators*, World Scientific.

Guterrez L.B.; Lewis F.L. & Lowe J.A. (1998). Implementation of neural network tracking for a single flexible link: comparison with PD and PID controllers, *IEEE Transactions on Industrial Electronics*, Vol.45, No.2, pp. 307-318.

Haykin, S (1994). *Neural Networks: A Comprehensive Foundation*, McMillan.

Hui D.; Fuchun S. & Zengqi S. (2002). Observer-based adaptive controller design of flexible manipulators using time-delay neuro-fuzzy networks, *Journal of Intelligent and Robotic Systems*, Springer, Vol. 34, No.4, pp. 453-466.

Lewis F.L.; Jagannathan S. & Yesildirek A. (1999). *Neural Network control of Robot Manipulators and Nonlinear Systems*, Taylor and Francis.

Lin C.K. (2006). Nonsingular terminal sliding mode control of robot manipulators using fuzzy wavelet networks, *IEEE Transactions on Fuzzy Systems*, Vol. 14, No 6, pp. 849-859.

Lin J. & Lewis F.L. (2002). Fuzzy controller for flexible-link robot arm by reduced-order techniques, *IEE Proceedings on Control Theory and Applications*, Vol. 149, No.3, pp. 177-187.

De Luca A. & Siciliano B. (1993). Regualation of flexible arms under gravity, *IEEE Transactions on Robotics and Automation*, Vol.9, No 4, pp. 463-467.

Kanoh H.; Tzafestas S.G., Lee H.G. & Kalat J. (1986). Modelling and control of flexible robot arms, *Proc IEE CDC86, 25th Conference on Decision and Control*, pp. 1866-1870, Athens, Greece.

Karimi H.R. & Lohmann B. (2006). A Computational Method to Robust Vibration Control of Vehicle Engine-Body System using Haar Wavelets, *Proc. of the 2006 IEEE International Conference on Control Applications*, pp. 169-174, Munich, Germany.

Krzyzak A. & Sasiadek J. (1991). Flexible robot identification using nonparametric techniques. *Proc.IEEE CDC91, 30th IEEE Conference on Decision and Control*, Brighton UK.

Mallat S (1999). *A wavelet tour of signal processing*, 2nd Edition, Acedemic Press.

Nguyen V.B. & Morris A.S. (2007). Genetic algorithm tuned fuzzy logic controller for a robot arm with two-link flexibility and two-joint elasticity, *Journal of Intelligent and Robotic Systems*, Springer, Vol. 49, No.1, pp. 3-18.

Refregier A. (2003). Shapelets - I. A method for image analysis. *Mon. Not. R. Astron. Soc.*, Vol. 338, No 1, pp. 35-47.

Renno J.M. (2007). Inverse dynamics-based tuning of a fuzzy logic controller for a single-link flexible manipulator, *Journal of Vibrations Control*, Sage Publications, Vol.13, No.12, pp. 1741-1759, 2007.

Rigatos G.G. (2009). Model-based and model-free control of flexible-link robots: a comparison between representative methods, *Applied Mathematical Modelling*, Elsevier, Vol.33, No 10, pp. 3906-3925.

Rigatos G.G. (2006). Model-free control of flexible-link robots, *Proc. IC-SCCE 2006, 2nd Intl. Conference "From Scientific Computing to Computational Engineering*, Athens, Greece.

Rigatos G.G. & Tzafestas S.G. (2006). Feed-forward neural networks using Hermite polynomial activation functions. *Lecture Notes in Artificial Intelligence*, Springer, Vol. 3925, pp. 323-333.

Shi L.Z. & Trabia M.B. (2006). Design and tuning of importance-based fuzzy logic controller for a flexible-link manipulator, *Journal of Intelligent and Fuzzy Systems*, IOS press, Vol. 17, No.3, pp. 313-323.

Subudhi B. & Morris A.S. (2003). Fuzzy and neuro-fuzzy approaches to control a flexible single-link manipulator, *Proc. IMechE, Journal of Systems and Control Engineering*, Vol. 217, pp. 387-399.

Subudhi B. & Morris A.S. (2009). Soft computing methods applied to the control of a flexible robot manipulator, *Applied Soft Computing*, Elsevier, Vol 9, No. 1, pp. 149-158.

Sureshbabu N. & Farell J.A. (1999). Wavelet-based system identification for nonlinear control. *IEEE Transactions on Automatic Control*, Vol. 44, No, 2 , pp. 412-417.

Talebi H.A.; Khorasani K. & Patel R.V. (1998). Neural network based control schemes for flexible-link manipulators: simulations and experiments, *Neural Networks*, Elsevier, Vol.11, No. 7-8, pp. 1357-1377.

Tian L. & Collins C. (2005). Adaptive neuro-fuzzy control of a flexible manipulator, *Journal of Mechatronics*, Elsevier, Vol. 15, No. 10, pp. 1305-1320.

Tian L.; Wang J. & Mao Z. (2004). Constrained motion control of flexible robot manipulators based on recurrent neural networks, *IEEE Transactions on System, Man and Cybernetics - Part B: Cybernetics*, Vol. 34, No. 3, pp. 1541-1552.

Torrésani B (1995). *Analyse continue par ondelettes*, CNRS Editions.

Tzafestas S.G. & Rigatos G.G. (1998). Neural and Neuro-fuzzy FELA Adaptive Robot Control Using Feedforward and Counterpropagation Networks, *Journal of Intelligent and Robotic Systems*, Springer, Vol. 23, No 2-4, pp. 291-330.

Wai R.J. & Lee M.C. (2004). Intelligent optimal control of a single-link, flexible robot arm, *IEEE Transactions on Industrial Electronics*, Vol. 51, No.1, pp. 201-220.

Wang F.Y. & Gao Y (2005). *Advanced Studies of Flexible Robotic Manipulators*, World Scientific.

Yang S.S. & Cheng C.S. (1996). An orthogonal neural network for function approximation. *IEEE Transactions on Systems, Man and Cybernetics - Part B: Cybernetics*, Vol. 26 No. 5, 779-784.

Zhang Q. & Benveniste A. (1993). Wavelet Networks. *IEEE Transactions on Neural Networks*, Vol. 3, No. 6, pp. 889-898.

Zuo W. ; Zhu Y. & Cai L. (2009). Fourier-Neural-Network-Based Learning control for a class of nonlinear systems with flexible components. *IEEE Transactions on Neural Networks*, Vol. 20, No. 1, pp. 139-151.

26

Dimensional optimization of completely restrained positioning cable driven parallel manipulator with large span

XiaoQiang Tang and Rui Yao
Tsinghua University People's Republic of China

1. Introduction

Cable driven parallel manipulator develops rapidly from 1980s which is also called cable robot, cable driven robot or wire driven parallel robot. Cable driven parallel manipulator has lots of desirable characteristics, including low inertial properties, high payload-to-weight ratios, potentially vast workspace, transportability, ease of disassembly/ reassembly, reconfigurability, and economical construction and maintenance(Alp & Agrawal, 2002). At present, giving full play to its advantage on large workspace, more and more large span cable driven parallel manipulator are used in Large radio telescopes or Wind tunnel support system(Lambert et al., 2007).

Since the cable driven parallel manipulator can only bear tension but not compression, a cable system with n end-effector Degrees of Freedom (DOFs) requires at least m=n+1 cables. When m=n+1, the cable driven parallel manipulator is Completely Restrained Positioning (CRP) (Verhoeven et al., 1998; Hiller & Fang, 2005). When a cable length is longer than 10m in a cable driven parallel manipulator, it is called a large span cable driven parallel manipulator. In this paper, we will take a six-cable driven parallel manipulator as an example to study the modeling and dimensional optimization method of a CPR cable driven parallel manipulator with large span.

The optimization target of the previous work on dimensional optimization design is workspace requirement and constraint condition is cable tension, stiffness or motion Index (Hassan & Khajepour, 2008; Yang et al., 2006; Fang et al., 2004), which will greatly influence the accuracy and vibration of a cable driven parallel manipulator. So, the purpose of this paper is to optimize dimensions of the six-cable driven parallel manipulator to meet the workspace requirement of constraint condition in terms of cable tension. So far, great progress has been made in this field. However, two difficulties arose during the dimensional optimization of cable driven parallel manipulator: cable catenary algorithm (Yao et al., 2007) and optimization target of a cable driven parallel manipulator. Firstly, the weight of cable itself results in catenary. But most of previous research neglected the cable's catenary because they thought the weight of cable is inconsiderable, and did not consider the deformation of cable (Duan et al., 2008). For a CPR cable driven parallel manipulator with large span, the deformation and catenary greatly influence the modeling and control accuracy (Kozak et al., 2006). Many of the current study on this topic can be introduced from bridge and construction fields (Krishna & Prem. 1978; Yang & Chen, 2003). Based on previous improvements, modeling method for a CPR cable driven parallel manipulator is adopted in this paper.

Secondly, cable tension is always used as the optimization target in a cable driven parallel manipulator because it is a tension redundant mechanism. If one of the cable tensions is negative or small, cable driven parallel manipulator will be uncontrollable. So it is necessary to optimize the geometric parameters by adopting a cable tension as a constraint condition in the required workspace.

In China, a Five Hundred meter Aperture Spherical radio Telescope (FAST) employs a cable driven parallel manipulator with large span (more than 300m) as the first adjustable feed support system (Tang, et al.). For the design of the cable driven parallel manipulator in FAST, several design proposals are made, one of which is the six-cable driven parallel manipulator (Nan, 2006). This chapter will study the modeling and dimensional optimization method of this six-cable driven parallel manipulator.

In this chapter, section 2 introduces the classification of cable driven parallel manipulator. Section 3 gives the modeling method for a large cable drive parallel manipulator. The tension static equilibrium equation is set up by cable catenary equation with cable deformation. In Section 4, the six-cable driven parallel manipulator in FAST is taken as an example. The description and modeling method of the six-cable driven parallel manipulator is presented, and Tilt-and-Torsion Angles is adopted for modeling. Section 5 gives a dimensional optimization method. Taking the six-cable driven parallel manipulator in FAST as an example, the optimization target is pose angle of the six-cable driven parallel manipulator and constraint condition is cable tensions. Finally, a set of optimized dimensional parameters of the six-cable driven parallel manipulator is available for building the feed support system in FAST.

The result indicates that the methods in this chapter are feasible for modeling and optimizing a CPR cable driven parallel manipulator with large span. More importantly, it provides a theoretical basis for further study.

2. Classification of a cable driven parallel manipulator

A cable driven parallel manipulator, according to its number of cables (m) and degrees of freedom (DOFs) of end-effector (n), can be classified as follows (Verhoeven et al., 1998; Hiller & Fang, 2005):

IRPMs: Incompletely Restrained Positioning Mechanisms. The number of cables is less than or equal to the number of DOFs, namely,

$$m \leq n$$

CRPMs: Completely Restrained Positioning Mechanisms. There has a one extra cable for an n DOFs end-effector,

$$m = n + 1$$

RRPMs: Redundantly Restrained Positioning Mechanisms.

$$m > n + 1$$

Then, R stands for rotational, T for translational DOFs. The possible DOF classes are as follows,

DOFs	n	Type of motion
1T	1	Linear motion of a body
2T	2	Planar motion of a point
1R2T	3	Planar motion of a body
3T	3	Spatial motion of a point
2R3T	5	Spatial motion of a beam
3R3T	6	Spatial motion of a body

Table 1. possible DOF classes

3. Tension static equilibrium equation for a large cable drive parallel manipulator

3.1 cable static tension equilibrium equation with catenary and elastics deformation

In small cable driven mechanism, due to the relatively inconsiderable weight of the cable, a cable is often analyzed as a two-force member. However, the cable span of a large cable driven parallel manipulator is no less than 10m, so the cable weight and deformation are non-negligible. The catenary and deformation of the cable must be taken into account in the modeling of a large cable driven parallel manipulator.A meaningful work for analyzing a large cable driven parallel manipulator with catenary and elastic deformation is to describe the cable model mathematically.

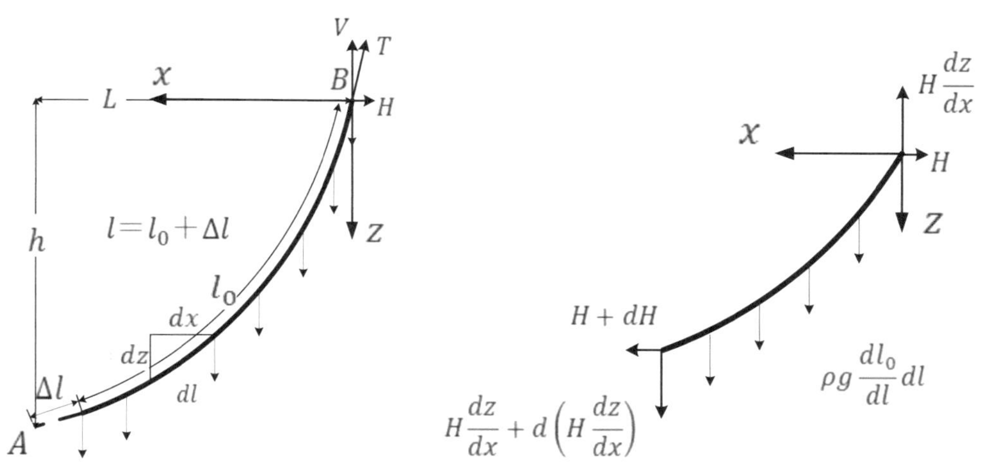

Fig. 1. catenary modelling of a cable

For setting up the cable model and cable static equilibrium equation, the symbols used in Fig. 1 are defined as:

l_0 is the unstrained length of the cables; A(the stain of the cable; T the force applied to the

fixed end of the cable; E the elastic modulus; A_0 the unstrained cross-sectional area. Using the variables and coordinate system above, we will briefly reproduce Irvine's derivation in this paper (Irivne, 1981).
To begin with, the cable must satisfy the geometric constraint:

$$\sum x = 0 \qquad H + dH - H = 0 \tag{1}$$

$$\sum y = 0 \qquad H\frac{dz}{dx} + d\left(H\frac{dz}{dx}\right) + \rho g dl_0 - H\frac{dz}{dx} = 0 \tag{2}$$

where,

$$\frac{dl}{dl_0} = \frac{T}{EA_0} + 1 \tag{3}$$

$$T = H\sqrt{1+\left(\frac{dz}{dx}\right)^2} \tag{4}$$

From $dl = dx\sqrt{1+\left(\frac{dz}{dx}\right)^2}$, Eq.(2) can be expressed as:

$$d\left(H\frac{dz}{dx}\right) + \rho g\frac{EA_0}{T+EA_0}ds = 0 \Rightarrow \frac{d^2z}{dx^2} + \frac{\rho g EA_0}{H}\frac{\sqrt{1+\left(\frac{dz}{dx}\right)^2}}{H\sqrt{1+\left(\frac{dz}{dx}\right)^2}+EA_0} = 0 \tag{5}$$

assuming $\frac{dz}{dx} = p$ Eq.(5) can be written as:

$$\frac{dp}{dx} + \frac{\rho g EA_0}{H}\frac{\sqrt{1+p^2}}{H\sqrt{1+p^2}+EA_0} = 0 \tag{6}$$

it yields:

$$\frac{dx}{dp} = -\frac{H^2}{\rho g EA_0} - \frac{H}{\rho g\sqrt{1+p^2}} \tag{7}$$

Therefore,

$$x = -\frac{H}{\rho g} sh^{-1}\left(\frac{dz}{dx}\right) - \frac{H^2}{\rho g E A_0}\frac{dz}{dx} + c_1 \tag{8}$$

where, $sh^{-1}(x) = \ln\left(x + \sqrt{1+x^2}\right), x \in (-\infty, +\infty)$

$$x = -\frac{H}{\rho g}\ln\left(\frac{dz}{dx} + \sqrt{1+\left(\frac{dz}{dx}\right)^2}\right) - \frac{H^2}{\rho g E A_0}\frac{dz}{dx} + c_1 \tag{9}$$

Integrating and applying the boundary conditions as follows:

$$x = 0, z = 0;$$

$$x = L, z = h;$$

The length of cable is l, the unstained length of the cable l_0, and Δl represents the stain of the cable, respectively. The relationship can be expressed as :$l = l_0 + \Delta l$.

$$l = \int_0^L \sqrt{1+\left(\frac{dz}{dx}\right)^2}\,dx \tag{10}$$

$$l_0 = \int_l \frac{1}{\dfrac{T}{EA_0}+1}\,dl \tag{11}$$

The numerical method is presented in the flow chart below:

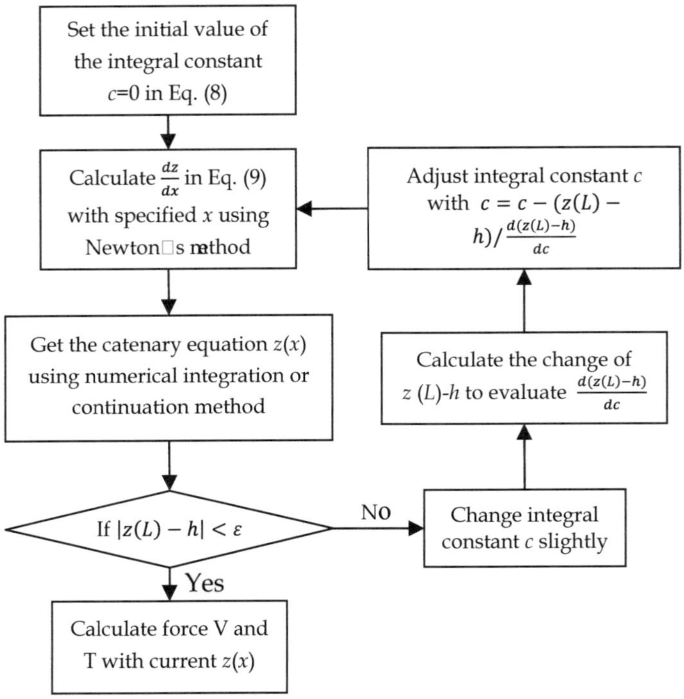

Fig. 2. Solution of a cable catenary with deformation

3.2 Static tension equilibrium equation for a cable driven parallel manipulator

In this paper, the CPR cable driven parallel manipulator with large span has following characteristics. The manipulator is static or moving slowly enough to be considered static. Gravity is considerable. The cables are assumed to be elastic. When cable catenary is not neglected, the static tension equation for a cable driven parallel manipulator becomes quite difficult to compute. In this section, the cable static tension equation is posed, and a solution method is presented for the CPR cable driven parallel manipulator.

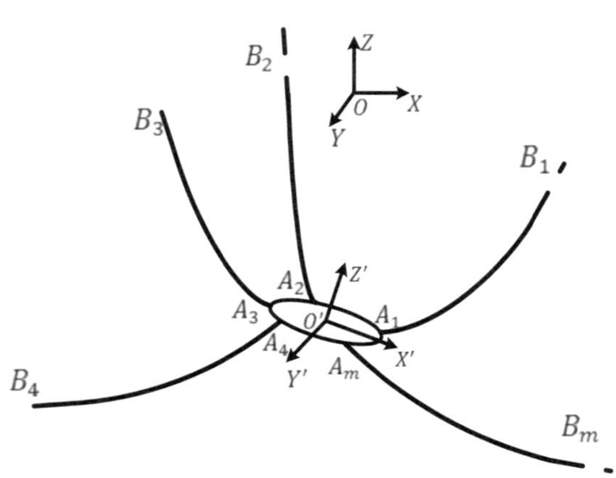

Fig. 3. A n DOFs cable driven parallel manipulator

Fig.3 is a n DOFs cable driven parallel manipulator, and two coordinates are set up: an inertial frame $\mathfrak{R}: O-XYZ$ is located at the center of the fixed platform. Another moving frame $\mathfrak{R}': O'-X'Y'Z'$ is located at the center of the moving platform. B_i (i= 1,2,...,m) are connected points of the cables and fixed platform, and A_i(i=1,2,...,m) are hinge center of the cables and moving platform.

For analysis, the symbols used in this section are defined as:

$O'^{\mathfrak{R}}$ is the O' expressed in the inertial frame; $B_i^{\mathfrak{R}}$ the vector B_i expressed in the inertial frame; $A_i^{\mathfrak{R}}$ the vector A_i expressed in the inertial frame; $A_i^{\mathfrak{R}'}$ the vector A_i expressed in the moving frame; The symbols with subscript i express the parameters of the $i-th$ cable.

According to Fig. 3, the vector of the cables can be defined as:

$$\boldsymbol{A}_i^{\mathfrak{R}} = R \cdot \boldsymbol{A}_i^{\mathfrak{R}'} + \boldsymbol{O}'^{\mathfrak{R}} \tag{12}$$

Where is R is the coordinate-axis rotation matrix.

Assuming $\boldsymbol{L}_i = \boldsymbol{B}_i^{\mathfrak{R}} \boldsymbol{A}_i^{\mathfrak{R}}, \boldsymbol{u}_i = \frac{L_i}{\|L_i\|}, \boldsymbol{r}_i = \boldsymbol{O}'^{\mathfrak{R}} \boldsymbol{A}_i^{\mathfrak{R}}$, static equilibrium equations of a cable driven parallel manipulator can be written as:

$$\boldsymbol{F} = J^T \boldsymbol{\sigma} \tag{13}$$

Where σ is the vector quantity of the cable tension (like T in Fig.1); J^T the tension transmission matrix of the cable driven parallel manipulator; $\boldsymbol{F} \in R^n$ the wrench of the moving platform.

$$\boldsymbol{\sigma} = [\sigma_1, \sigma_2, ..., \sigma_m]^T \tag{13}$$

$$J^T = \begin{bmatrix} \boldsymbol{u}_1 & \cdots & \boldsymbol{u}_m \\ \boldsymbol{r}_1 \times \boldsymbol{u}_1 & \cdots & \boldsymbol{r}_m \times \boldsymbol{u}_m \end{bmatrix}_{n \times m} \tag{14}$$

According to the section 3.1, the cable vertical tension can be expressed as:

$$V_i = H_i \frac{dz}{dx} \tag{15}$$

The static tension equilibrium equation for a cable driven parallel manipulator can be derived as:

$$\begin{cases} \sum_{i=1}^{m} \sigma_{ix} = 0; & \sum_{i=1}^{m} \sigma_{iy} = 0; & \sum_{i=1}^{m} \sigma_{iz} = 0; \\ \sum_{i=1}^{m} M_{ix} = 0; & \sum_{i=1}^{m} M_{iy} = 0; & \sum_{i=1}^{m} M_{iz} = 0; \end{cases} \tag{16}$$

So the solution flow chart can be shown in Fig.4.

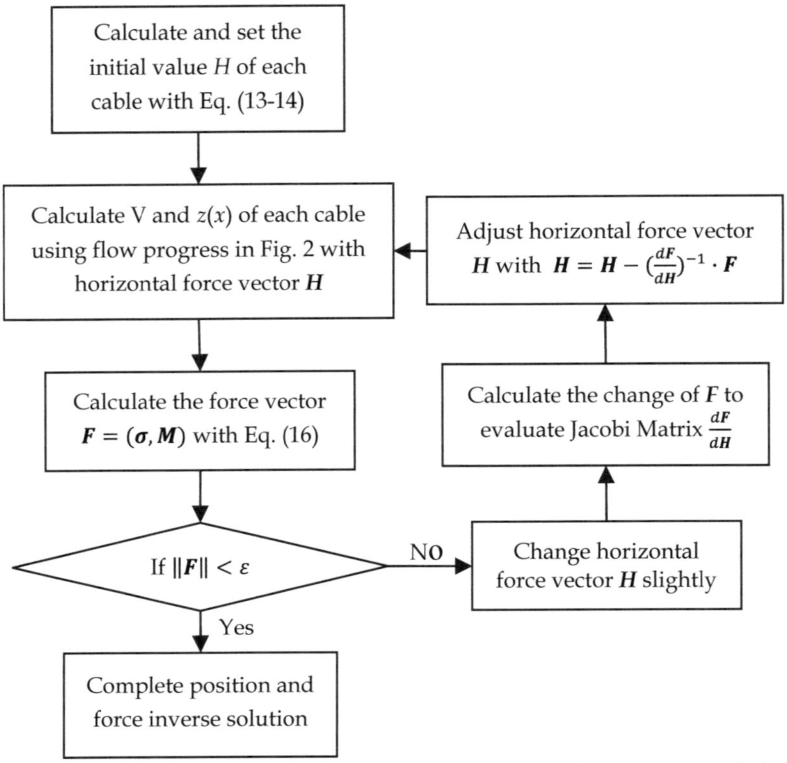

Fig. 4. Solution of a cable driven parallel manipulator with cable catenary and deformation

4. Description and modelling of the six-cable driven parallel manipulator in FAST

4.1 Description of the six-cable driven parallel manipulator

China began to build a Five-hundred meter Aperture Spherical radio Telescope (FAST). Due to the largeness in FAST, no solid support system is available for orientation of the feed. Therefore, a cable driven parallel manipulator is introduced for the first adjustable feed support system to provide a primary orientation. For the design of the cable driven parallel manipulator in FAST, several design proposals are made, one of which is the six-cable driven parallel manipulator.

In Figure 5, the feed support system of FAST is a hybrid manipulator which includes two parts: a cable driven parallel manipulator is the first adjustable feed support system and a feed mechanism is the second adjustable feed support system which includes an A-B platform and a Stewart platform with a feed on its moving platform.

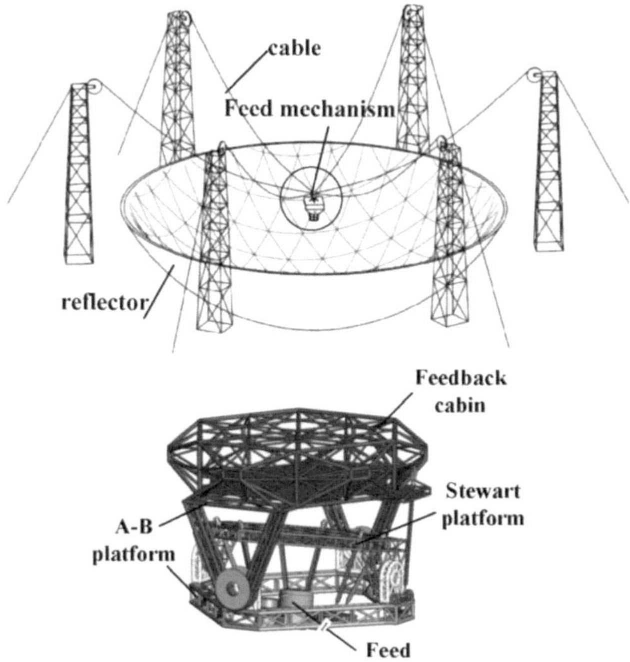

Fig. 5. Feed support system of FAST

The six-cable driven parallel manipulator in FAST draws the feed mechanism to primary orientation. The A-B platform, as the moving platform of the six-cable driven parallel manipulator, can provide accurate pose angle of feed to track celestial bodies. The Stewart platform, fixed on the A-B platform, is to improve the orientation accuracy of the feed. Due to the distribution of the cables, spin angle of this six-cable driven parallel manipulator is small, and the magnitude order of spin angel is 0.1°. The spin angle will influence the motion stability of the moving platform, so we restrict the spin angle. Therefore, we suppose the six-cable driven parallel manipulator is a CPR cable driven parallel manipulator with 5 DOFs.

4.2 The coordinate-axis rotation matrix based on Tilt-and-Torsion Angles

As shown in Fig.6, pose angle of the six-cable driven parallel manipulator is the tracking angle of feed. The tracking angle can be described as an azimuth angle and a tilt angle, which is difficult to be discriminated by Euler-angles description. Gosselin C.M (Gosselin, 2002) presented a T&T (Tilt-and-Torsion) angle to set up the coordinate-axis rotation matrix R, which can directly discriminate the azimuth and tilt angles.

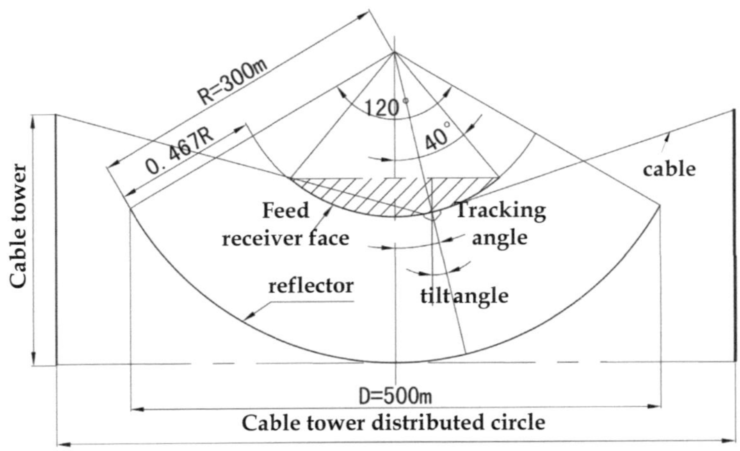

Fig. 6. Pose angle of the six-cable driven parallel manipulator

In Fig.7, three angles are defined: azimuth o, tilt 6 and torsion <x>. These new angles are easier to interpret geometrically and allow simple computation and representation of the 3D orientation workspace

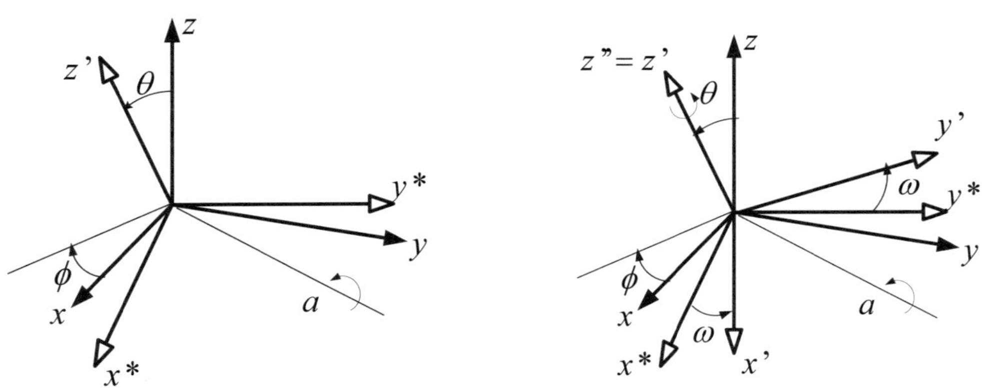

Fig. 7. Tilt-and-Torsion Angles The rotation matrix is derived as follows:

$$R_0 = R_a(\theta)R_z(\omega) = R_z(\phi)R_y(\theta)R_z(-\phi)R_z(\omega) = R_z(\phi)R_y(\theta)R_z\left(R_z(\phi)R_y(\theta)R_z(-\theta)-\phi\right)$$
$$= \begin{bmatrix} c\phi c\theta c(\omega-\phi)-s\phi s(\omega-\phi) & -c\phi c\theta s(\omega-\phi)-s\phi c(\omega-\phi) & c\phi s\theta \\ s\phi c\theta c(-\phi)+c\phi s(\omega-\phi) & -s\phi c\theta s(\omega-\phi)-c\phi c(\omega-\phi) & s\phi c\theta \\ -s\phi c(\omega-\phi) & s\phi c(\omega-\phi) & c\theta \end{bmatrix} \quad (17)$$

Where c(*) and s(*) correspond to the cos(*) and sin(*) , respectively.
The torsion angle of the six-cable driven parallel manipulator is restricted, the 0) = 0. The rotation matrix can be derived as:

$$R=\begin{bmatrix} c\phi c\theta c(-\phi)-s\phi s(-\phi) & -c\phi c\theta s(-\phi)-s\phi c(-\phi) & c\phi s\theta \\ s\phi c\theta c(-\phi)+c\phi s(-\phi) & -s\phi c\theta s(-\phi)-c\phi c(-\phi) & s\phi c\theta \\ -s\phi c(-\phi) & s\phi c(-\phi) & c\theta \end{bmatrix} \tag{18}$$

For the six-cable driven parallel manipulator, the azimuth o is the tracking azimuth and tilt θ is the pitching angle of the feed. For a cable driven parallel manipulator without torsion angle, it is useful to set up the coordinate-axis rotation matrix fi by the Tilt-and-Torsion Angles.

4.3 modeling of the six-cable driven parallel manipulator

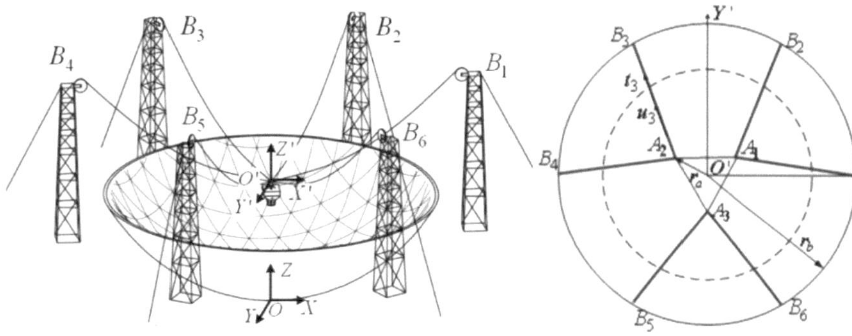

Fig. 8. Geometric parameter of the parallel mechanism

Fig.8 is the six-cable driven parallel manipulator in FAST, and two coordinates are set up: an inertial frame 5R: $0-XYZ$ is located at the center of the reflector's bottom. Another moving frame SR': $0'-X'Y'Z'$ is located at the center of the moving platform. ß; (i = 1,2, ...,6) are connected points of the cables and cable towers, and Aj (i = 1,2,3) are hinge center of the cables and moving platform.
For analysis, the symbols used in this section are defined as:
0'M is the O'expressed in the inertial frame; the vector ß; expressed in the inertial frame; A^ the vector Aj expressed in the inertial frame; the vector Aj expressed in the moving frame; r_b the radius of the cable towers' distributed circle; r_a the radius of the moving platform; D the aperture of the reflector; h the height of the cable tower; d the diameter of the cable; m the weight of the feed mechanism. It is defined as:

$$B_i^{\mathfrak{R}}=\left[r_b\cos(i-1)60°, r_b\cos(i-1)60°, h\right]^T \quad (i=1,2,...,6) \tag{19}$$

$$A_j^{\mathfrak{R}'}=\left[r_a\cos(4j-3)30°, r_a\sin(4j-3)30°, 0\right]^T \quad (j=1,2,3) \tag{20}$$

$$\boldsymbol{A}_j^{\mathfrak{R}}=R\cdot\boldsymbol{A}_j^{\mathfrak{R}'}+\boldsymbol{O}^{'\mathfrak{R}} \tag{21}$$

Where is R is the coordinate-axis rotation matrix based on Tilt-and-Torsion Angles. Then, according to the section 3, the static tension equilibrium equation for the six-cable driven parallel manipulator and its solution can be given.

5. Dimensional optimization of the six-cable driven parallel manipulator in FAST

5.1 Geometric parameters of the six-cable driven parallel manipulator

According to the reference report of FAST provided by National Astronomical Observatories Chinese Academy of Science (Nan, 2005), the related geometric parameters are given in Table 2.

Geometric Parameters	Symbol	dimension
Aperture of the reflector	D	500 (m)
Radius of the moving platform	r_a	5 (m)
Weight of the feed mechanism	m	30 (ton)
Young's modular	E	1.6×10^{11} (Pa)
Radius of cable tower's distributed circle	$r_{\prime\prime}$	To be optimized (m)
Height of cable tower	h	To be optimized (m)
Diameter of cable	d	To be optimized (mm)
Density of cable	P	To be optimized (Kg/m)

Table 2. related geometric parameters of the six-cable driven parallel manipulator

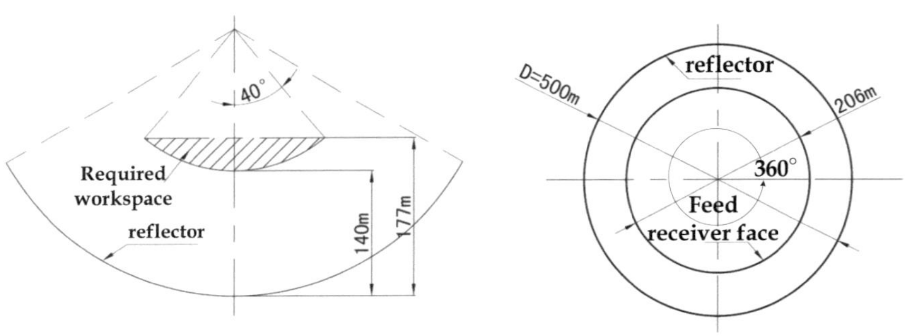

Fig. 9. required workspace of the feed

The required workspace of the feed is shown in Fig.9. It is a sphere crown with a radius of 160m. The center of the sphere and the reflector are concentric. Height of the required workspace is from 140m to 177m above the bottom of the reflector. The tracking angle of the feed is: azimuth ϕ: 0° - 360°, tilt θ: 0° -40°.

5.2 Dimensional optimization method for the six-cable driven parallel manipulator

Cable driven parallel manipulator is a tension redundant mechanism. If one of the cable tensions is negative or small, cable driven parallel manipulator will be uncontrollable. So it is necessary to optimize the geometric parameters by adopting a cable tension constraint condition in the required workspace.

The constraint conditions of cable tension are given as follows:

$$\sigma_{\min} \le \sigma_i \le \sigma_{\max} \quad (i = 1, 2, 3, \ldots, 6) \tag{22}$$

where σ_{min} is pretightening tension of cable, σ_{max} is the maximum tension of cable. Feed support system in FAST is a hybrid manipulator, and the six-cable driven parallel manipulator is used to primary orientation. The required tilt angle of feed is 0° —40°, and we hope the tilt angle of the six-cable driven parallel manipulator can closest to required title angle of feed. So, the optimization target is tilt angle of the six-cable driven parallel manipulator.

In the Six-cable driven parallel manipulator, assuming the tilt angle is θ and the required tilt angle is θ_{req}, the optimization target and conditions are as follows: Optimization target:

$$\min\left|\theta - \theta_{req}\right| \tag{23}$$

Optimization constraint conditions:

$$\sigma_{\min} \le \sigma_i \le \sigma_{\max} \quad (i = 1, 2, 3, ..., 6) \tag{24}$$

$$(x, y, z, \phi) \in \mathbb{R}_{required\ workspace} \tag{25}$$

5.3 Dimensional optimization for the six-cable driven parallel manipulator

The purpose of the dimensional design is to optimize the three important geometric parameters for the four-cable driven parallel manipulator: diameter of cable d, cable tower height h, and the radius of cable tower's distributed circle r_b. Considering the maximum observation scope, the cable tower height should be less than 290m (ref). A given route is the boundary of feed required workspace. The required pose angle in this route is azimuth angle ϕ:0° – 360°, tilt angle 6:40°. For the sake of safety, the cable tension is roughly calculated in an interval solution [80KN, 400KN].

So,

$$\theta_{req} = 40°; \sigma_{\min} = 80\text{KN}\ \sigma_{\max} = 400\text{KN};$$

According to the equation (22)-(25), the influence of the dimensional parameters on the tilt angle of the six-cable driven parallel manipulator is shown in Fig.10-13.

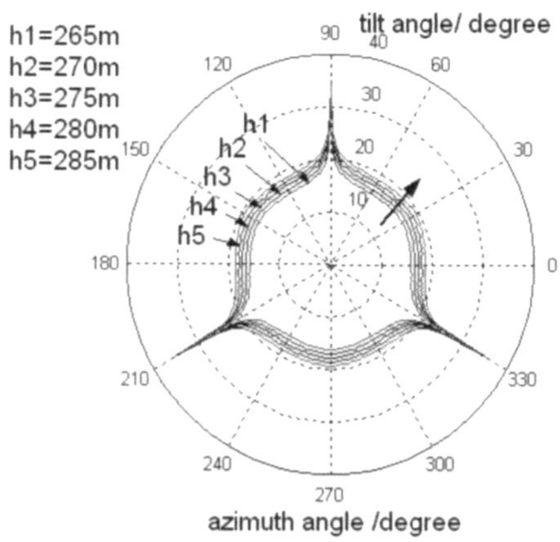

Fig. 10. Influence of cable tower height on tilt angle

In Fig.10, assuming the radius of the cable tower's distributed circle r_b is 290m and the cable's diameter d is 40mm.Given the selective cable tower heights, it shows that cable tower height has influence on workspace. Tilt angle of the six-cable driven parallel manipulator is larger by increasing the radius of the cable tower height h. So, the cable tower height h is designed as 285m.

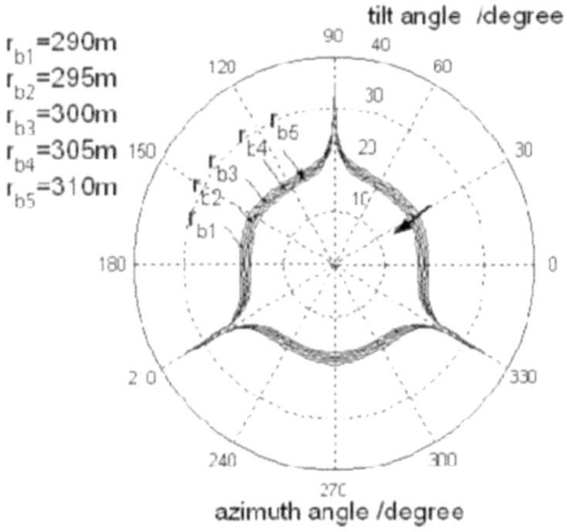

Fig. 11. Influence of radius of cable tower's distributed circle on tilt angle

As shown in Fig.11, the radius of the cable tower's distributed circle has not great influence on tilt angle. But, the increasing radius of the cable tower's distributed circle will slightly decrease the tilt angle. Therefore, the radius of the cable tower's distributed circle is designed r_b as 290m.

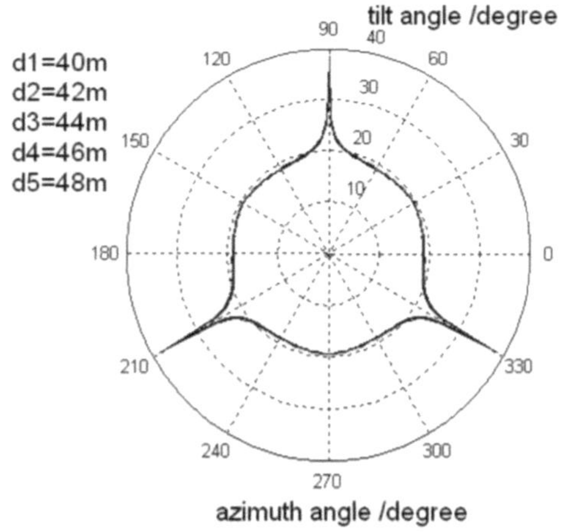

Fig. 12. Influence of diameter of cable on tilt angle

Fig.12 shows that the diameter of cable has almost no influence on tilt angle. It indicates that all the given diameter of cable can be used in FAST. However, For the sake of safety, cable's diameter d is designed as 42mm and its density is 7.35kg/m.

According to the dimensional optimization method, a set of final optimized dimensional parameters of the six-cable driven parallel manipulator are obtained in Table 3.This set of optimized dimensional parameters can be applied in the construction of the feed support system in FAST.

Geometric Parameters	Symbol	Optimized dimension
Aperture of the reflector	D	500 (m)
the radius of the moving platform	r_a	5 (m)
Weight of the feed mechanism	m	30 (ton)
Young's modular	E	1.6×10^{11} (Pa)
Radius of cable tower's distributed circle	r_h	290 (m)
Height of cable tower	h	285 (m)
Diameter of cable	d	42 (mm)
Density of cable	P	7.35 (Kg/m)

Table 3. optimized geometric parameters of the six-cable driven parallel manipulator

6. Conclusion

This paper addressed several important issues related to a large CPR cable driven parallel manipulator. In conclusion, we emphasize the following:

Firstly, in section 3 we have introduced an effective modeling method for a large CPR cable driven parallel manipulator. This method set up the effective catenary formula with cable elastic deformation into the tension equilibrium equation to work out the cable tensions. Secondly, according to the modeling method, a coordinate-axis rotation matrix based on Tilt-and-Torsion Angles is adopted to set up the tension equilibrium equation for the six-cable driven parallel manipulator in FAST. This rotation matrix can directly discriminate the tracking angle of feed.

Thirdly, based on the requirement workspace and tension condition, a set of optimization dimensional parameters is obtained to build the feed support system of FAST. The dimensional optimization method can avoid control invalidity for a cable driven parallel manipulator.

7. Acknowledge

This research is sponsored by National Natural Science Foundation of China (No.50605035,50975149), and the National 973 Program of China (No. 2007CB714007). and the Important National Science & Technology Specific Projects of China(No.2009ZX04001-042-02).

8. References

Alp, A.B.; Agrawal, S.K.(2002). Cable suspended robots : Deisgn, planning and control, Proceedings of the 2002 IEEE International Conference on Robotics and Automation, pp.4275-4280, May 2002,Washinton.

Lambert, C.; Nahon, M. & Chalmers, D. (2007). Implementation of an Aerostat Positioning System with Cable Control, IEEE/ASME Transactions on Mechatronics, vol. 12, (Feb. 2007) page numbers(32-40).

Verhoeven, R.; Hiller, M. & Tadoroko, S.(1998). Workspace, stiffness, singularities and classification of tendon driven Stewaet platform, Proceedings of the 6th International Symposium on Advances in Robot Kinematics, pp.105-114, 1998, Strobl, Austria.

Hiller, M.; Fang, S Q. (2005). Design, analysis and realization of tendon-based parallel manipulators, Mechanism and Machine Theory, vol.40, (2005) page numbers(429-445).

Hassan, M.; Khajepour, A. (2008). Optimization of Actuator Forces in Cable-Based Parallel Manipulators Using Convex Analysis, IEEE/ASME Transactions on Mechatronics, vol. 24, (June 2008) page numbers (736-740).

Yang, G L.; Pham, C B. & Yeo, S H.(2006). Workspace Performance Optimization of Fully Restrained Cable-Driven Parallel Manipulators, proceedings of the 2006 IEEE/RSJ International Conference on Intelligent Robots and Systems, pp.85-90, Jan 2006,Beijing, China.

Fang, S.; Franitza, D. & Torlo, M.(2004). Motion control of a tendon-based parallel manipulator using optimal tension distribution, IEEE/ASME Transactions on Mechatronics, vol. 9, (Sep. 2004) page number(561). Yao, R.; Tang, X Q. & Li, T M.(2007). Analysis and design of 3T cable-driven parallel manipulator for the feedback's orientation of the large radio telescope (in Chinese), Chinese Journal of Mechanical Engineering, vol.43, (Nov, 2007)page number (105-110). Duan, B Y.; Du, J L. & Bao, H.(2008). Dynamic modeling and active control of a cable-suspended parallel robot, Mechatronics, vol.18, (2008) page number (1-12).

Kozak, K.; Zhou, Q. & Wang, J S.(2006). Static Analysis of Cable-Driven Manipulators with Non-Negligible Cable Mass, IEEE transactions on robotics, vol.22, No.3,(Mar.2006), page number(425-433).

Krishna.; Prem.(1978). Cable-suspended roofs (Book style), New York McGraw-Hill, (1978), page number (27-65).

Yang, M G.; Chen, Z Q.(2003). Nonlinear analysis .of cable structure using a two-node curved cable element of high precision (in Chinese), China: Engineering Mechanics, vol.20, (2003), page number (42-47).

Tang, X Q.; Wang, J S. & Zhang, H. (2004). On the analysis of active reflector supporting mechanism for large spherical radio telescope, Mechatronics, vol.14, (Sep.2004) page numbers (1037-1053).

Nan, R D. (2006). Five hundred meter aperture spherical radio telescope (FAST), Science in China: Series G Physics, Mechanics & Astronomy, vol.49, (Feb.2006) page numbers (129-148).

Irivne, H.(1981). Cable structures, MA :MIT Press, Cambridge.

Gosselin, C M.(2002). Geometric analysis of parallel mechanisms.(2002) page numbers (75 – 81).

Nan, R D.(2005). Structure for supporting the feedback cabin of the FAST (Report style), National Astronomical Observatories Chinese Academy of Science, (2005).

27

Multi-Criteria Optimization Manipulator Trajectory Planning

E. J. Solteiro Pires and P. B. de Moura Oliveira
Universidade de Trás-os-Montes e Alto Douro
epires,oliveira@utad.pt
Portugal

J. A. Tenreiro Machado
Instituto Superior de Engenharia do Porto
`jtm@isep.ipp.pt`
Portugal

1. Introduction

In the last twenty years genetic algorithms (GAs) were applied in a plethora of fields such as: control, system identification, robotics, planning and scheduling, image processing, and pattern and speech recognition (Bäck et al., 1997). In robotics the problems of trajectory planning, collision avoidance and manipulator structure design considering a single criteria has been solved using several techniques (Alander, 2003).

Most engineering applications require the optimization of several criteria simultaneously. Often the problems are complex, include discrete and continuous variables and there is no prior knowledge about the search space. These kind of problems are very more complex, since they consider multiple design criteria simultaneously within the optimization procedure. This is known as a multi-criteria (or multi-objective) optimization, that has been addressed successfully through GAs (Deb, 2001). The overall aim of multi-criteria evolutionary algorithms is to achieve a set of non-dominated optimal solutions known as Pareto front. At the end of the optimization procedure, instead of a single optimal (or near optimal) solution, the decision maker can select a solution from the Pareto front. Some of the key issues in multi-criteria GAs are: i) the number of objectives, ii) to obtain a Pareto front as wide as possible and iii) to achieve a Pareto front uniformly spread.

Indeed, multi-objective techniques using GAs have been increasing in relevance as a research area. In 1989, Goldberg suggested the use of a GA to solve multi-objective problems and since then other researchers have been developing new methods, such as the multi-objective genetic algorithm (MOGA) (Fonseca & Fleming, 1995), the non-dominated sorted genetic algorithm (NSGA) (Deb, 2001), and the niched Pareto genetic algorithm (NPGA) (Horn et al., 1994), among several other variants (Coello, 1998).

In this work the trajectory planning problem considers: i) robots with 2 and 3 degrees of freedom (*dof*), ii) the inclusion of obstacles in the workspace and iii) up to five criteria that are used to qualify the evolving trajectory, namely the: joint traveling distance, joint velocity, end effector / Cartesian distance, end effector / Cartesian velocity and energy involved. These criteria are used to minimize the joint and end effector traveled distance, trajectory ripple and energy required by the manipulator to reach at destination point.

Bearing this ideas in mind, the paper addresses the planning of robot trajectories, meaning the development of an algorithm to find a continuous motion that takes the manipulator from a given starting configuration up to a desired end position without colliding with any obstacle in the workspace.

The chapter is organized as follows. Section 2 describes the trajectory planning and several approaches proposed in the literature. Section 3 formulates the problem, namely the representation adopted to solve the trajectory planning and the objectives considered in the optimization. Section 4 studies the algorithm convergence. Section 5 studies a $2R$ manipulator (*i.e.*, a robot with two rotational joints/links) when the optimization trajectory considers two and five objectives. Sections 6 and 7 show the results for the $3R$ redundant manipulator with five goals and for other complementary experiments are described, respectively. Finally, section 8 draws the main conclusions.

2. Trajectory planning

Trajectory planning for robotic manipulators is the process of creating trajectories free of collisions allowing the manipulators to perform the required task. The robotic planners substitute the human operator to specify explicitly the trajectory. Consequently, the operator is free to focus his attention on the task instead of worrying about the robot movement. Therefore, the operator only needs to specify the start and the end path points, leaving the planner to generate the appropriate trajectory that the manipulator must perform.

The trajectories are made by successive displacements of the robotic end effector and a trajectory can be seen as a sequence of points in which the end effector must pass. As a result of the end effector movement over the discrete points, it is obtained a continuous trajectory (Fig. 1). Optimizing the robot trajectory involves the identification of optimal points and the corresponding intermediate positions. Nevertheless, the trajectory optimization is difficult due to the non-linearity dynamics and the dimension of the trajectory search space.

The trajectory planning can be implemented using either the direct or the inverse kinematics. When it is adopted the direct kinematic, the problem is directly solved in the joint space. On the other hand, when the problem is solved through the inverse kinematics, it is determined the trajectory of the manipulator end effector in the operational space. Then, the values of the joints are evaluated using the inverse kinematics. The resulting values of the joints variables are then used by the planner to form the final trajectory. However, when the inverse kinematics is considered, due to the existence of singularities, some problems may arise (Duarte, 2002), such as:

- The mobility of the manipulator is reduced and it way be not possible to impose certain body movements to the end effector;
- The inverse kinematics problem can have multiple solutions;
- In the neighborhood of the kinematic singularities low speeds in operational space may require high speeds in joint space.

In order to solve the planning problem it is necessary to model the trajectory. Therefore, the representation of a manipulator trajectory can be seen as a string representing all the joint positions between the initial and final robot configurations. The trajectory is determined in order to satisfy some specifications with the best performance. Best performance may have a different meaning such as minimum energy consumption, trajectory duration, singularities avoidance, *etc.* Therefore, the trajectory optimization for robotic manipulators is a problem involving the use of multiple criteria. The resolution of such problems can benefit from the multi-objective optimization, particularly when the objectives are conflictous. This means that an improvement of one criterion leads, necessarily, to the degradation of another one. In these cases, the optimal compromise between them is possible considering the concepts of non-dominance theory proposed by Pareto and successfully integrated in evolutionary multi-objective (EMO). The major disadvantage of using EMO is the increasing of the computational time with the number of objectives under analysis, which is progressively reduced with advent of faster processors. This makes the techniques proposed in this chapter to have viability in the context of industrial applications.

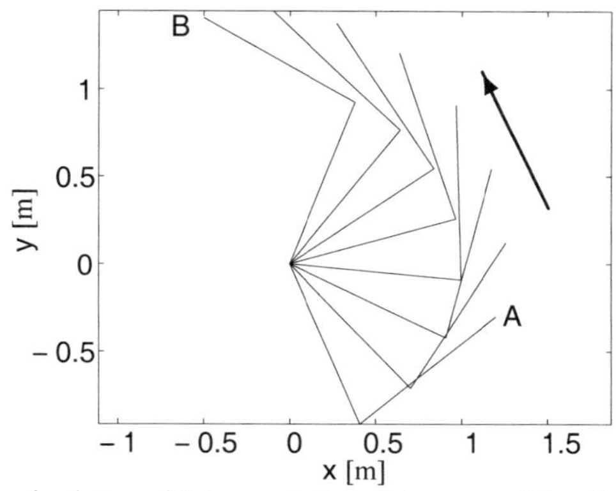

Fig. 1. Trajectory of a manipulator with two rotational degrees of freedom (2*R* robot)

2.1 Trajectory planning: A review

Trajectory planning is a fundamental research issue in robotics and, GAs that have been successfully adapted to solve many problems, also tracked this problem. This section describes GA-methods, proposed by several authors, to solve the planning of robot trajectories.

Davidor (1991) was the first researcher that applied GAs to robot path planning. He uses a string with a non fixed size to represent paths for a 3*R* robotic manipulator. The performance was evaluated considering the error between the desired path and the path yielded by the GA. To find the manipulator configurations he used the inverse kinematics. Other application similar was proposed by Nearchour and Aspragathos (1995) for a 7*R* robot. In this work a point-to-point movement was considered in a workspace with obstacles. The problems consisted in finding the configuration that the manipulator should adopt in order to put the end effector in the correct position. Another problem, using the inverse kinematics, was addressed by Lavoie and Boudreau (2001) for 7 and 10 *dof* manipulators. The end effector had to travel through several pre-defined points and the objective was to minimize the joint displacement without robot colliding with obstacles. Doyle and Jones (1996) proposed a GA to find the successive robot configurations with the algorithm searching the joint space in order to find the best path. Lee *et al.* (1999) proposed two applications based on evolutionary algorithms (EAs) for a 2*R* robot: One application using a GA with the penality method, and the second adopting a evolutionary strategy with a heuristic methods and using a time scheduling to lead with the joint motors torque constraints. A comparison with linear programming was included in the work.

Other algorithms inspired in nature have also been applied to robot trajectory planning. Kubota *et al.* (1997; 1996) proposed a hierarchical EA to finds the trajectory of a redundant manipulator without colliding with the workspace obstacles. The algorithm was based on the co-evolution of virus and host populations. One population calculates some manipulator collision-free positions and the other generates a collision-free trajectory by combining these intermediate positions. The objective consisted in minimizing an aggregate function including the distance traveled and the energy required to perform the task. Luo and Wei (2004) addressed the problem for a $3R$ manipulator based on a biological immune system. The problem was reduced to one objective, aggregating all the criteria together, and to one parameter (joint) making all the other joints dependent of that variable.

GAs can be used in trajectory generation to determine the intermediate points of the polynomial that describes the manipulator movement. In this kind of application the velocities and accelerations are considered null in the extreme points. Wei-Min and Yu-Geng (1996) developed a method for the $2R$ and $3R$ manipulators incorporating kinematic, dynamical and control constraints. The method uses polynomials of 4th and 5th degree and optimizing the time duration of the trajectory. Wang and Zalzala (1996) also developed an algorithm for an industrial robot to provide a short travel time duration. The algorithm divided the joint space through a grid, using a 6th degree polynomial and a binary GA with heuristic search techniques. Another application was proposed by Tian and Collins (2004), adopting a binary GA and a cubic polynomial for the trajectory generation of a $2R$ manipulator in a workspace containing point obstacles.

GAs are also applied in cooperative manipulator systems that is having more than one manipulator operating in the same workspace. Rana and Zazala (1996) proposed a technique evolving two manipulators. This algorithm minimized the trajectory duration while avoiding collisions between the manipulators. The planning was carried out in the joint space and the trajectory was represented as a string of via-points connected through cubic splines.

Another method based on a hybrid EA was described by Ridao *et al.* (2001). The GA had to determine the sequence of synchronization points where the length of the path is minimal. For this purpose, the algorithm used predetermined manipulators paths (given by other algorithms like local search). Ali *et al.* (2002) proposed a method, in a three-dimensional space, for two $2R$ manipulators which consisted in minimizing the trajectory distance without colliding with obstacles. Another application, where the objective is the minimization of the energy required by the manipulator to perform the trajectory, was proposed by Garg and Kumar (2001; 2002) using both GAs and simulated annealing.

In the trajectory planning there are some approaches that optimize independently the objectives. Ortmann (2001) presented a multi-criteria scheme for a manipulator where the workspace has one obstacle. The algorithm tries to find multiple joint positions of the trajectory.

Chen and Zalzala (1997) proposed a GA to generate the position and the configuration of a mobile manipulator. In this report the inverse kinematics optimizes the least torque norm, the manipulability, the torque distribution and the obstacle avoidance.

In table 1 are depicted the main differences of the mentioned works. The column 'Kinematics' indicates if the work solves the planning trough the direct (D) or the inverse (I) kinematics. The manipulator type and the number of obstacles are described in the next columns. The column 'Objective' specifies the criteria used in the optimization. The space dimension and the information if the algorithm uses cooperative techniques are described in next columns. Column eight indicates if exists more than one manipulator, and its number, in the workspace. 'Dynamics' and 'Polynomial interpolation' indicates if the algorithm uses the dynamic equations and polynomial functions, or splines, respectively. Finally, the last two columns indicate if the algorithm adopts immune systems, hybrid GAs, and if it incorporate a mobile base.

3. Problem description

This work considers robotic manipulators that are required to move between two configurations taking into consideration several objectives. The objectives considered in this work are: the joint distance (O_q), the Cartesian / gripper distance (O_p), the joint ripple ($O_{\dot{q}}$), the Cartesian / gripper ripple ($O_{\dot{p}}$) and the energy required by the manipulator to make the movement (O_{E_a}). Thus, it is intended to determine

Authors	Kinematics	Manipulator	Obstacles	Objective	Workspace	Co-Evolutive	Cooperative	Dynamics	Polinomial interpolation	Immune systems	Hybrid GA	Mobile manipulator
Davidor	I	3*R*	-	Cartesian distance	2*D*	-	-	-	-	-	-	-
Nearchour and Aspragathos	I	7*R*	Y	Cartesian distance	2*D*	-	-	-	-	-	-	-
Lavoie and Boudreau	I	7;10*R*	Y	Joint distance	2*D*	-	-	-	-	-	-	-
Doyle and Jones		3*R*	Y	Joint distance	2*D*	-	-	-	-	-	-	-
Lee *et al.*		2*R*	-	Trajecty duration	2*D*	-	-	Y	-	-	-	-
Kubota *et al.*	D	7 gdl	Y	Distance and Power	3*D*	Y	-	Y	-	-	-	-
Luo and Wei	I	3*R*	-	Cartesian distance	2*D*	-	-	-	-	Y	-	-
Wei-Min and Yu-Geng	I	2;3*R*	-	Joint distance	2*D*	-	-	Y	Y	-	-	-
Wang and Zalzala	D	6*R*	-	Joint distance	3*D*	-	-	-	Y	-	-	-
Tian and Collins	I	2*R*	Y	Joint distance	2*D*	-	-	-	Y	-	-	-
Rana and Zazala	I	2;3*R*	-	Trajecty duration	2;3*D*	-	2	Y	Y	-	-	-
Ridao *et al.*		2;3*R*	Y	Trajecty duration	2;3*D*	-	2	-	-	-	Y	-
Ali *et al.*	D	2*R*	Y	Cartesian distance	3*D*	Y	2	Y	Y	-	Y	-
Garg and Kumar	D	2*R*	-	Energy	2*D*	-	2	Y	Y	-	Y	-
Ortmann	D	3*R*	-	Arm ripple	3*D*	-	-	-	Y	-	Y	-
Chen and Zalzala	I	-	Y	Torque and manipulability	3*D*	-	-	-	-	-	-	Y

Table 1. Genetic applications in manipulator trajectory planning

a set of non-dominated solutions from the Pareto optimal front. The final solution is then chosen by the decision maker taking into account the commitment of the objectives that he finds more appropriate. Two (see figure 2) and three *dof* planar manipulators are used in the experiments. However, this algorithm can be extended to hyper-redundant robots. The rotational joint of each link, is free to rotate 2π rad. To test a possible manipulator / obstacle collision, the end effector structure is analyzed to verify if it is inside of any obstacle.

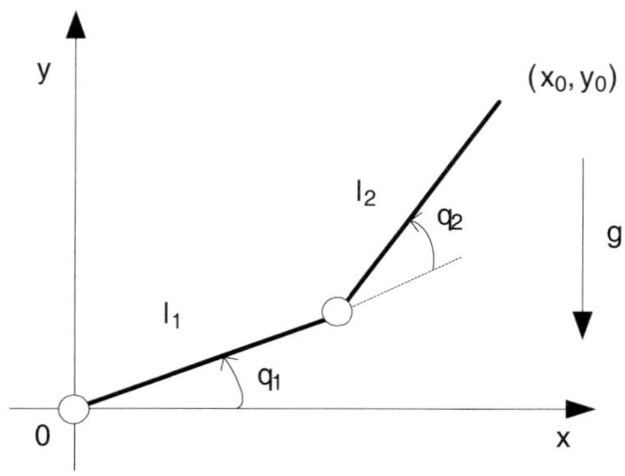

Fig. 2. Two joint (link) manipulator (2*R* robot) (g – gravity constant)

The path for a *iR* manipulator ($i = 2, 3$), at generation T, is directly encoded as a string in the joint space to be used by the GA. This string is represented by expression (1), where i represents the number of *dof* and Δt the sampling time between two consecutive configurations. Therefore, one string is codified as:

$$[\{q_1^{(\Delta t,T)}, q_2^{(\Delta t,T)}\}, \{q_1^{(2\Delta t,T)}, q_2^{(2\Delta t,T)}\}, \ldots, \{q_1^{((n-2)\Delta t,T)}, q_2^{((n-2)\Delta t,T)}\}] \quad (1)$$

where the joints values $q_l^{(j\Delta t,0)}$ ($j = 1, \ldots, n-2$; $l = 1, \ldots, i$) are randomly initialized in the range $]-\pi, +\pi]$ rad. The robot movement is described by $n = 8$ configurations. However, the initial and final configurations are not encoded into the string because they remain unchanged throughout the trajectory search. Without losing generality, for simplicity, it is adopted a normalized time of $\Delta t = 0.1$ s, but it is always possible to perform a time re-scaling.
Five indices $\{O_q, O_{\dot{q}}, O_p, O_{\dot{p}}, O_{E_a}\}$ presented in (2) quantify the qualify of the evolving trajectory robotic manipulators. The indices represents the joint distance, O_q, the joint ripple, $O_{\dot{q}}$, the Cartesian / gripper distance and ripple (O_p and $O_{\dot{p}}$), and the energy, O_{E_a}.

$$O_q = \sum_{j=1}^{n} \sum_{l=1}^{i} \left(\dot{q}_l^{(j\Delta t,T)}\right)^2 \quad (2a)$$

$$O_{\dot{q}} = \sum_{j=1}^{n} \sum_{l=1}^{i} \left(\ddot{q}_l^{(j\Delta t,T)}\right)^2 \quad (2b)$$

$$O_p = \sum_{j=2}^{n} d\left(p_j, p_{j-1}\right)^2 \quad (2c)$$

$$O_{\dot{p}} = \sum_{j=3}^{n} \left\{d\left(p_j, p_{j-1}\right) - d\left(p_{j-1}, p_{j-2}\right)\right\}^2 \quad (2d)$$

$$O_{E_a} = (n-1)T\,P_a = \sum_{j=1}^{n}\sum_{l=1}^{i} |\tau_l . \Delta q_l^{(j\Delta t,T)}| \tag{2e}$$

The function O_{E_a} evaluates the average mechanical energy during the trajectory whereas is assumed that power regeneration is not available by motors doing negative work, that is, by taking the absolute value of the power (Silva & Tenreiro Machado, 1999). The index i specifies the total number of manipulator joints. These criteria are minimized by the planner to find the non-dominated front. Before evaluating any solution, in order to remove virtual jumps, all the values such that $|q_l^{((j+1)\Delta t,T)} - q_l^{(j\Delta t,T)}| > \pi$ are readjusted, by adding or removing a multiple value of 2π in the strings.

The joint distance O_q represented in (2a) is used to minimize the manipulator joints traveling distance. In fact, for a function $y = g(x)$ the curve length is defined by equation (3) and, consequently, to minimize the curve length distance the simplified expression (4) is adopted.

$$\int\sqrt{1 + (dg/dx)^2}\,dx \tag{3}$$

$$\int (dg/dx)^2 dx = \int \dot{g}^2 dx \tag{4}$$

Throughout this chapter the results are only presented in the objective space, since the size of the space of attributes is of the order of $i(n-2)$ which makes the analysis difficult.

To solve the problem a multi-objective algorithm is used with selection operator based on the Pareto ranking proposed by Goldberg (1989) and a sharing scheme with $\sigma_{sharing} = 0.01$ and $\alpha = 2$ parameters (5), where $d_{k,j}$ is the distance between solution k and j. So, the fitness value is initially assigned according to the sorted rank of the solution in the population, and then this values is affected by the number of solutions in their neighborhood (niche count). To evaluate the niche count metric, all the neighborhood solutions of the population are considered regardless of their rank. After the crossover and the mutation operations, the elements that are selected to the next generation are the best of all between parents and their offspring.

$$f'^{(k)} = \frac{f^{(i)}}{n_{c_k}} \tag{5a}$$

$$n_{c_k} = \sum_{j=1} d_\sigma(d_{k,j}) \tag{5b}$$

$$d_\sigma(d_{k,j}) = \begin{cases} 1 - \left(\frac{d_{k,j}}{\sigma}\right)^\alpha, & d_{k,j} < \sigma \\ 0, & d_{k,j} \geq \sigma \end{cases} \tag{5c}$$

4. Algorithm convergence

This section analysis the convergence of the algorithm to the locals fronts when the trajectory is optimized. Several experiments consisting in moving a $2R$ manipulator from the point $A \equiv \{x_0, y_0\} = \{1.2, -0.3\}$ up to the point $B \equiv \{x_1, y_1\} = \{-0.5, 1.4\}$ are performed. The optimization considers the angular and the Cartesian end effector distance criteria, while the initial and final configurations are determined through the inverse kinematics. The simulations adopt the parameters (defined in advance): $pop_{dim} = 300$ population solutions, during $T_t = 1500$ generations, and probabilities of crossover and mutation of $p_c = 0.6$ and $p_m = 0.05$, respectively. Each link has a length of $l = 1$ m and mass of $m = 1$ kg.

The algorithm converges to the fronts illustrated in Fig. 3. One of the fronts is obtained when the planar manipulator moves around its base in a counterclockwise direction (Figure 4(a)). The corresponding optimal Pareto front is shown in Fig. 3(a) as f_P and is expanded separately illustrated in Fig. 3(b). The other local front is obtained when the manipulator circumvents the base in the clockwise direction (Fig. 4(b)). The front is denoted by f_l in Fig. 3(a) and is represented separately in Fig. 3(c).

Fig. 3(a) depicts the initial population I_{pop}, giving rise to the Pareto front f_P. In figure it is also illustrated the local front obtained from another experiment. In 80.1%of the tests carried out (in a set of 21 simulations) the algorithm converges to the optimal Pareto front f_P. In the rest of the cases the algorithm converges to the local front f_l.

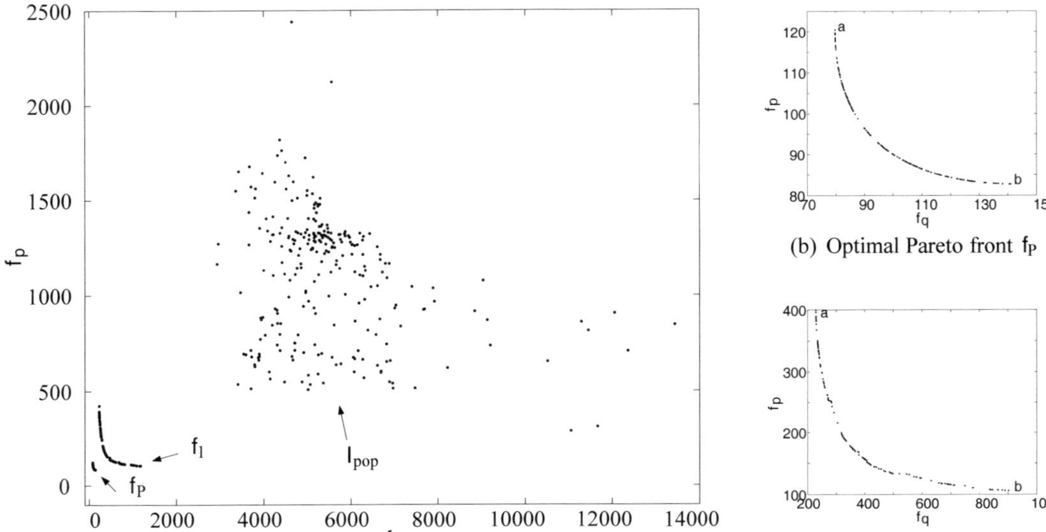

(a) Pareto front (f_P), local front (f_l) and initial population (I_{pop})

(b) Optimal Pareto front f_P

(c) Local front f_l

Fig. 3. The local fronts and the initial population

One of the problems associated with the use of EAs, in the context of optimization problems, is called premature convergence, which consists in the incapacity of the population to converge the global optimal (Bäck, 1996). This phenomenon arises when the objective function has an local optimal with a large base location or the local optimal is in a small spectrum. Therefore, the relationship between the convergence to the global optimum and the geometry function is very important. If the population of the EA is 'trapped' in a large local geometry, then it is difficult for the variation operators to produce an offspring with better performance than its parents. In the second case, if the global optimum is located in a geometry relatively small and the population did not found it until the moment, then variation operators have a low probability of producing descendants in these regions is very low. This happens in 19.9%of the cases, when the algorithm converges to local front getting trapped in it. Indeed, the optimal front solutions are so different from Pareto front, that operators have an extremely low probability to transform a solution of the local front in a solution of Pareto front.

In Fig. 5 is represented the configuration of the robotic manipulator and the joints displacement for extreme solutions of the Pareto optimal front, that are represented by solutions '*a*' and '*b*' in Fig. 3(b).

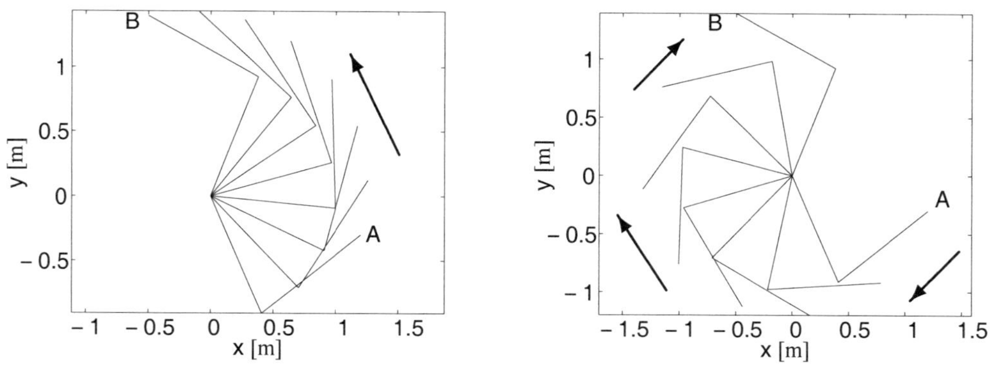

(a) A trajectory of Pareto optimal front f_P (b) A trajectory of local (non-optimal) front f_l

Fig. 4. Trajectories of the local fronts for the 2*R* manipulator

5. Experiments with the 2*R* manipulator

This section presents other experiments for the 2*R* manipulator, when are optimized five objectives, two by two at a time (2*D*), or the five criteria simultaneously (5*D*).

In Fig. 6 the Pareto optimal fronts are illustrated when two criteria are optimized a time, namely the ten cases: $O(q, \dot{q})$, $O(q, E_a)$, $O(q, p)$, $O(q, \dot{p})$, $O(\dot{q}, E_a)$, $O(\dot{q}, p)$, $O(\dot{q}, \dot{p})$, $O(E_a, p)$, $O(E_a, \dot{p})$ and $O(p, \dot{p})$. In the figures it can be seen a good distribution of the solutions along the Pareto fronts for all experiments.

In Fig. 7 are illustrated the results when the algorithm optimizes the joint distance and the joint velocity $O(q, \dot{q})$. The solution $s_{\dot{q},min}$ corresponds to the case where the $\dot{q}$ objective value is minimal. As expected the solution does not present any ripple (see Fig. 7(b)). On the other hand, in Fig. 8 is illustrated the results of the solution $s_{E_a\ min}$ for the optimization $O(q, E_a)$. It can be observed that the gripper describes a trajectory spending the lowest possible energy. Finally, Fig. 9 depicts the results when is minimized the joint distance and the Cartesian ripple. When is chosen an intermediate solution of a previous Pareto front, the behavior is a combination of the extreme solutions of that Pareto front.

In a second phase, the 2*R* manipulator motion is optimized when the five criteria (5*D*) are considered simultaneously. Figures 10 and 11 show the results for a number of generations $T_t = 50,000$ and a population size of $pop_{dim} = 1000$.

Fig. 10 shows the tradeoff of normalized values for the final population and the best solutions $s_i = \{s_1, s_2, s_3, s_4, s_5\}$ for the objectives $O_i = \{O_q, O_{\dot{q}}, O_{E_a}, O_p, O_{\dot{p}}\}$, respectively. Table 2 depicts the range of values obtained in the simulation. It can be observed that the algorithm 5*D* does not leads to results so good as those obtained with the algorithm 2*D*, due to the significant increase of the search space complexity. However, the solutions show a good distribution (Fig. 10(a)) and the results are close to those of the 2*D* fronts. It must be noted that, when it is considered only the 2*D* optimization it may happen that the 2*D* optimal Pareto front contains dominated solutions from the 5*D* optimization. These solutions are not taken into account when considering the 5*D* optimization.

From Fig. 10(b) it can be concluded that O_q and $O_{\dot{q}}$, or O_p and $O_{\dot{p}}$, are conflicting objectives with a small tradeoff between them. Moreover, the objective O_{E_a} presents the greatest compromise between the other objectives. Fig. 11 presents the best solution s_i, $i = \{1, ..., 5\}$, in the viewpoint of the objective i. For the trajectory under study, the results indicate that the closer the gripper moves near its base lower is the energy consumption (figures 11(e), 11(g) and 11(h)).

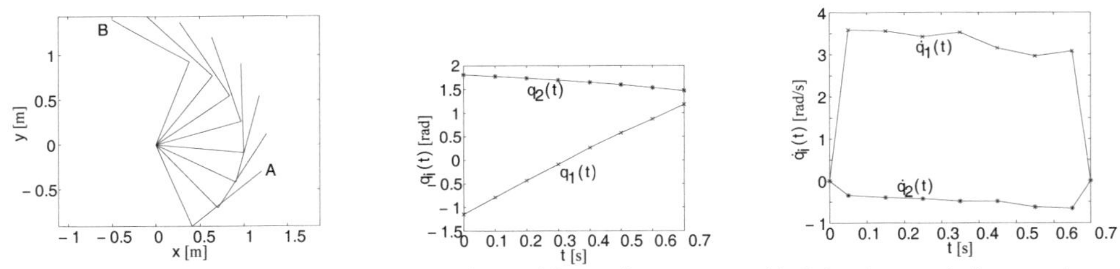

(a) Successive configurations, solution a of Fig. 3(b)) (b) Joint position trajectory vs. time, solution a of Fig. 3(b) (c) Joint time evolution vs. time, solution a of Fig. 3(b)

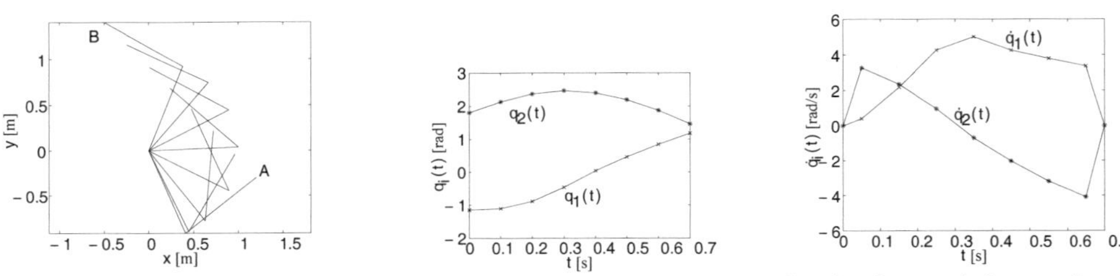

(d) Successive configurations, solution b of Fig. 3(b) (e) Joint position trajectory vs. time, solution a of Fig. 3(b) (f) Joint time evolution vs. time, solution b of Fig. 3(b)

Fig. 5. Trajectories of optimal the Pareto front for the $O(q, p)$ optimization and the $2R$ manipulator

	O_q (rad^2/s^2)	$O_{\dot{q}}$ (rad^4/s^4)	O_{E_a} (J)	O_p (m^2/s^2)	$O_{\dot{p}}$ (m^4/s^4)
Minimum	79.8	18.2	1056.7	83.5	15.0
Maximum	182.3	101.7	4602.7	121.8	56.4

Table 2. Range of the objectives with $5D$ optimization and the $2R$ manipulator

Fig. 12 illustrates the projections of the $5D$ front into the distinct $2D$ planes. In each plane, is also plotted the $2D$ non-dominated front obtained when the corresponding optimization is executed. It can be observed that the $5D$ optimization does not produce results as good as the $2D$ optimization. This is due to the increasing of the search space, reducing the ratio between the population size and the search space dimension. The distance between the fronts $2D$, $3D$, $4D$ and $5D$ increases with the number of objectives.

6. The 3*R* manipulator trajectory with 5*D* optimization

In this section the trajectory of a $3R$ manipulator is optimized considering simultaneously the five objectives listed in (2) and one circular obstacle with center at $c = (1.0, 0.4)$ and radius $\rho = 0.4$. Each link of the manipulator has a length of $l = 0.67$ m and a mass of $m = 0.66$ Kg. The initial and final configurations, $A \equiv \{q_1, q_2, q_3\} = \{-1.374, 1.129, 1.129\}$ and $B \equiv \{q_1, q_2, q_3\} = \{-1.050, 0.909, 0.909\}$ are comparable to the one adopted for the $2R$ manipulator.

Fig. 13 illustrates the normalized fitness values of the population and the solutions that have the best performance for each objective in the optimization. This simulation yields also a good distribution of the solutions. The objectives are also quarrelsome but the relative tradeoffs between them have changed. Table 3 depicts the range of objective values achieved by the manipulator in a single execution.

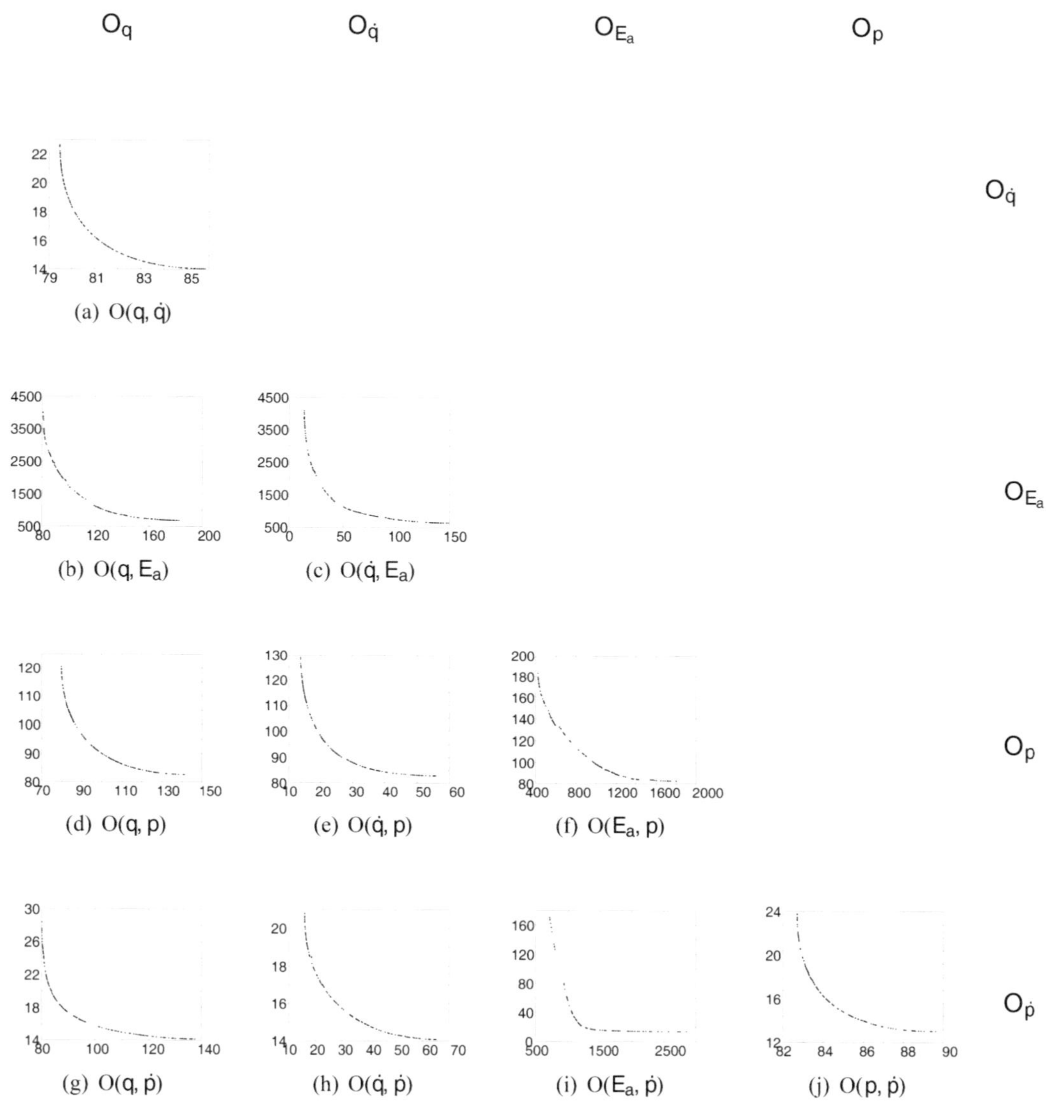

(a) $O(q, \dot{q})$

(b) $O(q, E_a)$ (c) $O(\dot{q}, E_a)$

(d) $O(q, p)$ (e) $O(\dot{q}, p)$ (f) $O(E_a, p)$

(g) $O(q, \dot{p})$ (h) $O(\dot{q}, \dot{p})$ (i) $O(E_a, \dot{p})$ (j) $O(p, \dot{p})$

Fig. 6. The *2D* Pareto optimal front for the ten cases: $O(q, \dot{q})$, $O(q, E_a)$, $O(q, p)$, $O(q, \dot{p})$, $O(\dot{q}, E_a)$, $O(\dot{q}, p)$, $O(\dot{q}, \dot{p})$, $O(E_a, p)$, $O(E_a, \dot{p})$ and $O(p, \dot{p})$ and the *2R* manipulator

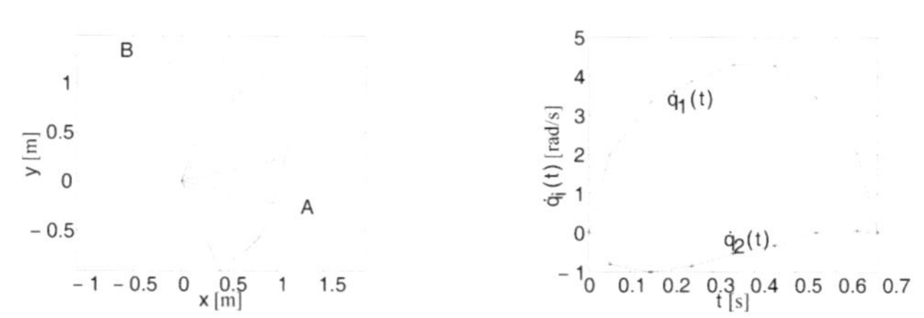

(a) Successive *2R* robot configurations (b) joint time evolution vs. time

Fig. 7. The solution $s_{\dot{q}\,min}$ of the $O(q, \dot{q})$ optimization for the *2R* manipulator

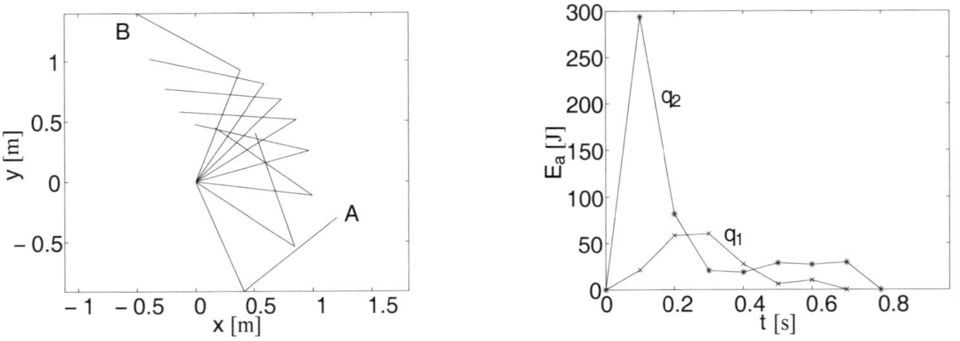

(a) Successive $2R$ robot configurations

(b) Energy required E_a

Fig. 8. The solution $s_{E_a\,min}$ of the $O(q, E_a)$ optimization for the $2R$ manipulator

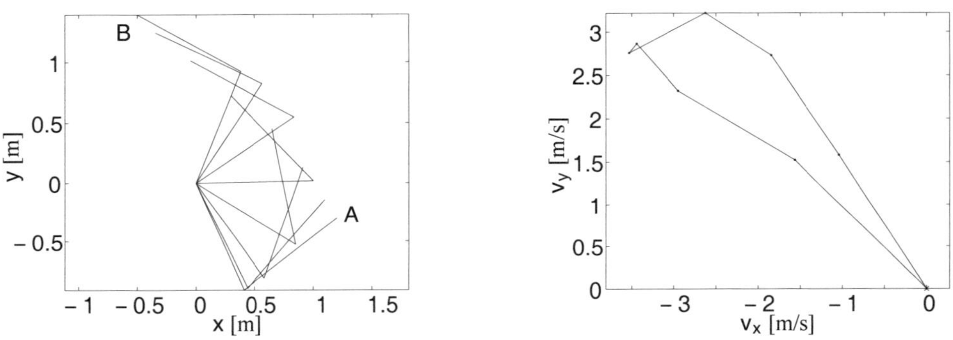

(a) Successive $2R$ robot configurations

(b) Cartesian time evolution v_x *vs.* v_y

Fig. 9. The solution $s_{\dot{p}\,min}$ of the $O(q, \dot{p})$ optimization for the $2R$ manipulator

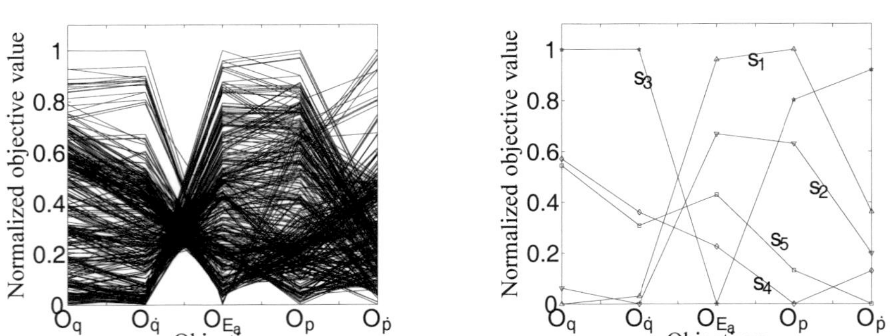

(a) Population tradeoffs

(b) Best solution of one objective tradeoffs

Fig. 10. Value path method representation of the O_q, $O_{\dot{q}}$, O_{E_a}, O_p and $O_{\dot{p}}$ objectives for the $2R$ manipulator

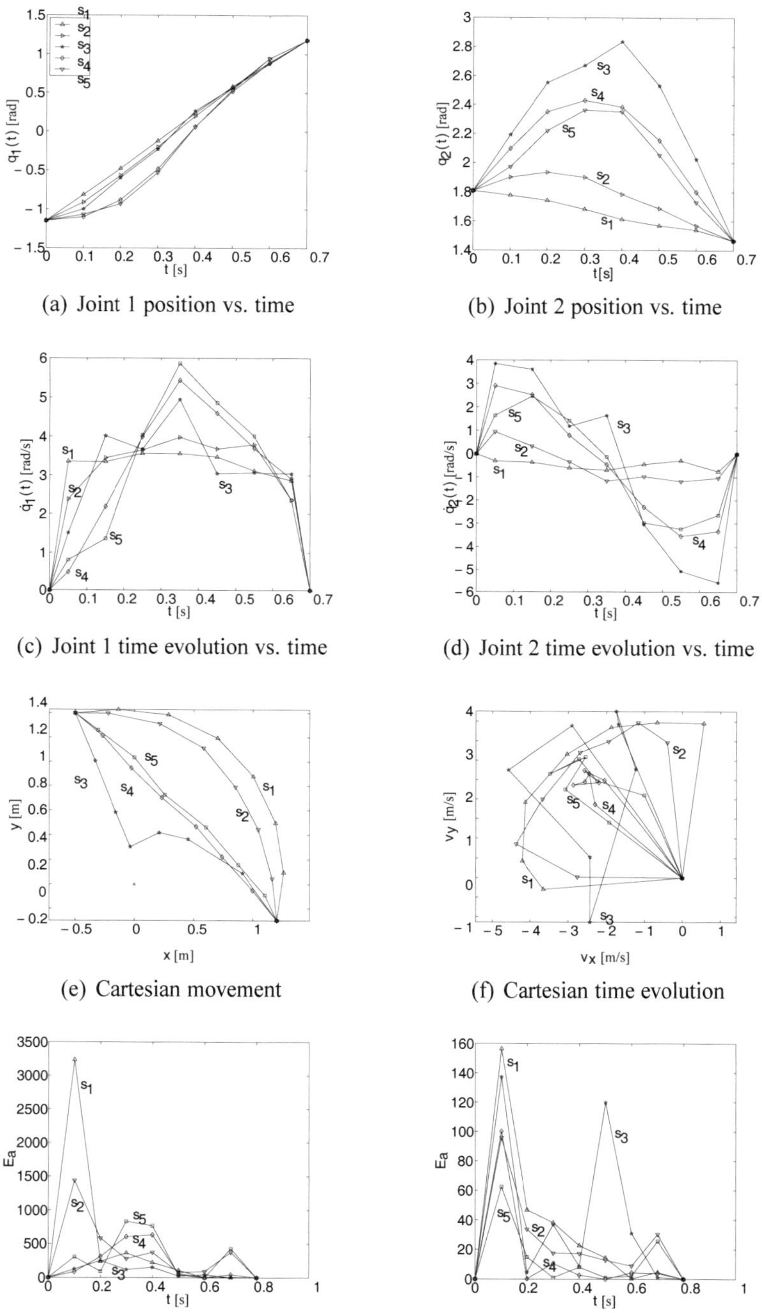

(a) Joint 1 position vs. time

(b) Joint 2 position vs. time

(c) Joint 1 time evolution vs. time

(d) Joint 2 time evolution vs. time

(e) Cartesian movement

(f) Cartesian time evolution

(g) Joint 1 energy required vs. time

(h) Joint 2 energy required vs. time

Fig. 11. Behavior of the best solutions obtained for the $2R$ robot with $5D$ optimization

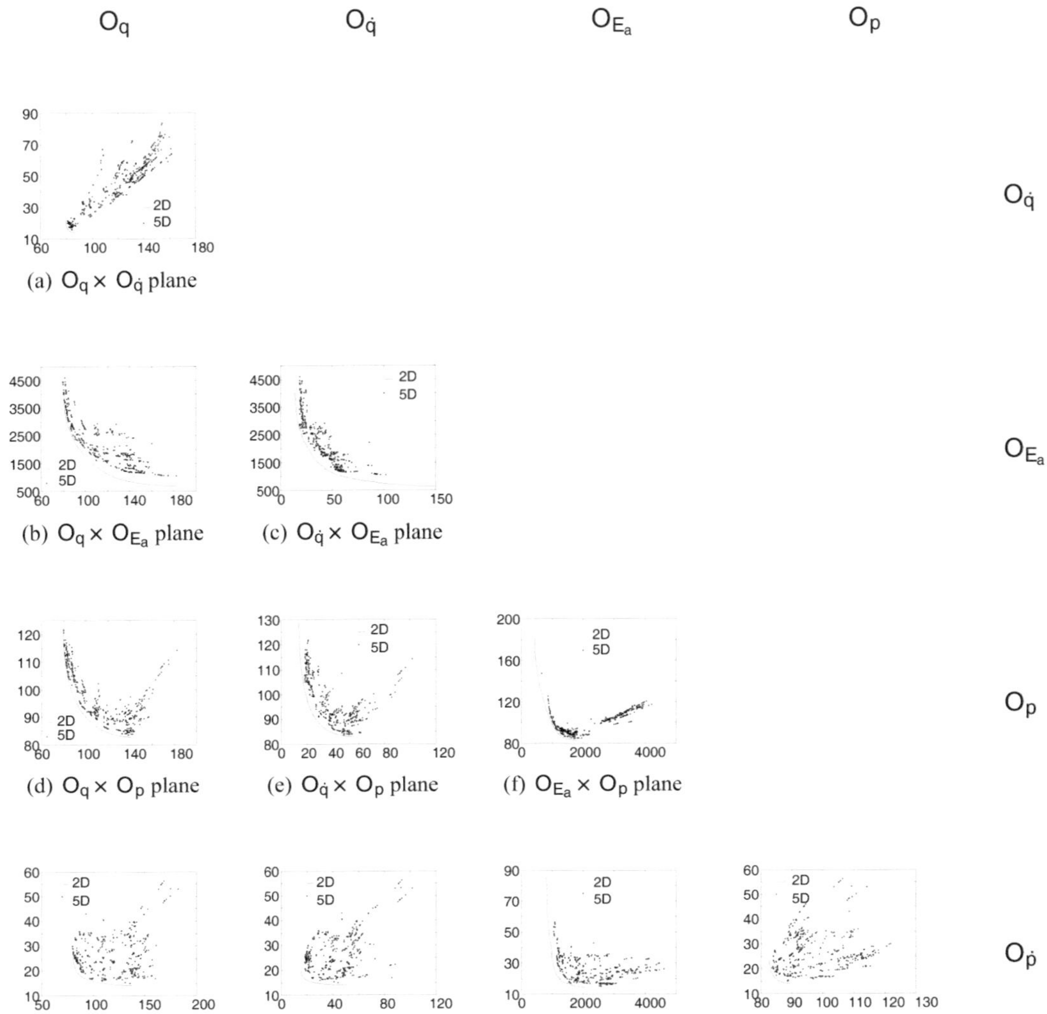

(a) $O_q \times O_{\dot{q}}$ plane

(b) $O_q \times O_{E_a}$ plane (c) $O_{\dot{q}} \times O_{E_a}$ plane

(d) $O_q \times O_p$ plane (e) $O_{\dot{q}} \times O_p$ plane (f) $O_{E_a} \times O_p$ plane

(g) $O_q \times O_{\dot{p}}$ plane (h) $O_{\dot{q}} \times O_{\dot{p}}$ plane (i) $O_{E_a} \times O_{\dot{p}}$ plane (j) $O_p \times O_{\dot{p}}$ plane

Fig. 12. Projections of the 5*D* Pareto optimal front for the 2*R* manipulator

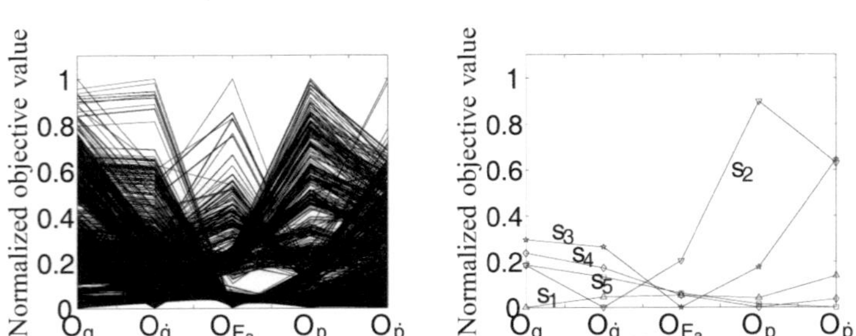

(a) Population tradeoffs (b) Best solutions tradeoffs

Fig. 13. Value path method representation of the O_q, $O_{\dot{q}}$, O_{E_a}, O_p and $O_{\dot{p}}$ objectives for the 3*R* manipulator

	O_q (rad^2/s^2)	$O_{\dot{q}}$ (rad^4/s^4)	O_{E_a} (J)	O_p (m^2/s^2)	$O_{\dot{p}}$ (m^4/s^4)
Minimum	155.1	52.8	446.9	74.4	12.9
Maximum	638.0	731.9	20531.3	333.7	108.3

Table 3. Range of the objectives with $5D$ optimization and the $3R$ manipulator

Fig. 14 shows the projections of $5D$ front into the different $2D$ planes. It is apparent that the solutions have a good distribution across the search space. From the figures it can be observed two optimal fronts. This 'discontinuity', of the optimal solutions, is due to all optimal solutions are not obtained when the manipulator moves in same direction.

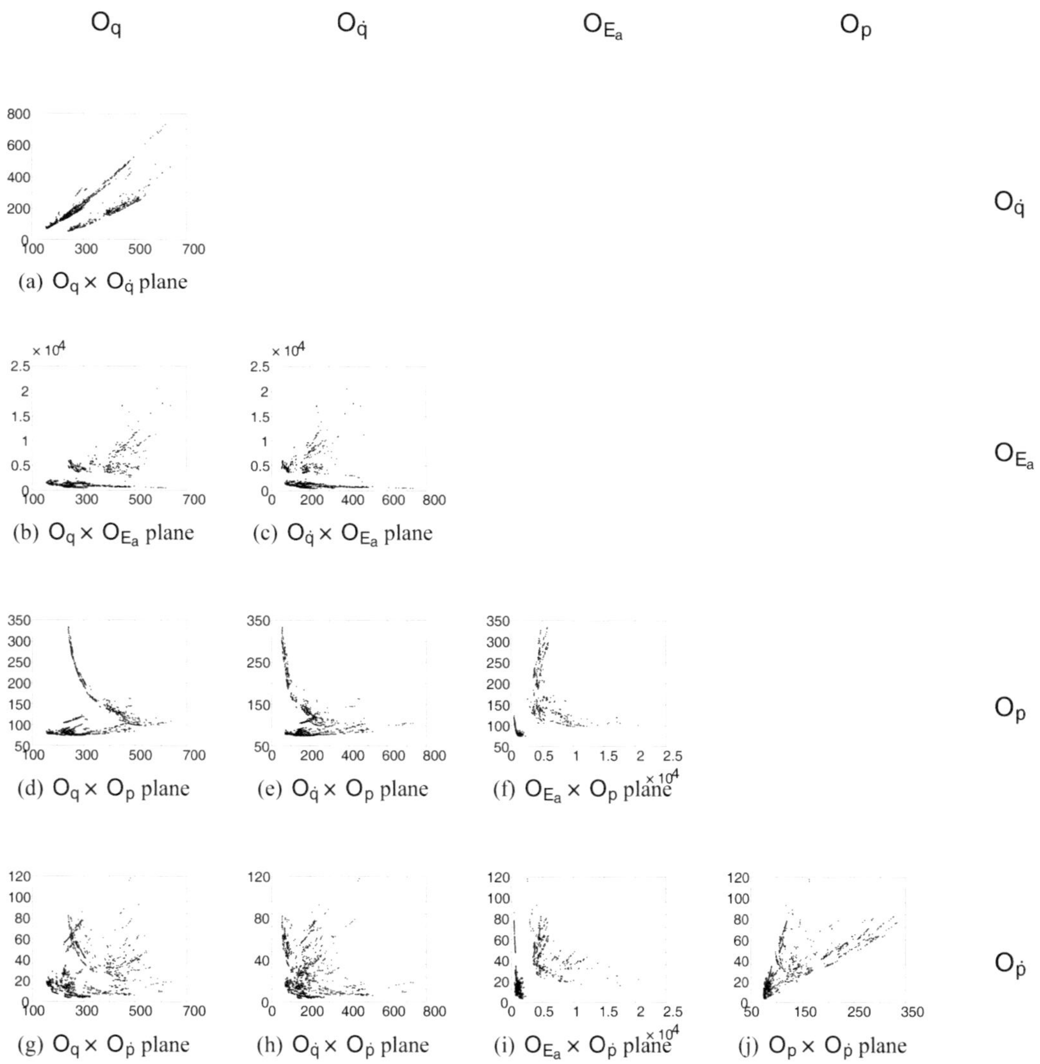

Fig. 14. The $5D$ Pareto optimal front projection for the $3R$ manipulator

Fig. 15 presents the best solutions for each objective. Most of them are obtained when the manipulator moves in the counterclockwise direction. However, the solution with lower $O_{\dot{q}}$ results from manipulator clockwise movement.

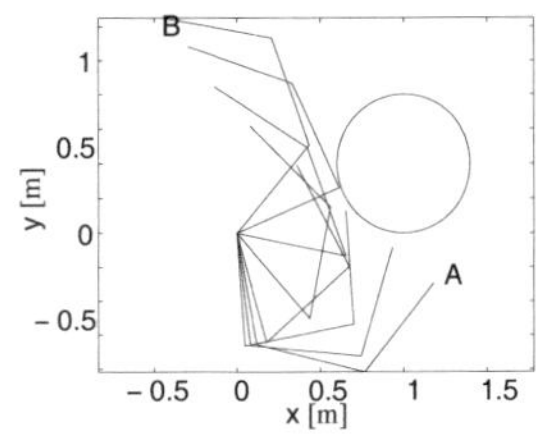

(a) Successive configurations for the achieved best solution in the viewpoint of the joint distance (O_q) objective

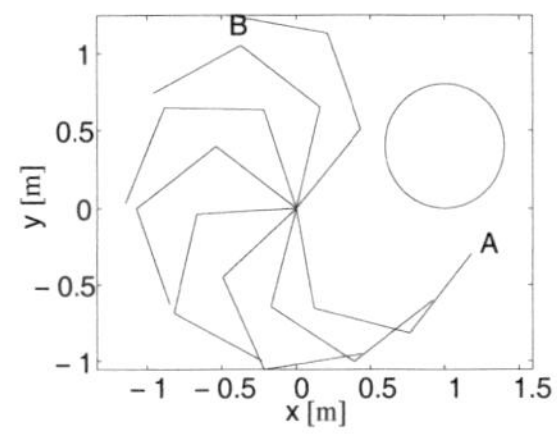

(b) Successive configurations for the achieved best solution in the viewpoint of the joint ripple ($O_{\dot{q}}$) objective

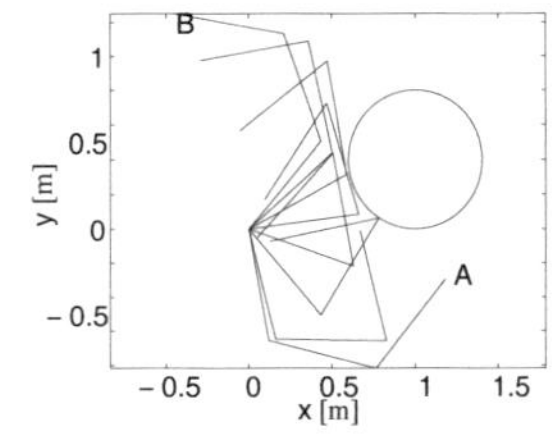

(c) Successive configurations for the achieved best solution in the viewpoint of the energy (O_{E_a}) objective

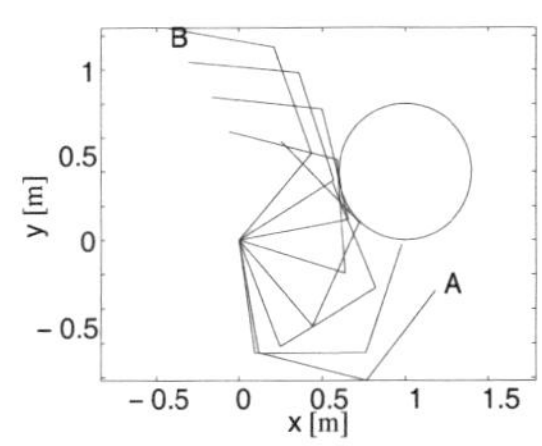

(d) Successive configurations for the achieved best solution in the viewpoint of the Cartesian distance (O_p) objective

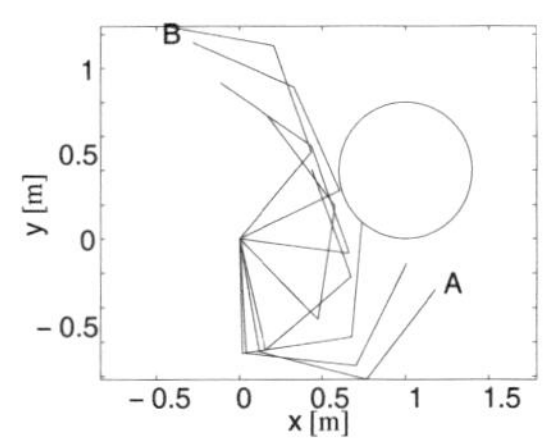

(e) Successive configurations for the achieved best solution in the viewpoint of the Cartesian ripple ($O_{\dot{p}}$) objective

Fig. 15. Results of the best solutions for $3R$ manipulator with one obstacle in the workspace, considering the five objectives optimization criteria

7. Additional experiments

In this section two tests are presented. The first experiment involves a $2R$ manipulator and three objectives. This simulation intends to emphasize the form of the optimal Pareto front for the $3D$ case. The second experiment involves a $3R$ manipulator and one obstacle, in order to study the effect of the obstacle in the non-dominated front.

During the experiments a 600 population elements is used over a total of 30000 generations. The other parameters remain the same as those earlier sections.

Fig. 16(a) depicts the results for the $2R$ manipulator when considering three objectives: the joint distance, the Cartesian distance and the energy required by the end effector to perform the task. The obtained front is continuous and has an 'Y' shape.

Figures 16(b)–16(d) show the Pareto optimal front projected into the different planes: $\{q, p\}$, $\{q, E_a\}$ and $\{p, E_a\}$. The chats are also overlapped the solutions obtained through the optimizations considering the two objectives under analysis. Comparing the results, it can be concluded that the algorithm with three objectives converges for the optimal front similarly to the Pareto front obtained with the two corresponding objectives.

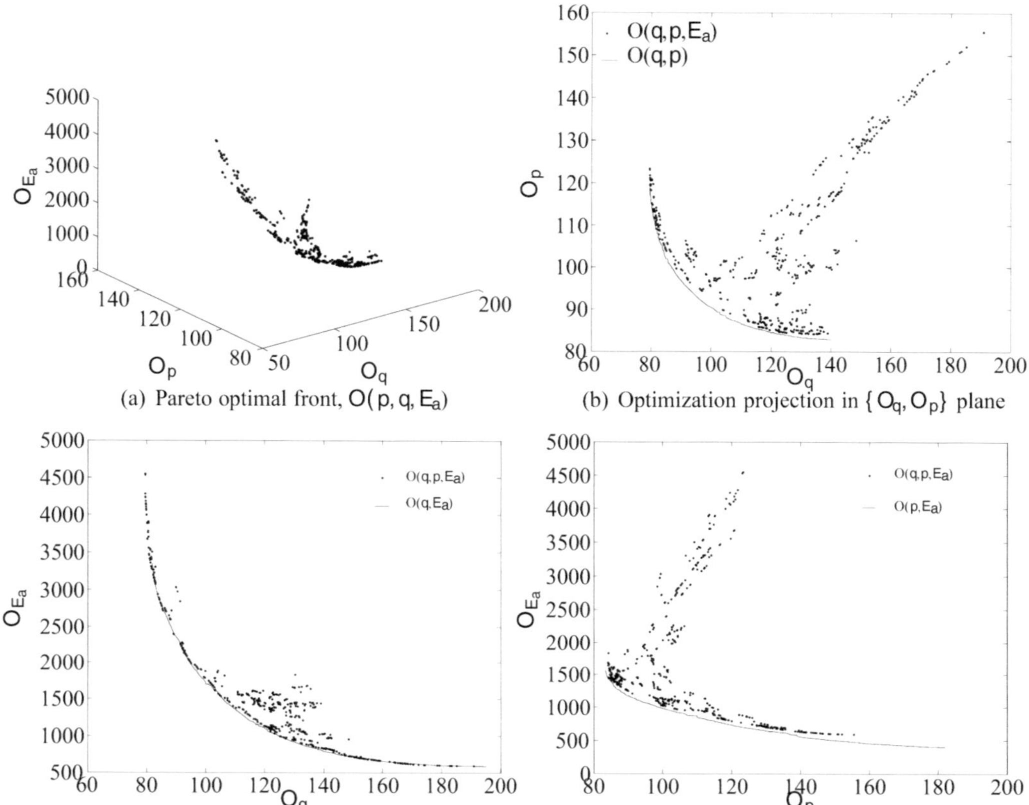

(a) Pareto optimal front, $O(p, q, E_a)$

(b) Optimization projection in $\{O_q, O_p\}$ plane

(c) Optimization projection in $\{O_q, O_{E_a}\}$ plane

(d) Optimization projection in $\{O_p, O_{E_a}\}$ plane

Fig. 16. The Pareto optimal front and its projections for the 2*R* manipulator and three objectives

In Fig. 17, a 3*R* manipulator is considered to move between two configurations $A \equiv \{q_1, q_2, q_3\} = \{-1.15, 1.81, -0.50\}$ and $B \equiv \{q_1, q_2, q_3\} = \{1.18, 1.47, 0.50\}$. The manipulator has link length $l_i = 1$ m length and mass $m_i = 1$ kg of mass, for $i = \{1, 2, 3\}$. The results concern environment with and without obstacles. The obstacle is a circle with center at $c = (2.0, 2.0)$ and radius $\rho = 1$. In Fig. 17(c) are represented the optimal Pareto fronts f_1 and f_2 for the workspace with and without obstacles, respectively. Thus, when the obstacle is introduced the optimal Pareto front is reduced. Consequently, the environment with obstacles does not lead to values with a cost as low as in the case of no obstacles, as can be seen by Figures 17(a), 17(b) and 17(c). At the other end of the front, the resulting solutions '*b*' and '*d*' are almost identical, therefore, it can be concluded that the optimization of the gripper Cartesian distance is not affected by inclusion of the obstacle in the workspace.

8. Conclusions

This chapter addresses the planning of robotic trajectories in a multiobjective optimization perspective, for finding a set of solutions belonging to the Pareto optimal front. The results demonstrate clearly that the algorithm finds the Pareto front or at least one very close.

Additionally, it is concluded that the algorithm leads to solutions with a good distribution along the front. The presence of obstacles in the environment can affect the optimal Pareto front but is not an

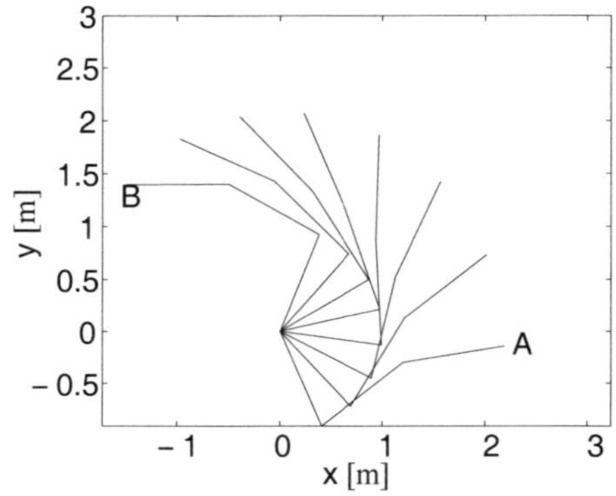

(a) Successive configurations, solution *a*

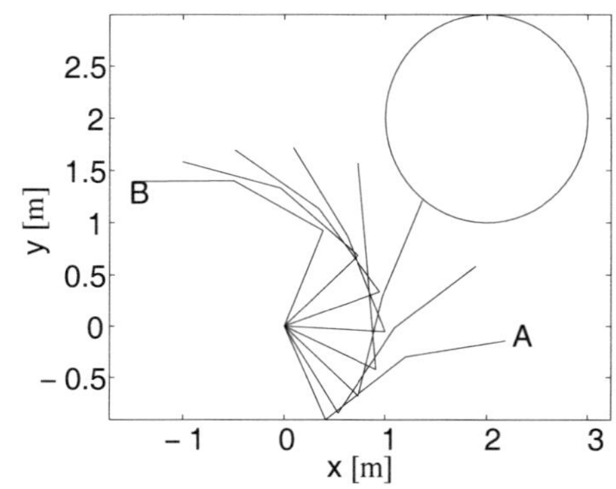

(b) Successive configurations, solution *c*

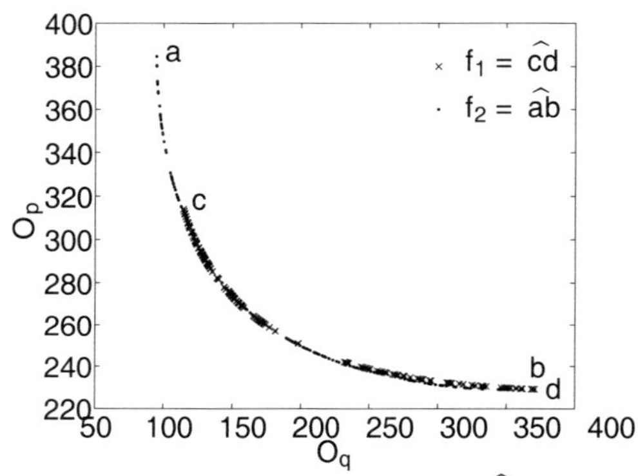

(c) Pareto optimal front for $O(q, p)$: $f_1 = \widehat{cd}$ (workspace with one obstacle), $f_2 = \widehat{ab}$ (workspace with one obstacle)

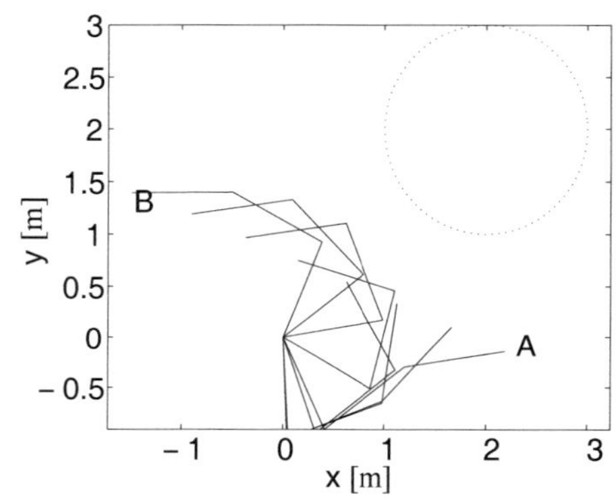

(d) Successive configurations, solution *b*

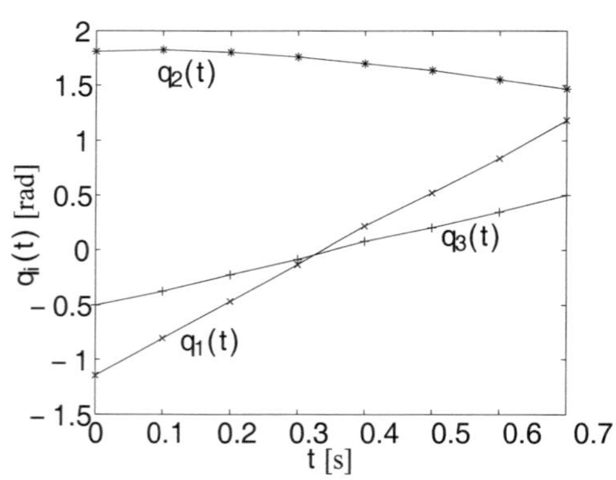

(e) Joint position trajectory, solution *a*

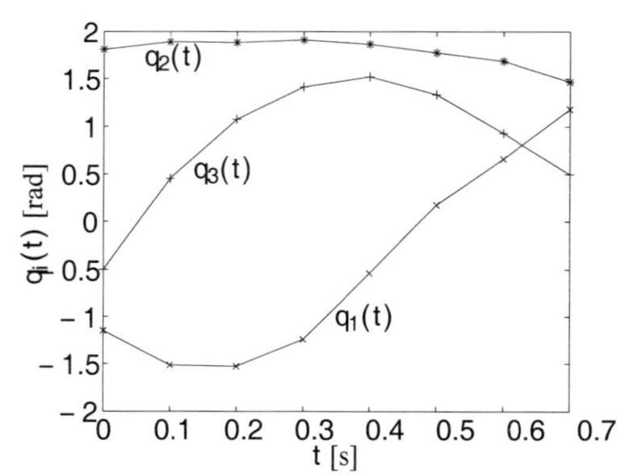

(f) Joint position trajectory, solution *b*

Fig. 17. Pareto optimal front, successive configurations and joint trajectory for the 3*R* manipulator obstacle to solve the problem. When it is increased the number of objectives, the burden and complexity of the algorithm increases and the quality of final solutions decreases. This disadvantage can be reduced by increasing population and the number of generations. However, the final population considering only the $2D$ optimization may include dominated solutions in the $5D$ optimization point of view. The proposed method helps the decision maker in choosing the best solution since the method gives a set of representative solutions belonging to the non-dominated front.

9. Acknowledgment

The authors would like to acknowledge the GECAD unit.

10. References

Alander, J. (2003). An indexed bibliography of genetic algorithms in robotics, *Technical report*, Department of Electrical Engineering and Production Economics, University of Vaasa.

Ali, A. D. M. S., Babu, N. R. & Varghese, K. (2002). Offline path planning of cooperative manipulators using co-evolutionary genetic algorithm, *Proceedings of International Symposium on Automation and Robotics in Construction, 19th (ISARC)*, National Institute of Standards and Technology, Gaithersburg, Maryland, pp. 415–424.

Bäck, T. (1996). *Evolutionary Algorithms in Theory and Practice: Evolutionary Strategies, Evolutionay Programming, Genetic Algorithms*, Oxford University Press, Oxford, New York.

Bäck, T., Hammel, U. & Schwefel, H.-P. (1997). Evolutionary computation: Comments on the history and current state, *IEEE Trans. on Evolutionary Computation* **1**(1): 3–17.

Chen, M. & Zalzala, A. M. S. (1997). A genetic approach to motion planning of redundant mobile manipulator systems considering safety and configuration, *Journal Robotic Systems* **14**(7): 529–544.

Coello, C. A. C. (1998). A comprehensive survey of evolutionary-based multiobjective optimization techniques, *Knowledge and Information Systems* **1**(3): 269–308.

Dae Lee, Y., Hee Lee, B. & Gyoo Kim, H. (1999). An evolutionary approach for time optimal trajectory planning of a robotic manipulator, *Information Sciences* **113**: 245–260.

Davidor, Y. (1991). *Genetic Algorithms and Robotics, a Heuristic Strategy for Optimization*, number 1 in *Series in Robotics and Automated Systems*, World Scientific Publishing Co. Pte Ltd.

Deb, K. (2001). *Multi-Objective Optimization Using Evolutionary Algorithms*, John Wiley & Sons, LTD.

Doyle, A. B. & Jones, D. (1996). Robot path planning with genetic algorithms, *2nd Portuguese Conf. on Automatic Control*, Porto, Potugal, pp. 312–318.

Duarte, F. B. M. (2002). *Anï£¡lise de Robots Redundantes*, Phd, Faculdade de Engenharia da Universidade do Porto.

Fonseca, C. M. & Fleming, P. J. (1995). An overview of evolutionary algorithms in multi-objective optimization, *Evolutionary Computation Journal* **3**(1): 1–16.

Garg, D. P. & Kumar, M. (2001). Optimal path planning and torque minimization via genetic algorithm applied to cooperating robotic manipulators, *IMECE – Congress of American Society of Mechanical Engineers*, Ney York.

Garg, D. P. & Kumar, M. (2002). Optimization techinques applied to multiple manipulators for path planning and torque minimization, *Engineering Applications of Artificial Intelligence* (15): 241–252.

Goldberg, D. E. (1989). *Genetic Algorithms in Search, Optimization, and Machine Learning*, Addison – Wesley.

Horn, J., Nafploitis, N. & Goldberg, D. (1994). A niched pareto genetic algorithm for multi-objective optimization, Proceedings of the First IEEE Conference on Evolutionary Computation, pp. 82–87.

Kubota, N., Arakawa, T. & Fukuda, T. (1997). Trajectory generation for redundant manipulator using virus evolutionary genetic algorithm, *IEEE International Conference on Robotics and Automation*, Albuquerque, New Mexico, pp. 205–210.

Kubota, N., Fukuda, T. & Shimojima, K. (1996). Trajectory planning of cellular manipulator system using virus-evolutionary genetic algorithm, *Robotics and Autonomous systems* **19**: 85–94.

Lavoie, M.-H. & Boudreau, R. (2001). Obstacle avoidance for redundant manipulators using a genetic algorithm, *Proc. of the 2001 CCToMM Symposium on Mechanisms, Machines, and Mechatronics*, Montréal.

Luo, X. & Wei, W. (2004). A new immune genetic algorithm and its application in redundant manipulator path planning, *Journal of Robotic Systems* **21**(3): 141–151.

Nearchou, A. C. & Aspragathos, N. A. (1995). Application of genetic algorithms to pointo-to-point motion of redundant manipulators, *Mecha. Mach. Theory, Pergamon* **31**(3): 261–270.

Ortmann, M. (2001). Multi-criterion optimization of robot trajectories with evolutionary strategies, *FACTA UNIVERSITATIS, Electronics and Energetics* **14**(1): 19–32.

Rana, A. & Zalzala, A. (1996). An evolutionary planner for near time-optimal collision-free motion of multi-arm robotic manipulators, *UKACC International Conference on Control*, Vol. 1, pp. 29–35.

Ridao, M. A., Camacho, E. F., Riquelme, J. & Toro, M. (2001). An evolutionary and local search algorithm for motion planning of two manipulators, *Journal of Robotic Systems* **18**(8): 463–476.

Silva, F. & Tenreiro Machado, J. (1999). Energy analysis during biped walking, *Proc. IEEE Int. Conf. Robotics and Automation*, Detroit, Michigan, pp. 59–64.

Tian, L. & Collins, C. (2004). An effective robot trajectory planning method using a genetic algorithm, mechatronics, *Mechatronics, In Press* **14**: 455–470.

Wang, Q. & Zalzala, A. M. S. (1996). Genetic control of near time-optimal motion for an industrial robot arm, *IEEE International Conference on Robotics and Automation*, Minneapolis, Minnesota, pp. 2592–2597.

Wei-Min, Y. & Yu-Geng, X. (1996). Optimum motion planning in joint space for robots using genetic algorithms, *Robotics and Autonomous Systems* **18**(4): 373–393.

On Designing Compliant Actuators Based On Dielectric Elastomers for Robotic Applications

Giovanni Berselli, Gabriele Vassura and Vincenzo Parenti Castelli
University of Bologna
Italy

Rocco Vertechy
PERCRO Laboratory, Scuola Superiore Sant'Anna
Italy

1. Introduction

This chapter reports on advances done in the design of compliant, Dielectric Elastomer based Linear Actuators (DELA).

Dielectric Elastomers (DE) are incompressible deformable dielectrics which can experience deviatoric (isochoric) finite deformations in response to applied large electric fields while, at the same time, alter the applied electric fields in response to the deformations undergone (Kofod et al., 2003; Pelrine et al., 1998; 2000; Toupin, 1956). Thanks to the strong electromechanical coupling, DE intrinsically offer great potentialities for conceiving novel solid-state mechatronic devices and in particular actuators. These devices can be profitably used for robotic applications (Bar-Cohen, 2004; Biddiss & Chaua, 2008; Kim & Tadokoro, 2007; Plante & Dubowsky, 2008). In fact, concerning conventional actuation technologies (such as electric motors), some basic problems remain unaddressed or, at least, improvable, such as:

- The need to increase flexibility and simplify design solutions. For an important number of applications, conventional systems are too heavy, inefficient (in terms of high power consumptions), too expensive and still relatively complex.
- The need to design human-friendly machines (Bicchi & Tonietti, 2004). In order to achieve highly precise position control, many industrial mechatronic systems (especially industrial robots) are designed to be very fast and stiff and thus potentially dangerous to humans. A possible way to greatly increase safety, is the design of robotic structures which are less precise but more compliant when compared to the traditional ones. To this respect, it is possible to concentrate the compliance in the actuated joints by using compliant (soft) actuators.
- The need for modern robots to interact with unstructured environments and to actively control force and stiffness at the contact interface. A promising approach is to rely on a mechanical system that is inherently compliant and to use active control strategies to vary this compliance (Biagiotti et al., 2004). The main advantages are a lower request both on actuator and controller bandwidth and improved stability (Paul & Shimano, 1976; Williamson, 1993).

- The need to develop alternative actuators to overcome design limitations which are imposed in few but very important specific applications. For instance, electromagnetic actuators cannot be used in the presence of the high magnetic fields generated by Magnetic Resonance Imaging (MRI) devices. Nevertheless, the possibility to accomplish manipulation tasks within an MRI environment would highly improve the diagnostic capabilities of this technology (Koseki et al., 2007).

Thanks to their intrinsic compliance, lightness, pliability, and low cost, actuators based on DE can be an explorable solution when trying to assess or improve the aforementioned issues.
DELA are usually composed of one or more DE shaped as a thin membrane (film) and a flexible supporting frame.
This chapter proposes a methodology that allows the modification of the available thrust as a function of the actuator length at will of the designer and, in particular, to obtain constant force actuators. The design procedure is divided in two steps: 1) optimization of the DE electromechanical parameters 2) design of the flexible frame. The supporting frame is conceived as a compliant mechanism (Howell, 2001) and it makes use of the stiffness characteristics of slider-crank mechanisms with elastic revolute pairs to be coupled (in symmetric or axis-symmetric configurations) with films of different geometries. Three actuator concepts are proposed which highlight the efficacy of the proposed method, in particular Rectangular DELA (1(a), 1(d)), Diamond DELA (1(b), 1(e)) and Conical DELA (1(c), 1(f)).

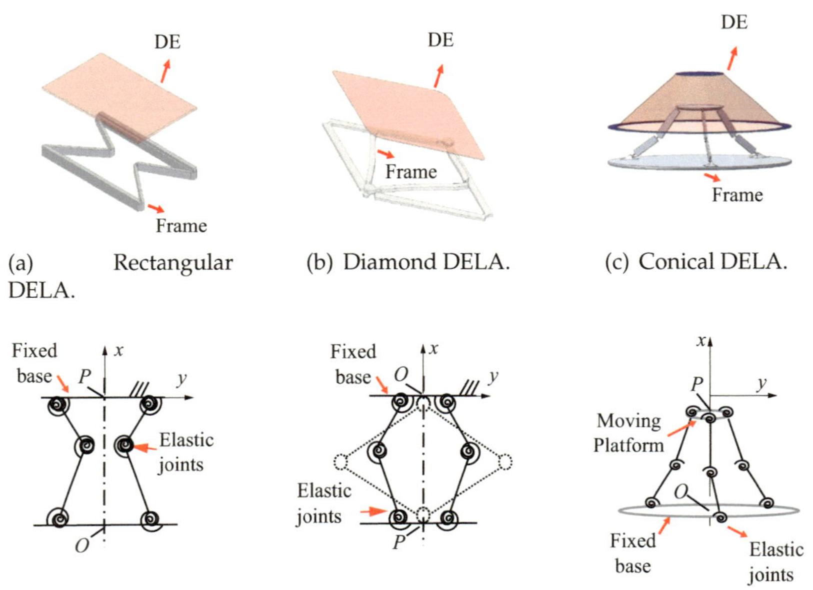

(a) Rectangular DELA. (b) Diamond DELA. (c) Conical DELA.

(d) Rectangular DELA frame schematic. (e) Diamond DELA frame schematic. (f) Conical DELA frame schematic.

Fig. 1. DELA concepts.

2. Background on Dielectric Elastomer Actuators

For actuation usage, DE can be shaped in thin films which are firstly prestretched then coated with compliant electrodes on both sides and piled one on the other to form an Electrically Deformable Film (EDF). An EDF can be mono or multi layered (Fig. 2). Activation of the EDF via the placement of an electric field, E_z, acting in the film thickness direction (z direction, Fig. 2) or, equivalently, of differential electric potentials (hereafter also called voltages, V) between the electrodes can induce a film area expansion and, thus, point displacements which can be used to produce useful mechanical work (whenever forces are applied to such points). The term EDF simply identifies a dielectric, hyperelastic, and incompressible isotropic membrane (the DE) placed between two compliant electrodes.

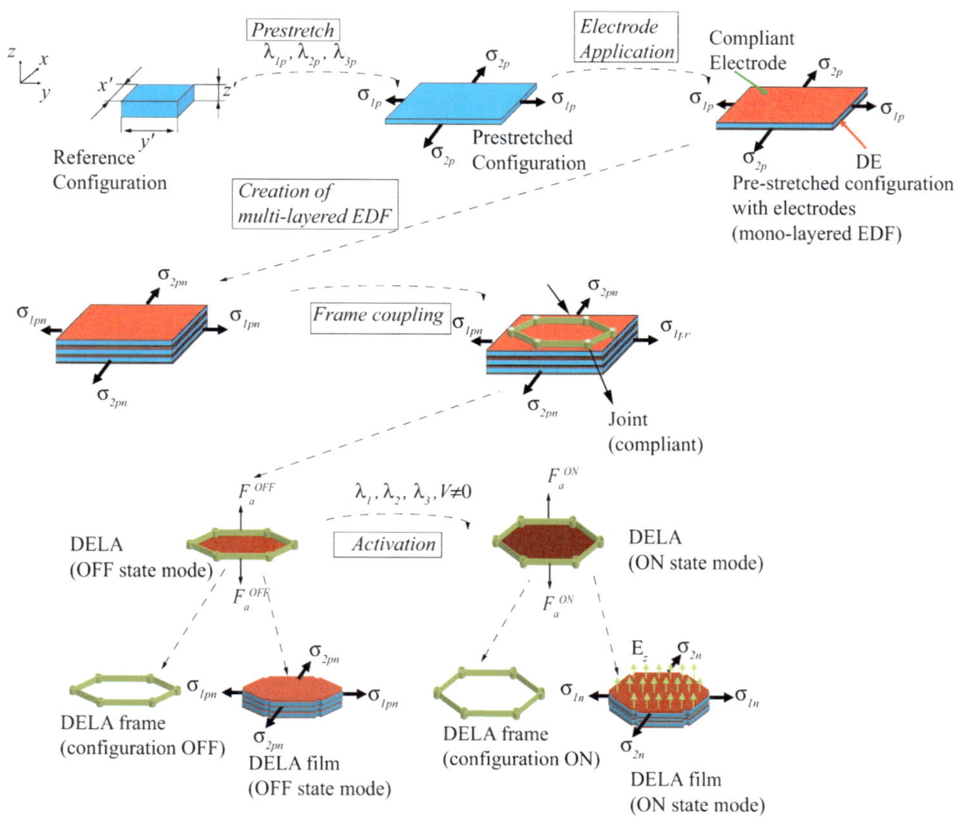

Fig. 2. DELA production steps.

Usually, DE-based actuators are obtained by first *uniformly* prestretching the DE (which is necessary since the film has negligible flexural rigidity) and then by coupling some segment of its boundary to some portion of a flexible supporting frame, either an elastic structural element (Kofod et al., 2006; Pei et al., 2003), for instance a helical spring, or a compliant mechanism (Berselli et al., 2008; 2009a; Plante, 2006; Vogan, 2004; Wingert et al., 2006), for instance a four-bar mechanism with elastic revolute joints. Figure 2 summerizes the conceptual steps carried

out for the production of DELA with generic geometry, that is DE prestretch, compliant electrode application (of single EDF layers), assembly of a multi-layered EDF and coupling with a frame of given geometry. In the following, the term *OFF state mode* and *ON state mode* will identify every actuator working condition where $V = 0$ and $V \neq 0$ respectively.
The major roles of the flexible polymeric frame are:

- to prevent the development of current arcs around the EDF border;
- to protect the EDF edges;
- to provide a firm support for the application of external forces;
- to coerce the EDF expansion in preferred directions;
- to maintain the EDF in a tensioned state so as to prevent wrinkling effects;
- to modify the overall actuator stiffness (EDF + frame) by using the frame own stiffness.

The actuator stiffness identifies the ratio among the variation of the actuator available thrust with respect to the variation of its length. In particular, specific compliant frames (Fig. 1) can be designed in order to achieve a constant force for a given range of actuator stroke and therefore a nearly naught stiffness. Constant force actuators can be beneficially employed in those applications that require constant but controllable output force over a given range of motion. For example, in grasping and manipulation applications, the desired grasp force might change with respect to the weight of the object to be manipulated. An actuator that provides a constant force would be capable of applying that given force even in presence of small deflections or positioning errors thus minimizing the control effort.
Different solutions for possible frame geometries have been proposed in the literature (Bolzmacher et al., 2004; Kofod & Sommer-Larsen, 2005; Lochmatter, 2007; Plante, 2006; Vogan, 2004) which make use of EDF in planar or spatial shapes (Fig. 3). Some of these shapes can be analytically solved in terms of EDF force vs EDF displacement (Plante (2006)) whereas most of them have to be studied through either Finite Element (FEM) analysis or experiments.

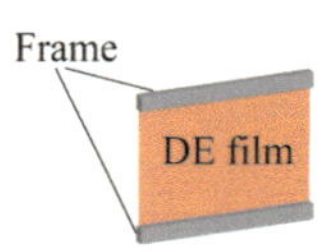

(a) Rectangular DELA (planar).

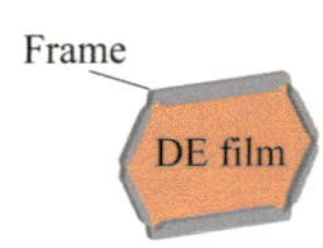

(b) Hexagonal shaped DELA (planar).

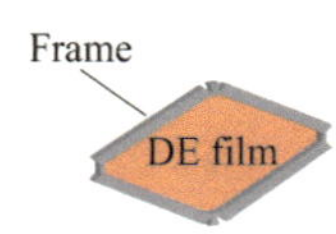

(c) Diamond DELA (planar).

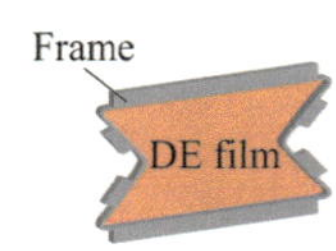

(d) Bow-tie DELA (planar).

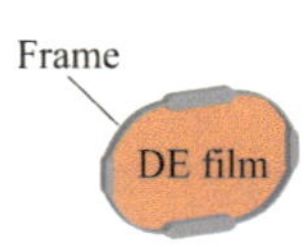

(e) Elliptical DELA (planar).

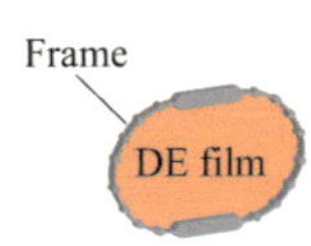

(f) Leaf spring DELA (planar).

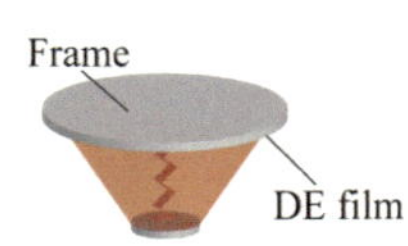

(g) Conical DELA (spatial).

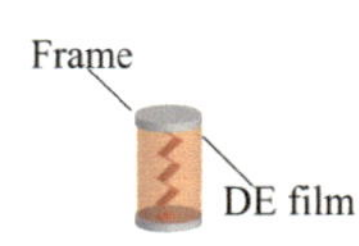

(h) Cylindrical DELA (spatial).

Fig. 3. DELA prototype for possible frame geometries.

3. General aspects of frame design

3.1 Compliant frame effect on the actuator performance

According to a simplified one-dimensional model, also adopted by Plante (2006), a DELA actuator can be considered as a set of two interacting springs: one, the EDF, operating under tension, the other, the frame, operating under compression. In general, the film deformation produces a variation of the actuator length $x = |(P - O)|$, where P and O are two points of the actuator (Figs. 1, 4(a), 4(b)), and a force having the same direction as that of the motion can be supplied to an external user.

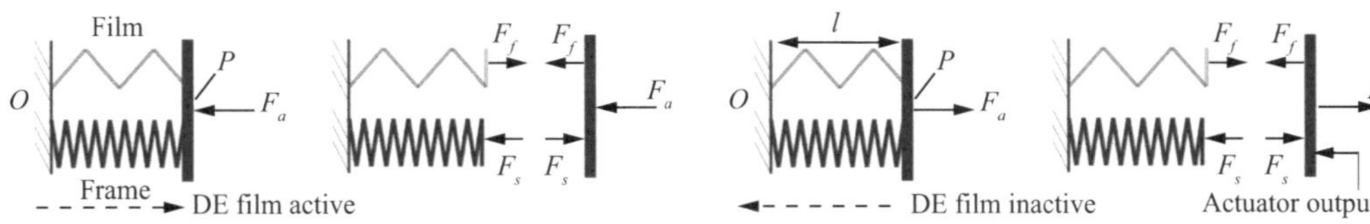

(a) Negative thrust: $|F_f| < |F_s|$ (b) Positive thrust: $|F_f| > |F_s|$

Fig. 4. Forces acting in the system in the ON state mode (a) and in the OFF state mode (b).

This force, called the actuator available thrust, F_a, can be represented as in Fig. 4 and is the resultant of two internal forces:

- The frame reaction force F_s due to the frame own stiffness, that is a function of the actual frame configuration. The compliant frame behaves, in general, like a non-linear compression spring coupled in parallel with the EDF.
- The EDF force F_f which represents the resultant force in the direction of the actuator motion due to the stress field arising in the EDF. This field depends on the amount of given prestretch, on the applied electric potential (voltage), and on the actuator configuration. In a one-dimensional model, where material viscoelasticity, creep and hysteresis are ignored, the film behaves, in general, as a non-linear tension spring. Under a step voltage variation the film force F_f undergoes a finite decrease, dependent on the magnitude of the applied voltage (see Fig. 5).

The available thrust F_a is therefore given by the difference between the film and the frame forces:

$$F_a(x, V) = F_f(x, V) - F_s(x) \quad (1)$$

Conventionally, F_a is the force that an external user supplies to the actuator.

Figure 5 shows qualitative diagrams of Force vs actuator Length (FL) curves concerning internal forces F_f and F_s, adopting a representation methodology widely used in the study of interacting elastic structures where the moduli of the forces are shown.

The *continuous* curve F_f^{off} represents the film force F_f in the inactive state whereas the *dotted* curve s represents the frame reaction force F_s.

In the OFF state mode (curve F_f^{off}), after the coupling of the prestretched film and the precompressed frame, the achieved equilibrium position is represented by point A. In this condition (no applied load and no applied voltage), the actuator initial length is x_0. The displacements imposed on the one-dimensional springs representing the film and the frame are given respectively by $|x_0 - x_{0f}|$ and $|x_{0s} - x_0|$, where x_{0f} is the free length of the EDF and x_{0s} is free length of the frame. Point A is taken as the reference point for the measurement of the

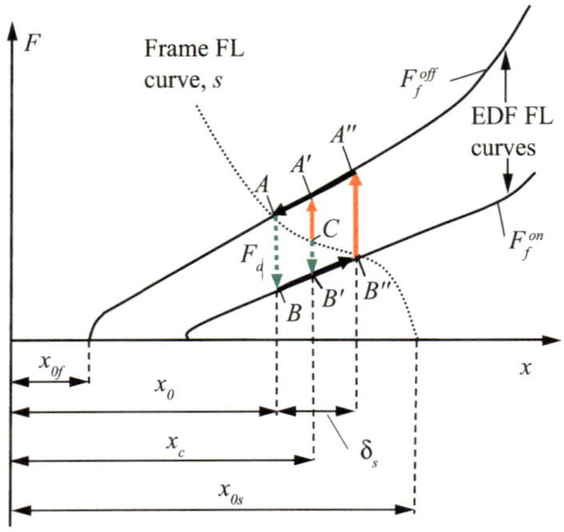

Fig. 5. FL relationship showing the effect of the film/frame coupling.

actuator stroke δ.
On the other hand, the *continuous* curve F_f^{on} represents the relationship between F_f and the actuator length, in the case of film activation under constant voltage. Point B represents the new value of F_f upon a step voltage rise, starting from point A. The distance between points A and B symbolizes the available thrust F_a. In this condition $|F_f| < |F_s|$ and the actuator output can move outward (Fig.4(a)). If the actuator output is free to move (no external load), the actuator can now reach a new equilibrium position represented by point B'' and δ_s represents the maximum stroke obtainable with a frame characterized by a FL profile like curve s.
In any intermediate position defined by x, points C and B' represent respectively forces F_s and F_f^{on}; in this condition, the available thrust is equal to the distance $\overline{B'C}$. If, at any point along the stroke, the excitation voltage is suddenly removed, the force acting on the film abruptly passes from point B' to point A', thus obtaining $|F_f| > |F_s|$ (Fig. 4(b)) and an available thrust acting in the opposite direction with absolute value equal to the distance $\overline{A'C}$. According to the adopted convention on the force sign, the available thrust can be plotted as in Fig. 6(a)

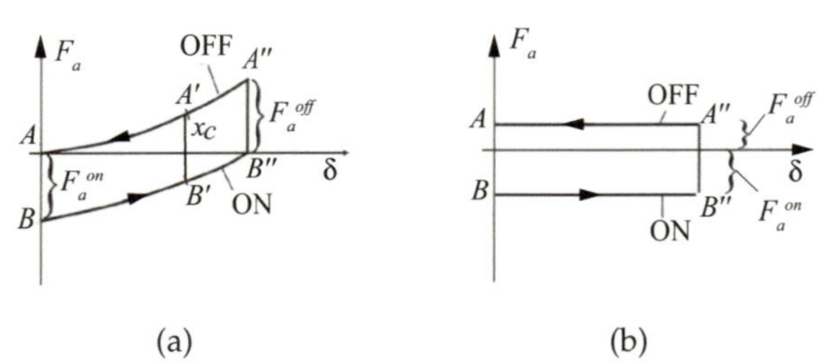

Fig. 6. Actuator force vs stroke relationship; real behavior (a) and ideal behavior (b).

The force profiles of F_a in the active and inactive states (F_a^{on} and $F_a{}^{off}$ respectively) can be modified by working on the frame design depending on the application requirements.
It can be seen from Fig. 6(a) that in general F_a is changing along the stroke δ, either when the actuator state is ON or OFF. However, it has been noticed (Pedersen et al., 2006) that a general purpose actuator should rather provide a constant thrust along its stroke (as in Fig. 6(b)).

3.2 Achievement of constant force actuators

Assuming the film electromechanical characteristics as given, the compliant frame stiffness can be designed in order to modify the curve s (Fig. 5) and obtain an actuator capable of providing a constant force over a given range of motion.
Let us suppose the EDF is coupled with a compliant mechanism whose elastic reaction force decreases as the actuator length x increases. In a one-dimensional model, such a frame can be conceived as a negative stiffness spring acting in parallel with the EDF. If this negative stiffness perfectly matches the EDF stiffness, a constant output force can be obtained.
The operating principle is illustrated in Fig. 7 where three different curves for F_s are represented assuming that the initial length of the actuator x_0 and the maximum obtainable stroke δ_s are invariant.

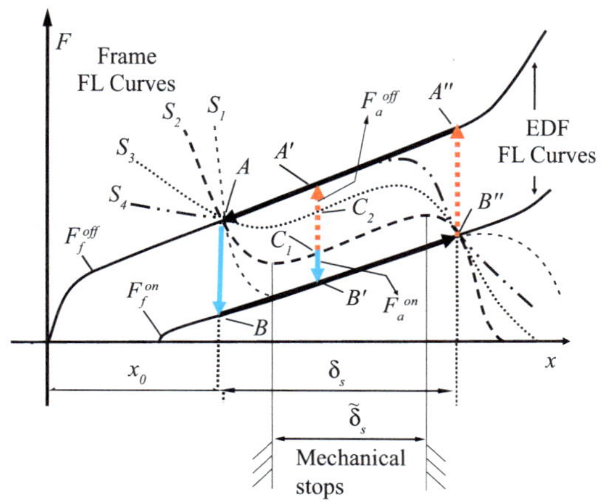

Fig. 7. FL relationship showing the "negative stiffness" effect.

Let us consider first, a frame FL curve depicted by the curve S_2, where the frame configuration is defined as the length $x = |P - O|$ (as previously defined). It can be seen that, for a large part of the stroke, the actuator available thrust, F_a, maintains a constant value, ${F_a}^{off}$, equal to the distance $\overline{A'C_1}$ if the EDF is deactivated.
In Fig. 7, ${F_f}^{off}$ and ${F_f}^{on}$ are approximated as parallel (as it happens for some actuator geometries (Berselli et al., 2008; 2009a; Wingert et al., 2006)). In such a case the same frame can be used also to obtain a constant available thrust ${F_a}^{on}$, equal to the distance $\overline{B'C_1}$ when the EDF is activated. If the actuator is required to supply a larger thrust when the EDF is active (actuator ON-state mode), a frame FL profile alike curve S_3 can be chosen, so as to increase ${F_a}^{on}$ from $\overline{B'C_1}$ to $\overline{B'C_2}$ (and consequently to decrease ${F_a}^{off}$). The OFF-state mode actuator thrust ${F_a}^{off}$ is maximized by designing a frame that provides a FL profile alike curve S_1 whereas the ON-state mode actuator thrust ${F_a}^{on}$ is maximized by designing a frame that provides a FL profile alike curve S_4. In this situation, however, no restoring force can pull back the actuator to its initial position when the voltage is switched off (actuator on OFF-state mode) and a returning device has to be provided. In practice, for applications requiring control via bidirectional forces, two identical actuators can be employed in an agonistic-antagonistic configuration.
Note that the actuator stroke can be limited to $\tilde{\delta}_s$ by purposely designed mechanical stops in order to prevent motion in undesired regions (e.g. regions where the thrust is not constant).

4. Possible frame designs

In order to achieve a constant force, provided that F_f monotonically increases as the actuator length increases, the frame should provide a negative force F_s whose value decreases along the same range (i.e the frame stiffness, $K_s = dF_s/dx$ should be negative). In general, this behavior is exhibited by mechanisms characterized by unstable equilibrium positions along their motion. For example, a bi-stable element presents an unstable equilibrium position (UEP) between two stable equilibrium positions (SEP) and can act as a negative stiffness spring over a given range of motion. Considering the FL profile of a bi-stable mechanism, the equilibrium positions occur at location where $F_s = 0$. In general, an equilibrium position is stable if the input force/moment increases as the generalized coordinate, x, increases (i.e. $K_s = dF_s/dx > 0$). On the other hand, the equilibrium position is unstable if the input force/moment decreases as the generalized coordinate increases (i.e. $K_s = dF_s/dx < 0$). Therefore a bi-stable element can act as a negative stiffness spring over a given range of motion and specifically around the UEP. Previously published solutions employed compound structure frames (i.e. additional mechanisms were employed in order to modify the FL profile of a given actuator). In particular, Plante (2006) and Vogan (2004) proposed the use of a bi-stable "snap through buckled beam" element or the use of an "over the center" device to be coupled with a "diamond" shaped supporting compliant mechanism with negligible joint stiffness. The critical aspects of such designs are related most of all to the use of non linear tension springs (Plante, 2006) or lateral instability due to beam buckling (Vogan, 2004).

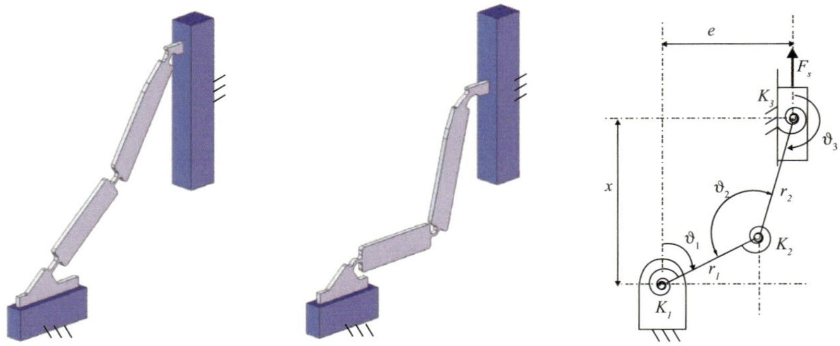

(a) (b) (c)

Fig. 8. Slider-Crank Compliant Mechanism (SCCM). Undeflected (a) and deflected (b) configuration. PRBM (c)

As an alternative design, here it is proposed to make use of the stiffness characteristic achievable with the Slider-Crank Compliant Mechanism (SCCM). A schematic of an SCCM in undeflected and deflected configuration is shown in Fig. 8(a) and Fig. 8(b). The Pseudo Rigid Body Model (PRBM) (Howell, 2001) of the same compliant mechanism is depicted in Fig. 8(c) where r_1 and r_2 are the crank and the connecting-rod lengths respectively, e is the slider-crank mechanism eccentricity, x is the mechanism length (i.e. the distance between the points O and P), K_1, K_2, K_3 are the spring constants of the compliant joints (Howell, 2001), ϑ_1 and ϑ_3 are the crank angle and connecting rod angular positions measured with respect to the actuator direction of motion ($\vartheta_2 = \vartheta_1 - \vartheta_3$), and ϑ_{10} ,ϑ_{20} ,ϑ_{30} are the undeflected angular positions of the flexural pivots ($\vartheta_{20} = \vartheta_{10} + \vartheta_{30}$).

Let us define the force F_s as the SCCM reaction force due to its own stiffness, that is a function of the actual mechanism configuration. Invoking the principle of the superimposition of the effects, the force F_s is given by:

$$F_s = F_1 + F_2 + F_3 \tag{2}$$

where the forces F_1, F_2 and F_3 are due to the deflection of the torsional springs with stiffness K_1, K_2 and K_3 respectively. Let us consider separately the contribution of each stiffness K_1, K_2 and K_3. At first, neglect the contribution of torsional springs with stiffnesses K_1 and K_2, such that $F_s = F_3$.

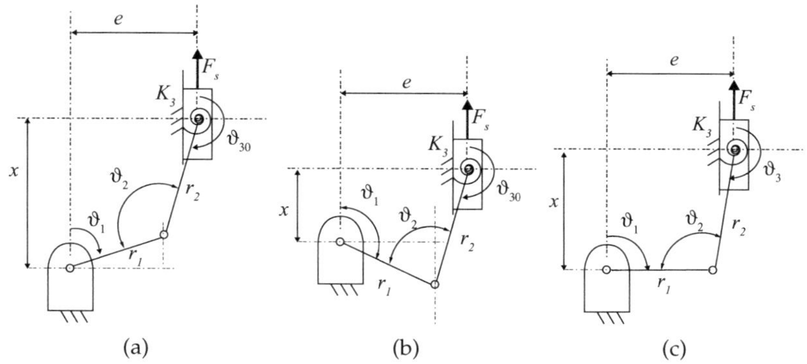

Fig. 9. SCCM. First stable position (a), second stable position (b), unstable position (c).

In such a situation, the resulting mechanism is depicted in Fig. 9(a) where the torsional spring with stiffness K_3 is represented in its undeflected position. Such mechanism is bi-stable. In fact two SEP are reached when the torsional spring is undeflected. If Figure 9(a) represents the first SEP, another SEP is reached when the connecting rod reaches a configuration parallel to its configuration at the first SEP. Therefore, Figure 9(b) represents the second SEP. An UEP is reached when the rocker arm is perpendicular to the slider direction of motion as depicted in Fig. 9(c).

Let us now neglect the contribution of the torsional spring with stiffness K_3 and, therefore, consider the contribution of the forces F_1 and F_2 only. Define $F_{12} = F_1 + F_2$. The resulting mechanism is a possible topology of a compliant mechanism that can supply a nearly constant-force if suitably dimensioned. The dimensions of the PRBM (i.e. the values of K_1, K_2, r_1, r_2, e, and the undeflected angular positions of the joints) resulting in a constant force may be found by minimizing the variation of the output force over the given input displacement as it will be shown in the next paragraph.

It is clear that, if coupled with DELA of different shapes, SCCM can be used to tailor the actuator stiffness to a given application and, in particular, design constant force actuators. Depending on the SCCM design, the actuator can work bidirectionally or monodirectionally (Berselli et al., 2009a) depending on whether or not the frame own stiffness provides a restoring force that brings back the actuator to an initial position when the EDF is deactivated. As a proof of concept, let us consider the mechanisms shown in Fig. 10(a) that will be referred to as Delta Element. Figure 10(b) represents a particular case where $e = 0$.

The Delta Element is in fact a parallel mechanism. However if the loads are directed along the x direction (shown in Fig. 10(a)) half mechanism can be modeled as an SCCM. The Delta Element can be coupled to every DE planar actuator to form a constant force actuator. Fig.

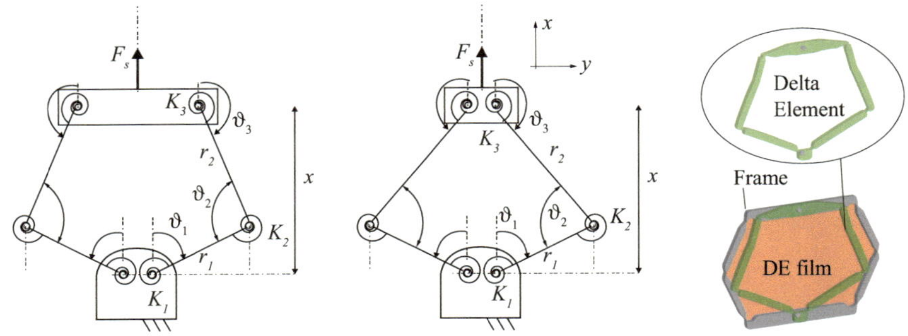

(a) Eccentric Delta Element (b) Delta Element (c) Delta element coupled with an hexagonal DELA

Fig. 10. Different possible configuration of compliant frames

10(c) shows a Delta Element coupled with a hexagonal DELA.
Qualitatively the behavior of the SCCM coupled with the EDF (Delta Element or other possible frame configuration based on the same concept) is shown in Fig. 11 where the contributions of the single forces F_3 and F_{12} are also depicted. The curve S_6 represents the total force $F_s = F_{12} + F_3$. Note that, given the desired stiffness of the actuator as a whole, that is EDF coupled with the SCCM (a null stiffness being represented in Fig. 11), the actuator thrust in the ON and OFF state modes can be adjusted by working on the force F_{12} only. In fact the curve S_6' which maximizes the thrust in the ON state mode has been obtained through an SCCM which provides a reaction force $F_s = F_3 + F_{12}'$.

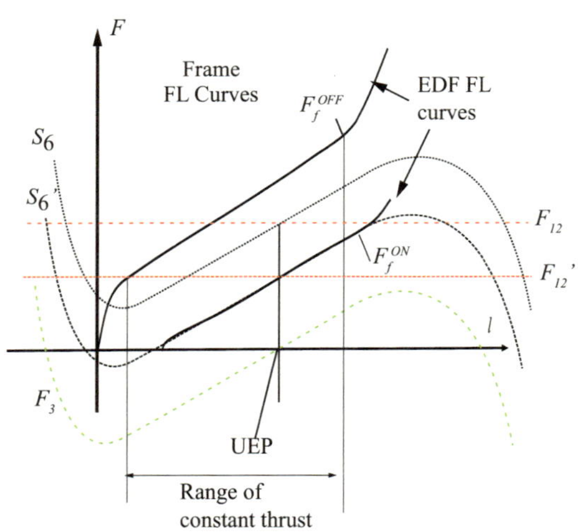

Fig. 11. Effect of the SCCM on the overall actuator stiffness.

5. Mathematical model of the Dielectric Elastomer film force

For design purposes DE can be considered as incompressible, hyper-elastic linear dielectrics whose electric polarization is fairly independent of material deformation (Berselli et al., 2008;

2009a; Kornbluh et al., 1995; Pelrine et al., 1998). For such elastomers, EDF activation generates an electric field, $E = V/z$ (V being the activation voltage applied between the EDF electrodes and z being the actual thickness of the DE film amid the EDF electrodes), and an electrically-induced Cauchy stress, $\sigma_{em} = \epsilon E^2$ (ϵ being the DE electric permittivity), both acting in the DE film thickness direction. As a consequence, the mechanical stress field in a stretched and activated DE, which is free to deform in its thickness direction, is given by the following relationships:

$$\sigma_1 = -p + \lambda_1 \frac{\partial \psi}{\partial \lambda_1}; \qquad \sigma_2 = -p + \lambda_2 \frac{\partial \psi}{\partial \lambda_2}; \qquad \sigma_3 = -p = -\epsilon E^2 = \frac{-\epsilon V^2}{z^2} = \frac{-\epsilon V^2}{{\lambda_3}^2 z'^2} \quad (3)$$

where λ_i and σ_i $(i = 1, 2, 3)$ are, respectively, the principal stretches and Cauchy stresses (the 3-rd principal direction coinciding with the film thickness direction), $\psi = \psi(\lambda_1, \lambda_2, \lambda_3)$ defines the DE strain-energy function (Ogden, 1972), and $z' = z/\lambda_3$ is the unstretched DE film thickness (in the reference configuration).

Considering an Ogden model for the constitutive behavior of incompressible rubber-like materials, it is postulated that the strain-energy function ψ has the form:

$$\psi = \psi(\lambda_1, \lambda_2) = \sum_{p=1}^{k} \frac{\mu_p}{\alpha_p} (\lambda_1^{\alpha_p} + \lambda_2^{\alpha_p} + \lambda_1^{-\alpha_p} \lambda_2^{-\alpha_p} - 3) \quad (4)$$

where k is the model order and μ_p, α_p are material parameters to be determined experimentally, that is curve fitted over experimental stress/stretch data. In his work, first order models are used ($k = 1, \alpha_1 = \alpha, \mu_1 = \mu$). Note that ψ is function of λ_1 and λ_2 only because incompressibility is assumed ($\lambda_3 = 1/\lambda_1\lambda_2$).

When $V = 0 \Rightarrow \sigma_3 = 0$, which is, in fact, the applied boundary condition for the EDF in the OFF state mode.

5.1 Rectangular actuators

Rectangular actuators are based on a rectangular mono-axially prestretched DE coupled to two rigid beams (Fig. 3(a)). Let us define (Fig. 12(a)) x' and y' as the EDF planar dimensions in the reference configuration (unstretched EDF) whereas x and y_p are EDF planar dimensions in the actual configuration. Note that y_p remains constant during actuator functioning. It is supposed that the DE deformation can be described by a pure shear deformation [1]. A principal prestretch $\lambda_{2p} = y_p/y'$ is applied in the y direction. The prestretch λ_{2p} is an independent design parameter. The points O and P are two points of the DELA frame placed, for instance, on its axis of symmetry and lying on the two opposite rigid beams.

As depicted in Fig. 12, in such actuators, activation of the EDF makes it possible to control the relative distance x (hereafter also called "DE length" or "actuator length") of the points O and P, which are supposed to be the points of application of the (given) external forces F_f acting on the actuator boundary. The DE deformation state (pure shear), (Ogden, 1972) is characterized by the following principal stretches:

$$\lambda_1 = \frac{x}{x'}; \qquad \lambda_2 = \lambda_{2,p}; \qquad \lambda_3 = \frac{1}{\lambda_1 \lambda_2} = \frac{z}{z'} \quad (5)$$

[1] According to the definition given by Ogden (1972), a pure shear deformation is characterized by the constancy of one principal stretch (for instance λ_2). A pure shear deformation can be achieved for infinitely wide EDF (i.e. for $y_p >> x \; \forall \; \Omega(t)$ where $\Omega(t)$ are the possible configurations of the EDF in working condition.

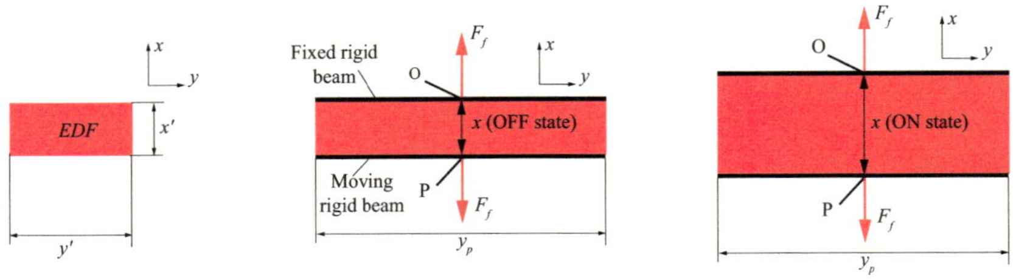

(a) Unstretched EDF.

(b) Rectangular DELA, schematic (OFF-state mode).

(c) Rectangular DELA, schematic (ON-state mode).

Fig. 12. Rectangular DELA.

Considering the xy plane, the principal stretch/stress directions are respectively aligned and orthogonal to the line joining the points O and P. Consequently, the mechanical stress field in a prestretched and activated DE, which is free to deform in its thickness direction, is given by Eq. 3.

Let us derive the expression of the external force $F_f = F_f(x, V)$ that must be supplied at O and P (and directed along the line joining these points) to balance the DE stress field at a given (fixed) generic configuration x of the actuator:

$$F_f(V,x) = zy_p\sigma_1(x,V) = y_p z'\lambda_3\lambda_1\frac{\partial\psi}{\partial\lambda_1} - \epsilon V^2\frac{y_p}{z} = \frac{y_p z'}{\lambda_{2,p}}\frac{\partial\psi}{\partial\lambda_1} - \epsilon V^2\frac{y_p\lambda_{2,p}}{z'x'}x \tag{6}$$

which, by convention, is positive if directed according to the arrows depicted in Fig. 12. Conventionally, F_f is the force that an external user supplies to the actuator.

It can be noted that $F_f(V,x)$ can be decomposed in two terms:

$$F_f^{off}(x,0) = F_f^{off}(x) = \frac{y_p z'}{\lambda_{2,p}}\frac{\partial\psi}{\partial\lambda_1} \tag{7}$$

and

$$F_f^{em}(x,V) = -\epsilon V^2\frac{y_p\lambda_{2,p}}{z'x'}x \tag{8}$$

The force F_f^{off} is the force supplied by an external user to the actuator when the voltage $V = 0$ (it has been termed as the DE *film force* in the OFF state mode). The force F_f^{on} is the force supplied by an external user to the actuator when the voltage $V \neq 0$. The DE *film force* in the ON state mode is given by:

$$F_f^{on}(x,V) = F_f^{off}(x) + F_f^{em}(x,V) \tag{9}$$

The "electrically induced" term F_f^{em} has the dimension of a force and is usually referred to as *Maxwell force* (Kofod & Sommer-Larsen, 2005; Plante, 2006) or *actuation force*.

Equation 8 shows that: 1) the "force" F_f^{em} does not depend on the strain energy function which is chosen to describe the material hyperelastic behavior 2) the "force" F_f^{em}, in case of rectangular actuators, is affected by prestretch (for the same undeformed DE geometry). In the following, the electrically induced force F_f^{em} will also be called *force difference* or *actuation force*.

5.2 Diamond actuators

Diamond actuators are based on a bi-axially prestretched lozenge shaped DE coupled to a frame made by a four-bar linkage mechanism having links with equal length, l_d (Fig. 13(c)). The DE is attached all over the frame border. Principal prestretches $\lambda_{1p} = x_p/x'$ and $\lambda_{2p} = y_p/y'$ are applied in the x and y directions and are independent design parameters. Let us define (Fig. 13(a)) x' and y' as the EDF planar dimensions in the reference configuration (unstretched EDF) whereas x_p and y_p are EDF planar dimensions in prestretched configuration. The coupling with the frame is done when the distance OP is equal to x_p (x_p can be chosen as desired) where O and P are the centers of two opposing revolute pairs of the four-bar mechanism (as shown in Fig. 13). In particular, it has been chosen $x = x_p$ for the EDF in the OFF state mode (Fig. 13(b)).

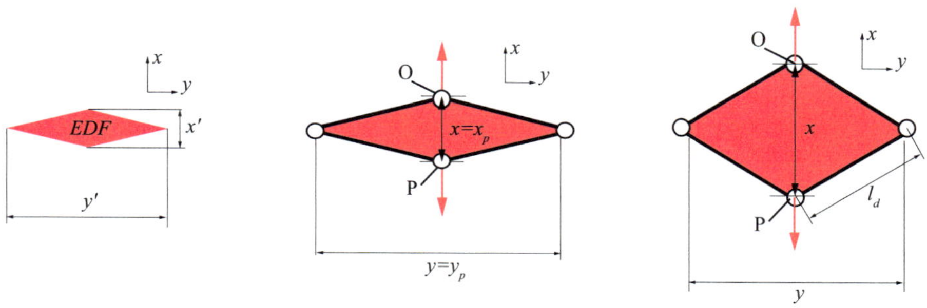

(a) Unstretched EDF. (b) Diamond DELA, schematic (OFF-state mode). (c) Diamond DELA, schematic (ON-state mode).

Fig. 13. Diamond DELA).

In such actuators, activation of the EDF makes it possible to control the relative distance x (hereafter also called "DE length" or "actuator length") of the points O and P, which are supposed to be the points of application of the (given) external forces F_f acting on the actuator boundary.

By construction, when coupled with a four-bar mechanism having links of equal length, lozenge-shaped EDF expand uniformly without changing their edge length l_d and principal stretch/stress directions. Thus, their deformation state is characterized by the following principal stretches:

$$\lambda_1 = \frac{x_p}{x'}\frac{x}{x_p} = \lambda_{1,p}\frac{x}{x_p}; \qquad \lambda_2 = \frac{y_p}{y'}\frac{y}{y_p} = \lambda_{2,p}\frac{y}{y_p} = \lambda_{2,p}\sqrt{(4l_d^2 - x^2)/(4l_d^2 - x_p{}^2)};$$
$$\lambda_3 = \frac{1}{\lambda_1\lambda_2} = \frac{z}{z'} \tag{10}$$

where the following kinematic relations can be easily found by the position analysis of the four-bar linkage machanism with links of equal length and observing that the displacements of both EDF boundary and frame must be identical, that is:

$$y = \sqrt{(4l_d^2 - x^2)} \qquad y_p = \sqrt{(4l_d^2 - x_p{}^2)} \tag{11}$$

Considering the xy plane, the principal stretch/stress directions are respectively aligned and orthogonal to the line joining the points O and P. Consequently, the mechanical stress field in

a prestretched and activated DE, which is free to deform in its thickness direction, is given by Eq. 3.
Let us now derive the expression of the external force that must be supplied at O and P, and directed along the line joining these points, to balance the DE stress field at a given (fixed) generic configuration x of the actuator. Because of symmetry, a quarter of the actuator can be schematized as in Fig. 14.

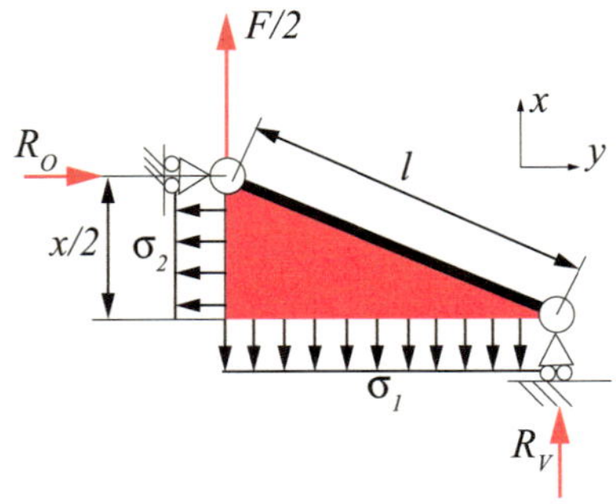

Fig. 14. Diamond DELA, force equilibrium.

The force F_f can be found using the equilibrium equations:

$$\frac{F_f}{2} + R_V = B_1; \qquad R_O = B_2; \qquad \frac{F_f}{2}\frac{y}{2} + R_o\frac{x}{2} = B_2\frac{x}{4} + B_1\frac{y}{4}; \tag{12}$$

where

$$B_1 = z\frac{y}{2}\sigma_1(x,V); \qquad B_2 = z\frac{x}{2}\sigma_2(x,V); \tag{13}$$

therefore

$$F_f = B_1 - B_2\frac{x}{y} = \frac{z'}{\lambda_1\lambda_2}\frac{y}{2}\lambda_1\frac{\partial\psi}{\partial\lambda_1} + \frac{z'}{\lambda_1\lambda_2}\frac{y}{2}\epsilon E^2 - \frac{z'}{\lambda_1\lambda_2}\frac{x}{2}\frac{x}{y}\lambda_2\frac{\partial\psi}{\partial\lambda_2} - \frac{z'}{\lambda_1\lambda_2}\frac{x}{2}\frac{x}{y}\epsilon E^2 \tag{14}$$

which, by convention, is positive if directed according to the arrows depicted in Fig. 13. It can be noted that $F_f(V,x)$ can be decomposed in two terms:

$$\begin{aligned} F_f^{off}(x,0) = F^{off}(x) &= z\lambda_1\frac{\partial\psi}{\partial\lambda_1}\frac{y^2}{4} - z\lambda_2\frac{\partial\psi}{\partial\lambda_2}\frac{x}{2}\frac{x}{y} \\ &= \frac{z'}{2}\left(\frac{\sqrt{(4l_d^2 - x_p{}^2)}}{\lambda_{2,p}}\frac{\partial\psi}{\partial\lambda_1} - \frac{x x_p}{\lambda_{1,p}\sqrt{(4l_d^2 - x^2)}}\frac{\partial\psi}{\partial\lambda_2}\right) \end{aligned} \tag{15}$$

and

$$F_f^{em}(x,V) = z\frac{y}{2}\left(\frac{-\epsilon V^2}{\lambda_3{}^2 z'^2}\right) + z\frac{x}{2}\frac{x}{y}\left(\frac{-\epsilon V^2}{\lambda_3{}^2 z'^2}\right) = \frac{-\epsilon V^2}{z'}\lambda_{1,p}\lambda_{2,p}\frac{x(2l_d^2 - x^2)}{x_p\sqrt{(4l_d^2 - x_p{}^2)}} \tag{16}$$

The force F_f^{off} is the film force in the OFF state mode whereas the film force in the ON state mode is given by:

$$F_f^{on}(x,V) = F_f^{off}(x) + F_f^{em}(x,V) \tag{17}$$

As stated for the rectangular actuators, the term expressed by F_f^{em} can be interpreted as an "electrically induced force" due to DE activation.

5.3 General remarks on the DE film models

Let us define: 1) the parameter $\zeta = x_b/x_f \leq 1$, where x_b is the initial actuator length and x_f is the final actuator length ($\delta = x_f - x_b = x_b(\zeta^{-1} - 1)$, being the actuator stroke); 2) the actuation force relative error as:[2]

$$e_T = [max(F_f^{em}(x))/min(F_f^{em}(x)) - 1] \quad \text{within } x_b \leq x \leq x_f \tag{18}$$

Different considerations can be drawn for the rectangular and the diamond actuators:

- Rectangular actuator. The actuation force is given by Eq. 8. Considering the DE parameters as fixed and given a maximum actuation voltage V_{max}:

$$e_T^{rectangular} = F_f^{em}(x_f)/F_f^{em}(x_b) - 1 = \frac{1}{\zeta} - 1 \tag{19}$$

  where it can be seen that the actuation force relative error depends on ζ and increases as ζ decreases being null for actuators presenting a null stroke. A possible way to keep $e_T = 0$ is by setting $V^2 = \frac{\Delta F_d}{C_{ps}x}$ where ΔF_d is the desired force difference, $C_{ps} = \frac{y_p \lambda_{2p} \epsilon}{z' x'}$. The information about the actual DE position x must be obtained with appropriate sensory systems or using the methods described in Jung et al. (2008) and fed back to a voltage controller. Obviously the actuation source should be capable of actively controlling the voltage.
- Diamond actuators. The actuation force is given by Eq. 16.
  Let us consider the adimensional parameter $\chi = x/l_d$ which uniquely identifies the lozenge configuration ($\chi_b = x_b/l_d, \chi_f = x_f/l_d, \zeta = \chi_b/\chi_f, \chi_b = \zeta\chi_f$). The actuation force in terms of χ can be written as:

$$F_f^{em}(x, V) = \frac{\epsilon V^2}{z'} \lambda_{1,p} \lambda_{2,p} \frac{l_d^3}{x_p \sqrt{(4l_d^2 - x_p{}^2)}} f_f^{em}(\chi) \tag{20}$$
$$f_f^{em}(\chi) = \chi(\chi^2 - 2)$$

  that shows how the actuation force becomes null when the lozenge shaped EDF degenerates into a square EDF (i.e. for $\chi = \tilde{\chi}_f = \sqrt{2}$ or $\tilde{x}_f = \sqrt{2}l_d$) and eventually changes sign for $\chi \geq \sqrt{2}$. The function $f_f^{em}(\chi)$ is plotted in Fig. 15 and has a minimum for $\chi = \sqrt{2/3}$.
  Let us consider configurations of the EDF such that $\chi < \tilde{\chi}_f$. Considering the EDF parameters as fixed and given a maximum actuation voltage V_{max}, then:

$$e_T^{diamond} = \frac{max_{\chi_b,\chi_f}(f_{em}(\chi))}{min_{\chi_b,\chi_f}(f_{em}(\chi))} \tag{21}$$

[2] In the following $max[f(x)]$within $x_b \leq x \leq x_f$ will be indicated as $max_{x_b,x_f}[f(x)]$

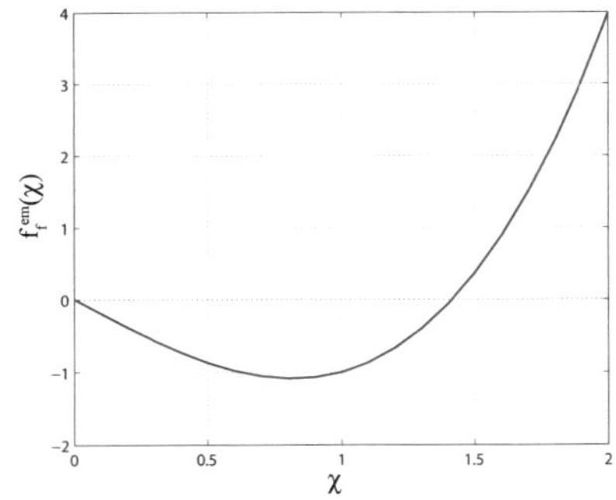

Fig. 15. Plot of $f_f^{em}(\chi)$

which is minimum if $f_{em}(\chi_b) = f_{em}(\chi_f)$, that is:

$$\chi_f = \sqrt{\frac{2}{\zeta^2 + \zeta + 1}} \tag{22}$$

The resulting force difference relative error being

$$e_T = \frac{\sqrt{8/3} - (2/3)^{\frac{3}{2}}}{2\chi_f - \chi_f^3} - 1 \tag{23}$$

Therefore, in order to minimize the force difference relative error, given x_b and x_f, l_d should be chosen such that:

$$l_d = \frac{x_f}{\sqrt{2}}\sqrt{\zeta^2 + \zeta + 1} \tag{24}$$

6. DE design constraints

In this section, three different types of design constraints or failure modes that can affect EDF design are described. These failure modes do not take into account the effect of localized material flaws, electric field concentrations or stress concentrations.

- *Mechanical failure.* This condition occurs when the mechanical strength of the material is exceeded. Experimental activities have shown that mechanical failure for hyperelastic polymers is primarily a function of stretch and not of stress and it takes place when folded polymer chains are straightened beyond their unfolded length. Plante (2006) reports a mechanical failure criterion based on DE film area expansion stating that failure is prevented if $A_{final}/A_{initial} < c$. The term $A_{initial}$ is the initial DE area before prestretch, A_{final} is the DE area at breaking and c is a characteristic constant. However, it has been shown (Vertechy et al., 2009) that also the Kawabata's failure criterion is suited for the study of DE materials and simpler to use when designing. This criterion (Hamdi et al., 2006) postulates that the mechanical failure of polymers under any loading path occurs when any principal stretch equals or exceeds the value of the stretch at break measured under uniaxial tension, that is:

$$max[\lambda_1, \lambda_2] \geq \lambda_{ut} \tag{25}$$

where λ_{ut} is the principal stretch at break achieved in an uniaxial test.

- *Electric breakdown.* This type of failure occurs when the electric field in a material becomes greater than its dielectric strength. In this situation the electric field may mobilize charges within the DE, producing a path of electric conduction. After electric breakdown, the DE will present a permanent defect preventing its usage for actuation. Electric breakdown occurs when:
$$E \geq E_{br} \tag{26}$$
where E_{br} is the electric field at break that is usually determined experimentally. A theoretical prediction of electric breakdown can be found in Whithead (1953). For actuation usage, it is useful to activate the DE electric fields which are as close as possible to the electric field at break (indeed an higher E signifies higher F_f^{em}). Recent experiments have shown that DE prestretching increases the DE dielectric strength. For this reason, in the following design procedure DE prestretch is maximized.
- *Loss of tension.* This condition occurs when the applied voltage induces deformations which may remove the tensile prestress. In fact, EDF have negligible flexural rigidity. This thin membrane can wrinkle out of its plane under slight compressive stresses which arise if the applied voltage is too high and exceed the given prestretch. Loss of tension is avoided if:
$$\sigma_1 > 0 \; \forall \; \Omega(t); \qquad \sigma_2 > 0 \; \forall \; \Omega(t) \tag{27}$$
where $\Omega(t)$ are the possible configurations of the DE film in working condition.
Another cause of DE failure is electromechanical instability or *pull-in* (Stark & Garton, 1955) and was identified as a mean of dielectric failure in insulators in 1950 (Mason, 1959). Pull-in is not properly a failure mode but a phenomenon that can eventually lead to either mechanical failure or electric breakdown. In fact, a voltage application causes DE expansion and subsequent reduction of thickness. A reduction in thickness signifies higher electric fields. Therefore there exists a positive feedback between a thinner elastomer and a higher electric field. An unrestricted area expansion of the material may lead to mechanical failure whereas higher electric fields may lead to electric breakdown. As reported by Lochmatter (2007), however, this hypothesis has not yet been proven experimentally and the condition of Eqs. 25, 26, 27 are considered sufficient for design purposes.

7. Analytical model development for the Slider Crank Compliant Mechanism

The FL curve concerning a compliant mechanism can be found by the PRBM using either the principle of virtual work or the free-body diagram approach Howell (2001).
Supposing the pin joints being torsional linear springs, the torques due to the deflection of the springs are given by:
$$T_i = -K_i \Psi_i \tag{28}$$
where, with reference to Fig. 8(c), K_i, $i = 1, 2, 3$ are the pivot torsional stiffnessess to be designed and $\Psi_1 = \vartheta_1 - \vartheta_{10}$, $\Psi_2 = \vartheta_3 - \vartheta_{30} - \vartheta_1 + \vartheta_{10}$, $\Psi_3 = \vartheta_3 - \vartheta_{30}$. The following relationships are found from the position analysis of the mechanism:
$$\vartheta_3 = \pi - asin\Big(\frac{r_1 sin(\vartheta_1) - e}{r_2}\Big); \qquad x = r_1 cos(\vartheta_1) - r_2 cos(\vartheta_3); \qquad \alpha = atan(\frac{e}{x}) \tag{29}$$

If the value of the eccentricity e is such that $e = 0$, the law of cosines can be used leading to the following expressions:

$$\vartheta_1 = acos\left(\frac{x^2 + r_1{}^2 - r_2{}^2}{2xr_1}\right); \qquad \vartheta_3 = acos\left(\frac{x^2 + r_2{}^2 + r_1{}^2}{2xr_2}\right)$$

Note that, if the compliant mechanism is formed form a monolithic piece, then:

$$\vartheta_{30} = \pi - asin\left(\frac{r_1 sin(\vartheta_{10}) - e}{r_2}\right) \tag{30}$$

From the static analysis of the mechanism, the following FL relationship can be obtained:

$$F_s = F_1 + F_2 + F_3 \tag{31}$$

where

$$F_1 = \frac{K_1\Psi_1 cos(\vartheta_3)}{r_1 sin(\vartheta_3 - \vartheta_1)}; \qquad F_2 = \frac{K_2\Psi_2 cos(\alpha)}{r_1 sin(\vartheta_1 - \alpha)}; \qquad F_3 = \frac{K_3\Psi_3 cos(\vartheta_1)}{x\ sin(\vartheta_1) - e\ cos(\vartheta_1)} \tag{32}$$

The same expressions holds when $e = 0$.
Let us define the variable $K_{12} = K_1/K_2$ and the function $\Xi = \Xi(K_{12}, r_1, r_2, e, \theta_{10})$ such that:

$$F_{12} = K_1\Xi \tag{33}$$

where

$$\Xi = \frac{\Psi_1 cos(\vartheta_3)}{r_1 sin(\vartheta_3 - \vartheta_1)} + \frac{K_{12}\Psi_2 cos(\alpha)}{r_1 sin(\vartheta_1 - \alpha)} \tag{34}$$

This expression will find a use when designing the SCCM such that F_{12} is quasi constant along a given range of motion (see section 9.2)

8. Design procedure and actuator optimization

Let us derive a general design methodology, that can be used to optimize DELA whose analytical model is available. Nevertheless, in case the geometry of the DELA does not make it possible to derive simple mathematical models, the considerations which are drawn concerning the frame stiffness remain valid.

8.1 Design variables

The actuator available thrust, F_a, is given by Eq. 1. The maximum thrust in the OFF state mode is $F_{max}^{off} = F_a(x, 0)$. The maximum thrust in the ON state mode is $F_{max}^{on} = F_a(x, V_{max})$. The overall design of a DELA depends on numerous parameters. In practical applications some of these parameters are defined by the application requirements whereas some others are left free to the designer.
First of all, when a DE material is chosen for applications (silicone or acrylic DE (Kofod & Sommer-Larsen, 2005; Plante, 2006)), the material electromechanical properties are given, that is the dielectric constant ϵ_r, the constants related to the material constitutive equation, the electric field at breaking E_{br} and the ultimate stretch at breaking λ_{br}. Supposing to use an Ogden model for the DE constitutive equation and electrodes, the constants μ_p, α_p are given.
It is supposed that the actuator size is given along with the maximum encumbrance of the

actuator is determined by the application requirements. Moreover, the maximum actuation voltage V_{max} which can be supplied by the circuitry is given along with the DELA initial and final positions x_b, x_f ($\delta = x_b - x_f$ being the desired actuator stroke).

The designer can specify the maximum thrust profile in the OFF state mode $\overline{F}^{off}_{max}$ or the maximum thrust profile in the ON state mode $\overline{F}^{on}_{max}$, the thrust profile being approximated with a linear function with slope (stiffness) K_d. However it is wiser to specify the desired thrust profile in the OFF state mode because it depends on the DELA elastic properties only. The thrust in the ON state mode depends both on the elastic properties and on the applied voltage meaning that it can be controlled at will to a certain extent (using controllable actuation sources and sensory units). At last, the force difference $\Delta\overline{F}_a$ between $\overline{F}^{off}_{max}$ and $\overline{F}^{on}_{max}$ must be defined. Note that, as long as the frame is a passive elastic element, $\Delta F_a(x) = F_f{}^{off}(x) - F_f{}^{on}(x) = F_f^{em}(x)$. Variables which are unknown at this stage are:

- The initial DE film dimensions x', y', z'. Due to the production techniques of the DE films (which are either purchased as thin films or obtained by injection moulding), it is likely that the film thickness z' cannot be chosen at will. However a stack of insulating DE films can be used to form a single DE. Therefore it will be assumed that $z' \in \mathbb{I}$, $\mathbb{I}$ being a given set of integer number, whereas x', y' are completely left free to the designer.
- The DE prestretches in the planar directions $\lambda_{1p}, \lambda_{2p}$. It should be underlined again that prestretch in some direction is necessary for the DE film not to wrinkle under actuation. In addition, prestretch increases the breakdown strength of DE films, therefore improving actuator performance (Kofod et al., 2003; Pelrine et al., 2000; Plante & Dubowsky, 2006). At last, the effect of prestretch is to alter the DE film dimensions making it thinner and wider and therefore increasing ΔF_f for a given voltage, V (for instance see Eqs. 8 and 16). Therefore prestretch should be kept as high as possible.
- The number of film layers N_{layers}.
- Concerning the SCCM used to correct the DELA stiffness every kinematic and structural variable is still unknown.

In summary:

- *Given data*:
    - **Material properties**.
        - DE mechanical properties: μ, α
        - DE stretch at break: λ_{br}
        - DE electrical properties: ϵ_r
        - Electric field at break: E_{br}
    - **Application requirements**.
        - Actuator planar dimensions (maximum allowable by application constraints);
        - Actuator initial and final position (and desired stroke): x_b,x_f;
        - Desired thrust profile: $\overline{F}^{off}_{max}$ (with approximately constant stiffness K_d);
        - Desired actuation force: $\Delta\overline{F}_a = \Delta\overline{F}_f$.
    - **Circuitry parameters**.
        - Maximum actuation voltage: V_{max};
- *Design variables*:
    - **DE film parameters**
        - DE film initial dimensions: x', y', z' where $z' \in \mathbb{I}$;

  - Amount of prestretch: $\lambda_{1p}, \lambda_{2p}$;
  - Number of film layers: N_{layers}.
- **Frame parameters.**
  - Links lengths and dimensions: r_1, r_2, e;
  - Flexural pivot dimensions: $K_i,\ i = 1,3$ and θ_{10}.

8.2 The design procedure

The design procedure comprises two steps: first the determination of the DE geometrical parameters, second the design of the flexible frame.

- **Determination of the DE geometrical parameters**

1. Choose a suitable DE geometry and define the actuator planar dimensions which are compatible with the application constraints.
   Recall that, for the diamond actuators, given the desired initial and final actuator lengths, x_b and x_f, it is possible to choose the lozenge length that minimizes the force difference error e_T. Minimizing this error means that F_f^{off} and F_f^{on} are close to parallel in the $x_b - x_f$ range. The same result cannot be achieved with rectangular actuators, where e_T is independent of the actuator geometrical parameter y_p.
2. Given x_b, x_f (and therefore y_b, y_f), the ultimate stretch at break λ_{br} and a suitable safety factor ϕ_λ to avoid mechanical break, find the initial DE film planar dimensions, x', y':

$$\lambda_{1max} = \frac{x_f}{x'} = \phi_\lambda \lambda_{br} \Rightarrow x' = \frac{x_f}{\phi_\lambda \lambda_{br}} \qquad \lambda_{2max} = \frac{y_b}{y'} = \phi_\lambda \lambda_{br} \Rightarrow y' = \frac{y_b}{\phi_\lambda \lambda_{br}} \tag{35}$$

3. Given x', y', find $\lambda_{1p}, \lambda_{2p}$:

$$\lambda_{1,p} = \frac{x_b}{x'}; \qquad \lambda_{2,p} = \frac{y_b}{y'} \tag{36}$$

4. Given $x', y', x_f, y_f, V_{max}$, the electric field at break E_{br} and a suitable safety factor to avoid electrical break ϕ_{el}, find z':

$$E_{max} = \phi_{el} E_{br} = \frac{V_{max}}{z_{min}} \Rightarrow z_{min} = \frac{V_{max}}{\phi_{el} E_{br}} = z_f \tag{37}$$

$$\lambda_{3min} = \frac{z_f}{z'} \Rightarrow z' = \frac{z_f}{\lambda_{3min}} = \frac{V_{max}}{\phi_{el} E_{br}} max_{x_b, x_f}(\lambda_1 \lambda_2)$$

   Choose $\tilde{z}' \in \mathbb{I}$ such that $\tilde{z}' \geq z'$
5. Given the aforementioned quantities, the quantity $min\ F_f^{em}(x) = \Delta F_f^{min}$ within $x_b \leq x \leq x_f$ can be analytically computed for one layer of film. Knowing the desired $\Delta \overline{F}_f$, find N_{layer} such that:

$$N_{layer} \Delta F_f^{min} \geq \Delta \overline{F}_f \tag{38}$$

6. Verify that $\sigma_i > 0\ \forall\ \Omega(t),\ i = 1, ...2$. If the condition is not verified, decrease V_{max} and, in case, increase N_{layer} in order to achieve the desired thrust.

- **Design of the flexible frame**

1. Impose the UEP at a point $\overline{x}$ such that $x_b \leq \overline{x} \leq x_f$. The imposition of the UEP at a given $\overline{x}$ constrains the dimension of either r_1 or r_2, for instance:

$$r_2 = \sqrt{r_1^2 + \overline{x}^2} \tag{39}$$

2. Considering Eq. 39, the frame force due to the torsional springs K_1, K_2 is given by

$$F_{1,2} = K_1 \Xi(K_{12}, r_1, e, \theta_{10}) \tag{40}$$

where $K_{12} = K_2/K_1$.
Use multivariable optimization (Howell, 2001) to find $K_{12}, r_1, e, \theta_{10}$ which minimize:

$$\Im = \frac{max_{x_b, x_f} \Xi}{min_{x_b, x_f} \Xi} \tag{41}$$

subjected to:

$$K_{12} \geq 0; \quad 0 < r_1^{min} < r_1 < r_1^{max}; \quad 0 \leq e^{min} < e < e^{max}; \quad \theta_{10}^{min} \leq \theta_{10} \leq \theta_{10}^{max} \tag{42}$$

The variable K_{12} and the connecting rod length r_1 are constrained to be positive. In addition the maximum and minimum values for r_1 and for the eccentricity e might be imposed by the application constraints. At last θ_{10} is allowed to vary in the range $[\theta_{10}^{min}, \theta_{10}^{max}]$ only, in order to avoid excessive deflections of the elastic joints.

3. Given the desired thrust profile $\overline{F}_{max}^{off}$ and therefore $\overline{F}_f^{off}(\overline{x})$, find K_1 such that

$$F_{1,2} - \overline{F}_f^{off}(\overline{x}) \approx 0 \tag{43}$$

4. Given K_{12} and K_1 then $K_2 = K_{12}K_1$

5. Given the DELA desired stiffness $K_d = \frac{dF_f^{off}(x)}{dx} \approx const$ find K_3 such that

$$K_3 + K_d \approx 0 \tag{44}$$

6. Given the values of K_i, $i = 1, 2, 3$ the designer can find the flexure dimensions. Supposing, for instance, the flexures are straight beam hinges with rectangular cross section then $K_i = \frac{EI_{a_i}}{L_i}$ where E is the frame material Young modulus, L_i is the length of the small-length flexural pivot, and $I_{a_i} = \frac{h_i^3 b_i}{12}$ is the moment of inertia of the pivot cross sectional area with respect to the axis a_i (h_i and b_i denotes the pivot thickness and width respectively, whereas a_i is the barycentric axis parallel to the width direction).

9. Case studies

9.1 Single-acting constant-force actuator of rectangular geometry

The objective of the present case study is to design a single-acting actuator capable of supplying a positive constant force over a given range of motion. The EDF is a Silicone DE (Whacker Elastosil RTV 625) coated with silver grease electrode (CW7100) (Kofod & Sommer-Larsen, 2005) and coupled to the rigid links containing the points O and P of the compliant frame schematized in Fig. 1(d). Displacement along the y direction (or, alternatively, rotation) of the rigid link containing the point O is prevented by the symmetry of both the compliant frame geometry and the EDF stress distribution.

- *Given data*:
    - **Material properties** (Kofod & Sommer-Larsen, 2005):
      $\mu = 596kPa$, $\alpha = 0.85$, $\lambda_{br} = 2$, $\phi_\lambda = 0.75$, $\epsilon_r = 2.7$, $E_{br} = 140MV/m$, $\phi_{el} = 0.7$
    - **Application requirements and Circuitry parameters**:
      $x_b = 16mm$, $x_f = 22mm$, $y_p = 43mm$ (Fig. 13(c)), $\overline{F}^{off}_{max} = 0.25N$, $\Delta\overline{F}_a = 0.25N$, $V_{max} = 2.5kV$
- *Design variables*:
    - impose $e = 0$, $\overline{x} = x_b$
    - **DE film parameters**:
      $x' = 14.6mm$, $y' = 28.6mm$, $z' = 0.5mm$, $\lambda_{1p} = 1.09$, $\lambda_{2p} = 1.5$, $N_{layers} = 2$;
    - **Frame parameters.**:
      $r_1 = 21.5mm$, $r_2 = 26.8$, $e = 0mm$, $\theta_{10} = 22°$
      $K_1 = 0.002Nm/rad$, $K_2 = 0.0034Nm/rad$, $K_3 = 0.067Nm/rad$

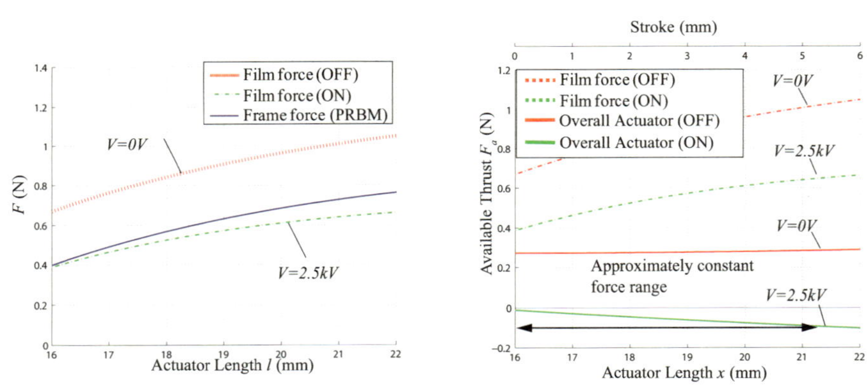

(a) (b)

Fig. 16. Analytical FL relationship showing film force F_f and frame force absolute value $|F_s|$ (a), DE actuator FL curves when coupled with the delta element (final design) (b).

Figure 16(a) shows the frame force $|F_s|$ and the theoretical film forces F_f^{off}, F_f^{on} as functions of the actuator length x. The frame behavior is as expected. Figure 16(b) shows the film force F_f compared to the overall actuator available thrust F_a. The actuator thrust in the OFF state mode is approximately constant over the range 16-22 mm with value 0.27 (in this range a maximal deviation by 0.01 N is admitted).

9.2 Bidirectional constant-force actuator of diamond shape

Objective of the present case study is the design of a bidirectional constant-force actuator of diamond shape using a compound-structure flexible frame. The EDF is an acrylic DE (VHB 4905) coated with silver grease electrodes (CW7100) (Plante, 2006) and coupled with the four bar linkage mechanism schematized with a dashed line in Fig. 1(e).

- *Given data*:
  - **Material properties** (Kofod & Sommer-Larsen, 2005; Vogan, 2004): $\mu = 60kPA$, $\alpha = 1.8$, $\lambda_{br} = 8$, $\phi_\lambda = 0.63$, $\epsilon_r = 4.7$, $E_{br} = 150MV/m$, $\phi_{el} = 0.8$
  - **Application requirements and Circuitry parameters**: $x_b = 20mm$, $x_f = 30mm$, $l_d = \frac{x_f}{\sqrt{2}}\sqrt{\zeta^2+\zeta+1} = 30.8mm$ (Fig. 13(c)), $\overline{F}^{off}_{max} = 0.25N$, $\Delta\overline{F}_a = 0.5N$, $V_{max} = 7kV$
- *Design variables*:
  - impose $e = 0$, $\overline{x} = x_m = (x_f + x_b)/2$
  - **DE film parameters**: $x' = 6mm$, $y' = 31.74mm$, $z' = 1.5mm$, $\lambda_{1p} = 3.3$, $\lambda_{2p} = 5$, $N_{layers} = 1$
  - **Frame parameters.**: $r_1 = 21.5mm$, $r_2 = 33$, $e = 0mm$, $\theta_{10} = 10°$ $K_1 = 0.002Nm/rad$, $K_2 = 0.008Nm/rad$, $K_3 = 0.050Nm/rad$

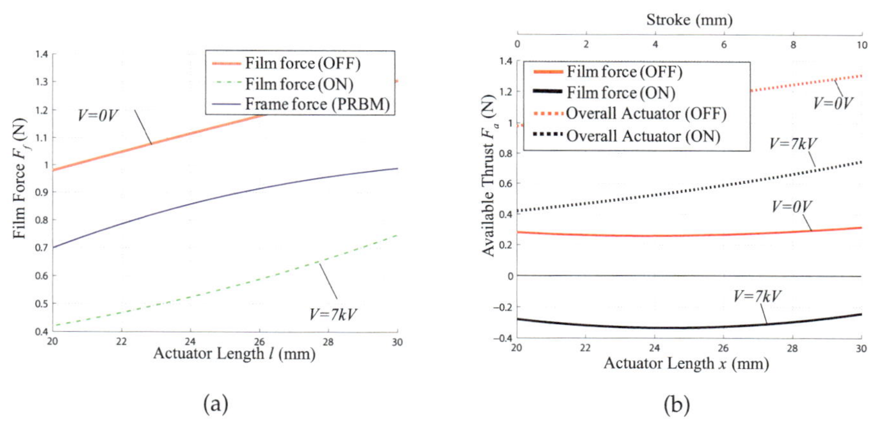

Fig. 17. Analytical FL relationship showing film force F_f and frame force absolute value $|F_s|$ (a), DE actuator FL curves when coupled with the delta element (final design) (b).

Figure 17(a) shows the frame force $|F_s|$ and the theoretical film forces F_f^{off}, F_f^{on} as functions of the actuator length x. The frame behavior is as expected. Figure 17(b) shows the film force F_f compared to the overall actuator FL curve (compound-structure frame coupled with the DE film). The available thrust in the OFF state mode keeps a value close to 0.25N over the range 20-30 mm (in this range a maximal deviation by 0.008 N is admitted). In order to prevent the actuator from working in the non-linear range, mechanical stops can be provided.

9.3 DELA of Conical Shape with predetermined stiffness

The properties of the SCCM can also be used to modify the behavior of axialsymmetric actuators. As an example consider the conical actuator depicted in Fig. 18

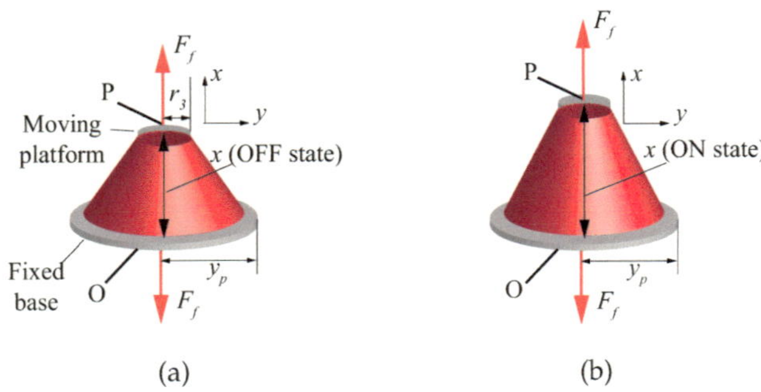

Fig. 18. Conical Actuator. OFF state mode (a), ON state mode (b).

Alike the other DELA geometries, the conical actuator supplies an available thrust that heavily changes along the stroke. This behavior is hereafter modified by coupling the conical EDF with the compliant frame shown in Fig. 1(f). The active film is shaped as a truncated cone. A planar circular DE film with initial radius of y' is first subjected to an equibiaxial prestretch up to a final radius denoted as y_p. Then, the application of an external force in the z direction (which is supplied by the moving platform of the compliant frame, see Fig. 1(f)) causes the DE film to gain a shape which is approximately conical. In this case, a simple mathematical model for the EDF is not available therefore the first part of the design procedure (concerning DE film design) cannot be employed. It should be stated, however, that a numerical solution of the conical DE has been proposed very recently by He et al. (2008). This solution relies on a set of differential equations to be solved numerically. Alternatively, FEM analysis can be used. In this work, the EDF FL curves have been determined experimentally using the procedure outlined in Berselli et al. (2009b). The EDF is an acrylic DE (VHB4905) coated with silver grease electrodes (CW7100). The objective of the present case study is to design a DELA capable of returning to an initial rest position when deactivated, that is to present a positive given stiffness K_d in the OFF state mode.

- *Given data*:
  - **DE film FL curve are determined experimentally; cone dimensions given below (Fig. 18)** (Berselli et al., 2009b)
    $y' = 20mm, y_p = 80mm, ' = 1.5mm, r_3 = 12mm$
  - **Application requirements and Circuitry parameters**:
    $x_b = 20mm, x_f = 30mm$ (Fig. 13(c)), $\overline{F}^{off}(x) = 0.07x, \Delta\overline{F}_a = 1.5N, V_{max} = 5kV$
- *Design variables*:
  - impose $e = 28, \overline{x} = x_b$
  - **DE film parameters**:
    $x' = 6mm, y' = 31.74mm, z' = 1.5mm, \lambda_{1p} = 3.3, \lambda_{2p} = 5, N_{layers} = 1$
  - **Frame parameters.**:
    $r_1 = 20.9mm, r_2 = 21.2, e = 28mm, \theta_{10} = 42°$
    $K_1 = 0.013Nm/rad, K_2 = 0.006Nm/rad, K_3 = 0.036Nm/rad$

In Figure 19(a), the modulus of the frame force $|F_s|$ and the film force F_f are plotted as functions of the actuator length x. The frame behavior is as expected. Figure 19(b) shows the the overall actuator available thrust F_a. The actuator thrust in the OFF state mode is a linear curve

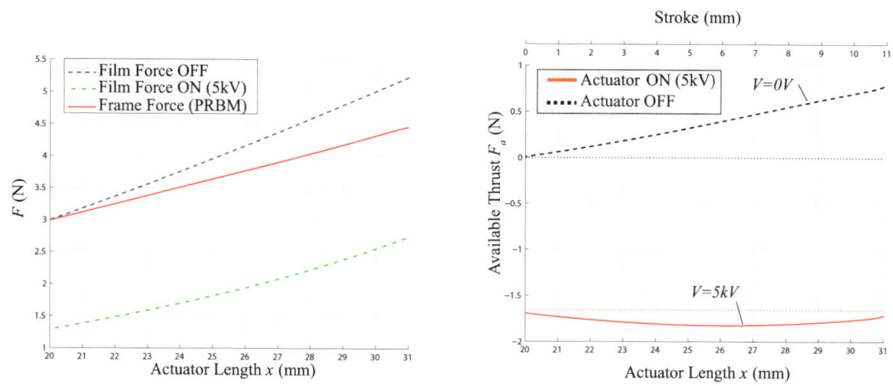

(a) (b)

Fig. 19. Analytical FL relationship showing film force F_f and frame force absolute value $|F_s|$ (a), DE actuator FL curves when coupled with the delta element (final design) (b).

vanishing at an actuator length of 20mm whereas the actuator thrust in the ON state mode is approximately constant (about 1.7N) over the range 20-30mm (in this range a maximal deviation by 0.1 N is admitted). A positive slope of the available thrust in the OFF state mode enables the actuator to come back to its initial rest position when deactivated.

10. Conclusions

The study of compliant actuators based on Dielectric Elastomers has been presented in a general framework which takes into account the interaction between the EDF and the film supporting frame. The key motivation of this work is based on the observation that a DELA presents an available thrust profile which can be heavily improved in terms of stiffness characteristics. Therefore, an easy methodology is needed to tailor the actuator stiffness to the application requirements.
In conclusion, the main contributions of this chapter can be summarized as it follows:

- A novel concept for the design of compliant frames has been proposed. The concept makes use of the stiffness properties of the slider-crank compliant mechanism. If suitably coupled with EDF of different geometries, such mechanism permits to adjust the DELA available thrust profile at the will of the designer.
- A novel design methodology has been presented which allows to tailor the stiffness of the actuator to the application requirements, defining its structural and geometrical properties. The procedure is composed of two main sub-procedures, one allowing the design of the EDF, the other one allowing the design of the compliant frame. In particular, the first part of the procedure can be employed to size EDF which are shaped as lozenges or rectangles for which analytical models are available. It is not applicable to general DE geometries (which require resorting to FEM analysis or experiments). The second part of the procedure is general and can be used to size the compliant frames even if a mathematical model of the EDF is not directly available.
- Three case studies have been presented concerning rectangular shaped EDF, lozenge shaped EDF, and conically shaped EDF. The response of the rectangular-shaped and

lozenge-shaped EDF have been determined analytically. The response of the conical-shaped EDF have been determined experimentally. Every geometry is coupled with a suitable compliant frame.

11. Acknowledgment

This research has been partially funded by Mectron Laboratory, Regione Emilia Romagna.

12. References

Bar-Cohen, Y. (2004). *Electroactive Polymer (EAP) Actuators as Artificial Muscles: Reality, Potential and Challenges*, Vol. PM136 of 2, SPIE Press.

Berselli, G., Vertechy, R., Vassura, G. & Parenti Castelli, V. (2008). A compound-structure frame for improving the performance of a dielectric elastomer actuator, *Springer, Advances in Robot Kinematics* **11**: 391–398.

Berselli, G., Vertechy, R., Vassura, G. & Parenti Castelli, V. (2009a). Design of a single-acting constant-force actuator based on dielectric elastomers, *ASME Journal of Mechanisms and Robotics, to be published* .

Berselli, G., Vertechy, R., Vassura, G. & Parenti Castelli, V. (2009b). Experimental evaluation of optimal conically-shaped dielectric elastomer linear actuators, *Intelligent Robots and Systems, 2009. IROS 2008. IEEE/RSJ International Conference on* .

Biagiotti, L., Tiezzi, P., Melchiorri, C. & Vassura, G. (2004). Modelling and controlling the compliance of a robotic hand with soft finger-pads, *IEEE Int. Conf. on Robotics and Automation, ICRA, Workshop on Multi-point Interaction in Robotics and Visrtual Reality*, New Orleans, LA.

Bicchi, A. & Tonietti, G. (2004). Fast and soft arm tactics: Dealing with the safety-performance tradeoff in robot arms design and control., *IEEE Rob. Aut. Mag.* **11**: 22–33.

Biddiss, E. & Chaua, T. (2008). Dielectric elastomers as actuators for upper limb prosthetics: Challenges and opportunities, *Medical Engineering and Physics* **30**(4): 403–418.

Bolzmacher, C., Hafez, M., Khoudja, M. B., Bernardoni, P. & Dubowsky, S. (2004). Polymer based actuators for virtual reality devices and rehabilitation applications, *Proceedings of the SPIE*, Vol. 5385, pp. 281–289.

Hamdi, A., Nait Abdelaziz, M., Ait Hocine, N., Heuillet, P. & Benseddiq, N. (2006). A fracture criterion of rubber-like materials under plane stress conditions, *Polymer Testing* **25(8)**: 994–1005.

He, T., Zhao, X. & Suo, Z. (2008). Equilibrium and stability of dielectric elastomer membranes undergoing inhomogeneous deformation, *Available online* .

Howell, L. (2001). *Compliant Mechanisms*, John Wiley and Sons.

Jung, K., Kima, K. J. & Choi, H. R. (2008). A self-sensing dielectric elastomer actuator., *Sensors and Actuators A* **143**: 343–351.

Kim, L. K. & Tadokoro, S. (2007). *Electroactive Polymers for Robotic Applications: Artificial Muscles and Sensors*, Eds. Springer.

Kofod, G., Paajanen, M. & Bauer, S. (2006). Self-organized minimum-energy structures for dielectric elastomer actuators, *Applied Physics A: Materials Science and Processing* **85(2)**: 141–143.

Kofod, G. & Sommer-Larsen, P. (2005). Silicone dielectric elastomer actuators: Finite-elasticity model of actuation, *Sensor and Actuators* **122**: 273–283.

Kofod, G., Sommer-Larsen, P., Kornbluh, R. & Pelrine, R. (2003). Actuation response of polyacrylate dielectric elastomers, *Journal of Intelligent Material Systems and Structures* **14**: 787–793.

Kornbluh, R. ., Pelrine, R. & Joseph, J. (1995). Dielectric artificial muscle actuators for small robots, *Proceedings of the Third IASTED International Conference on Robotics and Manufacturing, Cancun, Mexicoe.*

Koseki, Y., Tanikawa, T. & Chinzei, K. (2007). MRI-compatible micromanipulator-design and implementation and MRI-compatibility tests, *Proc. of EMBC 2007*, pp. 465–468.

Lochmatter, P. (2007). *Development of a Shell-like Electroactive Polymer (EAP) Actuator*, PhD thesis, Federal Institute of Technology, Zurich.

Mason, J. (1959). *Dielectric Breakdown in Solid Insulation*, Progress in Dielectrics I, Heywood and Company Ltd.

Ogden, R. W. (1972). Large deformation isotropic elasticity: on the correlation of theory and experiment for incompressible rubber-like solids, *Proc. Roy. Soc. London* **A-326**: 565–584.

Paul, R. & Shimano, B. (1976). Compliance control., *Proceedings of the JACC.*

Pedersen, C. B. W., Fleck, N. A. & Ananthasuresh, G. K. (2006). Design of a compliant mechanism to modify an actuator characteristic to deliver a constant output force, *Journal of Mechanical Design* **128**(5): 1101–1112.

Pei, Q., Pelrine, R., Standford, S., Kornbluh, R. & Rosenthal, M. (2003). Electroelastomer rolls and their application for biomimetic walking robots, *Synth. Met.* **135/136**: 129–131.

Pelrine, R., Kornbluh, R. & Joseph, J. (1998). Electrostriction of polymer dielectrics with compliant electrodes as a means of actuation, *Sensors Actuators A* **64**: 77–85.

Pelrine, R., Kornbluh, R. & Kofod, G. (2000). High-strain actuator materials based on dielectric elastomers, *Advanced Materials* **12(16)**: 1223–1225.

Plante, J. S. (2006). *Dielectric elastomer actuators for binary robotics and mechatronics*, PhD thesis, Department of Mechanical Engineering, Massachusetts Institute of Technology, Cambridge, MA.

Plante, J. S. & Dubowsky, S. (2006). Large-scale failure modes of dielectric elastomer actuators, *Int. J. Solids Struct.* **43**.

Plante, J. S. & Dubowsky, S. (2008). MRI compatible needle manipulator for robotic assisted interventions to prostate cancer, *in* F. Carpi & E. Smela (eds), *Biomedical applications of electroactive polymer actuators*, Wiley.

Stark, K. H. & Garton, C. G. (1955). Electric strength of irradiated polythene, *Nature* **176**: 1225–1226.

Toupin, R. A. (1956). The elastic dielectrics, *J. Rational Mech. Anal.* **5**: 849–915.

Vertechy, R., Berselli, G., Vassura, G. & Parenti Castelli, V. (2009). A new procedure for the optimization of a dielectric elastomer actuator, *Springer, Computational Kinematics* **1**: 391–398.

Vogan, J. (2004). *Development of dielectric elastomer actuators for MRI devices*, Master's thesis, Department of Mechanical Engineering, Massachusetts Institute of Technology, Cambridge, MA.

Whithead, S. (1953). *Dielectric Breakdown of Solids*, Oxford University Pres.

Williamson, M. M. (1993). *Series elastic actuators*, Master's thesis, Department of Electrical Engineering and Computer Science, Massachusetts Institute of Technology, Cambridge, MA.

Wingert, A., Lichter, M. D. & Dubowsky, S. (2006). On the design of large degree-of-freedom digital mechatronic devices based on bistable dielectric elastomer actuators, *IEEE/ASME Transactions on Mechatronics* **11**(4): 448–456.

29

Hybrid Control Techniques for Static and Dynamic Environments: a Step towards Robot-Environment Interaction

Fabrizio Romanelli
Comau Robotics S.p.A.
Italy

1. Introduction

Nowadays industrial robots have to perform complex tasks at high speeds and have to be capable of carrying out extremely precise and repeatable operations in an industrial environment; however, robot manipulators can perform just a small set of interactions with structures in their surrounding environment and with human operators which can be eventually present in the working area. The great gap between the capability of performing tasks with a high precision and speed, and the ability to perceive the environment and to act with it, claims the need of a smart and versatile control system which must guarantee a high degree of interaction between the robot manipulator and its world, whilst assuring the same performances required by the industry processes. In this context the ability to perceive its environment forms a crucial characteristic of such control system, which has to cope with problems of incompleteness of data and uncertainty. In order to achieve a high level of interaction, the control has to provide the system with the capability of robustly perceive the robot surroundings and to promptly react to any change in the state of its environment. A fundamental aspect of the robot-environment interaction is related to the capability of the control paradigm to model the structured and unstructured environment, such as its static and dynamic features. In this chapter, the main effort is that of describing new control architectures capable of taking into account both the static and dynamic characteristics of the robot world. In order to correctly introduce the problem, a particular focus has to be done on the static and dynamic aspects of the environment, considering at a first glance the two sides as different and separated and then, the last effort has to be done to merge the control techniques together in order to give a general solution to the modelling of environment and control of the robot. As an introduction for the first aspect, and as stated before, one of the most innovative and important problem in the nowadays industrial and service robotics is that one of completely controlling, not just the robot itself with its kinematics and dynamics, but even its interaction with the space where it works (Van Wichert & Lawitzky, 2001; Kraiss, 2006). It happens very often that the manipulators have to work in spaces shared with human operators or with other robots (eventually in some interchange zones) or they have to move and operate in places with static and/or dynamic facilities (Barraquand et al., 1989). Under this point of view, the use of a system that can control speed in a safe way (Winkler, 2007), allows to control operative areas, assuring a greater safety to human operators and between robots and machines, and it increases the efficiency of the used space as it is possible to concentrate more industrial devices in the same space with a great economical and cycle-time savings. The robots indeed can interact between them covering shorter paths, with the certainty that they will not collide, as far as the algorithm is active; at the same time, the presented paradigm (Romanelli & Tampalini, 2008a) allows the interaction of robots with other moving machines sharing the same spaces, increasing the efficiency of the used space. In addition to this, the system is capable of identifying the presence of the robot end effector inside the controlled zone or inside a larger zone called warning zone, using configurable outputs in order to communicate to the other devices inside the robot cell or to communicate to the other robots as well, if a warning or controlled zone is violated. This first paradigm gives a reference for the modelling of static industrial environment (such as interlock areas, machinery and other industrial devices) and for the control of the robot to interact with these predefined spaces. In order to model the robot surroundings, a system has been developed to manage a set of different geometrical shapes which define zones where the movement and access is forbidden or allowed; for the control system, the warning zone has been introduced. It is defined as a thickness from the controlled zone which is the core of the control system as it intrinsically defines the control for the general speed override of the robot end effector: an opportune control law has been studied in order to cope with particular geometrical conditions where more than one different shapes have to be controlled simultaneously. The proposed model and control has been applied and extended to dynamic objects moving in the robot working area (such as conveyors and rails which have a known trajectory) as well: the spatial checks and control in this case are performed on one or more moving geometrical zones. This aspect links the static model of the environment to the dynamic one, and puts the basis for a more general control paradigm. The dynamic and unstructured aspect of the problem instead has been analyzed in a slightly different way, taking into account the behaviour of the obstacles moving inside the robot working area without any prior knowledge, such as human operators working in tight connection to the robot; occupancy grids (Moravec, 1988; Elfes, 1989) have been taken into account in order to face this problem. They are used in order to tessellate the space (i.e. the operating area of the robot) in regular cells, and to store in each cell fine grained, quantitative information. With Bayesian occupancy grids (Collins et al., 2007), the idea is to extend the meaning of the value contained in each cell to the probability of that cell being occupied by an object. The nature of the decomposed space may be Euclidean space or a higher dimension state-space which could take into account velocities, accelerations and orientations as well. Such maps are extremely useful for robotic applications, such as obstacle/collision avoidance. In this kind of applications, the problem of the uncertainty of the information given by the sensors (proprioceptive or exteroceptive) is one of the biggest in this field. Such paradigm, using the Bayesian occupancy grids, face the problem in a very efficient way, as it models the unreliability of the measurements with probability. Another advantage of the use of occupancy grids is that they allow sensor fusion to be performed in a flexible way even if the system presents different typologies of sensors (even with very heterogeneous sensor models). Fuzzy logic (Zadeh, 1965, Klir & Yuan, 1965; Mamdani & Assilian, 1993) control has been chosen in order to take into account the behavioural aspect of the interaction between robots and human operators. Fuzzy logic controllers of particular interest are those used in Antilock Braking Systems (Mauer, 1995), in camera applications and where robots or automatic systems have to carry out behavioural tasks, such as collision avoidance and path planning (Kim et al., 1999). The paradigm concerning the dynamic aspect of the environment is based on the decisional and behavioural component on a fully reactive system based on Fuzzy logic controllers. The information about obstacle position around the working area related to the robot end effector, are computed in order to establish which kind of behaviour has to be taken and how it has to be applied. It is therefore possible after a proper adjustment of the control, to synthesize a system capable of acting with complex strategies, based on a simple set of behaviours; the result of this paradigm is a control which expresses precisely qualitative concepts, defined formally in terms of mathematical functions, called membership functions. Late in this chapter a new approach to deal with collision avoidance in dynamic environments is proposed. In the industry sphere the problem of collision and obstacle avoidance is relevant as the interactions between humans and machines are closer and closer. This is an important aspect which is matter of studies in the field of robotics and automation. In this context the basis idea of this chapter is to give a first step towards integration between the work of humans and robots; this integration can't be set aside of security which is the most relevant aspect of the problem and which has been taken into account during this study as a first requirement. The obstacle avoidance algorithm proposed (Romanelli & Tampalini, 2008b) is based both on a probabilistic framework, such to make the connection between the sensorial perception and the control of the robot, and on a polyvalent logic framework. There are no particular restrictions to the exteroceptive sensorial input model to the system, as the uncertainty of position of the obstacles given from the sensors can coexist in the same system, as the probabilistic framework also gives a good instrument to obtain sensor fusion. This method, which is efficient for medium-low distances obstacles, was combined with a fuzzy logic engine, which is very efficient for a medium-high distances obstacles and it is well adaptable to define politics to decide the reference override speed in function of heading. The advantages of utilizing a combination of the two approaches is that the robot override speed can be controlled, acting with both the controls in a continuous and smooth way. This control law takes into account both the trajectory of the obstacles moving around the robot area and the behaviour of the moving objects. The advantages of the aforementioned algorithms for the management of both static and dynamic environments have been merged in a hybrid control system, to make it capable of interacting with static objects (such as industrial facilities), objects with structured dynamics (i.e. objects bound to move on predefined paths, such as rails) and objects with unstructured dynamics (such as humans moving around the robot area). The results of the proposed control technique have been tested over a simulated system, for both the static and dynamic aspects; experiments on a real system have also been carried out, limited to the interaction between robot and structures, due to security reasons. Late in the chapter it will be showed how the presented approach can be extended in order to take into account other cognitive features.

2. Robot-Environment Interaction

Creating autonomous robots that can learn to act in unpredictable environments has been a long standing goal of robotics, artificial intelligence, and cognitive sciences. Robots are meant to become part of everyday life, as our appliances, assistants at home, and in particular in industrial environments, co-workers at the workplace. Nevertheless, to get robots operating outside research centres or universities and beyond the supervision of engineers or experts, it is necessary to face different technological challenges, amongst them, the development of strategies that allow robots to learn from their own experiences and interaction with the environment. This would provide robots with certain level of independence and dynamic behaviour. As a first step to face the problem of robot interaction with the environment it is important to understand and observe the interaction of humans with the environment in order to make the robot acting in a similar manner to what people do while moving in the same environment. A fundamental step to make the robot acting as a human is to define the typology of sensors it needs to perform a complex task, with enough accuracy and with robustness to unpredictability. The robot interacts with the environment in different ways: it acquires information from the environment through its sensors to provide the necessary input signals to the controller and it performs actions in its surroundings in order to achieve the desired tasks. The fundamental of interaction here is that sensing and acting are coupled dynamically and can not be analyzed independently since the perception of the robot influences the actions of the robot and the actions of the robot influence how the robot perceives the environment. This interaction exhibits complex and unpredictable characteristics and it is very difficult to identify the whole system using a generic method. In order to give the basis for a simplified theory, the analysis can start with the following assumptions:

- The robot controller is reactive where the output of the controller does not depend on the internal states of the controller, but only on the current input signals provided to the controller.
- The robot works and operates in a controlled environment with no other externals factors which influence the environment. Therefore it can be assumed that, when the robot performs actions in the environment, the change in the perception of the robot will be only dependent on the actions of the robot.

The entire robot-environment interaction can be described in a complete form using two models, under the previous assumptions: the robot controller model which computes the desired motor responses of the manipulator according to its perception and the perception model which emulates how the perception of the robot is affected by its own actions. In this chapter the attention will be focused on the robot controller model. There is also an important social aspect that has to be taken into account when developing new theories for automation and robotics in the industrial society; in the last century the growth in automation inside the factories and industries was exponential and this has led to a rapid change in the conditions of human operators. In particular for muscular fatigue technology has substituted tension and mental effort; for the more advanced automated plants, the transformation of physical energy into technical and mental skills is even emphasized (Marcuse, 1964). Another product of the increase in automation is the sense of alienation which has to be faced by the human operators who have to work in tight connection with robots or other mechanical instruments, being actors of repetitive tasks without complex interactions with their "mechanical co-workers". In this work, the interaction between robot and environment has been studied in order to also take taking into account the role of the human operator in the manoeuvring of robots, making his role more integrated to the productive process; this is possible, considering the overall increase in the interaction between robot and environment which is the basis idea of the present chapter. Furthermore the operation of a robot, as robot-environment interaction, is governed by three major components: the *robot* itself, its sensors, actuators and general hardware morphology, the *environment*, its perceptual properties and environmental conditions, and the *task*, the control program being executed on the robot. Given this, the behaviour of the robot emerges through the interaction of these three aspects. Any theory of robot-environment interaction will be dependent upon quantitative descriptions of the robot's behaviour. In order to assert that the robot-environment interaction is more influenced by the control program (the task) than the environment, here the chaos theory will be applied to describe the robot's behaviour quantitatively based on the considerations and results made in (Nehmzow & Walker, 2005). As a first approximation, it can be assumed that the robot's trajectory encapsulates the important aspects of the robot's behaviour; so this theory focuses the attention on the application of dynamical systems theory to the analysis of robot trajectories. One of the most distinctive characteristics of a chaotic system is its sensitivity to a variation in the system's variables, so that if two trajectories that started close each other will diverge from one another as time progresses, the more chaotic system, the greater the divergence. The Lyapunov exponent λ in (1) represents a measurement of chaos, so the larger the positive Lyapunov exponent, the quicker knowledge about the system is lost.

$$\lambda = \lim_{n \to \infty} \lim_{E_0 \to 0} \frac{1}{n} \sum_{k=1}^{n} \log \left| \frac{E_k}{E_{k-1}} \right| \tag{1}$$

Where E_0 is the initial error, and E_k, E_{k-1} is respectively the error at time *k* and *k-1*. From the experiments conducted by Nehmzow and Walker, the robot was firstly asked to conduct two different tasks in the same environment, and the computed Lyapunov exponent differed between the two tasks, signifying that the overall behaviour of the robot differed between the two experiments, and that this must have been due to the changes in control program. On the contrary the other experiment was conducted varying the environment and keeping the same task active. In this condition the Lyapunov exponent was the same in any of the changed environments, showing that the robot-environment interaction is far more influenced by the control program than the environment. This is an important result for the study of robot-environment interaction in the context of industrial robotics where the tasks are very often repetitive and where the environmental conditions can vary over time (as the presence of activated or deactivated controlled zones or the presence of human operators in the cell): this means that the obtained results greatly depend on the particular task the robot has to accomplish more than by the changes in its environment (e.g. changes in temperature which can affect motor dynamics, etc.).

3. Static environment modelling

In order to define a correct and effective model of the static features inside an industrial environment it is very important to define the possible set of structured objects inside a robot cell (which can be both static and dynamic). After this first step, where a general model can be proposed and analyzed in order to cover all the possible scenarios inside a real environment, it is necessary to synthesize a control system capable of taking into account these predefined geometrical areas, which can be both forbidden and allowed, in order to provide the robot with the fundamental tools to face an entry level of interaction; the study of this subject is then a good start to realize a more intelligent integrated robotized cell, where the strong interaction between robots and humans becomes closer and closer (taking into account this first modelling together with the dynamic environment modelling). In literature this topic is still at a basis level, and there are a lot of starting points of study: in the industrial automation background, PILZ developed a brand new control system for areas crowded with robots, machines and humans (Schulz, 2007). With this system, made of a safety camera capable of identifying access to forbidden areas, it is possible to send an output to safety devices in order to immediately stop the machines operating in those areas, avoiding harmful situations. Another application of this feature is the one made by ABB which implements a world zone management system made as follows: it is possible to define volumes where the robot presence is avoided. If the robot end effector is inside the allowed working space, the robot keeps working; if the robot end effector ends in an off-limits area, previously defined by the programmer, the control cuts the power and immediately stops the robot. This system lets the user program world zone software which is especially useful when two robots are working in close proximity to prevent collision and establish working protocols (Rooks, 2005). In literature there are several different approaches to the study of space occupancy and cooperation and to cope with collision avoidance problem. In particular, path planning is strongly associated to the problem of forbidden zones (Red et al., 1987; Brooks, 2003; Roy & Pratihar, 2003). The management of operative space is a matter of study and development in the field of telemanipulation and robot assisted tasks (Matinfar et al., 2007), where security of avoiding forbidden zones is the main objective of the work. Amongst the previous studies in collision detection, there are some works to be mentioned as the one (Canny, 1984) where given two general polyhedra of complexity n, one of which is moving while translating or rotating about a fixed axis, determine the first collision, if any, between the two objects. Another important aspect of collision detection and control of forbidden areas is presented in several works (Barak & Witkin, 1992; Bouma & Vanecek, 1992), where the dynamics of complex bodies is simulated over a system equipped with a collision detection algorithm. In the industrial field, other approaches aimed to reach a greater level of automation in robot-environment interaction. Fuzzy logic allows controlling a system in order to avoid access to dangerous areas (Shahrokhi & Bernard, 2004). There is a further approach that will be presented later on this chapter, where a system uses Bayesian occupancy grid and a fuzzy logic controller in order to avoid the collision between robot and other objects or humans moving around the cell (Romanelli & Tampalini, 2008b). In the following paragraphs, a new method to synthesize a control system capable of managing a set of predefined geometrical areas will be showed; with this paradigm, the advantages of taking into account a space model will be showed as well.

3.1 Management and control of multiple geometrical areas

The main problem for managing the robot working space, in particular when this space is shared with other robots or machinery, is the modelling of the robot surroundings. The first step to give the basis for a system capable of managing and control the interaction with the environment has to be the definition of primitive geometrical areas in order to cover all the possible configuration of the objects inside an industrial cell. The elementary objects are defined as parallelepipeds, cylinders and spheres (as depicted in Figure 1).

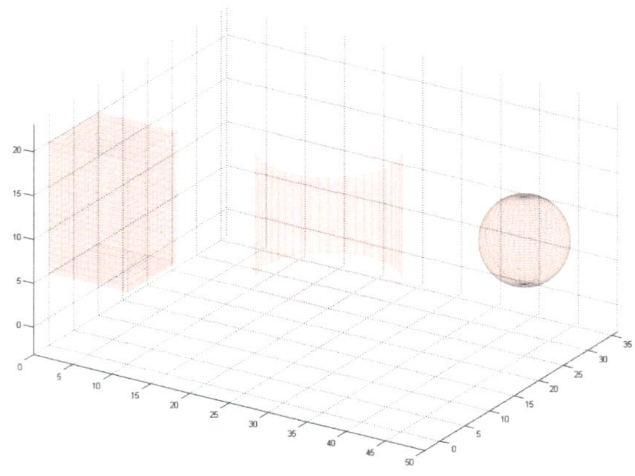

Fig. 1. Elementary geometrical shapes to model robot environment

With this simple modelling the system can be provided with the capability of defining multiple geometrical areas of these types in order to cover almost every object inside the cell (such as working tables, machinery or moving objects such as rails and conveyors). Starting from this definition of elementary geometrical area, in the following paragraph a control system will be showed. When this control system is active, it is possible to move the robot around the operating area with the certainty that, if the forbidden areas were previously defined and activated, if the robot end effector enters that zone, it will be immediately stopped (or similarly a digital output can be raised).

3.1.1 Control system

Starting from the elementary geometrical areas previously defined, the system allows the programmer to define multiple elementary zones which can be integrated inside the system and which represent the database of spatial forbidden areas which are used in order to control both the position and speed of the robot end effector. The areas can be easily defined by the programmer, considering that for the parallelepiped it is sufficient to define two points (the lower left corner at the base of the shape and the upper right corner); with these two points declared, the first shape can be integrated into the forbidden zone database. The cylinder on the other hand is defined considering the centre of the base circumference, the radius and its height: with this convention the base of the cylinder is parallel to the XY plane of the world frame reference. The sphere is instead defined considering its centre and radius. Thanks to the possibility of defining several zones in the same operating area, it is clear that a great part of the industrial applications can be covered by the use of this control paradigm. An important feature of this system is the possibility to consider the areas with two different features, concerning their *life*; they can be either constant or temporary. The first typology is programmable and modifiable from a particular class of users and they can not be ignored or modified by user programs: these zones are active during all the cycle of the robot. These can be used, for instance, in order to define zones that can not be covered by the robot end effector as they are occupied by fixed structures, such as pillars or other irremovable facilities. The second typology is temporary as it can be activated or deactivated from each user program and it is a very useful function in order to manage interlocks for exchange zones between robots: when a robot is inside an elementary zone, it is compulsory that the other robot is avoided to access the zone. This feature can be extended to those systems where a network of robot controllers is present and where the information about the elementary areas present in the robots cell is shared. In this case the control system of each robot can be supervised by another controller in order to update the information of the position of the robots in respect to the position of the multiple elementary areas defined on the cell. Another important feature of the presented system is that one which allows the user to define a further area bigger than the elementary one, called warning zone: in this zone, the robot can keep working but with a safe control system that checks the distance between the surface of the elementary area and the robot end effector, forcing the robot speed override to a value proportional to that distance. With this method, the robot end effector speed is reduced when the control system realizes that the robot violates the warning zone; with this control active, it is also possible to avoid mechanical solicitations due to hard brakes and to allow human operator to better perceive the enabled elementary zone around the robot. This control law is applied to each geometrical area and the resulting speed overrides (one for each declared elementary area) are computed in order to find the minimal value of them and to apply it to the robot. Another important innovation of the presented method is that concerning the implementation of dynamic management of elementary areas; in particular, with this system is possible to program areas which can change their position over time. This allows the elementary zones to be linked to moving objects (such as moving machines or to end effector of other robots inside the cell); in order to use this position it is necessary that the current position of the objects, to which the dynamic zone has to be attached, has to be known as the robot controller must know this information, or it must similarly share it with other cooperative robots inside the cell. In order to fully describe this feature, a space where several robots operate can be considered: in this configuration, it would be useful to define a dynamic zone on each robot end effector (linked to it). In this condition each robot knows exactly and instantly the position of other robots end-effector: if a robot draws too much to another one, the presented control is able to prevent damage between them. This allows the robot programmer to be released from the need of defining interlocks with some useless waits, while with this management it is possible to define digital outputs when a robot accesses a specified zone (not when specified in a user program); this control is parallel and acts in real-time, despite of the classic management of interlocks. Thus is very important during the productive cycle when robot programmer has to develop applications where more robots and machines share the same working space, and the presented method helps the user to exactly bound the working areas. The definition of a shared information on the state of the elementary zones (if they are occupied by a robot or not) is very useful and innovative for what concerns monitoring a dynamical geometrical area (e.g. with conveyors). With this system it is possible to link a dynamic zone to a moving object and this allows defining dynamical interlocks, which can be shared through a network between robots, giving a global visibility in the whole cell. The kinematics information about the robot joints and the tool allows the definition of useful information avoiding collision between robots cooperating in the same cell; this system, on the other hand, is not safe for the interaction between robots and human operators, but it is thorough in order to protect and prevent damage between robots and facilities without the need of further devices.

3.1.2 Proposed integrated solution

In this section an accurate description of the architecture of the control system will be shown. With the proposed solution each controlled elementary zone can be programmed and defined using both the programming language (PDL2 for Comau robots), describing the complete geometry, shape, thickness of the warning zone and its static or dynamic typology. With this method it is possible to program elementary areas to be controlled, in a precise way and this solution can be very suited for off-line programming. A second method to perform the definition of an elementary zone is to use the robot in order to teach the distinctive points of a geometrical area (as shown in the previous section). With this method, the user will be asked to move the robot around the working area and to locate the distinctive points of the geometrical shapes which have to be defined: teaching these points will bring the advantage of having a direct comparison with the taught elementary controlled zone and the real obstacle inside the working area. These methods provide the user with simple tools in order to create the database of the controlled elementary zones which allows the control system to perform complex operations of the spatial checks on the robot working area. With the proposed solution it is also possible to program and define for each declared zone, *channels* of shared information which can be activated automatically whenever the robot end-effector enters a controlled zone or a warning zone; this also allows to have a quantization of the working area. The operator who uses the proposed solution has the possibility of tuning a set of parameters which makes the system extremely flexible and modular. It is also possible for example to define a dynamic controlled zone, linked to the end-effector of another robot, in order to check the possible collisions between the robots. The control scheme is depicted in Figure 2.

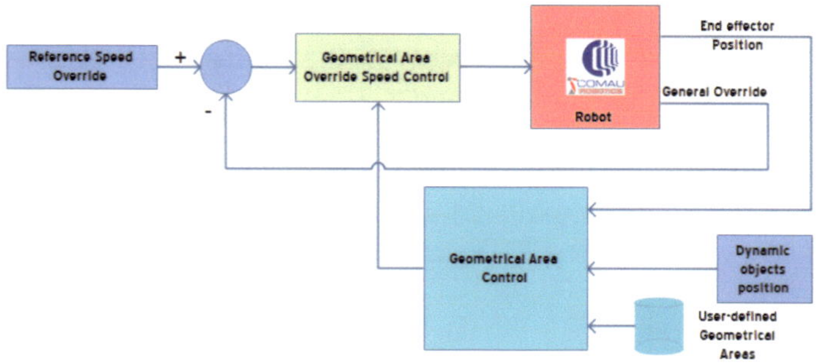

Fig. 2. Architecture of the multiple geometrical areas control system

As shown in the scheme the geometrical control algorithm checks if the robot end-effector is inside a controlled zone or, analogously a warning zone. This check is done on the basis of the database of geometrical areas, previously defined by the user; in this context, the dynamic objects position provides the control system with the possibility to link the geometrical areas to arbitrary moving points (as conveyors or rails), which can be read from external sensors like encoders. The speed control is performed by the geometrical area control block which detects the typology of the shape and selects the correct control law to be applied in order to modify the robot override, preventing collisions with the user-defined zones. The speed override is changed smoothly when the robot end-effector comes up against a spherical elementary zone, according to the following control law:

$$\begin{cases} v = v_o \cdot \dfrac{d-r}{\delta}, & r \le d \le r+\delta \\ v = v_o, & d > r+\delta \\ v = 0, & d < r \end{cases} \tag{2}$$

Where v is the actual speed override of the robot end-effector, v_o is the past override, d is the distance between the robot end-effector and the centre of the elementary spherical area, δ is the thickness of the warning zone and r is the radius of the sphere (the area is depicted in Figure 3.a). When the robot encounters a cylindrical elementary zone its speed override is subject to the following control law:

$$\begin{cases} v = v_o \cdot \dfrac{d_1 - r}{R-r}, & 0 \le z \le h \\ v = v_o \cdot \dfrac{d_2}{\delta}, & h < z \le h+\delta, -\delta \le z < 0, p \in cyl \\ v = v_o \cdot \dfrac{d_3}{\delta}, & h < z \le h+\delta, -\delta \le z < 0, p \notin cyl \end{cases} \tag{3}$$

Where h is the height of the cylinder, p the position of the robot end-effector and R is r+δ. The distances d_i represent: the distance between the centre of the cylinder and the robot position (d_1), the distance between the cylinder top or bottom base and the robot position, when it belongs to the top/bottom cylinder with thickness δ (d_2) and the minimal distance between the robot position and the points on the circumference of the top/bottom cylinder base (d_3). The robot speed override coincides to the old speed override when the robot end-effector is outside the warning zone; the cylindrical elementary area is depicted in Figure 3.b with its warning zone.

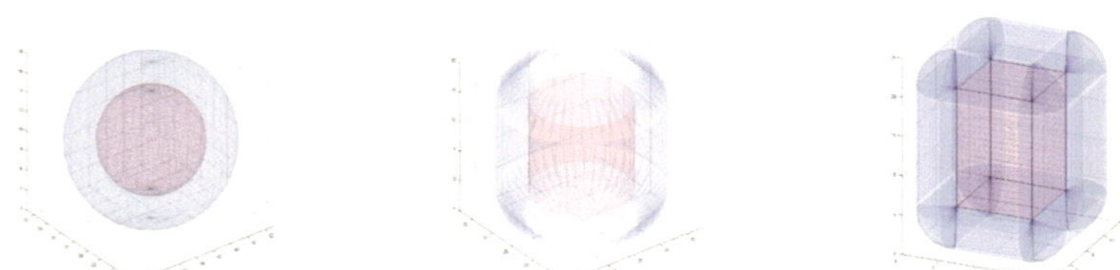

3.(a) 3.(b) 3.(c)

Fig. 3. Elementary shapes a) sphere, b) cylinder, c) parallelepiped (red) and their warning zones (blue)

The last modelled elementary geometrical area is represented by the parallelepiped; its control law is quite complex as its warning zone is composed by 8 half lunes, 12 quarters of cylinder and 6 planes. Given that, the mathematical treatment of the control law for the parallelepiped warning area is not reported here, but it is enough to consider that this control law aims at smoothly covering the whole warning zone area, with an appropriate speed override for each sector of it. Each elementary zone declared inside the working space has its own control law, also depending on the thickness of the warning zone; it is fundamental when the control system has to fix the controlled speed override that the correct value will be chosen in an efficient way. It is chosen according to the following:

$$\min(v_1, v_2, \dots, v_M) \tag{4}$$

Where M is the number of elementary geometrical areas defined inside the robot working space. This solution, notwithstanding its easiness, assures that the selected controlled speed override v_i follows a smooth trend when several different elementary zones are defined in the robot environment, even if they overlap. The last important feature of the presented paradigm is the possibility to manage dynamic geometrical areas, linking the position of moving objects to distinctive points belonging to the previously defined elementary shapes: for the spherical area, this point is characterized by the centre of the sphere. The cylinder will have its characteristic point on the centre of its bottom base; finally the bottom base centre of the parallelepiped will represent its characteristic point.

4. Dynamic environment modelling

In this section an effective and robust method to model the dynamic features in the industrial environment is described. As a first assumption, this model needs as an input the position over time or the trajectory of the objects which move around the robot working area. These can be identified as the inputs coming from different sensors (laser or camera scan and so on) or coming from more sophisticated devices like trackers (Harville & Li, 2004). The position of the multiple tracked objects is passed over time to the control algorithm which computes the correct speed override according to the combination of the Bayesian occupancy grid controller (Moravec & Elfes, 1985; Fulgenzi et al., 2007; Vasquez et al., 2006) and the fuzzy logic filter (Dong et al., 2005; Yen & Pfluger, 1995; Malhotra & Sarkar, 2005). In order to model the uncertainty coming from the sensors, a valid framework has to be taken into account; the Bayesian framework has suited models to cope with the uncertainty on the position of the obstacles but it also has an intrinsic capability to perform sensor fusion. In this context, a powerful instrument to face the problem of dynamic modelling of the space surrounding the robot is the Bayesian occupancy grid, a tessellated 2D grid in which each cell stores its probability of occupation. The behavioural side of the approach, which is necessary in order to model complex and unpredictable trajectories (like that of humans), is given by the fuzzy logic control which is very efficient for obstacles at medium-high distances and it is well adaptable to define politics to decide the reference speed override in function of heading. In this control system, sensor observations are processed from both the Bayesian occupancy grid algorithm and the fuzzy filter and the results of the computation are given as input to the collision avoidance algorithm. The combination of the two methods is effective as the robot speed override can be changed acting with both the controls in a continuous and smooth way; this control law in fact takes into account both the trajectory of the obstacles moving around the robot area and the behaviour of the obstacles. In the following paragraphs the two frameworks will be shown, examining the characteristics of both methods and comparing the advantages of the integrated solution in respect of the single solutions.

4.1 Bayesian control technique

The occupancy grid framework is based on the division of space (both Cartesian and multidimensional, taking for instance into account speed, acceleration and orientation as well) into cells. The probabilistic approach applied to the occupancy grid paradigm gives the possibility to extend the concept of cell value: if applied to obstacle or collision avoidance this value can fit well with the probability that the cell of the grid is occupied by an obstacle. Given as input for the algorithm the position $X = [x,y]^T$ of each obstacle, or likewise (ρ,θ), Bayes' theorem states:

$$P_c(Occ \mid X) \propto P_c(X \mid Occ) \cdot \hat{P}(Occ) \tag{5}$$

Where $P_c(Occ \mid X)$ is the probability that the cell of the grid is occupied by an obstacle, given the measurement, and the right side member of (5) is a distribution of probability, and it is shaped as Gaussian multimodal distribution as shown in the following:

$$P_c(X \mid Occ) \cdot \hat{P}(Occ) \propto \aleph(\mu, \Sigma) \tag{6}$$

Where $\mu = [\mu_1, \ldots, \mu_N]^T$ and Σ is the covariance matrix (positive-definite real N by N matrix). The probability density function is defined as follows:

$$f_X(x_1, \ldots, x_N) = \frac{1}{(2\pi)^{N/2} \left|\Sigma\right|^{1/2}} \cdot e^{-\frac{1}{2}(x-\mu)^T \Sigma^{-1} (x-\mu)} \tag{7}$$

The formula in (7) describes the probability density of an obstacle, in each point of the space $\mathbb{R}^N$ where ($N = 2$). In order to extend Bayes' theorem to more than one obstacle, assuming that all the events (obstacles) are independents, the generalized union probability theorem can be used:

$$\begin{aligned} P\left(\bigcup_{i=1}^{n} A_i\right) = & \sum_i P(A_i) - \sum_{ij}^{'} P(A_i \cap A_j) + \\ & + \sum_{ijk}^{''} P(A_i \cap A_j \cap A_k) - \ldots + \\ & + (-1)^{n-1} P\left(\bigcap_{i=1}^{n} A_i\right) \end{aligned} \tag{8}$$

This theorem states that if the probability that an obstacle is occupying a cell is independent from the others (which is reasonable for the problem), the union probability can be expressed in a closed form. Under this point of view, this method is a good approach to the obstacle avoidance problem since, besides the possibility to solve the problem of modelling multi-object space occupancy, it also faces the problem of sensor fusion, as the structure of Bayesian occupancy grid is well suited for the integration of different typologies of sensor measurements (Stepan et al., 2005). The algorithm is structured as follows:

1. At the beginning, the occupancy grid is initialized with a 0.5 probability of occupation (when no information is yet available from the field);
2. As a new measurement is available, the grid is updated following the Bayes' rule described in (5);
3. The grid is further updated using the generalized union probability theorem, in order to merge together all the obstacles in the robot area;
4. Back to step 2.

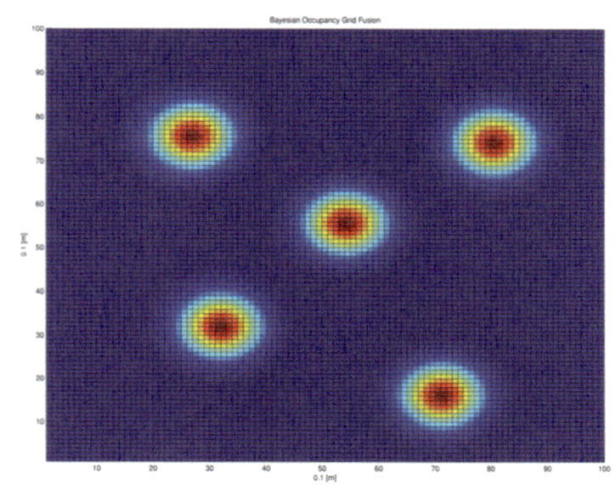

Fig. 4. Bayesian Occupancy Grid: the probability of a cell occupation grows from blue to red

Figure 4 shows an example taken from a simulation of a Bayesian occupancy grid, where 5 obstacles are moving around the robot area.

4.2 Fuzzy logic control technique

In the context of collision avoidance problem and robot-environment interaction the pure reactive systems could be a good solution to face these problems, even because they require few computational resources. Other advantages of purely reactive systems are the following:

- Emphasis on the importance of a tight relationship between perception and action;
- Absence of abstract knowledge and symbolic reasoning;
- Vertical decomposition of the problem into sub-problems to be executed in parallel;
- Modularity of the software;
- Architectures are often inspired by theory from several disciplines.

Following the classis outline (Brooks, 1986), the presented fuzzy logic controller is composed of two components: the functional module and the behavioural module. The functional component, depicted in Figure 5, acquires information which will be then used as an input to the engine.

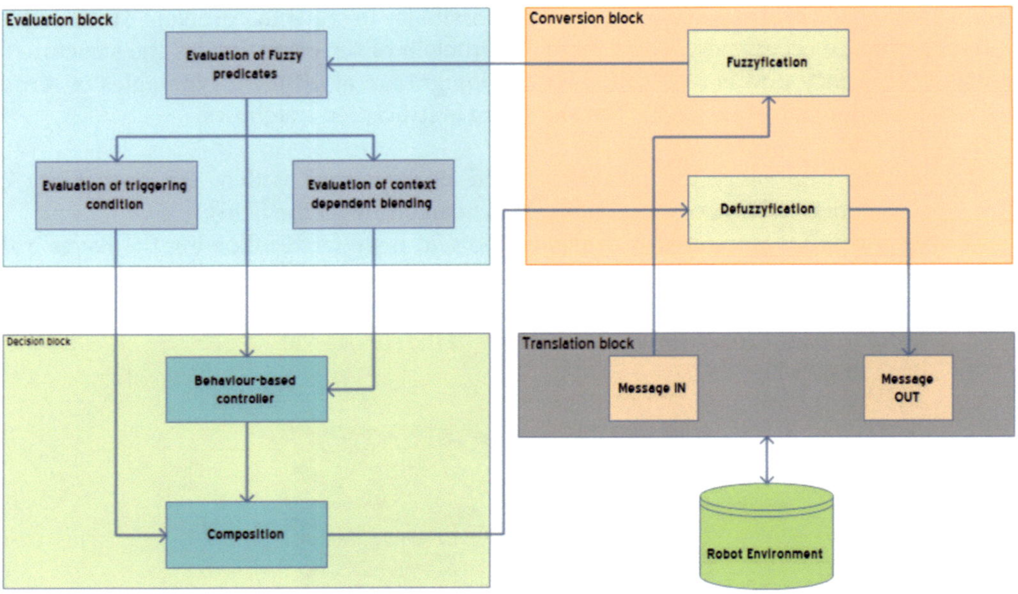

Fig. 5. Functional view of the Fuzzy logic controller

As these data have been computed, the functional component blends resulting actions and transmits values to the actuators and is composed of:

- Translational block: it is an interface between the environment surrounding the robot and data processed inside the engine. It transforms all the information from world frame to robot frame coordinates;
- Conversion layer: information acquired from translation layer is here transformed into fuzzy values (fuzzyfication process) and, after computation, output fuzzy values are transformed again into crisp values (defuzzyfication process);
- Evaluation layer: it is composed by three sub-modules, each one managing a sub-tree. Predicates sub-tree, behaviour triggering conditions sub-tree and behaviour evaluation sub-tree;
- Decision layer: decides actions to be carried out on the basis of environment information that is provided by previous layer. Behaviours are entities totally independent from each other and from the environment, describing activities to be carried out.

In respect of the classical implementation of the Fuzzy logic controller, the proposed solution does not have the purpose of determining type and position of the obstacles in the environment. The information about distance [*mm*] and heading [°] should be generated by the Bayesian occupancy grid controller. Fuzzy rules are the basis where the operative knowledge of the robot can be built from a human heuristic knowledge. A template of a Fuzzy rule can be the following: *IF <antecedent> THEN <consequent>*. Where the antecedent could consist of an arbitrary large number of precondition combined through logic operators AND, OR and NOT. So for instance, the following rule: *IF ((obstacle* $\notin$ *North) AND* *(obstacle* $\in$ *Far)) THEN (speed* $\in$ *Fast).* This Fuzzy rule states that if an obstacle is present in the working area and if it is far and north of the robot, then the robot must advance pretty fast. Obviously, while in the antecedent all the aforementioned logic operation can be used, in the consequent only the AND operator is acceptable. Moreover, credibility value (i.e. the membership degree of a variable to the membership function) range between 0 and 1, both included. This implements that *T-norm* and *T-conorm* are the AND and OR operators of classic logic, where *T-norm* is:

$$\begin{aligned} &\min(x, y) \\ &x \cdot y \\ &\max(x + y - 1, 0) \end{aligned} \tag{9}$$

And where *T-conorm* is:

$$\begin{aligned} &\max(x, y) \\ &x + y - x \cdot y \\ &\min(x + y, 1) \end{aligned} \tag{10}$$

The fuzzyfication, blending and defuzzyfication blocks in the functional engine scheme are depicted in Figure 6. The effective engine component is labelled *Inference*. This is the scheme chosen for the Fuzzy logic controller component, in which the block that evaluates the triggering condition is canned before the behavioural sub-tree (in order to avoid wasting computational resources) and there is a blending block for each behaviour. The purpose of activation threshold is to state the effective possibility that the robot behaviour is not going to change. In this way the computational load is decreased because the engine is not forced to scan all the rules. Another important component is the blending block that fuses the outputs of the basic behaviours: this block allows the coexistence of behaviours even if there are conflicting tasks to be performed.

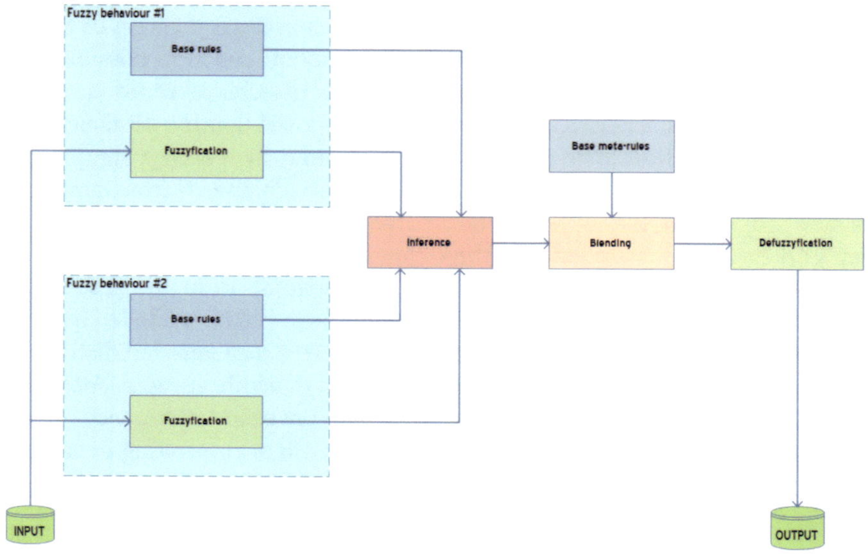

Fig. 6. Fuzzy inferential engine scheme

In order to implement the arbitration two different strategies can be chosen: a strategy in which there is for every fuzzy rule a blending block or the strategy that has a blending block for each basic behaviour.
Blending rules are contained in the component labelled meta-rules base in Figure 6. Formally:

$$Des_{B_i}\left(\underline{x},\underline{c}\right)=\max_k\left\{\left\{\min\begin{bmatrix}d_1^k\cdot\mu_{A_1}^k(x_1),d_2^k\cdot\mu_{A_2}^k(x_2),\ldots\\ \ldots,d_N^k\cdot\mu_{A_N}^k(x_N)\end{bmatrix}\right\},\mu_C^k(\underline{c})\right\}$$

$$Des_{B_i}\left(\underline{x},\underline{c}\right)=\max_k\left\{\min\left\{\min\begin{bmatrix}\mu_{A_1}^k(x_1),\mu_{A_2}^k(x_2),\ldots\\ \ldots,\mu_{A_N}^k(x_N)\end{bmatrix}\right\},d^k\cdot\mu_C^k(\underline{c})\right\} \quad (11)$$

Where Des_{Bi} is the deliverability function (Ruspini, 1991) and where $d_1^k, d_2^k, \ldots, d_N^k$ is the relative weight of the k-th meta-rule.

4.3 The integrated solution

In this paragraph the developed algorithm and the proposed solution to interact with dynamic unstructured environments will be showed. The Bayesian occupancy grid is a probabilistic method that models the space occupied by obstacles on an environment. The input of the occupancy grid is the position of each object (given either as a Cartesian position or as a ρ, θ representation); this position is considered as two-dimensional, in order to better fit with the readings from the sensors (such as camera trackers or laser scanners). The grid is relative to the space surrounding the robot, as the industrial robot manipulator is supposed to be always in a fixed position; a simple extension of the proposed control paradigm can be also taken into account, considering a non-fixed position of the robot. The grid is then mapped on the real working area of the robot, and it is possible to choose a different resolution of thickness of the grid, in order to achieve more accuracy on the possibility to have an obstacle in a cell. As it can be supposed that the obstacles are humans (i.e. workers that can interact with the manipulator around its working area) and that they can be seen as input from cameras, as an (x, y) position on a plane, or from laser scanners, as distances and angles, to the Bayesian occupancy grid algorithm, the two-dimensional occupancy grid has to be extended to a three-dimensional occupancy grid, since the robot end-effector is given each instant as a position and orientation in the space. In order to extend the 2D Bayesian occupancy grid to 3D, each obstacle can be modelled as a cylinder of probabilities, where the centre is given by the mean value of the Gaussian distribution of the obstacle. At the same time, the Bayesian occupancy grid override speed is computed by the algorithm: this is a quite simple operation, as the Bayesian occupancy grid gives a probability framework which is well suited for the problem of controlling override speed of the robot end-effector.

$$O_{BOG}=1-P_c(occ\mid X)\cdot K_s \quad (12)$$

The Bayesian occupancy grid override speed O_{BOG} is computed as the inverse of the probability that a cell is occupied given the measurement, multiplied by a constant K_s, which is the weight given to the probability that an obstacle is occupying the cell. The algorithm also gives as input for the Fuzzy logic controller, the distance and angle of the nearest obstacle to the robot. The Fuzzy logic controller takes the distance and angle computed from the Bayesian occupancy grid algorithm and, taking into account the behaviours of the nearest obstacle, computes the override speed according to the defuzzyfication process. The O_{FLC} is computed taking advantage of the behavioural nature of fuzzy logic filter. This override speed is then used by the control in order to compute the override value which has to be assigned to the robot. This value is computed considering the general override at time $t-1$, which is the value used to weight the Bayesian occupancy grid and Fuzzy logic controllers overrides.

$$\begin{aligned} O_{gen}(t) &= O_{gen}(t-1) \cdot O_{FLC}(t) + \\ &+ (1 - O_{gen}(t-1)) \cdot O_{BOG}(t) \end{aligned} \tag{13}$$

The meaning of using the past general override $O_{gen}(t-1)$, as shown in (13), as a weight is that the control law gives more importance to the Fuzzy logic control algorithm when the general override speed is high (i.e. the obstacle is far from the robot), and more importance to the Bayesian occupancy grid algorithm when the general override speed is low (i.e. the obstacle is near). This typology of control has two main behaviours: when the obstacle is far, and the general speed override is high, the Fuzzy logic controller acts as the main control, since it allows behaviour based on the heading of the robot end-effector towards the obstacle. In this state the decisional policy is submitted to the Fuzzy logic controller as the distance from the obstacle is high: for instance while the robot end-effector is moving and an obstacle is moving away, behind the heading of the end-effector, the Fuzzy logic override speed will be fixed to the maximum speed, if there is no obstacle in front of the heading of the robot end-effector. On the other hand when the obstacle is near, and the general speed override is low, the Bayesian occupancy grid algorithm acts as the main control, as it allows a behaviour based on the probability of encountering an obstacle in a portion of space, then adjusting the Bayesian occupancy grid override with an appropriate control law. In this state the decisional policy is submitted to the Bayesian algorithm as the distance from the obstacle is low, and a good accuracy in the choice of the correct speed override is needed by the control constraints. It is then also important to emphasize the way the control behaves, as the switch between the two typologies of control techniques is entirely smooth. This means that both the Bayesian occupancy grid and the Fuzzy logic controllers act simultaneously in every condition; the advantages of this technique is that it merges together the advantages of each control law and that, taking advantage of feedback on general speed override, the control is fast, robust and ready.

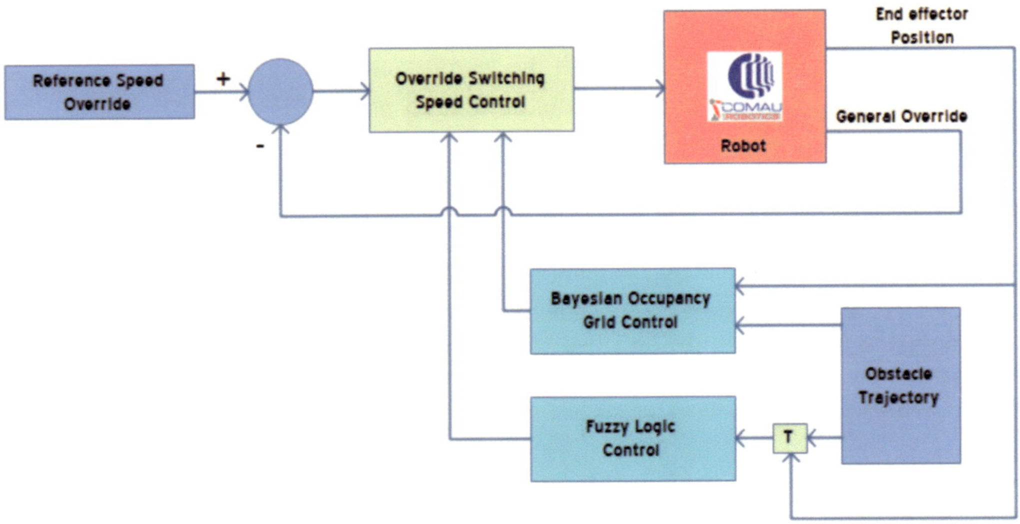

Fig. 7. Architecture of the control system for dynamic environments: robot is controlled by the feedback controller from the output (general speed override)

The proposed algorithm acts as follows (Figure 7):

1. At the beginning, the Bayesian occupancy grid and Fuzzy logic controller variables are initialized;
2. As a new measurement is available, the occupancy grid is updated;
3. The grid is further updated using the generalized union probability theorem, in order to merge together all the obstacles in the robot area;
4. The Bayesian occupancy grid override, the distance and angle of the nearest obstacle is then computed and passed to the Fuzzy logic controller;
5. The Fuzzy logic controller computes the speed override according to the defuzzyfication process;
6. The Bayesian occupancy grid and Fuzzy logic controller partial speed overrides are used to compute the general speed override according to (13);
7. Back to step 2.

5. Hybrid control paradigm

The features of the control systems, for the management of robot behaviour in static and dynamic environments, have been described as the capability to control the robot speed override as different changes in the environment arise. The architectural scheme of the proposed hybrid control system is depicted in Figure 8.
From the scheme it is possible to locate the three blocks where the decisions about the overrides related to static and dynamic models are taken; in particular the geometrical area control block takes as input the position of the robot end-effector, the dynamic objects (to be eventually connected to the elementary geometrical areas) and the database of user-defined geometrical areas.

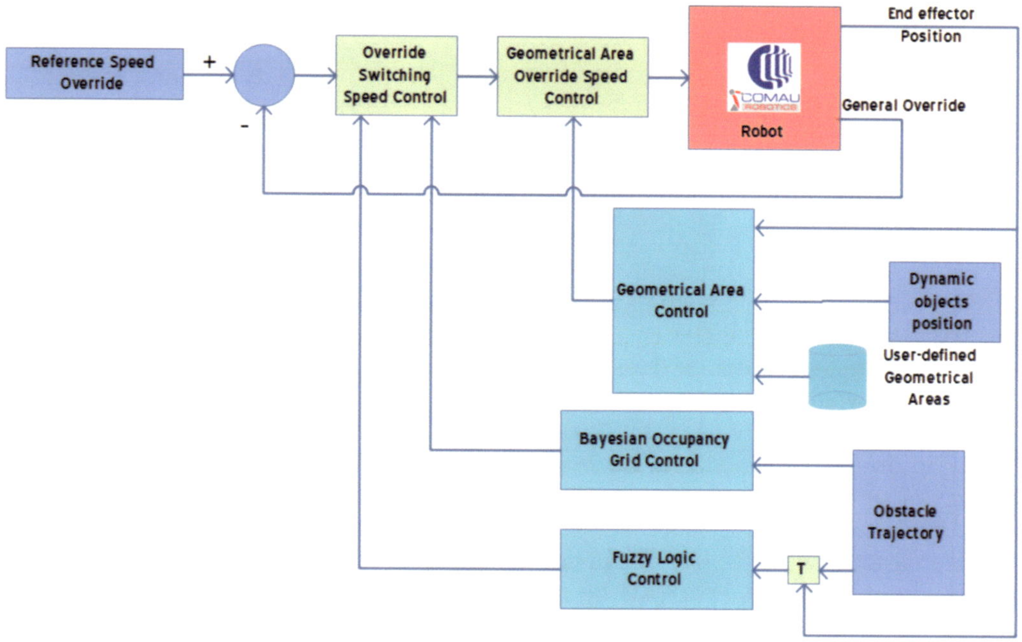

Fig. 8. Architecture of the hybrid control system

With this approach an override is computed for the static model of the system and then used by the geometrical area override speed control in order to select the best value to be taken into account for the regulation together with the values coming from the override switching speed control. The dynamic model of the objects are taken into account within the Bayesian occupancy grid and fuzzy logic control blocks which take as input the trajectory of the obstacles and the robot end-effector trajectory as well. The presented hybrid control gives the robot the possibility to react to changes both in the dynamic and static environment, preserving the characteristics of both controls; in the following sections the hybrid control will be tested in a simulation environment and in a real industrial robotized cell in order to prove the effectiveness of the studied control methods.

6. Simulation results

In this section a set of simulation results will be showed for both the static and dynamic control algorithms which have been analysed and developed in the previous sections. In particular, the simulation environment developed in Matlab and VRML will be showed, in order to prove the effectiveness of the presented paradigms and the hybrid system, and to validate the results before the application of the control methods on a real industrial robotic system. Given this, the simulation 3D environment has been developed with the aid of VRML in order to give a better representation of the virtual robot, while it is moving around its working area.
It is important to note that the control algorithms have been implemented into the real robot control, with the generation of real targets for the motors, but where the motors has not been attached to the controller unit.

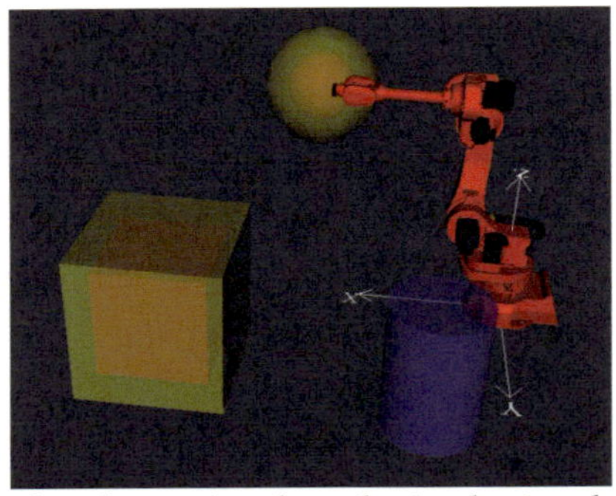

Fig. 9. 3D environment - The dynamic obstacle is depicted in blue as a cylinder; the controlled geometrical areas are depicted with their warning zones in yellow and red

Therefore the targets have been processed by the robot controller and then transferred via TCP/IP to the Matlab simulation environment through a client/server connection; in this typology of simulation, the motors target is updated real-time as in the real robot, and the VRML model of the manipulator acts in the same way as the robot should do in reality. Moreover a framework of the elementary geometrical areas and the dynamic obstacles moving around the robot area has been integrated in the simulation in order to better perceive the effectiveness of the hybrid control system while the robot is moving, changing its speed according to the control laws, when the manipulator end-effector encounters an obstacle in its trajectory or proximity or, in general, when it has to interact with its structured or unstructured environment. The obstacle avoidance control algorithm is capable of prevent impacts on the objects moving around the robot area; in this test, five obstacles moving around the robot surroundings were taken into account. The control system takes as input the override reference signal (which is considered constant), and it produces a controlled speed override value in order to control robot movements. The robot end-effector and obstacles are represented as points in space and the relative trajectories are defined a priori inside the simulator. The obstacles speeds are constant while the end-effector speed value is modified by the feedback control system output.

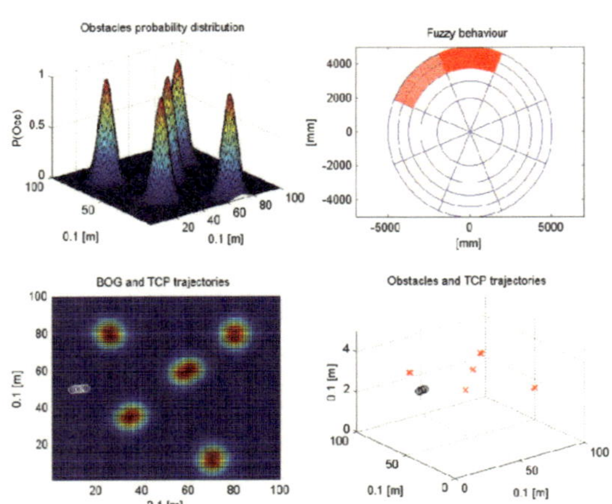

Fig. 10. General overview of simulation results for five obstacles moving around the robot surroundings

A general overview of the computed outputs is depicted in Figure 10 for five obstacles moving in the robot area; from this view it is possible to see the distribution of probabilities of the obstacles produced by the Bayesian occupancy grid filter in the upper left side. In the bottom left side the trajectories of the robot end-effector (represented as white circles) in 2D representation of the Bayesian occupancy grid are depicted. In upper right side, the Fuzzy logic controller behaviour is depicted for the closest obstacle. Figure 11 shows the behaviour of control system for five obstacles.

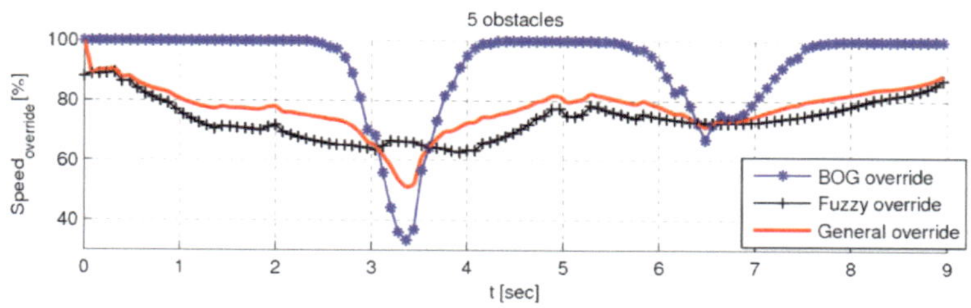

Fig. 11. General overview of simulation results for five obstacles moving around the robot surroundings

The trend shows that the Fuzzy logic controller reacts taking into account all the obstacles in the robot environment, and this can be seen considering the slow dynamics in the Figure. On the other hand Bayesian occupancy grid controller is more sensible to close obstacles and influences the general override in a significant way, when general override is low.

7. Experimental results

The second session of tests has been executed on the real Comau C4G robot controller and SMART NS16 manipulator (Figure 12); the SMART NS16 is a 6-axis industrial manipulator with a maximum load at wrist of 16 kilograms and a high repeatability of 0.05 mm.

Fig. 12. The SMART NS16 industrial robot manipulator

It is worth noting that for security reason, the algorithm to avoid collisions with humans has not been tested on the real robot as this experiment should include an appropriate hardware module to make the system absolutely safe, with redundancy controls on the real position of the robot end-effector. In this section will be then showed the results of the control algorithm, while the robot is moving at its maximum speed and performing technological tasks in a real industrial cell. During this test phase, the SMART NS16 has been programmed, as usual in industrial robotics, as if it was moving over a working path (technological move of welding process) with one active cylindrical controlled area. In Figure 13.a the robot end-effector position (*X*,*Y*) is plotted for two movements: the first one where the control is not active (the blue pattern) and the second one where the control algorithm is enabled (the red pattern). The manipulator moves from the starting position, where it is calibrated at home position, to a first point A. The following move is a Cartesian linear move, parallel to *Y* axis as far as the second point B. In the figure, two circles are depicted, highlighting the warning zone in yellow and the forbidden zone in red. When the forbidden zone is deactivated, on the blue trajectory, the end-effector moves towards the start position after reaching point B; with the red trajectory, the robot end-effector stops inside the warning zone at the boundaries of the forbidden zone when the control is enabled.

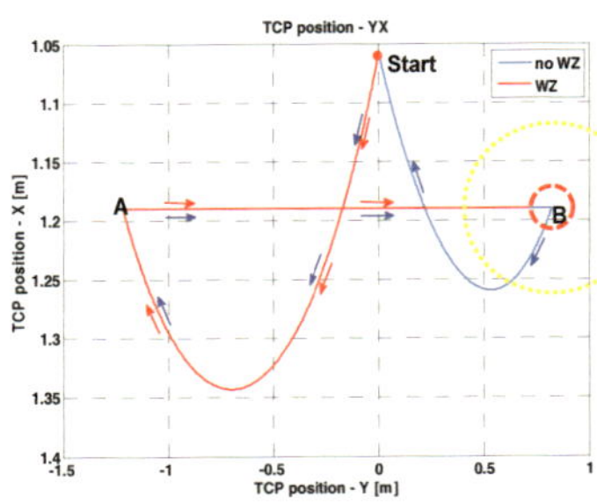

13.(a)

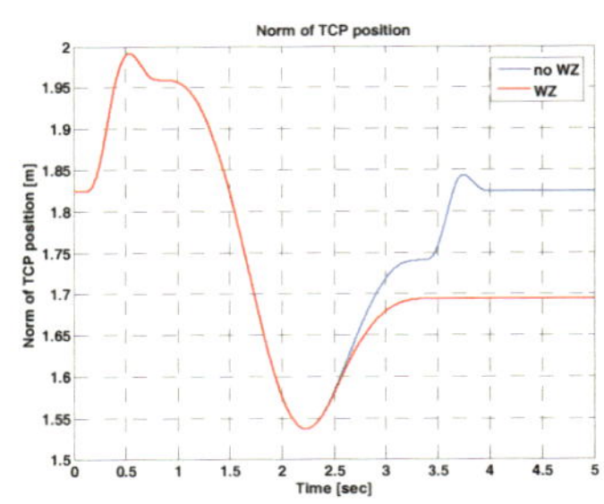

13.(b)

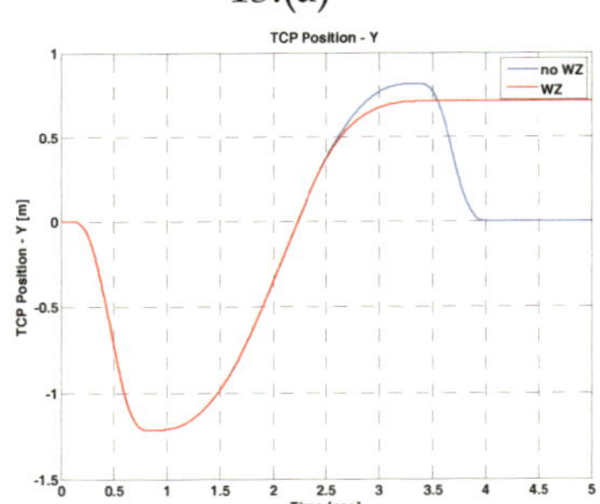

13.(c)

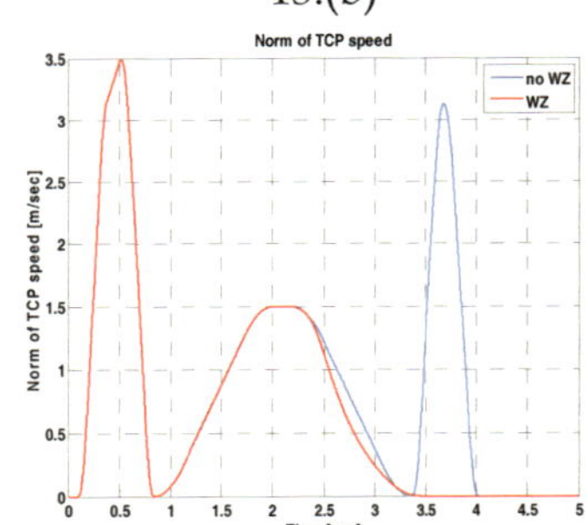

13.(d)

Fig. 13. a) Robot position (*X*,*Y*), with enabled/disabled control, b) Norm of the end-effector position over time, c) Trend of the *Y* end-effector position over time, d) Norm of the robot speed over time

The position trend over time is depicted in Figure 13.b and 13.c. From Figure 13.b it is possible to see the different trends of the norm position of the end-effector when the control is enabled (in red) and when it is deactivated (in blue). The robot starts to move along the trajectory parallel to *Y* at 1.5 seconds and after almost one second it enters the warning zone. In Figure 13.c it is possible to see how, without enabled warning zone control algorithm, the trend along the *Y* axis continues as far as the next movement (at about 3.5 seconds); when the control algorithm is enabled, it is shown that the Y position of the robot end-effector becomes constant when the end-effector encounters the forbidden zone. In Figure 13.d is then possible to see how the end-effector speed changes when it comes in contact with the warning zone at about 2.5 seconds. From a first analysis of these results it is clear how the reduction of speed is in inverse relation to the distance between the end-effector and the geometrical elementary zone surface. A similar analysis can be done concerning the acceleration trend shown in Figure 14.

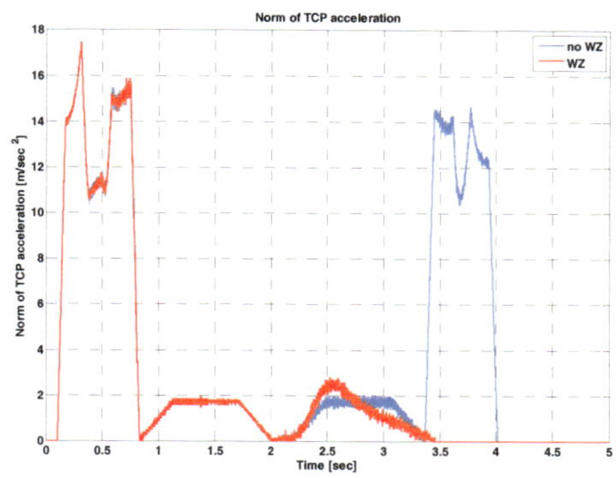

Fig. 14. Norm of the robot acceleration over time

These results can be easily extended to a more complex environment where several different geometrical areas are present and where the control algorithm has to manage the override control for each area, even if they overlap. Moreover the effectiveness of this method in presence of dynamic geometrical areas can be proved as well, when the system can be provided with sensors like external encoders which can be linked to a geometrical area, as stated in Section 3.1.2.

8. Conclusion

In this chapter new hybrid control techniques for the modelling of static and dynamic environments have been analyzed and developed, in order to make the designed controller capable to cope with both static and dynamic features; from the dynamic side of the problem, a framework which models both behavioural and probabilistic characteristics of the world surrounding the robot has been taken into account. A further control paradigm has been presented in order to interact with static environment, developing a solution with which it is possible to define several areas around the robot where its movements are allowed or likewise forbidden. It has also been showed as the presented hybrid control paradigm can be used as the basis for an overall increase of perception and interaction between the robot and its surroundings, distinguished by both environmental structures and human operators. The hybrid control system has been implemented both in simulation and on a real system, providing proofs of its feasibility, robustness and effective increase in robot-environment interaction. Moreover the modularity of the system allows extending its characteristics also taking into account other cognitive features.

9. Acknowledgments

This work has been supported by Comau Robotics S.p.A.; the author would like to thank the company and the Control Engineering team for the support and help on the experimental work and on the state of the art research.

10. References

Barak, D. & Witkin, A. (1992). Dynamic simulation of non-penetrating flexible bodies, *Computer Graphics*, vol. 26, no. 2, 1992, pp. 303-308

Barraquand, J.; Langlois, B. & Latombe, J.-C. (1989). Robot motion planning with many degrees of freedom and dynamic constraints, *Proceedings of the 5th international symposium on Robotics research,* pp. 435-444, 0-262-13253-2, 1991, MIT Press, Cambridge, MA, USA

Bouma, W. & Vanecek, J.G. (1992). Collision detection and analysis in a physically based simulation, *Proceedings of the 2nd Eurographics Workshop on Animation and Simulation*, 1992, Vienna, Austria

Brooks, R. A. (1986). A robust layered control system for a mobile robot, *IEEE Journal of Robotics and Automation*, vol. 2, no. 1, March 1986, pp. 14-23

Brooks, R. (2003). Planning collision-free motions for pick-and-place operations, *International Journal of Robotics Research*, vol. 2, no. 4, 2003, pp. 19-44

Canny, J. (1984). Collision detection for moving polyhedra, 1984, Cambridge, MA, USA

Collins, T.; Collins, J. & Ryan, C. (2007). Occupancy grid mapping: an empirical evaluation, *Proceedings of the 15th Mediterranean Conference on Control and Automation*, Athens, Greece, July, 2007

Dong, T.; Liao, X.; Zhang, R.; Sun, Z. & Song, Y. (2005). Path Tracking and Obstacles Avoidance of UAVs - Fuzzy Logic Approach, *Proceedings of the 14th IEEE International Conference on Fuzzy Systems. FUZZ apos; 05*, pp. 43-48, 0-7803-9159-4, May, 2005

Elfes, A. (1989). Using occupancy grids for mobile robot perception and navigation, *Computer*, vol. 22, June 1989, pp. 46-57

Fulgenzi, C.; Spalanzani, A. & Laugier, C. (2007). Dynamic obstacle avoidance in uncertain environment combining PVOs and occupancy grid, *Proceedings of the IEEE International Conference on Robotics and Automation*, Rome, Italy, 2007

Harville, M. & Li, D. (2004). Fast, Integrated Person Tracking and Activity Recognition with Plan-View Templates from a Single Stereo Camera, *Proceedings of IEEE Conference on Vision and Pattern Recognition*, pp. 398-405, 2004

Kim, B.; Kwon, O.; Kim, K.; Lee, E. & Hong, S. (1999). A study on path planning for mobile robot based on fuzzy logic controller, *Proceedings of IEEE TENCON'99*, pp. 1002-1005, December, 1999

Klir, G. & Yuan B. (1995). *Fuzzy Sets and Fuzzy Logic: Theory and Applications*, Upper Saddle River, NJ, Prentice Hall

Kraiss, K.-F. (2006). *Advanced Man-Machine Interaction: Fundamentals and Implementation*, Springer

Malhotra, R. & Sarkar, A. (2005). Development of a fuzzy logic based mobile robot for dynamic obstacle avoidance and goal acquisition in an unstructured environment, *Proceedings of Advanced Intelligent Mechatronics 2005 IEEE/ASME*, pp. 1198-1203, 0-7803-9047-4, 2005

Mamdami, E.H. & Assilian, S. (1993). An experiment in linguistic synthesis with a fuzzy logic controller, *in Readings in Fuzzy Sets for Intelligent Systems*, pp. 283-289, 1993, B. Dubois, H. Prade, and R.R. Yager Eds., San Mateo, CA

Marcuse, H. (1964). One-Dimensional Man, *Studies in the ideology of advanced industrial society*, 1964.

Matinfar, M.; Baird, C.; Batouli, A.; Clatterbuck, R. & Kazanzides, P. (2007). Robot-assisted skull base surgery, *Intelligent Robots and Systems - IROS 2007 IEEE/RSJ International Conference*, 2007, pp. 865-870

Mauer, G.F. (1995). A fuzzy logic controller for an abs braking system, *IEEE FS*, vol. 3, November, 1995, pp. 381-388

Moravec, H. & Elfes, A. (1985). High resolution maps from wide angle sonar, *Proceedings of the IEEE International Conference on Robotics and Automation*, pp. 116-121, 1985, St. Louis, Missouri: IEEE Computer Society Press

Moravec, H. (1988). Sensor fusion in certainty grids for mobile robots, *AI Magazine*, vol. 9, no. 2, July August 1988, pp. 61-74

Nehmzow, U. & Walker, K. (2005). Quantitative description of robot-environment interaction using chaos theory, *Robotics and Autonomous Systems*, vol. 53, Issues 3-4, 31 December 2005, pp. 177-193

Red, W.; Troung, C. & Kim, K. (1987). Robot path planning in three-dimensions using direct subspaces, *ASME Journal of Dynamics Systems, Measurement and Control*, vol. 109, 1987, pp. 238-244

Romanelli, F. & Tampalini, F. (2008a). A control algorithm for the management of multiple dynamical geometrical areas for industrial manipulators, *Proceedings of the RAAD 2008 17th International Workshop on Robotics in Alpe-Adria-Danube Region,* Ancona, Italy, September 15-17, 2008

Romanelli, F. & Tampalini, F. (2008b). BOG and fuzzy filters based multimodal collision avoidance for industrial manipulators, *Proceedings of the IEEE Conference on Robotics, Automation and Mechatronics*, pp. 779-784, 978-1-4244-1675-2, Chengdu, China, 21-24 September, 2008

Rooks, B. (2005). Precision robot welding for world's largest particle physics apparatus, *Industrial Robot: An International Journal*, vol. 32, no. 4, 2005

Roy, S. & Pratihar, D. (2003). A genetic-fuzzy approach for optimal path-planning of a robotic manipulator among static obstacles, *IE(I) Journal.CP*, vol. 84, 2003, pp. 86-89

Ruspini, E. H. (1991). On the semantics of fuzzy logic, *International Journal of Approximate Reasoning*, vol. 5, no. 1, January 1991, pp. 45-88

Schulz, B. (2007). The gift of sight, *Austrlian Manufacturing Technology*, 2007, pp. 86-89

Shahrokhi, M. & Bernard, A. (2004). A fuzzy approach for definition of dangerous zone in industrial systems, *SMC(7)*, 2004, pp. 6318-6324

Stepan, P.; Kulich, M. & Preucil, L. (2005). Robust data fusion with occupancy grid, *IEEE Transactions on Systems, Man and Cybernetics, Part C: Applications and Reviews*, vol. 35, Feb 2005, pp. 160-115

Van Wichert, G. & Lawitzky, G. (2001). Man-machine interaction for robot applications in everyday environments, *Proceedings of the 10th IEEE International Workshop on Robot and Human Interactive Communication*, pp. 343-346, 2001

Vasquez, D.; Romanelli, F.; Fraichard, T. & Laugier, C. (2006). Fast object extraction from Bayesian occupancy grids using self organizing networks, *Proceedings of the 9th International Conference on Control, Automation, Robotics and Vision (ICARCV)*, pp. 1-6, 2006

Winkler, B. (2007). Safe space sharing human-robot cooperation using a 3D time-of-flight camera, *Techical Conference - Robots and Vision Show*, 2007

Yen, J. & Pfluger, N. (1995). A fuzzy logic based extension to Payton and Rosenblatt's command fusion method for mobile robot navigation, *IEEE Transactions on Systems, Man and Cybernetics*, vol. 25, no. 6, 1995, pp. 971-978

Zadeh, L. (1965). Fuzzy sets, *Information and Control*, vol.8, June 1965, pp. 338-353

30

Maximal Operational Workspace of Parallel Manipulators

E. Macho, O. Altuzarra and A. Hernandez
University of Basque Country.
Faculty of Engineering in Bilbao.
Department of Mechanical Engineering
Spain

Nomenclature

Symbol	Meaning
MP	Mobile platform
KC	Kinematic chain(s)
TCP	Tool center point
DOF	Degree(s) of freedom
IKP	Inverse kinematic problem
DKP	Direct kinematic problem
$\lvert \mathbf{J}_{IKP} \rvert$	Inverse Jacobian
$\lvert \mathbf{J}_{DKP} \rvert$	Direct Jacobian

The term *kinematic chain* is used in the sense of limb, or *leg*, of the parallel manipulator.

1. Introduction

Parallel manipulators are an interesting alternative to serial robots given the important mechanical and kinematic advantages offered. Nevertheless, they often present more complex and smaller workspaces with internal singularities (Altuzarra et al., 2004; Gosselin & Angeles, 1990). Thus, the workspace size, shape and quality are considered some of the main design criteria of these robots (Merlet et al., 1998).
These robots often present multiple solutions for both the DKP and the IKP. The workspace singularity-free region where the manipulator is initially configured, i.e., the set of postures that a manipulator can reach in the same direct and inverse configuration, has been traditionally considered its operational workspace. This is due to the fact that it is widely extended the idea that to perform a transition between different kinematic solutions, the robot must cross a singular position where the control is lost, and that must be avoided (Hunt & Primrose, 1993). This idea leads to very limited operational workspaces.
In this chapter, a general methodology for obtaining the maximal operational workspace where a parallel manipulator can move in a controllable way will be presented. The basis for enlarging the operational workspace consists in superimposing all the singularity-free regions of the workspace associated with the same assembly mode for all different robot work-

ing modes. This work is the generalization of the methodology developed in (Macho et al., 2008b) for a planar two-*DOF* parallel manipulator, the 5*R* robot.
According to this, the first step is to develop a methodology capable of obtaining the complete workspace, i.e., the entire set of positions that a point of interest of the *MP*, the *TCP*, can reach. There are three main families of methodologies used to obtain the workspace of a manipulator, namely, discretization methods (Dash et al., 2002; Masory & Wang, 1995), geometrical methods (Bonev & Ryu, 1999; Gosselin, 1990) and analytical methods (Agrawal, 1991; Jo & Haug, 1998). In this case, the general purpose hybrid analytical-discrete procedure applicable to fully-parallel manipulators described in (Macho et al., 2009) will be used. Secondly, a complete singularity analysis must be done to carry out an efficient path planning. Singularity maps are traced for target manipulators making a kinematic analysis of the positions comprising the calculated workspace. To do this a systematic method to obtain the corresponding Jacobian matrices is introduced. This methodology is based on the mathematical disassembling of the manipulator into a *MP* and a set of serial *KC*.
Two main types of kinematic singularities are obtained. On the one hand, the *IKP* singularities which separate the different working modes of the robot and define the workspace boundaries, since they occur whenever a *KC* reaches an extreme position. At these postures a dependence relation among the output velocities appears, so the output capabilities of the *MP* are restricted, which is equivalent to an instantaneous reduction in the number of *DOF*. However, the manipulator can reach *IKP* singularities without compromising its controllability. On the other hand, *DKP* singularities, which are different for each working mode, occur inside the workspace and separate the different assembly modes of the robot (Li et al., 2007). At direct kinematic singularities a dependence relation among the input velocities occurs, which implies the robot becoming uncontrollable. The result of the whole process developed is the computation of all singularity-free workspace regions where the robot is fully controllable, associated with certain working and assembly modes.
This research shows the possibility of enlarging the operational workspace joining the different working modes through the *IKP* singularities, maintaining at all times the same assembly mode, that is, avoiding reaching a *DKP* singularity. Thus, the maximal operational workspace associated with a certain assembly mode of the robot is the union of the different singularity-free regions associated with that assembly mode from all working modes. In order to make a path planning inside this enlarged operational workspace, the strategies to identify the *IKP* singularities which connect the different regions joined will be provided. To illustrate all these general purpose procedures and strategies, an example of application will be proposed. An spatial complex parallel manipulator is the most appropriate candidate to show the potentiality, effectiveness and interest of this methodology.
The different solutions of the *IKP* of a parallel manipulator are known as working modes. In the same way, the different solutions of the *DKP* have been traditionally known as assembly modes. Nevertheless, in this chapter the concept of assembly mode will have a different meaning. It is well known that some parallel manipulators are able of moving from one *DKP* solution to another in a fully controlled way, i.e., without crossing any *DKP* singularity. These are the so-called cuspidal robots (Innocenti & Parenti-Castelli, 1992; McAree & Daniel, 1999). The notion of assembly mode as *DKP* solution is confusing in cuspidal robots, because different *DKP* solutions can be joined without loosing control and in fact, joined solutions are geometrically indistinguishable. In this work, those *DKP* solutions between which the manipulator can alternate without crossing a *DKP* singularity will not be considered different assembly modes. Assembly modes will be considered those kinematic configurations sepa-

rated by *DKP* singularities, always characterized by different signs of $|\mathbf{J}_{DKP}|$, or which require the physical disassembling of the manipulator to be reached (Macho et al., 2008a).

2. A case study of translational manipulator

The manipulator shown in Fig. 1 is a three-*DOF* spatial parallel robot. It has two similar *KC* actuated each one by means of a prismatic joint along a fixed sliding direction and a third *KC* actuated by a fixed revolute joint. This is a fully-parallel manipulator since it has one actuated joint per *KC*. In fully-parallel manipulators the number of *KC* and, therefore, the number of input variables coincides also the number of *DOF* (output variables). The kinematic structure of these three *KC*, containing articulated parallelograms, causes a manipulator having just three translational *DOF*. These are known as *Delta-like KC*. There are several well known mechanisms based on this kind of kinematic structure, like the DELTA Robot. The singular configurations of these types of robots is analyzed in (Gregorio, 2004; Lopez et al., 2005) and the workspace in (Laribi et al., 2007; Liu et al., 2004). This example is suitable to illustrate the operational workspace enlargement general strategies due to the complexity of its singularity loci. Probably, the robot in Fig. 1 is not the most adequate at a practical level, but it has been found interesting for research purposes. Motion limitations in the kinematic joints will not be considered. Neither interferences or collisions among elements.

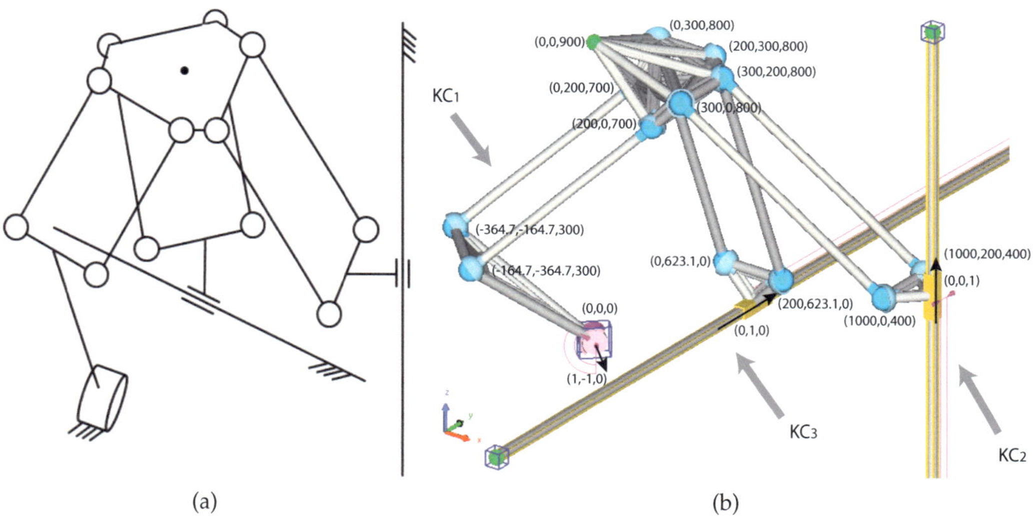

Fig. 1. *Delta-like* translational manipulator

An equivalent mechanism which will provide the same workspace and the same singularity loci will be analyzed. This is the manipulator shown in Fig. 3. Each *KC* of the original mechanism in Fig. 1 is split in two, breaking the parallelogram. The *KC* actuated by the revolute joint (KC_1 in Fig. 1) is split in two *R̲SS KC*, like shown in Fig. 2-(a), and each of the two *KC* actuated by a prismatic joint (KC_2 and KC_3 in Fig. 1) split in two *P̲SS KC*, Fig. 2-(b). The resulting fully-parallel manipulator has six-uncoupled-*DOF*. The *MP* can change its position and also spatial orientation. The kinematics of manipulators in Fig. 1 and Fig. 3 will be similar if the resulting couples of *KC* (Fig. 2) are actuated imposing for both *KC* the same input variable. Therefore the workspace and singularity loci of the manipulator in Fig.

1 can be obtained computing the constant orientation workspace of the manipulator in Fig. 3. The reason for analyzing the auxiliary manipulator instead of the real one is because it will be used a method capable of solving the kinematics of manipulators with $\underline{R}SS$ and $\underline{R}SS$ *KC*.

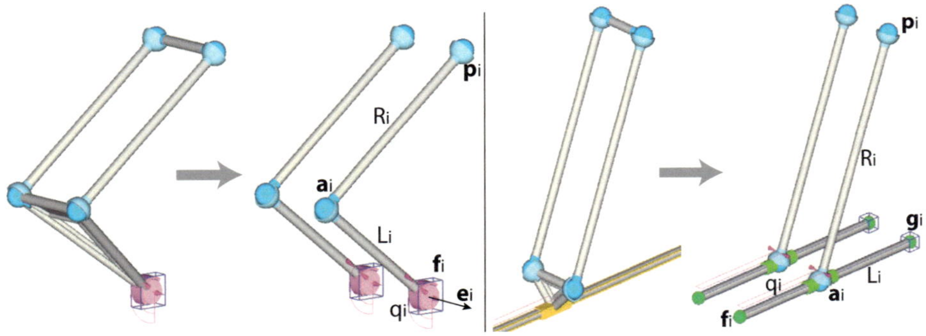

(a) Couple of $\underline{R}SS$ *KC* (b) Couple of $\underline{P}SS$ *KC*

Fig. 2. Auxiliary equivalent *KC*

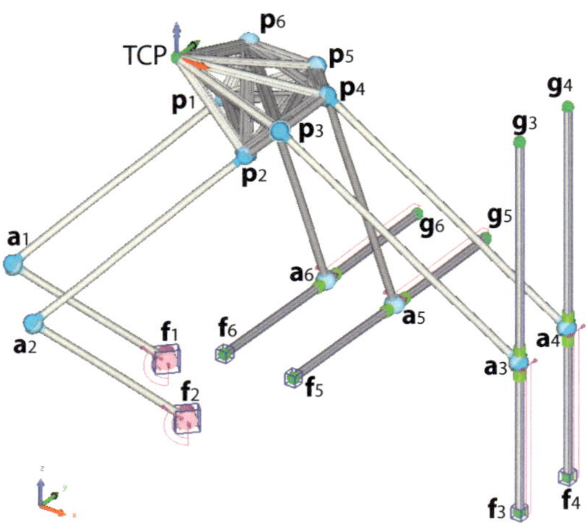

Fig. 3. Auxiliary equivalent 4-$\underline{P}SS$-2-$\underline{R}SS$ manipulator

In the original manipulator each of the three *KC* has two *IKP* configurations. Thus, the whole manipulator has a total of eight (2^3) working modes. In the auxiliary manipulator, each of the six equivalent *KC* as also two *IKP* configurations. Thus, the auxiliary manipulator has a total of sixty four (2^6) working modes, but just the eight coincident with those of the original manipulator will be considered. These are shown in Fig. 4. The nomenclature used is $WM_{c_1c_2c_3}$, where c_1, c_2 and c_3 stand for configuration of KC_1, KC_2 and KC_3 respectively, being either p (positive) or n (negative). For the analyzed manipulator, when the *IKP* has solution, each *KC* can have two different configurations. Mathematically those come from a quadratic equation where the two distinct solutions correspond to the use of a positive or negative sign.

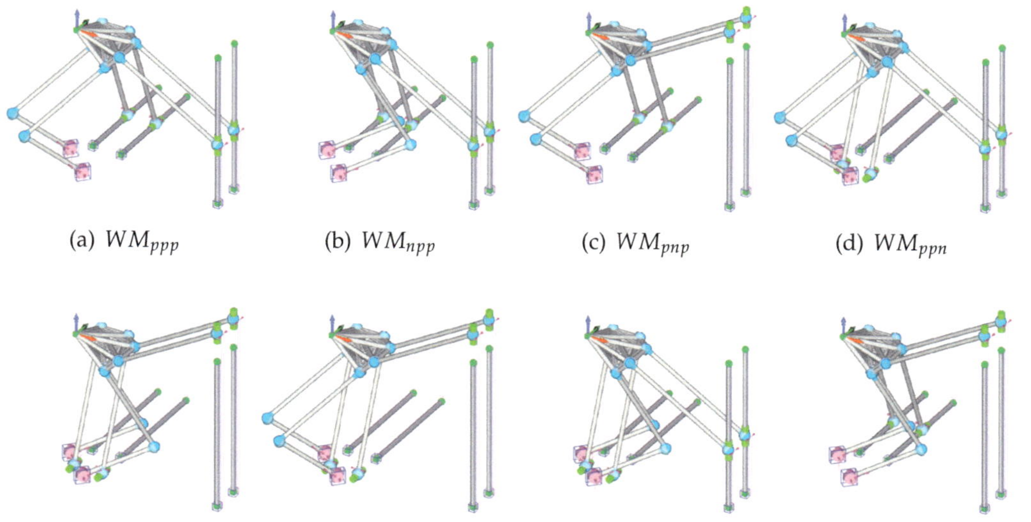

(a) WM_{ppp} (b) WM_{npp} (c) WM_{pnp} (d) WM_{ppn}

(e) WM_{nnn} (f) WM_{pnn} (g) WM_{npn} (h) WM_{nnp}

Fig. 4. Working modes considered in the auxiliary manipulator

3. Workspace computation

In order to compute the workspace, as well as to make the kinematic analysis to determine the singularity surfaces crossing and dividing the workspace into the singularity-free regions, a mathematical model of the robotic device must be done. In parallel manipulators, a MP is attached to a fixed frame through a set of KC. Therefore, the mathematical model developed to compute the workspace and the singularities consists in first separating the MP from each CK. The following notion for these basic entities of the model will be used.

The MP is a rigid body located in a reference frame $(O, \mathbf{i}, \mathbf{j}, \mathbf{k})$, by virtue of a moving frame $(TCP, \mathbf{u}, \mathbf{v}, \mathbf{w})$ attached to it, as shown in Fig. 5. The coordinates of the origin, position of the TCP, are the translational output variables (X, Y, Z). In a six-DOF manipulator, the spatial orientation of such a system is given by the rotational output variables of the MP, the three Euler angles (φ, θ, ψ), in their ZYZ version. In this moving frame, the position of the nodes where KC are attached to the MP are given by constant coordinates (x_{pi}, y_{pi}, z_{pi}).

A KC can be considered as an open limb with a large variety of topologies. In this example only $\underline{R}SS$ and $\underline{P}SS$ cases, shown in Fig. 2, appear. In fully-parallel manipulators, the number of KC is equal to the number of DOF and thus, each KC has a single actuated joint. Actuated joints are underlined and define input variables, denoted as q_i. Nodes where KC are attached to the MP are denoted by $\mathbf{p}_i = [X_{pi}, Y_{pi}, Z_{pi}]$. Nodes fixed to the base frame are $\mathbf{f}_i = [X_{fi}, Y_{fi}, Z_{fi}]$. For the $\underline{P}SS$ KC, to define the fixed sliding direction two nodes are used, $\mathbf{f}_i$ and $\mathbf{g}_i = [X_{gi}, Y_{gi}, Z_{gi}]$. Finally, intermediate nodes, those whose position is different according to the IKP configuration, are given by $\mathbf{a}_i = [X_{ai}, Y_{ai}, Z_{ai}]$. Further constant parameters are also considered, e.g. magnitudes like R_i, L_i...

The workspace is computed using a hybrid discrete-analytical procedure. The complete workspace of the manipulator depicted in Fig. 3 is a six-dimensional continuum entity. The method presented in (Macho et al., 2009) makes an approximation of the actual continuum by

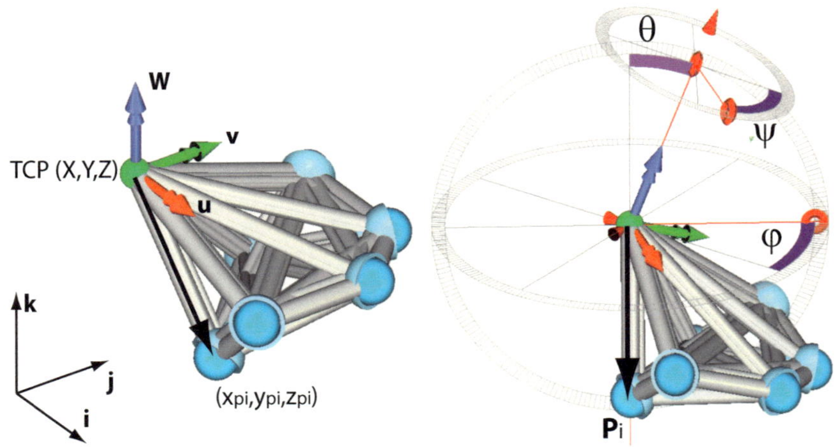

(a) Relative reference system (b) Euler angles

Fig. 5. Output variables defining the pose of the *MP*

a discretization of this real domain. On each point of such a discretized workspace, however, all calculations are done analytically.

The grid of discrete positions can be configured to the desired number of dimensions. Each dimension of such a grid contains an output variable to be incremented. All the remaining output variables, those not considered in the grid, will maintain a constant value. For example in Fig. 3, the case under analysis, a three-dimensional grid will be configured. Each dimension will increment one of the translation output variables (X, Y, Z). This way, since all the rotational output variables (φ, θ, ψ) will maintain a constant value, the constant orientation workspace of the manipulator will be computed. The constant orientation workspace is a three-dimensional subspace of the complete six-dimensional workspace. But the reader must remember that the constant orientation workspace of the manipulator in Fig. 3 is the total workspace of the original manipulator depicted in Fig. 1.

The most efficient method for providing the discrete candidate poses to the workspace is based on the propagation of a wave front. Starting from the pose where the manipulator is initially assembled, since this pose evidently belongs to the workspace. New posses, potentially belonging to the workspace, will be generated in the surroundings of those which have already provided satisfactory results, as shown schematically in Fig. 6 for a two-dimensional case.

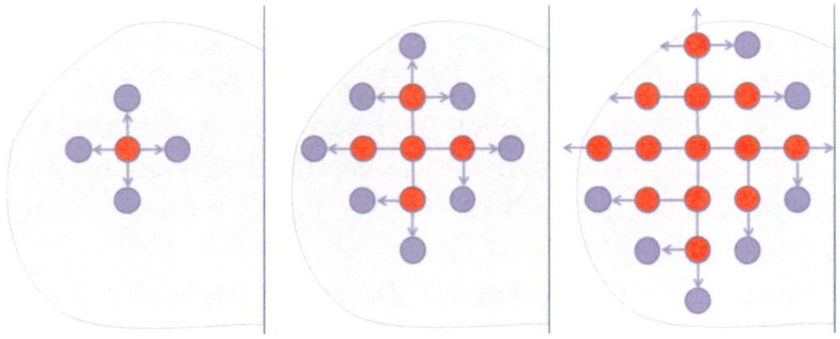

Fig. 6. Schematic advance of a wave front

The increment step for each *DOF* of the *MP* considered in the grid has to be defined. New candidate poses will be generated by incrementing each of the output variables considered, separately and in both senses, increasing and decreasing. This provides propagating capability in all directions of the domain comprising the subset of the workspace being computed. Each candidate pose will be tested to know if it has to be added to the workspace or not. The final result will be the workspace connected to the starting pose.
To check if a candidate pose belongs or not to the workspace, a verification of the *IKP* solution existence is performed. As this has to be done on a large number of poses, the most efficient method, the analytical one, has been chosen. Once the *MP* has been positioned according to the values of the output variables given by a point on the discrete grid, such a pose belongs to the workspace if and only if all *KC* can be physically assembled. As the *IKP* for parallel robots is an uncoupled problem, each *CK* is now considered as an independent sub-mechanism. The whole manipulator can be assembled when all *KC* can be assembled individually.
Once the *MP* is positioned in a specific location, the coordinates of the nodes at the end-joints $\mathbf{p}_i$ of each *KC* are defined. These positions and the *KC* dimensions are the data necessary to first check the existence of solution, and afterwards solve the *IKP*. As each *KC* is treated separately, next will be shown the algorithms applied to the two types of *KC* involved in the present manipulator,shown in Fig. 2.

- *PSS KC*: node $\mathbf{a}_i$ is located at the two possible intersections of a sphere centered at $\mathbf{p}_i$, with radius R_i and a line between points $\mathbf{f}_i$ and $\mathbf{g}_i$.
- *RSS KC*: node $\mathbf{a}_i$ is located at the two possible intersections of a sphere centered at $\mathbf{p}_i$, with radius R_i and a circumference centered at $\mathbf{f}_i$, with axis $\mathbf{e}_i$ and radius L_i.

In Fig. 7-(a) is shown the result of this computation.

4. Singularity analysis

4.1 Velocity problem

The previous result still has not all the information required to plan a safe motion. Singularity maps will be traced by carrying out a kinematic analysis of the positions obtained in the previous step. A systematic method to obtain the corresponding *DKP* and *IKP* Jacobian matrices has been developed. These matrices come from performing the derivatives with respect to time on the position equations. In fully-parallel manipulators, since the number of *DOF* coincides with the number of *KC*, a position equation will be posed for each *KC*.
This position equation is specific for each *KC* topology and it is called characteristic equation. It is posed always in the same systematic way, in function of three types of parameters, the coordinates of the node attached to the *MP* (X_{pi}, Y_{pi}, Z_{pi}) the input variable or actuator position (q_i) and the dimensional parameters, $(R_i, Li, ...)$, including here also the coordinates of fixed nodes $(X_{fi}, Y_{fi}, Z_{fi}, ...)$. This way, each *KC* is initially considered an independent sub-mechanism with its own position equation. In the case of $\underline{R}SS$ and $\underline{P}SS$ *KC*, the characteristic equation, $f_i = 0$, is given by Equation 1. The difference between both types of *KC* lies in the expressions $\mathbf{a}_i(q_i)$, the coordinates of the node $\mathbf{a}_i$ in function of the input variable.

$$f_i = (X_{pi} - X_{ai}(q_i))^2 + (Y_{pi} - Y_{ai}(q_i))^2 + (Z_{pi} - Z_{ai}(q_i))^2 - R_i^2 = 0 \tag{1}$$

Next step consists in performing the assembly of these equations to the output variables, which is the application of the physical assembly of *KC* to the *MP*. This mathematical assembly is performed by substituting in the characteristic equations f_i, the end joint coordinates

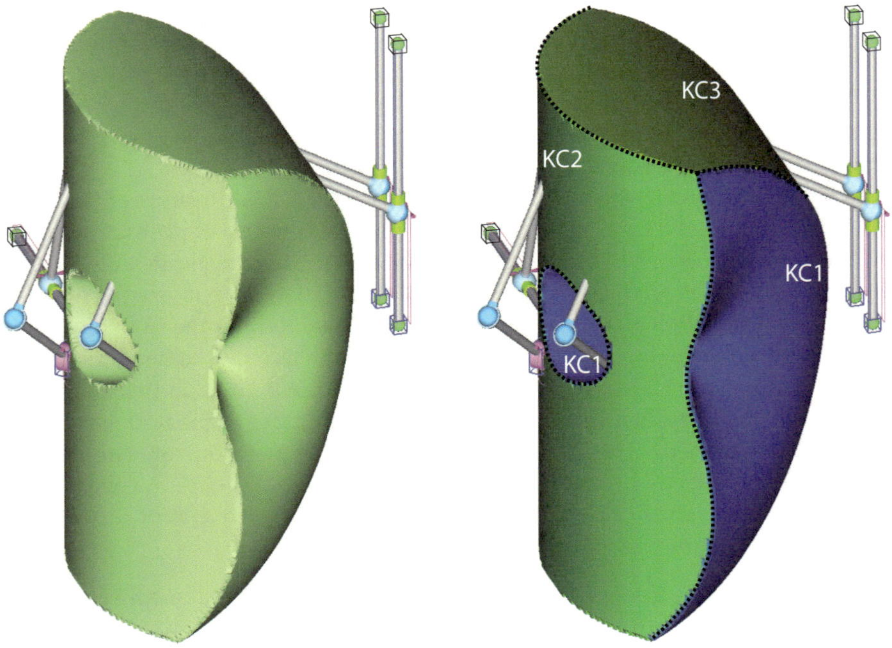

(a) (b)

Fig. 7. Workspace

(X_{pi}, Y_{pi}, Z_{pi}), as functions of the output variables $\mathbf{X} = (X, y, Z, \varphi, \theta, \psi)$. Taking into account Fig. 5, these functions are given in Equation 2.

$$\begin{Bmatrix} X_{pi} \\ Y_{pi} \\ Z_{pi} \end{Bmatrix} = \begin{Bmatrix} X \\ Y \\ Z \end{Bmatrix} + \begin{bmatrix} c\varphi & -s\varphi & 0 \\ s\varphi & c\varphi & 0 \\ 0 & 0 & 1 \end{bmatrix} \begin{bmatrix} c\theta & 0 & s\theta \\ 0 & 1 & 0 \\ -s\theta & 0 & c\theta \end{bmatrix} \begin{bmatrix} c\psi & -s\psi & 0 \\ s\psi & c\psi & 0 \\ 0 & 0 & 1 \end{bmatrix} \begin{Bmatrix} x_{pi} \\ y_{pi} \\ z_{pi} \end{Bmatrix} \tag{2}$$

The resulting system of six non-linear equations is dependent on the six output variables X_i and the six input variables q_i. By differentiating this system with respect to time, the velocity equations are obtained. This linear system of equations can be expressed in the matrix form given by Equation 3, being those matrices the *DKP* and *IKP* Jacobian ($\mathbf{J}_{DKP}$ and $\mathbf{J}_{IKP}$ respectively).

$$\left[\frac{\partial \mathbf{f}}{\partial \mathbf{X}}\right] \dot{\mathbf{X}} = \left[-\frac{\partial \mathbf{f}}{\partial \mathbf{q}}\right] \dot{\mathbf{q}} \tag{3}$$

4.2 DKP Jacobian matrix

To pose $\mathbf{J}_{DKP}$, the derivatives with respect to all output variables must be considered, independently from those chosen to compute a subset of the complete workspace. Since the actual auxiliary manipulator is a six *DOF* robot, the derivatives of the position equations with respect to the three translational output variables (X, Y, Z) and with respect to the three angular

ones (φ, θ, ψ) have to be performed, although just the translational variables have been incremented in the discrete grid to compute the constant orientation workspace.
Each element a_{ij} of $\mathbf{J}_{DKP}$ in Equation 3 is given by:

$$a_{ij} = \frac{\partial f_i}{\partial X_j} = \frac{\partial f_i}{\partial X_{pi}} \frac{\partial X_{pi}}{\partial X_j} + \frac{\partial f_i}{\partial Y_{pi}} \frac{\partial Y_{pi}}{\partial X_j} + \frac{\partial f_i}{\partial Z_{pi}} \frac{\partial Z_{pi}}{\partial X_j} \tag{4}$$

On the one hand appear the partial derivatives of the position equations f_i with respect to the end-joint coordinates (X_{pi}, Y_{pi}, Z_{pi}), which can be directly obtained from Equation 1. On the other hand, the partial derivatives of such coordinates with respect to the output variables X_j are independent from f_i and they can be obtained from Equation 2. For translational output variables (X, Y, Z) those expressions are immediate:

$$\frac{\partial X_{pi}}{\partial X} = 1 \qquad \frac{\partial X_{pi}}{\partial Y} = 0 \qquad \frac{\partial X_{pi}}{\partial Z} = 0 \tag{5}$$

$$\frac{\partial Y_{pi}}{\partial X} = 0 \qquad \frac{\partial Y_{pi}}{\partial Y} = 1 \qquad \frac{\partial Y_{pi}}{\partial Z} = 0 \tag{6}$$

$$\frac{\partial Z_{pi}}{\partial X} = 0 \qquad \frac{\partial Z_{pi}}{\partial Y} = 0 \qquad \frac{\partial Z_{pi}}{\partial Z} = 1 \tag{7}$$

>From Equation 1, considering Equation 5, Equation 6 and Equation 7, the Equation 4 for the three translational variables leads to:

$$\frac{\partial f_i}{\partial X} = X_{pi} - X_{ai} \tag{8}$$

$$\frac{\partial f_i}{\partial Y} = Y_{pi} - Y_{ai} \tag{9}$$

$$\frac{\partial f_i}{\partial Z} = Z_{pi} - Z_{ai} \tag{10}$$

However for the three rotational output variables (φ, θ, ψ) more complex expressions will be generated. As these expressions have to be evaluated in each of the numerous postures comprising the discretized workspace, they must be optimized as much as possible, regarding the computational cost. Firstly, applying Equation 5, Equation 6 and Equation 7, the Equation 4 for orientation variables is transformed into:

$$\frac{\partial f_i}{\partial \varphi} = \frac{\partial f_i}{\partial X} \frac{\partial X_{pi}}{\partial \varphi} + \frac{\partial f_i}{\partial Y} \frac{\partial Y_{pi}}{\partial \varphi} + \frac{\partial f_i}{\partial Z} \frac{\partial Z_{pi}}{\partial \varphi} \tag{11}$$

$$\frac{\partial f_i}{\partial \theta} = \frac{\partial f_i}{\partial X} \frac{\partial X_{pi}}{\partial \theta} + \frac{\partial f_i}{\partial Y} \frac{\partial Y_{pi}}{\partial \theta} + \frac{\partial f_i}{\partial Z} \frac{\partial Z_{pi}}{\partial \theta} \tag{12}$$

$$\frac{\partial f_i}{\partial \psi} = \frac{\partial f_i}{\partial X} \frac{\partial X_{pi}}{\partial \psi} + \frac{\partial f_i}{\partial Y} \frac{\partial Y_{pi}}{\partial \psi} + \frac{\partial f_i}{\partial Z} \frac{\partial Z_{pi}}{\partial \psi} \tag{13}$$

In (Macho et al., 2009) are obtained the following expressions:

$$\frac{\partial X_{pi}}{\partial \varphi} = -(Y_{pi} - Y) \tag{14}$$
$$\frac{\partial Y_{pi}}{\partial \varphi} = X_{pi} - X \tag{15}$$
$$\frac{\partial Z_{pi}}{\partial \varphi} = 0 \tag{16}$$

$$\frac{\partial X_{pi}}{\partial \theta} = \cos\varphi(Z_{pi} - Z) \tag{17}$$
$$\frac{\partial Y_{pi}}{\partial \theta} = \sin\varphi(Z_{pi} - Z) \tag{18}$$
$$\frac{\partial Z_{pi}}{\partial \theta} = -\sin\varphi(Y_{pi} - Y) - \cos\varphi(X_{pi} - X) \tag{19}$$

$$\frac{\partial X_{pi}}{\partial \psi} = -\cos\theta(Y_{pi} - Y) + \sin\varphi\sin\theta(Z_{pi} - Z) \tag{20}$$
$$\frac{\partial Y_{pi}}{\partial \psi} = \cos\theta(X_{pi} - X) - \cos\varphi\sin\theta(Z_{pi} - Z) \tag{21}$$
$$\frac{\partial Z_{pi}}{\partial \psi} = \sin\theta[\cos\varphi(Y_{pi} - Y) - \sin\varphi(X_{pi} - X)] \tag{22}$$

4.3 IKP Jacobian

For fully-parallel manipulators, $\mathbf{J}_{IKP}$ is diagonal. Each term of such diagonal is associated with one *KC* and it is given by the derivative of the characteristic equation f_i with respect to to its input variable q_i. Therefore, $|\mathbf{J}_{IKP}|$ vanishes, producing an *IKP* singularity, whenever any of the diagonal terms does. This means that all *KC* can be considered independent sub-mechanisms capable of separately causing *IKP* singularities.
It would be useful if the values of different Jacobian terms were comparable among them. Thus, given a posture of the manipulator, which *KC* is closer to produce an *IKP* singularity could be determined. To do this a normalization of these terms has been implemented.
Instead of the actual value of the derivative $\partial f_i/\partial q_i$, another parameter, called normalized term of the $\mathbf{J}_{IKP}$, $\|\partial f_i/\partial q_i\|$, will be used. This parameter has a specific expression for each type of *KC* and its value is directly proportional to the corresponding term of the $|\mathbf{J}_{IKP}|$, and thus, vanishes at the *IKP* singular configuration of the *KC*. But its value is limited to 1 at the furthest position from the singularity. Next will be shown the expressions of $\|\partial f_i/\partial q_i\|$ for the *KC* comprising the manipulator under analysis:

- $\underline{P}SS$ *KC*:

$$\left\|\frac{\partial f_i}{\partial q_i}\right\| = \frac{(\mathbf{p}_i - \mathbf{a}_i)\cdot(\mathbf{f}_i - \mathbf{g}_i)}{R_i L_i} \tag{23}$$

- $\underline{R}SS$ *KC*:

$$\left\|\frac{\partial f_i}{\partial q_i}\right\| = \frac{[(\mathbf{a}_i - \mathbf{f}_i)\times\mathbf{e}_i]\cdot(\mathbf{p}_i - \mathbf{a}_i)}{R_i L_i} \tag{24}$$

IKP singularities occur whenever any *KC* reaches an extreme position, thus they define the workspace boundary. Taking this into account, it is easy to check all discrete postures added to the workspace in order to know which of them are closer to one of these singularities. If one posture is surrounded completely by neighboring positions in the discrete grid, it is inside the workspace, whereas if it lacks some neighbor it is in a border and hence it is an approximate *IKP* singularity.
Once one of these postures has been identified, comparing the values of the normalized elements corresponding to all *KC*, it is possible to know which *KC* has produced the singularity, that one with the smaller value. This result is depicted in Fig. 7-(b). This information will be necessary later on to plan working mode transitions in the enlarged operational workspace.

4.4 DKP singularity maps and workspace singularity-free regions

The mapping of $|\mathbf{J}_{DKP}|$ on the workspace provides an approximate singularity map. In fact, the change in the sign of $|\mathbf{J}_{DKP}|$ is the best way to detect a transition over a singular posture. This phenomenon has been used to obtain *DKP* singular postures within the computed workspace. Whenever two contiguous postures in the discrete grid with different signs are found, they are considered approximated singularities. Once all approximated singular postures are found, there is an easy way to refine the singular locus. Taking every two neighboring singular postures with opposite signs, by means of a linear interpolation that makes use of the actual values of the $|\mathbf{J}_{DKP}|$, it can be placed an intermediate posture at the presumed null value of the determinant, as shown schematically in Fig. 8.

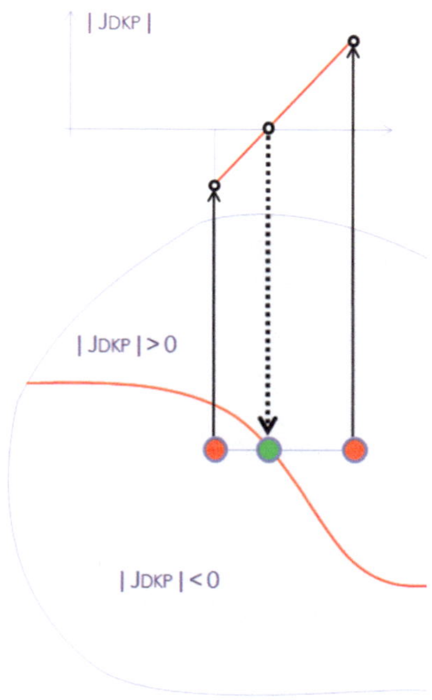

Fig. 8. *DKP* singularity refinement

DKP singularity locus divides the workspace into a set of regions free of internal singularities. For a three-dimensional workspace, like the one depicted in Fig. 7, the loci of postures

where the $|\mathbf{J}_{DKP}|$ vanishes defines a spatial surface which completely splits the aforementioned workspace into regions associated with different signs of such a kinematic indicator. These correspond with different assembly modes of the manipulator.
In $\mathbf{J}_{DKP}$ appear the coordinates of nodes $\mathbf{a}_i$. For the same pose of the *MP*, the coordinates of such points are different for each *KC IKP* configuration. Therefore, each working mode of the manipulator has its own *DKP* singularity loci. This result is depicted in Fig. 9. The resulting workspace singularity-free regions associated to $|\mathbf{J}_{DKP}|$ positive and negative are the volumes depicted in Fig. 10.

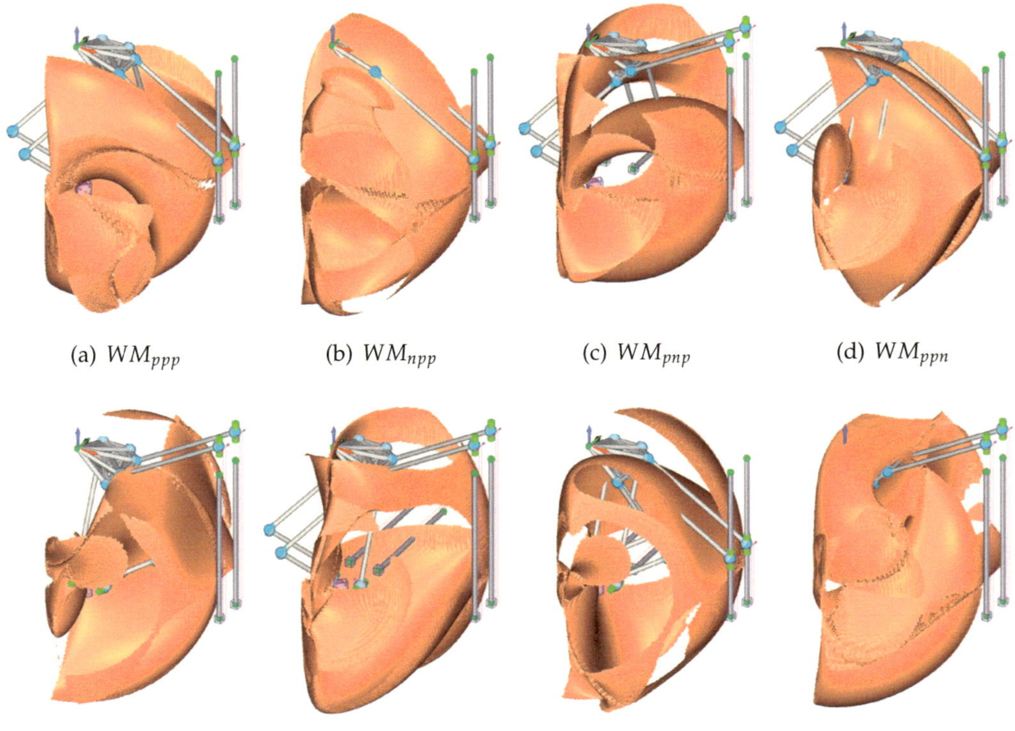

(a) WM_{ppp} (b) WM_{npp} (c) WM_{pnp} (d) WM_{ppn}

(e) WM_{nnn} (f) WM_{pnn} (g) WM_{npn} (h) WM_{nnp}

Fig. 9. *DKP* singularity surfaces for each working mode. Loci of postures with $|\mathbf{J}_{DKP}| = 0$

5. Enlarged operational workspace

As mentioned before, it is a common practice to define as the basic operational workspace one of these singularity-free regions, i.e., the region of the workspace associated with a working mode, in which the manipulator keeps the same assembly mode configuration. The robot will have its home posture in such a region and will be kept inside it all the time. However, it is also possible to consider the union of singularity-free regions associated with the same assembly mode. This requires a path planning implementing transitions between working modes, which will lead to an enlarged operational workspace.
Therefore, finally, there is a workspace, shown in Fig. 7, crossed by *DKP* singularities that must be avoided, shown in Fig. 9, which divide that workspace into a set of singularity-free

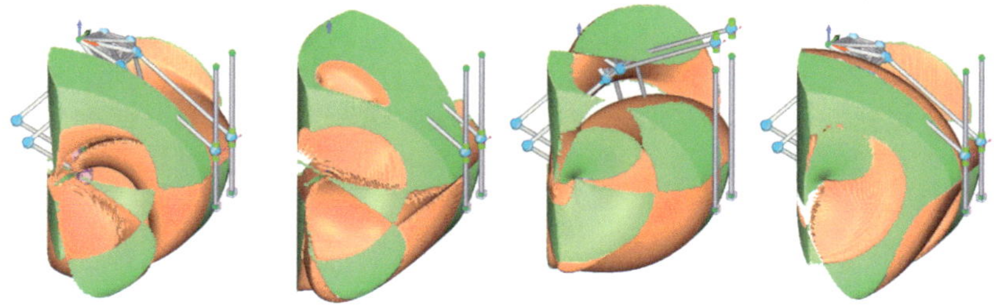

(a) WM_{ppp} (b) WM_{npp} (c) WM_{pnp} (d) WM_{ppn}

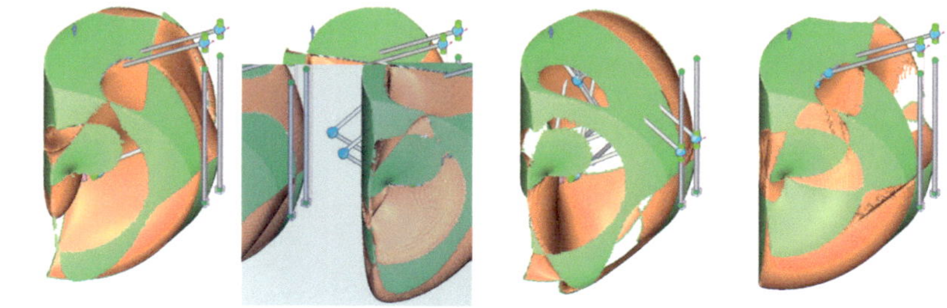

(e) WM_{nnn} (f) WM_{pnn} (g) WM_{npn} (h) WM_{nnp}

$|\mathbf{J}_{DKP}| \geq 0$

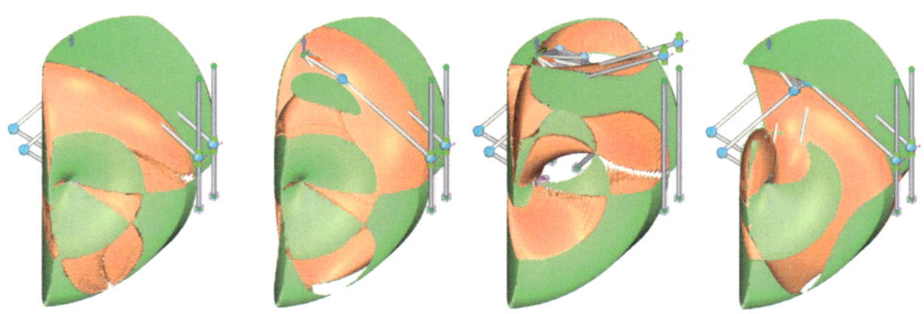

(i) WM_{ppp} (j) WM_{npp} (k) WM_{pnp} (l) WM_{ppn}

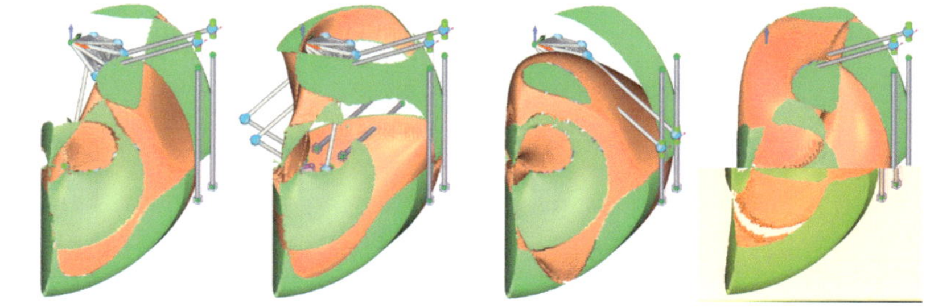

(m) WM_{nnn} (n) WM_{pnn} (o) WM_{npn} (p) WM_{nnp}

$|\mathbf{J}_{DKP}| \leq 0$

Fig. 10. Workspace singularity-free regions

regions, Fig. 10, that are obviously smaller than the complete workspace. Alternatively, the enlarged operational workspace associated with an assembly mode, also being usually smaller than the total workspace, since comes from the union of several singularity-free regions, is normally larger than any of such composing regions, and thus, can be defined as the largest set of postures that the manipulator can reach without the blockade of the actuators.
The analysis of the operational workspace done in the obtained three-dimensional workspace may result confusing. In order to understand these concepts clearly, only a plane section of the workspace will be considered. A bi-dimensional analysis will be much more illustrative and the explained ideas can be immediately extrapolated to the three-dimensional entity. Remember that according to the procedure described, just the desired output variables are incremented in the discrete grid. So, configuring the workspace computation grid only in variables (Y, Z), a planar slice, for a constant value of the output variable $X = 0$, of the whole workspace will be obtained, as shown in Fig. 11.

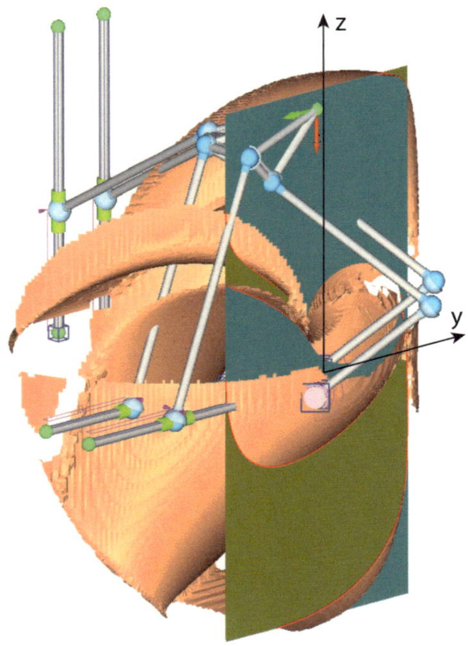

Fig. 11. Planar slice of the workspace for a constant value of X

Thus, without losing generality, it will be assumed that the robot moves in such a way that the TCP is always on that plane. In this planar case, singularity loci are defined by curves. As shown in Fig. 11, the singularity curves in the bidimensional case are the intersections between the considered plane and the general three-dimensional singularity surfaces. In Fig. 12 are shown the workspace singularity-free regions associated with different signs of $|\mathbf{J}_{DKP}|$ for the eight existing working modes.
The enlarged, or maximal, operational workspace associated to an assembly mode will be obtained overlapping the singularity-free regions, corresponding to all existing working modes, associated with that assembly mode. This result is shown in Fig. 13. As it can be observed, obtained enlarged operational workspaces do not fill completely the complete workspace, but the not reachable areas are just quite small corners (less than a 5% of the whole surface).

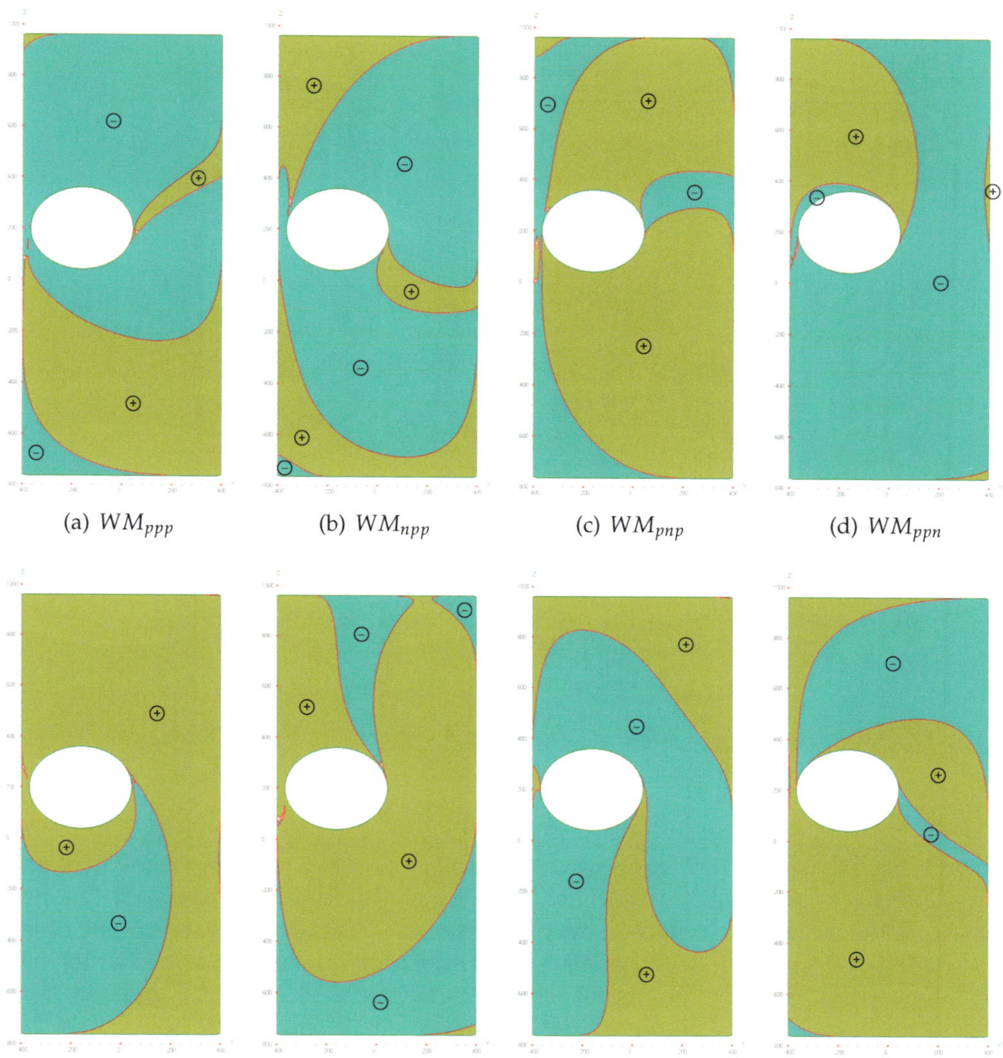

(a) WM_{ppp} (b) WM_{npp} (c) WM_{pnp} (d) WM_{ppn}

(e) WM_{nnn} (f) WM_{pnn} (g) WM_{npn} (h) WM_{nnp}

Fig. 12. Workspace singularity-free regions in a bi-dimensional section

5.1 Path planning inside the maximal operational workspace

After having obtained the singularity-free regions of the workspace associated with all working and assembly modes, the strategies to enlarge the accessible space are easier to plan and implement. To do this, it is necessary to know and use, as said before, *IKP* singularities, because they enable transitions between regions associated to different working modes. By virtue of this knowledge, appropriate paths can be defined to fulfill with desired motion planning tasks.

As an example let's suppose that the manipulator is initially configured in the WM_{ppp} working mode, with the *TCP* located in the initial position, p_i, shown in Fig. 14-(a). This posture

(a) $|\mathbf{J}_{DKP}| \leq 0$ (b) $|\mathbf{J}_{DKP}| \geq 0$

Fig. 13. Maximal operational workspaces associated with existing assembly modes. Superimposition of all singularity-free regions associated with the same sign of $|\mathbf{J}_{DKP}|$

is located in a singularity-free region associated with $|\mathbf{J}_{DKP}| \leq 0$ assembly mode. According to this, in Fig. 14-(b) all regions associated with $|\mathbf{J}_{DKP}| \geq 0$ have been removed. But the remaining workspace is still composed by several disjoint singularity-free regions, all of them associated with $|\mathbf{J}_{DKP}| \leq 0$. Although all of them have been initially considered as belonging to the same assembly mode, the current elemental operational workspace is given by the the region that contains the *TCP* location, i.e., that one depicted in Fig. 14-(c).
>From this initial region, on reaching the *IKP* singularity curve highlighted with a dashed line in Fig. 15-(a), the KC_3 in Fig. 1 can change its *IKP* configuration. Thus, the manipulator will change its working mode from WM_{ppp} to WM_{ppn}. At that moment *DKP* singularity curves, which are specific for each working mode, change and the workspace becomes divided into new singularity-free regions. The new region where the robot can move without losing control is depicted in Fig. 15-(b). The transition from one region to another is be performed always via the *IKP* singularity locus due to the *KC* that changes its configuration. In fact those curves (surfaces in the case of the general three-dimensional workspace) are completely shared by the regions connected, as shown in Fig. 15-(c). Once the new region has been reached, new postures in the workspace are attainable, for example p_f in Fig. 15-(b). Also some postures that were attainable in the starting region cannot be directly reached now, like p_i.
The overlapping of the two regions corresponding to WM_{ppp} and WM_{ppn} working modes, Fig. 15-(c), almost fill the enlarged operational workspace corresponding to $|\mathbf{J}_{DKP}| \leq 0$, indicated in Fig. 13-(a). But lacking postures can be reached making another working mode

(a) (b) (c)

Fig. 14. Initial posture and corresponding singularity-free region

change since they are inside the region corresponding to the WM_{npp} working mode. Thus, starting from the same initial region, associated with the WM_{ppp} working mode, on reaching a posture over the IKP singularity curve highlighted with dashed line in Fig. 15-(d), the KC_1 will change its configuration form p to n to perform the desired transition between working modes. Doing this, the region depicted in Fig. 15-(e) is reached, enlarging the initial operational workspace as shown in Fig. 15-(f).

In this example, just overlapping three singularity-free regions, those corresponding to WM_{ppp}, WM_{ppn} and WM_{npp} working modes, the maximal enlarged operational workspace associated with $|\mathbf{J}_{DKP}| \leq 0$ is fulfilled. Nevertheless, additional transitions connecting all other working mode regions are feasible. In Fig. 16 are shown all the possibilities starting from the initial region depicted in Fig. 14-(c). Next to the connection lines displayed is indicated which KC changes its configuration to make the transition between connected regions. Note that there are some regions in Fig. 16 which cannot be reached (not colored regions) and some working mode changes that are not possible (dashed connection lines) because the corresponding initial and final regions do not share the adequate IKP singularity curve.

In consequence, operational workspace enlargement capability may depend on the region where the manipulator is initially configured. For example, if the starting region was the one marked in Fig. 18, only the color filled regions will be accessible, resulting on an operational workspace, shown in Fig. 19 smaller than the one achieved in Fig. 17. This fact has to be taken into account when making the path planning.

A transition scheme, like the one shown in Fig. 16, is very useful to perform the path planning task. For example, let's suppose that the manipulator has to from the initial posture p_i, to the final posture p_f shown in Fig. 20-(a). Let's suppose also that the manipulator will work according to the $|\mathbf{J}_{DKP}| \leq 0$ assembly mode. Since both postures belong tho the enlarged operational workspace, it is known in advance that the robot will be able to join them in a controlled way. And finally, let's suppose that there are some technical hypothetical rea-

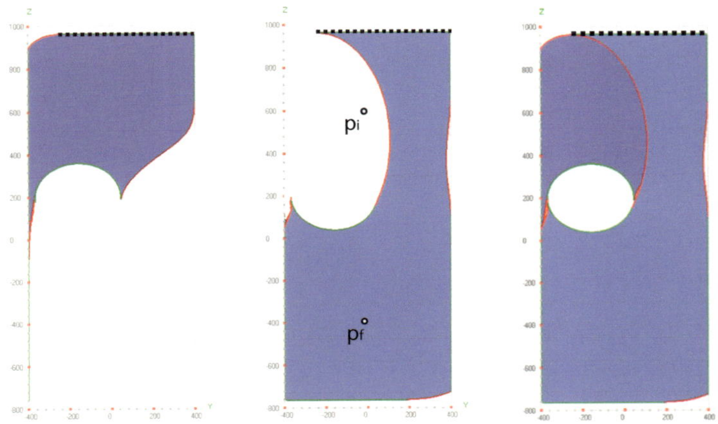

(a) (b) (c)

Transition from WM_{ppp} to WM_{ppn} (KC_3 changes from p to n)

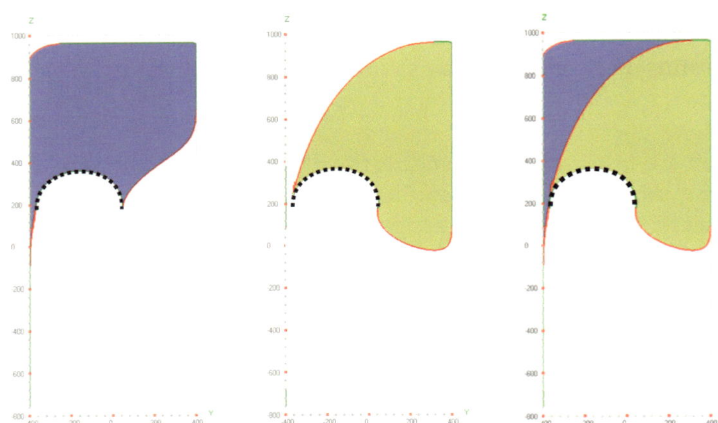

(d) (e) (f)

Transition from WM_{ppp} to WM_{npp} (KC_1 changes from p to n)

Fig. 15. Transitions between singularity-free regions trough IKP singularities

sons which impose that the the manipulator must start in the initial posture configured in the WM_{ppp} working mode and must reach the final posture configured in the WM_{nnn} working mode. In Fig. 20-(b) and (c) are shown the initial and final singularity-free regions, containing the initial and final postures, which must be joined making working mode changes.

The map in Fig. 16 allows to know easily how to go from p_i to p_f verifying also the prescribed initial and final working modes. Note that as the three KC have to change their configuration, at least three transitions will be required. There are several possibilities, or ways, for joining the initial and final regions. Just one will be depicted in Fig. 21:

- Starting form the initial position p_i in the WM_{ppp} working mode, the TCP of the manipulator goes to the intermediate position p_{g1} over the IKP singularity curve caused by KC_3 to make transition to the WM_{ppn} working mode, Fig. 21-(a).

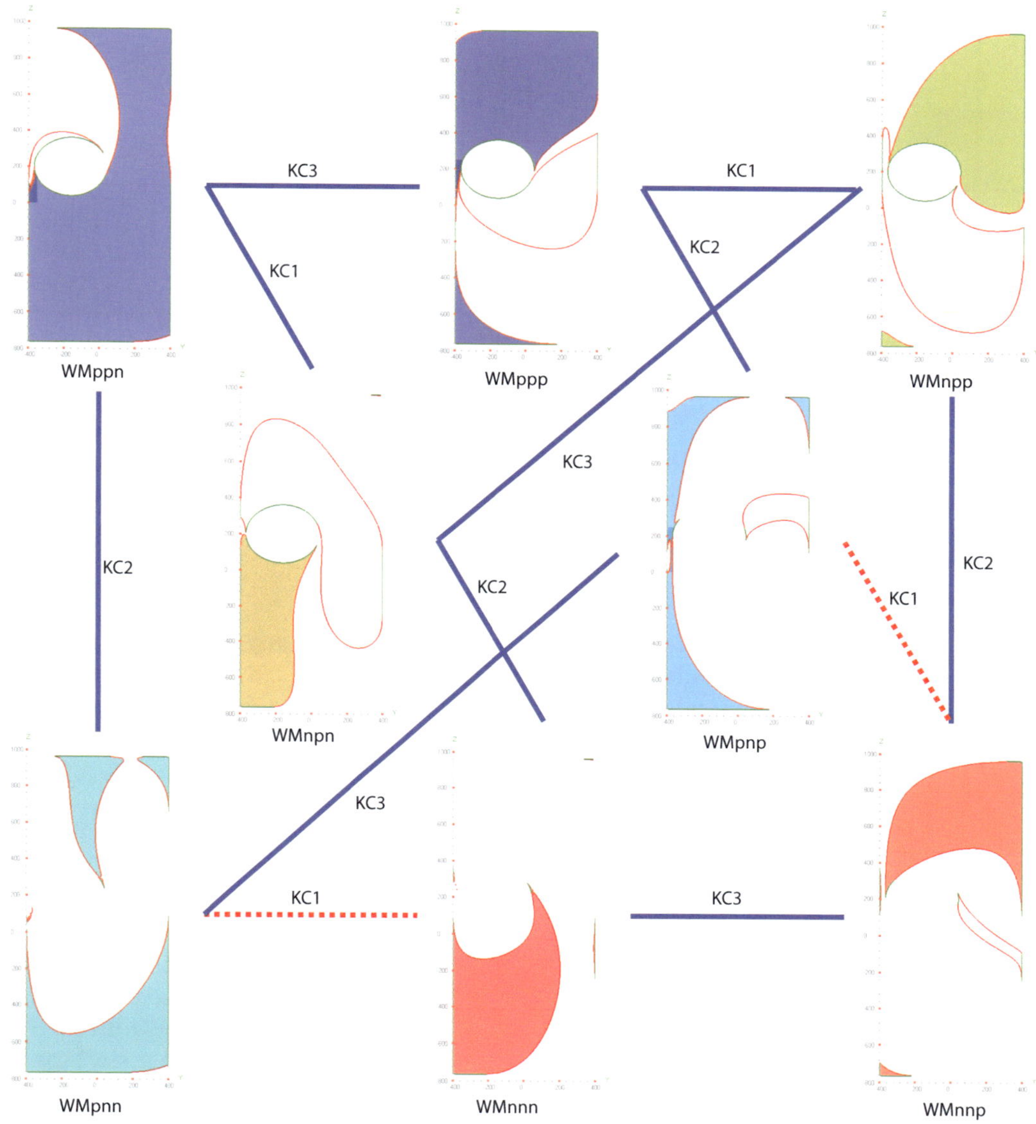

Fig. 16. Transitions among singularity-free regions associated with different working modes

- On reaching the new region, the motion has to continue inside it from p_{g1}, to the second intermediate position p_{g2} over the IKP singularity curve caused by KC_2 to make transition to the WM_{pnn} working mode, Fig. 21-(b).
- On reaching the third region, the motion has to continue inside it from p_{g2}, to the third intermediate position p_{g3} over the IKP singularity curve caused by KC_1 to make transition to the WM_{nnn} working mode, Fig. 21-(c).
- Finally, the motion can evolve inside the final region to the final posture p_f, Fig. 21-(d).

All motions have been done without reaching any DKP singularity.

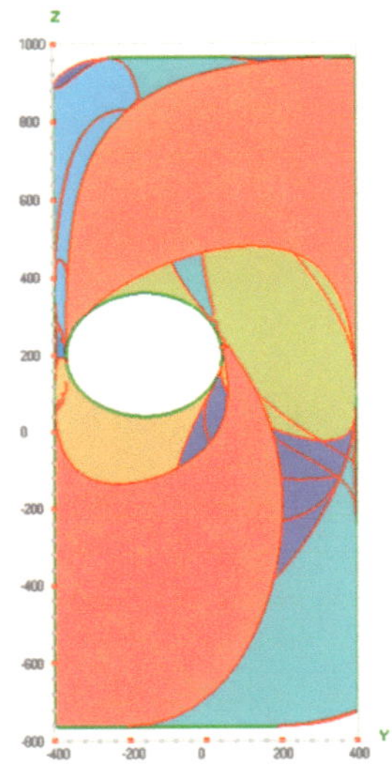

Fig. 17. Overlapping of all joined singularity-free regions. Enlarged operational workspace

Generalizing the described process, to go from a singularity-free region associated with a working mode $WM_{i,j,..,k}$, to another region associated with $WM_{l,m,..,n}$, they must be done consecutively as many elemental transitions as kinematic chains have to change their configuration, namely:

- from $WM_{\mathbf{i},j,..,k}$ to $WM_{\mathbf{l},j,..,k}$
- from $WM_{l,\mathbf{j},..,k}$ to $WM_{l,\mathbf{m},..,k}$
- ...
- from $WM_{l,m,..,\mathbf{k}}$ to $WM_{l,m,..,\mathbf{n}}$

This way any pose inside the operational workspace can be reached without blockade of actuators. All these procedures can be generalized directly to the complete three-dimensional workspace. In Fig. 22 are shown the enlarged operational workspaces associated with the two existing assembly modes. A size comparison (in volume) among the complete workspace, Fig. 7, the largest singularity-free region (WM_{npn}, Fig. 10-(o)) and the enlarged operational workspace associated with the assembly mode $|\mathbf{J}_{DKP}| \leq 0$, Fig. 22-(b) can be done resulting on:

- Complete workspace: 100%
- Largest singularity-free region: 75%
- Enlarged operational workspace: 97%

6. Conclusions

This chapter has described and illustrated the strategy of obtaining enlarged workspaces connecting all working modes and keeping one of the parallel manipulator assembly modes. It has been used a procedure which is able to obtain all the singularity-free regions of the

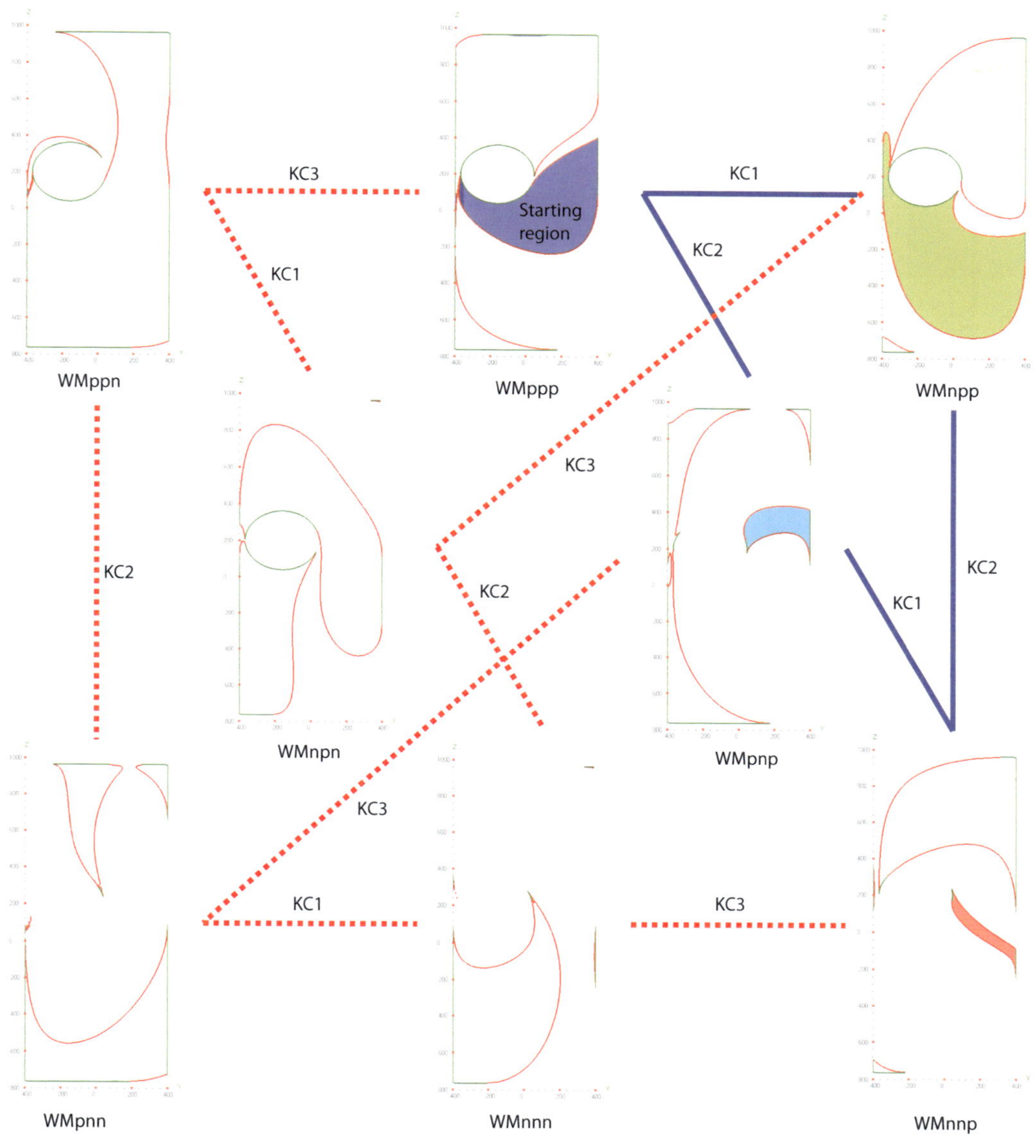

Fig. 18. Feasible transitions starting from another region

robot workspace, taking into account that each working mode presents a different direct kinematic problem singularity locus. The direct singularities imply that a dependency relation appears among actuator velocities and separate assembly modes, while the inverse singularities, which define the workspace boundaries and separate working modes, can be reached without loss of control. Therefore, it is possible to make use of the inverse singularities to transit between different singularity-free regions associated with the same assembly mode. Joining all the regions associated with the same assembly mode for all working modes, the largest set of postures that the manipulator can reach is obtained. In order to make an efficient path planning within this enlarged workspace, the workspace borders allowing the transi-

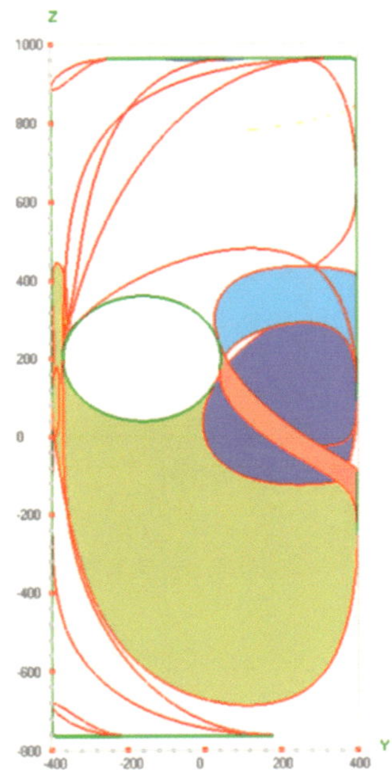

Fig. 19. Joined singularity-free regions. Enlarged operational workspace starting from another region

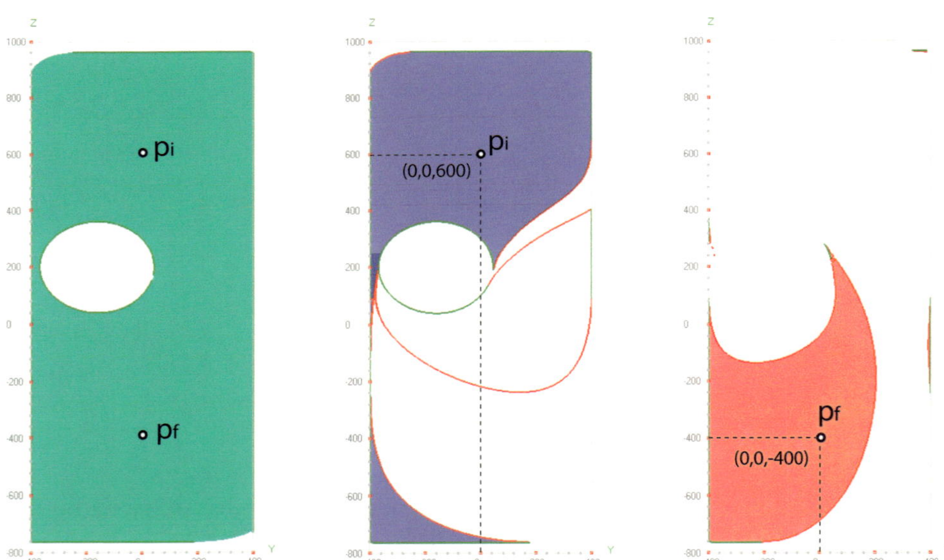

(a) (b) (c)

Fig. 20. Initial and final postures in the enlarged operational workspace and in their singularity-free regions

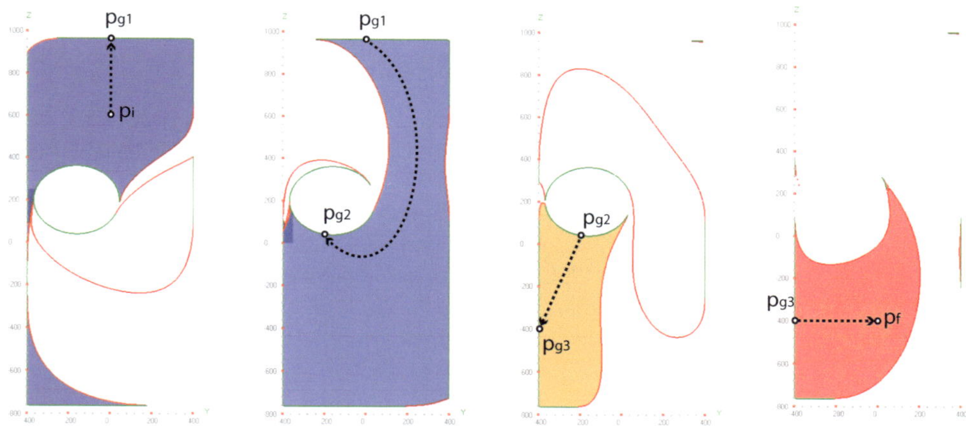

(a) (b) (c) (d)

Fig. 21. Path planning

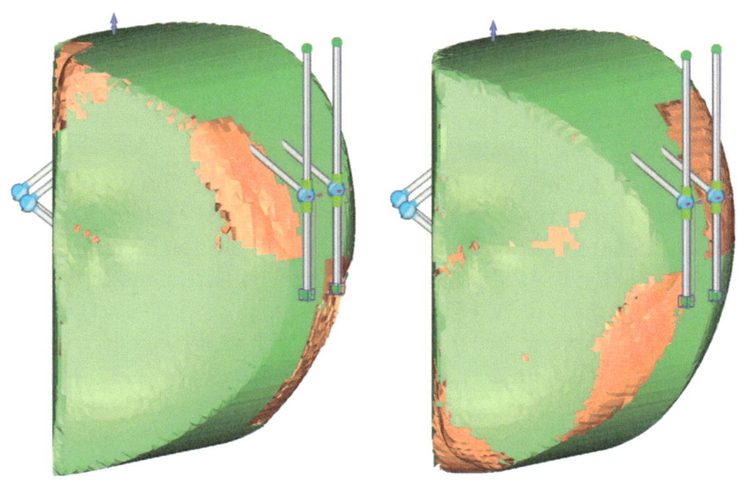

(a) $|\mathbf{J}_{DKP}| \leq 0$ (b) $|\mathbf{J}_{DKP}| \geq 0$

Fig. 22. Enlarged three-dimensional workspace

tions between the different working modes need to be known and used. This is the first time that the systematic general purpose procedure has been applied to a *Delta like* manipulator by means of an auxiliary equivalent robot.

Acknowledgment

This research work was supported in part by the Spanish Ministerio de Ciencia y Tecnologia (Project DPI2008-00159), the FEDER funds of the European Union, the University of the Basque Country (Project GIU07/78) and the Grant (BFI104.148.R2) from the Basque Government.

7. References

Agrawal, S. (1991). Workspace boundaries of in-parallel manipulator systems, *IEEE Journal on Robotics and Automation* **7(2)**: 94–99.

Altuzarra, O., Pinto, C., Aviles, R. & Hernandez, A. (2004). A practical procedure to analyze singular configurations in closed kinematic chains, *IEEE Transactions on Robotics* **20(6)**: 929–940.

Bonev, I. & Ryu, J. (1999). Workspace analysis of 6-prrs parallel manipulators based on the vertex space concept, *ASME Design Engineering Technical Conference* .

Dash, A., Yeo, S., Yang, G. & Chen, I. (2002). Workspace analysis and singularity representation of three-legged parallel manipulators, *ICARCV 02* pp. 962–967.

Gosselin, C. (1990). Determination of workspace of 6-dof parallel manipulators, *ASME Journal of Mechanical Design* **112**: 331–336.

Gosselin, C. & Angeles, J. (1990). Singularity analysis of closed loop kinematic chains, *IEEE Transactions on Robotics and Automation* **6(3)**: 281–290.

Gregorio, R. D. (2004). Determination of singularities in delta-like manipulators, *The International Journal of Robotics Research* **23(1)**: 89–96.

Hunt, K. & Primrose, E. (1993). Assembly configurations of some in-parallel-actuated manipulators, *Mechanism and Machine Theory* **28(1)**: 31–42.

Innocenti, C. & Parenti-Castelli, V. (1992). Singularity free evolution from one configuration to another in serial and fully parallel manipulators, *ASME Journal of Mechanical Design* **120**: 73–99.

Jo, D. & Haug, E. (1998). Workspace analysis of closed-loop mechanisms with unilateral constraints, *ASME Design Automation Conference* pp. 53–60.

Laribi, M., Romdhane, L. & Zeghloul, S. (2007). Analysis and dimensional synthesis of the delta robot for a prescribed workspace, *Mechanism and Machine Theory* **42**: 859Ű870.

Li, H., Gosselin, C. & Richard, M. (2007). Determination of the maximal singularity-free zones in the six-dimensional workspace of the general goughŰstewart platform, *Mechanism and Machine Theory* **42(4)**: 497–511.

Liu, X., Wang, J., Oh, K. & Kim, J. (2004). A new approach to the design of a delta robot with a desired workspace, *Journal of Intelligent and Robotic Systems* **39**: 209Ű225.

Lopez, M., Castillo, E., Garcia, G. & Bashir, A. (2005). Delta robot: inverse, direct, and intermediate jacobians.

Macho, E., Altuzarra, O., Amezua, E. & Hernandez, A. (2009). Obtaining configuration space and singularity maps for parallel manipulators, *Mechanism and Machine Theory* **44(11)**: 2110–2125.

Macho, E., Altuzarra, O., Pinto, C. & Hernandez, A. (2008a). Transitions between solutions of the direct kinematic problem, *Advances in Robot Kinematics* .

Macho, E., Altuzarra, O., Pinto, C. & Hernandez, A. (2008b). Workspaces associated to assembly modes of the 5r planar parallel manipulator, *Robotica* **26(3)**: 395–403.

Masory, O. & Wang, J. (1995). Workspace evaluation of stewart platforms, *Advanced Robotics* **9(4)**: 443–46.

McAree, P. & Daniel, R. (1999). An explanation of the never-spacial assembly changing motions for 3-3 parallel manipulators, *The International Journal of Robotics Research* **18(6)**: 556–574.

Merlet, J., Gosselin, C. & Mouly, N. (1998). Workspace of planar parallel manipulators, *Mechanism and Machine Theory* **33(1-2)**: 7–20.

31

Kinematical and Dynamical Models of KR 6 KUKA Robot, including the kinematic control in a parallel processing platform

John Faber Archila Díaz
Universidad Industrial de Santander. UIS
Colombia
Max Suell Dutra and Fernando Augusto de Noronha Castro Pinto
Universidade Federal do Rio de Janeiro, UFRJ
Brazil

1. Introduction

This chapter presents the study and modelling of KR 6 KUKA Robot, of the Robotics Laboratory, Federal University of Rio de Janeiro, see fig 1. The chapter shows the CAD model (Computer Aided Design), the direct kinematics, the inverse kinematics and the inverse dynamical model. The direct kinematic is based in the use of homogeneous matrix. The inverse kinematics uses the quadratic equations model. The dynamical model is based on the use of Euler-Lagrange equations, using the D-H (Denavit-Hartenberg) algorithm and taking into account the inertia tensor, which was found with help of CAE tools (Computer Aided Engineering), On the other hand the Jacobian of robot manipulator is present, it's necessary for the kinematic control. The chapter finishes with the implementation of the inverse kinematic in one parallel processing platform and analyzes its performance.

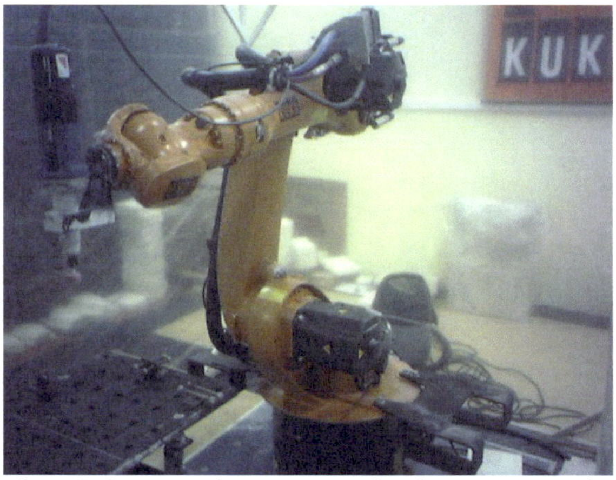

Fig. 1. KR 6 KUKA Robot, Robotics Laboratory, Federal University of Rio de Janeiro UFRJ.

Space movement representation
For the representation of space movements there are several methods such as rotation matrix, vectors, quaternions, roll pitch and yaw, Euler angles, homogenous matrix, among others (Barrientos, 1997). The selected method used for the developing of the direct kinematic model in this work is the homogeneous matrix. The basic concepts for mathematical models formulations are:
Homogeneous Matrix
Homogeneous matrices are 4X4 matrixes, which can represent rotations, translations, scales and perspectives (Ollero, 2001). In general, the homogeneous matrices represent linear transformations. The general form is presented in equation (1)

$$A = \begin{bmatrix} [R(3\times3)] & [T(3\times1)] \\ [P(1\times3)] & [E(1\times1)] \end{bmatrix} \quad (1)$$

$R(3\times3)$ Corresponds to a matrix of three rows by three columns, representing rotations.

$T(3\times1)$ Corresponds to an array of three rows by a column that represents translation.

$P(3\times1)$ Represents a vector of a row of three columns representing the perspective.

$E(1\times1)$ Corresponds to a scalar that represents the scale of the transformation.

For this case $\vec{P} = \vec{0}$ and $E = 1$

Principal homogeneous matrix
Rotation around the Z axis, figure 2.

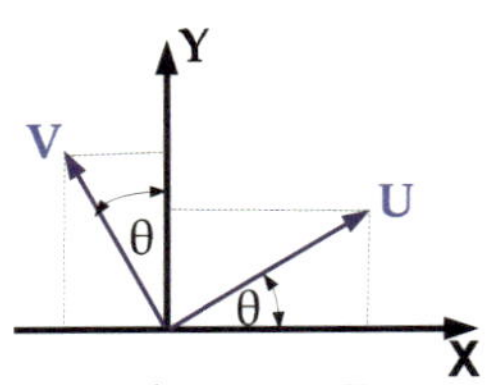

$$\begin{bmatrix} Cos(\theta) & -Sin(\theta) & 0 & 0 \\ Sin(\theta) & Cos(\theta) & 0 & 0 \\ 0 & 0 & 1 & 0 \\ 0 & 0 & 0 & 1 \end{bmatrix}$$

Fig. 2. Rotation around to axis Z.

Translation Px, Py, Pz, figure 3.

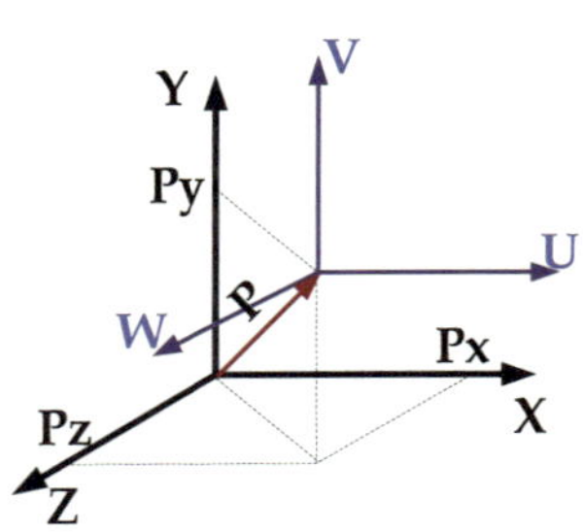

$$\begin{bmatrix} 1 & 0 & 0 & P_x \\ 0 & 1 & 0 & P_y \\ 0 & 0 & 1 & P_z \\ 0 & 0 & 0 & 1 \end{bmatrix}$$

Fig. 3.Translation

The movements in the space are represented by a series of rotations and translations, these rotations and translations, can be represented as a homogeneous matrix multiplication.

Direct Kinematics

The direct kinematics is the robot kinematic model. In this model, the movements of the robot (coordinates of degrees of freedom) are given and the final positions are found. See Figure 4.

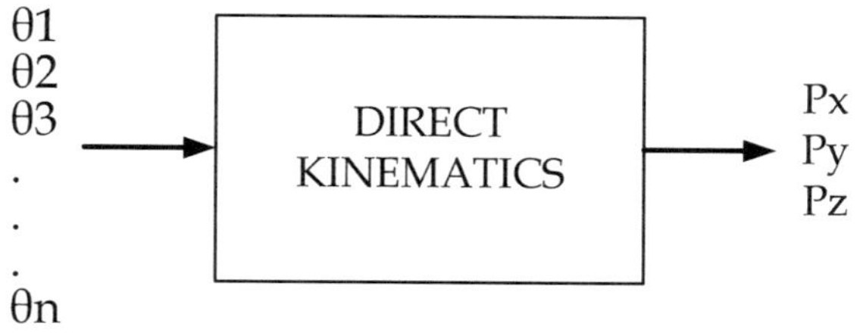

Fig. 4. Direct Kinematics

To find the direct kinematic model, using the homogeneous matrix method, is necessary to make the moves of coordinated system from the fixed base until the last link. For each movement, homogeneous matrices are obtained and the final result is the product of these matrices.

Inverse kinematics

The inverse kinematics seeks the coordinates of each degree of freedom based on the final position of the robot. Figure 4.

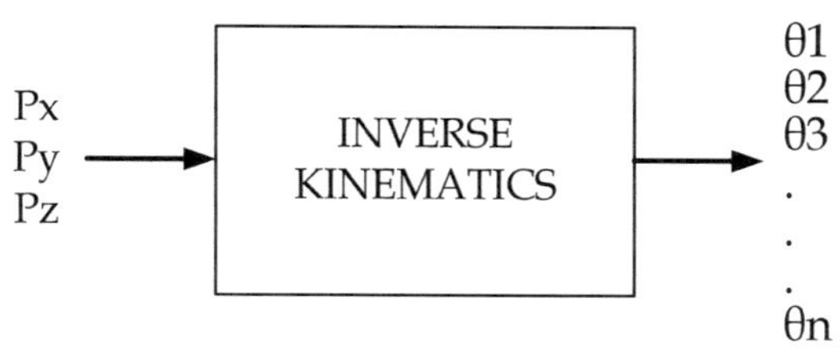

Fig. 5. Inverse Kinematics

The methods used are: the geometric method using the quadratic equation (Dutra, 2006) and the gradient method

Path planning

For the robot operation is necessary to identify different working positions, coordinates and the paths that allow performing a specific task. This work is makes with path planning methods.

Dynamic

In order to the robot control is necessary to know the dynamic model and formulate the dynamic control strategies. To find the dynamic model of the KUKA KR6 the Euler Lagrange model was used. (Kurfles, 2005).

2. Modelling

For the Robot modelling, CAD models and mathematical models were developed. The mathematical models are for the kinematics and dynamics of the manipulator and are presented below.

2.1 CAD Model

The Figure 6 shows the model developed in the CAD software Solid Edge®, based on the KR6 KUKA Robot of Robotics Laboratory at UFRJ, taking into account its main geometric characteristics.

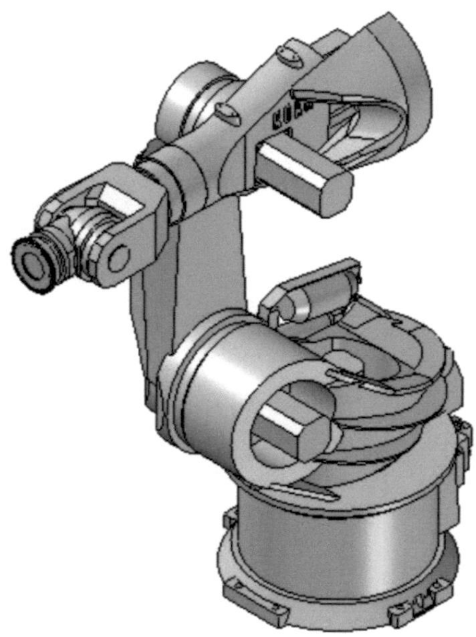

Fig. 6. CAD model of the KR6 KUKA Robot

2.2 Direct Kinematics

The manipulator kinematics model is based on the use of homogeneous matrix for this purpose; coordinated systems are located in a convention proposed by the authors. Supported by recommendations of the Denavit-Hartenberg algorithm (Denavit & Hartenberg, 1955). The convention adopted is:

- On the spin axis of each joint, locate the z-axis direction such that positive rotations are counter clockwise. See Figure 7.
- The x-axis is located parallel to each link, oriented to the follow coordinate system. See Figure 7.

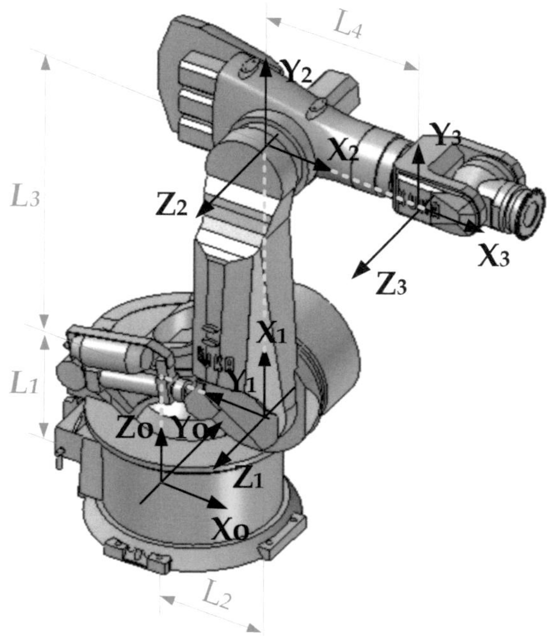

Fig. 7. Reference coordinate systems, for the KR6 KUKA Robot.

The proposed convention, seeks to define all positive rotations of joints when the rotation direction is counter clockwise and to make the translations always on the x-axis in positive direction.

The generated movements for going from one frame to another are mathematically represented by homogeneous matrix transformations and follow the particular geometry of the robot link to link:

1. R(Zo,θo)*T(Zo,L1)
2. T(Xo´,L2)*R(Xo´,π/2)*R(Zo´, π/2)
3. R(Z1,θ1)*T(X1,L3) *R(Z1´, -π/2)
4. R(Z2,θ2)*T(X2,L4)

The full kinematic model is presented in equation (2):

$$\begin{aligned} T = {} & R(Z_o,\theta_0)\times T(Z_o,L_1)\times T(X'_o,L_2) \\ & \times R\left(X'_o,\frac{\pi}{2}\right)\times R\left(Z'_o,\frac{\pi}{2}\right)\times R(Z_1,\theta_1) \\ & \times T(X_1,L_3)\times R\left(Z'_1,-\frac{\pi}{2}\right)\times R(Z_2,\theta_2) \\ & \times T(X_2,L_4) \end{aligned} \tag{2}$$

2.2 Inverse Kinematics

To obtain the inverse kinematic model the geometry of the robot is used, using the quadratic equation method (Dutra, 2006), obtaining the solution for different configurations of the robot in closed form. The first step is to characterize the robot in a vectorial model, for any position, see Figure 8.

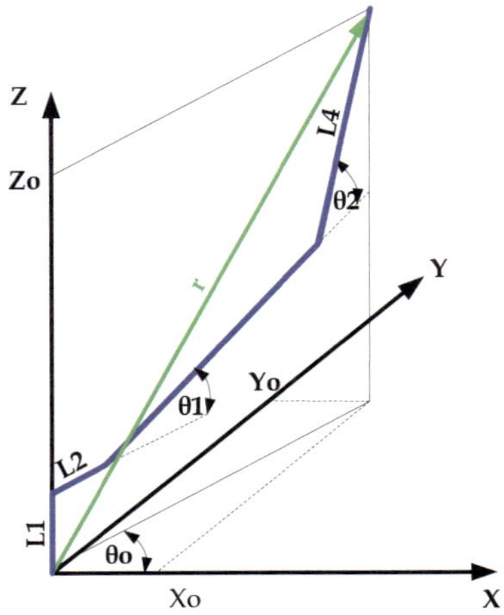

Fig. 8. Vectorial model for inverse kinematics.

Based on the figure 8, the vector sum is developed and the algebraic equations for each joint are obtained:
For the joint θ_1 equation (3)

$$\theta_1 = Sin^{-1}\left(\frac{-B \pm \sqrt{B^2 - 4AC}}{2A}\right) \tag{3}$$

Were

$$\begin{aligned} D &= L_4{}^2 - L_3{}^2 - X^2 - Z^2 \\ A &= 4X^2L_3{}^2 + 4Z^2L_3{}^2 \\ B &= 4ZL_3D \\ C &= D^2 - 4X^2L_3{}^2 \end{aligned} \tag{4}$$

For the joint θ_2 equation (5)

$$\theta_2 = Sin^{-1}\left(\frac{-b \pm \sqrt{b^2 - 4ac}}{2a}\right) - \theta_1 \tag{5}$$

were:

$$\begin{aligned} d &= L_3{}^2 - L_4{}^2 - X^2 - Z^2 \\ a &= 4X^2L_4{}^2 + 4Z^2L_4{}^2 \\ b &= 4ZL_4D \\ c &= d^2 - 4X^2L_4{}^2 \end{aligned} \tag{6}$$

The above equations, equation (3) to equation (6), represent the inverse kinematic model in closed form and providing solutions for different settings of KR 6 KUKA Robot in the workspace.

2.3 Numerical method to the inverse kinematics

For the implementation of inverse kinematic model in a parallel processing platform, the gradient method was proposed in order to observe the platform potential for the application in the control of the manipulator. The general form of the method corresponds to equation (7)

$$\theta_{n+1} = \theta_n - J^{-1}(\theta_n) \times f(\theta_n) \tag{7}$$

Were $J^{-1}(\theta_n)$ correspond to inverse jacobian of the robot, which can be seen in equation (8) and the functions $f(\theta_n)$ depends of kinematical direc model, equation (9)

$$J(\theta_n) = \left[\frac{\partial f_i(\theta)}{\partial \theta_j}\right] = \begin{bmatrix} \frac{\partial f_1(\theta)}{\partial \theta_1} & \frac{\partial f_1(\theta)}{\partial \theta_2} & \frac{\partial f_1(\theta)}{\partial \theta_3} \\ \frac{\partial f_2(\theta)}{\partial \theta_1} & \frac{\partial f_2(\theta)}{\partial \theta_2} & \frac{\partial f_2(\theta)}{\partial \theta_3} \\ \frac{\partial f_3(\theta)}{\partial \theta_1} & \frac{\partial f_3(\theta)}{\partial \theta_2} & \frac{\partial f_3(\theta)}{\partial \theta_3} \end{bmatrix} \tag{8}$$

$$\begin{aligned} f_1(\theta_1, \theta_2, \theta_3) &= a_1 \\ f_2(\theta_1, \theta_2, \theta_3) &= a_2 \\ f_3(\theta_1, \theta_2, \theta_3) &= a_3 \end{aligned} \tag{9}$$

According to the equations (8) and (9), it is necessary to determine the Jacobian of the robot and develop the algorithm to find the corresponding angles for each position of the manipulator, the functions of equations (9) are made explicit in the set of equations (10):

$$\begin{aligned} f_1 &= L_2 \times Cos(\theta_0) \\ &\quad + L_3 \times Cos(\theta_1) \times Cos(\theta_0) \\ &\quad + L_4 \times (Cos(\theta_1 + \theta_2) \times Cos(\theta_0) - X_o \\ f_2 &= L_2 \times Seno(\theta_0) \\ &\quad + L_3 \times Cos(\theta_1) \times Seno(\theta_0) \\ &\quad + L_4 Cos(\theta_1 + \theta_2) \times Seno(\theta_0) - Y_o \\ f_3 &= L_1 + L_3 \times Seno(\theta_1) \\ &\quad + L_4 \times Seno(\theta_1 + \theta_2) - Z_o \end{aligned} \tag{10}$$

The equations (10), correspond to the solution for the direct kinematics of the robot.

2.4 Inverse Dynamic

The dynamic model is based on the Euler Lagrange equations (11). Specifically in the model presented by Fu (1987), it considers the inertial effects of the manipulator by means of the inertia tensor. The use of the transformation matrices is an advantage by the fact that its derivatives can be obtained as a linear combination of a constant matrix multiplied by the original matrix.

$$T_i = \frac{d}{dt}\left(\frac{\partial L}{\partial \dot{\theta}_i}\right) - \frac{\partial L}{\partial \theta_i} \quad (11)$$

were:
L corresponds to the Lagrangian (kinetic energy less potential energy) equation (12)

$$L\left(\theta(t),\dot{\theta}(t)\right) = K\left(\theta(t),\dot{\theta}(t)\right)\text{-U}\left(\theta(t)\right) \quad (12)$$

The model proposed by Fu (1987) is presented in equation (13)

$$T_i = \sum_{k=1}^{n} D_{ik}\ddot{\theta}_k + \sum_{k=1}^{n}\sum_{m=1}^{n} h_{ikm}\dot{\theta}_k\dot{\theta}_m + c_i \quad (13)$$

The model represented in a matrix form is shown in equation (14)

$$T(t) = D(\theta(t))\ddot{\theta}(t) + h(\theta(t),\dot{\theta}(t)) + c(\theta(t)) \quad (14)$$

were:

$T(t) = \left[T_1(t) \quad T_2(t) \quad \ldots\ldots \quad T_n(t)\right]^T$ Vector torque, size nx1.

$\theta(t) = \left[\theta_1(t) \quad \theta_2(t) \quad \ldots\ldots \quad \theta_n(t)\right]^T$ Vector of joint positions, size nx1.

$\dot{\theta}(t) = \left[\dot{\theta}_1(t) \quad \dot{\theta}_2(t) \quad \ldots\ldots \quad \dot{\theta}_n(t)\right]^T$ Vector of angular Velocities, size nx1.

$\ddot{\theta}(t) = \left[\ddot{\theta}_1(t) \quad \ddot{\theta}_2(t) \quad \ldots\ldots \quad \ddot{\theta}_n(t)\right]^T$ Vector of angular acceleration, size nx1.

$D_{ik} = \sum_{j=\max(i,k)}^{n} Tr\left(U_{jk}J_jU_{ji}^T\right) \qquad i,k = 1,2,\ldots.,n$ Inertia matrix, size nxn

$$J_i = \begin{bmatrix} \frac{-I_{xx}+I_{yy}+I_{zz}}{2} & I_{xy} & I_{xz} & m_i\bar{x}_i \\ I_{xy} & \frac{I_{xx}-I_{yy}+I_{zz}}{2} & I_{yz} & m_i\bar{y}_i \\ I_{xz} & I_{yz} & \frac{I_{xx}+I_{yy}-I_{zz}}{2} & m_i\bar{z}_i \\ m_i\bar{x}_i & m_i\bar{y}_i & m_i\bar{z}_i & m_i \end{bmatrix}$$ (15) Inertia Tensor, size 4x4.

$h(\theta,\dot{\theta}) = \left[h_1 \quad h_2 \quad \ldots\ldots \quad h_n\right]^T$ Vector of Coriolis and centrifugal force, Size nx1

$h_i = \sum_{k=1}^{n}\sum_{m=1}^{n} h_{ikm}\dot{\theta}_k\dot{\theta}_m \qquad i = 1,2,\ldots.,n$

$h_{ikm} = \sum_{j=\max(i,k,m)}^{n} Tr\left(U_{jkm}J_jU_{ji}^T\right) \quad i,k,m = 1,2,\ldots.,n$

$h_i = \sum_{k=1}^{n}\sum_{m=1}^{n} h_{ikm}\dot{\theta}_k\dot{\theta}_m \qquad i = 1,2,\ldots.,n$

$c(\theta) = \begin{bmatrix} c_1 & c_2 & \ldots\ldots & c_n \end{bmatrix}^T$ Gravity forces vector size nx1

$$c_i = \sum_{j=i}^{n} \left(-m_j g U_{ji} {}^j r_j \right) \qquad i = 1,2,\ldots,n$$

From the direct kinematic model, presented in equation (2), is applied the model of Fu , to which is necessary determine Ujk matrices, the inertia tensor Ji for each link, the inertia effects D, the matrix hi and hijk of Coriolis and centrifugal acceleration, the position vector R and the gravitational vectors force C.
To calculate the matrix U_{jk} is used the canonical equation (16):

$$U_{jk} = \frac{\partial\, {}^0 A_j}{\partial \theta_k} = {}^0 A_{j-1} Q_i\, {}^{j-1} A_k \tag{16}$$

To determine the inertia tensor of each link the CAD model was used, obtaining the inertia moments around of the reference system used in the assembly module of the CAD software. As an example is presented the case of link 2, the software presents the inertia in the way shown in Figure 8.

Fig. 8. Inertia obtained in Solid Edge®.

The Table 1 was obtained using the software solid edge®, showing data of inertia and centroids.

m_2	38,767	Kg
I_{xx2}	112,0126	Kg-m2
I_{yy2}	107,2872	Kg-m2
I_{zz2}	16	Kg-m2

I_{xy2}	7,1702	Kg-m^2
I_{xz2}	-22,7151	Kg-m^2
I_{yz2}	-30,4911	Kg-m^2
X_c	-0,3661	M
Y_c	-0,505	M
Z_c	1,55	m

Table 1. Inertial and centroidal data for link two.

To obtain the inertia tensor, equation (15), it is necessary to make two changes to the inertia tensor data obtained with solid edge®. The first change is a translation of the reference base system to the reference of the joint and the second change is to orient the base reference system according to the reference system selected in the kinematic model. Figure 9

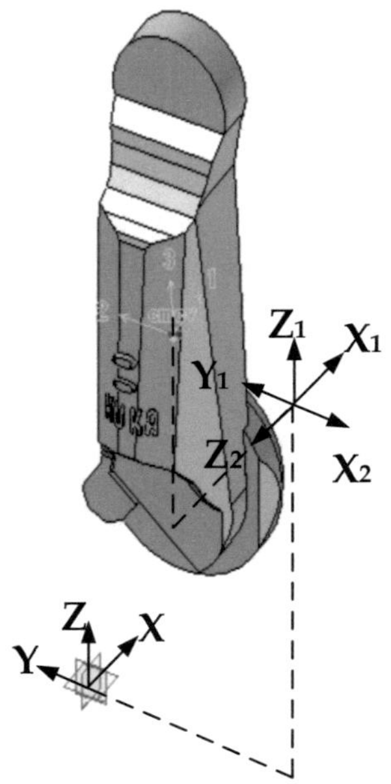

Fig. 9. Reference system for link two

To the translation of the references systems it is necessary to apply the Steiner theorem (Beer, 1980), the distances to apply the theorem are presented in Table 2

X_{axis2}	0	M
Y_{axis2}	-0,505	M
Z_{axis2}	1,35	M

Table 2. Distances to Steiner theorem application.

The inertia tensor for the link two after making changes based on the equation (15) is presented in Table 3

2,295	4,062	7,170	0
4,062	53,840	22,715	7,753
7,170	22,715	1,111	14,195
0	7,753	14,195	38,767

Table 3. Inertia tensor of link two.

For the determination of the matrix D (matrix of inertial effects). It is necessary to use the Equation (17)

$$D(\theta) = \begin{bmatrix} D_{11} & D_{12} & D_{13} \\ D_{21} & D_{22} & D_{23} \\ D_{31} & D_{34} & D_{33} \end{bmatrix} \tag{17}$$

were:

$$D_{11} = Tr\left(U_{11}J_1U_{11}^T\right) + Tr\left(U_{21}J_2U_{21}^T\right) + Tr\left(U_{31}J_3U_{31}^T\right)$$

$$D_{12} = D_{21} = Tr\left(U_{22}J_2U_{21}^T\right) + Tr\left(U_{32}J_3U_{31}^T\right)$$

$$D_{13} = D_{31} = Tr\left(U_{33}J_3U_{31}^T\right)$$

$$D_{22} = Tr\left(U_{22}J_2U_{22}^T\right) + Tr\left(U_{32}J_3U_{32}^T\right)$$

$$D_{23} = D_{32} = Tr\left(U_{33}J_3U_{32}^T\right)$$

$$D_{33} = Tr\left(U_{33}J_3U_{33}^T\right)$$

For the determination of vector h, vector of Coriolis and centrifugal forces, the equation (18) is proposed. This equation presents the angular velocities independently through the matrix *Hi,v*, for purposes of calculation and simulation.

$$h_i = \dot{\theta}^T H_{i,v} \dot{\theta} \tag{18}$$

were:

$$H_{i,v} = \begin{bmatrix} h_{i11} & h_{i12} & h_{i13} \\ h_{i21} & h_{i22} & h_{i23} \\ h_{i31} & h_{i32} & h_{i22} \end{bmatrix}$$

$$h_{ikm} = Tr\left(U_{jkm}J_jU_{ji}^T\right) \quad j = \max(i,k,m)$$

$$U_{jkm} = \frac{\partial U_{jk}}{\partial \theta_m} = {}^0A_{k-1}Q_k\,{}^{k-1}A_{m-1}Q_m\,{}^{m-1}A_i \quad j \geq k \geq m$$

$$U_{jkm} = \frac{\partial U_{jk}}{\partial \theta_m} = {}^0A_{m-1}Q_m\,{}^{m-1}A_{k-1}Q_k\,{}^{k-1}A_i \quad j \geq m \geq k$$

$$U_{jkm} = \frac{\partial U_{jk}}{\partial \theta_m} = 0 \qquad \text{se} \quad j < k \quad \text{ou} \quad j < m$$

$$U_{jkm} = \frac{\partial U_{jk}}{\partial \theta_m} = \frac{\partial}{\partial \theta_m}\frac{\partial^0 A_j}{\partial \theta_k}$$

Thus the vector h of centrifugal and Coriolis forces is:

$$h_i = \dot{\theta}^T H_{i,v} \dot{\theta}$$

$$\begin{bmatrix} h_1 \\ h_2 \\ h_3 \end{bmatrix} = \begin{bmatrix} \begin{bmatrix} \dot{\theta}_1 & \dot{\theta}_2 & \dot{\theta}_2 \end{bmatrix} \begin{bmatrix} h_{111} & h_{112} & h_{113} \\ h_{121} & h_{122} & h_{123} \\ h_{131} & h_{132} & h_{122} \end{bmatrix} \begin{bmatrix} \dot{\theta}_1 \\ \dot{\theta}_2 \\ \dot{\theta}_2 \end{bmatrix} \\ \begin{bmatrix} \dot{\theta}_1 & \dot{\theta}_2 & \dot{\theta}_2 \end{bmatrix} \begin{bmatrix} h_{211} & h_{212} & h_{213} \\ h_{221} & h_{222} & h_{223} \\ h_{231} & h_{232} & h_{222} \end{bmatrix} \begin{bmatrix} \dot{\theta}_1 \\ \dot{\theta}_2 \\ \dot{\theta}_2 \end{bmatrix} \\ \begin{bmatrix} \dot{\theta}_1 & \dot{\theta}_2 & \dot{\theta}_2 \end{bmatrix} \begin{bmatrix} h_{311} & h_{312} & h_{313} \\ h_{321} & h_{322} & h_{323} \\ h_{331} & h_{332} & h_{322} \end{bmatrix} \begin{bmatrix} \dot{\theta}_1 \\ \dot{\theta}_2 \\ \dot{\theta}_2 \end{bmatrix} \end{bmatrix}$$

For the Determination of the gravity force vector C:

$$c(\theta) = \begin{bmatrix} c_1 & c_2 & c_3 \end{bmatrix}^T$$

$$c_i = \sum_{j=i}^{3} \left(-m_j g U_{ji} \, {}^j\overline{r}_j \right) \qquad i = 1,2,3$$

$$c_1 = -m_1 g U_{11} \, {}^1\overline{r}_1 - m_2 g U_{21} \, {}^2\overline{r}_2 - m_3 g U_{31} \, {}^3\overline{r}_3$$

$$c_2 = -m_2 g U_{22} \, {}^2\overline{r}_2 - m_3 g U_{32} \, {}^3\overline{r}_3$$

$$c_3 = -m_3 g U_{33} \, {}^3\overline{r}_3$$

$$\begin{bmatrix} c_1 \\ c_2 \\ c_3 \end{bmatrix} = \begin{bmatrix} -m_1 g U_{11} \, {}^1\overline{r}_1 - m_2 g U_{21} \, {}^2\overline{r}_2 - m_3 g U_{31} \, {}^3\overline{r}_3 \\ -m_2 g U_{22} \, {}^2\overline{r}_2 - m_3 g U_{32} \, {}^3\overline{r}_3 \\ -m_3 g U_{33} \, {}^3\overline{r} \end{bmatrix}$$

The vector r in the reference system of rotation axes is:

$${}^1r_1 = \begin{bmatrix} -0.0052 \\ 0.0026 \\ 0.369 \\ 1 \end{bmatrix} \quad {}^2r_2 = \begin{bmatrix} 0 \\ 0.2 \\ 0.366 \\ 1 \end{bmatrix} \quad {}^3r_3 = \begin{bmatrix} 0 \\ 0.454 \\ -0.00015 \\ 1 \end{bmatrix}$$

$$g = \begin{bmatrix} 0 & 0 & -g & 0 \end{bmatrix}$$

The total system is then as follows:

$$T(t) = D(\theta(t))\ddot{\theta}(t) + h(\theta(t), \dot{\theta}(t)) + c(\theta(t))$$

$$\begin{bmatrix} T_1 \\ T_2 \\ T_3 \end{bmatrix} = \begin{bmatrix} D_{11} & D_{12} & D_{13} \\ D_{21} & D_{22} & D_{23} \\ D_{31} & D_{34} & D_{33} \end{bmatrix} \begin{bmatrix} \ddot{\theta}_1 \\ \ddot{\theta}_2 \\ \ddot{\theta}_3 \end{bmatrix}$$

$$\begin{bmatrix} \begin{bmatrix} \dot{\theta}_1 & \dot{\theta}_2 & \dot{\theta}_2 \end{bmatrix} \begin{bmatrix} h_{111} & h_{112} & h_{113} \\ h_{121} & h_{122} & h_{123} \\ h_{131} & h_{132} & h_{122} \end{bmatrix} \begin{bmatrix} \dot{\theta}_1 \\ \dot{\theta}_2 \\ \dot{\theta}_2 \end{bmatrix} \\ + \begin{bmatrix} \dot{\theta}_1 & \dot{\theta}_2 & \dot{\theta}_2 \end{bmatrix} \begin{bmatrix} h_{211} & h_{212} & h_{213} \\ h_{221} & h_{222} & h_{223} \\ h_{231} & h_{232} & h_{222} \end{bmatrix} \begin{bmatrix} \dot{\theta}_1 \\ \dot{\theta}_2 \\ \dot{\theta}_2 \end{bmatrix} \\ \begin{bmatrix} \dot{\theta}_1 & \dot{\theta}_2 & \dot{\theta}_2 \end{bmatrix} \begin{bmatrix} h_{311} & h_{312} & h_{313} \\ h_{321} & h_{322} & h_{323} \\ h_{331} & h_{332} & h_{322} \end{bmatrix} \begin{bmatrix} \dot{\theta}_1 \\ \dot{\theta}_2 \\ \dot{\theta}_2 \end{bmatrix} \end{bmatrix}$$

$$+ \begin{bmatrix} -m_1 g U_{11}\,{}^{1}\overline{r}_1 - m_2 g U_{21}\,{}^{2}\overline{r}_2 - m_3 g U_{31}\,{}^{3}\overline{r}_3 \\ -m_2 g U_{22}\,{}^{2}\overline{r}_2 - m_3 g U_{32}\,{}^{3}\overline{r}_3 \\ -m_3 g U_{33}\,{}^{3}\overline{r} \end{bmatrix}$$

3. Control

3.1 Kinematic Control

For the robot control, the first stage is to make a kinematical control. The different types of Kinematical control are shown below:

Control of position in open loop

This control is appropriated in the case of low operation velocities or in case that the dynamics of the robot does not perturb the behavior of the manipulator. The figure 10 presents the open loop kinematics control.

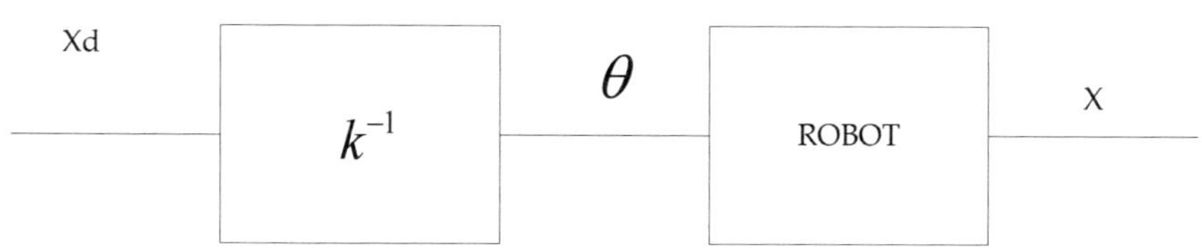

Fig. 10. Block Diagram to open loop kinematics control

In the figure 10 the operator K^{-1} corresponds to inverse kinematics model.

Control of position in closed loop

To make the kinematical position control iterative in a closed loop, it is required to work whit the inverse jacobian J^{-1} as is shown in the figure 11. The operator f(.) corresponds to the inverse kinematic model.

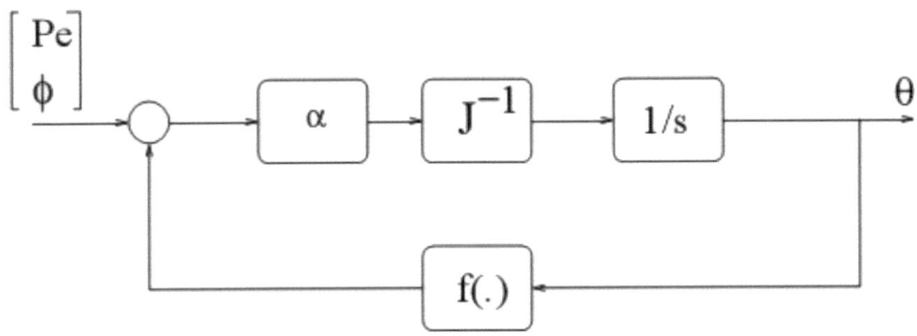

Fig. 11. Iterative control position

Joint Control
Other option is to make a joint level control. In this case, the control is represented in the equation (19) in state variables.

$$\begin{bmatrix} v \\ w \end{bmatrix} = J(\theta)\dot{\theta} \tag{19}$$

Considering that most of the robots usually have a speed control loop at the level of joints: where for one input $u = \dot{\theta}_d$ and a high-gain control $k \longrightarrow \infty$ is found, the error tends to zero $e \longrightarrow 0$ and consequently $u \approx \dot{\theta}$, the desired angular velocity. Figure 12.

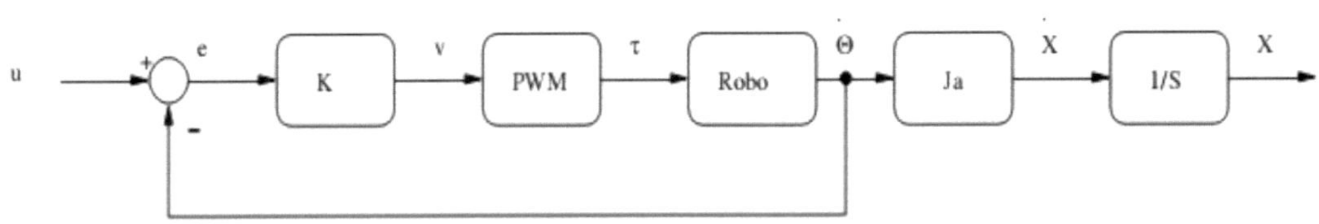

Fig. 12. Joint velocity control

Therefore, the manipulator motion can be described by $\dot{\theta}_i = u_i$, $i = 1,\ldots,n$, were u_i is a signal applied to the motor speed of the i-th joint. Then the control system to be considered is presented in the equation (20):

$$\begin{bmatrix} v \\ w \end{bmatrix} \approx J(\theta)u \tag{20}$$

For the problem of trajectory tracking and to avoid errors in steady state, the control law u can be chosen as a proportional control adding one feed forward term.
Equation (21) and figure 13

$$\overline{u} = \dot{x}_d + K\left(x_d - x\right) \tag{21}$$

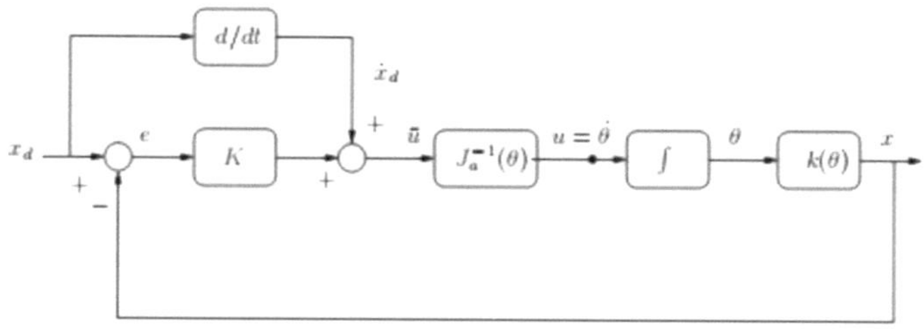

Fig. 13. Control law.

3.2 Control in a parallel processing platform

The Parallel Processing is the ability of an entity to carry out multiple operations or tasks simultaneously, obtaining a better performance and achieving lower response time. There are different platforms of parallel processing like DSPs (Digital Signal processor), FPGAs (Field Programmable Gate Array), among others. In this work the Inverse kinematic model was implemented into a DSP and a microprocessor used to compare its performance. To implement the inverse kinematics model an algorithm based on the gradient method or Newton's method were used (Tsai, 1999). In this case the algorithm calculates the position error and the number of iterations needed to get the result. It is notable that the bigger the number of iterations, the bigger the calculation time will be for the control system, (Archila & Dutra, 2007), which is the response speed limit allowed by the robot and/or the task to realize. An overview of the algorithm is presented in Figure 14.

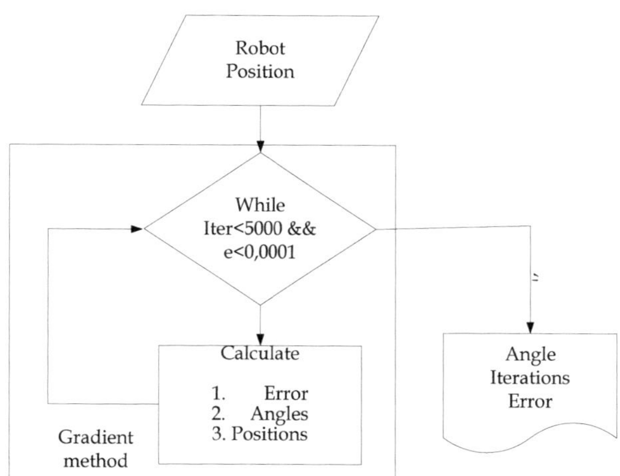

Fig. 14. Algorithm for inverse kinematics.

4. Numeric Simulation

To validate the kinematic and dynamic models Matlab 7.0® was used evaluating the behaviour of each mathematical model to be applied in the KUKA robot controller. The validation of the models works with oblique trajectories passing through two singularities of the robot geometry.

4.1 Computational model to validate the direct kinematics model

In the validations of this model, the input data are angular positions for each joint and the positions for terminal elements are obtained, Figure 15. In this case the joint 2 and 3 takes angular increments up to a 90 degrees rotation in each link.

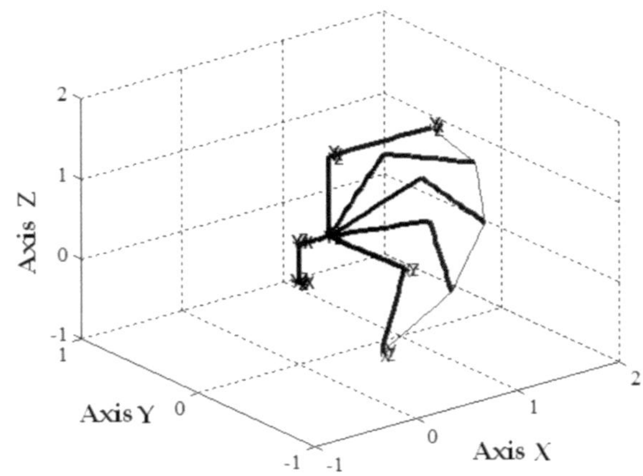

Fig. 15. Direct kinematic model.

4.2 Computational model to validate the inverse kinematics model

The validation of this model requires knowledge of the spatial positions for the robot, the validation used an oblique path obtaining the following results Figure 16.

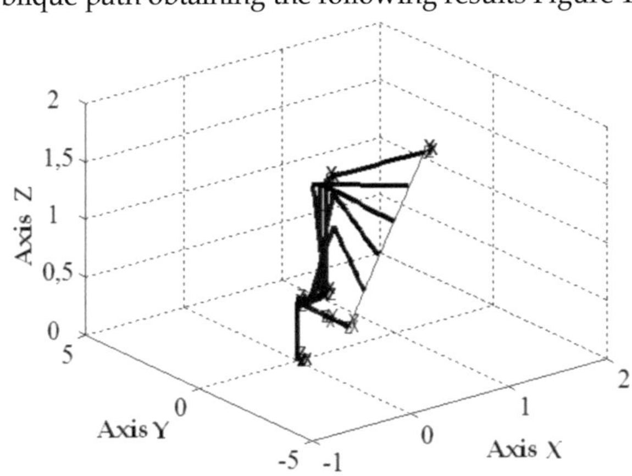

Fig. 16. Inverse kinematic model.

Figure 16 shows that the inverse kinematics model follows the path requested, providing the angular positions needed to achieve the desired positions even for singular points in the workspace of the manipulator.

4.3 Computational model to validate the dynamical model

To validate the dynamic model, the input data were oblique trajectories and tasks that require the motion of the terminal element with constant speed, working load of 16 kg which corresponds to the maximum load recommended by KUKA Roboter (2005). Through the validated inverse kinematics algorithm the angular positions are calculated. Once obtained, the angular positions, it is calculated the angular velocities using the Jacobian inverse, obtaining the following joint speeds Figure 17.

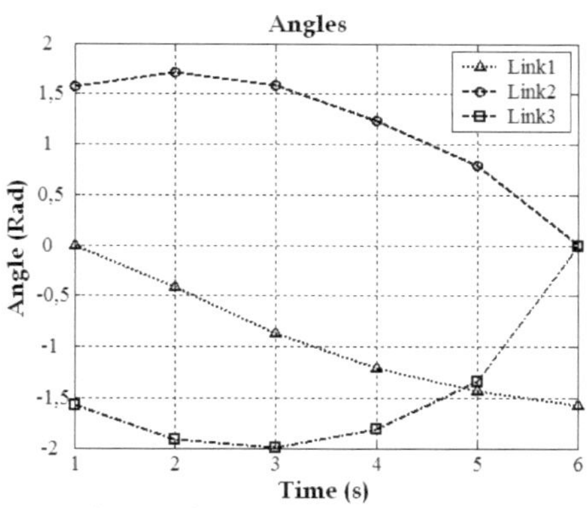

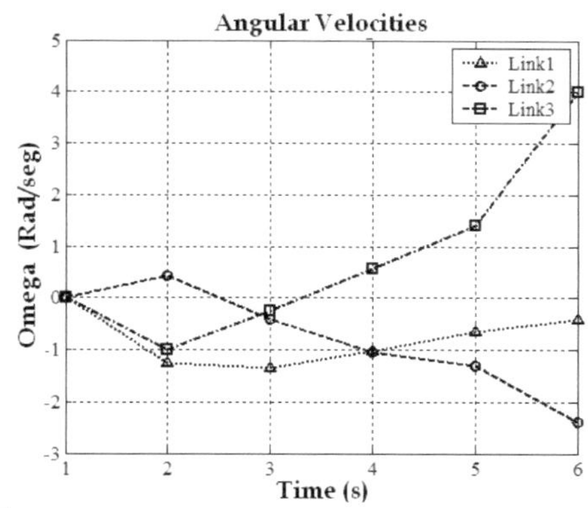

Fig. 17. Angles and angular velocity for the joints.

The angular accelerations required to perform the requested trajectory were calculated and the values are presented in Figure 18.

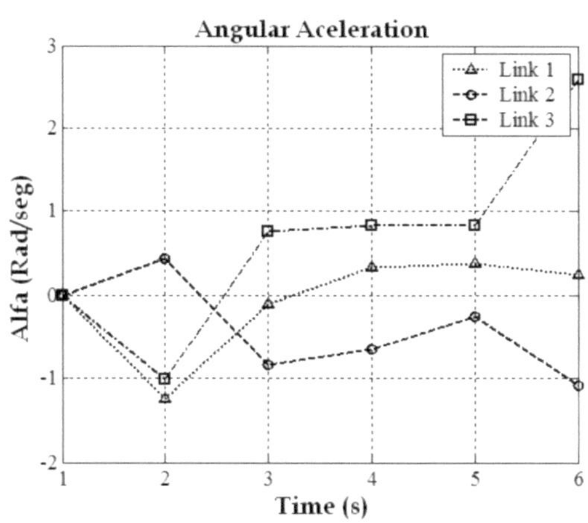

Fig. 18. Angular accelerations obtained in the joints.

With the data of angular positions, angular velocities, angular accelerations and loads applied on the dynamic model of equation (14), the following values of the torques of the robot actuators are obtained, Figure 19.

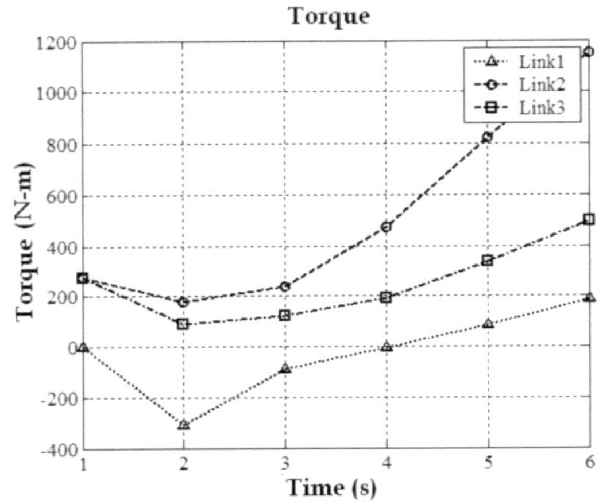

Fig. 19. Torques obtained in the joints.

The data provided by KUKA Roboter (2005) are presented in Table 4, It shows the operational limits of the robot.

	KUKA KR 6
Maximum Torque	3400 Nm
Total mass	206 Kg + 16 Kg
Maximum Velocity	600 Degrees/sec
Repeatability	+/- 0,1 mm

Table 4. Technical data to KR6 KUKA Robot

4.4 Implementation in a parallel processing platform

The inverse kinematics algorithm was implemented in VisualDSP®, Matlab® and C® to observe its performance, evaluating the processing time and the position error.

Processing Time

To evaluate the processing time, oblique trajectories were used in the same way as in the case of the dynamic model validation. The processing times obtained are presented in Figure 20 which clearly shows that the DSP has the shortest processing time.

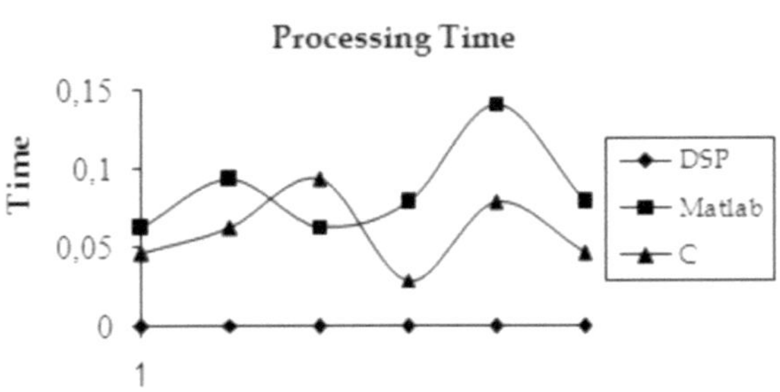

Fig. 20. Comparation of processing time

Evaluation of processing time

The Figure 20 shows that the DSP processing time is the shortest. However, we must assess whether it is appropriate to implement it as a Possible driver. For this purpose it is necessary to review the technical data of the robot to know its response speed. This data can be seen in Table 5, where the main information are: the encoder pulses and maximum angular velocity to which the robot can move, due to technical characteristics of the servo motors used to its operation.

Encoder	512	Pulses/Rev
	1,422	Pulses/Degree
W Max	600	Degrees/sec
Frequency	853,33	Pulses/sec
Period	0,00117	sec/pulse
	0,703	Degrees/pulse
	0,0123	Radians/pulse

Table 5. Technical data to calculus of response speed of KR6 KUKA Robot

The Figure 21 shows that the DSP processing time is appropriated to be implemented as a controller because it offers a maximum response of 0.55 ms, and the robot required 1.17 ms.

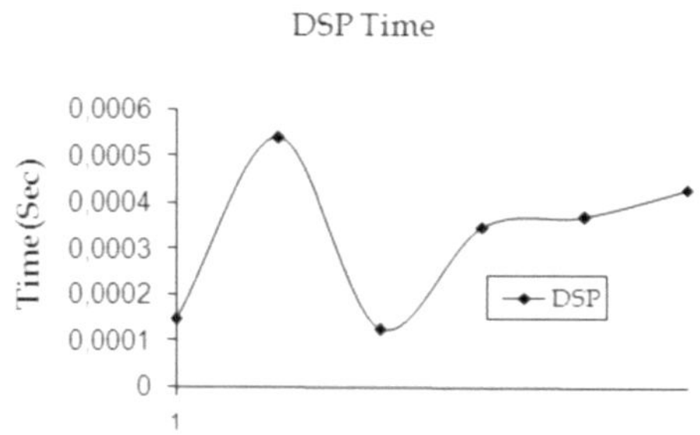

Fig. 21. DSP Processing time

Position Error

Another important characteristic to evaluate, is the position error which corresponds to two main factors: the first one is the quantization error and the second one the error due to the calculation method. The Figure 25 shows the position errors. The comparison parameter corresponds to the repeatability of the robot, which for this case is 0.1 mm in accordance **to** Figure 22.

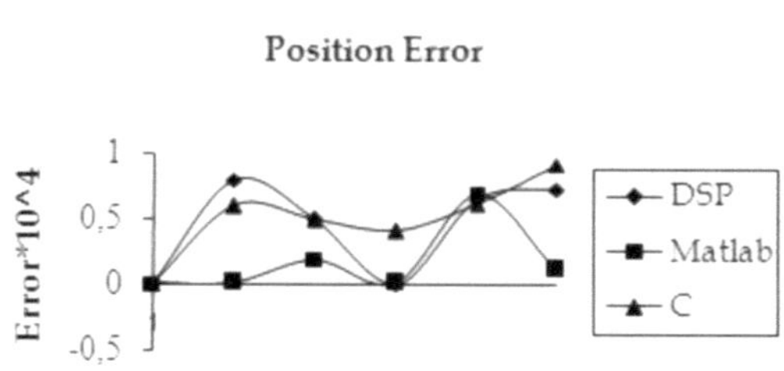

Fig. 22. Position error

In the figure 22, the maximum position error of the DSP is 0.08 mm.

5. Conclusions

The work presents the direct kinematics model of the robot KUKA KR 6, which was evaluated, obtaining suitable results for the kinematic control implemented.
The inverse kinematics model using the quadratic equation was shown to be an appropriated method. It presents the possible configurations of the robot in order to achieve the desired position to follow a trajectory.
The calculation of the inertia tensor was performed with the aid of CAD software, which gives us the principal inertia moments, products of inertia, and the centroid location of each link of the robot
Is Important to show that the inertia tensor obtained needs to be understood and translated to the appropriate reference system for the dynamic model.
The dynamic model was evaluated, obtaining values of 1250 Nm of torque to the oblique trajectory at constant speed work-rate. Obtaining as a maximum speed 12.7 RPM in the joint two, working well within the parameters of the KUKA Roboter.
The performance evaluation of the DSP was adequate, obtaining a processing time of 0.55 ms, appropriate for the given operation.
The position error found corresponds to the calculation method and the quantization error The maximum error value found was 0.08 mm which is below the repeatability value given by KUKA Roboter.

6. References

Archila J. F, Dutra M. S., (2007). Design and construction of a SCARA Type Manipulator, implementing a control system, International Congress of Mechanical Engineering COBEM 2007. Brasilia.

Barrientos, A., (2001). Fundamentos de robótica, McGraw Hill, pp 15 - 38

Beer F., (1980) Mecânica Vetorial para Engenheiros, McGraw H, pp 399 - 447.

Denavit, J., Hartenberg R. S., (1955) A Kinematic Notation for Lower- Pair Mechanism Based on Matrices, Journal of Applied Mechanics 22: 215 -221.

Dutra, M. S., (2006). Notas de Aula Mecanismos, Universidade Federal do Rio de Janeiro UFRJ.

Fu K. S., (1987) Control, Sensing, Vision, and Intelligence, McGraw Hill, pp 82 - 102.

KUKA Roboter. (2005)Technical specifications Manual.

Kurfles, T., (2005). Robotics and Automation Handbook. CRC Press, pp 26 - 84

Ollero, A., (2001). Robótica, Manipuladores y robots móviles, Primera edición, Alfaomega, Barcelona, pp 43 - 80

Tsai L. W., (1999) Robot Analysis. The Mechanics of Serial and Parallel Manipulators. Editorial John Wiley & Sons, Inc, pp 55 - 72.

Visual DSP ++ 4.5, C/C++ Compiler and library Manual for sharc, Analog Device.

Manipulator Design Strategy for a Specified Task Based on Human-Robot Collaboration

Seungnam Yu[a], Seungwhan Suh[a], Woonghee Son[b,c], Youngsoo Kim[b,d] and Changsoo Han[a]
[a] *Dept. of Mechanical Eng., Hanyang University, 1271 Sa-3 dong, Sangnok-gu, Ansan, Kyeonggi-do, Korea*
[b] *Dept. of Mechatronics Eng., Hanyang University, 1271 Sa-3 dong, Sangnok-gu, Ansan, Kyeonggi-do, Korea*
[c] *Korea Institute of Industrial Technology, 1271-18 Sa-3 dong, Sangnok-gu, Ansan, Kyeonggi-do, Korea*
[d] *Ministry of Education, Science and Technology, 77-6, Sejong-No, Jongno-Gu, Seoul, Korea*

1. Introduction

1.1 Overview and Motivation

This study starts from the improvement of developed curtain-wall installation robot.[1] Especially the robot was designed to handle a heavy weight curtain wall by integrating commercial excavator with custom-made manipulator system. Developed prototype system was tested in the real construction site and several considerations were deduced. First, separation of control source to operate the entire system is not good idea.

Fig. 1. Curtain-wall for skyscraper and developed curtain-wall installation robot system [1]

Second, hydraulic excavator can't exhibit the ideal performance on the curtain wall installation process because of its intrinsic jerking and shaking characteristics while handling the heavy weight materials.[2]

New research project is raised to solve those problems and to develop the useful automation system in real construction site. As a first step, the design process of scale down manipulator is proposed to ease the kinematical robot design. This scaled down robot and corresponding assumptions are applied to the simulation and comparison works of each robot design candidate. Then, the target task is selected as curtain wall installation which is required 6-DOF motion to cover the whole installation process. As abovementioned, this study based on the modification of prototype system hence end-effector of prototype system is utilized continuously and focused to develop the arm part except the wrist part. Therefore, the final goal of this study is to propose the proper design of robot manipulator handled by human passively as man-machine cooperation system specialized to curtain wall installation process. This idea is not only applied to passive system but also active controlled system because newly modified design makes to promote the physical performance of manipulator system intrinsically.

This work primarily focuses on selection of optimal type of joint type and link length considering task type and motion requirements. The developments of this work provide an open and objective design and analysis framework for a serial robot. This framework may be applied at any stage in the design process. As abovementioned, this study applies this frame work to prototype of our curtain wall installation robot and modified its kinematic design.

1.2 Experimental analysis of prototype system

Originally developed curtain wall installation system is combined by two modules, one is general commercial excavator and the other is newly designed 3-DOF manipulator. That system is tested in the real construction site by a new installation task scenario and compared with conventional installation task operated by human. Fig.2 shows the task results of curtain wall installation task. As the result, Robot requires more time than human for same task! Throughout this test, several considerations for the prototype system are deduced

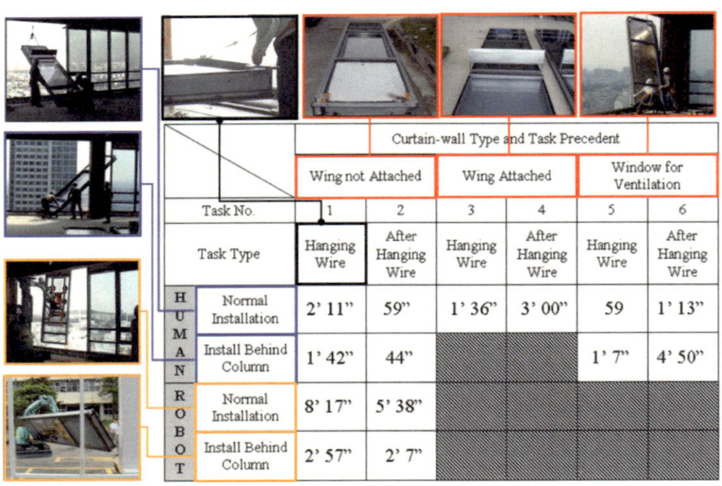

		Curtain-wall Type and Task Precedent					
		Wing not Attached		Wing Attached		Window for Ventilation	
	Task No.	1	2	3	4	5	6
	Task Type	Hanging Wire	After Hanging Wire	Hanging Wire	After Hanging Wire	Hanging Wire	After Hanging Wire
HUMAN	Normal Installation	2' 11"	59"	1' 36"	3' 00"	59	1' 13"
HUMAN	Install Behind Column	1' 42"	44"			1' 7"	4' 50"
ROBOT	Normal Installation	8' 17"	5' 38"				
ROBOT	Install Behind Column	2' 57"	2' 7"				

Not Applicable

Fig. 2. Field test of curtain wall installation robot and its elapsed time for each process [3]

First, 2-way control strategy of Excavator and 3-DOF manipulator is not efficient because it is hard to operate excavator boom and 3-DOF manipulator simultaneously through whole installation process. That is, the tasks of the two machines could not be completed independently hence the two workers had to work simultaneously. (Fig.3)
Second, using excavator for delicate motion is not appropriate. Excavator is not a machine for material handling but digging a ground. That hydraulic system has the leakage and drooping characteristics. Therefore, it is difficult to handle the heavy curtain wall and assembly it on the slot gently.
Based on those considerations, new concept for manipulator design for man-machine collaboration is proposed.

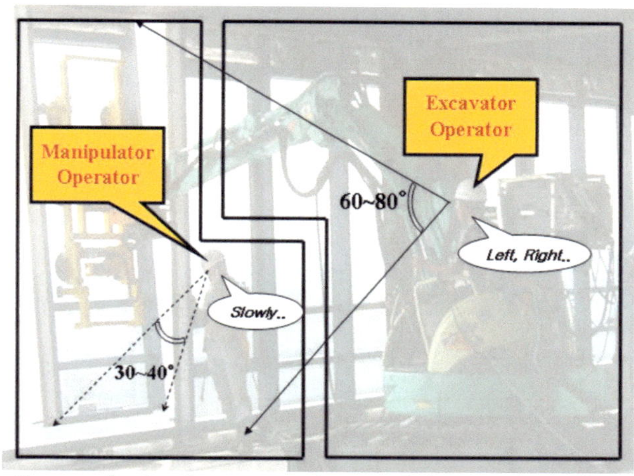

Fig. 3. Communication confusion in cooperative work in curtain wall installation

1.3 Conceptual design of new system

Final goal of this research is dedicated to man-machine cooperation system. That is the hybrid system that mixes the sensitivity of human and high performance of machine. This system is powerful at a field area. There is, unlike factory, not arranged and materials which handled by machine are relatively low standard one. Most of all, those materials are very heavy and large! Therefore, it is hard to automate the whole task using the robot and automation system.
Human sense is very useful in that case. Through this entire research is dedicated to the development of manipulator which handled by human and especially the major issue is to decide the adequate kinematical combination of manipulator components through the simulation of scale down manipulator of real one. Fig.4 shows the considerations for new manipulator design and Fig.5 describes concept of the intuitive curtain wall installation robot derived by upper considerations and its simulation strategy.

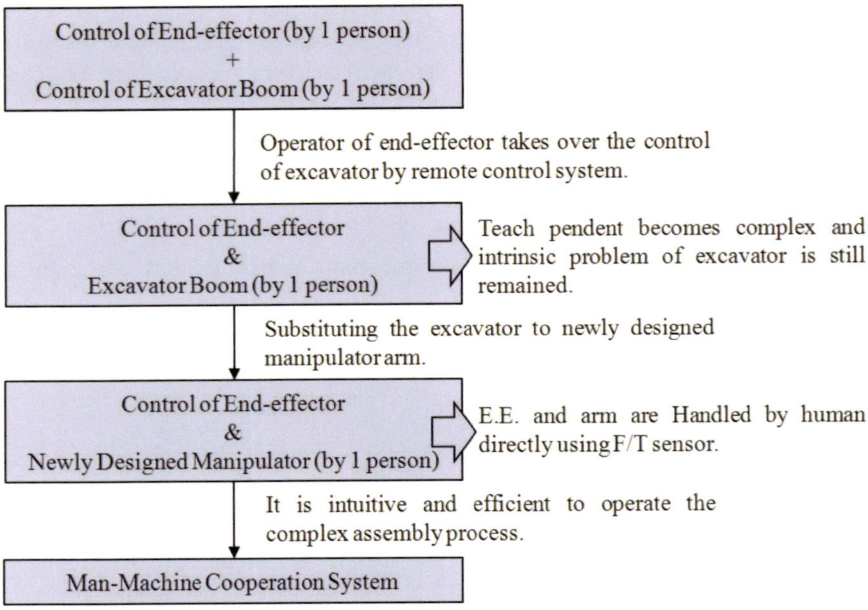

Fig. 4. Considerations for new mechanism

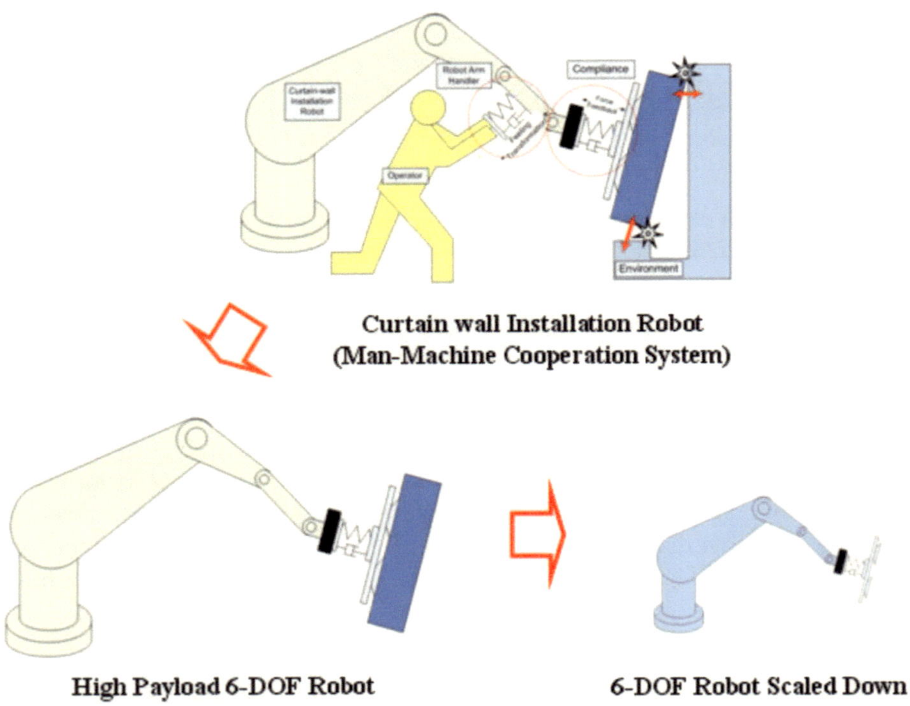

Fig. 5. Scaled Down Manipulator for Specified Robot Manipulator Design

1.4 Basic Assumptions for Specified Manipulator Design

1.4.1 Invariant part of proposed manipulator system

To design the modified manipulator, the several invariant parts are considered. First part is DOF of end-effecter. It is designed by analysis of conventional process of curtain wall installation (Fig.6(a)) and as shown in Fig.6(b), R-R-P type end-effector is developed. New design for manipulator system is focused to arm part integrating with this end-effector.

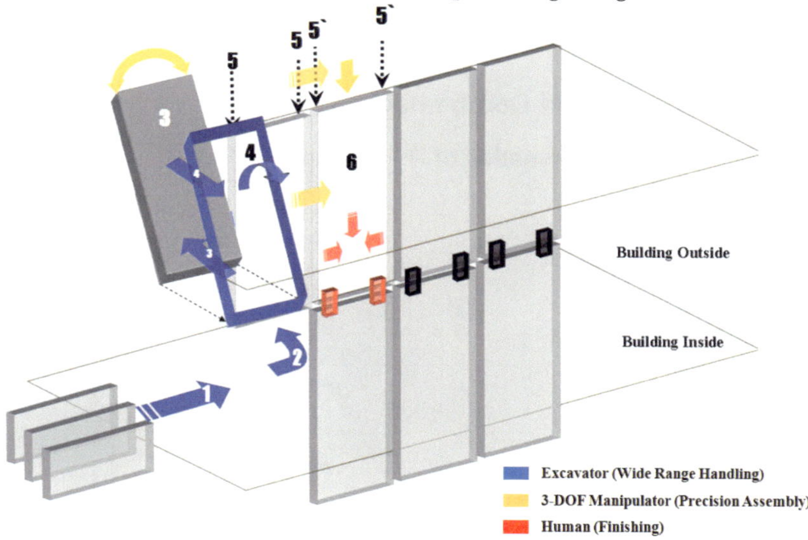

(a) Curtain wall Installation Process

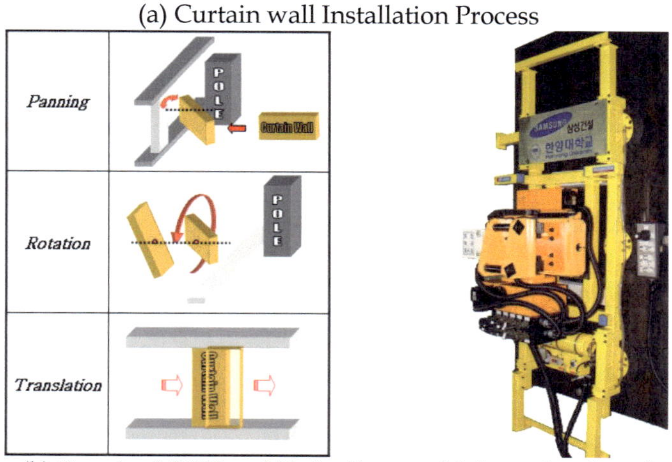

(b) Required DOF and End-effecter which we developed

Fig. 6. Invariant Condition - DOF of End effecter [1]

1.4.2 Variant part of proposed manipulator system

Variant part of the newly designed manipulator system is mainly arm part. Finally, joint of base frame, shoulder joint and elbow joint are remained except the invariant elements. Therefore, this research process can be simplified to find a combination of joint type of three joints and link ratio of upper and lower arm. Detailed assumptions for this study are as follows.

1. Each joint of manipulator is actuated by its own general actuators.
2. To cover the working area, one revolute joint which axis is perpendicular to the ground is should be included.
3. Each joint has one DOF and revolute joint and prismatic joint are considered only.
4. Target task is performed by nearly full stretched manipulator (generally, this task is performed in the skyscraper hence safety is most important issue) and manipulator is handled by human operator directly.

2. Selection of Adequate Joint Combination

2.1 Comparison of the basic kinematics of R-R type and R-P type

2.1.1 Simulation Conditions

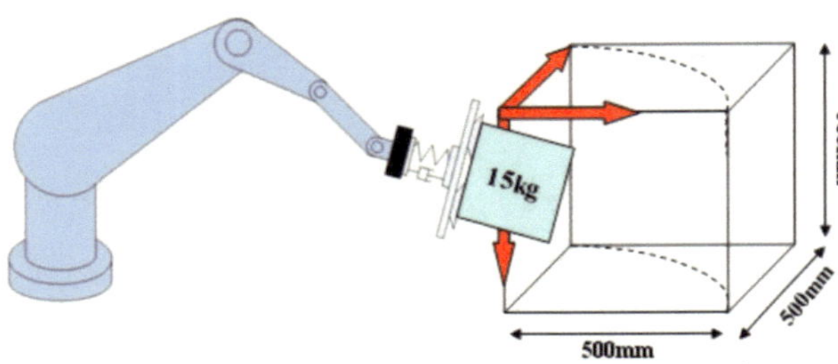

Fig. 7. Motion Range of 6-DOF Scaled Down

First, to select the part of shoulder joint and elbow joint, the motion range of manipulators is considered to 500mm×500mm×500mm, and link length of each manipulator is fixed by $l_1 = 300mm,\ l_2 = 330mm$. This process is dedicated to compare the performance of R-P and R-R type manipulator. Table.1 shows the several conditions for this simulation.

R-R Type	Contents	R-P Type
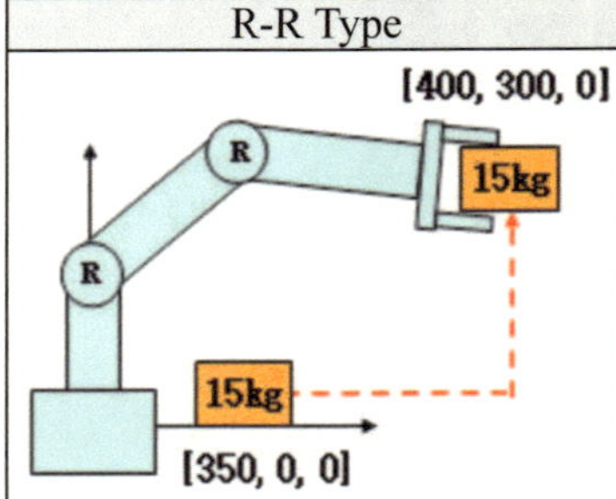	[Task definition] : In X-Y plane, 15kg weight box is moved initial pt. to final pt. through same trajectory	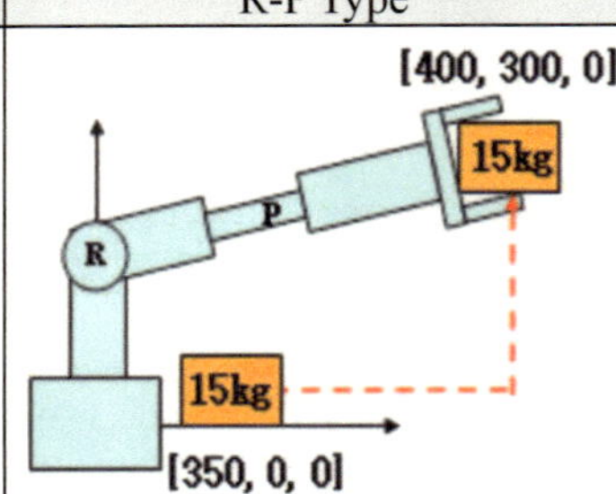
[350,0,0]	Initial Position	[350,0,0]
[400,300,0]	Final Position	[400,300,0]
$l_1 = 300mm,\ l_2 = 330mm$	Link Length	$l_1 = 300mm$ $0 \le d_2 \le 350mm$

Table. 1. Experimental Condition of two joint types of manipulators

2.1.2 Basic kinematics

2.1.2.1 R-R Type Manipulator

Table 2 shows the DH table[4] of R-R type manipulator.

	α_i	a_i	d_i	θ_i
1	0	0	0	θ_1
2	0	l_1	0	θ_2
3	0	l_2	0	0

Table. 2. DH table of R-R type manipulator

Using the upper information, transformation matrix and position vector of end effecter can be calculated. (l_1, l_2 is link length, and $s_1 = \sin\theta_1$, $c_1 = \cos\theta_1$)

Transformation matrix,

$$ {}^{0}_{E}T = \begin{bmatrix} c_{12} & -s_{12} & 0 & l_1c_1 + l_2c_{12} \\ s_{12} & c_{12} & 0 & l_1s_1 + l_2s_{12} \\ 0 & 0 & 1 & 0 \\ 0 & 0 & 0 & 1 \end{bmatrix} \tag{1} $$

Position vector of End effecter

$$ {}^{0}P_{E-E} = \begin{bmatrix} l_1c_1 + l_2c_{12} \\ l_1s_1 + l_2s_{12} \\ 0 \end{bmatrix} \tag{2} $$

Using Cramer rule,

$$ c_2 = \frac{x^2 + y^2 - l_1^2 - l_2^2}{2l_1l_2}, \quad s_2 = \pm\sqrt{1 - c_2^2} \tag{3} $$

Thus, manipulator pose at the initial position is $\theta_2 = -112.70, \theta_1 = 60.44$, and pose at the final position is $\theta_2 = -75.04, \theta_1 = 76.48$.

Jacobian and Hessian matrix is as follows,

$$ \begin{bmatrix} \dot{x} \\ \dot{y} \end{bmatrix} = \begin{bmatrix} -l_1s_1 - l_2s_{12} & -l_2s_{12} \\ l_1c_1 + l_2c_{12} & l_2c_{12} \end{bmatrix} \begin{bmatrix} \dot{\theta}_1 \\ \dot{\theta}_2 \end{bmatrix} \tag{4} $$

$$\left[H_{\phi\phi}^{u}\right]_{1;;}=\begin{bmatrix}\dfrac{\partial}{\partial\phi_1}\left[G_{\phi}^{u}\right]^{x}\\ \dfrac{\partial}{\partial\phi_2}\left[G_{\phi}^{u}\right]^{x}\end{bmatrix}=\begin{bmatrix}-l_1c_1-l_2c_{12} & -l_2c_{12}\\ -l_2c_{12} & -l_2c_{12}\end{bmatrix}$$

$$\left[H_{\phi\phi}^{u}\right]_{2;;}=\begin{bmatrix}\dfrac{\partial}{\partial\phi_1}\left[G_{\phi}^{u}\right]^{y}\\ \dfrac{\partial}{\partial\phi_2}\left[G_{\phi}^{u}\right]^{y}\end{bmatrix}=\begin{bmatrix}-l_1s_1-l_2s_{12} & -l_2s_{12}\\ -l_2s_{12} & -l_2s_{12}\end{bmatrix} \tag{5}$$

2.1.2.2 R-P Type Manipulator

Table.3 represents the DH table of R-P type manipulator.

	α_i	a_i	d_i	θ_i
1	0	0	0	$\theta_1+\frac{\pi}{2}$
2	$\frac{\pi}{2}$	0	d_2	θ_2
3	0	0	l_1	0

Table. 3. DH table of R-P type manipulator

Likewise, transformation matrix, position vector of End effecter, Jacobian and Hessian of R-P type can be derived.

Transformation matrix is,

$${}_{E}^{0}T=\begin{bmatrix}-s_1 & 0 & c_1 & (l_1+d)c_1\\ c_1 & 0 & s_1 & (l_1+d)s_1\\ 0 & 1 & 0 & 0\\ 0 & 0 & 0 & 1\end{bmatrix} \tag{6}$$

Position vector of End effecter

$${}^{0}P_{E-E}=\begin{bmatrix}(l_1+d)c_1\\ (l_1+d)s_1\\ 0\end{bmatrix} \tag{7}$$

Jacobian and Hessian of R-P type are as follows

$$\begin{bmatrix}\dot{x}\\ \dot{y}\end{bmatrix}=\begin{bmatrix}-(l_1+d)s_1 & c_1\\ (l_1+d)c_1 & s_1\end{bmatrix}\begin{bmatrix}\dot{\theta}_1\\ \dot{d}\end{bmatrix} \tag{8}$$

$$
\begin{aligned}
\left[H^u_{\phi\phi}\right]_{1;;} &= \begin{bmatrix} \dfrac{\partial}{\partial \phi_1}\left[G^u_\phi\right]^x \\ \dfrac{\partial}{\partial \phi_2}\left[G^u_\phi\right]^x \end{bmatrix} = \begin{bmatrix} -(l_1+d)c_1 & -s_1 \\ -c_1 & 0 \end{bmatrix} \\
\left[H^u_{\phi\phi}\right]_{2;;} &= \begin{bmatrix} \dfrac{\partial}{\partial \phi_1}\left[G^u_\phi\right]^y \\ \dfrac{\partial}{\partial \phi_2}\left[G^u_\phi\right]^y \end{bmatrix} = \begin{bmatrix} -(l_1+d)s_1 & c_1 \\ c_1 & 0 \end{bmatrix}
\end{aligned} \tag{9}
$$

2.1.3 Performance Analysis

2.1.3.1 Required torque

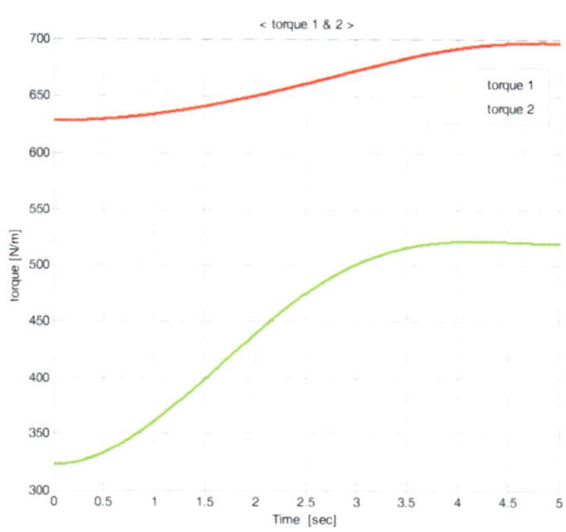 RR type	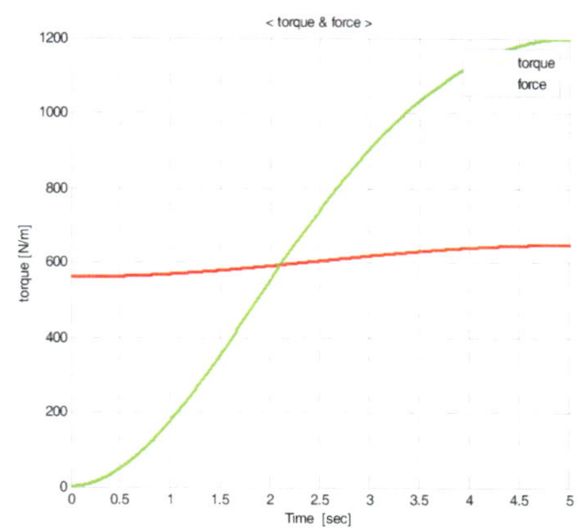 RP type
Min. Torque 1: 628.2526 Min. Torque2: 322.8128 Max. Torque 1: 697.2908 Max. Torque 2: 521.3266	Min. Torque 1: 560.9 Min. Force 2: 2.4696 Max. Torque 1: 649.8 Max. Force 2: 1199.2

Table. 4. Required torque of each type of manipulator

Required torque of each joint combination is calculated as shown in Table.4. Required torques of each joint is almost same but performance of prismatic joint of R-P type changes more rapidly.

2.1.3.2 Force Transmission Ratio (FTR)

Force transmission is important index to verify manipulator performance as changes of manipulator pose. The following equation represents force transmission ratio.[5]

$$\text{Force transmission ratio} = \frac{1}{(\text{Min. eigenvalue of Jacobian})} \tag{10}$$

$$\tau = J^T f$$

In this simulation, the reach of manipulator is increasing by time. As shown in Fig.8, R-P type has better performance when the reach of manipulator is increased. It can be observed from the simulation results that when the reach of R-R type manipulator becomes larger, the force transmission ratios become smaller. Specifically, manipulator pose are placed on last position, the force transmission ratio of R-P type becomes three times of FTR of R-R type.

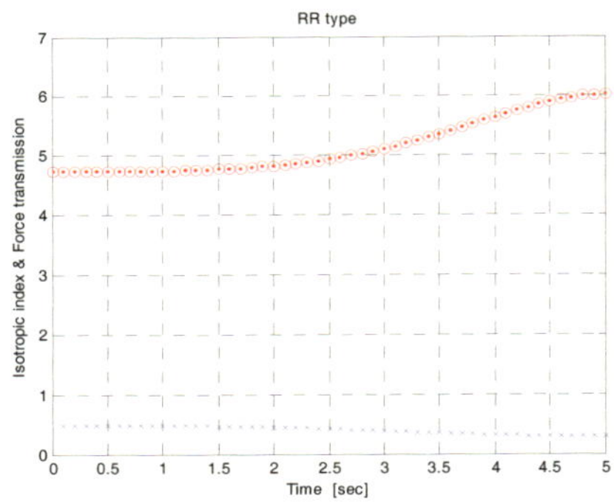

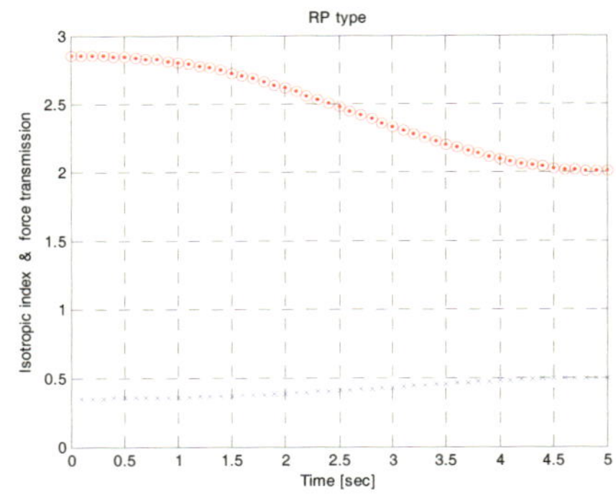

(a) R-R type (b) R-P type

Fig. 8. Force Transmission Ratio of R-R type and R-P type
(Isotropic Index: Upper line, FTR: Lower Line)

2.2 Comparison of Computation Complexities

2.2.1 Three DOF Planar manipulator

Expanding the considerations of previous chapter, combination of three joints including base frame is considered. This concept is illustrated with the 3-DOF planar and spatial robots with two revolute and one prismatic joint. This provides the most economic computational counts.

Based on the model given by eq. (11) and (12), computational complexities for the GIM and MCI of an n-link, n-revolute joint serial robot can be computed. GIM means complexities Index for Generalized Inertia Matrix and MCI means Matrix of Convective Inertia. [6]

$$GIM:(3.5n^2+11.5n-7)M(2n^2+9n-7)A \tag{11}$$

$$MCI:(7n^2+13n+4)M(4n^2+13n+2)A \tag{12}$$

Table.5 represents computation complexities of 3-DOF planar manipulator calculated by eq. (11) and (12). As shown in Table.5, R-P-R type has minimum value for the GIM

Computation and P-R-R type has minimum value for the MCI computation.

Type	GIM Computation	MCI Computation	Total
R-R-P	46M; 29A	76M; 42A	122M; 71A
R-P-R	34M; 22A	88M; 39A	122M; 61A
P-R-R	42M; 27A	53M; 29A	96M; 56A

Table. 5. Computation Complexities of 3-DOF Planar manipulator

2.2.2 Three DOF Spatial manipulator

3-DOF spatial manipulator has an eq. of GIM and MCI as follows.

$$GIM:(11n^2+34n-18)M\ (7n^2+37n-18)A \tag{13}$$

$$MCI:(14n^2+22n+4)M\ (13.5n^2+55.5n-65.5)A \tag{14}$$

Note that the GIM complexity is less than the value reported by Walker and Orin (1982),i.e. $(12n^2+56n-27)M \quad (7n^2+67n-53)A$, whereas for MCI complexity value is not available for comparison.[6] Three-DOF spatial robot arms with two revolute and one prismatic joint are shown in Fig.9(a)-(c).

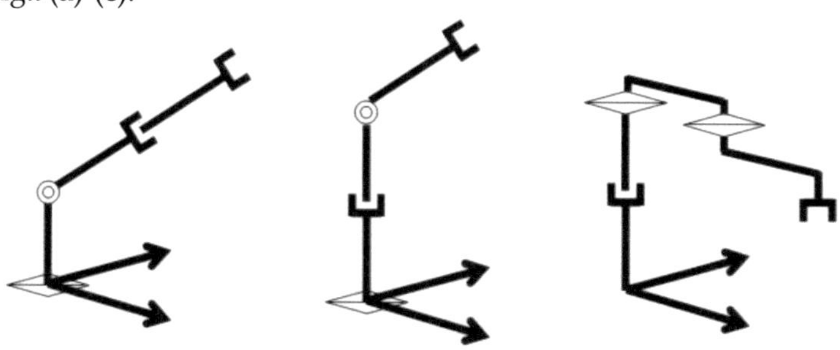

(a) R-R-P (b) R-P-R (c) P-R-R

Fig. 9. Three Types of Spatial manipulator

Two of which, namely, those in Fig.9(a) and (c) are the Stanford and RTX robot arms, respectively. While the Stanford arm has spherical workspace, RTX is of SCARA type. Computational complexities for the Generalized Inertia Matrix (GIM) and Matrix of Convective Inertia (MCI) terms for all these robot arms are tabulated in Table.6 which are based on eq. (13) and (14). Similar to the planar case, simplicity due to the orthogonal positions of the joint axes is also taken into account. Based on the total minimum computational complexity P-R-R configuration (RTX Robot), shown in Fig.9(c) has minimum computation complexities.

Type	GIM Computation	MCI Computation	Total
R-R-P	170M 154A	522M 352A	692M 506A
R-P-R	153M 148A	556M 415A	709M 583A
P-R-R	142M 138A	389M 317A	531M 455A

Table. 6. Computation Complexities of 3-DOF Spatial manipulator [6]

2.3 Comparison of Manipulability
2.3.1 Basic Formulations
Manipulability is the ability to reach a certain position or set of positions, and to change the position or orientation at a given configuration. This performance index is most general one, and it's possible to map on the workspace of a manipulator. Commonly, manipulability can be expressed following equations.[7]

$$W = \sqrt{\det(JJ^T)} \tag{15}$$

$$\begin{aligned} W &= \left|\det(J)\right| \\ &= \sqrt{\lambda_1 \lambda_2 \ldots\ldots \lambda_m} = \sigma_1 \sigma_2 \ldots\ldots \sigma_m \end{aligned} \tag{16}$$

Where, $\lambda_1 \geq \lambda_2 \geq \ldots\ldots \geq \lambda_m \geq 0$ are Eigen values of JJ^T and $\sigma_1 \geq \sigma_2 \geq \ldots\ldots \geq \sigma_m \geq 0$ $(\sigma_i = \sqrt{\lambda_i})$ are singular values of Jacobian matrix. Manipulability becomes equal to zero if and only if manipulability Space becomes less than m. That is singular point.

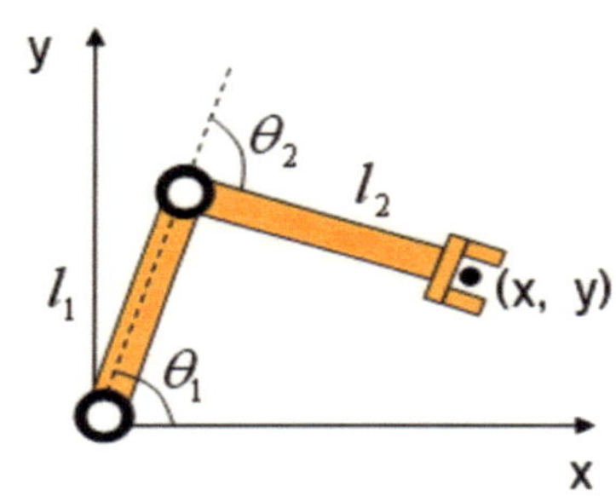

Fig. 10. Generalized 2-DOF R-R type Manipulator

Jacobian matrix of manipulator of Fig.10 is like eq. (17).

$$J(\theta) = \begin{pmatrix} l_1 c_1 + l_2 c_{12} & l_2 c_{12} \\ -l_1 s_1 - l_2 s_{12} & -l_2 s_{12} \end{pmatrix} \tag{17}$$

$$W = \left|\det J(\theta)\right| = l_1 l_2 \left|\sin\theta_2\right| \tag{18}$$

When $\theta_2 = \pm\frac{\pi}{2}$, the configuration of the Manipulator is Optimal ($l_1 + l_2 = C$; C is constant).
As a result, the optimal design of link length is $l_1 = l_2$. And optimal configuration is

$$\theta_2 = \pm\frac{\pi}{2}.$$

2.3.2 Manipulability Ellipsoid Analysis

It is natural to use the determinant of the Jacobian in a measure of manipulator dexterity. Manipulator Ellipsoid shows the manipulability of each manipulator by a shape of ellipsoid. That is, more closer to a complete circle, the manipulator has better manipulability. To calculate the manipulability ellipsoid, institute new matrices.

$$V = J(\theta)\dot{\theta} \tag{19}$$

$$J = Q_1 \Sigma Q_2^T \tag{20}$$

To verify the performance of manipulator in a workspace, generally this index is plotted to the workspace, and considers the phase of change of this index. In this study, the case which optimal link length is $l_1 = l_2$ is considered, and compared with the case which link length is different each other.

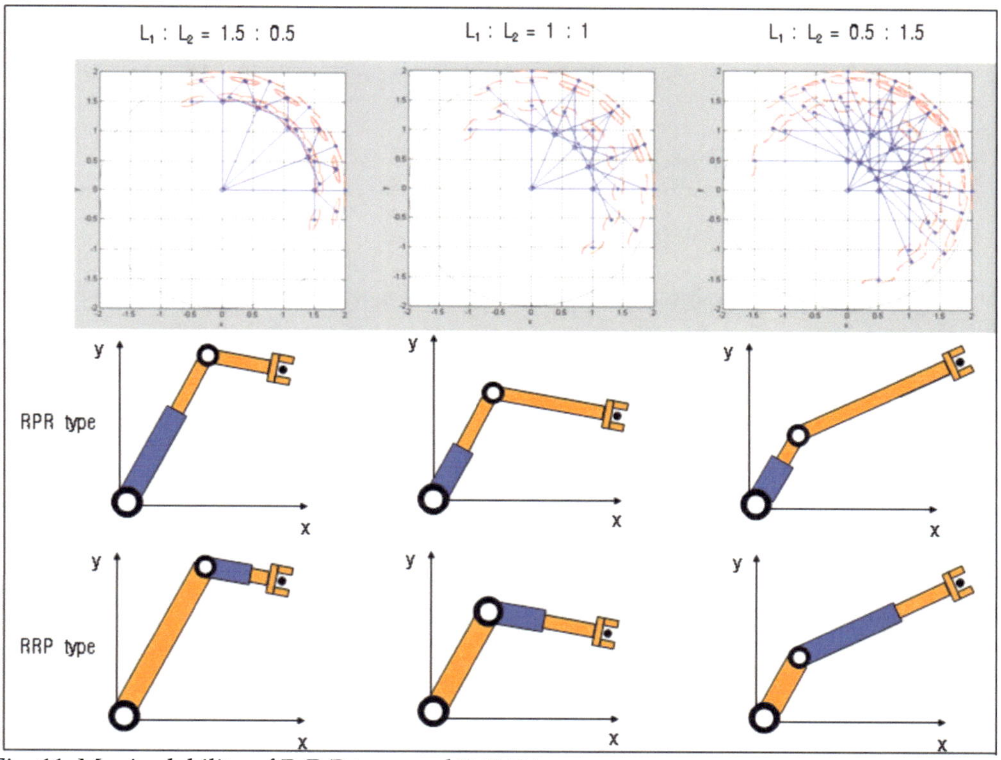

Fig. 11. Manipulability of R-P-R type and R-R-P type

Fig.11 represents manipulability ellipsoid of R-P-R and R-R-P type. First column is the case of link length 3:1 ratio, second column is the case of 1:1 ratio and last case shows the case of 1:3 ratio. This figure shows link length of lower arm becomes larger, and then the area of ellipsoid of near the outer line becomes larger. This means better design that the length of lower arm becomes larger than upper arm for the specific task which this study concerns.

2.4 Comparison of Motion Capability

2.4.1 Motion Capability of R-R-R type

The motion capability of the manipulator is defined to be the measure or volume of subset of all possible rotations and translations of rigid body. The technique that will be used is similar to that of calculating the volume of a region in R^3 by integrating the volume element $dV = dxdydz$ over the region of concern. The first application of the volume element to robotics appears in Karger (1989). It is also briefly mentioned as a possible area of study in Loncaric (1985). [8]

First, applying the volume element to a manipulator capable of both rotational and translation motion, it may be instructive to investigate the simple case of planar motion.

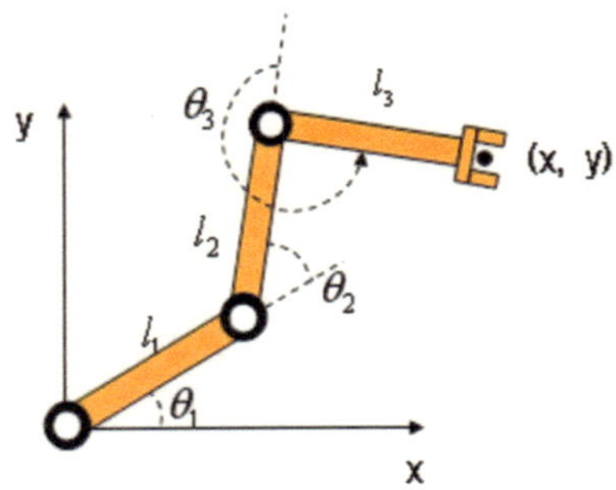

Fig. 12. R-R-R Type Planar Manipulator

If the planar R-R-R type manipulator shown in Fig.12 is considered, the problem reduces to reparameterizing the manifold $R^2 \times 2\pi$ in terms of the joint angles. This is accomplished via the forward analysis for the mechanism and yields

$$\begin{aligned} x &= l_1 \cos\theta_1 + l_2 \cos(\theta_1 + \theta_2) + l_3 \cos(\theta_1 + \theta_2 + \theta_3) \\ y &= l_1 \sin\theta_1 + l_2 \sin(\theta_1 + \theta_2) + l_3 \sin(\theta_1 + \theta_2 + \theta_3) \\ \theta &= \theta_1 + \theta_2 + \theta_3 \end{aligned} \tag{21}$$

The coordinate-induced tangent vectors are then given by

$$\begin{bmatrix} t_{\theta_1} \\ t_{\theta_2} \\ t_{\theta_3} \end{bmatrix} = \begin{bmatrix} \dfrac{\partial x}{\partial \theta_1} & \dfrac{\partial y}{\partial \theta_1} & \dfrac{\partial \theta}{\partial \theta_1} \\ \dfrac{\partial x}{\partial \theta_2} & \dfrac{\partial y}{\partial \theta_2} & \dfrac{\partial \theta}{\partial \theta_2} \\ \dfrac{\partial x}{\partial \theta_3} & \dfrac{\partial y}{\partial \theta_3} & \dfrac{\partial \theta}{\partial \theta_3} \end{bmatrix} \begin{bmatrix} t_x \\ t_y \\ t_\theta \end{bmatrix} \tag{22}$$

This equation can be described by following equation.

$$t_{\theta_1} = \begin{bmatrix} -l_1\sin(\theta_1) - l_2\sin(\theta_1+\theta_2) - l_3\sin(\theta_1+\theta_2+\theta_3) \\ l_1\cos(\theta_1) + l_2\cos(\theta_1+\theta_2) + l_3\cos(\theta_1+\theta_2+\theta_3) \\ 1 \end{bmatrix} \tag{23}$$

$$t_{\theta_2} = \begin{bmatrix} -l_2\sin(\theta_1+\theta_2) - l_3\sin(\theta_1+\theta_2+\theta_3) \\ l_2\cos(\theta_1+\theta_2) + l_3\cos(\theta_1+\theta_2+\theta_3) \\ 1 \end{bmatrix} \tag{24}$$

$$t_{\theta_3} = \begin{bmatrix} -l_3\sin(\theta_1+\theta_2+\theta_3) \\ l_3\cos(\theta_1+\theta_2+\theta_3) \\ 1 \end{bmatrix} \tag{25}$$

The magnitude of the volume of the parallelepiped formed by these tangent vectors is given by

$$\begin{vmatrix} t_{\theta_1}\cdot t_{\theta_1} & t_{\theta_1}\cdot t_{\theta_2} & t_{\theta_1}\cdot t_{\theta_3} \\ t_{\theta_2}\cdot t_{\theta_1} & t_{\theta_2}\cdot t_{\theta_2} & t_{\theta_2}\cdot t_{\theta_3} \\ t_{\theta_3}\cdot t_{\theta_1} & t_{\theta_3}\cdot t_{\theta_2} & t_{\theta_3}\cdot t_{\theta_3} \end{vmatrix}^{\frac{1}{2}} = l_1 l_2 \left|\sin\theta_2\right| \tag{26}$$

Each element of this matrix represents the standard dot product which is given by $v \cdot w = |v||w|\cos\theta$. Hence, the volume element, in terms of the joint angle parameterization, is given by

$$dV = l_1 l_2 \left|\sin\theta_2\right| d\theta_1 d\theta_2 d\theta_3 \tag{27}$$

The tangent vector represents the instantaneous motions of the end effecter. Hence the loss of one or more of these vectors represents a loss in a degree of freedom, and the volume element degenerates when the robot is in a singularity configuration. The volume, or what will be referred to as the motion capability, achieved by the robot, assuming all joints can rotate $2\pi\ radian$, is given by

$$\begin{aligned} V &= \int_{\theta_1=0}^{2\pi}\int_{\theta_2=0}^{2\pi}\int_{\theta_3=0}^{2\pi} l_1 l_2 \left|\sin\theta_2\right| d\theta_1 d\theta_2 d\theta_3 \\ &= 16\pi^2 l_1 l_2 \end{aligned} \tag{28}$$

$$V = 16\pi^2 l_1 l_2 = 4\pi \times [\pi(l_1+l_2)^2 - \pi(l_1-l_2)^2] \tag{29}$$

In the case of R-R-R type manipulator, motion capability is 4π times of physical workspace. Adding to this, consider the other types of manipulator. In addition to this type, we will consider other type of manipulator. Second type is R-P-R type (Fig.13).

2.4.2 Motion Capability of R-P-R type

Likewise, volume element and motion capability can be calculated.

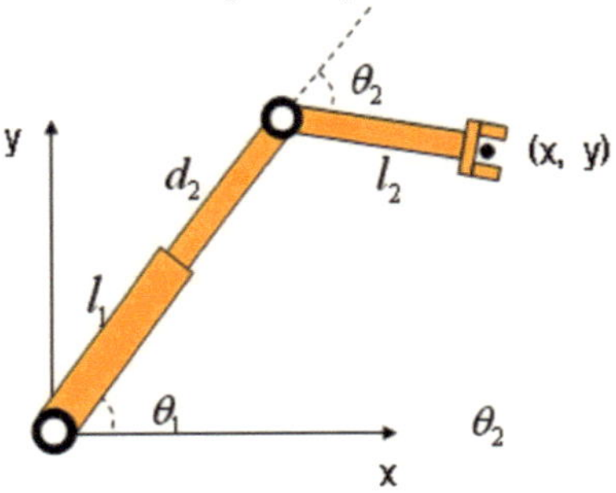

Fig. 13. R-P-R type manipulator

$$\begin{aligned} x &= l_1 \cos\theta_1 + d_1 \cos\theta_1 + l_2 \cos\theta_2 \\ y &= l_1 \sin\theta_1 + d_1 \sin\theta_1 + l_2 \sin\theta_2 \\ \theta &= \theta_1 + \theta_2 \end{aligned} \tag{30}$$

$$t_{\theta_1} = \begin{bmatrix} -(l_1 + d_1)\sin(\theta_1) \\ (l_1 + d_1)\cos(\theta_1) \\ 1 \end{bmatrix} \tag{31}$$

$$t_{\theta_2} = \begin{bmatrix} -l_2 \sin(\theta_2) \\ l_2 \cos(\theta_2) \\ 1 \end{bmatrix} \tag{32}$$

$$t_{d_1} = \begin{bmatrix} \cos\theta_1 \\ \sin\theta_1 \\ 0 \end{bmatrix} \tag{33}$$

To assume the length of each link and prismatic part is 1, the magnitude of the elements of the volume of these tangent vectors are given by

$$\begin{aligned} t_{\theta_1} \cdot t_{\theta_1} &= 4\sin^2\theta_1 + 4\cos^2\theta_1 + 1 = 5 \\ t_{\theta_1} \cdot t_{\theta_2} &= 2\sin\theta_1 \sin\theta_2 + 2\cos\theta_1 \cos\theta_2 + 1 = 2\cos(\theta_1 - \theta_2) + 1 \\ t_{\theta_1} \cdot t_{\theta_3} &= -2\sin\theta_1 \cos\theta_1 + 2\cos\theta_1 \sin\theta_1 = 0 \\ t_{\theta_2} \cdot t_{\theta_1} &= 2\sin\theta_1 \sin\theta_2 + 2\cos\theta_1 \cos\theta_2 + 1 = 2\cos(\theta_1 - \theta_2) + 1 \\ t_{\theta_2} \cdot t_{\theta_2} &= \sin^2\theta_2 + \cos^2\theta_2 + 1 = 2 \\ t_{\theta_2} \cdot t_{\theta_3} &= -\sin\theta_2 \cos\theta_1 + \cos\theta_2 \sin\theta_1 = \sin(\theta_1 - \theta_2) \end{aligned}$$

$$
\begin{aligned}
t_{\theta_3} \cdot t_{\theta_1} &= -2\sin\theta_1 \cos\theta_1 + 2\cos\theta_1 \sin\theta_1 = 0 \\
t_{\theta_3} \cdot t_{\theta_2} &= -\sin\theta_2 \cos\theta_1 + \cos\theta_2 \sin\theta_1 = \sin(\theta_1 - \theta_2) \\
t_{\theta_3} \cdot t_{\theta_3} &= \cos^2\theta_1 + \sin^2\theta_1 = 1
\end{aligned} \tag{34}
$$

The magnitude of the volume by these vectors are given by

$$
\begin{vmatrix}
t_{\theta_1} \cdot t_{\theta_1} & t_{\theta_1} \cdot t_{\theta_2} & t_{\theta_1} \cdot t_{\theta_3} \\
t_{\theta_2} \cdot t_{\theta_1} & t_{\theta_2} \cdot t_{\theta_2} & t_{\theta_2} \cdot t_{\theta_3} \\
t_{\theta_3} \cdot t_{\theta_1} & t_{\theta_3} \cdot t_{\theta_2} & t_{\theta_3} \cdot t_{\theta_3}
\end{vmatrix}^{\frac{1}{2}}
$$

$$
= \begin{vmatrix}
5 & 2\cos(\theta_1 - \theta_2) + 1 & 0 \\
2\cos(\theta_1 - \theta_2) + 1 & 2 & \sin(\theta_1 - \theta_2) \\
0 & (\theta_1 - \theta_2) & 1
\end{vmatrix}^{\frac{1}{2}} \tag{35}
$$

If $x = \cos(\theta_1 - \theta_2)$, eq.(35) can be expressed as following equations

$$
\begin{aligned}
&\sqrt{10 - 5\sin^2(\theta_1 - \theta_2) - (2\cos(\theta_1 - \theta_2) + 1)^2} \\
&= \sqrt{10 - 5(1 - x^2) - (2x + 1)^2} \\
&= \sqrt{x^2 - 4x + 4} \\
&= |x - 2| \\
&= |\cos(\theta_1 - \theta_2) - 2|
\end{aligned} \tag{36}
$$

$$
\begin{aligned}
V &= \int_0^{2\pi} \left(\int_0^{2\pi} \left(\int_0^{d1} |\cos(\theta_1 - \theta_2) - 2| \, dd1 \right) d\theta_1 \right) d\theta_2 \\
&= 8\pi^2
\end{aligned} \tag{37}
$$

2.4.3 Motion Capability of P-R-P type

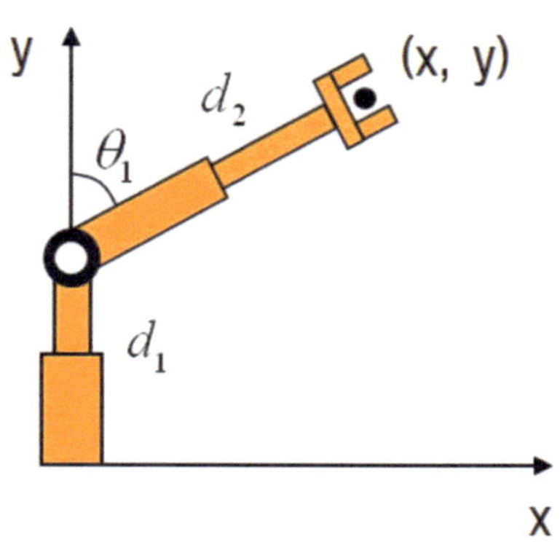

Fig. 14. P-R-P type manipulator

Finally, we consider the P-R-P type manipulator (Fig.14). This manipulator induces a parameterization of the manifold $R^2 \times 2\pi$ via the forward analysis. This yield.

$$\begin{aligned} x &= d_2 + \cos\theta_1 \\ y &= d_1 + d_2 \sin\theta_1 \\ \theta &= \theta_1 \end{aligned} \tag{38}$$

Assuming the revolute joint can rotate 2π radians, the volume element and motion capability of this manipulator is given by

$$V = \int_0^{d_1} \int_0^{d_2} \int_0^{2\pi} |\sin\theta_1| d\theta_1 ds_1 ds_2 = 4d_1 d_2$$

We can use this value of motion capabilities of each type to compare the average degree of joint. That is the relationship between this motion capability and the area of the manipulator's physical workspace. For a convenience of calculation, we assume that link length of each time is 1. First, motion capability of R-R-R type is, as stated above, 16π, and the physical workspace covers a circle of radius $l_1 + l_2 = 2$. Dividing this result by the area of the physical workspace of the robot yields

$$\frac{V_{RRR}}{\pi(2\cdot 1)^2} = 4\pi \tag{39}$$

Eq(39) shows that the R-R-R manipulator achieves a double covering its physical workspace and is able to rotate a body 2π radians about any point in this workspace. Similarly, we can calculate this value about R-P-R and P-R-P type.

$$\text{R-P-R type : } \frac{V_{RPR}}{\pi(l_1 + d_1)^2} = \frac{V_{RPR}}{\pi\cdot 2^2} = \frac{8\pi^2}{4\pi} = 2\pi \tag{40}$$

$$\text{P-R-P type : } \frac{V_{PRP}}{2l_1 l_2 + \pi l_2^2} = \frac{4l_1 l_2}{2l_1 l_2 + \pi l_2^2} = \frac{4}{2+\pi} \approx 0.78 \tag{41}$$

As a result, increasing the prismatic joint means decline of the motion capability of a manipulator.

3. Selection of Adequate Link Length Combinaion

Non-redundant planar manipulator is considered as shown in Fig.15.

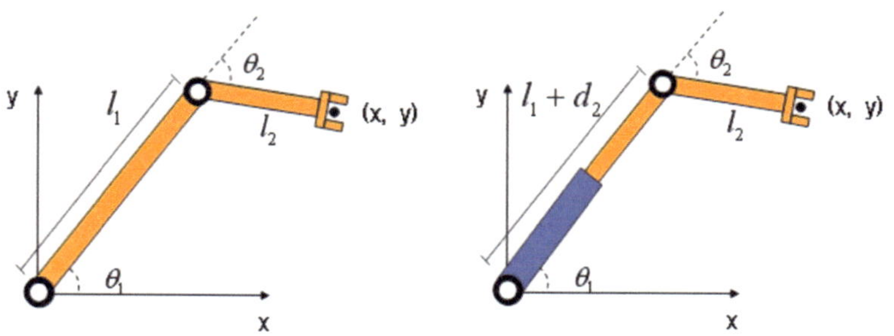

Fig. 15. Generalized planar manipulator

To deal with the simplest symbolic expressions, it is convenient to consider the Jacobian ${}^{2}J$ that correspond to task coordinates expressed in the second coordinate frame located at the manipulator end-effecter.

$$J \equiv {}^{2}J = \begin{bmatrix} l_1 \sin_2 & 0 \\ l_2 + l_1 \cos_2 & l_2 \end{bmatrix} \tag{42}$$

For this manipulator value decomposition (Golub and Van Loan 1989) of the Jacobian matrix have been derived in symbolic form (Kircanski and Boric 1993). Thus, the singular values σ_1 and σ_2 are given by

$$\sigma_{1,2} = \sqrt{\frac{l_1^2 + 2l_2^2 + 2l_1 l_2 c_2 \pm \sqrt{(l_1^2 + 2l_2^2 + 2l_1 l_2 c_2 - 4l_1^2 l_2^2 s_2^2)}}{2}} \tag{43}$$

It is obvious that $\sigma_1 \geq \sigma_2$ is always satisfied. One of the best measures of robot dexterity- condition number- is defined as the ratio of the maximum and minimum singular values. [9] Once the singular values have been derived symbolically, the condition number is also obtained symbolically as a function of the ratio between the link lengths k and the joint angle θ_2.

$$k = \frac{l_1}{l_2} \tag{44}$$

$$\mu(k, \theta_2) = \frac{\sigma_1}{\sigma_2} = \sqrt{\frac{k^2 + 2 + 2kc_2 + \sqrt{(k^2 + 2 + 2kc_2)^2 - 4k^2 s_2^2}}{k^2 + 2 + 2kc_2 - \sqrt{(k^2 + 2 + 2kc_2)^2 - 4k^2 s_2^2}}} \tag{45}$$

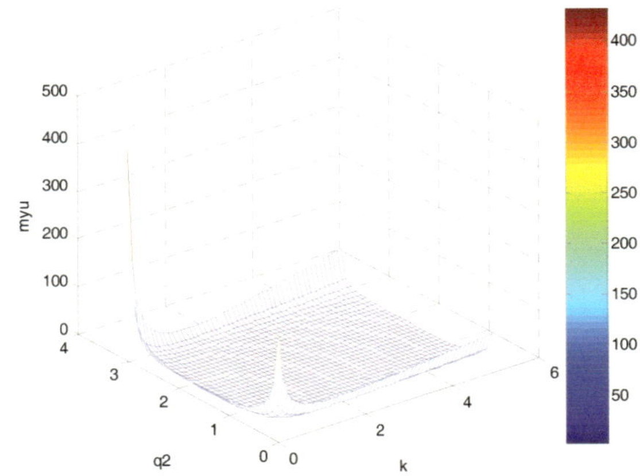

Fig. 16. Condition number versus joint angle and link length ratio

Fig.16 shows this relationship. The shaded zone at the contour graph designates the region where the condition number is less than 1.3. The isotropy condition means that the condition number has the optimal value of unity, that the two singular values are identical,

$$(k^2+2+2kc_2)^2 = 4k^2 s_2^2 \tag{46}$$

It can be satisfied only if $l_1 = l_2\sqrt{2}, \sin\theta_2 = \frac{\sqrt{2}}{2}, \cos\theta_2 = -\frac{\sqrt{2}}{2}$ (47)

The same result is obtained from the conditions that the rows of J must be mutually orthogonal and of equal lengths. Fig.17 shows the minimum condition number versus the link lengths ratio $k = l_1 / l_2$ for a family of 2-DOF planar manipulator.

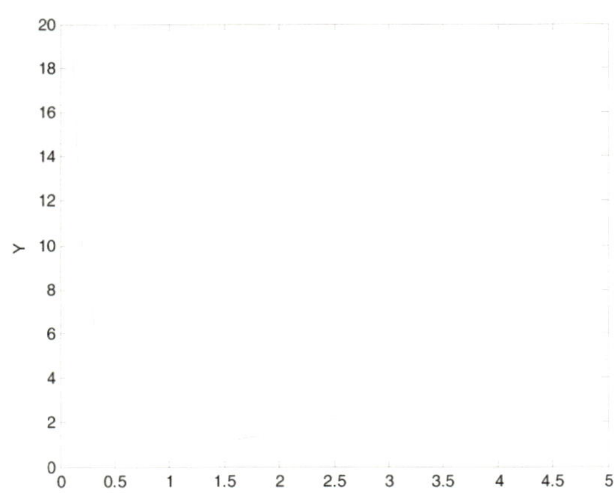

Fig. 17. The minimum condition number vs. the link lengths ratio $k = l_1 / l_2$

Fig. 17 is plotted by eq.(48)

$$\mu_{\min}(k) = \min_{q_2} \mu(k,q_2) = \sqrt{\frac{k^4+4+\left|k^2-2\right|\sqrt{k^4+4}}{k^4+4-\left|k^2-2\right|\sqrt{k^4+4}}} \quad (48)$$

Fig.18 shows the variation of the condition number with the joint angle θ_2 for several values of the link lengths ratio k. It is obvious not only that the optimal condition number of unity is achievable only if $k=\sqrt{2}$, but also that the condition number increases with θ_2 in vicinity of the isotropic configuration. Therefore, $k=\sqrt{2}$ can be a proclaimed value as the optimal 2-DOF planar manipulator design. This result, however, is opposed to the result of manipulability analysis of previous chapter. Therefore, robot designer has to choose the appropriate ratio of link lengths for his/her own purpose.

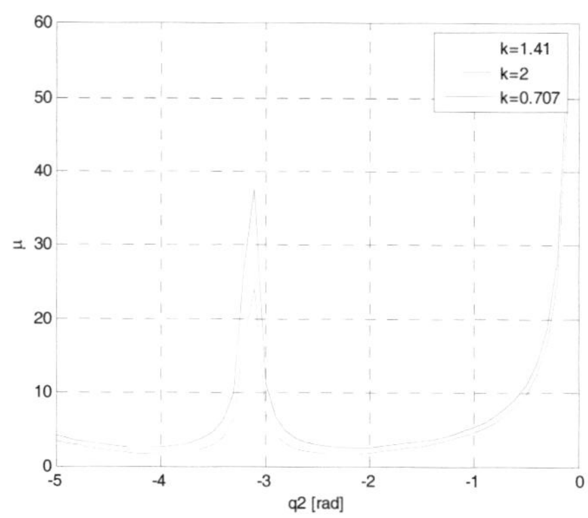

Fig. 18. The condition number with the joint angle θ_2 for several values of the link lengths ratio k

4. Conclusion

This paper introduces a design concept for a robot manipulator specific to the curtain wall installation task or any other similar task that is in a dominant task area with a full stretch, high manipulability and motion capability, etc. Fig.19 shows the results of this study. This design concept is based on the following assumptions.

1. The task was assumed to be curtain wall installation during the construction of a tall building.
2. The required DOF of the manipulator is six, with an unchangeable rotational base frame.
3. The end-effecter of the manipulator has a 3-DOF R-P-R type joint, which is also invariant.
4. The previous assumption makes this problem rather simple thus it was handled as a planar manipulator problem.

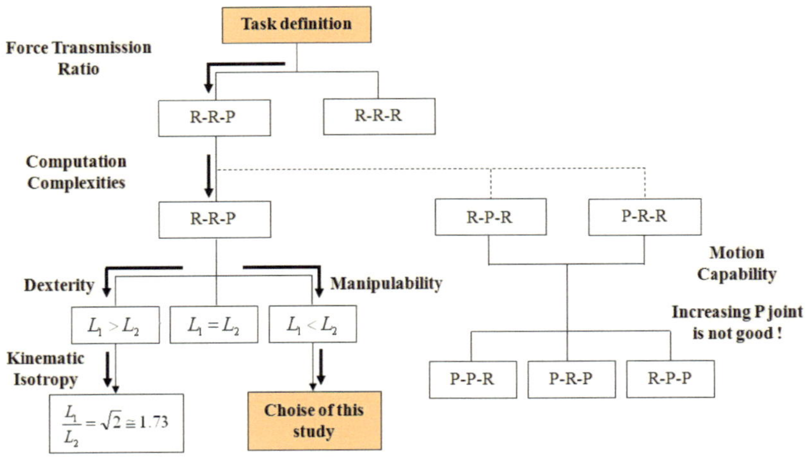

Fig. 19. Flowchart showing the decision process for determining the optimal design of a manipulator for the specified task.

The following figure illustrates the concept of a man-machine cooperation system for curtain wall installation based on the results of this study.

Fig. 20. Concept of an assistant manipulator for curtain wall installation

Acknowledgement

This work was supported by the Research Fund of Hanyang University (HYU-2009-T) and a grant from Construction Technology Innovation Program(CTIP) funded by Ministry of Land, Transportation and Maritime Affairs (MLTM),SRC/ERC program of MOST (grant #R11-2005-056-03003-0).

6. Reference

[1] SeungNam Yu, Seung Yeol Lee, Chang Soo Han, Kye Young Lee, Sang Heon Lee, "Development of the curtain wall installation robot: Performance and efficiency tests at a construction site", Autonomous Robots, Vol.22, No.3, pp.281~291, 2007

[2] Jin-u Lee, Chang-Baek Son, "An Analysis of Work Ratio for construction Curtain wall by Work Sampling Technique", The Korean Construction Society 2003, pp.519~522
[3] Seung-Nam Yu, Jong-Ho Choe, Seung-Yel Lee, Chang-Soo. Han, Kye-Young Lee, Sang-Heon Lee, "The Analysis of the Curtain Wall Installation Robot: Based on the Test in the Construction Processes", ISARC 2005, Ferrara, Italy, pp76, 2005
[4] John J. Craig, "Introduction to Robotics - Mechanics and Control, 3rd Edition", Addison Wesley, 2005
[5] Stephen L. Chiu, "Task Compatibility of Manipulator Postures", The International Journal of Robotics Research, Vol.7, No.5, 13-21, 1988
[6] Prasadkumar P. Bhangale, S. K. Saha, V. P. Agrawal, "Dynamic Model Based Selection Criterion for Robot Manipulators", Mutibody Dynamics 2003, Jorge A.C. Ambrósio(Ed.), IDMEC/IST, Lisbon, Portugal, July 1-4, 2003
[7] Tsuneo Yoshikawa, "Manipulability of Robotic Mechanisms", The international journal of Robotics Research, 1985, pp.3~4
[8] Udai Basavaraj, Joseph Duffy, "End-Effector Motion Capabilities of Serial Manipulators, The international journal of Robotics Research, 1993, pp132~137
[9] Manja Kircanski, "Kinematic Isotropy and Optimal Kinematic Design of Planar Manipulators and a 3-DOF Spatial Manipulator", The Int. Journal of Robotics Research", Vol.15, No.1, pp.61-77, 1996

33

P·SPR·D and P·SPR·D+I Control of Robot Manipulators and Redundant Manipulators

Kiyotaka Shimizu
Keio University
Japan

1. Introduction

This chapter is concerned with a PID-like control scheme for robot manipulators. We propose P·SPR·D control and P·SPR·D+I control for a set-point servo problem of the robot manipulators which are passive systems. P·SPR·D control consists of Proportional (P) action + Strict Positive Real (SPR) action + Derivative (D) action. Such control can asymptotically stabilize multi-joint robot manipulators. Stability analysis of the P·SPR·D control is made, based on the passivity theory and LaSalle's invariance principle. The L_2-gain disturbance attenuation problem is also investigated. The effectiveness of the proposed method is demonstrated by the simulation results for a two-link manipulator.

Let $\boldsymbol{u} \in R^r$ be the control input, $\boldsymbol{y} \in R^m$ the output, $\boldsymbol{r} \in R^m$ the desired value and $\boldsymbol{e} = \boldsymbol{r} - \boldsymbol{y}$ the error, then PID control is expressed as follows.

$$
\begin{aligned}
&\boldsymbol{u} = K_P\boldsymbol{e} + K_I\int_0^t \boldsymbol{e}(\tau)d\tau + K_D\dot{\boldsymbol{e}} + \boldsymbol{m}_0 \\
&i.e. \\
&\dot{\boldsymbol{z}} = \boldsymbol{e}, \ \ \boldsymbol{z}(0) = \boldsymbol{0} \\
&\boldsymbol{u} = K_P\boldsymbol{e} + K_I\boldsymbol{z} + K_D\dot{\boldsymbol{e}} + \boldsymbol{m}_0
\end{aligned}
$$

where $K_P, K_I, K_D \in R^{r\times m}$ are gain matrices corresponding to Proportional, Integral and Derivative action, respectively, and $\boldsymbol{m}_0$ denotes the so-called manual reset quantity.

We propose the following P·SPR·D control (Shimizu, 2009a) in which a SPR (strict positive real) element is used instead of Integral element:

$$
\begin{aligned}
&\dot{\boldsymbol{\xi}} = D\boldsymbol{\xi} + \boldsymbol{e} + \dot{\boldsymbol{e}}, \ \ \boldsymbol{\xi}(0) = \boldsymbol{0}, \ \ D < 0 \\
&\boldsymbol{u} = K_P\boldsymbol{e} + K_S\boldsymbol{\xi} + K_D\dot{\boldsymbol{e}} + \boldsymbol{m}_0
\end{aligned}
$$

In Section 2 we study stability analysis of the P·SPR·D control imitating the PID control for a set-point servo problem of multi-joint manipulator systems.

In regard to PD and PID control for robot manipulators, there exist many papers including (Arimoto, 1996; Arimoto & Miyazaki, 1984; Spong et al., 1992), etc.. So the feature of our method is to apply the P·SPR·D control instead of PID. By introducing the SPR element it has a merit that a design of passivity-based control becomes very simplified.

When the P·SPR·D control is applied to a plant possessing the Kalman-Yakubovich-Popov (K-Y-P) property (Byrnes et al., 1991), we can prove that the closed-loop system becomes asymptotically stable by the P·SPR·D control, applying the passivity theory and LaSalle's theorem (LaSalle & Lefschetz, 1961). The reader may refer to (Shimizu, 2008a;b; 2009b) with respect to the researches of P·SPR·D control for affine nonlinear systems and mechanical systems. SPR stabilization of mechanical system is discussed in (Lozano et al., 2000) also. The reader may refer to (Byrnes et al., 1991; Khalil, 2002; Lozano et al., 2000; van der Schaft, 2000) about passivity-based control theory in general.
By the way, static state feedback control law may be obtained by the passivity based design (Sepulchre et al., 1997; Shen, 2004) of the cascaded system also. Generally speaking, however, the control law using a storage function is complex. Besides, an advantage of the P·SPR·D control is of output feedback of simple structure.
In Section 3 an extension to redundant manipulators is investigated. Control of the redundant manipulator in the task space was studied in (Arimoto, 1996; Galicki, 2008; Khatib, 1987; Murakami et al., 2008; Shibata, 2007; Spong et al., 1992).
Section 4 investigates L_2-gain disturbance attenuation problem (γ-dissipativity (van der Schaft, 2000)) under the existince of disturbances. It is easy to solve the problem by applying the P·SPR·D control.
The simulation results is presented in Section 5 to demonstrate the effectiveness of the proposed methods.

2. P·SPR·D Control of Robot Manipulators

We consider a set-point servo problem for robot manipulators. An equation of motion of robot manipulator with n joints can be obtained by the Euler-Lagrange formulation. Let $\boldsymbol{q}$ be the position (angles of each link) of the manipulator, $\boldsymbol{\tau}$ the input torque, $\frac{1}{2}\dot{\boldsymbol{q}}^T M(\boldsymbol{q})\dot{\boldsymbol{q}}$ the kinetic energy and $U(\boldsymbol{q})$ the potential energy. Then it can be represented as (Arimoto, 1996; Spong et al., 1992)

$$M(\boldsymbol{q})\ddot{\boldsymbol{q}} + \frac{1}{2}\dot{M}(\boldsymbol{q})\dot{\boldsymbol{q}} + S(\boldsymbol{q},\dot{\boldsymbol{q}})\dot{\boldsymbol{q}} + \boldsymbol{g}(\boldsymbol{q}) = \boldsymbol{\tau} \tag{1}$$

where $M(\boldsymbol{q})$ denotes the inertia matrix which is positive definite and bounded, $\boldsymbol{g}(\boldsymbol{q}) \stackrel{\triangle}{=} U_q(\boldsymbol{q})^T$ is the gradient of the gravity potential energy and $S(\boldsymbol{q},\dot{\boldsymbol{q}})$ denotes

$$S(\boldsymbol{q},\dot{\boldsymbol{q}})\dot{\boldsymbol{q}} = \frac{1}{2}\left\{\dot{M}(\boldsymbol{q})\dot{\boldsymbol{q}} - \left[\frac{\partial}{\partial \boldsymbol{q}}\boldsymbol{q}^T M(\boldsymbol{q})\dot{\boldsymbol{q}}\right]^T\right\}$$

, which is a skew-symmetric matrix. Denoting $\boldsymbol{x}_1 = \boldsymbol{q} \in R^n$, $\boldsymbol{x}_2 = \dot{\boldsymbol{q}} \in R^n$, $\boldsymbol{x} = (\boldsymbol{x}_1^T, \boldsymbol{x}_2^T)^T$, and letting the output by $\boldsymbol{y} = \boldsymbol{x}_2 \in R^n$, and the control input by $\boldsymbol{\tau} \in R^n$, state space representation of (1) becomes as follows.

$$\dot{\boldsymbol{x}}_1 = \boldsymbol{x}_2 \tag{2a}$$

$$\begin{aligned}\dot{\boldsymbol{x}}_2 &= -M(\boldsymbol{x}_1)^{-1}\left\{\frac{1}{2}\dot{M}(\boldsymbol{x}_1)\boldsymbol{x}_2 + S(\boldsymbol{x}_1,\boldsymbol{x}_2)\boldsymbol{x}_2 + \boldsymbol{g}(\boldsymbol{x}_1)\right\} + M(\boldsymbol{x}_1)^{-1}\boldsymbol{\tau} \\ &\stackrel{\triangle}{=} \boldsymbol{f}_2(\boldsymbol{x}_1,\boldsymbol{x}_2) + G_2(\boldsymbol{x}_1)\boldsymbol{\tau}\end{aligned} \tag{2b}$$

$$y = x_2 \tag{3}$$

Now taking a storage function equal to the kinetic energy + the potential energy as

$$W(x) = \frac{1}{2}x_2^T M(x_1)x_2 + U(x_1) - U(x_1^*) \tag{4}$$

we calculate its time derivative with use of skew-symmetricity of $S(x_1, x_2)$ to obtain

$$\begin{aligned}\dot{W}(x) &= \frac{\partial}{\partial x_1}\left\{\frac{1}{2}x_2^T M(x_1)x_2\right\}\dot{x}_1 + \frac{\partial}{\partial x_2}\left\{\frac{1}{2}x_2^T M(x_1)x_2\right\}\dot{x}_2 + \frac{\partial U(x_1)}{\partial x_1}\dot{x}_1 \\ &= \frac{1}{2}x_2^T \dot{M}(x_1)x_2 + x_2^T\{-\frac{1}{2}\dot{M}(x_1)x_2 - S(x_1,x_2)x_2 - g(x_1) + \tau\} + g(x_1)^T x_2 \\ &\le y^T\tau \end{aligned} \tag{5}$$

Therefore, the robot manipulator is passive with respect to the input τ and the output $y = x_2$ (Arimoto, 1996). Thus, the so-called K-Y-P property holds :

$$W_{x_1}(x)x_2 + W_{x_2}(x)f_2(x_1, x_2) \le 0 \tag{6a}$$
$$W_{x_2}(x)G_2(x_1) = y^T \tag{6b}$$

Here let us consider a set-point servo problem (a set-point tracking control) with the desired value $(x_1, x_2) = (x_1^*, \mathbf{0})$. For that we consider the following system which consists of the robot manipulator (2),(3) and a SPR (strict positive real) element (8).

$$\dot{x}_1 = x_2 \tag{7a}$$

$$\begin{aligned}\dot{x}_2 &= -M(x_1)^{-1}\left\{\frac{1}{2}\dot{M}(x_1)x_2 + S(x_1,x_2)x_2 + g(x_1)\right\} + M(x_1)^{-1}\tau \\ &\triangleq f_2(x_1,x_2) + G_2(x_1)\tau\end{aligned} \tag{7b}$$

$$\dot{\xi} = D\xi + (x_1^* - x_1) - x_2, \;\; \xi(0) = \mathbf{0}, \;\; D < 0 \tag{8}$$

$$y = x_2 \tag{9}$$

And set up a feedback compensator (P·SPR·D controller) :

$$\tau = K_P(x_1^* - x_1) + K_S\xi - K_D x_2 + g(x_1^*) \tag{10}$$

where $K_P, K_S, K_D \in R^{n\times n}$ are all gain matrices being positive definite and diagonal. Here $g(x_1^*)$, gravity force compensation at the desired value x_1^*, corresponds to the so-called manual reset quantity of PID control.
We have the following theorem.
[Theorem 1] The closed-loop system (7)∼(10) of the robot manipulator with the P·SPR·D control is asymptotically stable at the equilibrium $(x_{1e}, x_{2e}, \xi_e) = (x_1^*, \mathbf{0}, \mathbf{0})$, provided that positive definite diagonal matrices $K_P, K_S, K_D \in R^{n\times n}$ and negative definite diagonal $D \in R^{n\times n}$ are appropriately chosen.

(Proof) At the equilibrium of system (7),(8),(10) hold the following relations.

$$\begin{aligned} \mathbf{0} &= x_{2e} \\ \mathbf{0} &= -g(x_{1e}) + \tau_e \\ \mathbf{0} &= D\xi_e + (x_1^* - x_{1e}) \end{aligned}$$

Thus it follows that $(x_{1e}, x_{2e}, \xi_e) = (x_1^*, \mathbf{0}, \mathbf{0})$ is an equilibrium point, provided that $\tau_e = g(x_1^*)$.
Now let us consider a Lyapunov function candidate

$$V(x,\xi) = W(x) + g(x_1^*)^T(x_1^* - x_1) + \frac{1}{2}\begin{bmatrix}(x_1^* - x_1)\\ \xi\end{bmatrix}^T\begin{bmatrix}K_P - \overline{K} & \overline{K}\\ \overline{K}^T & K_S - \overline{K}\end{bmatrix}\begin{bmatrix}(x_1^* - x_1)\\ \xi\end{bmatrix} \quad (11)$$

where $K_P - \overline{K} > 0,\ K_S - \overline{K} > 0$ and $\begin{bmatrix}K_P - \overline{K} & \overline{K}\\ \overline{K}^T & K_S - \overline{K}\end{bmatrix}$ is a positive definite matrix. The first term in the right-hand side of (11) is a semi-positive definite function. Since the second term plus the third one is a quadrtic function of $\begin{bmatrix}(x_1^* - x_1)\\ \xi\end{bmatrix}$ whose quadratic term is with the positive definite matrix, it has the minimum. Accordingly, $V(x,\xi)$ is a function bounded below.
Next calculate its time derivative along (7)∼(10) using the K-Y-P property (6) to get

$$\begin{aligned}
\dot{V}(x,\xi) &= W_{x_1}(x)x_2 + W_{x_2}(x)f_2(x_1,x_2) + W_{x_2}(x)G_2(x_1)\tau - g(x_1^*)^T x_2 \\
&\quad + \begin{bmatrix}(x_1^* - x_1)\\ \xi\end{bmatrix}^T\begin{bmatrix}K_P - \overline{K} & \overline{K}\\ \overline{K}^T & K_S - \overline{K}\end{bmatrix}\begin{bmatrix}-\dot{x}_1\\ \dot{\xi}\end{bmatrix} \\
&\le y^T\tau - g(x_1^*)^T x_2 + \begin{bmatrix}(x_1^* - x_1)\\ \xi\end{bmatrix}^T\begin{bmatrix}K_P - \overline{K} & \overline{K}\\ \overline{K}^T & K_S - \overline{K}\end{bmatrix}\begin{bmatrix}-x_2\\ D\xi + (x_1^* - x_1) - x_2\end{bmatrix} \\
&= x_2^T(K_P(x_1^* - x_1) + K_S\xi - K_D x_2 + g(x_1^*)) - g(x_1^*)^T x_2 \\
&\quad + \begin{bmatrix}(x_1^* - x_1)\\ \xi\end{bmatrix}^T\begin{bmatrix}-(K_P - \overline{K})x_2 + \overline{K}D\xi + \overline{K}(x_1^* - x_1) - \overline{K}x_2\\ -\overline{K}^T x_2 + (K_S - \overline{K})D\xi + (K_S - \overline{K})(x_1^* - x_1) - (K_S - \overline{K})x_2\end{bmatrix} \\
&= x_2^T(K_P(x_1^* - x_1) + K_S\xi - K_D x_2) + \begin{bmatrix}(x_1^* - x_1)\\ \xi\end{bmatrix}^T\begin{bmatrix}\overline{K} & \overline{K}D\\ (K_S - \overline{K}) & (K_S - \overline{K})D\end{bmatrix}\begin{bmatrix}(x_1^* - x_1)\\ \xi\end{bmatrix} \\
&\quad - (x_1^* - x_1)^T K_P x_2 - \xi^T K_S x_2 \\
&= -x_2^T K_D x_2 + \begin{bmatrix}(x_1^* - x_1)\\ \xi\end{bmatrix}^T\begin{bmatrix}\overline{K} & \overline{K}D\\ (K_S - \overline{K}) & (K_S - \overline{K})D\end{bmatrix}\begin{bmatrix}(x_1^* - x_1)\\ \xi\end{bmatrix} \qquad (12)
\end{aligned}$$

Here we try to make

$$\begin{bmatrix}\overline{K} & \overline{K}D\\ (K_S - \overline{K}) & (K_S - \overline{K})D\end{bmatrix}$$

be negative definite. For that purpose, set $\overline{K} < 0,\ K_S - \overline{K} = (\overline{K}D)^T$ and $D < -I$ such that we have $K_S = (I + D)\overline{K} > 0$. Then the above matrix becomes

$$\begin{bmatrix}\overline{K} & \overline{K}D\\ (\overline{K}D)^T & \overline{K}D^2\end{bmatrix}$$

Since the (1,1) element and the (2,2) element are $\overline{K} < 0$, $\overline{K}D^2 < 0$, respectively, we can choose $\overline{K} < 0$ and diagonal $D < 0$ such that the above matrix becomes negative definite. This can be concluded from the Schur complement also.
Consequently, $\dot{V}(x, \xi)$ becomes semi-negative definite, and it follows that the P·SPR·D control is stable in the sense of Lyapunov, but it is unknown if asymptotically stable. So we apply LaSalle's invariance principle (LaSalle & Lefschetz, 1961) as below.
Let $\Omega_c = \{(x, \xi) \mid V(x, \xi) \leq c\}$ and suppose that Ω_c is bounded and $\dot{V}(x, \xi) \leq 0$ in Ω_c (c is a positive number such that $\dot{V}(x, \xi) \leq 0$). Here define Ω_E as a set of all points of Ω_c satisfying $\dot{V}(x, \xi) = 0$ and put

$$\Omega_E = \{(x, \xi) \mid \dot{V}(x, \xi) = 0, \ (x, \xi) \in \Omega_c\}$$

From (12) (x, ξ) satisfying $\dot{V}(x, \xi) = 0$ is given as $\boldsymbol{x}_2 = \boldsymbol{0}, \ \boldsymbol{x}_1^* - \boldsymbol{x}_1 = \boldsymbol{0}, \ \boldsymbol{\xi} = \boldsymbol{0}$. So we have

$$\Omega_E = \{(x, \xi) | \boldsymbol{x}_1 = \boldsymbol{x}_1^*, \boldsymbol{x}_2 = \boldsymbol{0}, \boldsymbol{\xi} = \boldsymbol{0}, \ (x, \xi) \in \Omega_c\}$$

Accordingly, we know from (7),(8),(10) that (x, ξ) in Ω_E consists of only the equilibrium point $(\boldsymbol{x}_{1e}, \boldsymbol{x}_{2e}, \boldsymbol{\xi}_e) = (\boldsymbol{x}_1^*, \boldsymbol{0}, \boldsymbol{0})$ with $\boldsymbol{\tau}_e = \boldsymbol{g}(\boldsymbol{x}_1^*)$. Thus the largest invariance set Ω_M in Ω_E consists of the equilibrium point $(\boldsymbol{x}_{1e}, \boldsymbol{x}_{2e}, \boldsymbol{\xi}_e) = (\boldsymbol{x}_1^*, \boldsymbol{0}, \boldsymbol{0})$. Therefore, by LaSalle's invariance principle all trajectories in Ω_c converges to Ω_M as $t \to \infty$. Thus $(\boldsymbol{x}_1^*, \boldsymbol{0}, \boldsymbol{0})$ is aymptotically stable. Namely, it is achieved that $\boldsymbol{x}_1(t) \to \boldsymbol{x}_1^*, \ \boldsymbol{x}_2(t) \to \boldsymbol{0}, \ \boldsymbol{\xi}(t) \to \boldsymbol{0}$, as $t \to \infty$. □

[Remark 1] It is well-known (Khalil, 2002) that if affine nonlinear system is passive and zero state detectable, then the output feedback control $\boldsymbol{u} = -K\boldsymbol{y}, \ K > 0$ asymptotically stabilizes an equilibrium point $\boldsymbol{x}_e = \boldsymbol{0}$. Since the robot manipulator is not zero state detectable, however, one cannot apply this well-known fact to asymptotical stabilization to the origin. In order to stabilize the origin $(\boldsymbol{x}_1, \boldsymbol{x}_2) = (\boldsymbol{0}, \boldsymbol{0})$, one must apply Theorem 1 letting $\boldsymbol{x}_1^* = \boldsymbol{0}$.

[Remark 2] P·SPR·D control of affine nonlinear systems is investigated in (Shimizu, 2008a; 2009b). Its asymptotical stability is proved under the assumption of passivity and zero state detectability.

[Remark 3] Although we consider only a rigid robot manipulator in this chapter, elastic joint robot arm is studied in (Shimizu, 2009b; Spong et al., 1992).

Local asymptotical stability of PID control for the robot manipulator was first proved by (Arimoto, 1996; Arimoto & Miyazaki, 1984). For comparison with the P·SPR·D control, its proof based on the K-Y-P property is given in Appendix.

It is well-known (Arimoto, 1996) that PD control + gravity force compensation yields superior control performance. However, in case where the gravity force compensation $\boldsymbol{g}(\boldsymbol{x}_1^*)$ at the desired value $\boldsymbol{x}_1^*$ is not available, we can consider the following P·SPR·D+I control instead of (10).

$$\boldsymbol{\tau} = K_P(\boldsymbol{x}_1^* - \boldsymbol{x}_1) + K_S\boldsymbol{\xi} - K_D\boldsymbol{x}_2 + K_I \int_0^t (\boldsymbol{x}_1^* - \boldsymbol{x}_1(\tau))d\tau \tag{13}$$

i.e.

$$\dot{\boldsymbol{\xi}} = D\boldsymbol{\xi} + (\boldsymbol{x}_1^* - \boldsymbol{x}_1) - \boldsymbol{x}_2, \ \boldsymbol{\xi}(0) = \boldsymbol{0}, \ D < 0$$
$$\dot{\boldsymbol{z}} = \boldsymbol{x}_1^* - \boldsymbol{x}_1, \ \boldsymbol{z}(0) = \boldsymbol{0}$$
$$\boldsymbol{u} = K_P(\boldsymbol{x}_1^* - \boldsymbol{x}_1) + K_S\boldsymbol{\xi} - K_D\dot{\boldsymbol{x}}_2 + K_I\boldsymbol{z}$$

Since the stability of transient state is sufficiently guaranteed by the P·SPR·D control, we devise here only a counterplan to remove a steady state error (an off-set). Of course, the P·SPR·D+I control is inferior to the P·SPR·D control with the gravity force compensation. Yet sufficiently satisfactory control performance can be obtained.

3. P·SPR·D Control of Redundant Manipulators

Robot manipulators with multi-freedom, the so-called redundant manipulators, can perform complex and flexible operation utilizing the redundancy. Stability analysis of robot manipulators should be made basically in the joint-space coordinates. But actual robot manipulators aim to control direct motion in the task-space. Therefore, it is more convinient for the robot task to represent a model in the task-space showing a manipulator end-point position rather than a model in the joint-space. If joint angles $\boldsymbol{q}^*$ corresponding to the desired target position $\boldsymbol{z}^o$ in the task-space can be accurately calculated from inverse kinematics, one may consider stabilization only in the joint-space.
For the redundant manipulators, however, the joint angles $\boldsymbol{q}^*$ corresponding to the target position $\boldsymbol{z}^o$ can not be determined uniquely and in addition calculation of inverse kinematics is usually complex and inaccurate. Thus, from a view-point of practice a stable control scheme based on the task-space plus joint-space coodinates is very desirable.
Now let us consider a multi-joint redundant manipulator with n links. Let $\boldsymbol{z} \in R^p$ $(p < n)$ be the end-point position vector in the task-space. Then one has a relation from the kinematics as (Arimoto, 1996; Khatib, 1987; Spong et al., 1992)

$$\boldsymbol{z} = \boldsymbol{f}(\boldsymbol{q}) = \boldsymbol{f}(\boldsymbol{x}_1) \tag{14}$$

$$\dot{\boldsymbol{z}} = \frac{\partial \boldsymbol{f}(\boldsymbol{x}_1)}{\partial \boldsymbol{x}_1}\dot{\boldsymbol{x}}_1 \stackrel{\triangle}{=} J(\boldsymbol{x}_1)\boldsymbol{x}_2 \tag{15}$$

It is easy to calculate forward kinematics $\boldsymbol{q} \mapsto \boldsymbol{z}$, but hard to calculate inverse one $\boldsymbol{z} \mapsto \boldsymbol{q}$. Namely, given the desired position coordinates $\boldsymbol{z}^o$, it is very difficult to determine the joint coordinates $\boldsymbol{x}_1^* = \boldsymbol{q}^*$ realizing $\boldsymbol{z}^o$, as the degree of freedom of redundancy is large.
In order to achieve accurate end-point position control, it is desired to obtain a control method for realizing $\boldsymbol{z}^o$, even when the inverse transformation $\boldsymbol{x}_1^* = \boldsymbol{f}^{-1}(\boldsymbol{z}^o)$ may not be attained correctly. For that purpose we propose a stabilizing control method for the end-point position setting, combining the P·SPR·D control in the joint-space and the P·SPR one in the task-space. Namely, we add the following P·SPR control in the task-space to the P·SPR·D one in the joint-space.

$$\dot{\boldsymbol{z}} = J(\boldsymbol{x}_1)\boldsymbol{x}_2 \tag{16}$$

$$\dot{\boldsymbol{\eta}} = D'\boldsymbol{\eta} + (\boldsymbol{z}^o - \boldsymbol{z}) - J(\boldsymbol{x}_1)\boldsymbol{x}_2,\ \boldsymbol{\eta}(0) = \boldsymbol{0},\ D' < 0 \tag{17}$$

$$\boldsymbol{\tau}' = J(\boldsymbol{x}_1)^T K_P'(\boldsymbol{z}^o - \boldsymbol{z}) + J(\boldsymbol{x}_1)^T K_S'\boldsymbol{\eta} \tag{18}$$

where $K_P', K_S' \in R^{p\times p}$ are positive definite diagonal gain mtrices. Therefore, actual control input becomes in consideration of task-space coordinates as follows.

$$\begin{aligned}\boldsymbol{\tau} &= K_P(\boldsymbol{x}_1^* - \boldsymbol{x}_1) + K_S\boldsymbol{\xi} - K_D\boldsymbol{x}_2 + \boldsymbol{g}(\boldsymbol{x}_1^*) + \boldsymbol{\tau}' \\ &= K_P(\boldsymbol{x}_1^* - \boldsymbol{x}_1) + K_S\boldsymbol{\xi} - K_D\boldsymbol{x}_2 + \boldsymbol{g}(\boldsymbol{x}_1^*) + J(\boldsymbol{x}_1)^T K_P'(\boldsymbol{z}^o - \boldsymbol{z}) + J(\boldsymbol{x}_1)^T K_S'\boldsymbol{\eta}\end{aligned} \tag{19}$$

where the set-point x_1^* denotes the joint angles corresponding to the desired position z^o, which is determined from the inverse kinematics. It is not unique, however. Then the following theorem holds.

[Theorem 2] The closed-loop system (7)∼(9),(16),(17),(19) of the redundant manipulator with P·SPR·D control is asymptotically stable at the equilibrium $(x_{1e}, x_{2e}, \xi_e, z_e, \eta_e) = (x_1^*, 0, 0, z^o, 0)$, provided that positive definite diagonal matrices $K_P, K_S, K_D \in R^{n\times n}, K_P', K_S' \in R^{p\times p}$ and negative definite diagonal matrices $D \in R^{n\times n}, D' \in R^{p\times p}$ are appropriately chosen.

(Proof) For simplicity of description, we will state only a part to be added to the proof of Theorem 1.

Consider a Lyapunov functiion candidate for the overall system

$$V_{total}(x,\xi,z,\eta) = V(x,\xi) + V'(z,\eta) \tag{20}$$

where $V(x,\xi)$ is given by (11) and $V'(z,\eta)$ denotes a Lyapunov function candidate corresponding to the additional part (16) ∼ (18) ;

$$V'(z,\eta) = \frac{1}{2}\begin{bmatrix}(z^o - z)\\ \eta\end{bmatrix}^T \begin{bmatrix}K_P' - \overline{K}' & \overline{K}'\\ \overline{K}'^T & K_S' - \overline{K}'\end{bmatrix}\begin{bmatrix}(z^o - z)\\ \eta\end{bmatrix} \tag{21}$$

where $K_P' - \overline{K}' > 0$, $K_S' - \overline{K}' > 0$ and $\begin{bmatrix}K_P' - \overline{K}' & \overline{K}'\\ \overline{K}'^T & K_S' - \overline{K}'\end{bmatrix}$ is a positive definite matrix.

Next calculate a time derivative of (20) along (7)∼ (9),(16),(17),(19) to get

$$\dot{V}_{total}(x,\xi,z,\eta) = \dot{V}(x,\xi) + \dot{V}'(z,\eta) \tag{22}$$

But $\dot{V}(x,\xi)$ has been evaluated by (12) except for a part of $y^T\tau'$, already, so we calculate only the remained part as follows.

$$\begin{aligned}
& y^T\tau' + \dot{V}'(z,\eta)\\
& \le y^T\tau' + \begin{bmatrix}(z^o - z)\\ \eta\end{bmatrix}^T \begin{bmatrix}K_P' - \overline{K}' & \overline{K}'\\ \overline{K}'^T & K_S' - \overline{K}'\end{bmatrix}\begin{bmatrix}(\dot{z}^o - \dot{z})\\ \dot{\eta}\end{bmatrix}\\
& = x_2^T\tau' + \begin{bmatrix}(z^o - z)\\ \eta\end{bmatrix}^T \begin{bmatrix}K_P' - \overline{K}' & \overline{K}'\\ \overline{K}'^T & K_S' - \overline{K}'\end{bmatrix}\begin{bmatrix}-J(x_2)x_2\\ D'\eta + (z^o - z) - J(x_1)x_2\end{bmatrix}\\
& = x_2^T(J(x_1)^T K_P'(z^o - z) + J(x_1)^T K_S'\eta)\\
& \quad + \begin{bmatrix}(z^o - z)\\ \eta\end{bmatrix}^T \begin{bmatrix}-(K_P' - \overline{K}')J(x_1)x_2 + \overline{K}'D'\eta + \overline{K}'(z^o - z) - \overline{K}'J(x_1)x_2\\ -\overline{K}'^T J(x_1)x_2 + (K_S' - \overline{K}')D'\eta + (K_S' - \overline{K}')(z^o - z) - (K_S' - \overline{K}')J(x_1)x_2\end{bmatrix}\\
& = x_2^T(J(x_1)^T K_P'(z^o - z) + J(x_1)^T K_S'\eta) + \begin{bmatrix}(z^o - z)\\ \eta\end{bmatrix}^T \begin{bmatrix}\overline{K}' & \overline{K}'D'\\ (K_S' - \overline{K}') & (K_S' - \overline{K}')D'\end{bmatrix}\begin{bmatrix}(z^o - z)\\ \eta\end{bmatrix}\\
& -(z^o - z)^T K_P' J(x_1)x_2 - \eta^T K_S' J(x_1)x_2\\
& = \begin{bmatrix}(z^o - z)\\ \eta\end{bmatrix}^T \begin{bmatrix}\overline{K}' & \overline{K}'D'\\ (K_S' - \overline{K}') & (K_S' - \overline{K}')D'\end{bmatrix}\begin{bmatrix}(z^o - z)\\ \eta\end{bmatrix}
\end{aligned} \tag{23}$$

Therefore, combining (12) and (23), we have

$$\dot{V}_{total}(x,\xi,z,\eta) \leq -x_2^T K_D x_2 + \begin{bmatrix} (x_1^* - x_1) \\ \xi \end{bmatrix}^T \begin{bmatrix} \overline{K} & \overline{K}D \\ (K_S - \overline{K}) & (K_S - \overline{K})D \end{bmatrix} \begin{bmatrix} (x_1^* - x_1) \\ \xi \end{bmatrix}$$
$$+ \begin{bmatrix} (z^o - z) \\ \eta \end{bmatrix}^T \begin{bmatrix} \overline{K}' & \overline{K}'D' \\ (K_S' - \overline{K}') & (K_S' - \overline{K}')D' \end{bmatrix} \begin{bmatrix} (z^o - z) \\ \eta \end{bmatrix} \tag{24}$$

The third term in the right-hand side can be made negative definite by the similar argument in Theorem 1. Hence the function (24) is semi-negative definite.
By the way, when $\tau = \tau_e = g(x_1^*)$, it is obvious that an equilibrium of (7),(8),(16),(17),(19) becomes $(x_{1e}, x_{2e}, \xi_e, z_e, \eta_e) = (x_1^*, \mathbf{0}, \mathbf{0}, z^o, \mathbf{0})$. Therefore, by the similar argument in Theorem 1, we can show that the equilibrium $(x_1^*, \mathbf{0}, \mathbf{0}, z^o, \mathbf{0})$ is asymptotically stable by LaSalle's invariance principle. Namely, it is achieved that $x_1(t) \to x_1^*$, $x_2(t) \to \mathbf{0}$, $\xi(t) \to \mathbf{0}$, $z(t) \to z^o$, $\eta(t) \to \mathbf{0}$, as $t \to \infty$. □
AyCollorary 1Az Theorem 2 holds also with setting $K_S' = 0$ and $\tau' = J(x_1)^T K_P'(z^o - z)$.

Meanwhile, when $n > p$ in the redundant manipulator, we can set some joint angles $q_i, i \in I$ at arbitrary values $x_{1i}^* = q_i^*, i \in I$ within the possible freedom (the number of elements of I is less than $n - p$). In this case define a vector $\hat{x}_1 \in R^n$ as $\{\hat{x}_{1i} = x_{1i},\ i \in I,\ \hat{x}_{1i} = 0,\ i \notin I\}$ and $\{\hat{x}_{1i}^* = x_{1i}^*,\ i \in I,\ \hat{x}_{1i}^* = 0,\ i \notin I\}$and let us modify the control law (16),(17),(19) as follows.

$$\dot{\xi} = D\xi + (\hat{x}_1^* - \hat{x}_1) - \hat{x}_2,\ \ \xi(0) = \mathbf{0},\ \ D < 0 \tag{25}$$
$$\dot{z} = J(x_1)x_2 \tag{26}$$
$$\dot{\eta} = D'\eta + (z^o - z) - J(x_1)x_2,\ \eta(0) = \mathbf{0},\ \ D' < 0 \tag{27}$$
$$\tau = K_P(\hat{x}_1^* - \hat{x}_1) + K_S\xi - K_D\hat{x}_2 + g(x_1^*) + J(x_1)^T K_P'(z^o - z) + J(x_1)^T K_S'\eta \tag{28}$$

It is then noted that elements of K_P and K_S corresponding to $x_{1i},\ i \notin I$ do not give any effect. In this case, asymptotical stability in the subspace of joint coodinates $x_i,\ i \in I$ is guaranteed such that $x_{1i}(t) \to x_{1i}^*,\ i \in I$ as $t \to \infty$. But joint angles $x_{1i},\ i \notin I$ are not known where to converge, although stable in the sense of Lyapunov.
Theorem 2 does not take damping (Derivative action) in the task-space in consideration. However, damping $-K_D x_2$ in the joint-space contributes to it indirectly.
In order to add the damping in the task-space to control input τ', we can add Derivative action term $-XK_D'\dot{z}$ to τ'. Although we can calculate the matrix X theoretically, however, it is not of practical use because of too much complexity.
On the other hand, under the situation of $\ddot{q} = \dot{x}_2 \approx \mathbf{0}$ one can prove asymptotical stability, even if $-J(x_1)^T K_D'\dot{z}$ is added to (18). But we do not know whether the damping is effective or not, as $\dot{x}_2 \approx \mathbf{0}$ is not assumed.

4. L_2-Gain Disturbance Attenuation Problem

In this section we study L_2-gain disturbance attenuation problem under the existence of disturbance w. Let us consider again the following cascaded system of the robot manipulator and the SPR element.

$$\dot{x}_1 = x_2 \tag{29a}$$

$$\begin{aligned}\dot{x}_2 &= -M(x_1)^{-1}\left\{\frac{1}{2}\dot{M}(x_1)x_2 + S(x_1,x_2)x_2 + g(x_1)\right\} + M(x_1)^{-1}\tau + L(x)w \\ &\stackrel{\triangle}{=} f_2(x_1,x_2) + G_2(x_1)\tau + L(x)w\end{aligned} \tag{29b}$$

$$\dot{\xi} = D\xi + (x_1^* - x_1) - x_2, \;\; D < 0 \tag{30}$$

$$y = x_2 \tag{31}$$

where $w \in R^l$ is the disturbance vector.
And set up a feedback compensator (P·SPR·D controller) :

$$\tau = K_P(x_1^* - x_1) + K_S\xi - K_D x_2 + g(x_1^*) \tag{32}$$

where $K_P, K_S, K_D \in R^{n\times n}$ are all positive definite diagonal matrices. Here $g(x_1^*)$, gravity force compensation at the desired value x_1^*, corresponds to the manual reset quantity $\boldsymbol{m}_0$.
The L_2-gain disturbance attenuation problem is defined to obtain the P·SPR·D control such that the closed-loop system satisfies the following two conditions under the given disturbance attenuation level $\gamma > 0$.

P1. When $w = \mathbf{0}$, the closed-loop system is asymptotically stable at the equilibrium $(x_{1e}, x_{2e}, \xi_e) = (x_1^*, \mathbf{0}, \mathbf{0})$.

p2. When $x(0) = \mathbf{0}$, the following inequality holds for arbitrarily given $T > 0$.

$$\int_0^T \|y(t)\|^2 dt \leq \gamma^2 \int_0^T \|w(t)\|^2 dt$$

It is noticed that P2 is equivallent to having L_2 gain below γ when $x(0) = \mathbf{0}$, that is, $\|y\|_2 \leq \gamma^2\|w\|_2$. It implies that for all $w \in L_2[0,T]$ and for the supply rate $s(y,w) = \frac{1}{2}\{\gamma^2 w^T w - y^T y\}$, the following γ-dissipation inequality holds (van der Schaft, 2000).

$$\dot{V}(x,\xi) \leq \frac{1}{2}\{\gamma^2 w^T w - y^T y\} \tag{33}$$

The following theorem solves the L_2-gain disturbance attenuation problem.
[Theorem 3] Suppose that $W(x)$ and $L(x)$ satisfy the matching condition

$$W_x(x)L(x) = y^T S(x)^T \tag{34}$$

where $S(x) \in R^{l\times n}$ denotes the function matrix and $S(x)^T S(x) = I_n$. Then the closed-loop system (29)∼(32) satisfies P2, provided that positive definite diagonal matrices $K_P, K_S, K_D \in R^{n\times n}$ and negative definite diagonal matrix $D \in R^{n\times n}$ are appropriately chosen and $K_D \geq \frac{1}{2}(1+\frac{1}{\gamma^2})I_n$. Namely, it possesses L_2 gain less than γ (i.e., γ-dissipation inequality holds.)
Furthermore, by the P·SPR·D control (32) the closed-loop system satisfies P1 so that $(x_{1e}, x_{2e}, \xi_e) = (x_1^*, \mathbf{0}, \mathbf{0})$ is asymptotically stable.

(**Proof**) To prove that the γ-dissipation inequality holds, make the following calculation for a storage function (11) (semi-positive definite function).

$$
\begin{aligned}
&\dot{V}(x,\xi) + \frac{1}{2}\{y^T y - \gamma^2 w^T w\} \\
&= W_{x_1}(x)x_2 + W_{x_2}(x)f_2(x_1,x_2) + W_{x_2}(x)G_2(x_1)\tau + W_{x_2}(x)L(x)w - g(x_1^*)^T x_2 \\
&\quad + \begin{bmatrix}(x_1^* - x_1) \\ \xi\end{bmatrix}^T \begin{bmatrix}K_P - \overline{K} & \overline{K} \\ \overline{K}^T & K_S - \overline{K}\end{bmatrix}\begin{bmatrix}(\dot{x}_1^* - \dot{x}_1) \\ \dot{\xi}\end{bmatrix} + \frac{1}{2}\{y^T y - \gamma^2 w^T w\}
\end{aligned}
$$

Here using the K-Y-P property (6) and the control (32) and the matching condition (34),

we have

$$
\begin{aligned}
&\dot{V}(x,\xi) + \frac{1}{2}\{y^T y - \gamma^2 w^T w\} \\
&\leq y^T\tau + y^T S(x)^T w - g(x_1^*)^T x_2 + \begin{bmatrix}(x_1^* - x_1) \\ \xi\end{bmatrix}^T \begin{bmatrix}K_P - \overline{K} & \overline{K} \\ \overline{K}^T & K_S - \overline{K}\end{bmatrix}\begin{bmatrix}-x_2 \\ D\xi + (x_1^* - x_1) - x_2\end{bmatrix} \\
&\quad + \frac{1}{2}\{y^T y - \gamma^2 w^T w\} \\
&= x_2^T(K_P(x_1^* - x_1) + K_S\xi - K_D x_2 + g(x_1^*)) + x_2^T S(x)^T w - g(x_1^*)^T x_2 \\
&\quad + \begin{bmatrix}(x_1^* - x_1) \\ \xi\end{bmatrix}^T \begin{bmatrix}-(K_P - \overline{K})x_2 + \overline{K}D\xi + \overline{K}(x_1^* - x_1) - \overline{K}x_2 \\ -\overline{K}^T x_2 + (K_S - \overline{K})D\xi + (K_S - \overline{K})(x_1^* - x_1) - (K_S - \overline{K})x_2\end{bmatrix} \\
&\quad + \frac{1}{2}\{y^T y - \gamma^2 w^T w\} \\
&= x_2^T(K_P(x_1^* - x_1) + K_S\xi - K_D x_2) + y^T S(x)^T w \\
&\quad + \begin{bmatrix}(x_1^* - x_1) \\ \xi\end{bmatrix}^T \begin{bmatrix}\overline{K} & \overline{K}D \\ (K_S - \overline{K}) & (K_S - \overline{K})D\end{bmatrix}\begin{bmatrix}(x_1^* - x_1) \\ \xi\end{bmatrix} - (x_1^* - x_1)^T K_P x_2 - \xi^T K_S x_2 \\
&\quad + \frac{1}{2}\{y^T y - \gamma^2 w^T w\} \\
&= -x_2^T K_D x_2 + y^T S(x)^T w + \begin{bmatrix}(x_1^* - x_1) \\ \xi\end{bmatrix}^T \begin{bmatrix}\overline{K} & \overline{K}D \\ (K_S - \overline{K}) & (K_S - \overline{K})D\end{bmatrix}\begin{bmatrix}(x_1^* - x_1) \\ \xi\end{bmatrix} \\
&\quad - \frac{1}{2}\left\{\frac{1}{\gamma}y^T S(x)^T - \gamma w^T\right\}\left\{\frac{1}{\gamma}S(x)y - \gamma w\right\} + \frac{1}{2}y^T y + \frac{1}{2}\frac{1}{\gamma^2}y^T S(x)^T S(x)y \\
&\quad - \frac{1}{2}w^T S(x)y - \frac{1}{2}y^T S(x)^T w \\
&= \begin{bmatrix}(x_1^* - x_1) \\ \xi\end{bmatrix}^T \begin{bmatrix}\overline{K} & \overline{K}D \\ (K_S - \overline{K}) & (K_S - \overline{K})D\end{bmatrix}\begin{bmatrix}(x_1^* - x_1) \\ \xi\end{bmatrix} - y^T\{K_D - \frac{1}{2}(1 + \frac{1}{\gamma^2})I_n\}y \\
&\quad - \frac{1}{2}\left\{\frac{1}{\gamma}S(x)y - \gamma w\right\}^T\left\{\frac{1}{\gamma}S(x)y - \gamma w\right\}
\end{aligned}
$$

The first term in the right-hand side is negative definite, as mentioned below (12). Hence, using $K_D \geq \frac{1}{2}(1+\frac{1}{\gamma^2})I_n$,

$$\dot{V}(x,\xi) + \frac{1}{2}\{y^T y - \gamma^2 w^T w\} \leq -\frac{1}{2}\left\{\frac{1}{\gamma}S(x)y - \gamma w\right\}^T \left\{\frac{1}{\gamma}S(x)y - \gamma w\right\} \leq 0 \tag{35}$$

Consequently, γ-dissipation inequality (33) holds, and so it follows that we have L_2 gain below γ.
When $w = \mathbf{0}$, P1 has been already concluded by Theorem 2. □

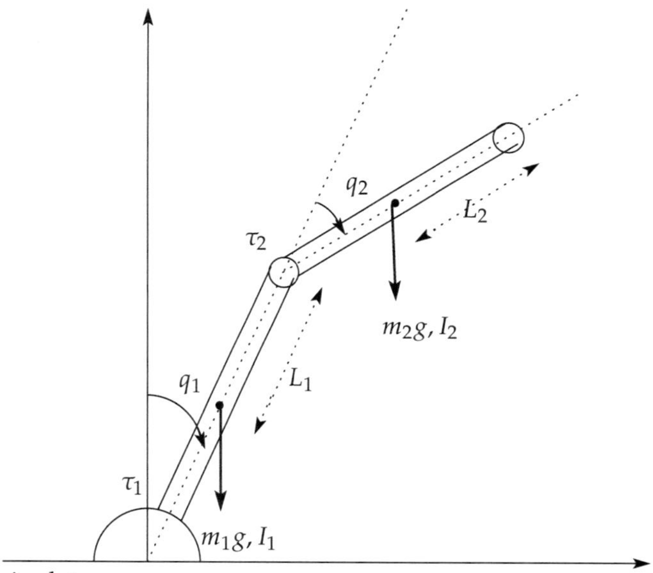

Fig. 1. 2-Link Manipulator

5. Simulation

Let us apply the P·SPR·D control of robot manipulator developed in Section 2 to a two-link manipulator depicted in Fig.1. Here generalized coordinates q_1, q_2 are relative joint angles, and $x_{11} \triangleq q_1$ denotes the perpendicular angle (angle from vertical line) of link 1 and $x_{12} \triangleq q_2$ relative angle of link 2 from link 1, τ_1 *and* τ_2 denote the torque of each joint acting clockwise. $L_1, L_2, m_1, m_2, I_1, I_2$ denote the length, the mass and the inertia moment of each link, respectively.
A numerical example of two-link manipulator is given as follows.

$$\begin{bmatrix} \dot{x}_{11} \\ \dot{x}_{12} \\ \dot{x}_{21} \\ \dot{x}_{22} \end{bmatrix} = \begin{bmatrix} x_{21} \\ x_{22} \\ f_{21}(\boldsymbol{x}_1,\boldsymbol{x}_2) + G_{211}(\boldsymbol{x}_1)\tau_1 + G_{212}(\boldsymbol{x}_1)\tau_2 \\ f_{22}(\boldsymbol{x}_1,\boldsymbol{x}_2) + G_{221}(\boldsymbol{x}_1)\tau_1 + G_{222}(\boldsymbol{x}_1)\tau_2 \end{bmatrix}$$

where

$$f_{21}(\boldsymbol{x}_1,\boldsymbol{x}_2) \stackrel{\triangle}{=} \frac{-1}{\det\ M}[1.05\{(-6x_{21}x_{22}-3x_{22}^2)\sin x_{12}+5x_{21}-117.6\sin x_{11}$$
$$-14.7\sin(x_{11}+x_{12})\}-(1+3\cos x_{12})(3x_{21}^2\sin x_{12}+5x_{22}-14.7\sin(x_{11}+x_{12}))]$$
$$f_{22}(\boldsymbol{x}_1,\boldsymbol{x}_2) \stackrel{\triangle}{=} \frac{-1}{\det\ M}[(-1-3\cos x_{12})\{(-6x_{21}x_{22}-3x_{22}^2)\sin x_{12}+5x_{21}-117.6\sin x_{11}$$
$$-14.7\sin(x_{11}+x_{12})\}+(21.2+6\cos x_{12})(3x_{21}^2\sin x_{12}+5x_{22}-14.7\sin(x_{11}+x_{12}))]$$
$$G_{211}(\boldsymbol{x}_1) \stackrel{\triangle}{=} \frac{1.05}{\det\ M},\quad G_{212}(\boldsymbol{x}_1) \stackrel{\triangle}{=} \frac{1}{\det\ M}(1-3\cos x_{12}),$$
$$G_{221}(\boldsymbol{x}_1) \stackrel{\triangle}{=} \frac{1}{\det\ M}(-1-3\cos x_{12}),\quad G_{222}(\boldsymbol{x}_1) \stackrel{\triangle}{=} \frac{1}{\det\ M}(21.2+6\cos x_{12})$$

and $\det M \stackrel{\triangle}{=} 21.26+0.3\cos x_{12}-9(\cos x_{12})^2$.
Further, $\boldsymbol{g}(\boldsymbol{x}_1)$ is also given as

$$\begin{bmatrix} g_1(\boldsymbol{x}_1) \\ g_2(\boldsymbol{x}_1) \end{bmatrix} = \begin{bmatrix} -117.6\sin x_{11}-14.7\sin(x_{11}+x_{12}) \\ -14.7\sin(x_{11}+x_{12}) \end{bmatrix}$$

Applying Theorem 1, let us solve a set-point servo problem with the desired value $\boldsymbol{x}_1^* = (1.5,\ 1)^T$. We set the SPR element as (8) and take an initial state as $(\boldsymbol{x}_1(0),\boldsymbol{x}_2(0)) = (\mathbf{0},\mathbf{0})$. The simulation results is shown in Fig.2, when $D = \begin{bmatrix} -2 & 0 \\ 0 & -2 \end{bmatrix}$, $K_P = \begin{bmatrix} 180 & 0 \\ 0 & 180 \end{bmatrix}$, $K_S = \begin{bmatrix} 40 & 0 \\ 0 & 40 \end{bmatrix}$, $K_D = \begin{bmatrix} 60 & 0 \\ 0 & 60 \end{bmatrix}$. We see that the convergence speed is very quick and the overshoot is very little.
Furthermore, as mentioned in Remark 1, the regulation problem (asymptotical stabilization to the origin) can be solved by setting $\boldsymbol{x}_1^* = \mathbf{0}$. At this time $\boldsymbol{m}_0 = \boldsymbol{g}(\boldsymbol{x}_1^*)$ is zero. The simulation results is shown in Fig.3.

Fig.4 shows the simulation results of P·SPR·D+I control in case where gravity force compensation $\boldsymbol{g}(\boldsymbol{x}_1^*)$ is not available. Here K_I is given as $\begin{bmatrix} 80 & 0 \\ 0 & 80 \end{bmatrix}$. Fig.5 shows the simulation results for the regulation problem.
It is seen that the P·SPR·D control is superior to the P·SPR·D+I control (Notice scales of y-axis). Nevertheless, the control performance by the P·SPR·D+I control is also satisfactory enough. Meanwhile, ordinary PID control(with $mv_0 = \mathbf{0}$) is represented as follows.

$$\boldsymbol{\tau} = K_P(\boldsymbol{x}_1^*-\boldsymbol{x}_1)+K_I\int_0^t(\boldsymbol{x}_1^*-\boldsymbol{x}_1(\tau))d\tau - K_D\boldsymbol{x}_2$$

i.e.

$$\dot{\boldsymbol{z}} = \boldsymbol{x}_1^*-\boldsymbol{x}_1,\quad \boldsymbol{z}(0)=\mathbf{0}$$
$$\boldsymbol{u} = K_P(\boldsymbol{x}_1^*-\boldsymbol{x}_1)+K_I\boldsymbol{z}-K_D\dot{\boldsymbol{x}_2}$$

The simulation results by the PID control with $K_P = \begin{bmatrix} 180 & 0 \\ 0 & 180 \end{bmatrix}$, $K_I = \begin{bmatrix} 40 & 0 \\ 0 & 40 \end{bmatrix}$, $K_D = \begin{bmatrix} 60 & 0 \\ 0 & 60 \end{bmatrix}$ are shown in Fig.6 and Fig.7 for the set-point servo problem and the regulation

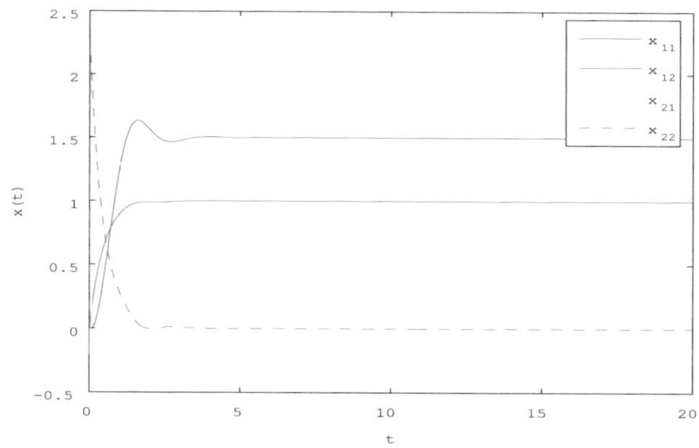

Fig. 2. P·SPR·D Control

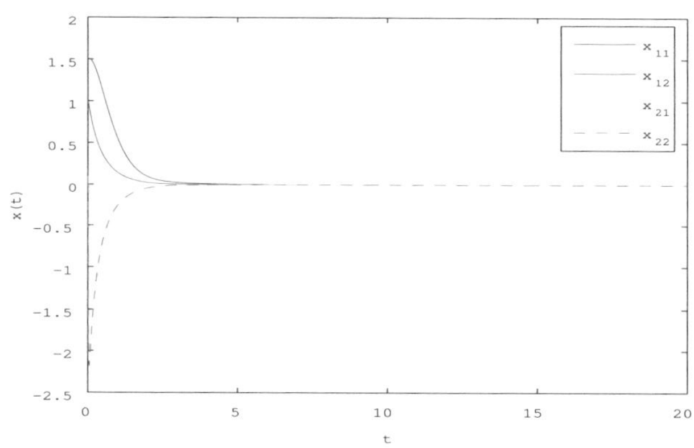

Fig. 3. P·SPR·D Control

one, respectively. It is observed that though the convergence was attained by the ordinary PID control also, the convergence speed is slower than the P·SPR·D+I control.

Of course control performance changes depending on controller parameters K_P, K_S, K_D, D. However, it was known that the proposed methods attained always much better performances than the ordinary PID control. Comparing these three cases, we can say that the P·SPR·D control is the best in regard to response speed, overshoot and steady state error. This indicates that the P·SPR·D or P·SPR·D+I control possesses a possivility of a new and promising control scheme.

Note that nothing has been mentioned in regard to controller parameter adjustment. The values of K_P, K_S, K_D used in the simulations are the same one with almost optimum values of K_P, K_I, K_D for the ordinary PID control which were obtained by trial and error. Of course the control performance depends on the parameter values, and so there is a room of argument for further improvement.

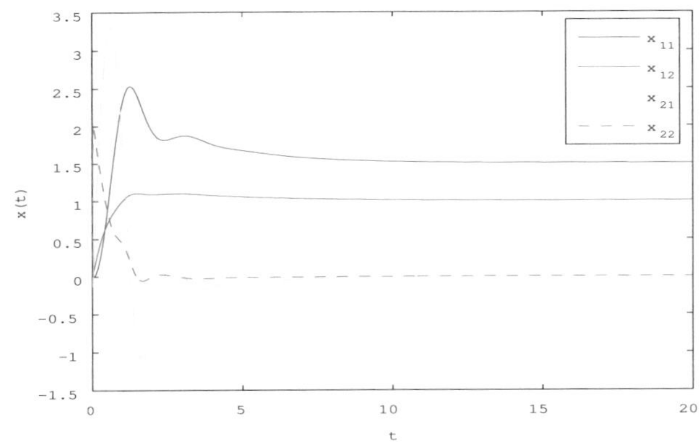

Fig. 4. P·SPR·D+I Control

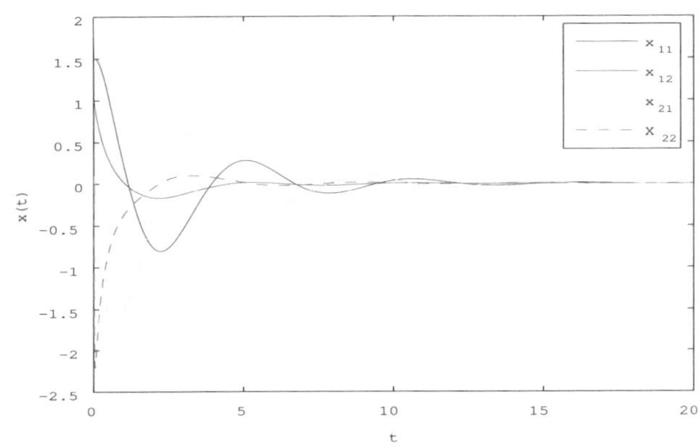

Fig. 5. P·SPR·D+I Control

6. Conclusion

Based on the passivity theory and LaSalle's invariance principle, we studied first the set-point servo problem for the robot manipulators by the P·SPR·D and P·SPR·D+I control. Next we investigated a stabilizing control method for the end-point position setting of redundant manipulator, combining the P·SPR·D control in the joint-space and the P·SPR one in the task-space. The effectiveness of the proposed methods are demonstrated with a two-link manipulator. We showed the simulation results of the P·SPR·D+$\boldsymbol{g}(\boldsymbol{x}_1^*)$ control and P·SPR·D+I control by which very excellent control performances were obtained. Further, the L_2-gain disturbance attenuation problem was studied also.
The P·SPR·D or P·SPR·D+I control can be said to be a new general control scheme and the use of SPR element as a part of controller possesses an advantage from a passivity-based design point of view. In particular the SPR element contributes powerfully to stabilization of the closed-loop system. They can be applied widely to linear systems and/or affine nonlinear sysems also. The optimum adjustment of controller parameters is left as a future topic.
Implementation of the P·SPR·D control is not difficult with a digital processor.

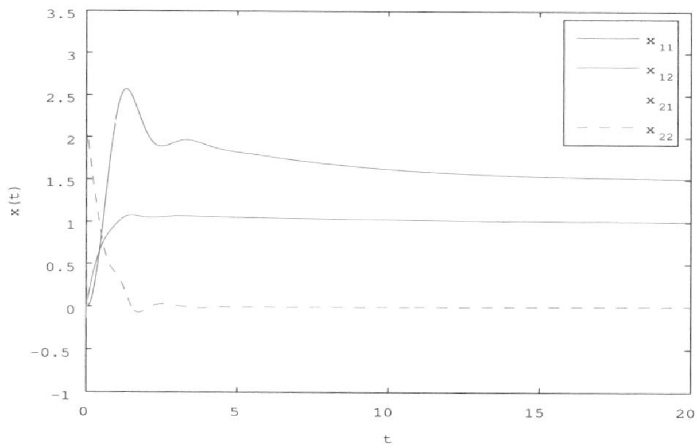

Fig. 6. Ordinary PID Control

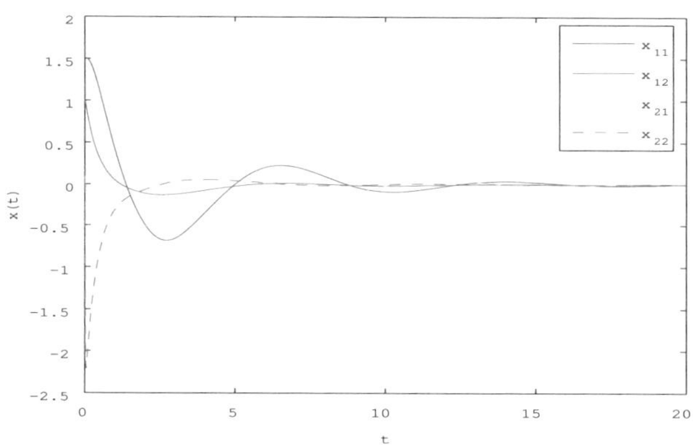

Fig. 7. Ordinary PID Control

7. References

Arimoto, S. (1996), *Control Theory of Nonlinear Mechanical Systems : A Passivity-Based and Circuit Theoretic Approach*, Oxford Univ. Press

Arimoto, T. & Miyazaki, F. (1984). Stability and Robustness of PID Feedback Control for Robot Manipulators of Sensory Capability, In: *Robotics Research*, M.Brady and R.Paul (Ed.), pp.783-799, MIT Press, Cambridge

Byrnes, C. I., ; Isidori, A. ; Willems, C. J. (1991). Passivity, Feedback Equivalence, and the Global Stabilization of Minimum Phase Nonlinear Systems. *IEEE Trans. Automat. Contr.*, Vol.AC36, No.11, 1228-1240

Galicki, M. (2008). An Adaptive Regulator of Robotic Mnipulators in the Task Space, *IEEE Trans. Automat. Contr.*, Vol.53,No.4, 1058-1062

Khalil, H. K. (2002). *Nonlinear Systems, 3rd ed.*, Prentice Hall

Khatib, O. (1987). A Unified Approach for Motion and Force Control of Robot Manipulators : The Operational Space Formulation. *IEEE J. Robot. Automat.*, Vol.RA-3, No.1, 43-53

LaSalle, J. & Lefschetz, S. (1961). *Stability by Liapunov's Direct Method*, Academic Press

Lozano, B. ; Brogliato, B. ; Egeland, O. & Maschke, B. (2000). *Dissipative Systems Analysis and Control*, Springer-Verlag
Murakami, T. ; Ohnishi, K. (1993). A Study of Stability and Work Space Decoupling Control Based on Robust Control in Multi-Degrees of Freedom Robot, *IEEJ. Ind. Appl.*, Vol.113-D, No.5, 639-646 (in Japanese)
Murakami, T. ; Kahlen, K. ; De Doncker, R. W. (2008). Robot Motin Control Based on Projection Plane in Redundant Manipulator, *IEEE Trans. Ind. Electron*, Vol.49, No.1, 248-255
Sepulchre, R. ; Jankovic, M. & Kokotovic, P. V. (1997) *Constructive Nonlinear Control*, Springer-Verlag
Shen, T. (2004). Review No.11, Basis for Passivity-Based Design, *J. of SICE*, Vol.43, No.5, 447-453
Shimizu, K. (2008a). P·SPR·D Control of Affine Nonlinear System and Its Application to Inverted Pendulum-Stability Theory Based on Passivity-. *Trans. SICE*, Vol.44, No.7, 575-582 (in Japanese)
Shimizu, K. (2008b). P·SPR·D Control for Affine Nonlinear System and Robot Manipulators–Stability Analysis Based on K-Y-P Property and LaSalle's Invariance Principle-, *Proceedings of the 47th IEEE Conference on Decision and Control*, pp.4332-4338, Cancun, Mexico
Shimizu, K. (2009a). P·SPR·D and P·SPR·D·I Control for Linear Multi-Variable Systems-Stabilization Theory Based on High Gain Output Feedback-, *Proceedings of 2009 American Control Conference*, pp.4666-4671, St.Louis, Missouri
Shimizu, K. (2009b). P·SPR·D and P·I·SPR·D Control for Affine Nonlinear Systems-Stabilization Theory Based on Passivity-, *Proceedings of 2009 American Control Conference*, pp.4680-4686, St.Louis, Missouri
Shibata, T. & Murakami, T. (2007). A Null Space Control Based on Passivity in Redundant Manipulator, *J.of the JPSP*, Vol.73, No.11, 1238-1245 (in Japanese)
Spong, M. ; Lewis, F. L. & Abdallah, C. T.(Eds.) (1992). *Robot Control -Dynamics, Motion Planning and Analysis*, Part 7 and Part 10, IEEE Press
Van der Schaft, A. (2000). *L_2-Gain and Passivity Techniques in Nonlinear Control, 2nd ed.*, Springer-Verlag

8. Appendix Case Where Ordinary PID Control

Let us consider the robot manipulator (2),(3). As is well-known, the robot manipulator is passive with respect to input τ and output y, and hence the K-Y-P property (6) holds.
We study here a set-point servo problem with the desired value $(x_1, x_2) = (x_1^*, 0)$, and set up an ordinary PID controller

$$\dot{z} = (x_1^* - x_1) \tag{36}$$
$$\tau = K_P(x_1^* - x_1) + K_I z - K_D x_2 \tag{37}$$

where $K_P, K_I, K_D \in R^{n\times n}$ are positive definite diagonal gain matrices.
Below we prove asymptotical stability of the closed-loop system, applying LaSalle's invariance principle.
An equilibrium of the closed-loop system (2),(36),(37) satisfies

$$\begin{aligned} 0 &= x_{2e} \\ 0 &= -g(x_{1e}) + K_P(x_1^* - x_{1e}) + K_I z_e \\ 0 &= (x_1^* - x_{1e}) \end{aligned} \tag{38}$$

, hence $(x_{1e}, x_{2e}, z_e) = (x_1^*, \mathbf{0}, \bar{z})$, $\bar{z} = K_I^{-1} g(x_1^*)$ becomes an equilibrium point.
Now consider a Lyapunov function candidate

$$\begin{aligned} V(x,z) &= W(x) + g(x_1^*)^T(x_1^* - x_1) + \frac{1}{2}(x_1^* - x_1)^T K_P (x_1^* - x_1) \\ &\quad +(x_1^* - x_1)^T K_I (z - \bar{z}) + \frac{1}{2}\alpha (z - \bar{z})^T K_I (z - \bar{z}) \\ &\quad -\alpha (x_1^* - x_1)^T M(x_1) x_2 \end{aligned} \tag{39}$$

where $W(x) = \frac{1}{2}x_2^T M(x_1)x_2 + U(x_1) - U(x_1^*)$, $\alpha > 0$.
It can be proved that $V(x,z)$ is a function bounded below in the neighborhood of $(x^*, \mathbf{0}, \bar{z})$.
Calculate a time derivative of $V(x,z)$ along (2),(3),(36),(37), using the K-Y-P property (6), to obtain

$$\begin{aligned} \dot{V}(x,z) &= W_{x_1}(x)x_2 + W_{x_2}(x)\{f_2(x_1,x_2) + G_2(x_1)\tau\} - g(x_1^*)^T x_2 \\ &\quad +(x_1^* - x_1)^T K_P(-x_2) - x_2^T K_I (z-\bar{z}) + (x_1^* - x_1)^T K_I \dot{z} + \alpha (z-\bar{z})^T K_I \dot{z} \\ &\quad +\alpha x_2^T M(x_1)x_2 - \alpha (x_1^* - x_1)^T \dot{M}(x_1)x_2 - \alpha (x_1^* - x_1)^T M(x_1)\dot{x}_2 \\ &\le y^T\tau - g(x_1^*)^T x_2 - (x_1^* - x_1)^T K_P x_2 - x_2^T K_I (z-\bar{z}) \\ &\quad +(x_1^* - x_1)^T K_I (x_1^* - x_1) + \alpha (z-\bar{z})^T K_I (x_1^* - x_1) + \alpha x_2^T M(x_1)x_2 \\ &\quad -\alpha (x_1^* - x_1)^T \dot{M}(x_1)x_2 - \alpha (x_1^* - x_1)^T M(x_1)\{f_2(x_1,x_2) + G_2(x_1)\tau\} \\ &= x_2^T (K_P(x_1^* - x_1) + K_I z - K_D x_2) - g(x_1^*)^T x_2 - (x_1^* - x_1)^T K_P x_2 \\ &\quad -x_2^T K_I (z-\bar{z}) + (x_1^* - x_1)^T K_I (x_1^* - x_1) + \alpha (z-\bar{z})^T K_I (x_1^* - x_1) \\ &\quad +\alpha x_2^T M(x_1)x_2 - \alpha (x_1^* - x_1)^T \dot{M}(x_1)x_2 \\ &\quad -\alpha (x_1^* - x_1)^T \{\frac{1}{2}\dot{M}(x_1)x_2 + S(x_1,x_2)x_2 + g(x_1)\} \\ &\quad -\alpha (x_1^* - x_1)^T (K_P(x_1^* - x_1) + K_I z - K_D x_2) \\ &= -x_2^T (K_D - \alpha M(x_1))x_2 - (x_1^* - x_1)^T (\alpha K_P - K_I)(x_1^* - x_1) \\ &\quad -\alpha g(x_1^*)^T (x_1^* - x_1) - \alpha (x_1^* - x_1)^T \dot{M}(x_1)x_2 \\ &\quad +\alpha (x_1^* - x_1)^T \{\frac{1}{2}\dot{M}(x_1)x_2 + S(x_1,x_2)x_2 + g(x_1)\} + \alpha (x_1^* - x_1)^T K_D x_2 \\ &= -x_2^T (K_D - \alpha M(x_1))x_2 - (x_1^* - x_1)^T (\alpha K_P - K_I)(x_1^* - x_1) \\ &\quad +\alpha (g(x_1) - g(x_1^*))^T (x_1^* - x_1) + \alpha (x_1^* - x_1)^T Q(x_1, x_2; K_D) x_2 \end{aligned}$$

where $Q(x_1,x_2;K_D) \overset{\triangle}{=} -\frac{1}{2}\dot{M}(x_1) + S(x_1,x_2) + K_D$.
Here assume for $\beta > 1$ that

$$(x_1^* - x_1)^T K_P (x_1^* - x_1) \ge \beta (g(x_1) - g(x_1^*))^T (x_1^* - x_1),$$

then we have

$$
\begin{aligned}
\dot{V}(x,z) &\leq -x_2^T(K_D-\alpha M(x_1))x_2 - (x_1^*-x_1)^T(\alpha K_P - K_I)(x_1^*-x_1) \\
&\quad + \frac{\alpha}{\beta}(x_1^*-x_1)^T K_P (x_1^*-x_1) + \alpha(x_1^*-x_1)^T Q(x_1,x_2;K_D)x_2 \\
&= -x_2^T(K_D-\alpha M(x_1))x_2 \\
&\quad - \begin{bmatrix}(x_1^*-x_1)\\ x_2\end{bmatrix}^T \begin{bmatrix}(\alpha-\frac{\alpha}{\beta})K_P - K_I & -\frac{1}{2}\alpha Q(x_1,x_2;K_D) \\ -\frac{1}{2}\alpha Q(x_1,x_2;K_D)^T & K_D - \alpha M(x_1)\end{bmatrix} \begin{bmatrix}x_1^*-x_1\\ x_2\end{bmatrix} \\
&\leq 0
\end{aligned} \tag{40}
$$

By supposing that x_2 exists in the neighborhood of $x_2 = \mathbf{0}$, spectral radius of $Q(x_1,x_2;K_D)$ can be considered whithin a certain value. When x_2 exists within that bounds, by taking α suffisiently small and $K_I > 0$ appropriatly small for the given β, we can make the matrix

$$
\begin{bmatrix}(\alpha-\frac{\alpha}{\beta})K_P - K_I & -\frac{1}{2}\alpha Q(x_1,x_2;K_D) \\ -\frac{1}{2}\alpha Q(x_1,x_2;K_D)^T & K_D - \alpha M(x_1)\end{bmatrix}
$$

and $K_D - \alpha M(x_1)$ be positive definite by choosing $K_P > 0$ and $K_D > 0$ large enough. In other words, if $K_P > 0$ and $K_D > 0$ are large enough and $K_I > 0$ is small, there exists α such that the above matrix and $K_D - \alpha M(x_1)$ become positive definite for the given β.
Let $\Omega_c = \{(x,z) \mid V(x,z) \leq c\}$ and suppose that Ω_c is bounded and $\dot{V}(x,z) \leq 0$ in Ω_c (c is a positive number such that $\dot{V}(x,z) \leq 0$). Here define Ω_E as a set of all points of Ω_c satisfying $\dot{V}(x,z) = 0$ and put

$$
\Omega_E = \{(x,z) \mid \dot{V}(x,z) = 0,\ (x,z) \in \Omega_c\} \tag{41}
$$

From (40),(2),(36) (x,z) satisfying $\dot{V}(x,z) = 0$ is given as $x_1^* - x_1 = \mathbf{0},\ x_2 = \mathbf{0},\ z = \bar{z}$, that is to say, it is a point $(x_1,x_2,z) = (x_1^*,\mathbf{0},\bar{z})$. Accordingly, we know from (40),(2),(36),(37) that (x,z) in Ω_E consists of only the equilibrium point $(x_{1e},x_{2e},z_e) = (x_1^*,\mathbf{0},\bar{z})$ when $\tau = K_I\bar{z} = g(x_1^*)$. Thus, the largest invariance set Ω_M in Ω_E consists of only the equilibrium point $(x_1^*,\mathbf{0},\bar{z})$. Therefore, by LaSalle's invariance principle all trajectories in Ω_c converges to Ω_M, i.e., to $(x_1^*,\mathbf{0},\bar{z})$ as $t \to \infty$. Thus $x = (x_1^*,\mathbf{0})$ is aymptotically stable. Q.E.D

34

3D Imaging System for Tele-Manipulation

Hideki Kakeya
The University of Tsukuba
Japan

1. Introduction

In accordance with the advancement of flat panel display technologies, recent researches of 3D display technologies are mostly focused on thin 3D displays using parallax barriers, lenticular lenses, or fry-eye lenses. Though some products based on these technologies have already been commercialized, they have not acquired many users. One of the main reasons why these 3D displays have not attained commercial successes is that most of these displays are developed for general purposes, not for a specialized purpose. In general, realization of 3D display has to sacrifice resolution. Also 3D displays give the viewer more eyestrains than 2D displays. In addition to these factors it should be noted that most 3D information can be obtained from 2D display with the intelligence of the viewer. Therefore 3D displays cannot be an attractive option unless they can offer substantial merits besides the impact at first impression.

Then where 3D displays become necessary and essential? To answer this question, let us consider how we perceive 3D information from 2D displays. There are several factors which can help depth perception from a simple 2D image. One of the most important factors is motion. Near objects tend to move faster than far objects in the 2D image. When the motion comes from the camera work, this relation always holds as long as the objects themselves are static. In this case it can be said that motion pictures on 2D display are presenting 3D information composed of x axis, y axis, and time axis. In other word, 2D displays cannot present 4D space composed of x axis, y axis, z axis, and time axis, which 3D displays showing motion pictures can present.

Then when 4D perception is necessary? One answer is real-time operation of 3D space. In real-time tasks such as tele-manipulation of robots or remote car driving, the operators or the drivers are required to grasp 3D location of the objects on the real-time basis. They cannot spare time to wait and see the change of image to perceive depth.

Real-time interaction systems where the viewer needs to perceive 3D space instantly include interactive visualization, 3D drawing, surgery simulations, etc. besides robot tele-operation and car driving. Since interactive systems can be used for various kinds of tasks, desired design of the system varies depending on the features of interaction. Therefore ready-made 3D displays cannot meet up with the requirement of the users. To increase the number of 3D display users for specialized purposes, it is important to present 3D displays designed for each use with low cost because we cannot expect cost reduction by mass production.

One ideal form of 3D interaction is that the viewer can work as if he or she were touching 3D objects directly with his or her hands and can feel that he or she is part of the 3D world. To achieve it, however, 3D space should be presented in the air so that the display may not interfere with the motion of the viewer. To present 3D image in the air conventional stereoscopic displays need to fabricate large artificial parallax. Unfortunately human depth perception does not rely only on the parallax but also on the focal accommodation of the eyes. Dependency on the focal accommodation of the eyes in depth perception increases when the viewer sees nearer objects. Therefore the contradiction between the binocular convergence of the eyes based on the parallax and the focal accommodation of the eyes (convergence-accommodation conflict) causes severe eyestrain or failure of depth perception especially when 3D image near the viewer is to be presented.

Besides convergence-accommodation conflict, motion parallax is also an important factor to realize physical 3D interaction. In the real world the objects are often located so that they hide one another. To give proper operation in the complex space, the operator has to move his or her head often to see what is hidden behind. Therefore realizing motion parallax is esential to let the operator work properly. Unfortunately, however, most conventional 3D displays cannot show motion parallax with wide viewing angle.

In the first half of this chapter, we measure the influence of motion parallax and convergence-accommodation conflict on depth perception of the operator. In the latter half of this chapter we introduce coarse integral volumetric imaging system, where both convergence-accommodation conflict and imperfection of motion parallax are overcome. This chapter is organized as follows. In Section 2 a couple of imaging systems which can show motion parallax are discussed and compared. In Section 3 effect of convergence-accommodation conflict on depth perception is discussed. In Section 4 a new 3D display system which can show motion parallax without convergence-accommodation conflict is explained.

2. Depth Perception by Motion Parallax

Clues of human depth perception are roughly categorized into two factors. One is the psychological factor. People can grasp rough 3D structure in the image even through 2D displays by perspective, motion, or occulusional relationship among the objects in the image. The other is the physiological factor.

The most famous and major physiological factor of depth perception is binocular parallax based on stereo disparity, which cannot be given by conventional 3D dispays. Generally the phrases "stereoscopic display" or "3D display" are used to the displays which can show binocular parallax to the viewer. As for parallax, motion parallax, the change of view in accordance with the viewer's motion, is also a physicological factor of depth perception.

To show motion parallax to the viewer who controls robots in a remote place, we can think of two major options. One is to use a master-slave camera system where the camera follows the viewer's motion. The other is to use multi-camera system.

In the former system, we measure motion of the viewer with a 3-D position sensor and send the information to the slave camera system, which is controlled so that it moves to the position where the necessary texture can be best taken. This mechanism is expected to work well when the viewer moves slowly. The viewer can observe exact motion parallax as long as the position of the slave camera is correct. When the viewer moves faster, however, it is

impossible to catch up with the motion of the viewer because of the delay of mechanical control, which becomes larger as the motion of the viewer becomes faster. When the delay is large, the image presented to the viewer has large geometrical distortion.
In the latter system, the images which correspond to the positions of the viewer's right and left eyes are selected and sent to the viewer. Since the image can be switched instantly, delay is negligible as long as broadband signal transmission system is available. The problem of this system is that it cannot express continuous motion parallax. Since images are switched as the viewer moves, motion parallax is discrete, which can give unnatural impression to the viewer.
To evaluate the influence of delay and discretization of motion parallax, we have made computer applications with 3D CG which simulate delayed and discretized motion parallax. We set 6 conditions for simulations as follows:

(1) Continuous motion and binocular parallax without delay,
(2) Continuous motion parallax without delay, but no binocular parallax,
(3) Discretized motion and binocular parallax (3cm) without delay,
(4) Discretized motion and binocular parallax (6cm) without delay,
(5) Continuous motion and binocular parallax with delay (1 second),
(6) Continuous motion and binocular parallax with delay (2 seconds).

As for the hardware for the experiments, we use a CRT monitor which has 120Hz refresh rate, a pair of LCD shutter glasses which synchronize with CRT refresh rate and deliver 60Hz image to each eye (odd frames for the right eye and even frame for the left eye), and a magnetic 3D position sensor as shown in Fig. 1. With the LCD shutter system binocular parallax can be given to the viewer, while motion parallax is generated by changing the image to follow the viewer's position detected by the magnetic sensor.

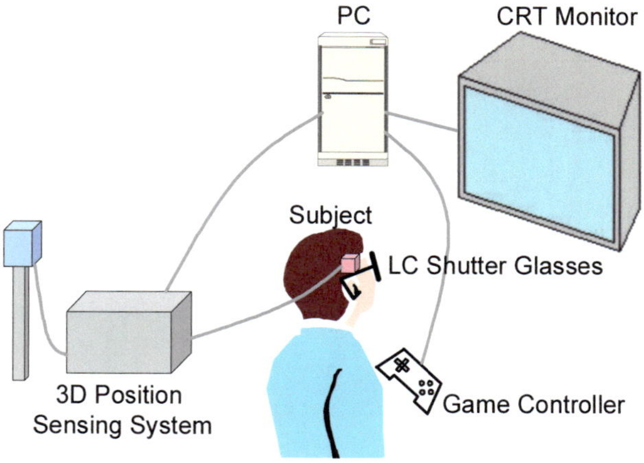

Fig. 1. Components of simulation system to test depth perception with delayed or discretized parallax.

To evaluate the effect of discretization and delay of motion parallax, we have made 3 simulators which require depth perception. The first simulator requires a simple task of hitting an approaching ball with a small pad bounded in a 2D plane parallel to the display

screen. The subject can move the pad in the first 8 seconds. Since the ball reaches the plane which includes the small pad 12 seconds after the ball is launched, the subject has to fix the position where he expects the ball reaches 4 seconds before it goes through the hitting plane. The time left for pad control is shown with the bars as shown in Fig. 2.
The result of the experiment is shown in Fig. 3. Here 7 subjects have tried 18 trials each. The ball moves inside the box whose size is 24cm (width) X 12cm (height) X 24cm (depth). As the figure shows, the ideal condition with no delay and discretization (condition 1) marked the best result. The difference between the ideal condition and the other conditions, however, are relatively small in this experiment.

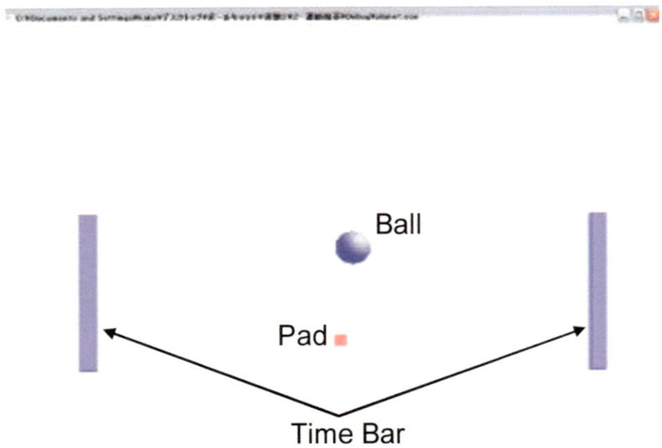

Fig. 2. Screen shot of the simulator to hit an approaching ball with a small pad.

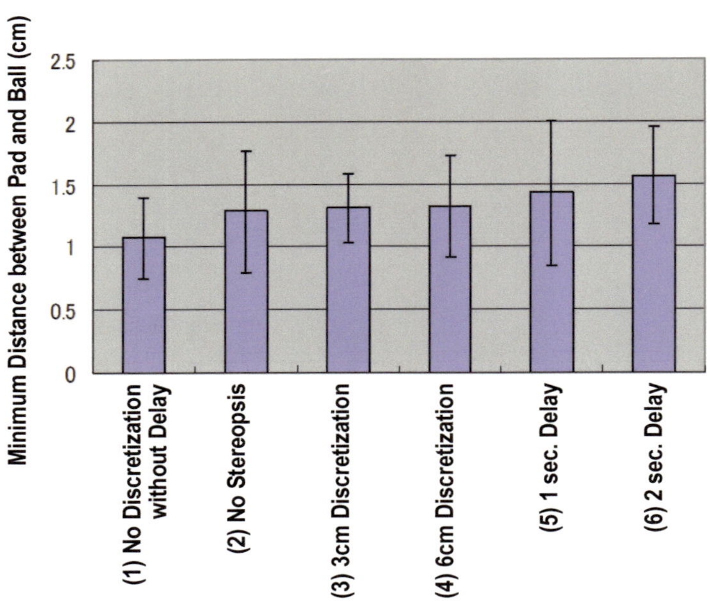

Fig. 3. Experimental results of subjects' performance in the ball-hitting simulator under various parallax conditions.

The second simulator is a crane game where the subject is required to pick up the target object. Since the target is static in this task, we can evaluate precision of depth perception more directly. While the subject is pushing the first button, the crane moves leftward. After the subject releases the button, the crane starts going forward. When the subject pushes the second button, the crane stops moving and it goes down to pick the target.
Here we have prepared two kinds of settings for the experiment. In the first setting, the target lies on the floor with textures (Fig. 4 (a)). In the second setting the floor does not exist and the target looks like an object floating in the air (Fig. 4 (b)). The former setting is expected to be easier because the subject can perceive depth from the perspective information also, while the subject is required to grasp depth only from the binocular and motion parallax in the latter setting, which is the best condition to test the effect of delay and discretization of parallax.
The result of the experiment is shown in Figs. 5 and 6. Here the number of subjects is 7 and the size of the workspace is 28cm (width) X 28cm (depth). The speed of the sliding motion of the crane is 2.5cm/s and the height of the crane from the target is 10.5 cm. 10 trials are given to each subjects and the average performances under different conditions are compared.
As Fig. 5 shows superiority of the condition without delay and discretization is weak when the floor is shown, while it becomes obvious when the floor is not shown. Without the floor, the subject has little psychological clues such as perspective to grasp the positional relationship among objects in the image. Therefore he or she has to rely on physiological clues such as binocular parallax and motion parallax to perceive depth. In this case lack of binocular parallax (condition 2), rough discretization of binocular and motion parallax (condition 4), and delay of motion parallax (conditions 5 and 6) can have strong influence on the performance of the operator.

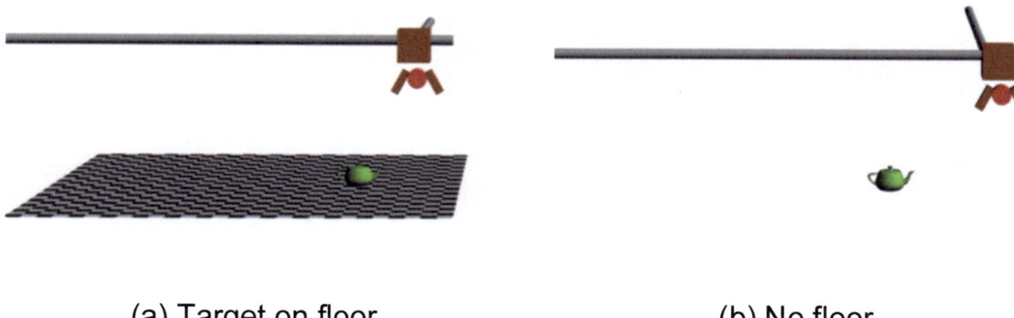

Fig. 4. Screen shots of crane game simulator with textured floor (a) and without floor (b).

The third and the last simulation is remote control of helicopter as shown in Fig. 7. The first and the second simulators explained above require only simple operations, while the third simulator requires more complex operations. The subject has to use 8 buttons as shown in Fig. 8 to control the helicopter. Also since this simulator is based on physical dynamics of the real world, the helicopter has inertia, which makes the control even harder.

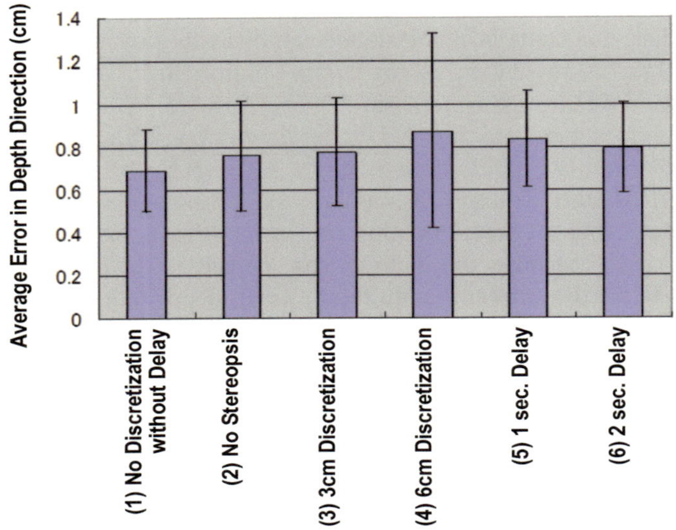

Fig. 5. Experimental results of subjects' performance in the crane game simulator with textured floor.

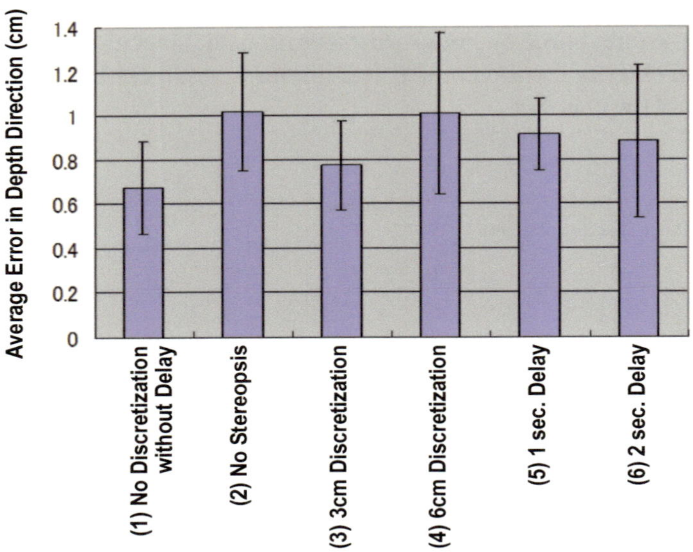

Fig. 6. Experimental results of subjects' performance in the crane game simulator without floor.

The subject is required to let the helicopter go through the ring floating in the air, which requires precise depth perception of the ring and the helicopter. The other difference of this simulation from the former simulations is emergence of occlusion. In the environment where the target ring and the helicopter can hide one another, motion of head to check the occluded area becomes important, where precise motion parallax can play an important role for depth perception.

The result of the experiment is shown in Fig. 9. Here 20 trials are given to each of the 7 subjects and the average failure rate of the task for each condition is calculated. The helicopter can move inside the box with the size of 24cm (width) X 12cm (height) X 24cm (depth). The maximum speed of the helicopter is 4cm/s and the subjects are required to let the helicopter go through the ring within 15 seconds.
In this experiment lack of stereopsis affects the performance of the operator most as shown in Fig. 9. Discretization and delay of motion parallax also affects the performance when the extent of discretization and delay is large.

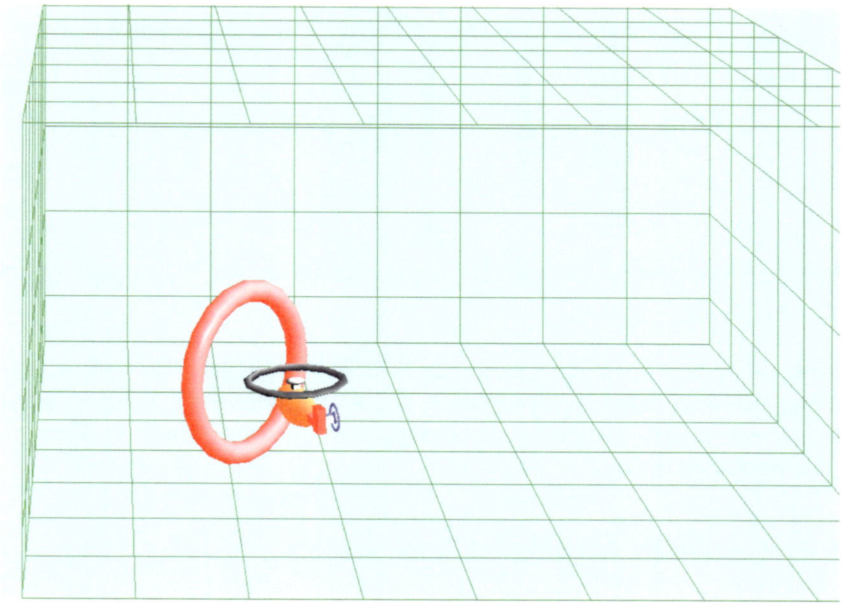

Fig. 7. Screen shot of helicopter remote-control simulator.

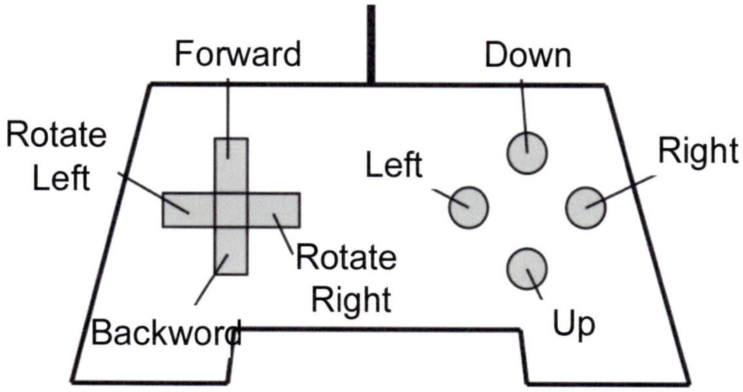

Fig. 8. Alignment of buttons in the pad to control helicopter remote-control simulator.

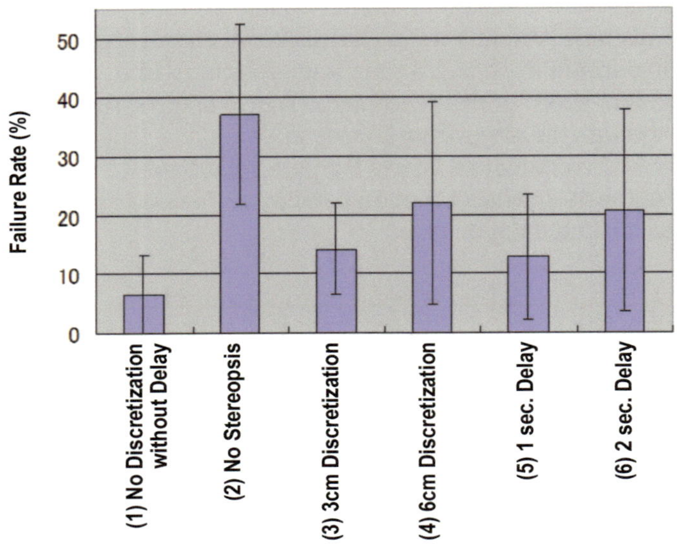

Fig. 9. Experimental results of subjects' performance in the helicopter remote-control simulator.

To sum up the results of these 3 experiments, we can say that smooth motion parallax with little discretization and delay is important in the tasks where the psychological depth cues are limited or the objects in the space occlude one another, which is often the case in complex tele-operation tasks.

3. Depth Perception by Accommodation

3.1 Convergence-Accommodation Conflict

For human depth perception, lack or imperfection of motion parallax is not the only problem of conventional 3D displays. Convergence-accommodation conflict can also affects depth perception of the observer.

When we watch a certian point in the scenery, we focus on that point so that it can be seen clearly. Then the points far from the focused depth are blured. In the natural scene we unconciously keep on controling focus of our eyes so that we can see the objects of interest clearly. When we see typical stereoscopic displays, however, our focus is always fixed on the screen because they rely only on the setereo disparity to show depth of the space. Usually stereoscopic displays emit light from one plane, where the focus of the viewer is always fixed, while binocular convergence of our eyes chnages depending on the stereo displarity of the object we look at.

Under natural circumstances the status of binocular convergence and focal accommodation has stable one-to-one correspondance. Since both convergence and accommodation are unconciously coordinated physiological processes, loss of correspondance has bad influences on physiology of human vision. Concretely the viewers of stereoscopic display often experience eyestrain or sickness peculiar to stereo vision. This loss of correspondance between binocular convergence and focal accommodation has been one of the major problems of stereoscopic display researchers.

Besides eyestrain and sickness, convergence-accommodation conflict can also affects depth perception, which can be a fatal problem for robot tele-operation systems. In the next subsection we show the results of the experiment to examine the effect of convergence-accommodation conflict on depth perception of the viewers.

3.2 Effect of Convergence-Accommodation Conflict on Depth Perception

As for the experiment we use the crane game used in the last section. To prepare the setting where the convergence-accommodation conflict is reduced, we produce the experimental system as shown in Fig. 10. In this system we use two displays and the points in the middle is depicted in both displays with DFD algorithm (Suyama 2000, Suyama 2002, Suyama 2004), where the points are depicted brighter on the nearer display. We trace the head position and use it to depict each point on both displays so that the points on both displays are overlapped and perceived as one point. The light from two displays can be combined by using a half mirror as shown in Fig. 11.

Here we have let 8 subjects try the crane game described in the previous section under 1 display condition and 2 display condition. The subjects have repeated 10 trials and the avergage performances are compared. The result of the experiment is shown in Fig. 12.

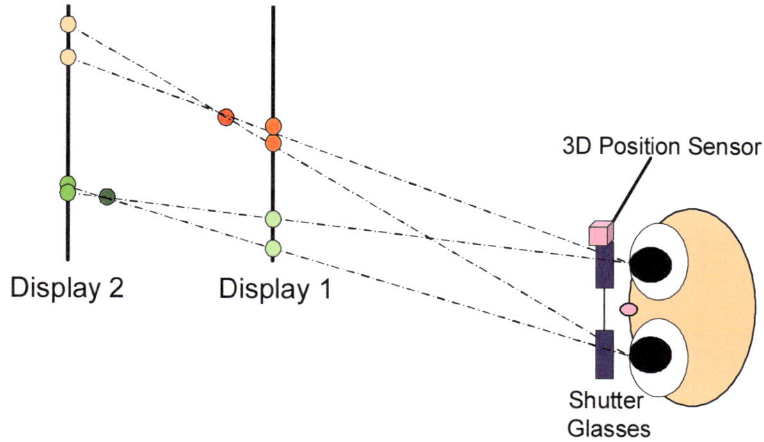

Fig. 10. Experimental system with 2 displays to ease convergence-accommodation conflict.

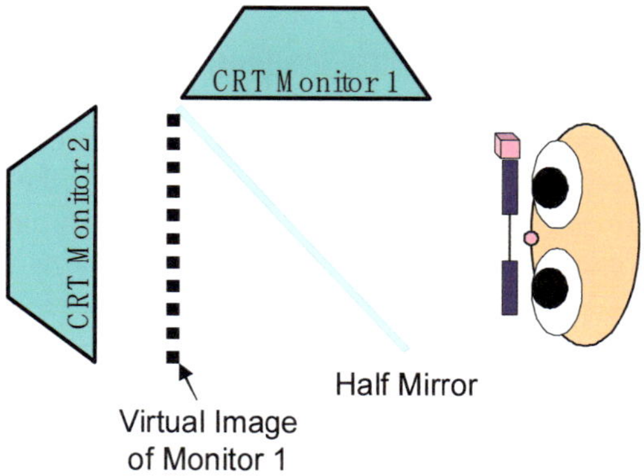

Fig. 11. Experimental instruments to merge images from 2 displays at different depths.

Though 2 display condition is better on average, the difference between two conditions are not very wide. It should be noted, however, that all of those who have performed poorly under 1 display condition have improved their performances under 2 display condition. It suggests that those who are not good at traditional stereopsis can perceive depth better when convergence-accommodation conflict is eased by inserting another display at different depth.

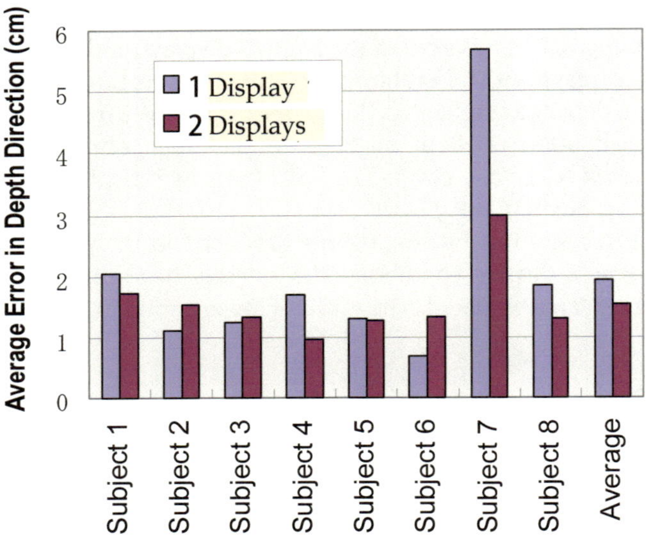

Fig. 12. Experimental results of subjects' performance in the crane game under 1-display and 2-display conditions.

4. Coarse Integral Volumetric Imaging

As described in the previous sections, to let the operator of robot grasp precise depth, a 3D display system with smooth motion parallax and little convergence-accommodation conflict is required. In this section we introduce a new 3D display system which can meet these requirements.

4.1 Concept of Coarse Integral Volumetric Imaging

Integral imaging, which combines fly-eye lenses and a high resolution flat display panel, is a prominent 3D display system in the sense that it can show not only horizontal parallax but also vertical parallax. In the conventional integral imaging, the number of pixels each component lens of the fly-eye lens sheet covers is usually the same as the number of views, which means that the viewer perceives each component lens as one pixel. Therefore the focus of the viewer's eyes is always fixed on the screen (fly-eye lens sheet), which makes it hard to show realistic images far beyond the screen or popping up from the screen.

Besides the orthodox integral imaging described above, we can also think of integral imaging where each component lens is large enough to cover pixels dozens of times more than the number of views. We have defined this type of integral imaging as coarse integral imaging (Kakeya 2008). In recent years coarse integral imaging has been studied by the research group lead by Prof. Byongho Lee (Lee 2002, Min 2005). The advantage of coarse integral imaging is that it can induce focal accommodation off the screen, for it generates a real image or a virtual image with the lenses. Thus we can show realistic images far beyond the screen or popping up from the screen. Yet it cannot overcome the problem of convergence-accommodation conflict because the eyes of the viewer are always focusing on the real image or the virtual image generated by the lens array.

To solve this problem, the author has proposed coarse integral volumetric display method, which combines volumetric solution with multiview solution based on coarse integral imaging (Kakeya 2008). In the proposed system layered transparent display panels are used instead of a single layer display panel for the coarse integral imaging. When we use multi-layered display panels, we can show volumetric real image or virtual image. To express pixels between image planes we can apply DFD approach, where 3D pixels are expressed with two adjacent panels, each of which emit light in inverse proportion to the distance between the 3D pixels and the panel. With this method we can overcome the shortcomings of multiview displays and volumetric displays at the same time.

Conventional volumetric displays can achieve natural 3D vision without contradiction between binocular convergence and focal accommodation, while they cannot express occlusion or gloss of the objects in the scene. On the contrary multiview displays can express the latter while it cannot achieve the former. The coarse integral volumetric display can realize both natural 3D vision and expression of occlusion and gloss.

4.2 Detail of Coarse Integral Volumetric Imaging

Before explaining detail of coarse integral volumetric imaging, we give a brief review of coarse integral imaging. As explained above, coarse integral volumetric imaging has real image version where the 3D image is popping up and the virtual image version where the 3D image is shown beyond the screen. Here we explain the real image version, for it can show 3D structure of remote spaces better because of the closeness.

In the real-image coarse integral imaging we usually keep the distance between the display panel and the lens sheet (convex lens array) the same as the focal distance of the component lenses of the lens array. With this configuration only, the light is just collimated and real image is not generated. To generate real image, we use a large aperture convex Fresnel lens as shown in Fig. 13. Then the real image with little aberration is generated at the focal distance of the Fresnel lens away from the Fresnel lens surface.

We can make a multiview system where the whole image observed in each eye switches alternately when we keep the distance between the lens array and the large aperture Fresnel lens long enough to generate the real image of the lens array, which corresponds to the viewing zone where the view for each eye changes alternately (Kakeya 2007).

The merit of this configuration is that the center of the image from all the viewpoints goes through the optical axis of each component lenses. If we try to converge light only with small component lenses, the light which goes to the center of the image is not perpendicular to the optical axis of each component lenses. Then the distance between the LCD panel and the lens array in the optical path becomes larger as the viewpoint becomes farther from the center, which makes the distance between the lens array and the real image shorter. With the configuration shown in Fig. 13, the distance between the center of the image on LCD and the center of the lens is constant regardless of the difference of viewpoints. Thus we can form real images almost on the same plane with the help of the large convex Fresnel lens.

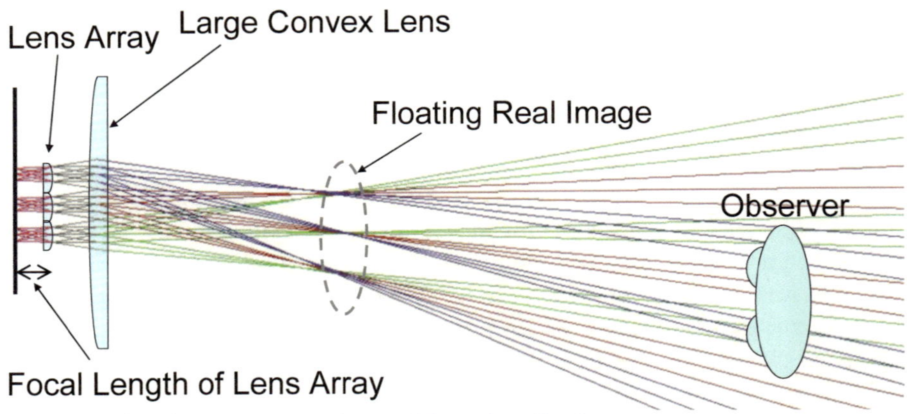

Fig. 13. Optics of real-image coarse integral imaging display.

The main differences between this system and the conventional integral imaging displays are the size of the component lenses of the convex lens array and the use of large aperture Fresnel lens. In this system each component lens of the lens sheet covers about hundred by hundred pixels. In the traditional integral imaging all the edges of the image are in the plane of lens sheet, because each component lens of the lens sheet corresponds to one pixel. In coarse integral imaging, however, the image through each lens includes large number of pixels, whose edges can induce the viewer to focus on the real image produced by the lenses. Thus the image produced with coarse integral imaging can be perceived as an image floating in the air.

Though coarse integral imaging can show images off the screen, the problem of convergence-accommodation conflict still exists, for it can only generate one image plane. This can deteriorate depth perception of the viewer as discussed in Section 3. Besides convergence-accommodation conflict, coarse integral imaging has another major problem which can severely damage the quality of the image. It is discretizaton of parallax as discussed in Section 2. When the distance between the lens array and the large aperture Fresnel lens is not far enough, multiple images from different component lenses are observed at the same time. In this case discontinuity of the images from different lenses becomes severe because of the parallax discretization when the 3D image to be shown has large depth. To show depth of the image, this system depends only on the parallax given by multiview principle. The parallax among the images from different lenses has to be larger as the depth of the 3D object to be shown becomes wider. Consequently discontinuity of the images on the boundaries of the lenses becomes apparent as shown in Fig. 14, which damages the image quality.

To solve the problem of convergence-accommodation conflict and discontinuity of image at the same time, we have proposed coarse integral volumetric imaging, which is based on the idea of introducing volumetric approach in addition to multiview approach (Yasui et al. 2006, Ebisu et al. 2007).

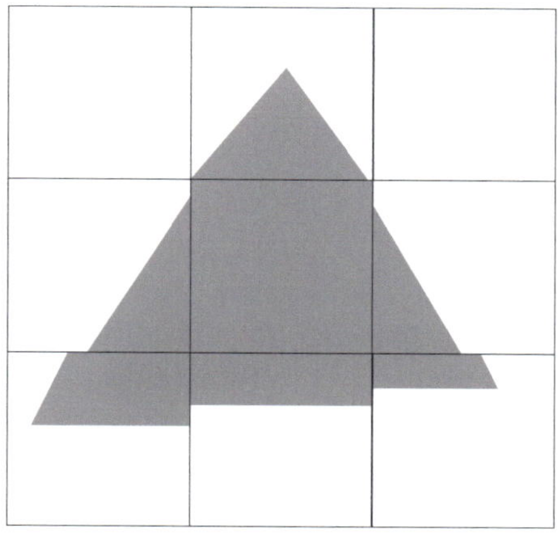

Fig. 14. Discontinuity of image in coarse integral imaging bacause of parallax discretization.

As shown in Fig. 15, multiple display panels are inserted to generate volumetric real image to keep the parallax between the images from two adjacent lenses small enough. Since artificial parallax is kept small, discontinuity between images from adjacent lenses are also kept small. Convergence-accommodation conflict is also reduced since each 3D pixel is displayed at the real-image layer near the right depth.

To express pixels between two panels we can use DFD approach, where 3D pixels are expressed with two adjacent panels, each of which emit light in inverse proportion to the distance between the 3D pixels and the panel. Thus natural continuity of depth is realized.

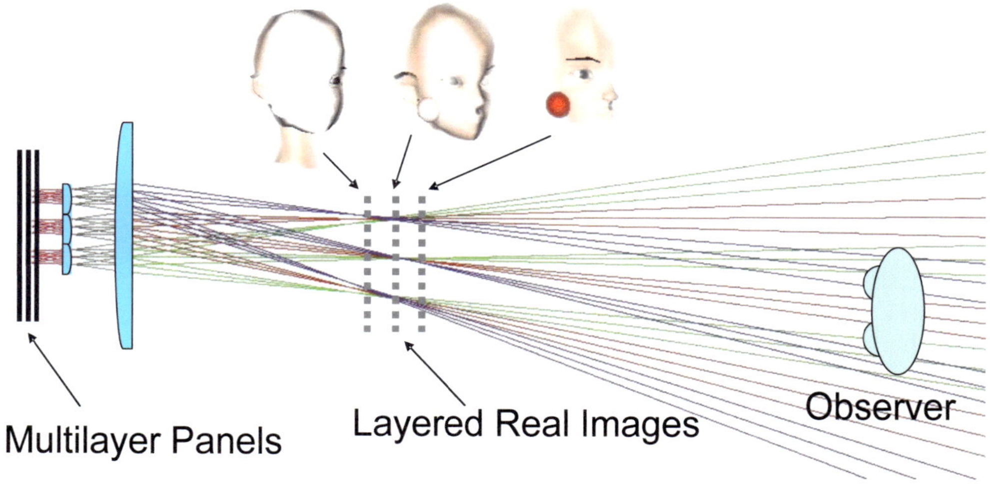

Fig. 15. Principle of coarse integral volumetric imaging.

4.3 Improvement and Application of Coarse Integral Volumetric Imaging

Fig. 15 approximates that the real-image planes are flat. In reality, however, the real image is curved and distorted as shown in Fig. 16. Not only the generated image plane is distorted, but also the image planes generated by component lenses of the lens array are not uniform. The image planes generated by the component lenses off the optical axis of the large aperture Fresnel lens do not have line symmetry about the optical axis, but are slanted toward the optical axis. The slant becomes greater as the component lens goes farther from the optical axis.

To compensate these distortions, each 3D pixel should be drawn on the adjacent two distorted image planes so that the brightness may be in inverse proportion to the distance to each plane as shown in Fig. 17. Since ways of distortion are different among component lenses, we apply this method for each component image by modifying the parameters so that each pixel is drawn at proper 3D positions (Kakeya 2009).

Fig. 17 shows a 3D still picture shown with the coarse integral volumetric display using DFD for distorted image plane. Here to increase connectivity between adjacent images, hexagon lenses are used for the component lenses of the lens array, for less artificial parallax is needed when the lens pitch becomes shorter.

To apply coarse integral volumetric imaging to tele-manipulation, we need electronic display and camera system for it. Currently it is hard to obtain inexpensive volumetric display with resolutions high enough to be applied to integral imaging. With the advancement of current display technologies, however, it is expected that we can obtain inexpensive solution to realize high resolution volumetric displays in a couple of years. With the current technology we have realized an electronic display for coarse integral volumetric imaging by merging images from different depths by half mirrors, though the hardware becomes bulky (Fig. 18).

As for camera system, coarse integral volumetric imaging needs not only images from multiple cameras, but also depth of pixles in the image from each camera. One way to realize it is stereo-matching. This algorithm, however, requires much calculation and frame rate becomes low unless we use high-spec computers. Further research is needed to overcome this problem.

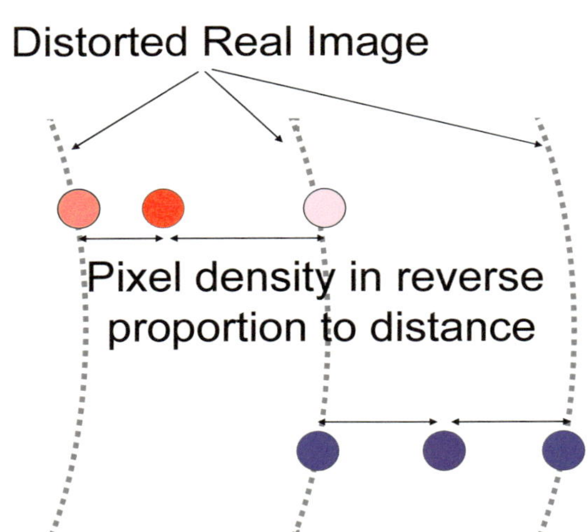

Fig. 16. Distortion of image palnes and DFD algorithm for distorted image planes

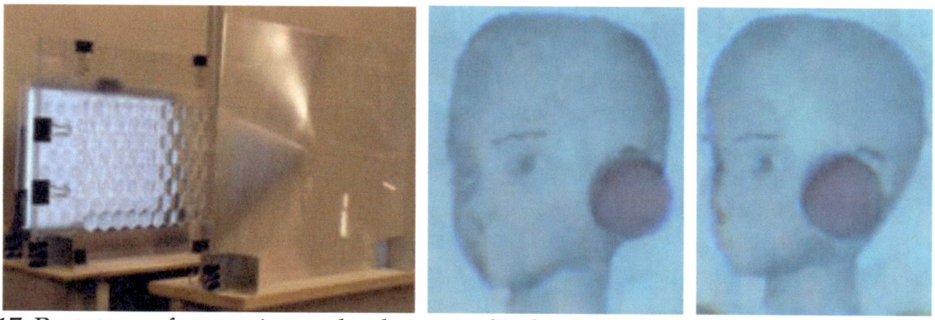

Fig. 17. Prototype of coarse integral volumetric display using DFD for distorted image plane and hexagon component lenses (left) and the images observed from two different viewpoints with the prototype system (middle and right).

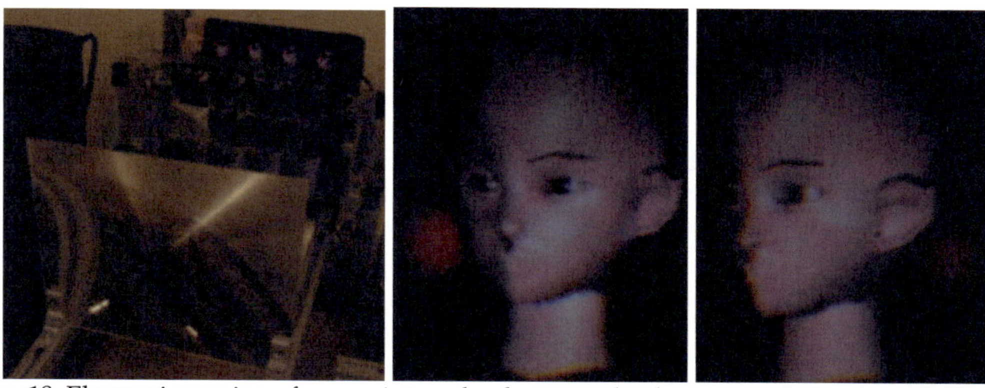

Fig. 18. Electronic version of coarse integral volumetric display using a half mirror (left) and the images observed from two different viewpoints with this display (middle and right).

7. Conclusion

In this chapter, we have discussed the influence of imperfect motion parallax and convergence-accommodation conflict on depth perception of the operator. Smooth motion parallax with little discretization and delay is important in the tasks where the psychological depth cues are limited or the objects in the space occlude one another. Also convergence-accommodation conflict has bad influence on depth perception.

To overcome these problems we have introduced coarse integral volumetric imaging, which can show smooth motion parallax and reduce convergence-accommodation conflict. In the coarse integral volumetric display, multi-layered display panels are used for each component image, which is refracted by a small component lens of the convex lens array and a large aperture Fresnel lens to generate volumetric real image or virtual image. To express pixels between two panels, 3D pixels are expressed with two adjacent panels, each of which emit light in inverse proportion to the distance between the 3D pixels and the panel. It has been confirmed with a prototype system that coarse integral volumetric imaging can realize smooth motion parallax.

8. References

Ebisu, H., Kimura, T., & Kakeya, H. (2007). Realization of electronic 3D display combining multiview and volumetric solutions, *SPIE proceeding* Volume 6490: *Stereoscopic Displays and Virtual Reality Systems XIV*, 64900Y.

Kakeya, H. (2007). MOEVision: simple multiview display with clear floating image, *SPIE proceeding* Volume 6490: *Stereoscopic Displays and Virtual Reality Systems XIV*, 64900J .

Kakeya, H. (2008). Coarse integral imaging and its applications, *SPIE proceeding* Volume 6803 : *Stereoscopic Displays and Virtual Reality Systms XV*, 680317.

Kakeya, H. (2009). Improving Image Quality of Coarse Integral Volumetric Display, *SPIE proceeding* Volume 7237, *Stereoscopic Displays and Virtual Reality Systems XVI* , 723726.

Lee, B., Jung, S, Park, J. & Min, S. (2002). Viewing-angle-enhanced integral imaging using lens switching, *SPIE proceeding* Volume 4660: *Stereoscopic Displays and Virtual Reality Systems IX*, pp.146-154.

Min, S., Kim, J., & Lee, B. (2005). Three-dimensional electro-floating display system based on integral imaging schem, *SPIE proceeding* Volume 5664: *Stereoscopic Displays and Virtual Reality Systems XII*, pp.332-339.

Suyama, S., Takada, H., Uehira, K., Sakai S. & Ohtsuka, S. (2000). A Novel Direct-Vision 3-D Display using Luminance-Modulated Two 2-D Images Displayed at Different Depths, *SID'00 Digest of Technical Papers*, 54.1, pp. 1208-1211.

Suyama, S., Takada, H. & Ohtsuka, S. (2002). A Direct-Vision 3-D Display Using a New Depth-fusing Perceptual Phenomenon in 2-D Displays with Different Depths, *IEICE Trans. on Electron.*, Vol. E85-C, No. 11, pp.1911-1915.

Suyama, S., Ohtsuka, S., Takada, H., Uehira, K. & Sakai, S. (2004). Apparent 3-D image perceived from luminance-modulated two 2-D images displayed at different depths, *Vision Research*, 44, pp. 785-793.

Yasui, R., Matsuda, I., & Kakeya, H. (2006). Combining volumetric edge display and multiview display for expression of natural 3D images, *SPIE proceeding* Volume 6055: *Stereoscopic Displays and Virtual Reality Systems XIII*,60550Y.

Experimental evaluation of output–feedback tracking controllers for robot manipulators

Javier Moreno–Valenzuela
Centro de Investigación y Desarrollo de Tecnología Digital del IPN
Mexico

Víctor Santibáñez and Ricardo Campa
Instituto Tecnológico de La Laguna
Mexico

1. Introduction

While the position of a robot link can be measured accurately, measurement of velocity and acceleration tends to result in noisy signals. In extreme cases, these signals could be so noisy that their use in the controller would no longer be feasible (Daly & Schwartz, 2006).
In order to overcome the problem of noisy velocity measurements and to guarantee that the error between the time–varying desired position and the actual position of the robot system goes asymptotically to zero for a set of initial conditions, a controller/observer scheme, based on position measurements, can be used. In this sort of schemes, the incorporated observer is used to estimate the velocity signal and sometimes the acceleration signal.
Another approach consists in using the Lyapunov theory to design a controller/filter to guarantee the tracking of the desired trajectory, no matter if an estimate of the velocity and acceleration can be possible with the obtained design.
In the perspective of control engineering, the approach of using only joint position measurements in either a controller/observer or a controller/filter to achieve tracking of a desired joint trajectory is denominated output–feedback tracking control of robot manipulators.
Recently, attention has been paid to the practical evaluation of output–feedback tracking controllers. In the paper by Arteaga & Kelly (2004) a comparison of several output–feedback tracking controllers is made, showing that those schemes which incorporate either an observer or filter are better than those which incorporate a numerical differentiation to obtain an estimation of the joint velocity.
The work by Daly & Schwartz (2006) reported the experimental results concerning three output–feedback tracking controllers. They showed the advantages and disadvantages of each control scheme that was tested.
On the other hand, saturation functions have been used in output–feedback tracking control schemes to guarantee that the control action is within the admissible actuator capability. See the papers by Loría & Nijmeijer (1998), Dixon et al. (1998), Dixon et al. (1999) and, more recently, Santibáñez & Kelly (2001). Although much effort was done to derive those controllers and very complex stability analyses were necessary, as far as we know, no experimental evalu-

ation of output–feedback tracking controllers that contain saturation functions in its structure has been reported.
Considering that output–feedback tracking controllers can be more efficient than full–state feedback tracking controllers, specially if noisy velocity measurements are present, and taking into account the philosophy of including saturation functions in an output–feedback tracking control design, the objective of this paper is to present an experimental comparison between output–feedback tracking controllers that do not have saturation functions in its structure and controllers that do have saturation functions. The experiments were carried out in a two degrees–of–freedom direct–drive robot, which is important from the control point of view, because the dynamics in this type of robots is highly nonlinear.
This chapter is organized as follows. The robot model and control problem formulation is presented in Section 2. Section 3 describes the experimental robot arm used in the experiments. Section 4 concerns to the description of the desired position trajectory and performance criterion. The controllers as well as the experiments on output–feedback tracking control are presented in Section 5, while Section 6 contains some discussions. Finally, concluding remarks are drawn in Section 7.

2. Robot dynamics and control goal

The dynamics in joint space of a serial–chain n-link robot manipulator considering the presence of friction at the robot joints can be written as (Canudas de Wit et al., 1996; Kelly et al., 2005; Ortega et al., 1998; Sciavicco & Siciliano, 2000):

$$M(q)\ddot{q} + C(q,\dot{q})\dot{q} + g(q) + F_v\dot{q} = \tau \tag{1}$$

where $M(q)$ is the $n \times n$ symmetric positive definite inertia matrix, $C(q,\dot{q})$ is the $n \times n$ vector of centripetal and Coriolis torques, $g(q)$ is the $n \times 1$ vector of gravitational torques, $F_v = diag\{f_{v1},\ldots,f_{vn}\}$ is the $n \times n$ positive definite diagonal matrix which contains the viscous friction coefficients of the robot joints, and τ is the $n \times 1$ vector of applied torque inputs.
Assume that only the robot joint displacements $q(t) \in \mathbb{R}^n$ are available for measurement. Then, the output–feedback tracking control problem is to design a control input $\tau(t)$ so that the joint displacements $q(t) \in \mathbb{R}^n$ converge asymptotically to the desired joint displacements $q_d(t) \in \mathbb{R}^n$, i.e.,

$$\lim_{t\to\infty} \tilde{q}(t) = 0, \tag{2}$$

where

$$\tilde{q}(t) = q_d(t) - q(t) \tag{3}$$

denotes the tracking error.

3. Experimental robot system

A direct–drive arm with two vertical rigid links —see Fig. 1— is available at the Mechatronics and Control Laboratory of the *Instituto Tecnológico de La Laguna*, which was designed and built at the Robotics Laboratory of CICESE Research Center. High–torque brushless direct–drive motors operating in torque mode are used to drive the joints without gear reduction.
A motion control board based on a TMS320C31 32–bit floating–point microprocessor from Texas Instruments is used to execute the control algorithm. The control program is written in C programming language and executed in the control board at $h = 2.5$ [ms] sampling

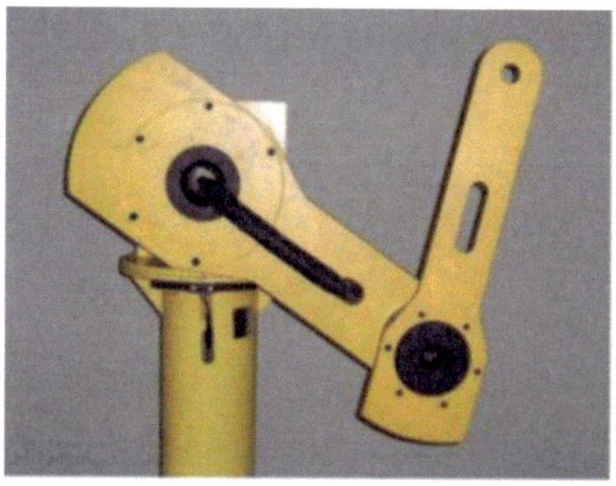

Fig. 1. Experimental robot arm

period. The maximum torque limits are τ_1^{Max}=150 [Nm] and τ_2^{Max}=15 [Nm] for motor 1 and 2, respectively.
The robot dynamics is described with details in (Reyes & Kelly, 1997; 2001). With reference to the symbols listed in Table 1, we present below the entries of the robot dynamics:

	notation	*value*	*unit*
Link 1 length	l_1	0.45	m
Link 2 length	l_2	0.45	m
Link 1 center of gravity	l_{c1}	0.091	m
Link 2 center of gravity	l_{c2}	0.048	m
Link 1 mass	m_1	23.90	kg
Link 2 mass	m_2	3.88	kg
Link 1 moment of inertia	I_1	1.27	kg m^2
Link 2 moment of inertia	I_2	0.09	kg m^2
Gravity acceleration	g	9.8	m/s^2

Table 1. Parameters of the manipulator

The elements $M_{ij}(q)$ $(i,j = 1,2)$ of the inertia matrix $M(q)$ are

$$\begin{aligned}
M_{11}(q) &= m_1 l_{c1}^2 + m_2\left(l_1^2 + l_{c2}^2 + 2 l_1 l_{c2}\cos(q_2)\right) \\
&\quad + I_1 + I_2, \\
M_{12}(q) &= m_2\left(l_{c2}^2 + l_1 l_{c2}\cos(q_2)\right) + I_2, \\
M_{21}(q) &= m_2\left(l_{c2}^2 + l_1 l_{c2}\cos(q_2)\right) + I_2, \\
M_{22}(q) &= m_2 l_{c2}^2 + I_2.
\end{aligned}$$

The elements $C_{ij}(q,\dot{q})$ $(i,j=1,2)$ of the centripetal and Coriolis matrix $C(q,\dot{q})$ are

$$\begin{aligned} C_{11}(q,\dot{q}) &= -m_2 l_1 l_{c2}\sin(q_2)\dot{q}_2, \\ C_{12}(q,\dot{q}) &= -m_2 l_1 l_{c2}\sin(q_2)\left(\dot{q}_1+\dot{q}_2\right), \\ C_{21}(q,\dot{q}) &= m_2 l_1 l_{c2}\sin(q_2)\dot{q}_1, \\ C_{22}(q,\dot{q}) &= 0. \end{aligned}$$

The entries of the gravitational torque vector $g(q)$ are given by

$$\begin{aligned} g_1(q) &= (m_1 l_{c1}+m_2 l_1)g\sin(q_1)+m_2 l_{c2} g\sin(q_1+q_2), \\ g_2(q) &= m_2 l_{c2} g\sin(q_1+q_2). \end{aligned}$$

The coefficients of the viscous friction are

$$F_v = diag\{2.288, 0.175\}\ [\text{N m s}].$$

Experiments showed that static and Coulomb friction at the motor joints are present and they depend in a complex manner on the joint position and velocity. We have decided to consider them as disturbances for the closed–loop system.

4. Desired position trajectory and performance criterion

The desired position trajectory $q_d(t)$ used in all experiments is given by

$$q_d(t) = \begin{bmatrix} 45[1-e^{-2.0t^3}]+10[1-e^{-2.0t^3}]\sin(7.50t) \\ 60[1-e^{-1.8t^3}]+125[1-e^{-1.8t^3}]\sin(1.75t) \end{bmatrix}\ [\text{degrees/s}]. \tag{4}$$

An important characteristic of the position trajectory $q_d(t)$ in (4) is that the desired velocity $\dot{q}_d(t)$ and acceleration $\ddot{q}_d(t)$ are null in $t=0$, then the closed–loop system trajectories will not present rude transients if the robot starts at rest. It is noteworthy that the execution of the proposed trajectory $q_d(t)$ in (4) demanded a 75% of the torque capabilities, which was estimated through numerical simulation and verified with the experiments.
The time evolution of the position error $\tilde{q}$ reflects how well the control system performance is. The performance criterion considered in this chapter was the Root Mean Square —RMS— value of the velocity error computed on a trip of time T, that is,

$$RMS[\dot{\tilde{q}}] = \sqrt{\frac{1}{T}\int_0^T \|\dot{\tilde{q}}(\sigma)\|^2 d\sigma}\ [\text{degrees/s}]. \tag{5}$$

In practice, the discrete implementation of the criterion (5) leads to

$$RMS[\dot{\tilde{q}}] = \sqrt{\frac{1}{T}\sum_{k=0}^{i} \|\dot{\tilde{q}}(kh)\|^2\, h}\quad [\text{degrees/s}],$$

where $h = 2.5$ [ms] is the sampling period and $T = 10$ [s].

5. Tested controllers

We have tested six controllers. Three of the tested controllers do not contain saturation functions, while the others do it. The main goal of the experimental evaluation was to assess the tracking performance of controllers that do not have saturation functions, i.e.,

- PD+ (Paden & Panja, 1988),
- Loría & Ortega (1995), and
- Lee & Khalil (1997),

with respect to those that have saturation functions,

- Loría & Nijmeijer (1998),
- New Design 1, and
- New Design 2.

The controllers denoted as New Design 1 and New Design 2, which are defined explicitly later, were proposed in (Moreno et al., 2008). Tools for analysis of singularly perturbed systems are used to show the local exponential stability of the closed–loop system given by those controllers and the robot dynamics.

Let us first describe the results concerning controllers without saturation functions. The first controller tested was the PD+ control (Paden & Panja, 1988), which is written as

$$\tau = M(q)\ddot{q}_d + C(q,\dot{q})\dot{q}_d + g(q) + F_v\dot{q} + K_d\dot{\tilde{q}} + K_p\tilde{q}, \tag{6}$$

where K_p and K_d are $n \times n$ symmetric positive definite matrices, $\tilde{q} = q_d - q$ denotes the tracking error and the joint velocity measurements $\dot{q}$ are approximated via simple numerical differentiation, i.e.,

$$\dot{q}_i(hk) = \frac{q_i(hk) - q_i(h[k-1])}{h} \tag{7}$$

where h is the sampling period and k is the discrete time. It is well known that the approach (7) is very common in many robot control platforms to obtain an estimation of the velocity measurements. The controller was tested using the following proportional and derivative control gains

$$\begin{aligned} K_p &= diag\{3500, 1000\}\ [1/\text{s}^2], \\ K_d &= diag\{45, 15\}\ [1/\text{s}], \end{aligned} \tag{8}$$

Let us notice that the gains (8) were obtained by trial and error until obtaining a reasonable performance in the tracking of the desired joint position $q_d(t)$, i.e., a relatively small bound of the maximum values of $\tilde{q}_1(t)$ and $\tilde{q}_2(t)$. Fig. 2 shows the time evolution of tracking errors $\tilde{q}_1(t)$, $\tilde{q}_2(t)$, and applied torques $\tau_1(t)$, $\tau_2(t)$.

Further improvement could have been obtained in the tracking performance, but paying the price of a noisy control action, which would excite other dynamics such as vibrating modes of the mechanical structure.

Since all of the tested controllers have a proportional–derivative structure, we have used the same numerical value of the gains in (8) for all of them, while the remaining gains in each controller were selected so that a reasonable performance were obtained, as we will explain later.

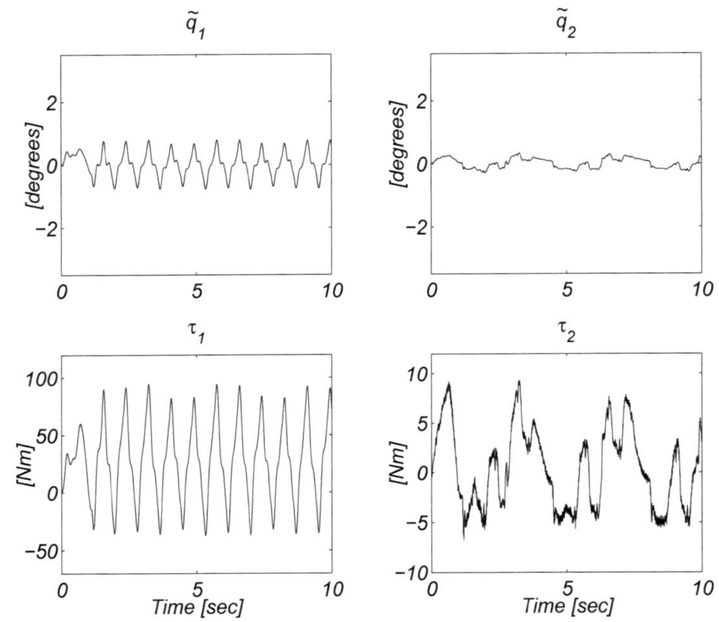

Fig. 2. PD+ controller : Tracking errors $\tilde{q}_1(t)$, $\tilde{q}_2(t)$, and applied torques $\tau_1(t)$, $\tau_2(t)$.

The other two control schemes that do not contain saturation functions correspond to the output–feedback tracking controllers by Loría & Ortega (1995) and Lee & Khalil (1997). The Loría & Ortega (1995) controller is written as

$$\tau = M(q)\ddot{q}_d + C(q,\dot{q}_d)\dot{q}_d + g(q) + F_v\dot{q} + K_d\tilde{\vartheta} + K_p\tilde{q}, \tag{9}$$

where $\tilde{\vartheta} \in \mathbb{R}^n$ is obtained with the linear filter

$$\dot{x} = -b_f\tilde{\vartheta},$$
$$\tilde{\vartheta} = x + b_f\tilde{q},$$

with $b_f > 0$. The controller (9) was implemented in our system with the control gains K_p and K_d in (8) and

$$b_f = 600.0\ [1/\mathrm{s}]. \tag{10}$$

The Lee & Khalil (1997) controller in its non–adaptive version can be written as

$$\begin{aligned}\tau = {} & M(q_d - x_1)\ddot{q}_d + C(q_d - x_1,\dot{q}_r)\dot{q}_d + g(q_d - x_1) \\ & + F_v[\dot{q}_d - x_2] + K_d x_2 + K_p x_1,\end{aligned} \tag{11}$$

where $\dot{q}_r = \dot{q}_d - x_2 + \lambda\tilde{q}$ and the second order high–gain observer is defined as

$$\dot{x}_1 = x_2 + \frac{L_1}{\epsilon}[\tilde{q} - x_1], \tag{12}$$

$$\begin{aligned}\dot{x}_2 = {} & \frac{L_2}{\epsilon^2}[\tilde{q} - x_1] + \ddot{q}_d \\ & + M(q_d - x_1)^{-1}[C(q_d - x_1,\dot{q}_r)\dot{q}_d \\ & + g(q - x_1) + F_v[\dot{q}_d - x_2] - \tau].\end{aligned} \tag{13}$$

The gains used for the implementation of this controller were K_p and K_v in (8), $\lambda = 1$ [1/s],

$$L_1 = diag\{5.0, 40.0\} \; [1/\mathrm{s}], \;\; L_2 = diag\{50.0, 400.0\} \; [1/\mathrm{s}^2],$$

and $\epsilon = 0.1$. We tried several sets of gains for the observer (12)–(13) until obtaining a good response in the tracking error. However, it was pretty difficult to find numerical values for which instability was avoided. As pointed out in (Daly & Schwartz, 2006), the real–time implementation of the controller/observer (11)–(13) may make estimations of the position and velocity errors inaccurate to the point of not being useful, which was confirmed during the experimental set up.
The obtained experimental results are given in Fig. 3, for the Loría and Ortega controller, and in Fig. 4, for the Lee and Khalil scheme.

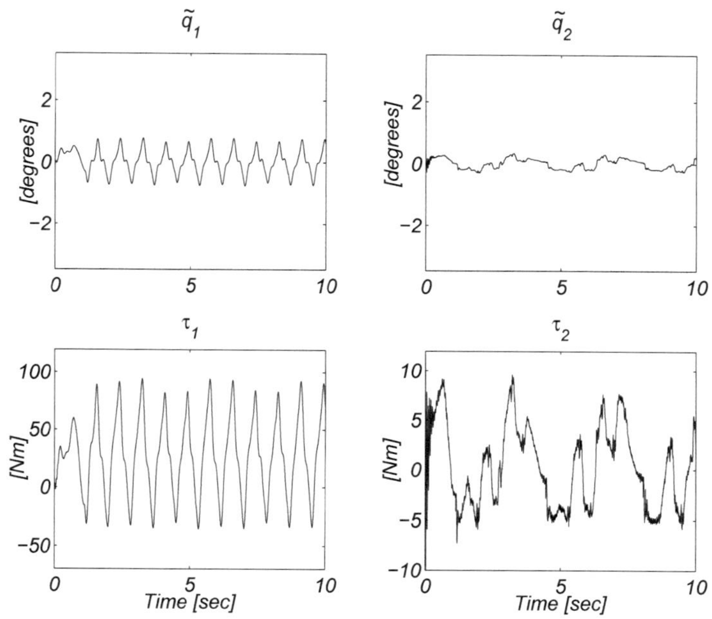

Fig. 3. Loría and Ortega controller: Tracking errors $\tilde{q}_1(t)$, $\tilde{q}_2(t)$, and applied torques $\tau_1(t)$, $\tau_2(t)$.

On other hand, concerning output–feedback tracking controllers that contain saturation functions in their structure, the first controller tested was the Loría & Nijmeijer (1998) approach, written as

$$\begin{aligned} \tau = {} & M(q)\ddot{q}_d + C(q, \dot{q}_d)\dot{q}_d + g(q) + F_v\dot{q} \\ & + K_d \, col\{\tanh(\tilde{\vartheta}_i)\} + K_p \, col\{\tanh(\tilde{q}_i)\}, \end{aligned} \tag{14}$$

where $col\{f(x_i)\} = [f(x_1) \;\cdots\; f(x_n)]^T \in \mathbb{R}^n$ for any scalar function f, used along with the saturated filter

$$\begin{aligned} \dot{x} &= -b_f \, col\{\tanh(\tilde{\vartheta}_i)\}, \\ \tilde{\vartheta} &= x + b_f \tilde{q}. \end{aligned}$$

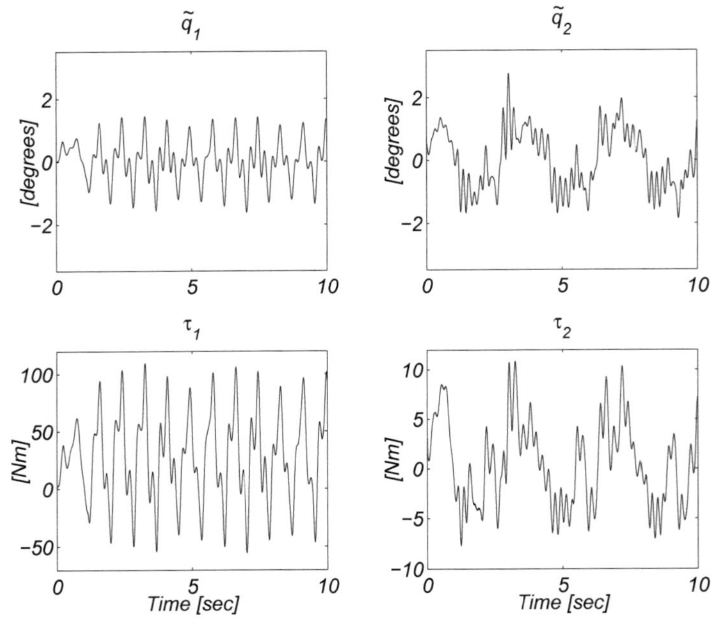

Fig. 4. Lee and Khalil controller: Tracking errors $\tilde{q}_1(t)$, $\tilde{q}_2(t)$, and applied torques $\tau_1(t)$, $\tau_2(t)$.

Once again, in order to keep a fair comparison with respect to the three previous controllers that do not use saturation functions, we used the numerical values of K_p and K_d in (8) and b_f in (10). The result of the experiment is depicted in Fig. 5.
Besides, we implemented the controller denoted as New Design 1 (Moreno et al., 2008)

$$\tau = M(q)\ddot{q}_d + C(q,\dot{q}_d)\dot{q}_d + g(q) + F_v\dot{q}_d$$
$$+ K_d\, col\left\{\frac{\tilde{\vartheta}_i}{\sqrt{\delta_{di} + \tilde{\vartheta}_i^2}}\right\} + K_p\, col\left\{\frac{\tilde{q}_i}{\sqrt{\delta_{pi} + \tilde{q}_i^2}}\right\}, \tag{15}$$

where δ_{pi} and δ_{di} are strictly positive constants, and the filter

$$\dot{x} = -b_f\, col\left\{\frac{\tilde{\vartheta}_i}{\sqrt{\delta_{di} + \tilde{\vartheta}_i^2}}\right\},$$
$$\tilde{\vartheta} = x + b_f\tilde{q},$$

with $b_f > 0$ is used. As previously, we chose the numerical values of K_p and K_d as in (8), and b_f as in (10). Parameters δ_{pi} and δ_{di} were

$$\delta_{p1} = 0.3,\ \delta_{p2} = 0.75,\ \delta_{d1} = 1.0,\ \delta_{d2} = 1.0.$$

Fig. 6 shows the results of the experiment.
Finally, the controller New Design 2 (Moreno et al., 2008)

$$\tau = M(q)\ddot{q}_d + C(q,\dot{q}_d)\dot{q}_d + g(q) + F_v\dot{q}_d$$

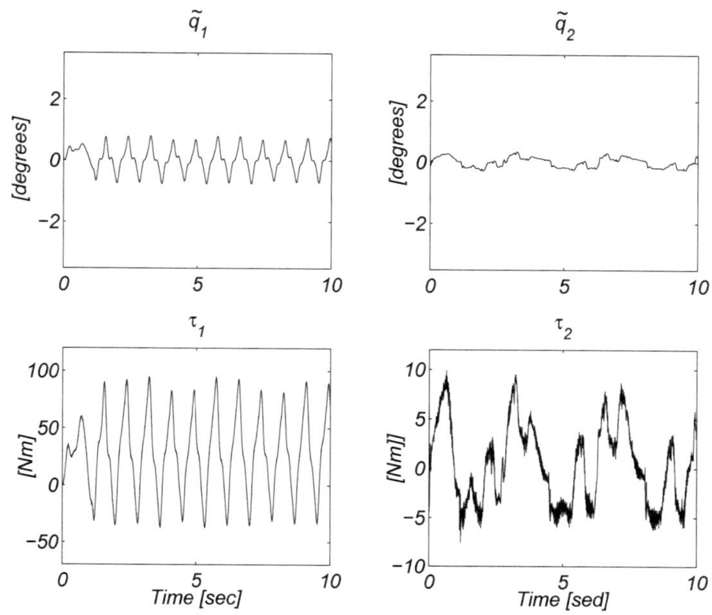

Fig. 5. Loría and Nijmeijer controller: Tracking errors $\tilde{q}_1(t)$, $\tilde{q}_2(t)$, and applied torques $\tau_1(t)$, $\tau_2(t)$.

$$+K_d \; col \left\{ \frac{\tilde{\vartheta}_i}{\delta_{di} + \ln(\cosh(\tilde{\vartheta}_i))} \right\}$$

$$+ K_p \; col \left\{ \frac{\tilde{q}_i}{\delta_{pi} + \ln(\cosh(\tilde{q}_i))} \right\}, \qquad (16)$$

which is used with the filter

$$\begin{aligned} \dot{x} &= -b_f \; col \left\{ \frac{\tilde{\vartheta}_i}{\delta_{di} + \ln(\cosh(\tilde{\vartheta}_i))} \right\}, \\ \tilde{\vartheta} &= x + b_f \tilde{q}, \end{aligned}$$

was tested under the same conditions that the New Design 1 in (15), while the parameters δ_{pi} and δ_{di} used in this case were

$$\delta_{p1} = 0.55, \; \delta_{p2} = 1.0, \; \delta_{d1} = 1.0, \; \delta_{d2} = 1.0.$$

The results are illustrated in Fig. 7.

6. Discussions

All the tested controllers assure theoretically that the position error $\tilde{q}(t)$ must vanish as time increases. In practice, Figures 3 to 8 reveal a steady-state oscillatory behavior. This is due to several factors such as the uncompensated Coulomb friction and the discrete implementation of the controller .

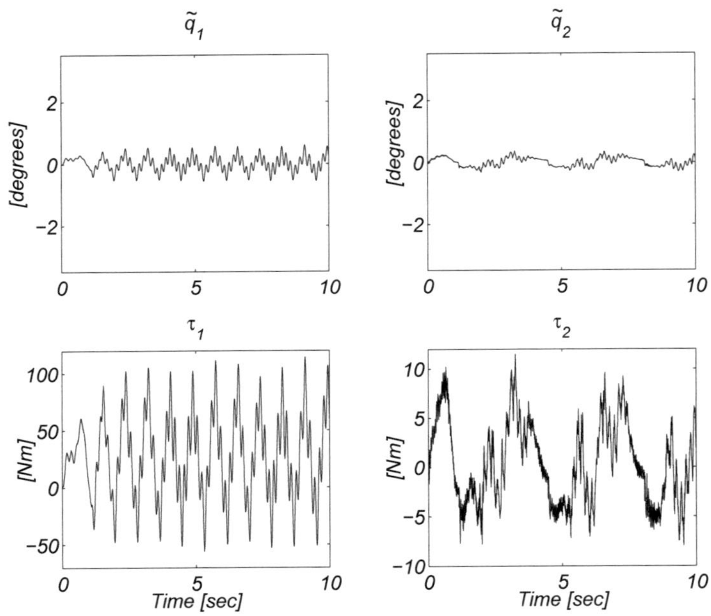

Fig. 6. New Design 1: Tracking errors $\tilde{q}_1(t)$, $\tilde{q}_2(t)$, and applied torques $\tau_1(t)$, $\tau_2(t)$.

	Controllers without saturation functions			Controllers with saturation functions		
	PD+	LO	LK	LN	New 1	New 2
$\max\{\lvert\tilde{q}_1(t)\rvert\}$ [deg]	0.78	0.75	1.62	0.79	0.59	0.61
$\max\{\lvert\tilde{q}_2(t)\rvert\}$ [deg]	0.34	0.34	2.76	0.34	0.36	0.39
$RMS[\tilde{q}]$ [deg]	0.418	0.417	1.134	0.419	0.284	0.299

Table 2. Performance of the controllers: PD+, Loría and Ortega (LO), Lee and Khalil (LK), Loría and Nijmeijer (LN), New Design 1, and New Design 2.

Table 2 summarizes the information about the tracking performance of the six schemes, remarking the difference between controllers that do not use saturation functions and controllers that do use them. In addition, Fig. 8 shows a bar chart of the $RMS[\tilde{q}]$ value computed for the six tested controllers. With respect to controllers without saturation function we can see that the performance of the PD+ controller (6) and the Loría and Ortega algorithm (9) is very similar, while the worst performance of the six controllers was obtained with the Lee and Khalil controller (11). On the other hand, concerning the experimental results using controllers with saturation function, we can see that the performance of the Loría and Nijmeijer scheme (14) is very similar to the one of the controllers PD+ (6) and Loría and Ortega (9). The reason is that in the situation of a very small tracking error $\tilde{q}$ the structure of the three controllers is very similar. In addition, the best performance of the six controllers was obtained with the New Design 1 and New Design 2, in equations (15) and (16), respectively, because they presented the lowest values of $\max\{|\tilde{q}_1(t)|\}$, $\max\{|\tilde{q}_2(t)|\}$ and $RMS[\tilde{q}]$.

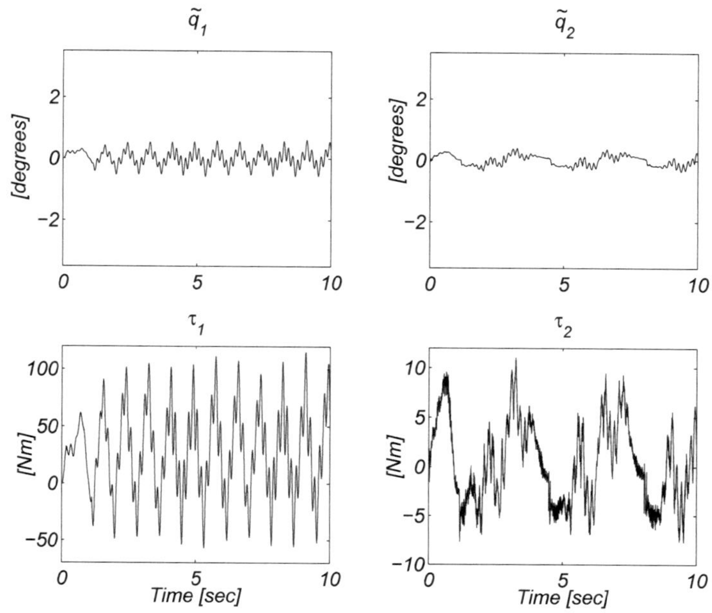

Fig. 7. New design 2: Tracking errors $\tilde{q}_1$, $\tilde{q}_2(t)$, and applied torques $\tau_1(t)$, $\tau_2(t)$.

The comparison reveals that all the controllers work efficiently since the tracking errors are relatively close to the performance of the PD+ controller (6), although the new controllers (15) and (16), which incorporate saturation functions, present a lower tracking error $\tilde{q}(t)$ than other output–feedback tracking controllers, including the PD+ control (6). The explanation of this is that the new controllers (15) and (16) incorporate the extra parameters δ_{pi}, δ_{di}, whose numerical value has effect in increasing the slope of the profile of the saturation function in the proximity of the origin. Different numerical values of δ_{pi} and δ_{di} lead to a similar behavior with respect to the PD+ (6), Loría and Ortega (9), and Loría and Nijmeijer (14) schemes.

7. Concluding remarks

In this work the output–feedback tracking control of robot manipulators was studied. An extensive experimental study in a two degrees–of–freedom direct–drive robot was presented, where it was shown that output–feedback tracking controllers having saturation functions in the corresponding proportional and derivative parts present better tracking performances than the controllers that have simple linear functions in the corresponding proportional and derivative parts.

To the best of the authors' knowledge, the output–feedback tracking controllers discussed in the experimental results have been tested in a real–time robot control system for the first time. The results obtained in practice suggest that output–feedback tracking controllers that incorporate saturation functions are reliable for application in industrial robots.

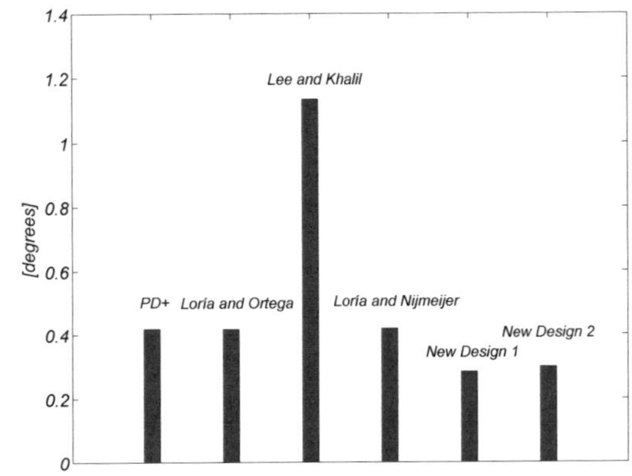

Fig. 8. Bar chart of the $RMS[\tilde{q}]$ value computed for the six tested controllers

8. Acknowledgments

This work was supported in part by CONACyT, Secretaría de Investigación y Posgrado–IPN, PROMEP, and DGEST, Mexico.

9. References

Arteaga, M. & Kelly, R. (2004). Robot control without velocity measurements: New theory and experimental results, *IEEE Transactions on Robotics and Automation* **39**(2): 297–308.

Daly, J. & Schwartz, H. (2006). Experimental results for output feedback adaptive robot control, *Robotica* **24**: 727–738.

Canudas de Wit, C., Siciliano, B. & Bastin, G. (1996). *Theory of Robot Control*, Springer-Verlag, London.

Dixon, W., de Queiroz, M., Dawson, D. & Zhang, F. (1998). Tracking control of robot manipulators with bounded torque inputs, *in Proc. of the 6th IASTED International Conference on Robotics and Manufacturing*, Banff, Canada, July, pp. 112–115.

Dixon, W., de Queiroz, M., Zhang, F. & Dawson, D. (1999). Tracking control of robot manipulators with bounded torque input, *Robotica* **17**(2): 121–129.

Kelly, R., Santibáñez, V. & Loría, A. (2005). *Control of Robot Manipulators in Joint Space*, Springer–Verlag, Berlin.

Lee, K. & Khalil, H. (1997). Adaptive output feedback control of robot manipulators using high–gain observer, *International Journal of Control* **67**(6): 869–886.

Loría, A. & Nijmeijer, H. (1998). Bounded output feedback tracking control of fully actuated Euler–Lagrange systems, *Systems and Control Letters* **33**: 151–161.

Loría, A. & Ortega, R. (1995). On tracking control of rigids and flexible joint robots, *Applied Mathematics and Computer Sciences* **5**(2): 329–341.

Moreno, J., Santibáñez, V. & Campa, R. (2008). A class of OFT controllers for torque–saturated robot manipulators: Lyapunov stability and experimental evaluation, *Journal of Intelligent and Robotic Systems* **55**(1): 65–88.

Ortega, R., Loría, A., Nicklasson, P. & Sira-Ramirez, H. (1998). *Pasivity–Based Control of Euler–Lagrange Systems*, Springer, London.

Paden, B. & Panja, R. (1988). Globally asymptotically stable PD+ controller for robot manipulators, *International Journal of Control* **7**(6): 1697–1712.

Reyes, F. & Kelly, R. (1997). Experimental evaluation of identification schemes on a direct drive robot, *Robotica* **15**: 563–571.

Reyes, F. & Kelly, R. (2001). Experimental evaluation of model–based controllers on a direct–drive robot arm, *Mechatronics* **11**: 267–282.

Santibáñez, V. & Kelly, R. (2001). Global asymptotic stability of bounded output feedback tracking control for robot manipulators, *in Proc. of the 40th IEEE Conf. Decision and Control*, Orlando, USA, December, pp. 1378–1379.

Sciavicco, L. & Siciliano, B. (2000). *Modelling and Control of Robot Manipulators*, Springer, London.

Higher Dimensional Spatial Expression of Upper Limb Manipulation Ability based on Human Joint Torque Characteristics

Makoto Sasaki[1], Takehiro Iwami[2], Kazuto Miyawaki[3], Ikuro Sato[4], Goro Obinata[5] and Ashish Dutta[6]
[1]Iwate University, [2]Akita University, [3]Akita National College of Technology
[4]Miyagi Cancer Center Research Institute, [5]Nagoya University
[6]Indian Institute of Technology Kanpur
[1-5]Japan, [6]India

1. Introduction

A basic ergonomic approach to biomechanical analysis of the human body for using assistive devises considers the individual's body functions, his physical load and the efficiency of the assistive devices. Biomechanical analysis using a rigid link model and the musculoskeletal model is a more objective technique. It replaces electromyogram analysis, and has been applied for the estimation of joint forces and torques, muscular tensions, and energy consumption during the use of various devices such as wheelchairs and assistive carts (Sasaki et al., 2005, 2008; Miyawaki et al., 2000, 2007). In this case both the simultaneous measurements of the external forces acting on a human body and the body movements are required for the analysis. On the other hand, analytical tools using computer generated mannequins, which need no measurement data of human motion, have also been developed (Jack; Siemens PLM Software and RAMSIS; Human Solution GmbH and Safework; Safework Inc., etc.). A computer generated mannequin--with standard body characteristics such as length, weight and joint range of motion--is useful for evaluating the compatibility of humans and different industrial products from the viewpoints of the workplace, work posture and also for reducing related design time and development costs. However, it is very difficult to standardize the body characteristics of elderly or physically handicapped people. Especially, the body functions and muscle characteristics of people with spinal cord injuries differ strikingly from those of normal people, depending on the lesion level and extent of paralysis. Therefore, accurate evaluation of an individual's proper body functions and development of assistive devices based on such evaluations are indispensable.

In order to quantify an individual's body functions, the concept of manipulability (Yoshikawa, 1984, 1985, 1990) that is used to analyze robot manipulators, was applied to evaluate the manipulability of the upper and lower limbs (Ohta et al., 1998; Hamada et al., 2000). All possible velocities, accelerations, and forces at the end-effector can be represented

as ellipsoids or polyhedra using the concept of manipulability. This evaluation method, which is commonly used in the field of robotics, provides effective knowledge for evaluation of the manipulability of upper and lower limbs considering both the kinematics and dynamics of the system.
However, this evaluation method is based on the assumption that the torque characteristic of a human joint is constant, and is hence not very accurate because the maximum torque generated at a human joint varies with the joint angle and direction of rotation. Moreover, although the manipulability of the upper and lower limbs in three-dimensional space is expressed as a six-dimensional hyperellipsoid at the maximum, it is difficult to imagine an ellipsoid in four or more dimensions. Even in the field of robotics, the visualization method of a higher-dimensional hyperellipsoid defined in a space with dimensions equal to the degrees of freedom of a hand, has not been clarified yet. A slack variable is generally introduced to search for the vertex of the polytope (Shim & Yoon, 1997; Chiacchio et al., 1997; Lee, 2001). It is extremely difficult to search for the region of a higher-dimensional polytope accurately using these methods because of its huge computational complexity, and it is also difficult to formulate an objective function in the case of linear programming.
In this chapter, we present a manipulating force ellipsoid and polytope reflecting an individual's joint torque characteristics as a new evaluation method for assessing the manipulability of an upper limb. We also present a visualization algorithm for a higher dimensional hyperellipsoid and a vertex search algorithm for a higher-dimensional polytope. As a practical application of the proposed methods, we also introduce a wheelchair adaptation problem. These methods, which were mainly developed to evaluate the upper limb manipulability, are also applicable to analyze a robot manipulator's manipulation capabilities.

2. Seven Degree of Freedom Upper-Limb Model

Figure 1 portrays a seven DOF-rigid-link model of the upper limb. In the model, θ_1, θ_2, and θ_3 represents the shoulder flexion(+)/extension(-), adduction(+)/abduction(-), and external rotation(+)/internal rotation(-) respectively. In addition, θ_4 is elbow flexion(+)/extension(-); θ_5, θ_6, θ_7 signify wrist supination(+)/pronation(-), palmar flexion(+)/dorsiflexion(-), and radial flexion(+)/ulnar flexion(-) respectively. The physical parameters of each link, e.g., mass, center of gravity, inertia matrix, are calculated using regression equations based on the body weight and link length (Ae et al., 1992). In general, the dynamic equation of motion of the upper limb is given as

$$\tau = M(\theta)\ddot{\theta} + h(\theta,\dot{\theta}) + g(\theta) + J^T F, \tag{1}$$

where $\tau \in R^l$ is a joint torque vector, $\theta \in R^l$ is a joint angle vector, $M(\theta) \in R^{l\times l}$ is an inertia matrix, $h(\theta,\dot{\theta}) \in R^l$ are centrifugal and Coriolis terms, $g(\theta) \in R^l$ is a gravity term, $J \in R^{n\times l}$ is a Jacobian matrix, $F \in R^n$ is a hand force vector, $l\,(=7)$ is the number of joints, and $n\,(n \le 6)$ represents the degrees of freedom of the hand. Because the net joint torque for generating the hand force can be written as

$$\tilde{\tau} = \tau - M(\theta)\ddot{\theta} - h(\theta,\dot{\theta}) - g(\theta), \tag{2}$$

the relation between the joint torque and the hand force is given as

$$\tilde{\boldsymbol{\tau}} = \boldsymbol{J}^T \boldsymbol{F} . \tag{3}$$

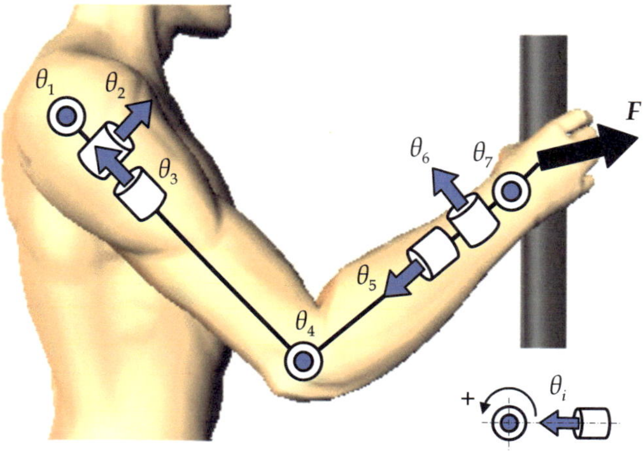

Fig. 1. Seven-rigid-link model of the upper limb.

This equation means that an individual's hand force characteristic is obtained by substituting measurable joint torque characteristics into τ . Because the joint torque characteristics vary according to the joint angle and direction of rotation, the maximum joint torque that can be generated at an arbitrary condition is given as

$$\tilde{\tau}_{imax} = \tau_{imax}(\theta_i) - M_i(\boldsymbol{\theta})\ddot{\boldsymbol{\theta}} - h_i(\boldsymbol{\theta},\dot{\boldsymbol{\theta}}) - g_i(\boldsymbol{\theta}) \tag{4}$$

$$\tilde{\tau}_{imin} = \tau_{imin}(\theta_i) - M_i(\boldsymbol{\theta})\ddot{\boldsymbol{\theta}} - h_i(\boldsymbol{\theta},\dot{\boldsymbol{\theta}}) - g_i(\boldsymbol{\theta}) , \tag{5}$$

where $\tau_{imax}(\theta_i)$ and $\tau_{imin}(\theta_i)$ signify the maximum joint torque that can be generated at joint angle θ_i in a positive direction or a negative direction, and $i\ (=1,2,\cdots,l)$ is the joint number. These joint torques can be quantified using a Cybex (Cybex Inc.) or Biodex machine (Medical Systems Inc.). Therefore, all the possible hand forces in a daily life motion is given by joint torque that satisfies the following conditions.

$$\tilde{\tau}_{imin} \le \tilde{\tau}_i \le \tilde{\tau}_{imax} \tag{6}$$

3. Manipulating Force Ellipsoid Considering Joint Torque Characteristics

3.1 Derivation of the Ellipsoid

The manipulating force ellipsoid presents an image in robot operational force space of the unit sphere defined in the actuated joint torque space:

$$\|\tilde{\tau}\|^2 = \tilde{\boldsymbol{\tau}}^T \tilde{\boldsymbol{\tau}} = \tilde{\tau}_1{}^2 + \tilde{\tau}_2{}^2 + \cdots + \tilde{\tau}_l{}^2 \le 1 . \tag{7}$$

It is described as

$$\boldsymbol{F}^T \boldsymbol{J}\boldsymbol{J}^T \boldsymbol{F} \le 1 \tag{8}$$

(Yoshikawa, 1990). However, in order to apply this method to evaluate the manipulability of the human upper limb, one must consider the differences of human joint torque characteristics on the joint rotational directions as defined in Eqs. (4) and (5). A simple way to do this evaluation is to map a unit sphere whose origin is the center position of joint torque described in Eq. (6) (Sasaki et al., 2004).

The center position of the set of all possible joint torque vectors can be written as

$$\tilde{\boldsymbol{\tau}}_{mean} = [\tilde{\tau}_{1mean}, \tilde{\tau}_{2mean}, \cdots, \tilde{\tau}_{lmean}]^T , \tag{9}$$

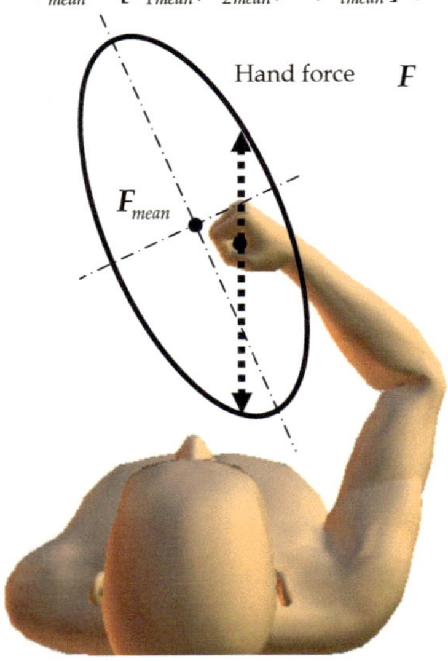

Fig. 2. Manipulating force ellipsoid based on human joint torque characteristics.

where

$$\tilde{\tau}_{imean} = \frac{\tilde{\tau}_{imax} + \tilde{\tau}_{imin}}{2} . \tag{10}$$

By assuming that $\tilde{\boldsymbol{\tau}}_{mean}$ is equal to the joint torque necessary to generate the hand force

$$\boldsymbol{F}_{mean} = [f_{1mean}, f_{2mean}, \cdots, f_{nmean}]^T , \tag{11}$$

the relation between $\tilde{\boldsymbol{\tau}}_{mean}$ and $\boldsymbol{F}_{mean}$ is given as

$$\tilde{\boldsymbol{\tau}}_{mean} = \boldsymbol{J}^T \boldsymbol{F}_{mean} , \tag{12}$$

where

$$f_{jmean} = \frac{f_{jmax} + f_{jmin}}{2} , \; j = 1,2,\cdots,n . \tag{13}$$

Therein, f_{jmax} and f_{jmin} respectively signify the maximum hand force in the positive and negative directions.

The joint torque space, with origin $\tilde{\boldsymbol{\tau}}_{mean}$ is given by:

$$\tilde{\boldsymbol{\tau}} - \tilde{\boldsymbol{\tau}}_{mean} = \boldsymbol{J}^T (\boldsymbol{F} - \boldsymbol{F}_{mean}) . \tag{14}$$

By scaling Eq. (14) with the diagonal matrix we have:

$$\boldsymbol{T}_\tau = \mathrm{diag}\left(\frac{1}{\tilde{\tau}_{1max} - \tilde{\tau}_{1mean}}, \frac{1}{\tilde{\tau}_{2max} - \tilde{\tau}_{2mean}}, \cdots, \frac{1}{\tilde{\tau}_{lmax} - \tilde{\tau}_{lmean}} \right) , \tag{15}$$

the normalized joint torque is given as

$$\hat{\boldsymbol{\tau}} = \hat{\boldsymbol{J}}^T (\boldsymbol{F} - \boldsymbol{F}_{mean}) , \tag{16}$$

where,

$$\hat{\boldsymbol{\tau}} = \boldsymbol{T}_\tau (\tilde{\boldsymbol{\tau}} - \tilde{\boldsymbol{\tau}}_{mean}) \tag{17}$$

$$\hat{J} = JT_{\tau} . \tag{18}$$

Therefore, the manipulating force ellipsoid, which reflects human joint torque characteristics maps the image in hand force spaces of a unit sphere in normalized joint torque space.

$$\|\hat{\tau}\|^2 = \hat{\tau}^T\hat{\tau} = \hat{\tau}_1^{\,2} + \hat{\tau}_2^{\,2} + \cdots + \hat{\tau}_l^{\,2} \le 1 ; \tag{19}$$

It is described as an n-dimensional hyperellipsoid

$$(F - F_{mean})^T \hat{J}\hat{J}^T (F - F_{mean}) \le 1 . \tag{20}$$

Figure 2 portrays the proposed ellipsoid; the distance between the hand position and the ellipsoid surface shows the ease of hand force manipulation. Additionally, the difference between the push force and the pull force that can be added to the object is expressed quantitatively by defining the center of the ellipsoid as F_{mean} . F_{mean} can be obtained directly from the relation between $\tilde{\tau}_{mean}$ and F_{mean} defined by Eq. (12) when J is a regular matrix $(l = n)$. However, the device which replaces F_{mean} with the center position of the manipulating force polytope introduced in section 4 is needed because F_{mean} , which balances $\tilde{\tau}_{mean}$, does not necessarily exist when J is not a regular matrix $(l \ne n)$.

3.2 Visualization Algorithm for Higher Dimensional Hyperellipsoid

Visualization of the ellipsoid in four or more dimensions is extremely difficult. We propose a visualization algorithm for an n-dimensional hyperellipsoid using two projection methods: orthogonal projection and section. The difference between the orthogonal projection and section is whether to consider the influences of all components of the hand force vector to visualize the n-dimensional hyperellipsoid (see Fig. 3).

First, we describe the orthogonal projection. The influences of all components of the hand force vector are included in the m-dimensional ellipsoid $(m \le n)$ visualized by the orthogonal projection of the n-dimensional hyperellipsoid.

The m-component of the hand force vector that we visualize can be set arbitrarily as F_1 :

$$F - F_{mean} = [F_1 \mid F_2]^T = [\bar{f}_1, \cdots, \bar{f}_m \mid \bar{f}_{m+1}, \cdots, \bar{f}_n]^T , \tag{21}$$

where,

$$\bar{f}_j = f_j - f_{jmean} . \tag{22}$$

To satisfy the relation of Eq. (20), each component of $\hat{J}\hat{J}^T$ is according to the definition of Eq. (21).

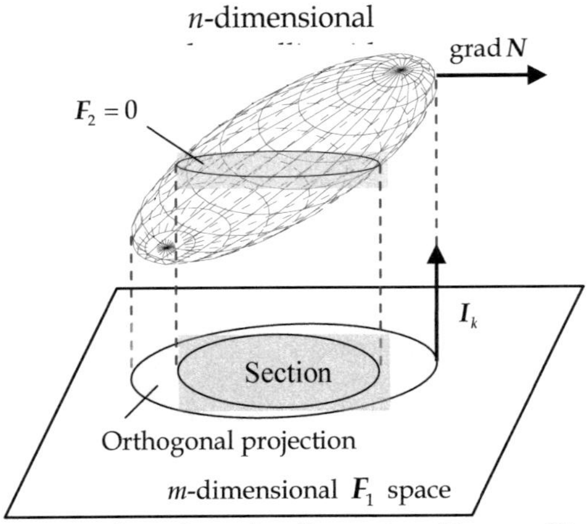

Fig. 3. Orthogonal projection and section of n-dimensional hyperellipsoid.

$$\hat{J}\hat{J}^T = \left[\begin{array}{cccc|cc} h_{1,1} & h_{1,2} & \cdots & & \cdots & h_{1,n} \\ h_{2,1} & \ddots & & & & \vdots \\ \vdots & & h_{m,m} & & & \vdots \\ \hline \vdots & & & & \ddots & \vdots \\ h_{n,1} & \cdots & \cdots & & \cdots & h_{n,n} \end{array}\right] = \left[\begin{array}{c|c} A & B \\ \hline C & D \end{array}\right], \tag{23}$$

where $A \in R^{m\times m}$, $B \in R^{m\times(n-m)}$, $C \in R^{(n-m)\times m}$ and $D \in R^{(n-m)\times(n-m)}$ are submatrices of real symmetric matrix $\hat{J}\hat{J}^T \in R^{n\times n}$.

The vertical vector to the n-dimensional hyperellipsoid is equal to the gradient vector

$$\operatorname{grad} N = \left[\frac{\partial N}{\partial \bar{f}_1}, \frac{\partial N}{\partial \bar{f}_2}, \cdots, \frac{\partial N}{\partial \bar{f}_n}\right]^T, \tag{24}$$

where

$$N = (F - F_{mean})^T \hat{J}\hat{J}^T (F - F_{mean}) - 1 . \tag{25}$$

The $(n-m)$ vertical unit vectors to the m-dimensional F_1 space are given as shown below.

$$\begin{aligned} I_1 &= [0,\cdots,0,1,0,\cdots,0]^T \\ &\vdots \\ I_{n-m} &= [\underbrace{0,\cdots,0}_{m},\underbrace{0,0,\cdots,1}_{n-m}]^T \end{aligned} \tag{26}$$

To express the orthogonal projection of the n-dimensional hyperellipsoid to the m dimensional F_1 space, the unit vector I_k $(k = 1,2,\cdots,n-m)$ must be an orthogonal relation to the vertical vector $\operatorname{grad} N$, as depicted in Fig. 3

$$\operatorname{grad} N \bullet I_k = 0 . \tag{27}$$

Substituting Eqs. (24) and (26) into Eq. (27), we obtain

$$\frac{\partial \boldsymbol{N}}{\partial \bar{f}_{m+1}} = h_{m+1,1}\bar{f}_1 + h_{m+1,2}\bar{f}_2 + \cdots + h_{m+1,n}\bar{f}_n = 0$$
$$\vdots \qquad . \tag{28}$$
$$\frac{\partial \boldsymbol{N}}{\partial \bar{f}_n} = h_{n,1}\bar{f}_1 + h_{n,2}\bar{f}_2 + \cdots + h_{n,n}\bar{f}_n = 0$$

Equation (28) expresses the orthogonal condition for the orthogonal projection. Using the definition described in Eqs. (21) and (23), the orthogonal condition can be written as

$$\boldsymbol{C}\boldsymbol{F}_1 + \boldsymbol{D}\boldsymbol{F}_2 = 0 \tag{29}$$

$$\boldsymbol{F}_2 = -\boldsymbol{D}^{-1}\boldsymbol{C}\boldsymbol{F}_1 \,. \tag{30}$$

Therefore, the general equations of the orthogonal projection are given as

$$(\boldsymbol{F}-\boldsymbol{F}_{mean})^T \hat{\boldsymbol{J}}\hat{\boldsymbol{J}}^T (\boldsymbol{F}-\boldsymbol{F}_{mean}) = [\boldsymbol{F}_1 \;\; \boldsymbol{F}_2]\begin{bmatrix} \boldsymbol{A} & \boldsymbol{B} \\ \boldsymbol{C} & \boldsymbol{D} \end{bmatrix}\begin{bmatrix} \boldsymbol{F}_1 \\ \boldsymbol{F}_2 \end{bmatrix} = \boldsymbol{F}_1^T(\boldsymbol{A}-\boldsymbol{B}\boldsymbol{D}^{-1}\boldsymbol{C})\boldsymbol{F}_1 \,. \tag{31}$$

Additionally, using the orthogonal transformation from $\boldsymbol{F}_1$ space to $\boldsymbol{E}$ space

$$\boldsymbol{F}_1 = \boldsymbol{P}\boldsymbol{E} \tag{32}$$

$$\boldsymbol{E} = [e_1, e_2, \cdots, e_m]^T \,, \tag{33}$$

Eq. (31) can be written in the following normal form

$$\boldsymbol{F}_1^T(\boldsymbol{A}-\boldsymbol{B}\boldsymbol{D}^{-1}\boldsymbol{C})\boldsymbol{F}_1 = \boldsymbol{E}^T\boldsymbol{P}^T(\boldsymbol{A}-\boldsymbol{B}\boldsymbol{D}^{-1}\boldsymbol{C})\boldsymbol{P}\boldsymbol{E} = \lambda_1 e_1{}^2 + \lambda_2 e_2{}^2 + \cdots + \lambda_m e_m{}^2 = 1 \,, \tag{34}$$

where λ_r $(r = 1,2,\cdots,m)$ are eigenvalues of $(\boldsymbol{A}-\boldsymbol{B}\boldsymbol{D}^{-1}\boldsymbol{C})$, $\boldsymbol{P} \in \boldsymbol{R}^{m\times m}$ signifies the orthogonal matrix whose columns are the eigenvectors $\boldsymbol{p}_r$ of $(\boldsymbol{A}-\boldsymbol{B}\boldsymbol{D}^{-1}\boldsymbol{C})$, the principal axes and directions of the m-dimensional ellipsoid are denoted, respectively, as $1/\sqrt{\lambda_r}$ and $\boldsymbol{p}_r$.

Next, we explain another visualization algorithm viz. section. The m-dimensional ellipsoid visualized by the section of the n-dimensional hyperellipsoid shows the set of generatable hand force vectors under $\boldsymbol{F}_2 = 0$ (see Fig. 3). Therefore, by substituting $\boldsymbol{F}_2 = 0$ into Eq. (31), the general equations of sectioning are given as

$$(\boldsymbol{F}-\boldsymbol{F}_{mean})^T \hat{\boldsymbol{J}}\hat{\boldsymbol{J}}^T (\boldsymbol{F}-\boldsymbol{F}_{mean}) = [\boldsymbol{F}_1 \;\; 0]\begin{bmatrix} \boldsymbol{A} & \boldsymbol{B} \\ \boldsymbol{C} & \boldsymbol{D} \end{bmatrix}\begin{bmatrix} \boldsymbol{F}_1 \\ 0 \end{bmatrix} = \boldsymbol{F}_1^T\boldsymbol{A}\boldsymbol{F}_1 \,. \tag{35}$$

Using the orthogonal transformation from $\boldsymbol{F}_1$ space to $\boldsymbol{W}$ space

$$\boldsymbol{F}_1 = \boldsymbol{Q}\boldsymbol{W} \tag{36}$$

$$\boldsymbol{W} = [w_1, w_2, \cdots, w_m]^T \,, \tag{37}$$

Eq. (35) can also be written as

$$\boldsymbol{F}_1^T\boldsymbol{A}\boldsymbol{F}_1 = \boldsymbol{W}^T\boldsymbol{Q}^T\boldsymbol{A}\boldsymbol{Q}\boldsymbol{W} = \mu_1 w_1{}^2 + \mu_2 w_2{}^2 + \cdots + \mu_m w_m{}^2 = 1 \,, \tag{38}$$

where μ_r $(r = 1,2,\cdots,m)$ are eigenvalues of $\boldsymbol{A}$, $\boldsymbol{Q} \in \boldsymbol{R}^{m\times m}$ signifies the orthogonal matrix whose columns are the eigenvectors $\boldsymbol{q}_r$ of $\boldsymbol{A}$, the principal axes and directions of the m dimensional ellipsoid are given, respectively, by $1/\sqrt{\mu_r}$ and $\boldsymbol{q}_r$.

Equations (34) and (38) show that an n-dimensional hyperellipsoid that cannot be viewed directly is expressible as a visible m-dimensional ellipsoid using orthogonal projection or sectioning. The proposed visualization algorithm is applicable to evaluation of the hand force characteristics suitable for the purpose because hand force $\boldsymbol{F}_1$ in Eq. (21) can be defined freely.

3.3 Experimental Validation

The proposed evaluation method is verified by comparing the manipulating force ellipsoid based on maximum joint torque with the measured hand force characteristics.

3.3.1 Measurement of Maximum Joint Torque

In order to calculate the manipulating force ellipsoid reflecting an individual's joint torque characteristics, it is indispensable to measure the maximum joint torque $\tau_{imax}(\theta_i)$ and $\tau_{imin}(\theta_i)$ in advance. For these studies, a Cybex machine (Cybex Inc.) was used for measuring them. The device can measure the maximum joint torque continuously at every joint angle in both the positive and negative directions.

For our experiment, we measured the maximum joint torque produced by the concentric contraction for every pair of movement directions; flexion and extension, adduction and abduction, and external rotation and internal rotation at the shoulder; flexion and extension at the elbow; and supination and pronation, palmar flexion and dorsiflexion, and radial flexion and ulnar flexion at the wrist (see Fig. 4). The subject in the experiment was a healthy young person (21 years old, 180 cm, and 60 kg). The purpose of this research was explained to the subject before the experiment.

3.3.2 Measurement of the Hand Force and Joint Angle

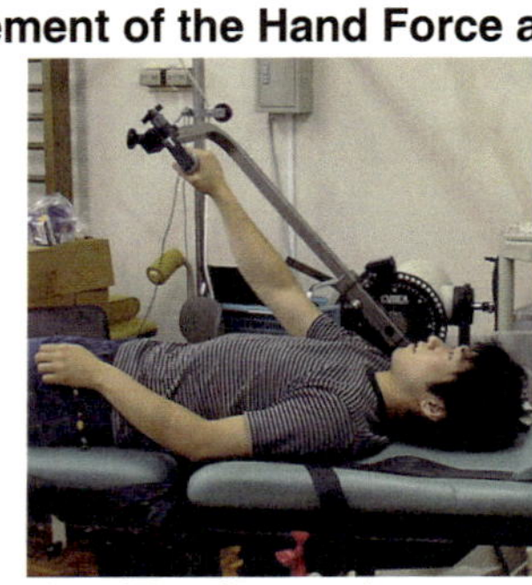

(a) Shoulder flexion/extension

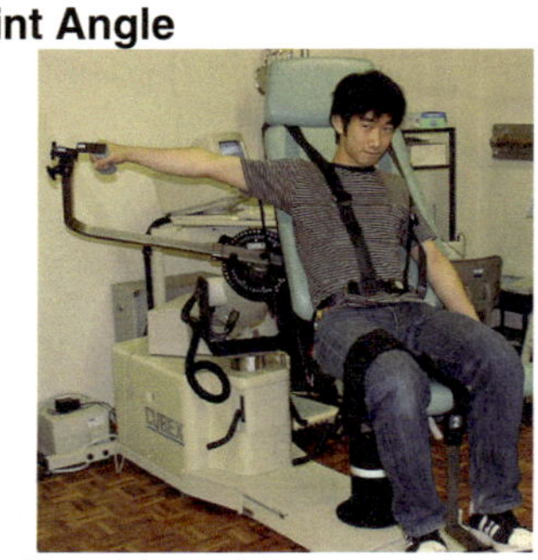

(b) Shoulder adduction/abduction

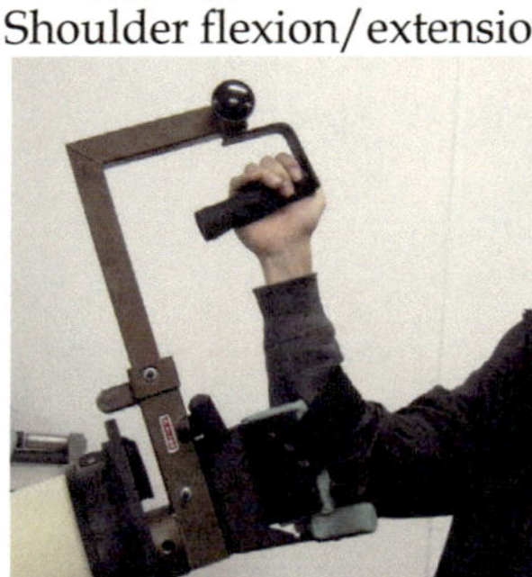

(c) Shoulder external/internal rotation

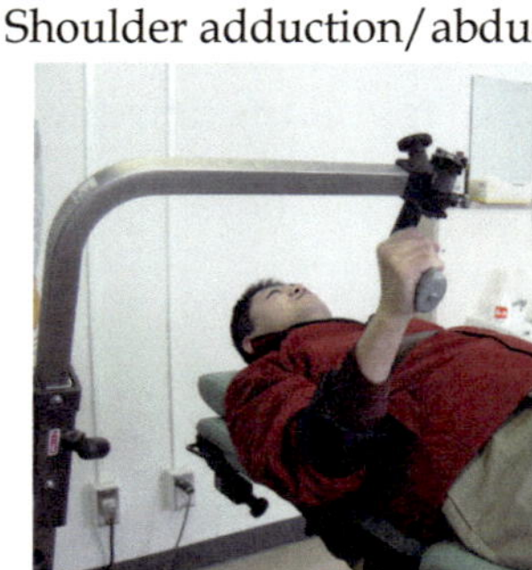

(d) Elbow flexion/extension

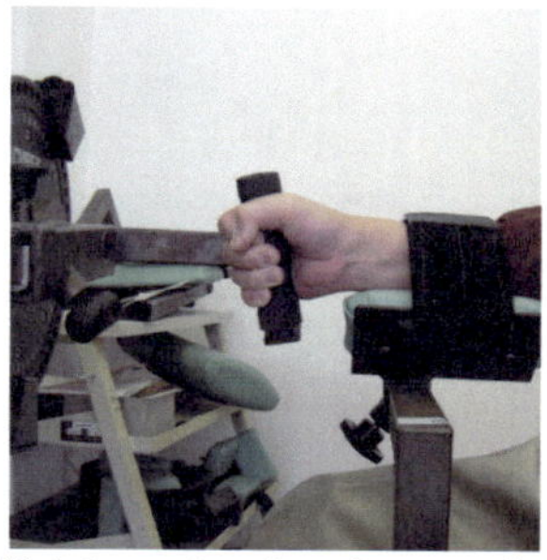

(e) Wrist supination/pronation

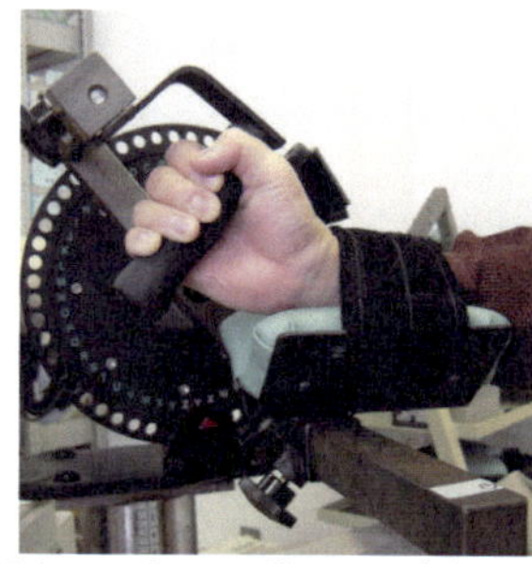

(f) Wrist palmar flexion/dorsiflexion

(g) Wrist radial flexion/ulnar flexion

Fig. 4. Measurement of maximum joint torque.

Figure 5 shows a measurement system comprising a six-axis force sensor (IFS-105M50A20-I63; Nitta Corp.) and a three-dimensional magnetic position and orientation sensor (Fastrak;Polhemus). The subject added the hand force in eight directions using maximum effort. The hand force applied to a grip was measured using the six-axis force sensor. The receiver to detect the position and posture was put to the hand and the humerus. The joint position and angle of the upper limb, which has 7 degrees of freedom, was calculated using the method developed by Oikawa and Fujita (2000). During this measurement, the subject's upper body was fixed to the back of a chair by belts to eliminate the influence of upper body movements on maximum hand force measurements. The measurements were taken at a sampling frequency of 20 Hz.

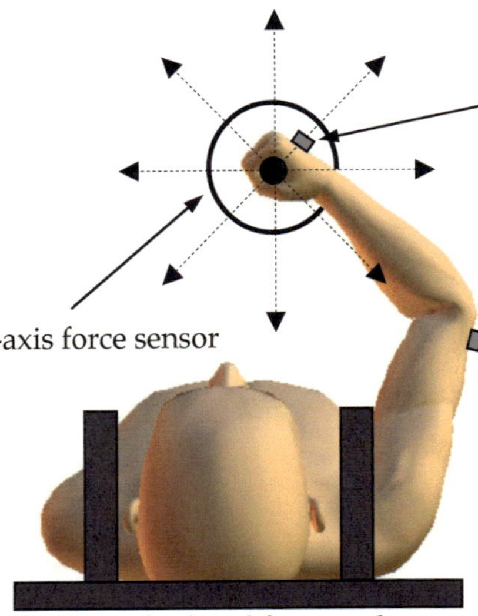

Fig. 5. Measurement system for hand force and upper limb posture.

3.3.3 Results

Figure 6 depicts a typical example of the maximum joint torque measured using the Cybex device. It is very clear that the ellipsoid described above, which does not consider human joint torque characteristics, is insufficient to evaluate the manipulability of upper limbs because even the joint torque characteristics of healthy person vary greatly according to the joint angle and rotational direction.

Figure 7 portrays a two-dimensional ellipsoid calculated from the maximum joint torque and the posture of the upper limb under the definition of

$$\boldsymbol{F} - \boldsymbol{F}_{mean} = [\boldsymbol{F}_1 \mid \boldsymbol{F}_2]^T = [f_x, f_y \mid f_z, m_x, m_y, m_z]^T , \tag{39}$$

where f , m , and subscripts signify the hand force, moment, and its direction respectively. A large ellipse represents the orthogonal projection of the six-dimensional hyperellipsoid and shows the set of all hand force vectors that can be generated under constraint conditions defined in Eq. (19). A small ellipse represents the section of the six-dimensional hyperellipsoid and means the set of generatable hand force vector under $\boldsymbol{F}_2 = 0$. On the other hand, circles in Fig. 7 show the measured hand force. Furthermore, the hand force is represented as a black circle when the joint torque to generate the measured hand force is satisfied as per the constraint condition of Eq. (19). A distribution map of the circle shows

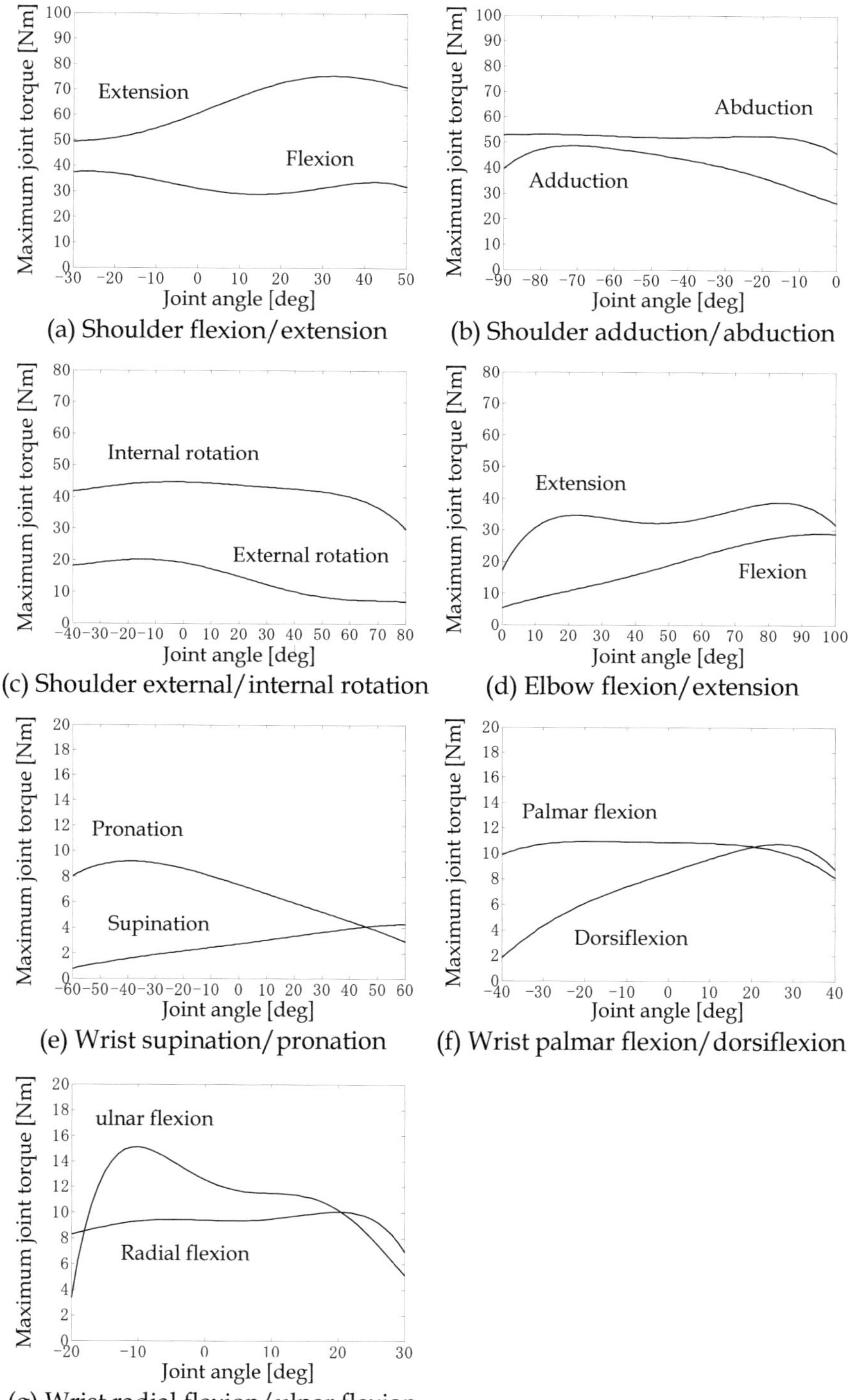

Fig. 6. Measurement results of maximum joint torque.

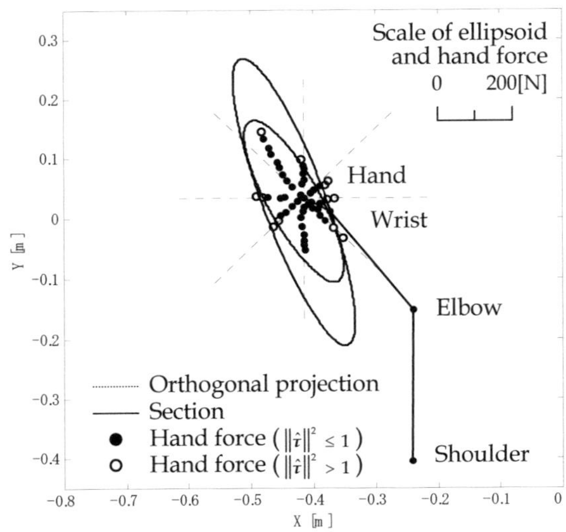

Fig. 7. Manipulating force ellipsoid and hand force.

that the magnitude of the push force and pull force in a certain direction is different. Also the ellipse visualized by the section is suitable for expression of the subject's manipulability of the upper limb in this experimental condition because the small ellipse and the black circle almost correspond to each other. One reason is that the influence of force $\boldsymbol{F}_2$ on the measured force appears only slightly because we instructed the subject to apply hand force simply on the horizontal plane.

The effectiveness of the new method proposed for quantitative evaluation of the individual's manipulability of the upper limb can be verified through the experimental results.

4. Manipulating Force Polytope Considering Joint Torque Characteristics

The manipulating force ellipsoid based on human joint torque characteristics represents the set of generatable hand force in the constraint condition of Eq. (19). It is effective for evaluation of the ease of hand force manipulation. However, to develop an assistive device that makes the best use of the remaining function, it is also important to evaluate the set of hand forces that can be generated using all joint torques defined in Eq. (6). We introduce the manipulating force polytope based on human joint torque characteristics, along with its vertex search algorithm (Sasaki et al., 2007a).

4.1 Derivation of the Polytope

All the hand forces that can be generated in a daily life motion is given by the joint torques satisfying the condition of Eq. (6). The set of all hand forces can be calculated using Eq. (6) and

$$\boldsymbol{F} = (\boldsymbol{J}^T)^{-1}\tilde{\boldsymbol{\tau}} . \tag{40}$$

This set of forces can be expressed as a convex polytope in n-dimensional hand force space.

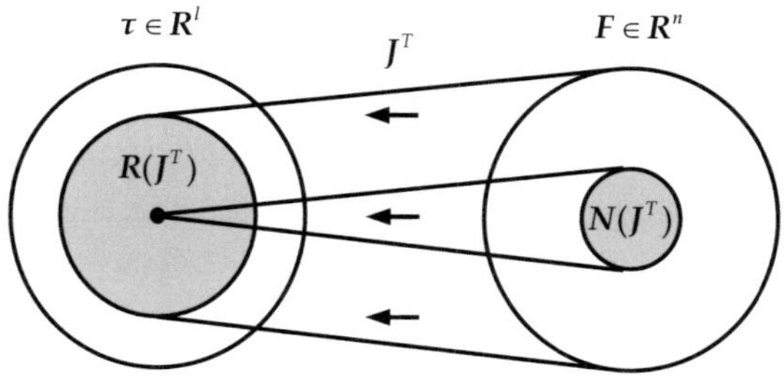

Fig. 8. Null space and range space of J^T.

The convex polytope can be called a manipulating force polytope. For a redundant manipulator such as the human upper limb $(l > n)$, in general, the set of hand forces cannot be calculated directly because J^T is not a regular matrix. The pseudo-inverse matrix $(J^T)^+$ is a general solution that minimizes the error norm $\|\tilde{\tau} - J^T F\|$ and it is introduced instead of $(J^T)^{-1}$

$$F = (J^T)^+ \tilde{\tau}, \tag{41}$$

(Chiacchio et al., 1997). However, Eq. (3) does not always have a solution for hand force because all joint torque space cannot be covered with the range space $R(J^T)$ of J^T, as shown in Fig. (8) (Asada & Slotine, 1986). In other words, a unique solution is not guaranteed and $\tilde{\tau}$ of both sides of the following equation is not always corresponding.

$$\tilde{\tau} = J^T F = J^T (J^T)^+ \tilde{\tau} = \tilde{\tau} \tag{42}$$

Therefore, to obtain the manipulating force polytope for a human upper limb, searching the subspace of the joint torque space given by $R(J^T)$ and projecting it to the hand force space are required.

Here, because the null space $N(J)$ of J is an orthogonal complement of $R(J^T)$, the following relation can be written

$$N(J) = \{R(J^T)\}^{\perp}. \tag{43}$$

In addition, the singular value decomposition of Jacobian matrix J is given as

$$J = U \Sigma V^T = [U_1 \ \ U_2] \begin{bmatrix} S & 0 \\ 0 & 0 \end{bmatrix} \begin{bmatrix} V_1^T \\ V_2^T \end{bmatrix}, \tag{44}$$

where $\Sigma \in R^{n \times l}$ is a diagonal matrix with arranged nonzero singular values of J such as $S = \mathrm{diag}(s_1, s_2, \cdots, s_r)$, $U \in R^{n \times n}$ is an orthogonal matrix, $U_1 \in R^{n \times r}$ and $U_2 \in R^{n \times (n-r)}$ are submatrices of U, $V \in R^{l \times l}$ is an orthogonal matrix, $V_1 \in R^{l \times r}$ and $V_2 \in R^{l \times (l-r)}$ are submatrices of V, and r is the rank of J. Because the column vector v_t $(t = 1, 2, \cdots, r)$ of V_1 is equal to the base vector of Eq. (43), $R(J^T)$ is represented as the space covered by r base vectors of dimension l. By projecting to the hand force space the joint torque's subspace given by $R(J^T)$, the manipulating force polytope is obtainable.

4.2 Vertex Search Algorithm for Higher Dimensional Polytope

To search for the vertex of the convex polytope, a slack variable is generally introduced. However, the vertex search using a linear programming method such as the simplex method engenders huge computational complexity and a complex definition of the objective function. Especially, it is extremely difficult to search for all vertexes of a high-dimensional polytope. Therefore, we propose a new vertex search algorithm.

The vertex search algorithm is based on the geometric characteristic that the points of intersection between the l-dimensional joint torque space and the space covered by r base vectors of dimension l exist in the $(l-r)$-dimensional face of joint torque space. The algorithm is explained using a three-dimensional rectangle in Fig. 9 for clarification.

For $l=3$ and $r=2$, the two-dimensional plane covered by two base vectors intersects with a side (= one dimension) of a three-dimensional rectangle (see Fig. 9(a)). Because its side is a common set of two planes, the number of joint torque components equal to the maximum joint torque $\tilde{\tau}_{imax}$ or the minimum joint torque $\tilde{\tau}_{imin}$ is equal to two. For $l=3$ and $r=1$, the one-dimensional straight line covered by a base vector intersects with a face (= two dimension) of a three-dimensional rectangle (see Fig. 9(b)). The number of joint torque components equal to $\tilde{\tau}_{imax}$ or $\tilde{\tau}_{imin}$ is then equal to one. Consequently, generalizing the geometric characteristics shows that the space covered by r base vectors of dimension l intersects with the $(l-r)$-dimensional face of the l-dimensional joint torque space. It also reveals that the number of joint torque components equal to $\tilde{\tau}_{imax}$ or $\tilde{\tau}_{imin}$ is equal to r.

By defining the points of intersection between the l-dimensional joint torque space and the range space $\boldsymbol{R}(\boldsymbol{J}^T)$ of $\boldsymbol{J}^T$ as

$$\boldsymbol{K}=[k_1,k_2,\cdots,k_r]^T, \tag{45}$$

the subspace of the joint torque space is given as

$$\boldsymbol{T}=k_1\boldsymbol{v}_1+k_2\boldsymbol{v}_2+\cdots+k_r\boldsymbol{v}_r=\boldsymbol{V}_1\boldsymbol{K}. \tag{46}$$

Equation (46) can also be written as

$$\begin{bmatrix}\boldsymbol{T}_1\\ \boldsymbol{T}_2\end{bmatrix}=\begin{bmatrix}\tau_1\\ \vdots\\ \tau_r\\ \hline \tau_{r+1}\\ \vdots\\ \tau_l\end{bmatrix}=\begin{bmatrix}k_1v_{11}+k_2v_{21}+\cdots+k_rv_{r,1}\\ \vdots\\ k_1v_{1,r}+k_2v_{2,r}+\cdots+k_rv_{r,r}\\ \hline k_1v_{1,r+1}+k_2v_{2,r+1}+\cdots+k_rv_{r,r+1}\\ \vdots\\ k_1v_{1,l}+k_2v_{2,l}+\cdots+k_rv_{r,l}\end{bmatrix}=\begin{bmatrix}\boldsymbol{V}_{11}\\ \boldsymbol{V}_{12}\end{bmatrix}\boldsymbol{K}, \tag{47}$$

where $\boldsymbol{T}_1\in\boldsymbol{R}^r$ and $\boldsymbol{T}_2\in\boldsymbol{R}^{l-r}$ are submatrices of $\boldsymbol{T}\in\boldsymbol{R}^l$, and where $\boldsymbol{V}_{11}\in\boldsymbol{R}^{r\times r}$ and $\boldsymbol{V}_{12}\in\boldsymbol{R}^{(l-r)\times r}$ are the submatrices of the base vector $\boldsymbol{V}_1$. From this equation, the relation between $\boldsymbol{T}_1$ and $\boldsymbol{T}_2$ is obtained as

$$\boldsymbol{T}_2=\boldsymbol{V}_{12}\boldsymbol{K}=\boldsymbol{V}_{12}\boldsymbol{V}_{11}^{-1}\boldsymbol{T}_1. \tag{48}$$

Because there are 'r' joint torque components equal to $\tilde{\tau}_{imax}$ or $\tilde{\tau}_{imin}$ in the intersection points, we can define $\boldsymbol{T}_1$ as shown below:

$$\boldsymbol{T}_{11}=\begin{bmatrix}\tau_1\\ \vdots\\ \tau_r\end{bmatrix}=\begin{bmatrix}\tilde{\tau}_{1max}\ or\ \tilde{\tau}_{1min}\\ \vdots\\ \tilde{\tau}_{rmax}\ or\ \tilde{\tau}_{rmin}\end{bmatrix} \tag{49}$$

The points of intersection K are obtained by judging whether the joint torque component of T_2 calculated from Eqs. (48) and (49) satisfies the condition of the joint torque in Eq. (6). Therefore only when T_2 satisfies this condition, the joint torque T is calculated from K,

$$K = V_{11}^{-1} T_1 = V_{12}^{-1} T_2 , \tag{50}$$

And it becomes the vertex of the l-dimensional convex polytope. Herein, the number of combinations which select the n equations from l equations in Eq. (47) and define V_{11} is ${}_lC_r$, while the number of combinations defining T_1 in Eq. (49) is 2^r. All vertexes of the l dimensional convex polytope can be found by calculating the intersection points in all combinations. The manipulating force polytope based on human joint torque characteristics is finally expressed by calculating the convex hulls of all the vertexes projected using Eq. (41) on the hand force space. This is done because the vertex of the l-dimensional convex polytope defined by the proposed algorithm always guarantees the unique solution shown in Eq. (42).

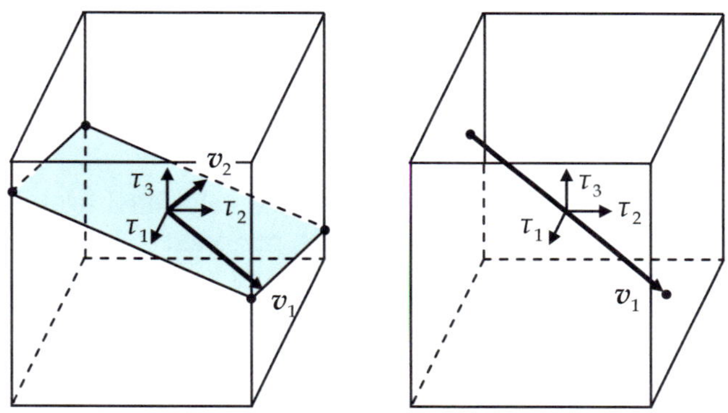

(a) $l = 3$, $r = 2$ (b) $l = 3$, $r = 1$

Fig. 9. Vertexes of l-dimensional convex polytopes.

4.3 Experimental Validation

To evaluate the effectiveness of the proposed method, an experiment to compare the manipulating force polytope based on maximum joint torque with the measured hand force characteristics was performed. The experimental methodology and the device were identical to those shown in subsection 3.3. The participant in the experiment was a person with a spinal cord injury (60 years old, 170 cm, 55 kg, and L2 lumbar injury).

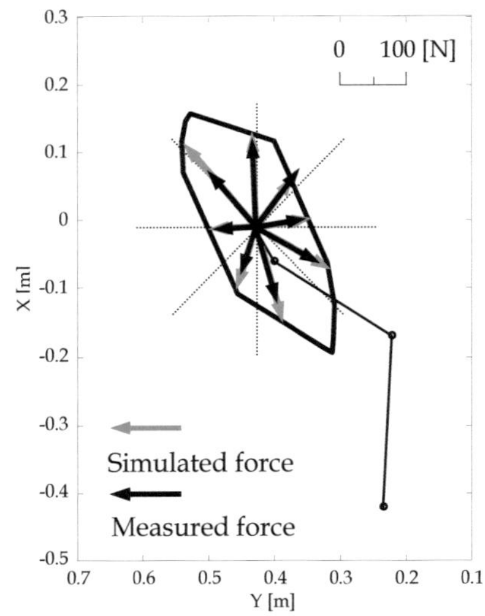

Fig. 10. Manipulating force polytope and hand force.

Figure 10 portrays the manipulating force polytope, as calculated from the maximum joint torque and posture of the upper limb. The polytope, projected to the two-dimensional plane, represents the set of all the possible hand force on the horizontal plane, demonstrating a hexagonal shape. This feature of shape of the manipulating force polytope agrees with findings of Oshima et al. (1999) that the distribution of the hand force vector in a two-dimensional plane is a hexagonal shape. In addition, the hand force vector (gray arrow) presumed from the manipulating force polytope approximately corresponds to the measured hand force (black arrow). The effectiveness of the method proposed for quantitative evaluation of the individual's manipulability of the upper limb can be confirmed through the presented experimental result, but it is necessary to perform further verification to achieve a more accurate evaluation.

5. Application to Wheelchair Propulsion

The proposed evaluation methods are useful for developing assistive devices, planning of rehabilitation, and improving living environments because an individual's manipulability of the upper limb can be evaluated quantitatively and visually. As a practical application of the proposed methods, we present the analysis of wheelchair propulsion.
Wheelchairs are commonly used mobility devices that support people for whom walking is difficult or impossible because of illness, injury, or disability. However, more than 50% of all manual wheelchair users experience upper limb pain or injury (Gellman et al., 1988; Sie et al., 1992; Pentland & Twomey, 1994). The most common problems of wheelchair users, who push on the handrim an average of 2000–3000 times a day, are shoulder, wrist, and hand injuries including carpal tunnel syndrome. For these reasons, studies of wheelchair propulsion have mainly addressed the physical load borne by wheelchair users, in addition to technical improvements, wheelchair configurations, and design optimization (Cooper,

1998; Engstrom, 2002; Sasaki et al., 2007b, 2008). A study related to optimal wheelchair design that we performed, (analytical results of wheelchair propulsion using the manipulating force ellipsoid) based on human joint torque characteristics is described here.

5.1 Experiments

The participants in the experiment were eight experienced wheelchair users with spinal cord injuries (55±15 years old, 167.5±7.5 cm, 63.5±11.5 kg, and injury levels T12–L2). In the experiment the maximum joint torque characteristics of the upper limb were measured using the Cybex machine portrayed in Fig. 4. The upper limb movement and hand force during wheelchair propulsion were measured using the measurement system presented in Fig. 11. The system comprises a three-dimensional magnetic position and orientation sensor (Fastrak; Polhemus) for measuring the joint angle and joint position of the upper limb. A six-axis force sensor (IFS-67M25A-I40; Nitta Corp.) for measuring the three-directional force and moment applied to the handrim, and a rotary encoder (OHI48-6000P4-L6-5V; Tamagawa Seiki Co. Ltd.) for measuring the rear wheel's rotation angle. Based on advice from an occupational therapist, the height of the rear wheel axis was adjusted to become equal with the subject's hand position when the hand was hung straight downward.

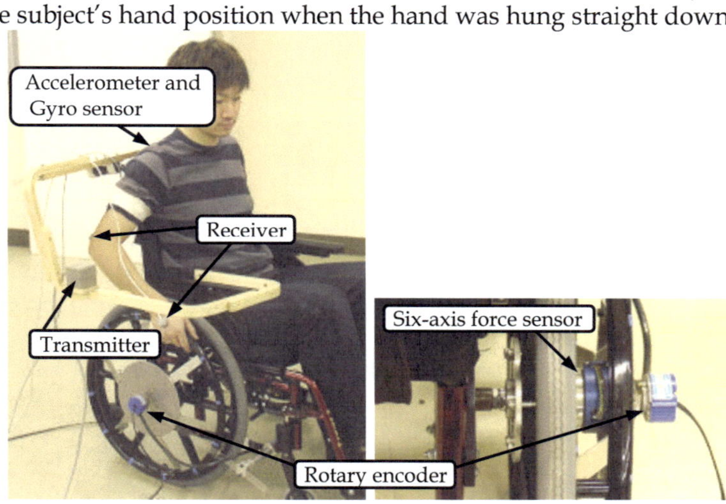

Fig. 11. Measurement system for hand force and upper limb posture during wheelchair propulsion.

5.2 Analysis of Wheelchair Maneuverability

Figure 12 portrays the measurement results of the upper limb postures and force vectors applied to the handrim, in addition to the calculation results of the manipulating force ellipsoid based on maximum joint torque characteristics as the stick diagram on the (a) sagittal plane and (b) frontal plane. Because most wheelchair users do not grasp the handrim during wheelchair propulsion, the ellipsoid was projected to each plane under the condition that all moment components are zero. In general, an increase of the hand force component toward the handrim tangential direction is necessary to achieve efficient wheelchair propulsion, because the handrim has only one degree of freedom which is rotation around the rear wheel axis. To analyze the wheelchair maneuverability quantitatively, the parameters presented in Fig. 13 were defined as $\boldsymbol{F}_m$ is the maximum generatable force, φ is the hand contact position, and α and β are angles on each plane between $\boldsymbol{F}_m$ and the most effective force direction d_{mef}.

Figures 14 and 15 respectively portray calculation results of the maximum generatable force $\boldsymbol{F}_m$, the angle α on the sagittal plane, and angle β on the frontal plane. The bands in the figure present the average value ± standard deviation among all subjects. Strong hand force is applicable to the handrim efficiently in the latter half of the propulsion cycle. This is because an increase in the maximum generatable force $\boldsymbol{F}_m$ and a decrease in angle α can be confirmed from the figure. However, angle α differs by more than 80 deg in the first half of the propulsion cycle. In this phase, the hand force applied in the direction $\boldsymbol{F}_m$ will not generate a sufficient hand force component in direction d_{mef}, which contributes directly to driving a wheelchair. Even if it is purely applied in the direction d_{mef}, the physical load of the upper limb will increase because d_{mef} is the direction in which wheelchair users have difficult applying hand force. Consequently, the fact that the users must start driving the wheelchair from an upper limb posture with bad maneuverability is inferred as one of the factors increasing the physical load of wheelchair users.

On the other hand, by looking at angle β on the frontal plain, it is very apparent that the direction to which the hand force can be generated easily, i.e., the direction of $\boldsymbol{F}_m$, differs from direction d_{mef} by about 5–10 deg. It shows that attaching the camber angle for the rear wheel is an efficient way to transmit hand force. The camber angle is known to be efficient for producing lateral stability of a wheelchair, reducing the downward turning tendency on side slopes, and preventing interference between the upper limb and the handrim (Trudel et al., 1995; Cooper, 1998). Furthermore, using the proposed evaluation methods, the effectiveness of the camber angle was proved from the new viewpoint of manipulability of the upper limb, bringing ease of hand force manipulation to wheelchair users.

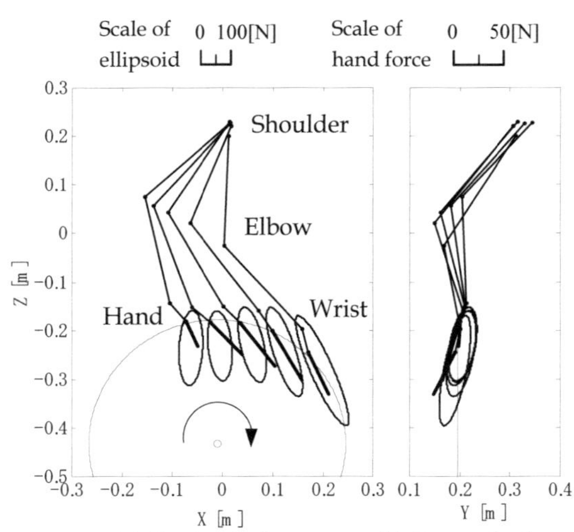

(a) Sagittal plane (b) Frontal plane

Fig. 12. Stick diagram of the upper limb, hand force, and manipulating force ellipsoid.

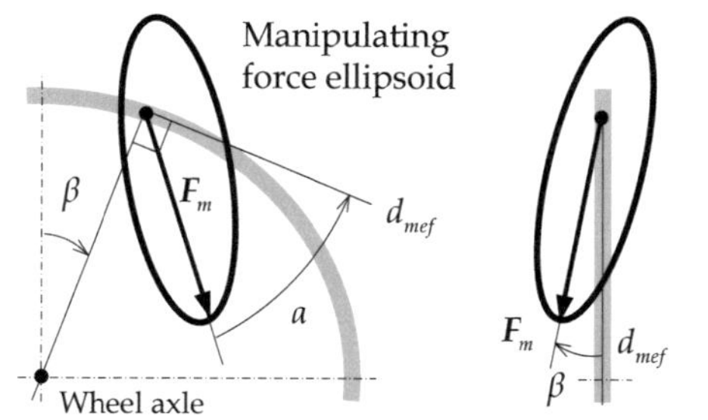

(a) Sagittal plane (b) Frontal plane

Fig. 13. Definition of component and angle of manipulating force ellipsoid.

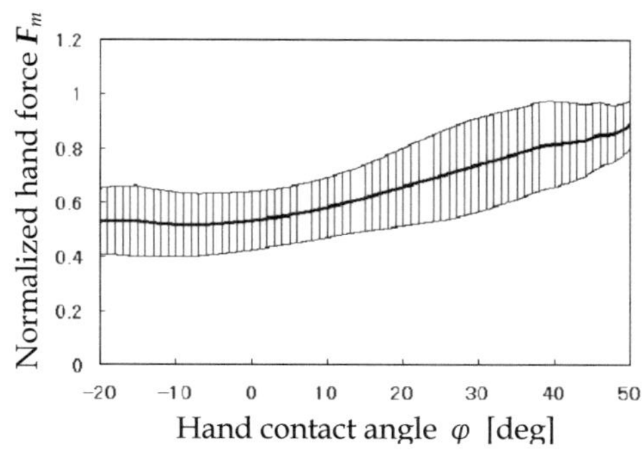

Fig. 14. Maximum generatable hand force F_m .

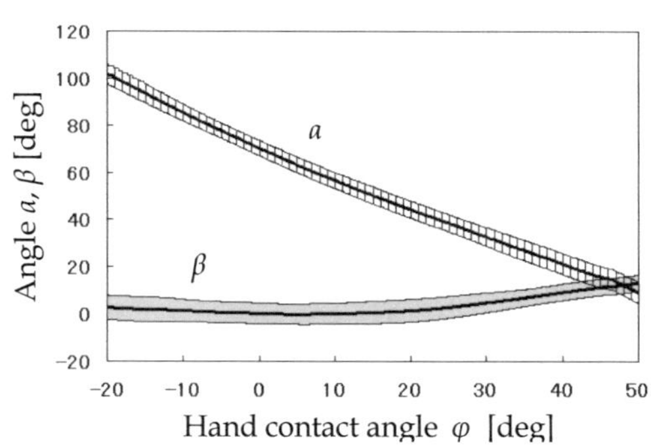

Fig. 15. Angle between maximum generatable hand force F_m and mechanically most effective force direction d_{mef} .

5.3 Analysis of Hand Force Patterns

Next, we analyze the hand force patterns applied to the handrim. The stick diagram of Fig. 12 shows that the direction of the measured hand force does not necessarily correspond to the direction in which the maximum hand force can be generated. One reason is that the hand force applied in direction of F_m does not necessarily engender an efficient driving force, as clearly depicted in the results portrayed in Fig. 15. However, it is readily inferred

that long-term wheelchair users perform efficient propulsion patterns. Therefore, we propose a new concept of the driving force contribution figure reflecting the driving efficiency to the manipulating force ellipsoid. Thereafter, we analyze hand force patterns used in wheelchair propulsion.
The driving force contribution figure is the set of driving forces obtainable using all hand force components of the manipulating force ellipsoid (see Fig. 16) and the driving force $\boldsymbol{F}_e$ is

$$\boldsymbol{F}_e = \frac{|\boldsymbol{F}_t|}{|\boldsymbol{F}_a|}\boldsymbol{F}_a, \tag{51}$$

where $\boldsymbol{F}_a$ is an arbitrary hand force vector in the manipulating force ellipsoid, and where $\boldsymbol{F}_t$ is a tangential component of $\boldsymbol{F}_a$ to the handrim, directly contributing to driving a wheelchair. In addition, driving force $\boldsymbol{F}_e$ has a direction equal to $\boldsymbol{F}_a$ and magnitude equal to $\boldsymbol{F}_t$. The distance between the boundary of the driving force contribution figure and the hand position on the handrim represents the contribution to driving the wheelchair. In other words, if the driving force contribution figure takes a large value along the driving force direction, the applied hand force efficiently supports wheelchair propulsion.
Figure 17 portrays a stick diagram, which is a product of the driving force contribution figure and Fig. 11. To analyze this numerical result more quantitatively, the parameters

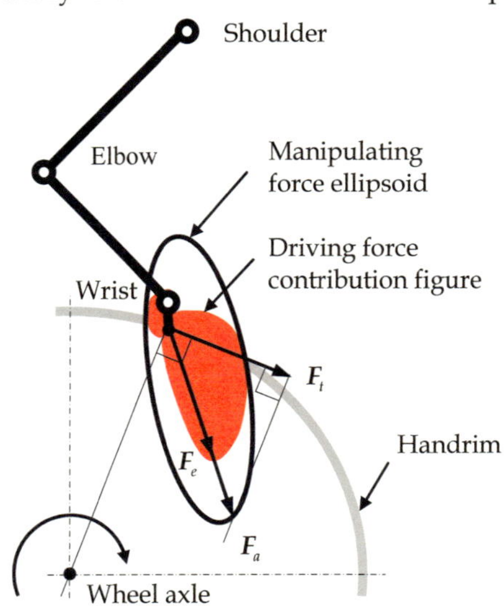

Fig. 16. Definition of driving force contribution figure.

presented in Fig. 18 were defined as follows: $\boldsymbol{F}_{em}$ signifies the maximum driving force, $\boldsymbol{F}_s$ denotes the hand force applied to the handrim, $\boldsymbol{F}_{ts}$ stands for a tangential component of $\boldsymbol{F}_s$ to the handrim, φ represents the hand contact position, and α' and β' are angles on each plane between $\boldsymbol{F}_{em}$ and the measured force $\boldsymbol{F}_s$.
Figures 19 and 20 respectively portray the calculation results of the maximum driving force $\boldsymbol{F}_{em}$, the tangential component of measured force $\boldsymbol{F}_{ts}$, the angle α' on the sagittal plane and

the angle β' on the frontal plane. The bands in the figure present average values ± standard deviation among all subjects, showing that possible hand forces expressed by the manipulating force ellipsoid can be converted efficiently into the driving force in the latter half of the propulsion cycle because the maximum generatable driving force F_{em} increases gradually. The results show that angle α' on the sagittal plain is about 10 degrees and angle β' on the frontal plain is 20 degrees, except for the time when the wheelchair starts to move. In addition, the force of the wheelchair users was applied to the direction in which the driving force can be generated easily. Based on the fact that most wheelchair users do not grasp the handrim during wheelchair propulsion, it can be understood that the hand force applied to the perpendicular direction to the handrim is also necessary to transmit the hand force to the tangential direction to the hand rim, although it does not contribute directly to driving. Especially at the time a wheelchair is moved from its halted state, the friction force between the hand and the handrim is required. That is thought to be the reason for the difference of the hand force direction, as presented in Fig. 20. Taken together, all these results reflect that the wheelchair users are considering both the efficiency and physical load of the upper limb, and are performing a very skillful operation for the task that they are given for wheelchair propulsion. This is confirmed to be true because, except for the time when a wheelchair must be moved initially, the direction in which the hand force is actually applied agrees mostly with the direction in which the driving force can be easily produced.

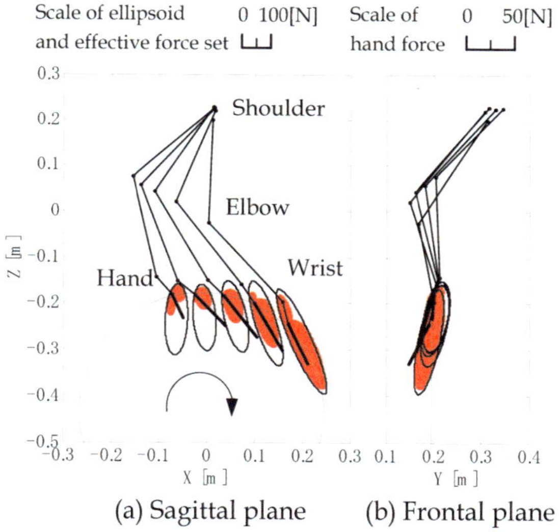

Fig. 17. Stick diagram of the upper limb, hand force, manipulating force ellipsoid, and driving force contribution figure.

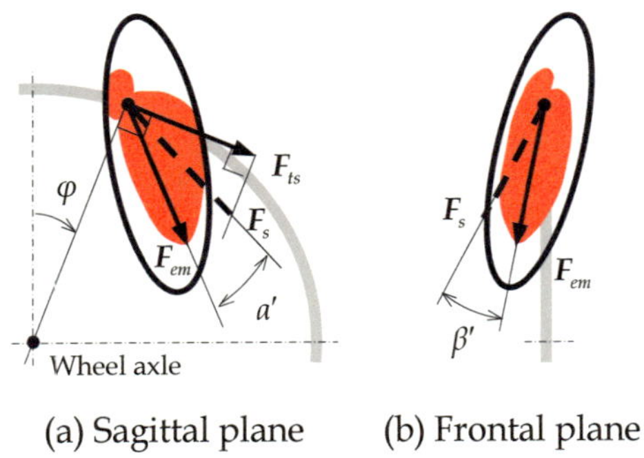

Fig. 18. Definition of component and angle of driving force contribution figure.

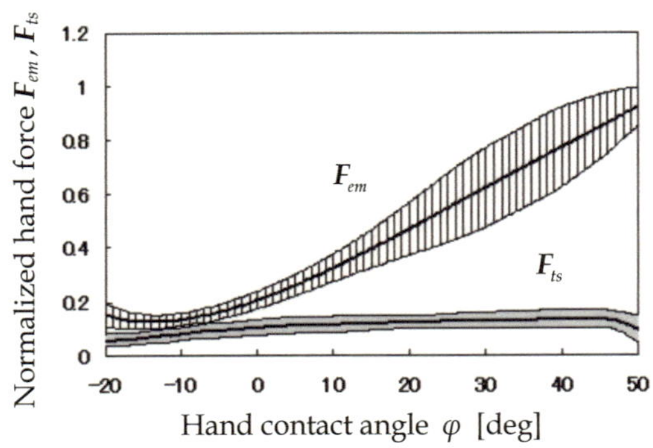

Fig. 19. Maximum driving force F_{em} and tangential component of measured hand force F_s to the handrim F_{ts}.

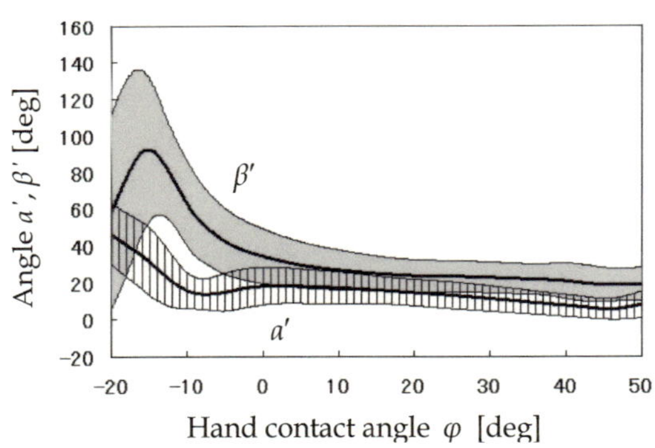

Fig. 20. Angle between F_{em} and measured hand force F_s.

5.4 Optimal Wheelchair Design

As described above, we performed analyses of wheelchair maneuverability quantitatively from the viewpoint of upper limb manipulability. The analytical results show that wheelchair users start driving the handrim in such a posture that it is difficult to generate the necessary hand force to drive the wheelchair. This might be a problem of wheelchairs,

and might be a cause of the increased physical load borne by wheelchair users. Using a new concept of the driving force contribution figure reflecting the driving efficiency to the manipulating force ellipsoid, the results accurately characterize wheelchair users driving the wheelchair, with consideration of the upper limb load and wheelchair propulsion efficiency.
The design and the adaptation of the wheelchair have generally been performed using trial and error based on experience and knowledge acquired over many years. However, their grounds and effects remain unclear. The wheelchair design criteria and evaluation of the adaptability between users and designed wheelchairs have not been established.
The proposed methods are useful not only for the quantitative evaluation of upper limb manipulability based on individuals' joint torque characteristics but also for prediction of the hand force pattern taken for the given task that the user must perform. In addition, for optimal wheelchair design, we have been developing other evaluation methods (Miura et al., 2004, 2006; Sasaki et al., 2008) including the estimation of physical loads using an upper limb musculoskeletal model, optimization of the driving form using genetic algorithms, and development of a wheelchair simulator that can freely adjust wheelchair dimensions according to the user's body functions. Therefore, using the evaluation methods proposed in this chapter or by combining them with other optimization methods we have developed, we can reasonably provide individually adjusted wheelchairs that reduce the physical load on users' upper limbs during wheelchair propulsion and which increase the wheelchair propulsion efficiency.

6. Conclusion

This chapter has presented a manipulating force ellipsoid and polytope based on human joint torque characteristics for evaluation of upper limb manipulability. As described in sections 3 and 4, the proposed methods are based on the relation between the joint torque space and the hand force space. Therefore, it is certain that more accurate evaluation can be achieved by expanding these concepts and by considering the relations among muscle space, joint torque space, and hand force space. However, the development of the three-dimensional musculoskeletal model of the human is a respected research area in the field of biomechanics. It is difficult to model the individual's muscle properties strictly, such as the maximum contraction force, the origin, the insertion and the length of each muscle. Because of this fact, the proposed evaluation method is a realistic technique by which the influence of the remaining muscle strength or paralysis can be modeled directly and easily as the individual's joint torque characteristics. Nevertheless, further improvements are necessary to achieve a more accurate evaluation because the bi-articular muscle characteristics cannot be reflected sufficiently using the method of separately measuring the maximum joint torque characteristics of each joint.
Through our investigations, we have solved three problems to express the manipulating force ellipsoid and polytope based on the measured maximum joint torque. The first is to have reflected the human joint torque characteristics depending on the joint angle and the rotational direction into the formulation of the manipulating force ellipsoid and polytope. Here, the peculiar feature of humans, that the region of maximum joint torque is not symmetric about the origin, was expressed by introducing the offset between the origin of the ellipsoid and the hand position. The second is to have derived two visualization methods of higher-dimensional hyperellipsoids such as the orthogonal projection and the

section, to evaluate the upper limb manipulability quantitatively and visually. Furthermore, the third is to have derived a new vertex search algorithm for higher-dimensional polytopes to search for all vertexes of convex polytopes without oversight by an easy calculating formula and few computational complexities. It is certain that the proposed methods are effective not only for evaluation of the manipulability of human upper limbs but also for the evaluation of a robot manipulator's manipulation capability because no reports, even in the robotics literature, have described solutions to these problems. Therefore, the proposed methods can probably contribute to progress in the field of robotics in a big way, offering useful new findings.

In this chapter, the analysis of the wheelchair propulsion was introduced as one example to evaluate the proposed methods' practical importance. In addition, the potential problems of wheelchairs and the wheelchair maneuverability were clarified quantitatively from the viewpoint of the upper limb manipulability. Results described herein show that the ease of hand force manipulation engenders improvement in all scenes of daily living and yields various new findings. Especially, it is important to evaluate upper limb manipulability of elderly and physically handicapped people quantitatively and visually for development of assistive devices, planning of rehabilitation, and improvement of living environments. Further applications of the proposed methods as a new evaluation index for the manipulability analysis of upper and lower limbs in various fields, including ergonomics and robotics, can be anticipated.

7. Acknowledgements

The authors gratefully acknowledge the support provided for this research by a Japan Society of Promotion of Science (JSPS) Grant-in-Aid for Young Scientists (Start-up 18800034 and B 20700463).

8. References

Ae, M.; Tang, H.P. & Yokoi, T. (1992). Estimation of inertia properties of the body segments in Japanese athletes, In: *Biomechanism*, Vol. 11, pp. 23–32, The Society of Biomechanisms Japan, ISBN 978-4-13-060132-0.

Asada, H. & Slotine, J.-J.E. (1986). *Robot Analysis and Control*, John Wiley and Sons, ISBN 0-471-83029-1.

Chiacchio, P.; Bouffard-Vercelli, Y. & Pierrot, F. (1997). Force polytope and force ellipsoid for redundant manipulators. *Journal of Robotic Systems*, Vol. 14, No. 8, pp. 613–620, ISSN 0741-2223.

Cooper, R.A. (1998). *Wheelchair selection and configuration*, Demos Medical Publishing, ISBN 1-888799-18-8.

Engstrom, B. (2002). *Ergonomic seating. A true challenge. Wheelchair seating & mobility principles*, Posturalis Books, ISBN 91-972379-3-0.

Gellman, H.; Chandler, D.R.; Petrasek, J.; Sie, I.; Adkins, R. & Waters, R.L. (1988). Carpal tunnel syndrome in paraplegic patients. *The Journal of Bone and Joint Surgery*, Vol. 70, No. 4, pp. 517–519, ISSN 0021-9355.

Hamada, Y.; Joo, S-W. & Miyazaki, F. (2000). Optimal design parameters for pedaling in man-machine systems. *Transactions of the Institute of Systems, Control and Information Engineers*, Vol. 13, No. 12, pp. 585–592, ISSN 1342-5668.

Lee, J. (2001). A structured algorithm for minimum l∞-norm solutions and its application to a robot velocity workspace analysis. *Robotica*, Vol. 19, pp. 343–352, ISSN 0263-5747.

Miura, H.; Sasaki, M.; Obinata, G.; Iwami, T.; Nakayama, A.; Doki, H. & Hase, K. (2004). Task based approach on trajectory planning with redundant manipulators, and its application to wheelchair propulsion, *Proceedings of the 2004 IEEE Conference on Robots, Automation and Mechatronics*, pp. 758–761, Singapore, ISBN 0-7803-8645-0.

Miura, H.; Sasaki, M.; Obinata, G.; Iwami, T. & Hase, K. (2006). Three-dimensional Motion Analysis of Upper limb for Optimal Design of Wheelchair, In: *Biomechanisms*, Vol. 18, pp. 89–100, The Society of Biomechanisms Japan, ISBN 4-7664-1305-9.

Miyawaki, K.; Iwami, T.; Obinata, G.; Kondo, Y.; Kutsuzawa K.; Ogasawara, Y. & Nishimura, S. (2000). Evaluation of the gait of elderly people using an assisting cart: gait on flat surface. *JSME International Journal, Series C*, Vol. 43, No. 4, pp. 966–974, ISSN 1344-7653.

Miyawaki, K.; Iwami, T.; Ogasawara, Y.; Obinata, G. & Shimada, Y. (2007). Evaluation and development of assistive cart for matching to user walking. *Journal of Robotics and Mechatronics*, Vol. 19, No. 6, pp. 637–645, ISSN 0915-3942.

Ohta, K.; Luo, Z. & Ito, M. (1988). Analysis of human movement under environmental constraints: Adaptability to environment during crank rotation tasks. *Transactions of the Institute of Electronics, Information and Communication Engineers*, Vol. J81-D-II, No. 6, pp. 1392–1401, ISSN 0915-1923.

Oikawa, K. & Fujita K. (2000). Algorithm for calculating seven joint angles of upper extremity from positions and Euler angles of upper arm and hand. *Journal of the Society of Biomechanisms*, Vol. 24, No. 1, pp. 53–60, ISSN 0285-0885.

Oshima, T.; Fujikawa, T. & Kumamoto, M. (1999). Functional evaluation of effective muscle strength based on a muscle coordinate system consisted of bi-articular and mono-articular muscles: contractile forces and output forces of human limbs. *Journal of the Japan Society for Precision Engineering*, Vol. 65, No. 12, pp. 1772–1777, ISSN 0912-0289.

Pentland, W.E. & Twomey, L.T. (1994). Upper limb function in persons with longterm paraplegia and implications for independence: part I. *Paraplegia*, Vol. 32, pp. 211–218, ISSN 0031-1758.

Sasaki, M.; Iwami, T.; Miyawaki, K.; Doki, H. & Obinata, G. (2004). Three-dimensional spatial expression of the manipulability of the upper limb considering asymmetry of maximum joint torque. *Transactions of the Japan Society of Mechanical Engineers, Series C*, Vol. 70, No. 697, pp. 2661–2667, ISSN 0387-5024.

Sasaki, M.; Iwami, T.; Obinata, G.; Doki, H.; Miyawaki, K. & Kinjo, M. (2005). Biomechanics analysis of the upper limb during wheelchair propulsion. *Transactions of the Japan Society of Mechanical Engineers, Series C*, Vol. 71, No. 702, pp. 654–660, ISSN 0387-5024.

Sasaki, M.; Iwami, T.; Miyawaki, K.; Obinata, G.; Sato, I.; Shimada, Y. & Kiguchi, K. (2007a). A study on evaluation of the manipulability of the upper limb using convex polyhedron: First report, new vertex search algorithm. *Transactions of the Japan Society of Mechanical Engineers, Series C*, Vol. 73, No. 729, pp. 1514–1521, ISSN 0387-5024.

Sasaki, M.; Iwami, T.; Obinata, G.; Miyawaki, K.; Miura, H.; Shimada, Y. & Kiguchi, K. (2007b). Analysis of wheelchair propulsion and hand force pattern based on manipulability of the upper limb. *Transactions of the Japan Society of Mechanical Engineers, Series C,* Vol. 73, No. 732, pp. 2279–2286, ISSN 0387-5024.

Sasaki, M.; Kimura, T.; Matsuo, K.; Obinata, G.; Iwami, T.; Miyawaki, K. & Kiguchi, K. (2008). Simulator for optimal wheelchair design. *Journal of Robotics and Mechatronics*, Vol. 20, No. 6, pp. 854–862, ISSN 0915-3942.

Shim, I.C. & Yoon, Y.S. (1997). Stabilization constraint method for torque optimization of a redundant manipulator, *Proceedings of the 1997 IEEE International Conference on Robotics and Automation*, pp. 2403–2408, ISBN 0-7803-3612-7, Albuquerque, NM, April, 1997.

Sie, I.H.; Waters, R.L.; Adkins, RH & Gellman, H. (1992). Upper extremity pain in the postrehabilitation spinal cord injured patient. *Archives of Physical Medicine and Rehabilitation*, Vol. 73, pp. 44–48, ISSN 0003-9993.

Trudel, G.; Kirby, R.L. & Bell, A.C. (1995). Mechanical effects of rear-wheel camber on wheelchairs. *Assistive Technology*, Vol. 7, No. 2, pp. 79–86, ISSN 1040-0435.

Yoshikawa, T. (1984). Analysis and control of robot manipulators with redundancy, In: *Robotics Research: The First International Symposium of Robotics Research*, Brady, M. & Paul, R. (Ed.), pp. 735–747, MIT Press, ISBN 0-262-52392-2.

Yoshikawa, T. (1985). Dynamic manipulability of robot manipulators. *Journal of Robotic Systems*, Vol. 2, pp. 113–124, ISSN 0741-2223.

Yoshikawa, T. (1990). *Foundations of Robotics: Analysis and Control,* MIT Press, ISBN 0-262-24028-9.